POLYNUCLEAR AROMATIC HYDROCARBONS

Third International Symposium on Chemistry and Biology— Carcinogenesis and Mutagenesis

POLYNUCLEAR AROMATIC HYDROCARBONS

Third International Symposium on Chemistry and Biology— Carcinogenesis and Mutagenesis

Sponsored by
Battelle-Columbus Laboratories
U.S. Environmental Protection Agency
Hoffmann-La Roche, Inc.
Battelle Memorial Institute,
Corporate Technical Development

Edited by

PETER W. JONES

Environmental Assessment Department
Electric Power Research Institute
Palo Alto, California

and

PHILIP LEBER

Toxicologist
Environmental Affairs Department
PPG Industries, Inc.
Pittsburgh, Pennsylvania

ANN ARBOR SCIENCE
PUBLISHERS INC /THE BUTTERWORTH GROUP
P.O. BOX 1425 ▪ ANN ARBOR, MICHIGAN 48106

Second Printing 1980

230 Collingwood, P. O. Box 1425, Ann Arbor, Michigan 48106

Library of Congress Catalog Card No. 79-642622
ISBN 0-250-40317-X

Manufactured in the United States of America

PREFACE

The decade of the 1970s will be remembered as the period in which societies and science became cognizant of the impact that technology and certain associated factors have on human health. In particular, many chemicals have undergone considerable scrutiny to determine their biological effects in terms of carcinogenicity, mutagenicity, teratogenicity or other toxicities which affect behavior, reproduction, longevity or general health.

One group of chemicals currently receiving attention is the polynuclear aromatic hydrocarbons (PAH), probably the most widespread of all chemical environmental contaminants. They occur as a result of combustion involving carbon and hydrogen and are naturally present in vegetation and fossil fuels. Since many PAH are known to be carcinogenic or cocarcinogenic, apparently as a result of oxidative reactions in the body, there is justifiable concern regarding their measurement and ultimate control.

A prerequisite for the control of PAH emission is an understanding of factors that influence PAH formation, in addition to high-sensitivity techniques for the measurement of these species. Significant advances in measurement methodology are reported in this volume, particularly in regard to high-performance capillary column gas chromatography and high-performance liquid chromatography separations of PAH isomers. State-of-the-art analytical and sampling methods are discussed by many authors, in addition to numerous studies of biological and chemical reactivity of PAH, their analogues and metabolites.

An understanding of metabolic conversion of PAH to biologically active metabolites is fundamental to the understanding of PAH carcinogenesis. Such an understanding is mandatory if we are ever to have the ability to control or prevent carcinogenesis; the attainment of this would be a decisive milestone in biological evolution. The biologically oriented papers provide some of the most recent advances in the mechanistic exploration of metabolism and carcinogenesis of PAH.

The Third International Symposium on Polynuclear Aromatic Hydrocarbons, on which this volume is based, was held at Battelle-Columbus Laboratories in October 1978. The proceedings of the first symposium were published under the title *Polynuclear Aromatic Hydrocarbons: Chemistry,*

Metabolism and Carcinogenesis (*Carcinogenesis–A Comprehensive Survey: Volume I*), 1976, Raven Press; the proceedings of the second symposium were published under the title *Polynuclear Aromatic Hydrocarbons: Second International Symposium on Analysis, Chemistry and Biology* (*Carcinogenesis –A Comprehensive Survey: Volume 3*), 1978, Raven Press. It is presently intended that subsequent symposia in this series will be held annually in October at Battelle-Columbus Laboratories.

This volume, with contributions from many of the world's leading scientists in the subject area, provides a valuable forum for discussion and examination of the most recent research findings in the area of analysis, chemistry and biology of PAH. It is our wish that this and subsequent volumes in this series, will encourage and motivate further research in order to provide clearer understanding of this complex and important area.

Peter W. Jones
Philip Leber

Peter W. Jones joined the Environmental Assessment Department of EPRI in June 1978, in order to develop and manage a program for the source assessment of processes associated with the generation of electric power from fossil fuels.

He was born and educated in England, and obtained a BSc in physics and chemistry from London University, and subsequently earned a D. Phil in physical and organic chemistry from York University in 1969. Prior to joining EPRI, he was a Program Manager for Analytical and Environmental Organic Chemistry at Battelle's Columbus Laboratories in Ohio.

Dr. Jones has authored and edited numerous technical publications; he was an IR-100 award winner in 1975, and is a Fellow of The Chemical Society (London) and member of the American Chemical Society. He was instrumental in establishing this PAH Symposium Series, and was co-chairman of the First, Second and Third International Symposia on Polynuclear Aromatic Hydrocarbons.

Dr. Philip Leber is a staff toxicologist in the Environmental Affairs Department of PPG Industries, Inc., Pittsburgh, and was formerly with Battelle-Columbus Laboratories. He received a PhD and an MS in Pharmacology-Toxicology from Purdue University and a BS in Chemistry from the University of California. He has studied at Ohio State University and North Dakota State University, and taught secondary education for two years while serving with the Peace Corps in India.

While at Battelle he directed toxicological evaluations of various pesticides, drugs and industrial chemicals in laboratory animals; coordinated and directed chemical metabolism and pharmacokinetic studies; acted as chemical carcinogen monitor for the Life Sciences Department and directed analytical work for determining chemical concentrations and stability in feeds, tissues and biological fluids for in-house toxicology projects.

Dr. Leber's research activities include drug and pesticide toxicology, development of *in vitro* cell techniques for examination of carcinogen mechanisms of heavy metals and asbestos; drug metabolism; and development of liquid chromatographic techniques for metabolic profiles.

He is presently a member of the American Association for the Advancement of Science, the Environmental Mutagen Society, the Genetic Toxicology Association and the New York Academy of Sciences.

DEDICATION

This volume is dedicated to the individuals of past and present times who have suffered from chemically induced diseases associated with exposure to polynuclear aromatic hydrocarbons in their occupational and living environments—and to the efforts of all who work to prevent and alleviate these afflictions through the applications of behavioral, biological, chemical and technological research.

CONTRIBUTORS

Thomas Alsberg
Department of Analytical Chemistry
University of Stockholm
Arrhenius Laboratory
Stockholm, Sweden

Shantilal Amin
Naylor Dana Institute for
Disease Prevention
American Health Foundation
Valhalla, New York 10595

Marilyn S. Arnott
Department of Biology
The University of Texas System
Cancer Center
M.D. Anderson Hospital and
Tumor Institute
6723 Bertner
Houston, Texas 77030

R. F. Arrendale
Tobacco and Health Research Laboratory
Science and Education Administration/
Federal Research
U.S. Department of Agriculture
Athens, Georgia 30604

William M. Baird
The Wistar Institute of Anatomy
and Biology
36th Street at Spruce
Philadelphia, Pennsylvania 19104

W. L. Banwart
Institute for Environmental Studies
and Department of Agronomy
University of Illinois
Urbana, Illinois 61801

Victoria Bedenko
Naylor Dana Institute for
Disease Prevention
American Health Foundation
Valhalla, New York 10595

Earl P. Benditt
Department of Pathology SM-30
University of Washington
School of Medicine
Seattle, Washington 98195

Roy L. Bennett
Environmental Sciences Research
Laboratory
U.S. Environmental Protection Agency
Research Triangle Park, North Carolina
27711

D. L. Berry
Biology Division
Oak Ridge National Laboratory
Oak Ridge, Tennessee 37830

E. Bingham
University of Cincinnati
College of Medicine
Department of Environmental Health
Kettering Laboratory
Cincinnati, Ohio 45267

Alf Bjørseth
Battelle–Columbus Laboratories
505 King Avenue
Columbus, Ohio 43201

Lars Blomberg
Department of Chemistry
University of Stockholm
Arrhenius Laboratory
Stockholm, Sweden

James A. Bond
Department of Pharmacology SJ-30
University of Washington
School of Medicine
Seattle, Washington 98195

J. W. Bridges
Department of Biochemistry
University of Surrey
Guildford, Surrey, England

J. M. Brown
National Bureau of Standards
Washington, D.C. 20234

R. L. Carpenter
Inhalation Toxicology Research Institute
Lovelace Biomedical and Environmental Research Institute
P.O. Box 4890
Albuquerque, New Mexico 87115

E. Cavalieri
Eppley Institute for Research in Cancer
University of Nebraska Medical Center
Omaha, Nebraska 68105

F. D. Cazer
Division of Medicinal Chemistry
College of Pharmacy
The Ohio State University
Columbus, Ohio 43210

Cecilia Chan
Centre for Research on Environmental Quality
York University
Downsview (Toronto), Ontario
Canada

R. Chang
Department of Biochemistry and Drug Metabolism
Hoffmann-La Roche, Inc.
Nutley, New Jersey 07110

Ruth Chemerys
The Wistar Institute of Anatomy and Biology
36th Street at Spruce
Philadelphia, Pennsylvania 19104

Ching Jer Chern
The Wistar Institute of Anatomy and Biology
36th Street and Spruce
Philadelphia, Pennsylvania 19104

S. N. Chesler
National Bureau of Standards
Washington, D.C. 20234

O. T. Chortyk
Tobacco and Health Research Laboratory
Science and Education Administration/ Federal Research
U.S. Department of Agriculture
Athens, Georgia 30604

G. M. Cohen
Department of Biochemistry
University of Surrey
Guildford, Surrey, England

Anders Colmsjö
Deparmtent of Analytical Chemistry
University of Stockholm
Arrhenius Laboratory
Stockholm, Sweden

A. H. Conney
Department of Biochemistry and Drug Metabolism
Hoffmann-La Roche, Inc.
Nutley, New Jersey 07110

James P. Crowley
Biomedical Sciences Section
Battelle–Columbus Laboratories
505 King Avenue
Columbus, Ohio 43201

J. M. Daisey
New York University Medical Center
Institute of Environmental Medicine
New York, New York 10016

S. M. D'Ambrosio
Departments of Pharmacology and Radiology
College of Medicine
The Ohio State University
Columbus, Ohio 43210

F. B. Daniel
Departments of Pharmacology
and Radiology
College of Medicine
Division of Medicinal Chemistry
College of Pharmacy
The Ohio State University
Columbus, Ohio 43210

Anthony J. Dennis
Biomedical Sciences Section
Battelle–Columbus Laboratories
505 King Avenue
Columbus, Ohio 43201

R. Depaus
Commission of the European Communities
Joint Research Centre
Petten Establishment
The Netherlands

Leila Diamond
The Wistar Institute of Anatomy
and Biology
36th Street at Spruce
Philadelphia, Pennsylvania 19104

J. DiGiovanni
Department of Pharmacology
School of Medicine
University of Washington
Seattle, Washington 98195

J. Joussot-Dubien
Laboratoire de Chimie Physique
Photophysique et Photochimie Moléculaire
Universite de Bordeaux
Bordeaux, France

E. Eisenstadt
Department of Physiology and
Department of Microbiology
Harvard School of Public Health
665 Huntington Avenue
Boston, Massachusetts 02115

F. Englund
Freshwater Biological and
Department of Biochemistry
University of Minnesota
P.O. Box 100
Navarre, Minnesota 55392

J. L. Epler
Biology Division
Oak Ridge National Laboratory
Oak Ridge, Tennessee 37830

Andrew A. Erickson
The Wistar Institute of Anatomy
and Biology
36th Street at Spruce
Philadelphia, Pennsylvania 19104

Thomas J. Facklam
Biomedical Sciences Section
Battelle–Columbus Laboratories
505 King Avenue
Columbus, Ohio 43201

Detlef Fechner
Institut für Wasser-, Boden- und
Lufthygiene des Bundesgesundheitsamtes
Berlin, Federal Republic of Germany

F. Ferris
Scientific Research Staff
Ford Motor Company
Scientific Research Laboratories
20000 Rotunda Drive
Dearborn, Michigan 48121

P. P. Fu
Ben May Laboratory for Cancer Research
University of Chicago
Chicago, Illinois 60637

N. B. Furlong
Department of Chemistry
The University of Texas System
Cancer Center
M.D. Anderson Hospital and
Tumor Institute
6723 Bertner
Houston, Texas 77030

R. B. Gammage
Health and Safety Research Division
Oak Ridge National Laboratory
Oak Ridge, Tennessee 37830

G. L. Gleason
Biology Division
Oak Ridge National Laboratory
Oak Ridge, Tennessee 37830

A. Gold
Department of Physiology and
Department of Microbiology
Harvard School of Public Health
665 Huntington Avenue
Boston, Massachusetts 02115

W. H. Griest
Analytical Chemistry Division
Oak Ridge National Laboratory
Oak Ridge, Tennessee 37830

G. D. Griffin
Health and Safety Research Division
Oak Ridge National Laboratory
Oak Ridge, Tennessee 37830

Zoe A. Grosser
Arthur D. Little, Inc.
Cambridge, Massachusetts 02140

F. Guenther
National Bureau of Standards
Washington, D.C. 20234

Hans Güsten
Kernforschungszentrum Karlsruhe
Institute für Radiochemie
Karlsruhe, Federal Republic of Germany

R. L. Hanson
Inhalation Toxicology Research Institute
Lovelace Biomedical and Environmental
Research Institute
P.O. Box 4890
Albuquerque, New Mexico 87115

Judith C. Harris
Arthur D. Little, Inc.
Cambridge, Massachusetts 02140

R. W. Hart
Departments of Pharmacology
and Radiology
College of Medicine
Division of Medicinal Chemistry
College of Pharmacy
The Ohio State University
Columbus, Ohio 43210

R. G. Harvey
Ben May Laboratory for Cancer Research
University of Chicago
Chicago, Illinois 60637

J. J. Hassett
Institute for Environmental Studies and
Department of Agronomy
University of Illinois
Urbana, Illinois 61801

A. R. Hawthorne
Health and Safety Research Division
Oak Ridge National Laboratory
Oak Ridge, Tennessee 37830

Stephen S. Hecht
Naylor Dana Institute for
Disease Prevention
American Health Foundation
Valhalla, New York 10595

Gunther Heinrich
Kernforschungszentrum Karlsruhe
Institute für Radiochemie
Karlsruhe, Federal Republic of Germany

H. S. Hertz
National Bureau of Standards
Washington, D.C. 20234

A. Hidmark
Freshwater Biological and
Department of Biochemistry
University of Minnesota
P.O. Box 100
Navarre, Minnesota 55392

H. C. Higman
Tobacco and Health Research Laboratory
Science and Education Administration/
Federal Research
U.S. Department of Agriculture
Athens, Georiga 30604

L. R. Hilpert
National Bureau of Standards
Washington, D.C. 20234

Norio Hirota
Naylor Dana Institute for
Disease Prevention
American Health Foundation
Valhalla, New York 10595

V. Hitchins
Department of Medicine
University of Kentucky Medical Center
Lexington, Kentucky 40506

Dietrich Hoffman
Naylor Dana Institute for
Disease Prevention
American Health Foundation
Valhalla, New York 10595

J. J. Hutton
Medical Service
Veterans Administration Medical Center
Lexington, Kentucky 40507

J. L. Hylden
Freshwater Biological Institute and
Department of Biochemistry
University of Minnesota
P.O. Box 100
Navarre, Minnesota 55392

Radhakrishnan Iyer
The University of Tennessee–
Oak Ridge Graduate School of
Biomedical Sciences
Oak Ridge, Tennessee 37803

J. Jacob
Biochemisches Institut für
Umweltcarcinogene
Ahrensburg,
Federal Republic of Germany

D. M. Jerina
Laboratory of Bioorganic Chemistry
National Institute of Arthritis,
Metabolism and Digestive Diseases
National Institutes of Health
Bethesda, Maryland 20014

Dennis A. Johnston
The Univerity of Texas Health
Science Center
Medical School at Houston
Texas Medical Center
Houston, Texas 77030

C. A. Jones
Department of Biochemistry
University of Surrey
Guildford, Surrey, England

T. D. Jones
Health and Safety Research Division
Oak Ridge National Laboratory
Oak Ridge, Tennessee 37830

Peter W. Jones
Electric Power Research Institute
3412 Hillview Avenue
P.O. Box 10412
Palo Alto, California 94303

M. R. Juchau
Department of Pharmacology
School of Medicine
University of Washington
Seattle, Washington 98195

W. Karcher
Commission of the European Communities
Joint Research Centre
Petten Establishment
The Netherlands

Niilo T. Karki
Departments of Pharmacology,
Microbiology, Pediatrics and
Obstetrics & Gynecology
University of Oulu
Oulu, Finland

Morris Katz
Centre for Research on Environmental
Quality
York University
Downsview (Toronto), Ontario
Canada

Antti Kauppila
Departments of Pharmacology, Microbiology, Pediatrics and Obstetrics & Gynecology
University of Oulu
Oulu, Finland

Kenneth T. Knapp
Environmental Sciences Research Laboratory
U.S. Environmental Protection Agency
Research Triangle Park, North Carolina 27711

T. J. Kneip
New York University Medical Center
Institute of Environmental Medicine
New York, New York 10016

Richard M. Kocan
Department of Pathology SM-30
University of Washington
School of Medicine
Seattle, Washington 98195

Maila Koivisto
Departments of Pharmacology, Microbiology, Pediatrics and Obstetrics & Gynecology
University of Oulu
Oulu, Finland

A. Kootstra
Biology Division
Oak Ridge National Laboratory
Oak Ridge, Tennessee 37830

W. A. Korfmacher
Department of Chemistry
University of Illinois
Urbana, Illinois 61801

Pekka Korhonen
Departments of Pharmacology, Microbiology, Pediatrics and Obstetrics & Gynecology
University of Oulu
Oulu, Finland

S. Kumar
Department of Chemistry
University of Oklahoma
Norman, Oklahoma 73019

B. Laine
Department of Medicine
University of Kentucky Medical Center
Lexington, Kentucky 40506

J. R. Lakowicz
Freshwater Biological Institute and Department of Biochemistry
University of Minnesota
P.O. Box 100
Navarre, Minnesota 55392

M. Lamotte
Laboratoire de Chimie Physique
Photophysique et Photochimie Moléculaire
Universite de Bordeaux
Bordeaux, France

R. C. Lao
Chemistry Division
Air Pollution Control Directorate
Air Pollution Technology Centre
Environment Canada
Ottawa, Canada

Edmond LaVoie
Naylor Dana Institute for Disease Prevention
American Health Foundation
Valhalla, New York 10595

Frank S. -C. Lee
Scientific Research Staff
Ford Motor Company
Scientific Research Laboratories
20000 Rotunda Drive
Dearborn, Michigan 48121

Milton L. Lee
Department of Chemistry
Brigham Young University
Provo, Utah 84602

R. E. Lehr
Department of Chemistry
University of Oklahoma
Norman, Oklahoma 73019

W. Levin
Department of Biochemistry and
Drug Metabolism
Hoffmann-La Roche, Inc.
Nutley, New Jersey 07110

Philip L. Levins
Arthur D. Little, Inc.
Cambridge, Massachusetts 02140

M. A. Leyko
New York University Medical Center
Institute of Environmental Medicine
New York, New York 10016

Michael C. MacLeod
Biology Division
Oak Ridge National Laboratory
Oak Ridge, Tennessee 37830

M. McNamara
Freshwater Biological and
Department of Biochemistry
University of Minnesota
P.O. Box 100
Navarre, Minnesota 55392

H. D. Mah
Laboratory of Bioorganic Chemistry
National Institute of Arthritis,
Metabolism and Digestive Diseases
National Institutes of Health
Bethesda, Maryland 20014

G. Mamantov
Department of Chemistry
University of Tennessee
Knoxville, Tennessee 37916

Werner L. Margard
Biomedical Sciences Section
Battelle–Columbus Laboratories
505 King Avenue
Columbus, Ohio 43201

W. E. May
National Bureau of Standards
Washington, D.C. 20234

Robert Mazzarese
Naylor Dana Institute for Disease
Prevention
American Health Foundation
Valhalla, New York 10595

J. C. Means
Institute for Environmental Studies and
Department of Agronomy
University of Illinois
Urbana, Illinois 61801

Antonio H. Miguel
Instituto de Quimica
Univeisidade Federale de Rio de Janeiro
Rio de Janeiro, Brazil

B. P. Moore
Department of Biochemistry
University of Surrey
Guilford, Surrey, England

D. F. S. Natusch
Department of Chemistry
Colorado State University
Fort Collins, Colorado 80523

G. J. Newton
Inhalation Toxicology Research Institute
Lovelace Biomedical and Environmental
Research Institute
P.O. Box 4890
Albuquerque, New Mexico 87115

R. Niemeier
University of Cincinnati
College of Medicine
Department of Environmental Health
Kettering Laboratory
Cincinnati, Ohio 45267

M. Nordqvist
Section on Oxidation Mechanism
Laboratory of Bioorganic Chemistry
National Institute of Arthritis
Metabolism and Digestive Diseases
National Institutes of Health
Bethesda, Maryland 20205

D. E. Olins
Biology Division
Oak Ridge National Laboratory
Oak Ridge, Tennessee 37830

C. T. Oravec
Department of Pharmacology
College of Medicine
The Ohio State University
Columbus, Ohio 43210

John Papaconstantinou
Biology Division
Oak Ridge National Laboratory
Oak Ridge, Tennessee 37830

Olavi Pelkonen
Departments of Pharmacology,
Microbiology, Pediatrics and
Obstetrics & Gynecology
University of Oulu
Oulu, Finland

J. Pereyre
Laboratoire de Chimie Physique
Photophysique et Photochimie
Moléculaire
Universite de Bordeaux
Bordeaux, France

John M. Pezzuto
Department of Biochemistry
CMDNJ-New Jersey Medical School
Newark, New Jersey 07103

T. J. Prater
Scientific Research Staff
Ford Motor Company
Scientific Research Laboratories
20000 Rotunda Drive
Dearborn, Michigan 48121

T. K. Rao
Biology Division
Oak Ridge National Laboratory
Oak Ridge, Tennessee 37830

S. Risemberg
Laboratoire de Chimie Physique
Photophysique et Photochimie
Moléculaire
Universite de Bordeaux
Bordeaux, France

E. Rogan
Eppley Institute for Research in Cancer
University of Nebraska Medical Center
Omaha, Nebraska 68105

R. Roth
Eppley Institute for Research in Cancer
University of Nebraska Medical Center
Omaha, Nebraska 68105

E. R. Royer
Inhalation Toxicology Research Institute
Lovelace Biomedical and Environmental
Research Institute
P.O. Box 4890
Albuquerque, New Mexico 87115

D. Ryan
Deparment of Biochemistry and
Drug Metabolism
Hoffmann-La Roche, Inc.
Nutley, New Jersey 07110

Takeo Sakuma
Centre for Research on Environmental
Quality
York University
Downsview (Toronto), Ontario Canada

Patricia F. Schlotzhauer
Environmental Resarch Laboratory
U.S. Environmental Protection Agency
College Station Road
Athens, Georgia 306605

William S. Schlotzhauer
Tobacco and Health Research Laboratory
Science and Education Administration/
Federal Research
U.S. Department of Agriculture
Athens, Georgia 30604

J. Schultz
Department of Physiology and
Department of Microbiology
Harvard School of Public Health
665 Huntington Avenue
Boston, Massachusetts 02115

Bernd Seifert
Institut für Wasser-, Boden-
und Lufthygiene des
Bundesgesundheitsamtes
Berlin, Federal Republic of Germany

James K. Selkirk
Biology Division
Oak Ridge National Laboratory
Oak Ridge, Tennessee 37830

Ray F. Severson
Tobacco and Health Research Laboratory
Science and Education Administration/
Federal Research
U.S. Department of Agriculture
Athens, Georgia 30604

A. G. Sharkey, Jr.
U.S. Department of Energy
Pittsburgh Energy Technology Center
Pittsburgh, Pennsylvania 15213

D. S. Shewach
Department of Biochemistry
The University of Texas System
Cancer Center
M.S. Anderson Hospital and
Tumor Institute
6723 Bertner
Houston, Texas 77030

Lester L. Shipman
Chemistry Division
Argonne National Laboratory
Argonne, Illinois 60439

T. J. Slaga
Biology Division
Oak Ridge National Laboratory
Oak Ridge, Tennessee 37830

Maurice E. Snook
Tobacco and Health Research Laboratory
Science and Education Administration/
Federal Research
U.S. Department of Agriculture
Athens, Georgia 40604

Ulf Stenberg
Department of Analytical Chemistry
University of Stockholm
Arrhenius Laboratory
Stockholm, Sweden

Paul E. Strup
Battelle–Columbus Laboratories
505 King Avenue
Columbus, Ohio 43201

K. B. Sukumaron
Ben May Laboratory for Cancer Research
University of Chicago
Chicago, Illinois 60637

C. W. Taylor
Department of Chemistry
University of Oklahoma
Norman, Oklahoma 73019

D. R. Taylor
Department of Chemistry
Colorado State University
Fort Collins, Colorado 80523

D. R. Thakker
Laboratory of Bioorganic Chemistry
National Institute of Arthritis,
Metabolism and Digestive Diseases
National Institutes of Health
Bethesda, Maryland 20014

D. M. Thomas
Department of Biochemistry
The University of Texas System
Cancer Center
M.D. Anderson Hospital and
Tumor Institute
6723 Bertner
Houston, Texas 77030

F. B. Thomas
Department of Biochemistry
The University of Texas System
Cancer Center
M.D. Anderson Hospital and
Tumor Institute
6723 Bertner
Houston, Texas 77030

Paul E. Thomas
Department of Biochemistry and
Drug Metabolism
Hoffmann-La Roche, Inc.
Nutley, New Jersey 07110

R. S. Thomas
Chemistry Division
Air Pollution Control Directorate
Air Pollution Technology Centre
Environment Canada
Ottawa, Canada

J. H. Thorngate
Health and Safety Reserach Division
Oak Ridge National Laboratory
Oak Ridge, Tennessee 37830

B. A. Tomkins
Analytical Chemistry Division
Oak Ridge National Laboratory
Oak Ridge, Tennessee 37830

Helle Tosine
Centre for Research on Environmental
Quality
York University
Downsview (Toronto), Ontario Canada

Larry L. Triplett
Biology Division
Oak Ridge National Laboratory
Oak Ridge, Tennessee 37830

Risto Tuimala
Departments of Pharmacology,
Microbiology, Pediatrics and
Obstetrics & Gynecology
University of Oulu
Oulu, Finland

J. Van Eijk
Commission of the European Communities
Joint Research Centre
Petten Establishment
The Netherlands

Daniel L. Vassilaros
Department of Chemistry
Brigham Young University
Provo, Utah 84602

T. Vo-Dinh
Health and Safety Research Division
Oak Ridge National Laboratory
Oak Ridge, Tennessee 37830

P. J. Walsh
Health and Safety Research Division
Oak Ridge National Laboratory
Oak Ridge, Tennessee 37830

C'L. A. Wang
Department of Pharmacology
College of Medicine
The Ohio State University
Columbus, Ohio 43210

Thomas Wannman
Department of Analytical Chemistry
University of Stockholm
Arrhenius Laboratory
Stockholm, Sweden

C. Warren
University of Cincinnati
College of Medicine
Department of Environmental Health
Kettering Laboratory
Cincinnati, Ohio 45267

D. Warshawsky
University of Cincinnati
College of Medicine
Department of Environmental Health
Kettering Laboratory
Cincinnati, Ohio 45267

E. L. Wehry
Department of Chemistry
University of Tennessee
Knoxville, Tennessee 37916

Curt M. White
U.S. Department of Energy
Pittsburgh Energy Technology Center
Pittsburgh, Pennsylvania 15213

JoAnn E. Wilkinson
Battelle–Columbus Laboratories
505 King Avenue
Columbus, Ohio 43201

S. A. Wise
National Bureau of Standards
Washington, D.C. 20234

D. T. Witiak
Division of Medicinal Chemistry
College of Pharmacy
The Ohio State University
Columbus, Ohio 43210

L. K. Wong
Department of Pharmacology
College of Medicine
The Ohio State University
Columbus, Ohio 43210

A. W. Wood
Department of Biochemistry and
Drug Metabolism
Hoffmann-La Roche, Inc.
Nutley, New Jersey 07110

S. G. Wood
Institute for Environmental Studies and
Department of Agronomy
University of Illinois
Urbana, Illinois 61801

H. Yagi
Laboratory of Bioorganic Chemistry
National Institute of Arthritis,
Metabolism and Digestive Diseases
National Institutes of Health
Bethesda, Maryland 20014

Toshio Yamauch
Department of Biomathematics
The University of Texas System
Cancer Center
M.D. Anderson Hospital and
Tumor Institute
6723 Bertner
Houston, Texas 77030

Chung S. Yang
Department of Biochemistry
CMDNJ-New Jersey Medical School
Newark, New Jersey 07103

Richard G. Zepp
Environmental Research Laboratory
U.S. Environmental Protection Agency
College Station Road
Athens, Georgia 30605

ACKNOWLEDGMENTS

This symposium was sponsored by the Environmental Protection Agency, (IERL and HERL, Research Triangle Park), Hoffmann-La Roche and Battelle Columbus Laboratories; this support is acknowledge with our gratitude. Furthermore, many Batelle staff members provided invaluable assistance, which helped to assure the coordination and success of the symposium. We particularly wish to thank Susan Armstrong, Tom Stewart, Marty Buller, Ruth Anne Gibson, Pete DeWalt, Peggy A. Olson and Diane Harris.

Contents

Chemistry and Biology

Chemistry

Biology

Chemistry and Biology

Session Chairman

P. W. Jones

Polynuclear Aromatic Hydrocarbons, edited by P. W. Jones and P. Leber. Ann Arbor Science Publishers, Inc. Ann Arbor, MI ©1979.

CHARACTERIZATION OF POTENTIAL ORGANIC EMISSIONS FROM A LOW-BTU GASIFIER FOR COAL CONVERSION

R. L. Hanson, R. E. Royer, R. L. Carpenter and G. J. Newton

Inhalation Toxicology Research Institute
Lovelace Biomedical and Environmental Research Institute
P.O. Box 4890
Albuquerque, New Mexico 87115

INTRODUCTION

Coal gasification is being developed as an alternative method for energy production from coal. Low-Btu coal gas is of particular interest for use in combined-cycle electrical generators and as a raw material for syngas production. Although coal gasification has been under development for a number of years, relatively little is known about the potential toxicity of respirable particles in the process streams or in aerosols produced when low-Btu coal gas is released from a hot process stream. In part, this is due to the high pressures and temperatures encountered in these processes, which complicate stream characterization. A cooperative program with METC has been established for the study of respirable-sized low-Btu coal-gas aerosols to provide data on the nature of this material.

When considering the data and observations made in this paper, it is imperative that the reader be aware of several limitations. First, a unit at METC is a small research unit designed to produce gas for studying cleanup systems and optimization of gas production. Since the unit does not operate in the same fashion or under the same conditions as a full-scale gas-producing facility, the effects of scaleup and changing operating conditions may mean the results cannot be directly extrapolated to larger units. Since the purpose of the gas cleanup system is to study the phenomena associated with gas cleanup, the data obtained on the cleaned gas may not be representative of similar process streams in production facilities.

Tar material from the disengagement chamber of this gasifier has been analyzed for volatile and semivolatile components by Pellizzari (1). The volatile organics that were found were about 170 compounds eluting between propane and C_3-alkyl naphthalene isomers. The semivolatile organics that were found in the tar were compounds eluting between acetone and fluorene.

Development work for sampling vapor phase components in process streams using Tenax adsorbent samplers was reported by Jones *et al.* (2). Tenax has also been used in sampling the polynuclear aromatic hydrocarbons (PAH) emissions from the Hygas pilot plant as reported by Philips (3). He found several 3- 4- and 5-ring PAH in the ambient samples from the Hygas pilot plant.

System Description

The Morgantown Energy Technology Center coal gasifier is a pressurized version of the McDowell-Wellman (Wellman-Galusha) atmospheric stirred-bed coal gasifier. The gasifier vessel is a large cylinder 7.3 m high and 1 m in diameter. A grate in the lower end of the vessel serves as a support for fuel material and the bottom of the unit is sealed with lock hoppers through which spent ash may be removed. The top of the gasifier vessel is closed with a hemispherical dome through which coal is fed and gas exits. The gasifier uses the Lurgi process for low-Btu coal gasification which requires air, steam and feed coal. The process produces, in addition to the low-Btu gas, both fly ash and bottom ash that must be removed from the gasifier. Coal is fed into the unit from lock hoppers at the top of the gasifier and ash is removed through a similar lock hopper arrangement at the bottom. This system permits introduction of fuel and removal of ash from high-pressure zones inside the gasifier. Once inside the gasifier vessel, the coal undergoes a complex series of reactions beginning with the volatilization of the coal material and ending with partial combustion of this material to carbon monoxide. The resulting mixture of gas, along with ash particles, exits the gasifier and both are passed through a large cyclone to remove large ash particles, as shown in Figure 1.

From the cyclone, the gas flow is split into two streams with approximately one-fourth of the gas being carried to an experimental wet-scrub cleanup system. The remaining three-fourths of the raw gas passes through a pressure-dropping nozzle and is burned at the top of the gasifier by two flares. Figure 1 schematically illustrates the gas cleanup system in operation during the first four ITRI sampling periods. The purpose of the gas cleanup system is to remove flyash, oils, tars and entrained metallic materials in an effort to produce low-Btu gas suitable for methanation or turbine combustion. After exiting the gas cleanup system, the gas returns to the flares at the top of the gasifier.

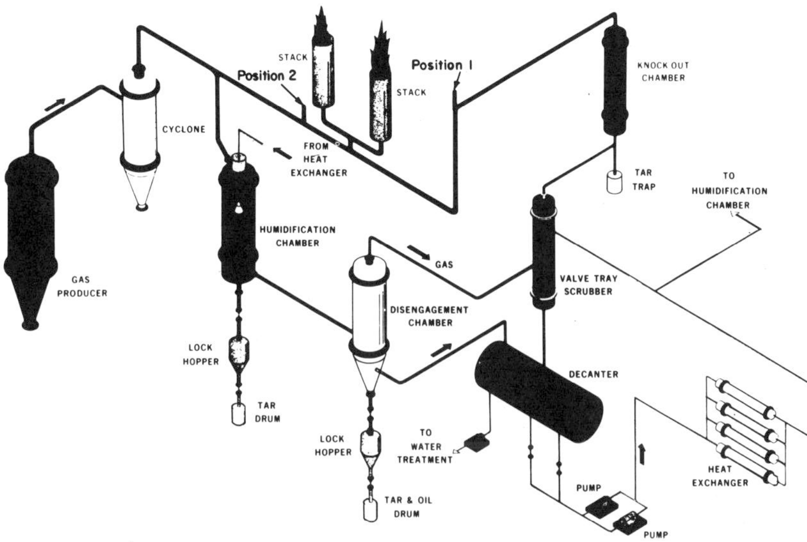

FIGURE 1. Low-Btu gasifier and gas cleanup system. Coal gas aerosol sampling positions for clean gas (1) and raw gas (2) are illustrated. The coal gas leaves the gas producer and first passes through a cyclone to remove large particles. About one-fourth of the gas goes to the humidification chamber and three-fourths of the gas is sent to the stack. Water from the heat exchanger is recycled to the humidification chamber. The coal gas leaves the humidification chamber and goes to the disengagement chamber at which liquids are separated from the gas. Liquids go to the decanter and then portions of the liquids go to water treatment and the remainder goes to the heat exchanger and then to the valve tray scrubber or the humidification chamber. The gas goes to the valve tray scrubber after leaving the disengagement chamber. After leaving the valve tray scrubber the gas goes to the knockout chamber and then recombines with the raw gas before it is burned in the stack.

In order to assess the physicochemical characteristics of low-Btu coal gas aerosols, samples of raw gas aerosol and cleaned gas aerosol and samples of scrubbed tar from the disengagement chamber were obtained. Two positions illustrated in Figure 1 were chosen as gas sampling points. Cleaned gas was sampled before (Position 1) it recombines with the raw gas stream prior to the flare. Position 2 was located before the flare and represents the raw gas stream. Gas exits the gasifier at temperatures on the order of 1,650°C and at pressures of 120-300 psi. At Position 2, just prior to the flare, the gas is essentially at ambient pressure and a temperature of about 430°C. At Position 1, the gas is at ambient temperature and pressure is 42 psi.

In order to mimic the effects of releasing process gas into the atmosphere, a sample of the gas was vented from the process stream and rapidly cooled and diluted in a sampling chamber from which the flow could be divided and routed to a variety of sampling instruments. It was necessary to dilute the coal gas with nitrogen since the coal gas forms flammable and explosive mixtures with air.

SAMPLING METHODS AND EQUIPMENT

Sampling System Design

Since the low-Btu coal gas contains appreciable quantities of hydrogen sulfide and other corrosive gases, the sampling system was constructed entirely of stainless steel. Space considerations limited the size of the sampling chamber to a cylinder approximately 1 m long by 34 cm in diameter. A similar sampling system has been used to collect effluent aerosols from a fluidized bed coal combuster (5).

A probe was used to sample the coal gas from Position 2 since the gas was near ambient pressure. Sample volume was determined by the difference between diluting nitrogen and exhaust vacuum flow. At Position 1, the gas pressure was 42 psi and a critical orifice was used to reduce pressure and establish sample flow. This orifice was placed immediately in front of a radial diluter.

The sampling system is illustrated in Figure 2. The sampled gas was mixed with dry nitrogen in the radially injected diluter and transported to the

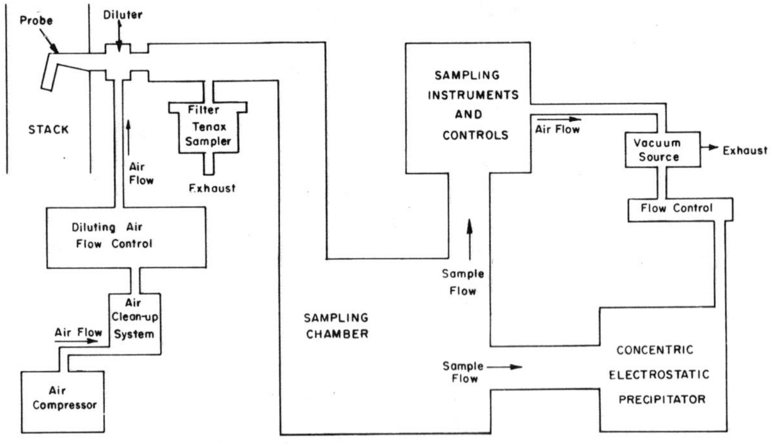

FIGURE 2. Sampling system schematic.

sampling chamber through 25.4-mm stainless steel pipes. Sampling and exhaust airflow controls were used to maintain a constant flow through the chamber at all times. Excess sample was routed through a concentric electrostatic precipitator and final filter prior to exhaust.

Following dilution of the sampled gas a portion of the aerosol formed was passed through a glass-fiber filter and then through a Tenax polymer adsorbent trap. The Tenax collected vapor-phase hydrocarbons above C_6. Samples of the tar from the disengagement chamber were also taken for organic analysis. This strategy provided aerosol samples produced from the raw stream and the clean stream in addition to tar removed from the stream by the disengagement chamber.

Methods of Analysis

Both filter particulate samples and Tenax samples were extracted with organic solvents to dissolve trapped organic material. Pentane was used in the Soxhlet extraction of the Tenax material while methylene chloride was used in the ultrasonic extraction of the particulate samples (3). Both the extraction and subsequent concentration by rotary evaporation were carried out at room temperature to minimize the loss of volatile material and possible thermal degradation of the samples. For gravimetric determinations of extracts the samples were concentrated to about 1 ml by rotary evaporation and then taken to dryness under a stream of dry nitrogen. After rotary evaporation, a portion of the extracted organic material from each sample was studied by gas chromatography (GC). A pair of filter and Tenax samples from both the cleaned and raw gas streams were analyzed by gas chromatography/mass spectrometry (GC/MS). A fraction of the disengagement chamber tar was also analyzed by GC/MS.

The disengagement chamber tar was first fractionated on Sephadex LH20 in order to obtain fractions for analysis by GC/MS and for bioassay using a modification of the procedure of Jones *et al.* (4). A 0.4-g sample of the tar was dissolved in 5 ml of tetrahydrofuran (THF) and filtered. The solution was placed on a 1 m x 15.5 mm i.d. column packed with 6 g of Sephadex LH20 which had been swelled overnight in THF. The sample was eluted with THF and the fractions listed in Table I were collected. The material remaining on the column after THF elution was eluted with 300 ml of methanol. The solvent was removed from these fractions by rotary evaporation to 5 ml followed by evaporation under a stream of dry nitrogen. The fractions were not exhaustively evaporated in order to avoid loss of volatile material. Fraction 5 was further separated into ether soluble bases, acids, neutrals, amphoteric compounds and ether insolubles. This ether soluble basic subfraction was analyzed by GC and GC/MS.

TABLE I. Sephadex LH-20 Fractionation of Disengagement Chamber Tar

Fraction Number	Fraction Volume (ml)	Sample Weight (mg)
0	70	0
1	30	66
2	15	67
3	15	127
4	15	88
5	95	128
6	300 (methanol)	115

For GC analysis a Varian Model 3700 gas chromatograph was used with a 30-m glass capillary column coated with Dexsil 300. Helium was used as carrier gas at 4 ml/min with a 1/10 split injection. A temperature program of 4°C/min from 60 to 270°C was used.

For GC/MS analysis a Finnigan Model 4023 GC/MS data system (GC/MS-DS) was used to obtain 70-eV electron impact mass spectra of components of the samples. A 30-m glass capillary column coated with Dexsil 300 was used with helium as carrier gas at 2.5 ml/min. Splitless injection of 1 μl was used. A temperature program of 2°C/min from 60 to 270°C was used.

In addition to the detailed gas-chromatographic analysis of the disengagement chamber tar, it was also analyzed by high-performance liquid chromatography (HPLC) using a Perkin-Elmer Model 601 in order to characterize some of its major components. A Perkin-Elmer reverse phase ODS-HC-SIL-X-I 0.26 x 25 cm column was used with a 40 to 100% methanol:water linear gradient at 45°C. Detection of the eluting compounds was accomplished with a Perkin-Elmer LC-55 ultraviolet absorption detector at 254 nm and a Perkin-Elmer 204A fluorescence detector with excitation at 260 nm and emission at 400 nm. Major peaks containing absorbing and fluorescent material were collected. Fluorescence excitation and emission spectra of these fractions were obtained with a Farrand Mark I spectrofluorometer. A number of aromatic compounds were tentatively identified by their fluorescence spectra and by coinjection of standards.

RESULTS AND DISCUSSION

Gravimetric analysis of the sample extracted from Tenax indicated a vapor-phase hydrocarbon concentration of 125 mg/m^3 in the raw gas stream and 82 mg/m^3 in the cleaned gas stream.

The extract of the filter sample of the raw gas aerosol contained many PAH with molecular weights from 152 to 252 amu. The reconstructed total ion chromatogram is shown in Figure 3. Table II contains the tentative identification of components found. The vapor-phase PAH collected by Tenax from the raw gas aerosol had molecular weights from 116 to 182 amu. Table III contains the tentative identification of components found in the extract of Tenax. Many PAH were found in both the vapor-phase sample and the filter sample. The reconstructed total ion chromatogram for a typical Tenax extract is shown in Figure 4. Table II illustrates that the filter sample of the raw gas aerosol contained many PAH with molecular weights above 182 amu. These compounds were removed from the vapor phase by cooling and diluting the raw gas.

The filter sample from the cleaned gas aerosol contained PAH from 178 to 216 amu. The reconstructed ion chromatogram of extract of this sample is shown in Figure 5. The tentative identification of components is contained in Table IV. The vapor-phase sample of the clean gas aerosol collected by Tenax contained PAH from 116 to 168 amu. The reconstructed ion chromatogram of the extract of the Tenax is shown in Figure 6. The tentative identification of components in this extract is given in Table V. Several of the major polynuclear aromatic components of the whole tar were determined by HPLC. These compounds are listed in Table VI. By comparing the chromatograms of the cleaned gas aerosol to the chromatograms of the raw gas aerosol, it can be seen that the cleanup process removes larger molecular weight PAH. Several of the larger molecular weight PAH found in the raw gas stream were also found in the whole tar from the disengagement chamber.

The ether-soluble basic compounds in fraction five from the Sephadex LH20 separation of the tar were selected for GC/MS analysis on the basis of initial bioassay results. The reconstructed ion chromatogram of this subfraction is shown in Figure 7. Mass spectral identification of components indicate that the ether-soluble bases consist of alkylated two-ring nitrogen heterocycles, three-ring nitrogen heterocycles and their alkylated derivatives, and some four-ring nitrogen heterocycles. The tentative identification of major components are listed in Table VII. Naphthalene and a trace of phenanthrene were present in this fraction, possibly due to their presence in large amounts in the tar.

CONCLUSIONS

Analysis of the aerosol samples and the cleanup system tar indicates that the heavy organic compounds produced by the low-Btu gasifier are primarily aromatic compounds and alkyl-substituted aromatics. The distribution of these compounds between vapor, liquid and solid phases is dependent on the physical state of the gas and the treatment to which it was subjected.

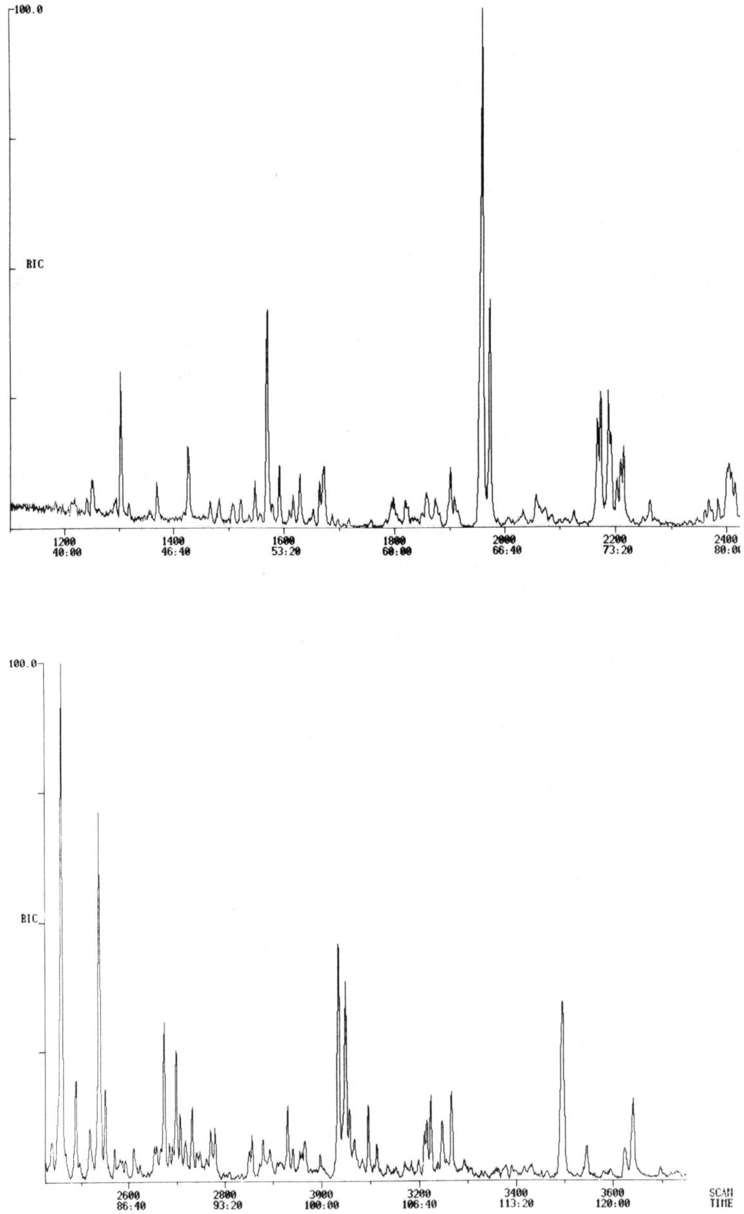

FIGURE 3. Reconstructed chromatogram of the extract of filter sample from raw gas stream. Chromatographic conditions: A 30-m Dexsil 300 glass capillary column, helium carrier gas at 30 cm/sec, temperature program rate of 2°C/min from 60 to 270°C, total ion current response of Finnigan Model 4023 GC/MS.

TABLE II. Tentative Identification of Components in Extract of Filter Sample From Raw Gas Stream

Scan	MW	Tentative Identification
1241	156	Dimethylnaphthanlene Isomer
1251	156	Dimethylnaphthalene Isomer
1304	152	Biphenylene or Acenaphthylene
1370	154	Acenaphthene or 1,4-Dihydro-1,4-Ethenonaphthalene
1427	168	Dibenzofuran
1467	170	Trimethylnaphthalene Isomer
1548	170	Trimethylnaphthalene Isomer
1572	166	Fluorene or Phenalene
1592	170	Trimethylnaphthalene Isomer
1630	182	Xanthene or Methyl Dibenzofuran Isomer
1673	182	Xanthene or Methyl Dibenzofuran Isomer
1778	180	Methylfluorene Isomer
1902	184	Dibenzothiophene
1961	178	Phenanthrene
1975	178	Anthracene
2168	192	Methylphenanthrene Isomer
2175	192	Methylphenanthrene Isomer
2189	192	Methylphenanthrene Isomer
2216	192	Methylphenanthrene Isomer
2406	206	Dimethylphenanthrene Isomer
2460	202	Fluoranthene
2539	202	Pyrene
2675	216	Benzofluorene or Methylpyrene Isomer
2734	216	Benzofluorene or Methylpyrene Isomer
3036	228	Benzo[c]phenanthrene or a Benzo Fluoranthene Isomer
3051	228	A Benzofluoranthene or Benzo[a]anthracene
3059	228	Benzo[a]anthracene or Chrysene
3096	228	Chrysene or Triphenylene
3225	242	Methylbenz[a]anthracene Isomer
3497	252	Benzofluoranthene Isomer
3546	252	Benzofluoranthene Isomer
3624	252	Benzo[e]pyrene or Benzofluoranthene Isomer
3642	252	Benzo[a]pyrene or Benzo[e]pyrene
3696	252	Perylene or Benzo[a]pyrene

TABLE III. Tentative Identification of Components in Extract of Tenax From Raw Gas Stream

Scan	MW	Tentative Identification
366	116	Indene or 1-Ethynyl-4-methylbenzene
537	122	Dimethylphenol Isomer
591	130	Methylindene Isomer
600	122	Dimethylphenol or Ethylphenol Isomer
673	128	Azulene or Napthalene
682	128	Napthalene or Azulene
902	117	Indole or Benzeneacetonitrile
940	142	Methylnaphthalene Isomer
948	148	Methylbenzothiophene Isomer
968	142	Methylnapthalene Isomer
1082	154	Acenapthene or Biphenyl
1140	156	Ethylnapthalene Isomer
1180	156	Dimethylnaphthalene Isomer
1200	156	Dimethylnaphthalene Isomer
1216	156	Dimethylnaphthelene Isomer
1221	154	1,4-Dihydro-1,4-Ethenonaphthalene or Acenapthene
1250	156	Dimethylnaphthalene Isomer
1260	152	Acenaphthylene or Biphenylene
1270	152	Acenaphthylene or Biphenylene
1277	156	Dimethylnaphthalene Isomer
1314	158	Methylbiphenyl Isomer
1326	154	Acenaphthene or 1,4-Dihydro-1,4-Ethenonaphthalene
1378	170	Trimethylnaphthalene Isomer
1388	168	Dibenzofuran
1393	170	Trimthylnaphthalene or Methylethylnapthalene
1407	170	Methylethylnapthalene Isomer
1420	170	Trimethylnapthalene Isomer
1438	170	Trimethylnapthalene Isomer
1462	170	Trimethylnapthalene Isomer
1475	170	Trimethylnapthalene Isomer
1503	170	Trimethylnapthalene Isomer
1511	166	Fluorene or Phenalene
1527	166	Fluorene or Phenalene
1547	170	Trimethylnaphthalene or Methylethylnaphthalene Isomer
1564	168	Propenylnaphthalene or Methylbiphenyl Isomer
1584	182	Xanthene or Methyldibenzofuran Isomer
1626	182	Biphenyl-4-carboxaldehyde, anthene or Methyldibenzofuran Isomer
1905	178	Phenanthrene
1918	178	Anthracene

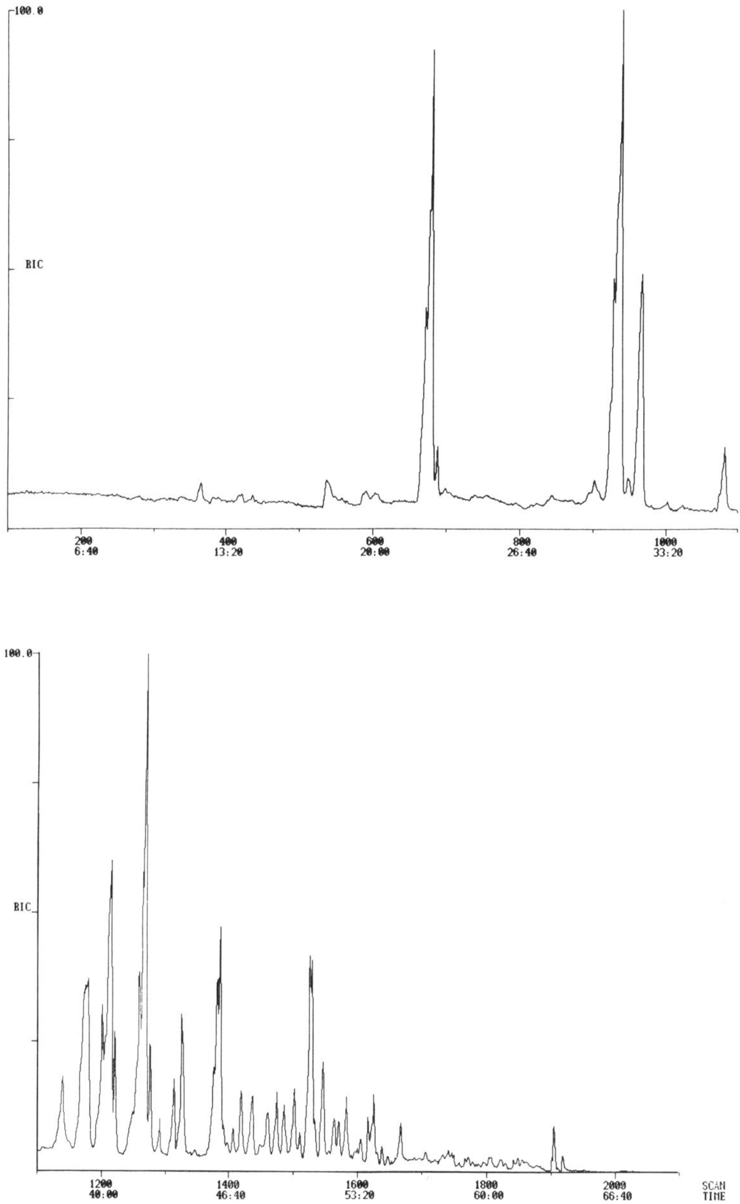

FIGURE 4. Reconstructed chromatogram of the extract of Tenax from raw gas stream. Chromatographic conditions: A 30-m Dexsil 300 glass capillary column, helium carrier gas at 30 cm/sec, temperature program rate of 2°C/min from 60 to 270°C, total ion current response of Finnigan Model 4023 GC/MS.

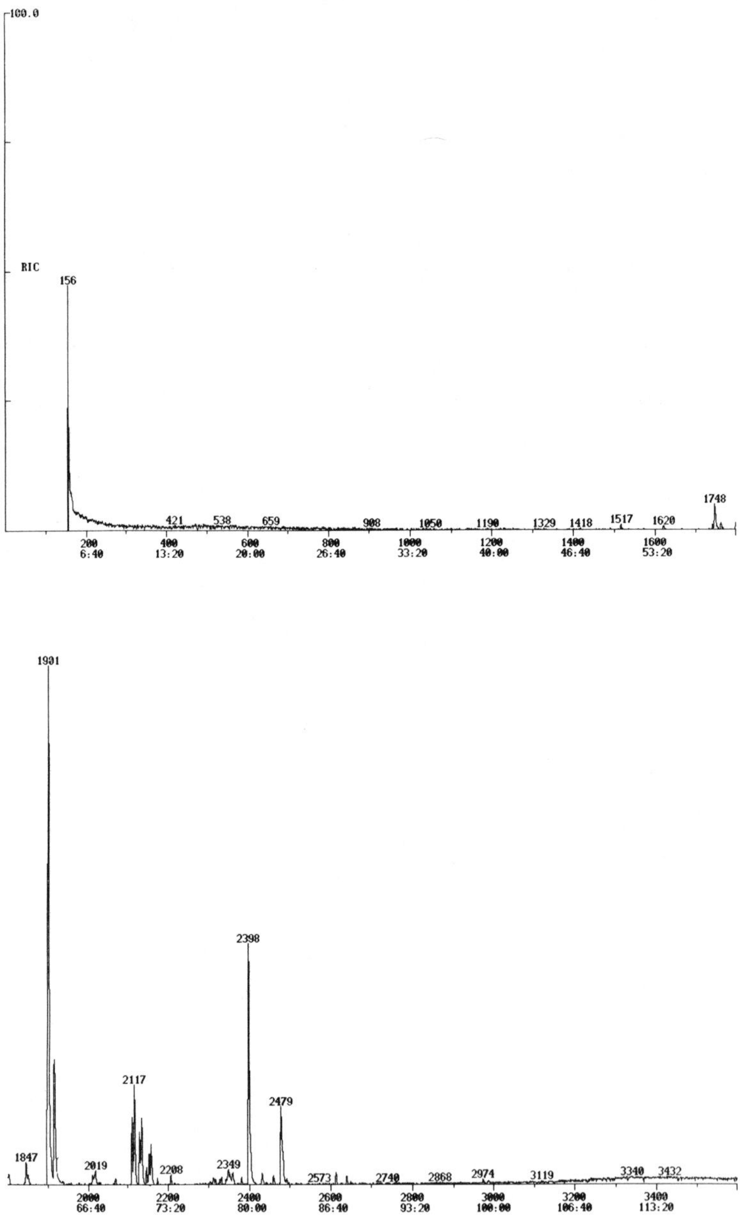

FIGURE 5. Reconstructed chromatogram of the extract of filter sample from cleaned gas stream. Chromatographic conditions: A 30-m Dexsil 300 glass capillary column, helium carrier gas at 30 cm/sec, temperature program rate of 2°C/min from 60 to 270°C, total ion current response of Finnigan Model 4023 GC/MS.

TABLE IV. Tentative Identifications of Components in Extract of Filter From Cleaned Gas Stream

Scan	MW	Tentative Identification
1901	178	Phenanthrene
1916	178	Anthracene
2111	192	Methylphenanthrene Isomer
2117	192	Methylphenanthrene Isomer
2349	206	Dimethylphenanthrene Isomer
2398	202	Fluoranthene
2479	202	Pyrene
2640	216	Benzofluorene or Methylpyrene Isomer

As may be seen by comparing Figures 3 and 4, aerosol formation due to rapid cooling and dilution of raw coal gas preferentially transfers PAH having molecular weights greater than 182 amu from the vapor phase to the particle phase. The gas cleanup system alters the material distribution in aerosols produced from the cleaned gas by preferentially removing high molecular weight PAH which are found in the disengagement chamber tar (Tables IV, V and VI). Note, however, that the particle phase of the cleaned gas aerosol contains compounds having molecular weights as high as 216 amu. This suggests that the cleanup is not totally effective in removing heavy PAH.

The gas cleanup system operates by transferring material from the vapor phase to the particulate phase (solid and liquid) and separating the two phases. In sampling the cleaned gas the gas was subjected to an isothermal expansion followed by dilution. This process should act to volatilize material, not condense it. The fact that we observed an aerosol containing particles composed of high molecular weight compounds suggests that these particles were generated by the cleanup system and escaped capture.

Examination of the disengagement chamber tar for minor constituents required prefractionation of the sample. Although many of the compound identifications are tentative, it is clear that many of these trace constituents are polynuclear aromatic hydrocarbons. In particular, the presence of aza-arenes should be noted. These compounds have become the subject of increasing chemical analysis and biological testing (6). The degree to which these materials transfer to the particle phase of coal gas aerosols is unknown, as is the potential environmental impact of these materials.

The data obtained provide a basis for development of a strategy for making further improvements in the gas cleanup system to reduce the presence of PAH. Although the biological properties of many of the compounds isolated have not been established, several are known or are considered to be carcinogens. Thus the data presented emphasizes the need for continued care in minimizing the release of emissions or effluents that may be a source of exposure to workers or the public.

ACKNOWLEDGMENTS

The authors acknowledge the excellent cooperation and assistance received from the staff of the Morgantown Energy Technology Center, especially W. Wallace, J. Kovach, J. Natestein, C. Pater and R. Rahfuse. The assistance of C. Hobbs, S. Weissman, R. Peele, R. Tamura and D. Horinek of the staff of the Inhalation Toxicology Research Institute is also acknowledged.

This research was performed under U.S. Department of Energy Contract Number EY-76-C-04-1013.

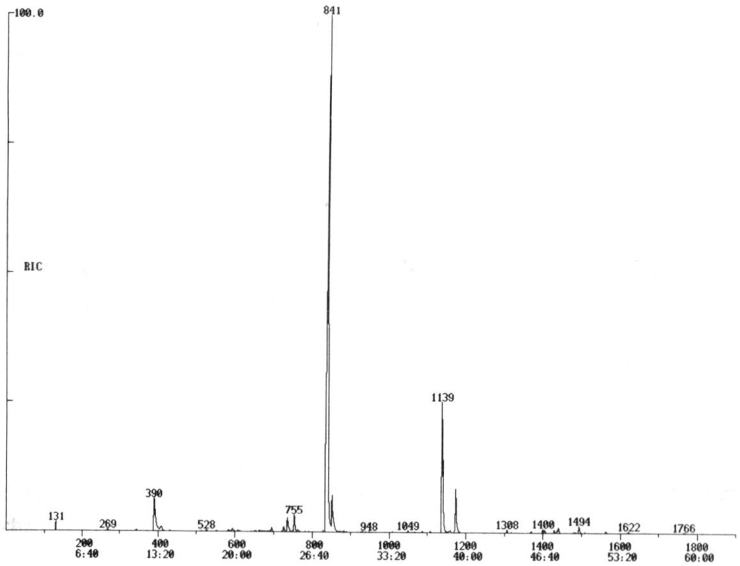

FIGURE 6. Reconstructed chromatogram of the extract of Tenax sample from cleaned gas stream. Chromatographic conditions: A 30-m Dexsil 300 glass capillary column, helium carrier gas at 30 cm/sec, temperature program rate of 2°C/min from 60 to 270°C, total ion current response of Finnigan Model 4023 GC/MS.

TABLE V. Tentative Identifications of Components in Extract of Tenax From Cleaned Gas Stream

Scan	MW	Tentative Identification
343	118	2,3-Dihydroindene
390	116	Indene
528	134	4-Carbon Alkylbenzene Isomer
594	132	Methylbenzofuran Isomer
696	132	Methyl-2,3-Dihydroindene Isomer
737	130	Methyl Indene Isomer
755	130	Methyl Indene Isomer
841	128	Naphthalene
1139	142	Methylnaphthalene Isomer
1175	142	Methylnaphthalene Isomer
1308	154	Biphenyl or 1,2-Dihydroacenaphthylene
1369	156	Ethylnaphthalene Isomer
1400	156	Dimethylnaphthalene Isomer
1404	156	Dimethylnaphthalene Isomer
1429	156	Dimethylnaphthalene Isomer
1438	156	Ethylnaphthalene Isomer
1441	156	Dimethylnaphthalene Isomer
1494	152	Acenaphthylene or Biphenylene
1562	154	Acenaphthene or 1,4-Dihydro-1,4-Ethenonaphthalene
1623	168	Dibenzofuran
1766	166	Fluorene

TABLE VI. Major Aromatic Compounds Found in Disengagement Chamber Tar

Naphthalene	Pyrene
Carbazole	Triphenylene
Acenaphthene	Benzo[a]fluorene
Fluorene	Chrysene
Phenanthrene	Perylene
Anthracene	Benzo[a]pyrene
Fluoranthene	

TABLE VII. Tentative Identifications of Major Components Found in Ether-Soluble Bases of Fraction 5

Scan	MW	Tentative Identification
103	128	Naphthalene
375	157	Dimethylquinoline Isomer
406	157	Dimethylquinoline Isomer
732	179	Benzoquinoline
746	179	Acridine
800	179	Phenanthridine
843	193	Methylbenzoquinoline Isomer
862	193	Methylbenzoquinoline Isomer
794	193	Methylphenanthridine Isomer
1276	229	Dibenzoquinoline

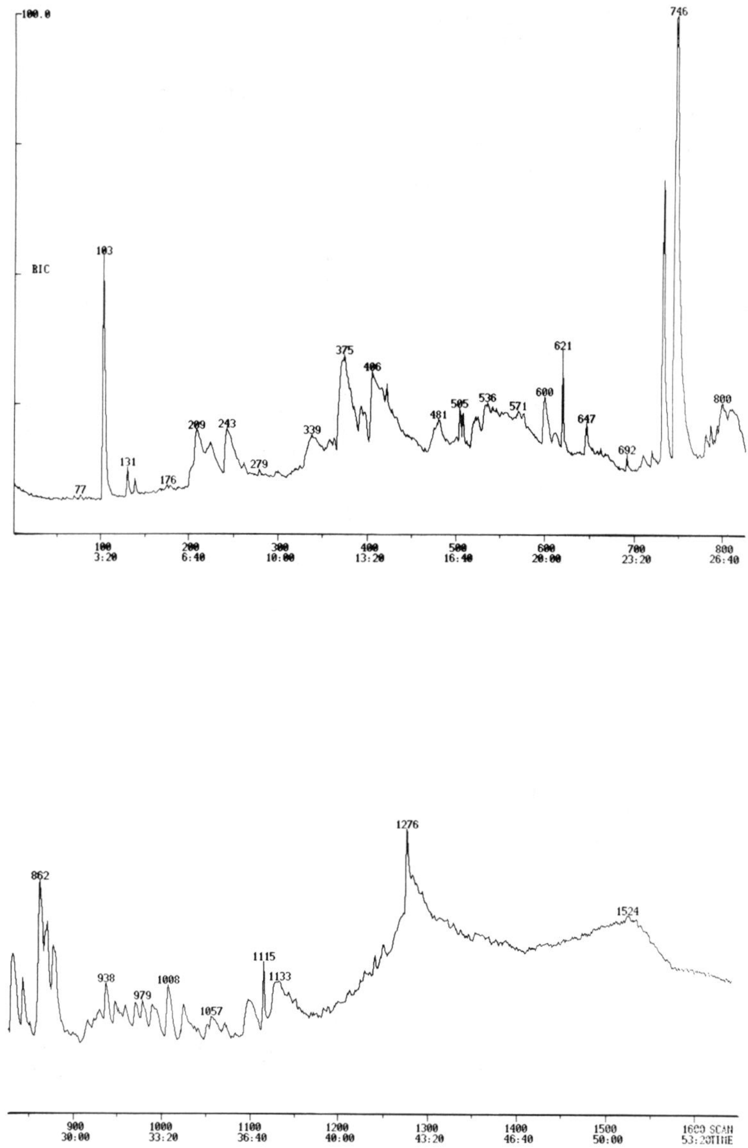

FIGURE 7. Reconstructed chromatogram of the ether-soluble bases of disengagement chamber tar fraction 5. Chromatographic conditions: A 30-m Dexsil 300 glass capillary column, helium carrier gas at 30 cm/sec, temperature program rate of 4 °C/min from 60 to 270 °C, total ion current response of Finnigan Model 4023 GC/MS.

REFERENCES

1. Pellizzari, E. D. (1978): Identification of components of energy-related wastes and effluents. EPA-600/7-78-004.
2. Jones, P. W., Graffeo, A. P., Detrick, R., Clarke, P. A., and Jakobsen, R. J. (1976): Technical manual for analysis of organic materials in process streams. EPA-600/2-76-072.
3. Phillips, R. D. (1977): A preliminary investigation of airborne PNA emissions from the Hygas pilot plant. In: *Proceedings of the Second ORNL Workshop on Exposure to Polynuclear Aromatic Hydrocarbons in Coal Conversion Processes.* CONF-770361, p. 49.
4. Jones, A. R., Guerin, M. R., and Clark, B. R. (1977): Preparative-scale liquid chromatographic fractionation of crude oils derived from coal and shale. *Anal. Chem.* 49:1766-1771.
5. Carpenter, R. L., Weissman, S., Heisler, T., Neton, G. J., Hanson, R. L., Peele, E. R.; Mazza, M. H., Kovach, J. J., Green, D. A., and Grimm, U. (1978): Characterization of Aerosols Produced by an Experimental Fluidized Bed Coal Combustor Operated With Sub-bituminous Coal. LF-57, available from NTIS.
6. Dong, M., Schmeltz, I., LaVoie, E., and Hoffmann, D. (1978): *Carcinogenesis, Vol. 3: Polynuclear Aromatic Hydrocarbons,* by P. W. Jones and R. I. Freudenthal, Eds., Raven Press, New York.

Polynuclear Aromatic Hydrocarbons, edited by P. W. Jones and P. Leber. Ann Arbor Science Publishers, Inc. Ann Arbor, MI ©1979.

METABOLISM AND MACROMOLECULAR BINDING OF BENZO[a]PYRENE AND ITS NONCARCINOGENIC ISOMER BENZO[e]PYRENE IN CELL CULTURE

James K. Selkirk and Michael C. MacLeod

Biology Division
Oak Ridge National Laboratory
Oak Ridge, Tennessee 37830

INTRODUCTION

The carcinogenicity of the polycyclic aromatic hydrocarbon benzo[a]-pyrene (B[a]P) (Figure 1) is thought to result from the metabolic formation of a highly reactive intermediate, 7,8-dihydrodiol-9,10-oxy-B[a]P which subsequently interacts with cellular macromolecules to produce neoplasia (10, 12,14,21,23,26,30,31). Evidence has been presented implicating similar "bay-region" diol-epoxides as the ultimate carcinogenic forms of benz[a]-anthracene (22,27), 7-methyl-benz[a]anthracene (3), dibenz[a,h]anthracene (29), and chrysene (28). Benzo[e]pyrene (B[e]P), on the other hand, is relatively inert when tested for carcinogenicity on mouse skin (16,25) and rat trachea (24) and when tested for mutagenicity in a mammalian system (6).

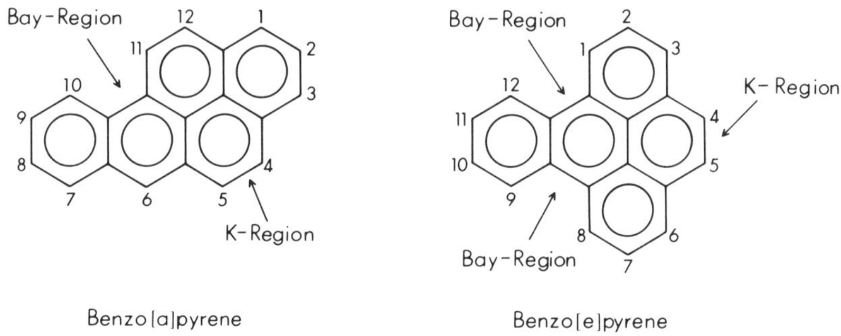

FIGURE 1. Structures of B[a]P and B[e]P, showing the locations of the K-regions and bay regions.

B[e]P contains two (equivalent) bay regions (Figure 1) and quantum mechanical model calculations suggest that 9,10-dihydrodiol-11, 12-oxy-B[e]P can easily form a reactive bay region-adjacent carbonium ion (11). In order to explain the noncarcinogenicity of B[e]P in the context of the bay-region theory, we have studied the metabolism of B[e]P by rat liver microsomes and by cultured hamster embryo fibroblasts. In both systems, the major metabolic pathways involve the K-region of B[e]P, with little or no attack on the isolated benzo ring. These results suggest that the lack of carcinogenicity of B[e]P may reflect its preferred mode of interaction with the cellular metabolic machinery.

MATERIALS AND METHODS

Chemicals

Tritium-labeled B[a] P (Amersham, Arlington Heights, IL, ⩾25 Ci/mmol) was diluted with unlabeled B[a]P (Aldrich, Milwaukee, WI, gold label) to a specific activity of 3.15 Ci/mmol, and stored at -20°C dissolved in dimethylsulfoxide (DMSO), at 0.9 m*M*. A stock solution of tritium-labeled B[e P (Midwest Research Institute, NCI Contract No. NO1-CP-33387, 5.02 Ci/mmol) was prepared and stored similarly, but at a final specific activity of 2.51 Ci/mmol. Ultrapure guanidine HC1 was obtained from Schwarz/Mann (Orangeburg, NY); density-gradient grade Cs_2SO_4 was from Atomergic Chemetals (Carle Place, NY). Glucurase (a preparation of β-glucuronidase containing very low aryl sulfatase activity) was obtained from Sigma Chemical Co. (St. Louis, MO).

Cells and Labeling

Hamster embryo fibroblasts (HEF) were grown in 100-mm dishes as described (20). All experiments were performed on confluent tertiary cultures which had been plated at a 1:2 dilution three days prior to the beginning of the labeling period. Medium was replaced with 10 ml of fresh medium containing labeled PAH at a final concentration of 4 μ*M* and incubation continued for 24 hr. The final concentration of DMSO was 0.44%, a level which gives no measurable cytotoxicity at 24 hr.

Analysis of Extracellular Metabolites

After 24-hr incubation of PAH with cells, the extracellular medium was aspirated, the cells were gently washed with 5 ml phosphate-buffered saline, and the wash and medium combined. Organic solvent-soluble metabolites were prepared by twice extracting the medium with 2.5 vol ethyl acetate. The distribution of ethyl acetate-soluble metabolites within the medium was analyzed by high-pressure liquid chromatography (HPLC) with a Spectra Physics 3500-B chromatograph fitted with a 1-m Zorbax ODS column (Dupont Inst., Wilmington, DE) and eluted with a linear 30-70% methanol and water gradient as previously described (18).

The aqueous phase was reextracted twice with 2.5 vol ethyl acetate. Metabolites which had been conjugated to glucuronic acid were released by a 16-hr treatment at 37°C with β-glucuronidase as described previously (4). The oxygenated metabolites released from the glucuronide conjugates were extracted twice with 2.5 vol ethyl acetate and analyzed by HPLC.

Subcellular Fractionation

The washed cell monolayers were harvested after a 15-min incubation at 37°C with 0.05% trypsin, 0.02% ethylenediamine tetraacetic acid (EDTA) in phosphate buffered saline, then washed twice at 0-4°C with buffer containing 10 m*M* NaCl, 10 m*M* tris (pH 7.0), 1.5 m*M* $MgCl_2$ and resuspended in the same buffer. The solution was adjusted to 0.5% Triton X-100 and 0.5% Na deoxycholate and lysis was completed by 10-15 strokes of a motor-driven Potter-Elvehjem homogenizer. Nuclei were deposited by centrifugation at 1,000 x g. Cytoplasmic organic solvent-soluble metabolites were determined by ethyl acetate extraction of the supernatant followed by HPLC. The nuclei were freed of residual cytoplasm by repeating the homogenization step.

Isopycnic Separation

The nuclear pellet was lysed and chromatin sheared by resuspension in 6 *M* guanidine·HCl, 10 m*M* EDTA (pH 7.0) and sonication. The lysate was extracted three times at room temperature with 3 vol ethyl acetate (water-saturated), and residual ethyl acetate was removed with a stream of N_2. The volume was adjusted to 2.2 ml with the same solution and layered over a 2.8-ml cushion of 2.2 *M* Cs_2SO_4, 10 m*M* EDTA (pH 7.0), 1.26 *M* DMSO in a 0.5 in. x 2 in. polyallomer tube. Centrifugation was carried out in a swinging-bucket rotor (SW 50.1) at 35,000 rpm (147,000 x g max) at 20°C for 40 ± 2 hr. The resulting gradients were fractionated by piercing the tubes near the bottom and collecting 5 or 6 drop fractions. Nucleic acids were monitored by A_{260}, proteins by reaction with fluorescamine. Density was determined by weighing 10-μl aliquots in glass capillaries. Bound hydrocarbon was determined by liquid scintillation counting using a Searle Mark III equipped with dpm facility. Using the known specific acitivity, dpm were converted to pmol hydrocarbon.

RESULTS AND DISCUSSION

Microsomal Metabolism of B[e]P

In experiments carried out in collaboration with the laboratories of D. Jerina (NIH, Bethesda, MD) and W. Levin (Hoffmann-La Roche, Nutley, NJ) microsomes were prepared from the livers of rats which had been pretreated with methylcholanthrene (MC) or phenobarbital (PB), or were untreated. The metabolites formed from ^{3}H-B[e]P after incubation with the three sets of microsomes, as assayed by HPLC, are shown in Figure 2. The mobilities of a set of synthetic standards are also indicated in the top panel. For all three sets of microsomes, the major metabolite cochromatographed with the synthetic K-region dihydrodiol, 4,5-dihydrodiol-B[e]P. The identification of this metabolite as the K-region dihydrodiol was confirmed by UV absorption

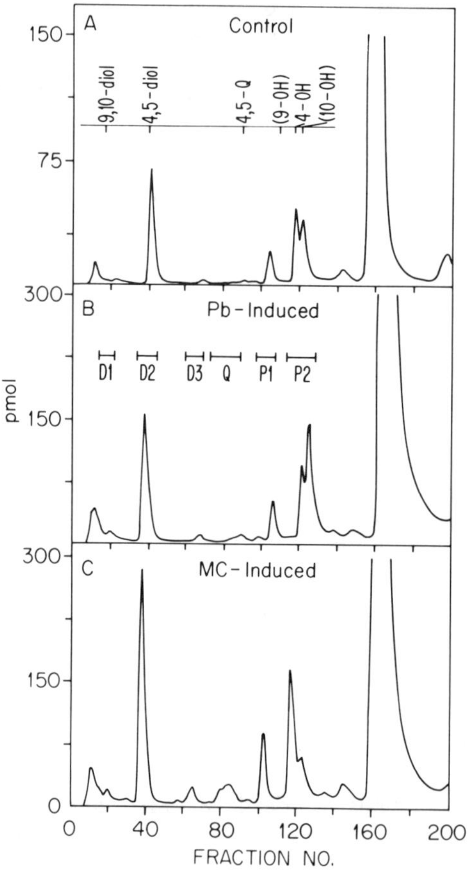

FIGURE 2. Pattern of B[e]P metabolites formed in incubations with liver microsomes from rats which were (A) untreated; (B) pretreated for three days with phenobarbital (75 mg/kg/day); or (C) pretreated for three days with methylcholanthrene (25 mg/kg/day). In each case, 100 nmol ^{3}H-B[e]P were incubated 10 min at 37° with from 0.1 to 1.0 mg microsomal protein in a total volume of 2.0 ml of a solution containing 0.1 *M* sodium phosphate (pH 7.4), 3 m*M* $MgCl_2$, 0.5 m*M* NADPH and 5% acetone. Reactions were terminated by extraction with 2.0 ml acetone and 4.0 ml ethyl acetate. Organic soluble metabolites were analyzed by HPLC. The relative mobilities of a series of synthetic standards are also shown in panel (A) and the regions of the chromatogram which contain the various metabolites are designated in panel (B) as D1, D2, D3, Q, P1 and P2.

spectra (Figure 3) and by fluorescence spectra (data not shown). Of particular interest was the absence of an identifiable peak which chromatographed with the same mobility as 9,10-dihydrodiol-B[e]P. This was true even in the incubation with MC-induced microsomes, where non-K-region metabolism should be favored (13,15).

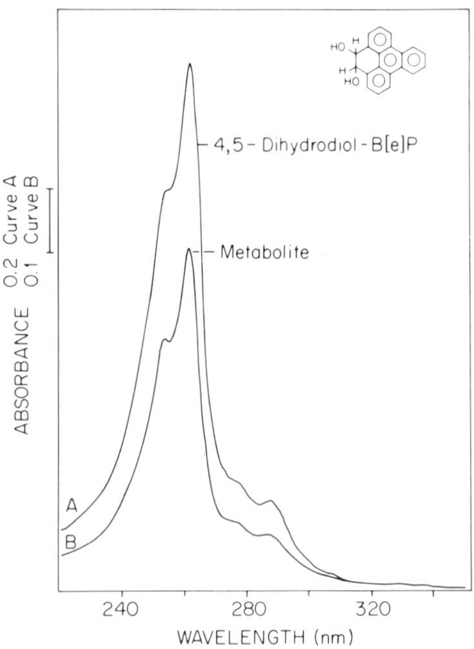

FIGURE 3. Comparison of the ultraviolet absorbance spectra of (A) synthetic 4,5-dihydriol-B[e]P and (B) the metabolite isolated from the chromatograph in region D2. Spectra were recorded with a Cary 118-C spectrophotometer.

This finding suggested the possibility of a metabolic explanation for the relative inactivity of B[e]P as a carcinogen (16,24,25). If vicinal diol-epoxides are indeed the ultimate carcinogens, then the major metabolite of B[e]P, 4,5-dihydrodiol-B[e]P is a carcinogenic "dead-end", since formation of a vicinal diol-epoxide in the K-region is impossible. The lack of carcinogenicity of B[e]P would therefore be due to the lack of formation of 9,10-dihydrodiol-B[e]P, the obligate precursor to the bay region diol epoxide 9,10-dihydrodiol-11,12-oxy-B[e]P. However, since previous work from this laboratory (20,17) has shown that *in vitro* metabolism is often very different from *in vivo* metabolism, and since the liver is not a target for PAH carcinogenesis, it was necessary to further test this suggestion in a situation which more closely resembled that found *in vivo*.

Metabolism of B[e]P by Hamster Embryo Fibroblasts

For these studies we chose tertiary cultures of HEF. These cells contain the full array of metabolic enzymes in an intact, nondisrupted orientation and can be transformed in culture by treatment with carcinogenic PAH (9). We incubated confluent cultures for 24 hr in medium containing ^{3}H-B[e]P at a concentration of 4 μM; this concentration was chosen to minimize cytotoxicity. The time period was chosen because we had previously found that the binding of ^{3}H-B[a]P to nuclear DNA reached a plateau level between 18 and 24 hr of incubation (unpublished data). Ethyl acetate-soluble metabolites found in the extracellular medium are shown in Figure 4A and those found in the cytoplasmic fraction are shown in Figure 4B. Chromatograms of

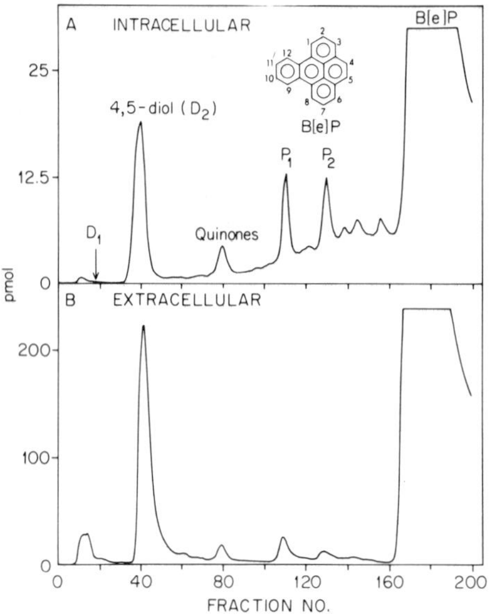

FIGURE 4. Pattern of B[e]P metabolites formed in a 24-hr incubation with HEF and found (A) in the intracellular medium or (B) in the extracellular fraction.

the small amount of ethyl acetate-soluble radioactivity found in the nuclear fraction (0.1-0.2% of the total input) demonstrated only unmetabolized B[e]P (data not shown). As seen previously in the microsomal incubations, the major metabolite of B[e]P formed by HEF was the K-region dihydrodiol. This metabolite was found in both the extracellular medium (Figure 4A) and in the cytoplasm (Figure 4B). Neither of these fractions contained demonstrable amounts of the non-K-region, 9,10-dihydrodiol-B[e]P, confirming the microsomal data. However, since intact cell systems generally produce appreciable amounts of water-soluble metabolites, primarily glucuronide conjugates (1), it was possible that the HEF cultures formed 9,10-dihydrodiol-B[e]P but rapidly conjugated it to the corresponding glucuronide. To test this, the extracellular water-soluble metabolites were hydrolysed with β-glucuronidase and then analyzed by HPLC. As shown in Figure 5, we detected by this method 4,5-dihydrodiol-B[e]P, and a phenolic derivative which cochromatographs with 4-hydroxy-B[e]P but no 9,10-dihydrodiol-B[e]P. Table I summarizes the overall recovery of metabolites in the various fractions we have assayed. It is clear that at least in terms of the dihydrodiols, K-region metabolism is favored to the virtual exclusion of metabolism at the 9,10 position. The small amount of radioactivity (0.64 to 1.07% of total metabolites) found in the 9,10-dihydrodiol-B[e]P region of the chromatogram may be due in part to triols, tetrols and other unidentified metabolites which contaminate this region, and thus represents an upper limit to 9,10-dihydrodiol-B[e]P formation in these cells.

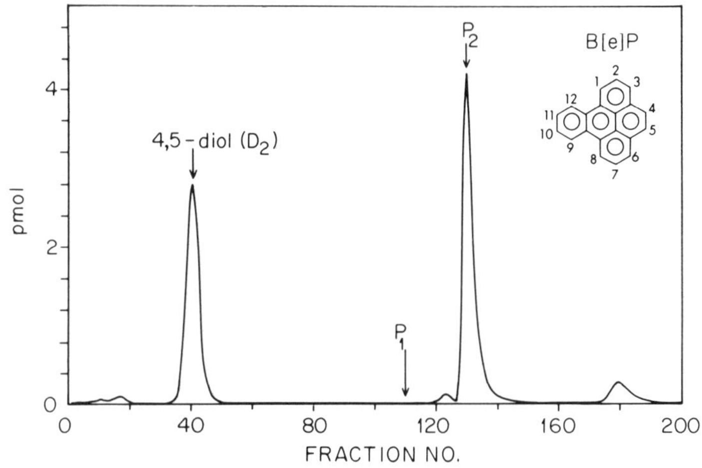

FIGURE 5. Identification by HPLC of B[e]P metabolites conjugated to glucuronides after a 24-hr incubation with HEF.

In contrast to this, metabolism of B[a]P by HEF produces large amounts of the non-K-region dihydrodiols, 7,8-dihydrodiol- and 9,10-dihydrodiol-B[a]P (Figure 6), but little or no K-region dihydrodiol which in this system chromatographs at approximately fraction 40. In addition, no dihydrodiols of B[a]P are conjugated to glucuronides by HEF (Figure 7). In terms of the phenols, it is likely that the P1 and P2 regions of the chromatogram (Figure 4) contain a number of different phenolic metaoblites of B[e]P in analogy to the clustering of the 12 monohydroxy-derivatives of B[a]P which are incompletely resolved in a single chromatographic run (7,19). It thus remains possible that HEF are capable of forming 9,10-oxy-B[e]P but that this

TABLE I. Distribution of Metabolites of B[e]P Formed by HEF

	% of Total Metabolites			
	Extracellular Organic Soluble	Extracellular Glucuronide	Intracellular Organic Soluble	Total
Expt. 1- 4,5-dihydrodiol	22.5	2.64	0.14	25.2
9,10-dihydrodiol	1.07	nd[a]	nd	1.07
monohydroxy compounds[b]	7.09	15.7	1.04	23.8
Expt. 2- 4,5-dihydrodiol	23.9	7.35	1.10	32.4
9,10-dihydrodiol	0.63	nd	0.01	0.64
monohydroxy compounds	3.91	9.63	1.06	14.6

[a]nd: not detectable

[b]Regions of the appropriate chromatograms corresponding to regions P1 and P2 in Figure 2 were summed.

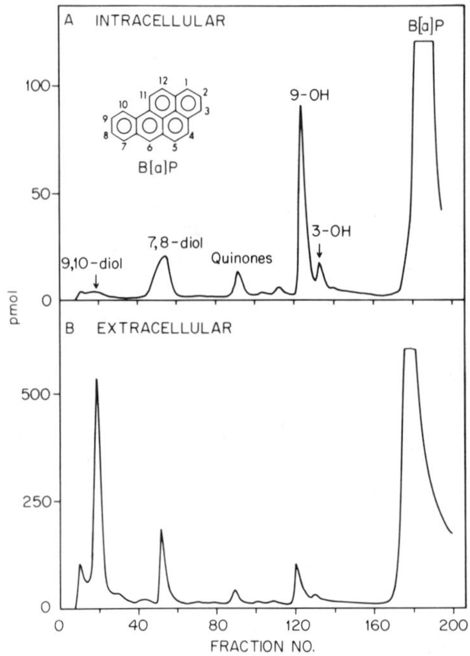

FIGURE 6. Pattern of B[a]P metabolites formed in a 24-hr incubation with HEF and found (A) in the intracellular medium, or (B) in the extracellular fraction.

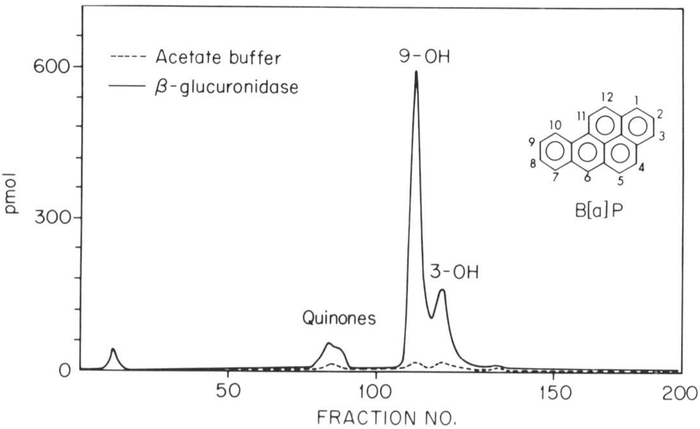

FIGURE 7. Identification by HPLC of B[a]P metabolites conjugated to glucuronides after a 24-hr incubation with HEF.

compound is very unstable, isomerizing to 9-hydroxy- and/or 10-hydroxy-B[e]P before it can be converted to the dihydrodiol by epoxide hydrase.

Binding to Nuclear Macromolecules

However, 9,10-oxy-B[e]P would be expected to have reasonably good alkylating ability as would the K-region oxide, 4,5-oxy-B[e]P. Considering the relative inactivity of B[e]P as a carcinogen it was of interest to determine if this was correlated with a low macromolecular binding. We chose to concentrate on nuclear macromolecules which we prepared by a novel method involving denaturation in 6 *M* guanidine followed by isopycnic separation in Cs_2SO_4. To demonstrate the separation which could be obtained with this method, nuclei were labeled with specific precursors of RNA, DNA and protein and then subjected to isopycnic centrifugation as described in Materials and Methods. In Figure 8A, the distribution of radioactivity after labeling RNA with ^{3}H-uridine and DNA with ^{14}C-thymidine is shown. The labeled RNA and DNA separated into two bands at densities of 1.65 g/cm^2 and 1.44 g/cm^2, respectively; these bands coincided with peaks of A_{260} (data not shown). Less than 2% of the ^{3}H-radioactivity was found in the DNA band and less than 0.1% of the ^{14}C-radioactivity was found in the RNA band. Figure 8B shows the distribution of radioactivity after labeling protein with ^{3}H-leucine and DNA with ^{14}C-thymidine. The ^{3}H-leucine was found in a broad band with densities less than about 1.30 g/cm^2. Determination of protein by reaction with fluorescamine yielded a profile which paralleled the radioactivity (data not shown) and similar results were obtained when calf

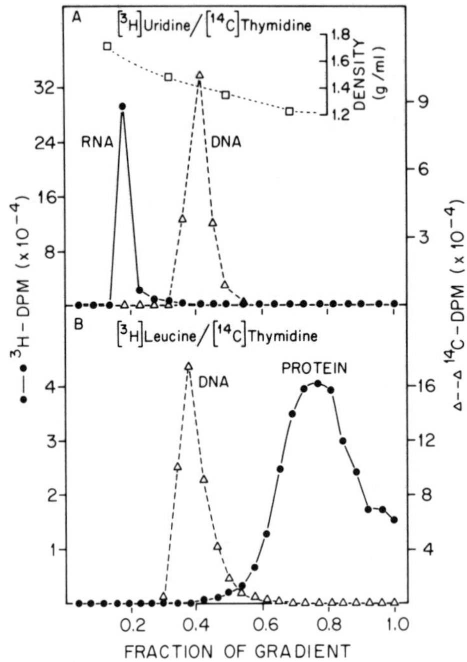

FIGURE 8. Separation of nuclear macromolecules by isopycnic centrifugation in Cs_2SO_4, guanidine·HCl gradients.

thymus histones (Sigma, St. Louis, MO) were applied to an isopycnic gradient. Less than 1% of the ^{3}H-radioactivity was found in the DNA band and less than 2% of the ^{14}C-radioactivity was found in the protein band.

In Figure 9, nuclei from cells labeled with ^{3}H-B[a]P (Figure 9A), or with ^{3}H-B[e]P (Figure 9B) were analyzed on isopycnic gradients. To permit comparisons, a fivefold scale difference has been introduced between the left and right sides of each panel. With B[a]P (Figure 9A) large peaks of radioactivity were found to band coincidentally with the RNA, DNA and protein bands. Table II gives the calculated specific activities attained in three experiments with B[e]P. For comparison, the average specific activities attained in a series of experiments with B[a]P were: RNA, 55.9 pmol/mg; DNA, 40.4 pmol/mg; protein, 169 pmol/mg. Thus, the binding of B[e]P to nuclear macromolecules is 80- to 150-fold lower than the binding of B[a]P.

SUMMARY

In summary, a plausible explanation for the difference in carcinogenicity between B[a]P and B[e]P is to be found in their respective metabolic fates in either microsomal incubations or HEF. With B[a]P the major site of metabolic attack is the isolated benzo ring, resulting in the formation of the

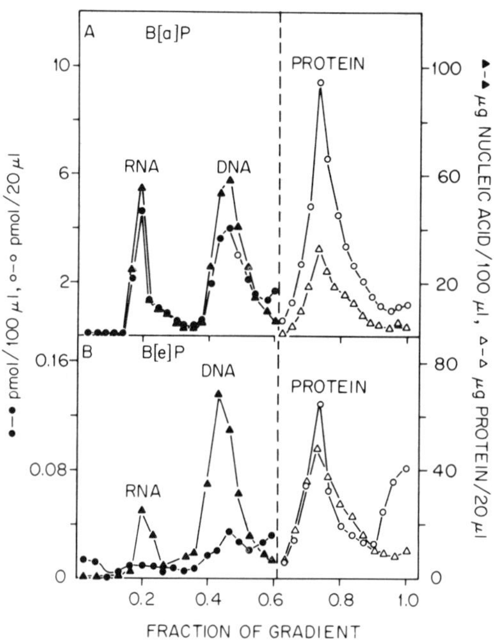

FIGURE 9. Isopycnic separation of nuclear macromolecules labeled in culture with (A) ^{3}H-B[a]P or (B) ^{3}H-B[e]P. Note the fivefold differences in the size of the aliquots of each fraction which were analyzed from the top (right) and bottom (left) of the gradients.

TABLE II. Binding of ^{3}H-B[e]P to Nuclear Macromolecules

	Specific Activity (pmol/mg)		
	RNA	DNA	Protein
Experiment 1	0.16	0.22	1.94
Experiment 2	0.20	0.19	1.63
Experiment 3	0.49	0.30	2.32
Average	0.28	0.24	1.96

non-K-region dihydrodiols, 7,8-dihydrodiol-B[a]P and 9,10-dihydrodiol-B[a]P. In particular, the 7,8-dihydrodiol, which is the precursor for the presumed ultimate carcinogen 7,8-dihydrodiol-9,10-oxy-B[a]P, is not conjugated to the corresponding glucuronide but is selectively retained by the cells (MacLeod, Cohen and Selkirk, submitted for publication). This favors formation of the bay-region diol epoxide and correlates with a relatively high level of binding to nuclear macromolecules. With B[e]P, the major attack is the K-region, producing 4,5-dihydrodiol-B[e]P which is found intracellularly, extracellularly and as the glucuronide conjugate. It has not been possible to demonstrate formation of 9,10-dihydrodiol-B[e]P in any of the fractions examined.

This absence of the precursor for bay-region diol-epoxide formation correlates with a very low level of binding of B[e]P to nuclear macromolecules.

Our data are in accord with predictions of the bay-region theory of chemical carcinogenesis and reduce the importance of K-region derivatives in the induction of neoplasia. The targets for reaction with the diol epoxide of B[a]P which are important in the production of cancer remain unknown as do the various steps in this pathway. We expect that continued study of a variety of noncarcinogenic PAH will provide us with cases in which the "block" occurs at later steps in this pathway. This will allow a dissection of this multistep process in a manner analogous to that used so successfully in elucidating cellular metabolic pathways by biochemical genetics (2,5,8).

ACKNOWLEDGMENTS

This research was supported by the National Cancer Institute (YO1-CP-50200) and the Division of Biomedical and Environmental Research, U.S. Department of Energy under contract with the Union Carbide Corporation, and by EPA-IAG-D5-E681. The microsomal experiments were done in collaboration with Dr. D. Jerina and Dr. W. Levin (manuscript in preparation).

REFERENCES

1. Baird, W. M., Chern, C. J., and Diamond, L. (1977): Formation of benzo[a]pyrene-glucuronic acid conjugates in hamster embryo cell cultures. *Cancer Res.* 37:3190-3197.
2. Beadle, G. W. (1959): Genes and chemical reactions in *Neurospora crassa. Science* 129:1715-1719.
3. Chouroulinkov, I., Gentil, A., Tierney, B., Grover, P., and Sims, P. (1977): The metabolic activation of 7-methylbenz[a]anthracene in mouse skin: high tumor initiating activity of the 3,4-dihydrodiol. *Cancer Lett.* 3:247-253.
4. Cohen, G. M., Haws, S. M., Moore, B. P., and Bridges, J. W. (1976): Benzo[a]pyren-3-yl hydrogen sulphate, a major ethyl acetate-extractable metabolite of benzo[a]pyrene in human, hamster and rat lung cultures. *Biochem. Pharmacol.* 25:2561-2570.
5. Ephrussi, B. (1942): Chemistry of "eye color hormones" of *Drosophila. Quart. Rev. Biol.* 17:327-338.
6. Harris, C. C., Hsu, I. C., Stoner, G. D., Trump, B. F., and Selkirk, J. K. (1978): Human pulmonary alveolar macrophages metabolize benzo[a]-pyrene to proximate and ultimate mutagens. *Nature* 252:633-634.

7. Holder, G., Yagi, H., Dansette, P., Jerina, D. M., Levin, W., Lu, A. Y. H., and Conney, A. H. (1974): Effects of inducers and epoxide hydrase on the metabolism of benzo[a]pyrene by liver microsomes and a reconstituted system: Analysis by high-pressure liquid chromatography. *Proc. Nat. Acad. Sci., U.S.* 71:4356-4369.
8. Horowitz, N. H., Bonner, D., Mitchell, H. K., Tatum, E. L., and Beadle, G. W. (1945): Genic control of biochemical reactions in *Neurospora. Am. Naturalist* 79:304-317.
9. Huberman, E., Kuroki, T., Marquardt, H., Selkirk, J. K., Heidelberger, C., Grover, P. L., and Sims, P. (1971): Transformation of hamster cells by epoxides and other derivatives of polycyclic hydrocarbons. *Cancer Res.* 31:2161-2167.
10. Jerina, D. M., Lehr, R., Schaefer-Ridder, M., Yagi, H., Karle, J. M., Thakker, D. R., Wood, A. W., Lu, A. Y. H., Ryan, D., West, S., Levin, W., and Conney, A. H. (1977): Bay region epoxides of dihydrodiols. A concept which explains the mutagenic and carcinogenic activity of benzo[a]pyrene and benzo[a]anthracene. In: *Origins of Human Cancer,* pp. 639-659, Cold Spring Harbor Laboratory, Cold Spring Harbor, NY.
11. Jerina, D. M., Lehr, R. E., Yagi, H., Hernandez, O., and Dansette, P. M. (1976): Mutagenicity of benzo[a]pyrene derivatives and the description of a quantum mechanical model which predicts the ease of carbonium ion formation from diol epoxides. In: *In Vitro Metabolic Activation in Mutagenesis Testing*, F. J. de Serres, J. R. Fouts, J. R. Bend, and R. M. Philpot, Eds., pp. 159-177, Elsevier/North Holland Press, Amsterdam.
12. King, H. W. S., Osborne, M. R., Beland, F. A., Harvey, R. G., and Brookes, P. (1976): (±)-7α, 8β-dihydroxy-9β, 10β-epoxy-7,8,9,10-tetrahydrobenzo[a]pyrene is an intermediate in the metabolism and binding to DNA of benzo[a]pyrene. *Proc. Nat. Acad. Sci., U.S.* 73: 2679-2681.
13. Kinoshita, N., Shears, B., and Gelboin, H. V. (1973): K-region and non-K-region metabolism of benzo[a]pyrene by rat liver microsomes. *Cancer Res.* 33:1936-1944.
14. Levin, E., Wood, A. W., Yagi, H., Dansette, P. M., Jerina, D. A., and Conney, A. H. (1976): Carcinogenicity of benzo[a]pyrene 4,5- 7,8-, and 9,10-oxides on mouse skin. *Proc. Nat. Acad. Sci., U.S.* 73:243-247.
15. Rasmussen, R. E., and Wang, I. Y. (1974): Dependence of specific metabolism of benzo[a]pyrene on the inducer of hydroxylase activity. *Cancer Res.* 34:2290-2295.
16. Scribner, J. (1973): Tumor initiation by apparently noncarcinogenic polycyclic aromatic hydrocarbons. *J. Nat. Cancer Inst.* 50:1717-1719.
17. Selkirk, J. K. (1977): Divergence of metabolic activation systems for short-term mutagenesis assay. *Nature* 270:604-607.
18. Selkirk, J. K., Croy, R. G., and Gelboin, H. V. (1974): Benzo[a]pyrene metabolites: Efficient and rapid separation by high-pressure liquid chromatography. *Science* 184:169-175.

19. Selkirk, J. K., Croy, R. G., and Gelboin, H. V. (1976): High-pressure liquid chromatographic separation of 10 benzo[a]pyrene phenols and the identification of 1-phenol and 7-phenol as new metabolites. *Cancer Res.* 36:922-926.
20. Selkirk, J. K., Croy, R. G., Wiebel, F. J., and Gelboin, H. V. (1976): Differences in benzo[a]pyrene metabolism between rodent liver microsomes and embryonic cells. *Cancer Res.* 36:4476-4479.
21. Sims, P., Grover, P. L., Swaisland, A., Pal, K., and Hewer, A. (1974): Metabolic activation of benzo[a]pyrene proceeds by a diol-epoxide. *Nature (Lond.)* 252:326-327.
22. Slaga, T. J., Huberman, E., Selkirk, J. K., Harvey, R. G., and Bracken, W. M. (1978): Carcinogenicity and mutagenicity of benz[a] anthracene diols and diol-epoxide. *Cancer Res.* 38:1699-1704.
23. Slaga, T. J., Viaje, A., Berry, D. L., Bracken, W., Buty, S. G., and Scribner, J. D. (1976): Skin tumor initiating ability of benzo[a]pyrene 4,5-, 7,8- and 7,8-diol-9,10-epoxides and 7,8-diol. *Cancer Lett.* 2: 115-122.
24. Topping, D. C., Pal, B. C., and Nettesheim, P. (1978): The interaction of benzo[a]pyrene and benzo[e]pyrene in respiratory tract carcinogenesis. *Proc., Assoc. Cancer Res.* 19:43.
25. Van Duuren, B. L., Sivak, A., Langseth, L., Goldschmidt, B. M., and Segal, A. (1968): Initiators and promoters in tobacco carcinogenesis. *J. Nat. Cancer Inst. Monograph* 28:173-180.
26. Weinstein, I. B., Jeffrey, A. M., Jennette, K. W., Blobstein, S. H., Harvey, R. G., Harris, C., Autrup, H., Kasai, H., and Nakanishi, K. (1976): Benzo[a]pyrene diol epoxides as intermediates in nucleic acid binding *in vitro* and *in vivo*. *Science* 193:592-595.
27. Wood, A. W., Levin, W., Chang, R. L., Lehr, R. E., Schaeffer-Ridder, M., Karle, J. M., Jerina, D. M., and Conney, A. H. (1977): Tumorigenicity of 5 dihydrodiols of benz[a]anthracene on mouse skin: Exceptional activity of benz[a]anthracene-3,4-dihydrodiol. *Proc. Nat. Acad. Sci., U.S.* 74:3176-3179.
28. Wood, A. W., Levin, W., Ryan, D., Thomas, P. E., Yagi, A., Mah, H. D., Thakker, D. P., Jerina, D. M., and Conney, A. H. (1977): High mutagenicity of metabolically activated chrysene 1,2-dihydrodiol: Evidence for bay region activation of chrysene. *Biochem. Biphys. Res. Comm.* 79:847-854.
29. Wood, A. W., Levin, W., Thomas, P. E., Ryan, D., Karle, J. M., Yagi, H., Jerina, D. M., and Conney, A. H. (1978): Metabolic activation of dibenzo[a,h]anthracene and its dihydrodiols to bacterial mutagens. *Cancer Res.* 38:1967-1973.
30. Wood, A. W., Wislocki, P. G., Chang, R. W., Levin, W., Lu, A. Y. H., Yagi, H., Hernandez, O., Jerina, D. M., and Conney, A. H. (1976): Mutagenicity and cytotoxicity of benzo[a]pyrene benzo-ring epoxides. *Cancer Res.* 36:3358-3366.

31. Yang, S. K., McCourt, D., Roller, P. O., and Gelboin, H. V. (1976): Enzymatic conversion of benzo[a]pyrene leading predominantly to the diol-epoxide r-7,6-8-dihydroxy-t,9,10-oxy-7,8,9,10-tetrahydrobenzo[a]-pyrene through a single enantiomer of r-7, t-8-dihydrodroxy-7,8-dihydrobenzo[a]pyrene. *Proc. Nat. Acad. Sci., U.S.* 73:2494-2598.

Polynuclear Aromatic Hydrocarbons, edited by P. W. Jones and P. Leber. Ann Arbor Science Publishers, Inc. Ann Arbor, MI © 1979.

DIFFERENCES IN METABOLISM PROVIDE A BASIS FOR THE LOW MUTAGENICITY AND CARCINOGENICITY OF BENZO[e]PYRENE COMPARED TO BENZO[a]PYRENE

R. E. Lehr, C. W. Taylor, S. Kumar, W. Levin,* R. Chang,* A. W. Wood,* A. H. Conney,* D. R. Thakker,† H. Yagi,† H. D. Mah† and D. M. Jerina†

Department of Chemistry
University of Oklahoma
Norman, Oklahoma 73019

*Department of Biochemistry and Drug Metabolism
Hoffmann-La Roche, Inc.
Nutley, New Jersey 07110

†Laboratory of Bioorganic Chemistry
National Institute of Arthritis, Metabolism and Digestive Diseases
National Institutes of Health
Bethesda, Maryland 20014

INTRODUCTION

A persistent problem facing research in the area of polycyclic aromatic hydrocarbon (PAH)[1] carcinogenicity has been the development of a basis for the significant differences in activity of structurally related compounds. A

[1]Abbreviations not defined in the text are: B[e]P 4,5-dihydrodiol, *trans*-4,5-dihydroxy-4,5-dihydro-B[e]P; B[e]P 9,10-dihydrodiol, *trans*-9,10-dihydroxy-9,10-dihydro B[e]P; B[e]P 9,10-diol-11,12-epoxides, (±)-9β, 10α-dihydroxy-11β, 12β (or 11α, 12α)-epoxy-9,10,11,12-tetrahydro B[e]P; 9,10-H_2B[e]P, 9,10-dihydro B[e]P; B[e]P H_4-9,10-diol, *trans*-9,10-dihydroxy-9,10,11,12-tetrahydro-B[e]P; B[e]P H_4-9,10-epoxide, 9,10-epoxy-9,10,11,12-tetrahydro-B[e]P: B[a]P 7,8-diol-9,10-epoxide (isomer 1), (±)-7β,8α-dihydroxy-9β,10β-epoxy-7,8,9,10-tetrahydro B[a]P; B[a]P 7,8-diol-9,10-epoxide (isomer 2), (±)-7β,8α-dihydroxy-9α,10α-epoxy-7,8,9,10-tetrahydro B[a]P; B[a]P H_4-7,8-epoxide, 7,8-epoxy-7,8,9,10-tetrahydro B[a]P; B[a]P H_4-9,10-epoxide, 9,10-epoxy-7,8,9,10-tetrahydro-B[a]P: B[a]P 9,10-diol-7,8-epoxide (isomer 1), (±)-9α, 10β-dihydroxy-7β,8β-epoxy-7,8,9,10-tetrahydro B[a]P; B[a]P 9,10-diol-7,8-epoxide (isomer 2), (±)-9α,10β-dihydroxy-7α,8α-epoxy-7,8,9,10-tetrahydro B[a]P.

good example is provided by benzo[a]pyrene (B[a]P) and benzo[e]pyrene (B[e]P), Figure 1. These two molecules are pentacyclic isomers which differ only by the position of benzo-ring annelation on pyrene. Although both B[a]P and B[e]P are widely disseminated throughout the biosphere, only B[a]P is of marked environmental concern, owing to its much greater carcinogenicity in animal tests.

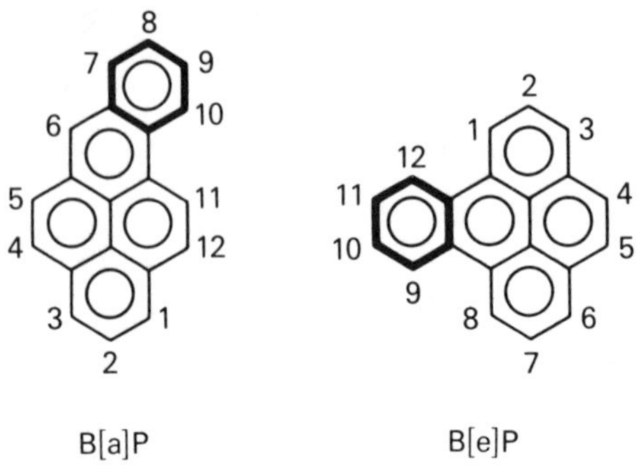

FIGURE 1. Structural formulas for benzo[a]pyrene and benzo[e]pyrene.

B[a]P has been the focus of intensive research that has greatly enhanced our understanding of the molecular pathway by which it is activated to metabolites which bind to critical cellular constituents such as DNA and RNA (1,13,15,20). Tumor studies have proved that a B[a]P 7,8-diol-9,10-epoxide is an important ultimate carcinogenic form of B[a]P (6) which is produced *via* a multistep pathway involving the conversions: arene → arene oxide → dihydrodiol → diol epoxide (Figure 2). An extensive review of the carcinogenicity and mutagenicity of B[a]P and its metabolites is available (11).

The strong evidence for the role of the diol-epoxide pathway in the carcinogenesis of B[a]P raised the question of whether similar diol-epoxide pathways were responsible for the carcinogenesis of other PAH (3). Perturbational molecular orbital (PMO) calculations predicted that significant differences in reactivity were to be expected for benzo-ring epoxides derived from different PAH and also for positional isomers of diol epoxides derived from a single PAH (4,5). Significantly, at least one of the potential benzo-ring positional isomers of the diol epoxides derived from any given highly carcinogenic PAH was invariably calculated to be highly reactive, and that diol epoxide was invariably the one in which the oxirane oxygen formed part of a "bay region" of the hydrocarbon. The prototype of a bay region is the sterically hindered region between C-4 and C-5 in phenanthrene. The bay regions in B[a]P and B[e]P diol epoxides are illustrated in Figure 2. Furthermore, the bay-region diol epoxides of weakly or noncarcinogenic PAH were generally calculated to be relatively less reactive compared to their analogs from the highly

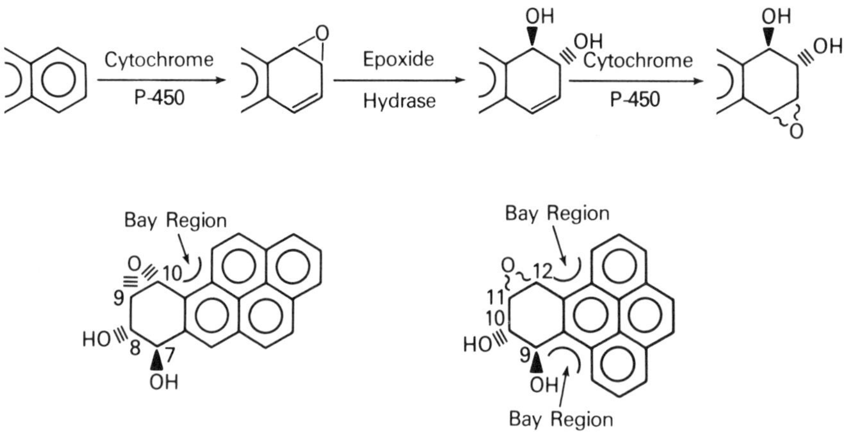

FIGURE 2. Generalized metabolic route for conversion of a PAH to a diol epoxide and the bay-region diol epoxides of B[e]P and B[a]P. The 7,8-diol-9,10-epoxide shown for B[a]P is isomer 2 while the isomer for the B[e]P 9,10-diol-11,12-epoxide is not specified.

carcinogenic PAH. The reactivity differences predicted for diol epoxides have been confirmed in the limited number of experiments reported (4). Rates of reaction of selected B[a]P, benzo[a]anthracene (BA) and napthalene diol epoxides with sodium *p*-nitrothiophenolate correlated well with the calculated PMO reactivity parameter. More recently, rates of solvolysis of selected chrysene, phenanthrene and B[a]P bay-region tetrahydroepoxides and diol epoxides were found to follow the predicted order of reactivity closely (21). The PMO reactivity parameter, $\Delta E_{deloc}/\beta$, provides an estimate of the ease of conversion of diol epoxides to benzylic triol carbonium ions. Larger values of $\Delta E_{deloc}/\beta$ correspond to predicted higher reactivities for the diol epoxides. Since chemical carcinogenesis and mutagenesis seem to generally involve the combination of critical cellular constituents such as DNA or RNA with reactive, electrophilic molecules or metabolites, chemical reactivity parameters calculated for the diol epoxides could be of value in predicting the relative mutagenicity and tumorigenicity of diol epoxides. An initial application of the calculations to BA and its derivatives was highly successful. As predicted, the "bay-region" 3,4-diol-1,2-epoxides of BA were considerably more potent, both in mutagenicity toward *S. typhimurium* strain TA 100 and Chinese hamster V79 cells (22) and in tumorigenicity in initiation-promotion experiments (9), than were other isomeric BA diol epoxides (14). More recently, evidence has been obtained which suggests that bay-region diol epoxides are important in the carcinogenicity of chrysene (10), dibenzo[a,h]-anthracene (25), 7-methylbenzo[a]anthracene (12,18), 7,12-dimethylbenzo[a]anthracene (2) and 3-methylcholanthrene (7,16,19).

BENZO[e]PYRENE DERIVATIVES

Synthesis

Benzo[e]pyrene can potentially form two dihydrodiol metabolites: the K-region dihydrodiol, B[e]P 4,5-dihydrodiol and the non-K-region dihydrodiol, B[e]P 9,10-dihydrodiol (Figure 3). Only the latter dihydrodiol can be metabolized to a benzo-ring diol epoxide. Details of the synthesis of these B[e]P dihydrodiols have been described (8). In the course of their preparation, a number of other compounds of biological interest were prepared. Their structures are also shown in Figure 3.

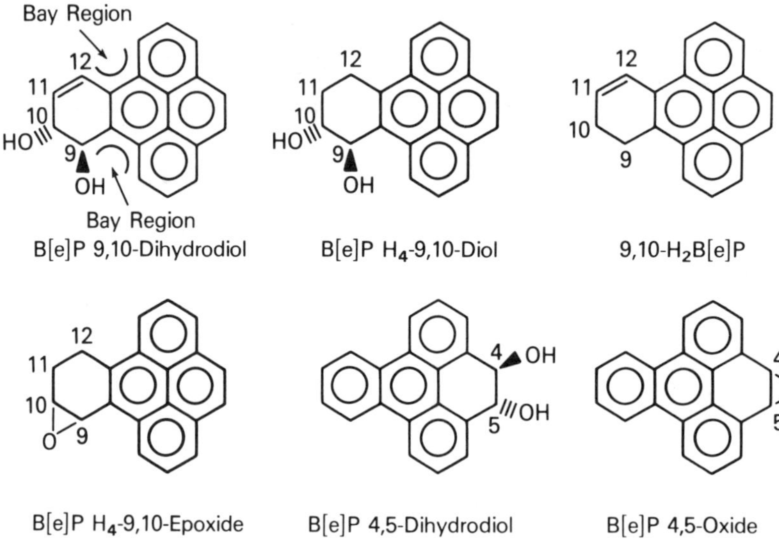

FIGURE 3. Synthetic derivatives of B[e]P.

Structural and Theoretical Properties

The non-K-region 9,10-dihydrodiol of B[e]P is unique among the non-K-region dihydrodiols whose syntheses have been reported. Both the double bond *and* the benzylic hydroxyl group of the benzo-ring form parts of bay regions in the molecule. As a consequence of nonbonded interactions between the C-9 hydroxyl group and the hydrogen atom at C-8, the hydroxyl groups in B[e]P 9,10-dihydrodiol adopt a predominantly quasidiaxial conformation as established from the very low coupling constant between the hydrogen atoms at C-9 and C-10 in the nuclear magnetic resonance spectrum of the dihydrodiol (8). Benzo-ring dihydrodiols in which only the double bond forms part of a bay region, such as B[a]P 7,8-dihydrodiol, have their hydroxyl groups in a predominantly quasidiequatorial conformation as shown in Figure 4. The effect of this conformational factor upon the ability of B[e]P 9,10-dihydrodiol to be metabolically activated to a bay-region diol epoxide was one of the major points of interest in this study.

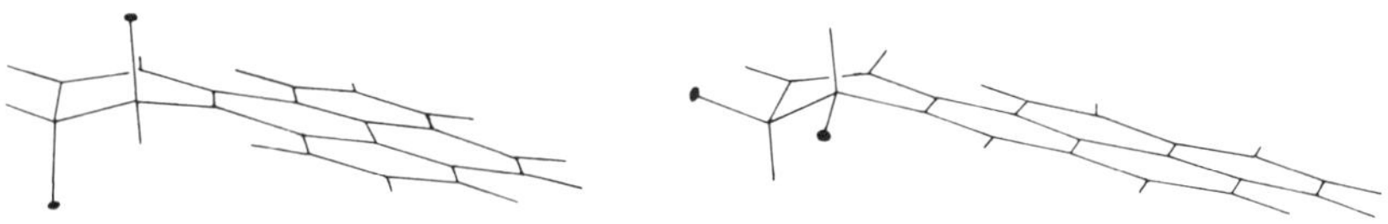

FIGURE 4. Principal conformations of B[e]P 9,10-dihydrodiol (left) and B[a]P 7,8-dihydrodiol (right). The hydroxyl groups, which are shown as darkened balls, adopt a quasiaxial conformation in the B[e]P 9,10-dihydrodiol due to steric hindrance at the bay region.

PMO calculations indicate that B[e]P 9,10-diol-11,12-epoxides (Figure 2), if formed, should be intermediate in reactivity between the corresponding chrysene and B[a]P bay-region diol epoxides. Values of $\Delta E_{deloc}/\beta$ are 0.639, 0.714 and 0.794 for the bay-region chrysene, B[e]P and B[a]P diol epoxides, respectively. Since the bay-region diol epoxides of chrysene have been implicated in the carcinogenicity of chrysene (10) and since the B[e]P diol epoxides are calculated to be substantially more reactive, considerable biological activity might be expected for the B[e]P 9,10-diol-11,12-epoxides.

METABOLIC ACTIVATION

The ability of B[e]P, 9,10-H_2B[e]P, B[e]P 4,5-dihydrodiol, B[e]P 9,10-dihydrodiol and B[e]P H_4-9,10-diol to be metabolically activated to mutagens was examined in *S. typhimurium* strains TA 98 and TA 100 with 10 nmol of each compound as substrate and varying amounts of hepatic microsomes from Aroclor-pretreated rats. None of the molecules were intrinsically mutagenic and only 9,10-H_2B[e]P was strongly metabolically activated. The relative ability of the substrates to be metabolically activated to products mutagenic toward *S. typhimurium* strain TA 98 at a cytochrome P-450 concentration of 0.6 nmol is shown in Figure 5. Similar results were obtained in strain TA 100, except that B[e]P 4,5-dihydrodiol was metabolically activated to moderately mutagenic products in this strain. When the same series of compounds was incubated with a highly purified monooxygenase system derived from rats pretreated with Aroclor 1243, only 9,10-H_2B[e]P was activated to compounds strongly mutagenic toward *S. typhimurium* strains TA 98 and TA 100.

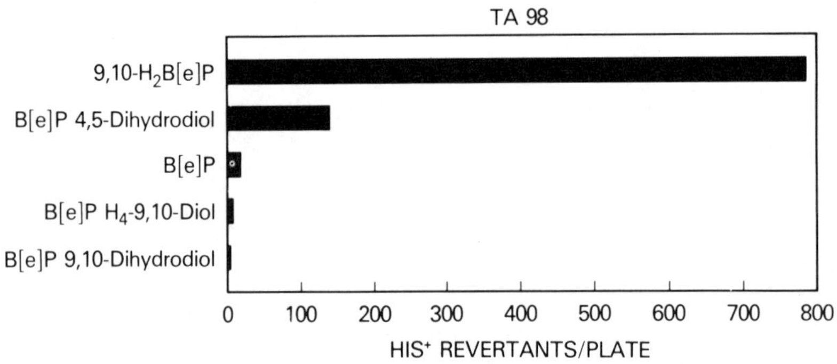

FIGURE 5. Relative ability of B[e]P and several B[e]P derivatives to be metabolically activated to compounds mutagenic toward *S. typhimurium* TA 98. See text for conditions.

There are three plausible explanations for the inability of B[e]P 9,10-dihydrodiol to be metabolically activated to potent mutagens. One possibility is that B[e]P 9,10-diol-11,12-epoxides are formed upon metabolic activation, but are nonmutagenic toward *S. typhimurium* strains TA 98 and TA 100, despite their relatively high calculated reactivity. A second possibility is that B[e]P 9,10-dihydrodiol is not metabolized to the bay-region diol epoxides, but that the diol epoxides, if formed, would be significantly mutagenic in *S. typhimurium* strains TA 98 and TA 100. Finally, it is possible that the bay-region diol epoxides are not formed and would not be nonmutagenic even if formed. That a bay-region tetrahydroepoxide of B[e]P would have strong mutagenicity was suggested by the ability of 9,10-H_2B[e]P to be metabolically activated to a highly mutagenic compound(s). Studies of the inherent mutagenicity of B[e]P derivatives as well as studies of the metabolism of B[e]P 9,10-dihydrodiol and 9,10-H_2B[e]P suggest that the bay region diol epoxides are not formed.

INHERENT MUTAGENICITY

To date, it has not been possible to synthesize the diastereomeric B[e]P 9,10-diol-11,12-epoxides in purity sufficient to permit a determination of their intrinsic mutagenicity. However, the closely related B[e]P H_4-9,10-epoxide has been prepared, and its inherent mutagenicity has been examined in *S. typhimurium* strains TA 98 and TA 100 as well as in Chinese hamster V79 cells (Figure 6). Data for the strongly mutagenic bay-region tetrahydro-epoxide of B[a]P (B[a]P H_4-9,10-epoxide) are given for comparison. The ratio of the mutagenicity of B[e]P H_4-9,10-epoxide to that of B[a]P H_4-9,10-epoxide is about 1.8, 0.4 and 1 in strains TA 98 and TA 100 of *S. typhimurium* and in V79 cells, respectively. The very high mutagenicity of B[e]P H_4-9,10-epoxide suggests that 9,10-H_2B[e]P is metabolically activated to B[e]P H_4-9,10-epoxide, that B[e]P 9,10-diol-11,12-epoxides would be highly mutagenic if formed, and also that B[e]P 9,10-dihydrodiol is not metabolically activated to the bay-region diol epoxides to a significant degree. In general tetrahydro-epoxides of PAH are known to have both chemical reactivity (21) and mutagenic activity (23) which are similar to their corresponding diol epoxides.

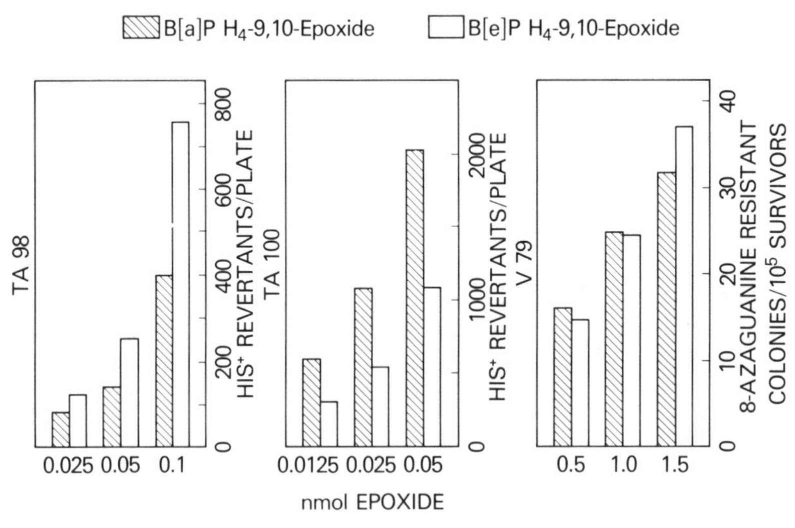

FIGURE 6. Relative mutagenicity of B[a]P H_4-9,10-epoxide and B[e]P H_4-9,10-epoxide toward *S. typhimurium* strains TA 98 and 100 and toward Chinese hamster V79 cells as a function of epoxide concentration.

The K-region B[e]P 4,5-oxide was found to be a weak mutagen with less than 10% of the activity of B[a]P 4,5-oxide toward *S. typhimurium* strains TA 98 and TA 100, and less than 20% of the activity of B[a]P 4,5-oxide toward Chinese hamster V79 cells. While the relative mutagenicities correlate with the greater predicted rate of ring opening to the zwitterion calculated for B[a]P 4,5-oxide ($\Delta E_{deloc}/\beta$ = 0.623 for opening at either C-4 or C-5) relative to B[e]P 4,5-oxide ($\Delta E_{deloc}/\beta$ = 0.586), such correlations are dependent on the mechanism of reaction of the arene oxide (5).

METABOLISM

The nature of the metabolites formed on incubation of B[e]P 9,10-dihydrodiol and 9,10-H_2B[e]P with hepatic microsomes from Aroclor-pretreated rats was examined. B[e]P 9,10-dihydrodiol was converted into four major metabolites, none of which was a tetraol such as would be expected to form by hydration of the benzo-ring diol epoxides derived from B[e]P 9,10-dihydrodiol. Rather, in each metabolite the benzo-ring double bond had resisted attack, and reaction had occurred at the aromatic nucleus (Figure 7), as shown by mass spectral and ultraviolet analysis of the products (24).

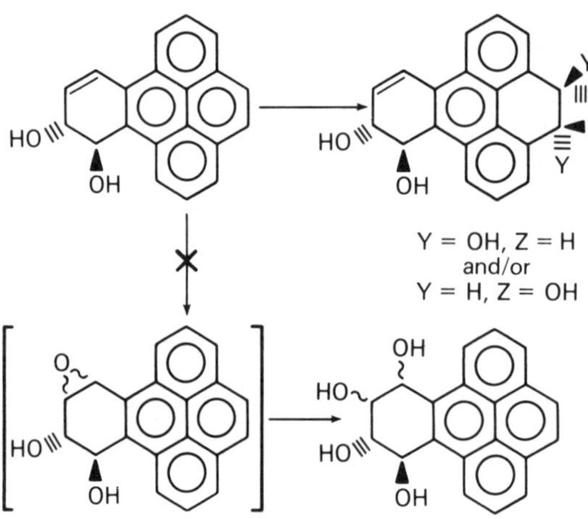

FIGURE 7. Metabolism of B[e]P 9,10-dihydrodiol results in the formation of a K-region diol and three phenolic metabolites.

9,10-H_2B[e]P was metabolized to a number of products, which included as major identified metabolites B[e]P, B[e]P H_4-9,10-diol, B[e]P 4,5-dihydrodiol, 9 (or 10)-hydroxy-9,10-H_2B[e]P and phenolic derivatives of 9,10-H_2B[e]P (24). Possible pathways for their formation are given in Figure 8. Most notable is the presence of B[e]P H_4-9,10-diol, which provides indirect, but strong, evidence for metabolism of 9,10-H_2B[e]P to B[e]P H_4-9,10-epoxide, since the epoxide would be expected to be rapidly hydrated spontaneously or enzymatically to this H_4-9,10-diol.

TUMORIGENICITY

The tumorigencity of B[e]P 9,10-dihydrodiol, B[e]P H_4-9,10-diol, 9,10-H_2B[e]P, B[e]P 4,5-dihydrodiol and B[e]P is being examined by initiation-promotion experiments on mouse skin. Preliminary data at 20 weeks of

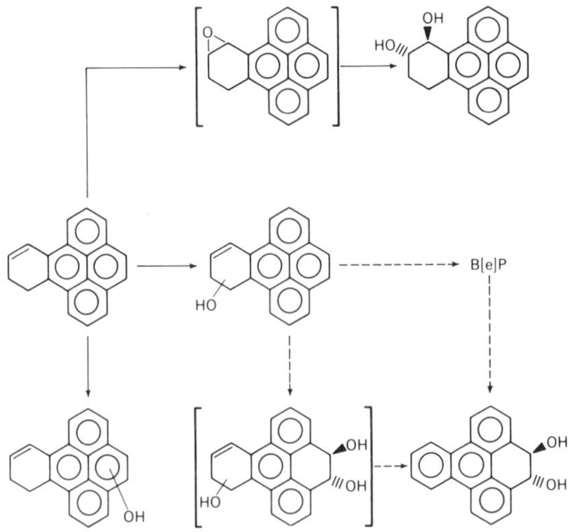

FIGURE 8. Metabolism of 9,10-H_2B[e]P.

promotion reveal that 9,10-H_2B[e]P is the only compound in the group with significant initiating activity. An initial application of 2.5 μmole of 9,10-H_2 B[e]P followed by twice weekly application of phorbol ester resulted in a 42% tumor incidence, with an average of 0.5 papillomas/mouse.

CONCLUSIONS

Our studies indicate that the basis for the very weak tumorigenicity of B[e]P when tested in rodents resides, at least in part, in the failure of B[e]P to be metabolized to the key benzo-ring bay-region diol epoxides. The lack of detectable metabolism of B[e]P 9,10-dihydrodiol to bay-region diol epoxides may be due to the predominantly quasidiaxial conformation of the hydroxyl groups in the molecule. A recent study of the metabolism of B[a]P 9,10-dihydrodiol (17) produced little or no evidence for its conversion to benzo-ring diol epoxides in contrast to the metabolism of B[a]P 7,8-dihydrodiol (Figure 9). Both B[e]P 9,10-dihydrodiol and B[a]P 9,10-dihydrodiol have their hydroxyl groups in the predominantly quasidiaxial conformation while the B[a]P 7,8-dihydrodiol has its hydroxyl groups predominantly quasiequatorial (Figure 4). This shift of metabolism away from the double bond of bay-region benzo-ring dihydrodiols may well prove to be a general metabolic phenomenon.

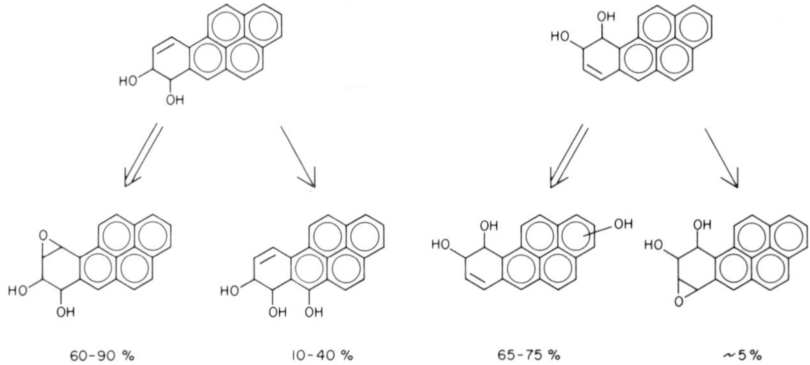

FIGURE 9. Comparative metabolism of B[a]P 7,8- and 9,10-dihydrodiols.

The results of this study emphasize the crucial role of metabolism in the carcinogenesis of PAH and the need for a better understanding of the factors that affect metabolism. Not only must a diol epoxide from a PAH be biologically active, it obviously must also be formed for the carcinogenic activity of the hydrocarbon to be expressed by this pathway. The B[e]P 9,10-diol-11,12-epoxides are structurally unique in that they must have their hydroxyl groups quasi-axial. They may or may not be biologically active since conformationally rigid diol epoxides of this type have yet to be tested.

ACKNOWLEDGMENTS

This investigation was supported in part by Grant No. 1 RO1 CA-22985-01, awarded to R.E.L. by the National Cancer Institute, Department of Health, Education and Welfare.

REFERENCES

1. Borgen, A., Darvey, H., Castagnoli, N., Crocker, T. T., Rasmussen, R. E., and Wang, I. Y. (1973): Metabolic conversion of benzo[a]pyrene by Syrian hamster liver microsomes and binding of metabolites to DNA. *J. Med. Chem.* 16:502-506.
2. Dipple, A., and Nebzydoskik, J. A. (1978): Evidence for the involvement of a diol epoxide in the binding of 7,12-dimethylbenzo[a]-anthracene to DNA in cells in culture. *Chem. -Biol. Interact.* 20:17-26.
3. Jerina, D. M., and Daly, J. W. (1976): Oxidation at Carbon. In: *Drug Metabolism,* D. V. Parke and R. L. Smith, Eds., pp. 13-32, Taylor and Francis, London.

4. Jerina, D. M., and Lehr, R. E. (1977): The bay region theory: A quantum mechanical approach to aromatic hydrocarbon-induced carcinogenicity. In: *Microsomes and Drug Oxidations,* V. Ullrich, I. Roots, A. G. Hildebrant, R. W. Estabrook and A. H. Conney, Eds., pp. 709-720, Pergamon Press, Oxford.
5. Jerina, D. M., Lehr, R. E., Yagi, H., Hernandez, O., Dansette, P. M., Wislocki, P. G., Wood, A. W., Chang, R. L., Levin, W., and Conney, A. H. (1976): Mutagenicity of benzo[a]pyrene derivatives and the description of a quantum mechanical model which predicts the ease of carbonium ion formation from diol epoxides. In: In Vitro *Metabolic Activation in Mutagenesis Testing,* F. J. de Serres, J. R. Fouts, J. R. Bend and R. M. Philpot, Eds., pp. 159-177, Elsevier, Amsterdam.
6. Kapitulnik, J., Wislocki, P. G., Levin, W., Yagi, H., Jerina, D. M., and Conney, A. H. (1978): Tumorigenicity studies with diol epoxides of benzo[a]pyrene which indicate that (±)-trans-7β, 8α-dihydroxy-9α,10α-epoxy-7,8,9,10-tetrahydrobenzo[a]pyrene is an ultimate carcinogen in newborn mice. *Cancer Res.* 38:354-358.
7. King, H. W. S., Osborne, M. R., and Brooke, P. (1977): The metabolism and DNA binding of 3-methylcholanthrene. *Int. J. Cancer* 20:564-571.
8. Lehr, R. E., Taylor, C. W., Kumar, S., Mah, H. D., and Jerina, D. M. (1978): Synthesis of the non-K-region and K-region trans-dihydrodiols of benzo[e]pyrene. *J. Org. Chem.* 43:3462-3466.
9. Levin, W., Thakker, D. R., Wood, A. W., Chang, R. L., Lehr, R. E., Jerina, D. M., and Conney, A. H. (1978): Evidence that benzo[a]-anthracene 3,4-diol-1,2-epoxide is an ultimate carcinogen on mouse skin. *Cancer Res.* 38:1705-1710.
10. Levin, W., Wood, A. W., Chang, R. L., Yagi, H., Mah, H. D., Jerina, D. M., and Conney, A. H. (1978): Evidence for bay region activation of chrysene 1,2-dihydrodiol to an ultimate carcinogen. *Cancer Res.* 38:1831-1834.
11. Levin, W., Wood, A. W., Wislocki, P. G., Chang, R. L., Kapitulnik, J., Mah, H. D., Yagi, H., Jerina, D. M., and Conney, A. H. (1978): Mutagenicity and carcinogenicity of benzo[a]pyrene and benzo[a]pyrene derivatives. In: *Polycyclic Hydrocarbons and Cancer: Environment, Chemistry, and Metabolism,* H. V. Gelboin and P. O. P. Ts'o, Eds, pp. 189-202, Academic Press, New York.
12. Malaveille, C., Tierney, B., Grover, P. L., Sims, P., and Bartsch, H. (1977): High microsome-mediated mutagenicity of the 3,4-dihydrodiol of 7-methylbenz[a]anthracene in *S. typhimurium* TA 98. *Biochem. Biophys. Res. Commun.* 75:427-433.
13. Moore, P. D., Koreeda, M., Wislocki, P. G., Levin, W., Conney, A. H., Yagi, H., and Jerina, D. M. (1977): *In vitro* reactions of the diastereomeric 9,10-epoxides of (+)- and (-)-*trans*-7,8-dihydroxy-7,8-dihydrobenzo[a]pyrene with polyguanylic acid and evidence for formation of an enantiomer of each diastereomeric 9,10-epoxide from benzo[a]pyrene in mouse skin. In: *Drug Metabolism Concepts,* ACS

Monograph Series 44, pp. 127-154. American Chemical Society, Washington, D.C.

14. Slaga, T. J., Huberman, E., Selkirk, J. K., Harvey, R. G., and Bracken, W. M. (1978): Carcinogenicity and mutagenicity of benz[a]anthracene diols and diol epoxides. *Cancer Res.* 38:1699-1704.
15. Sims, P., Grover, P. L., Swaisland, A., Pal, K., and Hewer, A. (1974): Metabolic activation of benzo[a]pyrene proceeds by a diol-epoxide. *Nature* 252:326-328.
16. Thakker, D. R., Levin, W., Wood, A. W., Conney, A. H., Stoming, T. A., and Jerina, D. M. (1978): Metabolic formation of 1,9,10-trihydroxy-9,10-dihydro-3-methylcholanthrene: A potential proximate carcinogen from 3-methylcholanthrene. *J. Am. Chem. Soc.* 100: 645-647.
17. Thakker, D. R., Yagi, H., Lehr, R. E., Levin, W., Buening, M., Lu, A. Y. H., Chang, R. L., Wood, A. W., Conney, A. H., and Jerina, D. M. (1978): Metabolism of trans-9,10-dihydroxy-9,10-dihydrobenzo[a]-pyrene occurs primarily by aryl-hydroxylation rather than formation of a diol epoxide. *Mol. Pharmacol.* 14:502-513.
18. Vigny, P., Duquesne, M., Coulomb, H., Lacombe, C., Tierney, B., Grover, P. L., and Sims, P. (1977): Metabolic activation of polycyclic hydrocarbons. Fluorescence spectral evidence is consistent with metabolism at the 1,2- and 3,4-double bond of 7-methylbenz[a]anthracene. *FEBS Lett.* 75:9-12.
19. Vigny, P., Duquesne, M., Coulomb, H., Tierney, B., Grover, P. L., and Sims, P. (1977): Fluorescence spectral studies on the metabolic activation of 3-methylcholanthrene and 7,12-dimethylbenz[a]anthracene on mouse skin. *FEBS Lett.* 82:278-282.
20. Weinstein, I. B., Jeffrey, A. M., Jennette, K. W., Blobstein, S. H., Harvey, R. G., Harris, C., Autrup, H., Kasai, H., and Nakanishi, N. (1976): Benzo[a]pyrene diol epoxides as intermediates in nucleic acid binding *in vitro* and *in vivo*. *Science* 193:592-595.
21. Whalen, D. L., Ross, A. M., Yagi, H., Karle, J. M., and Jerina, D. M. (1978): Stereoelectronic factors in the solvolysis of bay region diol epoxides of polycyclic aromatic hydrocarbons. *J. Am. Chem. Soc.* 100:5218-5221.
22. Wood, A. W., Chang, R. L., Levin, W., Lehr, R. E., Schaefer-Ridder, M., Karle, J. M., Jerina, D. M., and Conney, A. H. (1977): Mutagenicity and cytotoxicity of benzo[a]anthracene diol epoxides and tetrahydro-epoxides: exceptional activity of the bay region 1,2-epoxides. *Proc. Nat. Acad. Sci., USA* 74:2746-2750.
23. Wood, A. W., Levin, W., Chang, R. L., Yagi, H., Thakker, D. R., Lehr, R. E., Jerina, D. M., and Conney, A. H. (1979): Bay region activation of carcinogenic polycyclic hydrocarbons, this volume.
24. Wood, A. W., Levin, W., Thakker, D. R., Yagi, H., Chang, R. L., Ryan, D. E., Thomas, P. E., Dansette, P. M., Whittaker, N., Turujman, S., Lehr, R. E., Kumar, S., Jerina, D. M., and Conney, A. H. (1979):

Biological activity of benzo[e]pyrene: An assessment based on mutagenic activities and metabolic profiles of the polycyclic hydrocarbon and its derivatives. *J. Biol. Chem.*, in press.

25. Wood, A. W., Levin, W., Thomas, P. E., Ryan, D., Karle, J. M., Yagi, H., Jerina, D. M., and Conney, A. H. (1978): Metabolic activation of dibenzo[a,h]anthracene and its dihydrodiols to bacterial mutagens. *Cancer Res.* 38:1967-1973.

Polynuclear Aromatic Hydrocarbons, edited by P. W. Jones and P. Leber. Ann Arbor Science Publishers, Inc. Ann Arbor, MI ©1979.

POLYCYCLIC HYDROCARBONS AND SYRIAN HAMSTER EMBRYO CELLS: CELL TRANSFORMATION, ENHANCEMENT OF VIRAL TRANSFORMATION AND ANALYSIS OF DNA DAMAGE

Bruce C. Casto[1]

BioLabs, Inc.
2910 MacArthur Boulevard
Northbrook, Illinois 60062

INTRODUCTION

Short-term *in vitro* tests are being developed and several are currently in use for the identification of potential carcinogenic agents. Presently, the systems that appear to correlate best with activity *in vivo* are mammalian cell transformation (9,35,36), mutagenesis assays in mammalian (22) and microbial systems (29), analysis of DNA repair (40,41) or fragmentation (9,42), and enhancement of viral transformation (9,12). Many of the early investigations using *in vitro* assays for carcinogenic or mutagenic activity of polycyclic hydrocarbons failed, due to the fact that the systems used were incapable of metabolizing the parent compound into its ultimate active form. This complication has been resolved through the use of activation systems such as S-9 fractions of various tissue homogenates (2,27) or by using feeder layers of metabolically active cells (23,32) in conjunction with metabolically deficient indicator cells. Unlike the microbial systems and many of the mammalian cell systems, Syrian hamster embryo cells possess the necessary enzymes for activation of diverse classes of carcinogenic compounds (24). The following data show the outcome of the interaction between carcinogenic or noncarcinogenic polycyclic hydrocarbons when assayed in Syrian hamster embryo cells using four different *in vitro* systems.

---

[1]Present address: Northrop Services Inc., Environmental Sciences Center, Research Triangle Park, North Carolina, 27709.

CHEMICAL TRANSFORMATION

The transformation of hamster cells by polycyclic hydrocarbons using mass culture and colony assay techniques has been demonstrated in several laboratories (3,4,15,16,18,20). Recently, a quantitative focus assay in hamster cells has been described (8) using some of the methodology previously applied to assay in mouse cells (13, 38,39). With this assay, foci of transformed cells appear within 3-4 weeks as darkly stained clusters of cells against a lightly stained background of normal-appearing cells (Figure 1). The interior

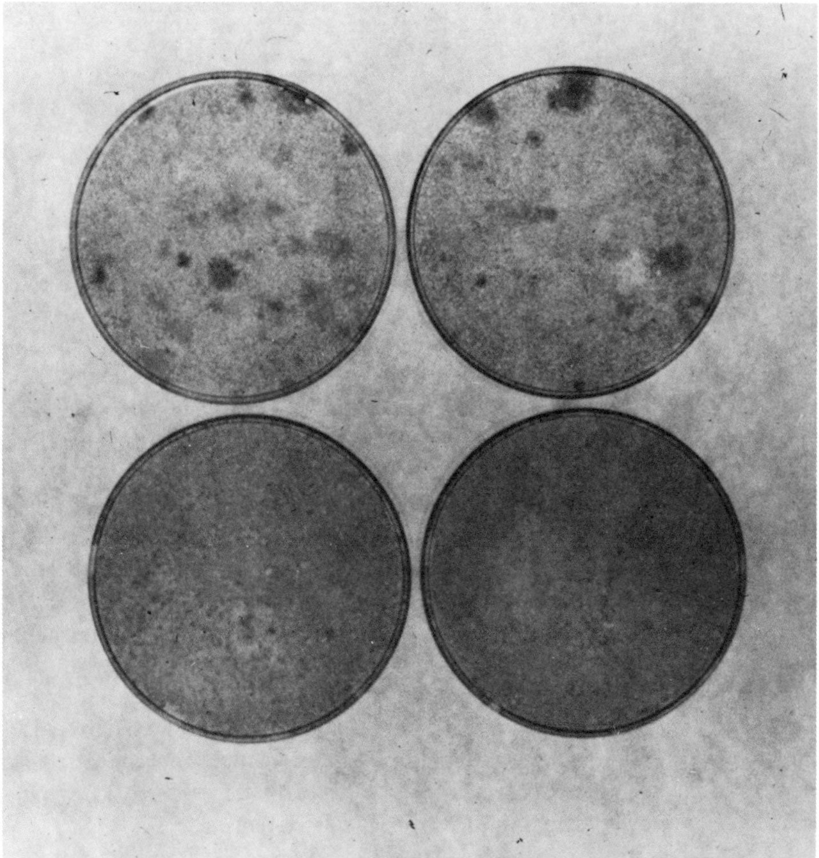

FIGURE 1. Focus assay for chemical transformation. Cultures of Syrian hamster embryo cells were treated with 2 μg/ml of B[a]P 24 hr after seeding 5 x 10^4 cells per 60-mm Petri dish. Chemical was removed after 6 days and the cultures incubated for an additional 19 days. Top, B[a]P; bottom, Control. (1X).

and the periphery of the foci contain fibroblastic-like cells showing random-orientation and a tendency to form multilayers (Figure 2). The polycyclic hydrocarbons benzo[a]pyrene (B[a]P), dibenz[a,h]anthracene (DB[a,h]A), 7,12-dimethylbenz[a]anthracene (DMBA) and 3-methylcholanthrene (MCA) induce transformed foci, whereas the noncarcinogens pyrene, phenanthrene and perylene are ineffective (Table I). The capacity to transform *in vitro* was equivalent to the *in vivo* activity when the dose required for transformation was taken into consideration; DMBA being the most potent followed by B[a]P, MCA and DB[a,h]A, respectively.

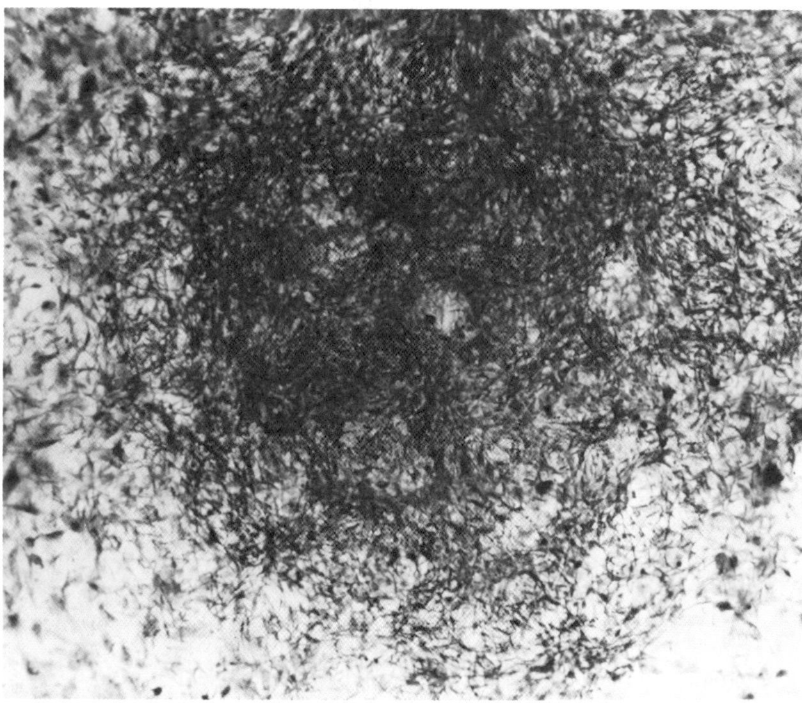

FIGURE 2. Focus of MCA-transformed hamster embryo cells showing the fibroblast morphology of the cells. Cultures were fixed and stained 19 days after treatment with 0.5 μg/ml of MCA. (32X).

Foci, selected from under a methylcellulose overlay, were tumorigenic when injected intradermally into weanling hamsters, however not all freshly derived foci formed tumors when first tested (Table II). Two of seven chemically transformed cell lines did not induce tumors in early passage (Passages

TABLE I. Transformation of Syrian Hamster Embryo Cells by Carcinogenic Polycyclic Hydrocarbons

Chemical[a]	μg/ml	Surviving Fraction	Foci/ Dishes	TF[b]
B(a)P	1.25	0.52	13/10	10.0
	0.62	0.55	11/10	8.0
DB(a,h)A	2.5	1.05	2/16	1.0
	1.2	1.07	1/10	0.5
DMBA	0.025	0.91	14/10	3.0
	0.0125	0.88	5/10	1.0
MCA	2.0	0.42	17/9	9.0
	1.0	0.48	5/10	2.1
Pyrene	20.0	0.98	0/9	0
	10.0	0.97	0/10	0
Acetone (0.5%)	—	1.00	0/50	0

[a]Chemicals were added 24 hr after plating 20,000 to 50,000 cells (transformation) or 1,000 cells (survival) per 60-mm dish. Transformation assay dishes were fixed and stained at 25 days, survival assays at 8 days.
[b]Transformation frequency. The number of cells seeded per dish was 2×10^4 (DB[a,h]A), 2.5×10^4 (B[a]P) or 5×10^4 (DMBA, MCA, pyrene and acetone). TF is expressed as the number of foci per 10^5 surviving cells.

TABLE II. Tumorigenicity of Hamster Embryo Cells Transformed *in vitro* by Polycyclic Hydrocarbons: Effect of Passage

Cell Line	Passage Level[a]	Cloning in Soft Agar[b]	Tumor Incidence (No.+/Total Inoc.)[c]
MCA-350	5	0	0/4
	28	+	6/6
MCA-354	6	0	0/4
	28	+	6/6
MCA-365	5	0	4/4
	20	+	6/6
MCA-390	4	±	3/3
	20	+	6/6
B(a)P-68	4	±	6/6
DMBA-36	5	+	6/6
DMBA-43	5	±	2/6

[a]Foci were picked from under methylcellulose with a Pasteur pipette, grown to confluency in a 60-mm dish and passaged at 1:5 split ratios.
[b]Capacity to clone in 0.3% agar: 0 = no colonies in 10^5 plated cells; ± = less than 1% cloning efficiency; + = greater than 10% cloning efficiency.
[c]Approximately 10^7 low-passage or 10^5 high-passage cells were inoculated i.d. on the abdomen or between the layers of the cheek pouch (low passage only).

5 and 6) but were highly tumorigenic after being subcultured *in vitro* more than 20 times. Similarly, the capacity to clone in soft agar was not always demonstrable with early-passage-transformed cell lines (Table II), but again this property was acquired after several transfers *in vitro*.

TABLE III. Transformation of Syrian Hamster Embryo Cells by 3-methylcholanthrene: Effect of Exposure for Varying Periods

	Foci/Dishes			TF[b]		
μg/ml[a]	1 day	3 days	6 days	1 day	3 days	6 days
2.00	2/10	3/7	7/10	0.96	3.5	16.0
0.50	0/5	2/9	2/5	0	1.1	4.0
0.12	0/10	0/5	2/8	0	0	1.6
Control	0/9	0/10	0/10	0	0	0

[a]Chemicals were added 24 hr after plating 50,000 cells (transformation) or 1,000 cells (survival); assays were fixed and stained after 25 or 8 days, respectively. Cells exposed for 6 days received fresh chemical after 3 days.
[b]Transformation frequency is expressed as the number of foci per 10^5 surviving cells.

With the polycyclic hydrocarbons, transformation was induced at lower doses and a higher transformation frequency was observed when exposure to the chemical was increased from 1 to 6 days (Table III). These data are in agreement with those reported earlier by Chen and Heidelberger (13) and Reznikoff *et al.* (38) with mouse fibroblasts, where an increase in transformation was observed following extended treatment periods. Analogous to these findings, Mondal *et al.* (33) have shown that application of a subeffective dose of MCA, followed one to three days later by the same concentration of B[a]P, DMBA or MCA, caused a significant increase in the formation of transformed foci in C3H1OT½ cells.

ENHANCEMENT OF VIRAL TRANSFORMATION

Pretreatment of hamster embryo cells with a variety of chemical carcinogens and mutagens will increase the frequency of transformation induced by oncogenic DNA viruses (6,7,10,11,12,14,26,31). The potency of the polycyclic hydrocarbons for induction of enhancement of viral transformation with a simian adenovirus (SA7) parallels that found for chemical transformation (Table IV). Treatment with 0.004 μg/ml of DMBA, 0.08 μg/ml of B[a]P or MCA or 0.4 μg/ml of DB[a,h]a significantly increased the viral transformation frequency, whereas pyrene was ineffective. Where the toxicity

TABLE IV. Enhancement of Viral Transformation Following Treatment of Syrian Hamster Embryo Cells with Polycyclic Hydrocarbons and Simian Adenovirus, SA7

Chemical[a]	μg/ml	Surviving Fraction	SA7 Foci[b] Treated	SA7 Foci[b] Control	Enhancement Ratio[c]
B[a]P	2.0	0.06	44	49	14.9
	0.4	0.24	50	49	4.3
	0.08	0.41	55	49	2.7
DMBA	0.02	0.21	81	60	6.4
	0.004	0.56	61	60	1.8
	0.0008	0.81	53	60	1.1
DB[a,h]A	10.0	0.40	103	60	4.3
	2.0	0.63	107	60	2.8
	0.4	0.71	86	60	2.0
MCA	2.0	0.04	24	60	10.0
	0.4	0.20	45	60	3.8
	0.08	0.62	88	60	2.4
Pyrene	20.0	0.85	37	40	0.9
	10.0	0.75	35	40	0.9
	5.0	0.75	34	40	0.9

[a]Chemical dilutions were added 24 hr prior to SA7 inoculation. Virus was adsorbed 3 hr and the cells transferred to 60-mm dishes at 200,000 (transformation) or 700 (survival) cells per dish. Survival assays were fixed and stained at 8 days, transformation assays at 21 days.
[b]Number of SA7 transformed foci in 5 dishes receiving 2×10^5 cells/dish.
[c]Enhancement was determined by dividing the transformation frequency (foci per 10^6 surviving cells) of chemical- and virus-treated cells by that obtained from cells treated only with virus.

was excessive, the enhancement was restricted to a relative increase in frequency, but an absolute increase in the numbers of SA7 foci was generally found at one or more concentrations of each chemical.

Maximal enhancement was observed after only 4-hr treatment with B[a]P and DMBA, although the toxicity continued to increase with longer exposure (Table V). However, after treatment with MCA, both the toxicity and the enhancement increased relative to the treatment time.

In contrast to the enhancement observed when cells were treated prior to viral inoculation, treatment with B[a]P, DMBA, or MCA after infection with virus resulted in a dose-dependent inhibition of viral transformation (Table VI). Concurrently, the cell lethality associated with chemical treatment was reduced or completely abolished. Although not shown in Table VI, addition of B[a]P or DMBA at times up to 24 hr after viral infection still

TABLE V. Effect of Pretreatment Time on the Enhancement of Viral Transformation by Polycyclic Hydrocarbons

		Hours of Pretreatment			
Chemical	**μg/ml**	**4**	**8**	**12**	**24**
B[a]P	0.4	2.6[a] (0.85)[b]	2.5 (0.58)	2.2 (0.56)	2.5 (0.43)
DMBA	0.025	4.5 (0.25)	4.3 (0.08)	2.4 (0.04)	3.8 (0.06)
MCA	1.25	2.6 (0.44)	3.2 (0.22)	3.9 (0.12)	6.3 (0.06)

[a]Enhancement ratio: determined by dividing the transformation frequency of chemical- and virus-treated cells by that obtained from cells treated only with virus.
[b]Surviving fractions were determined from dishes receiving 700 cells per 60-mm dish. Cloning efficiency ranged from 8 to 12%.

TABLE VI. Effect of Adding Polycyclic Hydrocarbons Before and After Virus Inoculation on the Enhancement of Viral Transformation

Chemical	**Time of Treatment[a]**	**μg/ml**	**Surviving Fraction[b]**	**SA7 Foci[c]**	**Enhancement Ratio[d]**
B[a]P	-4 hr	2.0	0.38	120	8.0
		0.4	0.86	87	2.5
	+4 hr	2.0	1.09	2	0.1
		0.4	0.80	9	0.6
DMBA	-4 hr	0.025	0.33	90	6.9
		0.012	0.89	77	2.2
	+4 hr	0.025	0.83	2	0.1
		0.012	1.01	13	0.7
Medium	-4 hr	0	1.00	40	1.0
Controls	+4 hr	0	1.00	20	1.0

[a]Chemicals were added 4 hr before (-4) or 4 hr after (+4) inoculation with SA7 virus.
[b]Determined from dishes seeded with 700 cells. Cloning efficiency ranged from 8 to 12%.
[c]Number of foci from 5 dishes, each seeded with 2×10^5 cells.
[d]Enhancement was determined by dividing the transformation frequency of chemical- and virus-treated cells by that obtained from cells treated only with SA7 virus.

caused a reduction in viral transformation (7,12). The inability to stimulate the viral transformation frequency when B[a]P, DMBA or MCA were added after virus may be related to the inhibition or delay of cell DNA synthesis often observed with certain polycyclic hydrocarbons (1). Casto (5) and Reed (37) have shown that early DNA synthesis after adenovirus infection is essential for maximal transformation of hamster cells to occur.

The *in vivo* or *in vitro* carcinogenicity and toxicity of certain polycyclic hydrocarbons can be modified by pretreatment with, or simultaneous addition of, 7,8-benzoflavone (ANF) or benz[a]anthracene (BA); the effects range from complete inhibition to enhancement of carcinogenicity, depending on the time of addition of ANF or BA relative to addition of the carcinogenic polycyclic hydrocarbon (17,21,28). Subsequently, Casto and DiPaolo (7) demonstrated an inhibition of viral enhancement by BA when added simultaneously with B[a]P; enhancement by MCA was not similarly affected. As was observed with B[a]P, treatment of hamster cells with BA in conjunction with DMBA resulted in a significant decrease in the enhancement caused by DMBA (Table VII). Although there was an overall loss in viral transformation frequency, there was a significant increase in the survivors as a result of the simultaneous treatment with DMBA or B[a]P and BA. Additionally, in cultures treated with B[a]P and BA (Table VII), there was an increase in the absolute numbers of SA7 foci per dish over that observed in cultures treated only with B[a]P. Similar patterns of protection from toxicity and reduction of enhancement of SA7 transformation were observed when ANF was substituted for BA in the above type of experiments.

ANALYSIS OF DNA REPAIR AND FRAGMENTATION

The capacity of many chemical carcinogens and mutagens to damage cellular DNA has prompted several investigators to explore the use of indirect or direct assays for such damage as a potential screen for carcinogens (25, 40,41,42). These assays are based upon the early studies showing breakage or repair of DNA following irradiation of *E. coli* and HeLa cells respectively (30,34).

Treatment with B[a]P, DMBA or MCA induced DNA repair synthesis in hamster embryo cells when monitored by autoradiography, but B[e]P, DB[a,h]A, pyrene, perylene and phenanthrene were ineffective (12). In contrast to the number of cells demonstrating repair synthesis when treated with UV-irradiation or alkylating agents (95-100%), only 40-70% of the cells treated with B[a]P, DMBA or MCA were observed to undergo DNA repair (12). Demonstration of low levels of repair synthesis was optimized by using two extended labeling (^{3}H-thymidine) periods after treatment; the first commencing immediately after treatment and continuing for 6 hr (0-6 hr)

TABLE VII. Inhibition of Enhancement of Viral Transformation by Simultaneous Treatment of Hamster Cells with Polycyclic Hydrocarbons and Benz[a]anthracene

μg/ml[a] (BA)		μg/ml (PAH)	Surviving Fraction	SA7 Foci[b]	Enhancement Ratio[c]
10	B[a]P	1.0	1.05	208	2.00
		0.5	1.06	133	1.27
2.5		1.0	0.42	177	4.26
		0.5	0.67	171	2.58
0		1.0	0.02	34	17.17
		0.5	0.27	90	3.36
0		0	1.00	99	1.00
10	DMBA	0.05	0.71	103	1.69
		0.025	0.93	85	1.06
2.5		0.05	0.32	200	7.27
		0.025	0.71	309	5.06
0		0.05	0.09	180	23.25
		0.025	0.41	235	6.66
0		0	1.00	86	1.00

[a]Benz[a]anthracene at 2X the indicated concentrations was added to hamster cells followed by 2X concentrations of B[a]P or DMBA, incubated for 24 hr, rinsed once, and SA7 adsorbed for 3 hr. Chemical- and virus-treated cells were then transferred for survival (700 cells) and transformation (200,000 cells) assays.

[b]Number of foci from 10 dishes, each seeded with 2×10^5 cells.

[c]Enhancement ratio was determined by dividing the transformation frequency (foci per 10^6 surviving cells) of chemical- and virus-treated cells by that obtained from cells treated only with virus.

and a second beginning 6 hr after treatment and continuing for 18 hr (6-24 hr). In the 0- to 6-hr period, there was an approximate twofold increase over control in the average grain count per nucleus following treatment with B[a]P or DMBA, but not with MCA (Table VIII). During the subsequent 18 hr, there was a three- to fivefold increase with all three chemicals; DB[a,h]A did not appear to induce repair since no significant increase in labeling occurred during either period. Cells treated with pyrene or phenanthrene had grain counts over the nucleus similar to that of solvent-treated cells (Table VIII).

When hamster cells were exposed overnight (18 hr) to B[a]P or DMBA, lysed and the DNA separated by centrifugation through an alkaline sucrose gradient (pH 12.5), evidence for damage to the DNA was obtained by alterations in the sedimentation patterns, indicative of DNA breakage. Treatment with 1 μg/ml of DMBA caused a shift in the major peak of labeled DNA to

TABLE VIII. Induction of DNA Repair Synthesis in Syrian Hamster Embryo Cells by Polycyclic Hydrocarbons

Chemical	Initial µg/ml	Labeling Period[a] (hr)	Chemical Dilution 1:1	1:2	1:4	1:8	Control
B[a]P	10.0	0-6	2.9[b]	3.1	6.3	4.1	1.5
		6-24	18.4	11.9	8.4	12.7	2.6
DB[a,h]A	20.0	0-6	0.7	1.1	0.9	1.3	1.6
		6-24	3.8	3.8	3.5	4.2	2.9
DMBA	0.5	0-6	2.9	4.1	4.6	5.2	1.6
		6-24	8.1	13.0	17.4	19.8	2.8
MCA	10.0	0-6	2.5	2.7	2.1	2.5	1.6
		6-24	9.5	9.0	11.4	9.1	2.9
Pyrene	20.0	0-6	0.3	0.3	0.4	0.3	0.2
		6-24	3.2	3.3	2.1	2.4	2.8
Phenanthrene	20.0	0-6	0.3	0.3	0.2	0.1	0.2
		6-24	2.6	3.2	4.6	2.8	3.0

[a]HEC in arginine-free medium were pulsed with 10 µCi/ml of ^{3}H-labeled thymidine for 6 hr immediately after 18-hr treatment (0-6 hr) and during the subsequent 18 hr (6-24 hr).
[b]Average grain count per nucleus (200 cells counted).

the lighter portion of the gradient when cells were examined immediately after treatment or 24 hr later. With 0.2 µg/ml of DMBA, similar changes in the sedimentation profile were apparent when the cells were tested at 24 hr after treatment, but not when tested at the time of chemical removal (12). Following B[a]P treatment, no changes in the sedimentation pattern were obvious when the cells were lysed immediately after B[a]P exposure, but breakage was demonstrated when the assay was run 24 hr later (Figure 3). Treatment with the carcinogens, MCA (Figure 4) and DB[a,h]A, or the noncarcinogens, pyrene and phenanthrene, resulted in no change in the DNA sedimentation pattern at either time period.

DISCUSSION

Carcinogenic, but not noncarcinogenic, polycyclic hydrocarbons are active in Syrian hamster embryo cells when assayed by certain selected *in vitro* tests. In two of these tests, the capacity to transform hamster cells and enhance viral transformation, the *in vitro* activity parallels their reported *in vivo* potency, with DMBA being the most active followed by B[a]P, MCA, DB[a,h]A, DB[a,c]A and B[e]P. In terms of the concentration of

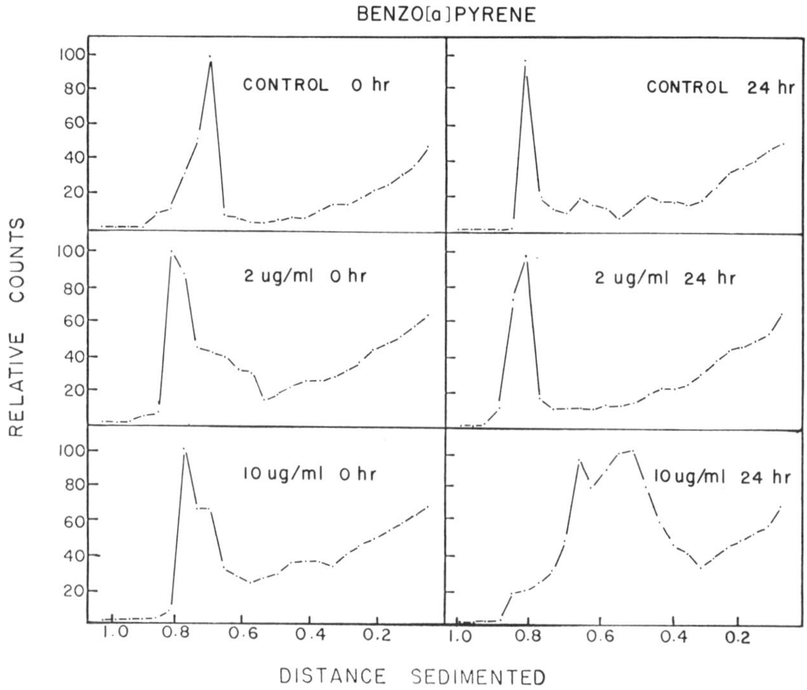

FIGURE 3. Alkaline-sucrose gradients of hamster embryo cells treated with B[a]P. Cells were prelabeled with ^{3}H-thymidine (0.5 μCi/ml for 24 hr), incubated in fresh medium for 24 hr, treated for 18 hr with B[a]P and removed from the dish with EDTA, immediately after treatment (left panels) or 24 hr later (right panels). Approximately 100,000 cells were added to the top of a 5-30% sucrose gradient (pH 12.5) into a lysing solution consisting of 1% sarkosyl in 0.05% EDTA. Cells were lysed for 1 hr at 25°, centrifuged for 1 hr at 30,000 rpm in an SW50 rotor at 20°. Three drop fractions were collected, neutralized, Bray's scintillation fluid added, and counts determined in Packard Tri-Carb scintillation spectrometer.

chemical required to demonstrate biologic activity, the enhancement of viral transformation appeared to be a more sensitive test than chemical transformation.

The qualitative and quantitative relationships that exist between the strong and weak carcinogenic polycyclic hydrocarbons in the above two tests were not as apparent when assays were performed in hamster cells for induction of DNA repair synthesis or DNA fragmentation. Although DMBA and B[a]P were positive in the latter two assays, the concentrations required were 10- to 100-fold greater than those necessary to demonstrate chemical transformation or enhancement of viral transformation; MCA was weakly positive

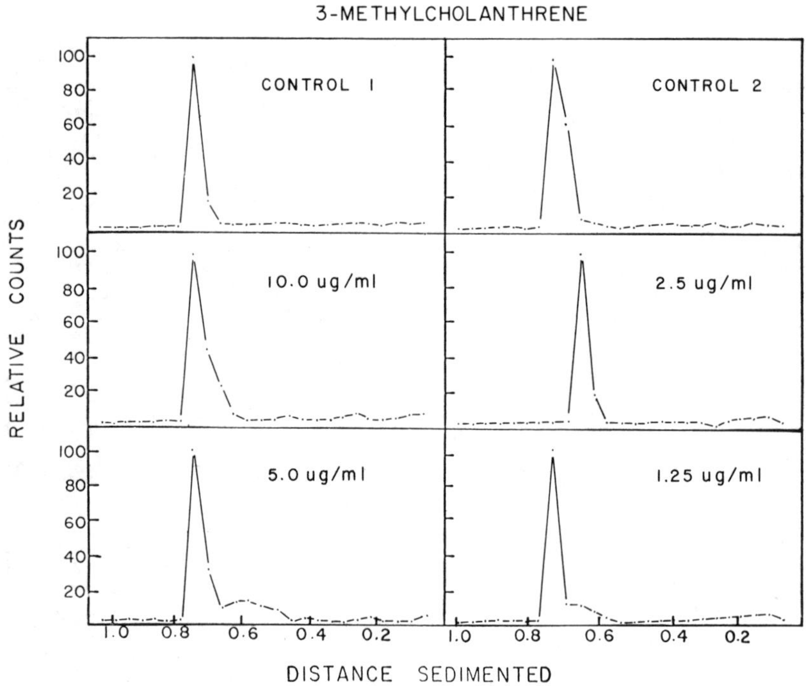

FIGURE 4. Alkaline-sucrose gradients of hamster embryo cells treated with MCA. Cells were prelabeled, treated and processed as described in Figure 3.

in the DNA repair assay and negative in the DNA fragmentation assay. The weak carcinogens DB[a,h]A, DB[a,c]A and B[e]P neither stimulated repair synthesis nor induced changes in the sedimentation of treated-cell DNA in alkaline sucrose gradients.

The sensitivity of the four *in vitro* assays, discussed here to detect activity of carcinogenic polycyclic hydrocarbons, agrees well with the results from the same assays using diverse classes of chemical carcinogens as reported by Casto *et al.* (9). Cell transformation and enhancement of viral transformation were equally sensitive for detection of carcinogens, accurately classifying as carcinogens or noncarcinogens 92% and 94%, respectively, of the chemicals tested; DNA fragmentation and induction of DNA repair assays correctly identified 72% and 50%. Some of the known carcinogens that were negative for cell transformation or enhancement of viral transformation are positive when tested in a host-mediated assay (19) or after activation by an S-9 fraction from hamster liver (Casto, unpublished data).

ACKNOWLEDGMENTS

The work upon which this publication is based was performed pursuant to NIH-NCI Contract NO1-CP-45615 with the National Cancer Institute, Department of Health, Education and Welfare.

REFERENCES

1. Alfred, L. J., and DiPaolo, J. A. (1968): Reversible inhibition of DNA synthesis in hamster embryo cells in culture: Action of 1,2-benzanthracene and 7,12-dimethyl-benz[a]anthracene. *Cancer Res.* 28:60-75.
2. Ames, B. N., Durston, W. E., Yamasaki, E., and Lee, R. D. (1973): Carcinogens are mutagens: A simple test system combining liver homogenates for activation and bacteria for detection. *Proc. Nat. Acad. Sci., USA* 70:2281-2285.
3. Berwald, Y., and Sachs, L. (1963): *In vitro* cell transformation with chemical carcinogens. *Nature, Lond.* 200:1182-1184.
4. Berwald, Y., and Sachs, L. (1965): *In vitro* transformation of normal cells to tumor cells by carcinogenic hydrocarbons. *J. Nat. Cancer Inst.* 35:641-661.
5. Casto, B. C. (1972): Biologic parameters of adenovirus transformation. *Prog. Exp. Tumor Res.* 18:166-198.
6. Casto, B. C. (1973): Enhancement of adenovirus transformation by treatment of hamster cells with UV, DNA base analogs, and dibenz-[a,h]anthracene. *Cancer Res.* 33:402-407.
7. Casto, B. C., and DiPaolo, J. A. (1974): Combined chemical and DNA viral carcinogenesis. *Excerpta Medica International Series No. 350, Vol. 2: Chemical and Viral Oncogenesis,* pp. 122-127, Excerpta Medica, Amsterdam.
8. Casto, B. C., Janosko, N., and DiPaolo, J. A. (1977): Development of a focus assay model for transformation of hamster cells *in vitro* by chemical carcinogens. *Cancer Res.* 37:3508-3515.
9. Casto, B. C., Janosko, N., Meyers, J., and DiPaolo, J. A. (1978): Comparison of *in vitro* tests in Syrian hamster cells for the detection of carcinogens. *Proc. Am. Assoc. Cancer Res.* 19:83.
10. Casto, B. C., Pieczynski, W. J., and DiPaolo, J. A. (1973): Enhancement of adenovirus transformation by pretreatment of hamster cells with carcinogenic polycyclic hydrocarbons. *Cancer Res.* 33:819-824.
11. Casto, B. C., Pieczynski, W. J., and DiPaolo, J. A. (1974): Enhancement of adenovirus transformation by treatment of hamster embryo cells with diverse chemical carcinogens. *Cancer Res.* 34:72-78.
12. Casto, B. C., Pieczynski, W. J., Janosko, N., and DiPaolo, J. A. (1976): Significance of treatment interval and DNA repair in the enhancement of viral transformation by chemical carcinogens and mutagens. *Chem.-Biol. Interact.* 13:105-125.

13. Chen, T. T., and Heidelberger, C. (1969): Quantitative studies on the malignant transformation of mouse prostate cells by carcinogenic hydrocarbons *in vitro*. *Int. J. Cancer* 4:166-178.
14. Coggin, J. H., Jr. (1969): Enhanced virus transformation of hamster embryo cells *in vitro*. *J. Virol.* 3:458-462.
15. DiPaolo, J. A., and Donovan, P. J. (1967): Properties of Syrian hamster cells transformed in the presence of carcinogenic hydrocarbons. *Exp. Cell Res.* 48:361–377.
16. DiPaolo, J. A., Donovan, P. J., and Nelson, R. (1969): Quantitative studies of *in vitro* transformation by chemical carcinogens. *J. Nat. Cancer Inst.* 42:867-874.
17. DiPaolo, J. A., Donovan, P. J., and Nelson, R. L. (1971): Transformation of hamster cells *in vitro* by polycyclic hydrocarbons without cytotoxicity. *Proc. Nat. Acad. Sci., Wash.* 68:2958-2961.
18. DiPaolo, J. A., Nelson, R. L., and Donovan, P. J. (1972): *In vitro* transformation of Syrian hamster embryo cells by diverse chemical carcinogens. *Nature* 235:278-280.
19. DiPaolo, J. A., Nelson, R. L., Donovan, P. J., and Evans, C. H. (1973): Host-mediated *in vivo-in vitro* assay for chemical carcinogens. *Arch. Pathol.* 95:380-385.
20. Freeman, A. E., Price, P. J., Bryan, R. J., Gordan, R. J., Gilden, R. V., Kelloff, G. J., and Huebner, R. J. (1971): Transformation of rat and hamster embryo cells by extracts of city smog. *Proc. Nat. Acad. Sci., Wash.* 68:445-449.
21. Gelboin, H. V., Wiebel, F., and Diamond, L. (1970): Dimethylbenzanthracene tumorigenesis and aryl hydrocarbon hydroxylase in mouse skin: Inhibition by 7,8-benzoflavone. *Science* 170:169-171.
22. Hsie, A. W., Couch, D. B., San Sebastian, J. R., Sun, W. N. C., Riddle, J. C., Brimer, P. A., Machanoff, R., Forbes, N. L., Kenney, F. T., and Hsie, M. H. (1978): Mutagenicity of carcinogens as determined in a quantitative mammalian cell mutation system. *Proc. Am. Assoc. Cancer Res.* 19:93.
23. Huberman, E., and Sachs, L. (1974): Cell-mediated mutagenesis of mammalian cells with chemical carcinogens. *Int. J. Cancer* 13:326-333.
24. Kouri, R. E., and Schechtman, L. M. (1977): State of the art: *In vitro* activation systems. In: *Short-term* In Vitro *Testing for Carcinogenesis, Mutagenesis and Toxicity,* J. Berky and C. Sherrod, Eds., pp. 423-438, The Franklin Institute Press, Philadelphia.
25. Laishes, B. A., and Stich, H. F. (1973): Repair synthesis and sedimentation analysis of DNA of human cells exposed to dimethylnitrosamine and activated dimethylnitrosamine. *Biochm. Biophys. Res. Commun.* 52:827-833.
26. Ledinko, N., and Evans, M. (1973): Enhancement of adeno-virus transformation of hamster cells by N-methyl-N′-nitro-N-nitrosoguanidine, caffeine, and hydroxylamine. *Cancer Res.* 33:2936-2938.

27. Malling, H. V. (1971): Dimethylnitrosamine: Formation of mutagenic compounds by interaction with mouse liver microsomes. *Mutation Res.* 13:425-429.
28. Marquardt, H., and Heidelberger, C. (1972): Influence of "feeder cells" and inducers and inhibitors of microsomal mixed-function oxidases on hydrocarbon-induced malignant transformation of cells derived from C3H mouse prostate. *Cancer Res.* 32:721-725.
29. McCann, J., Choi, E., Yamasaki, E., and Ames, B. N. (1975): Detection of carcinogens as mutagens in the *Salmonella*/ microsome test: Assay of 300 chemicals. *Proc. Nat. Acad. Sci., USA* 72:5135-5139.
30. McGrath, R. A., and Williams, R. W. (1966): Reconstruction *in vivo* of irradiated *Escherichia coli* deoxyribonucleic acid: The rejoining of broken pieces. *Nature, Lond.* 212:534-545.
31. Milo, G. E., Schaller, J. P., and Yohn, D. S. (1972): Hormonal modification of adenovirus transformation of hamster cells *in vitro. Cancer Res.* 32:2338-2347.
32. Mittelman, L. A., Sharovskaya, J. J., and Vasiliev, J. M. (1972): Toxic effect of 7,12-dimethylbenz[a]anthracene on neoplastic cells grown in mixed cultures with normal fibroblasts. *Int. J. Cancer* 10:667-674.
33. Mondal, S., Brankow, D. W., and Heidelberger, C. (1977): Syncarcinogenesis in cultured C3H/10T½ cells. *Proc. Am. Assoc. Cancer Res.* 18:8.
34. Painter, R. B., and Cleaver, J. E. (1967): Repair replication in HeLa cells after large doses of X-irradiation. *Nature* 216:369-370.
35. Pienta, R. J., Poiley, J. A., and Lebherz, W. B., III. (1977): Morphological transformation of early passage golden Syrian hamster embryo cells derived from cryopreserved primary cultures as a reliable *in vitro* bioassay for identifying diverse carcinogens. *Int. J. Cancer* 19:642-655.
36. Purchase, I. F. H., Longstaff, E., Ashby, J., Styles, J. A., Anderson, D., Lefevre, P. A., and Westwood, F. R. (1976): Evaluation of six short-term tests for detecting organic chemical carcinogens and recommendations for their use. *Nature* 264:624-627.
37. Reed, S. E. (1967): Transformation of hamster cells *in vitro* by adenovirus Type 12. *J. Gen. Virol.* 1:405-412.
38. Reznikoff, C. A., Bertram, J. S., Brankow, D. W.; and Heidelberger, C. (1973): Quantitative and qualitative studies of chemical transformation of cloned C3H mouse embryo cells sensitive to postconfluence inhibition of cell division. *Cancer Res.* 33:3239-3249.
39. Rhim, J. S., Creasy, B., and Huebner, R. J. (1971): Production of altered cell foci by 3-methylcholanthrene in mouse cells infected with AKR leukemia virus. *Proc. Nat. Acad. Sci., USA* 68:2212-2216.
40. San, R. H. C., and Stich, H. F. (1975): DNA repair synthesis of cultured human cells as a rapid bioassay for chemical carcinogens. *Int. J. Cancer* 16:284-291.
41. Stich, H. F., San, R. H. C., Lam, P. P. S., Koropatnick, D. J., Lo, L. W., and Laishes, B. A. (1976): DNA fragmentation or DNA repair as an

in vitro and *in vivo* assay for chemical procarcinogens, carcinogens, and carcinogenic nitrosation products. In: *Screening Tests in Chemical Carcinogenesis,* R. Montesano, H. Bartsch and L. Tomatis, Eds., pp. 617-636, IARC Scientific Pubications No. 12, Lyon.

42. Swenberg, J. A., Petzold, G. L.; and Harbach, P. R. (1976): *In vitro* DNA damage/alkaline elution assay for predicting carcinogenic potential. *Biochem. Biophys. Res. Comm.* 72:732-738.

Polynuclear Aromatic Hydrocarbons, edited by P. W. Jones and P. Leber. Ann Arbor Science Publishers, Inc. Ann Arbor, MI ©1979.

QUANTITATIVE EXTRACTION OF POLYCYCLIC AROMATIC HYDROCARBONS AND OTHER HAZARDOUS ORGANIC SPECIES FROM PROCESS STREAMS USING MACRORETICULAR RESINS

Zoe A. Grosser, Judith C. Harris and Philip L. Levins

Arthur D. Little, Inc.
Cambridge, Massachusetts 02140

INTRODUCTION

Qualitative and quantitative analysis of organic pollutants in aqueous industrial effluents has received increasing attention in environmental assessment activities. Recently promulgated EPA sampling and analysis procedures for organics, directed towards comprehensive characterization (7) or specific compound quantification (1), have called for collection and subsequent solvent extraction of rather large (1-10 liters) aqueous samples. This approach is capable of giving quantitative recoveries of a wide range of nonpolar organic materials. However, transport and liquid-liquid extraction of the large-volume aqueous samples, required to meet the detection-limit constraints, is cumbersome and frequently complicated by the formation of emulsions. In addition, quantitative recovery of very polar water-soluble organic species is not always achievable by solvent extraction. The decrease in the concentrations of organics in aqueous effluents considered to be of environmental concern and the increase in emphasis on polar organics, such as oxygenated and heterocyclic polynuclears, can be expected to accentuate the drawbacks of the solvent extraction approach.

An attractive alternative approach for collection of organics from large quantities of water is the use of an *in situ* continuous extractive sampler containing a solid sorbent material. One of the earlier devices of this type, once widely used in the field, was the EPA's carbon absorption module (CAM) (18) with generally good collection efficiency, but low recoveries, for organics. A number of previous workers (5,8,9,16,17,19,23) have done laboratory studies on the use of macroporous resins and have determined overall recoveries of selected organic compounds from water, under conditions designed to simulate sampling applications. The work of Simpson (15) and Van Rossum and Webb (22) compares the adsorption and recovery of organic compounds on macroreticular XAD resins with that on carbon. Other investigators (2,3,12) have explored the use of polyurethane foam plugs for collection of organics, especially polycyclic aromatic hydrocarbons.

Anion exchange resins have been used successfully in several instances for the concentration of polar or ionizable organic compounds (4,10). Anion exchange resins used by Pollio and Kunin (14) gave good adsorption for phenol.

The Ambersorb carbonaceous adsorbents have not yet been extensively researched as possible adsorbents for PAH or other compounds from water. Neely (13) shows Ambersorb XE-340 to be an effective adsorbent for trihalomethanes from water. Mullins *et al.* (11) compare the effectiveness of several activated carbons to XE-340.

The present work involves a systematic laboratory investigation of the applicability of macroreticular resins for general and compound specific sampling of organics, based on small-scale chromatographic experiments.

EXPERIMENTAL APPROACH

The basic approach used in this work was the generation and analysis of the frontal breakthrough curve produced by challenging a sorbent cartridge with an aqueous solution of organic analyte at constant flowrate. Figure 1 is a schematic representation of a frontal breakthrough curve, indicating the

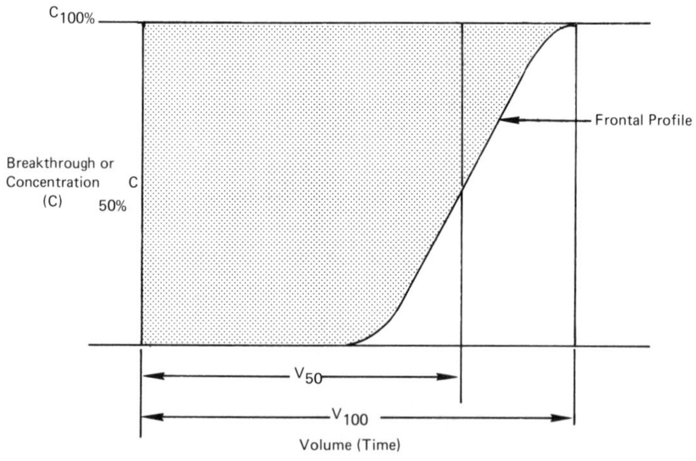

FIGURE 1. Schematic of frontal breakthrough curve.

50% and 100% breakthrough points. The entire shaded area in front of the curve corresponds to the total quantity of analyte taken up by the sorbent, up to 100% breakthrough. Interpretation of the frontal breakthrough curve according to established chromatographic principles (6) allows calculation of the adsorption isotherm.

The experimental apparatus employed a small-scale (6.0 cm x 0.5 cm i.d. or 3.2 cm x 0.5 cm i.d.) sorbent cartridge connected to an LDC UV monitor and Linear Instrument recorder. Test solutions (0.5-100 ppm) prepared using degassed distilled deionized water were introduced via a Milton Roy mini pump or Fluidmeter, Inc. pump. Prior to each test run, a system baseline was established by pumping distilled deionized water through the sorbent bed. The detector response corresponding to 100% breakthrough of each analyte was determined by shunting the test solution through a sorbent bypass loop. Figure 2 shows a typical experimental breakthrough curve and its associated 100% calibration.

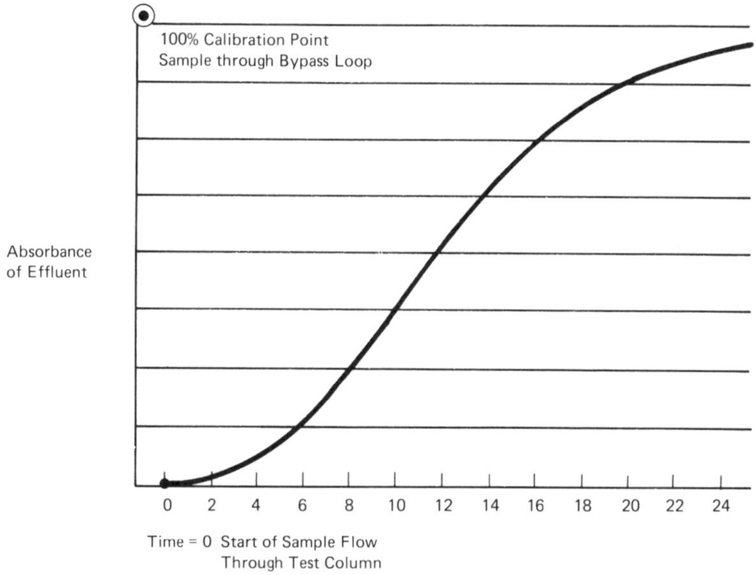

FIGURE 2. Experimental breakthrough curve for 100 ppm benzoic acid on Ambersorb XE-347.

In several experiments, the desorption branch of the curve was also generated by switching to a pure water feed after 100% breakthrough of analyte had been accomplished.

Regeneration of sorbents between breakthrough tests was accomplished by passing at least 20 bed volumes of methanol through the column and then reequilibrating with distilled deionized water.

The sorbent resins studied were obtained from Rohm & Haas and cleaned by sequential 24-hr Soxhlet extractions with distilled, deionized water, methanol and methylene chloride. The cleaned resins were stored in methanol. All other materials used were reagent-grade.

TREATMENT OF DATA

The frontal breakthrough curve generated in each experiment was analyzed to determine the characteristic 50% breakthrough parameters and the adsorption isotherm.

The specific retention volume corresponding to 50% breakthrough, V_{50}, is the number of sorbent bed volumes (BV) of aqueous solution that have passed through the bed up to the time when the exit concentration of organic species equals one half of the challenge concentration. The V_{50} value indicates the volumetric sampling capacity of a given sorbent cartridge for a particular organic material in water and is independent of challenge concentration as long as the latter is sufficiently low. At higher challenge concentrations, V_{50} will decrease because of overloading of the resin. The V_{50} value allows one to determine whether an organic compound is retained by a given volume of sorbent at a specified flowrate and sampling time.

The weight capacity of the sorbent for organic species, in mg/cc of resin, corresponding to 50% breakthrough is given by V_{50} x challenge concentration. The weight capacity is thus a function of challenge concentration, subject to an upper limit corresponding to saturation of the bed. The weight capacity, while useful for comparisons between laboratory experiments, is less valuable than V_{50} as a sampling-device design parameter.

The adsorption isotherm, a plot of uptake of solute in mg/cc of resin versus percent breakthrough, was calculated by integrating the area in front of the breakthrough curve at 5, 25, 50, 75 and 100% breakthrough. For example, the isotherm point corresponding to 50% breakthrough on the frontal chromatogram (shaded area of Figure 3) was calculated by a computer program using the algorithm:

$$\text{mg/cc of resin} = \frac{1}{\text{BV}} \times \frac{\text{Area abcd}}{\text{Area abce}} \times V_{100} \times C_{100}$$

Adsorption isotherms contain valuable information on the affinity of the resin for a sorbate, and allow the saturation capacity of the sorbent to be ascertained. If the breakthrough curve is generated under equilibrium sorption

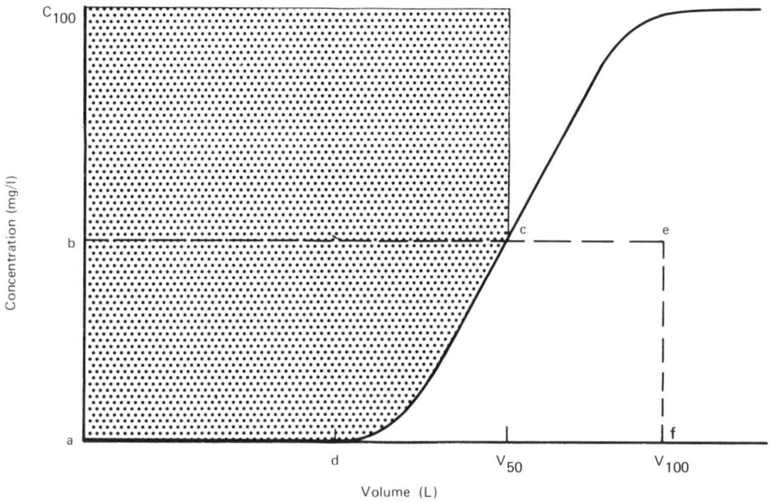

FIGURE 3. Illustration of isotherm calculation from breakthrough curve.

conditions, the isotherm is expected to show a Langmuir shape—linear in the low uptake region and reaching a plateau corresponding to saturation of the available adsorption sites. Deviation of the isotherm from the expected Langmuir shape is an indication of significant mass transfer effects in the frontal chromatography experiment.

RESULTS AND DISCUSSION

Effect of Flowrate

The initial studies in this program were done using XAD-2, the resin most extensively previously studied for collection of organic species from water. A series of experiments was performed to determine the effect of aqueous sample flowrate on the adsorption of a moderately polar organic species, acetophenone, on XAD-2. As shown in Figure 4, distinctly non-Langmuir isotherms were obtained at sample flowrates corresponding to linear velocities through the sorbent bed of 8 cm/min or higher. Note that the uptake at 50% breakthrough is also very much lower at the higher flowrates tested. These flowrate studies clearly show that mass transfer effects have an important impact on adsorption of acetophenone from water onto XAD-2, even at relatively low flowrates. It was necessary to reduce the flowrate to about 4 cm/min linear velocity in order to achieve the expected isotherm shape and maximum uptake. This linear velocity was maintained in subsequent breakthrough experiments with other analytes and sorbents.

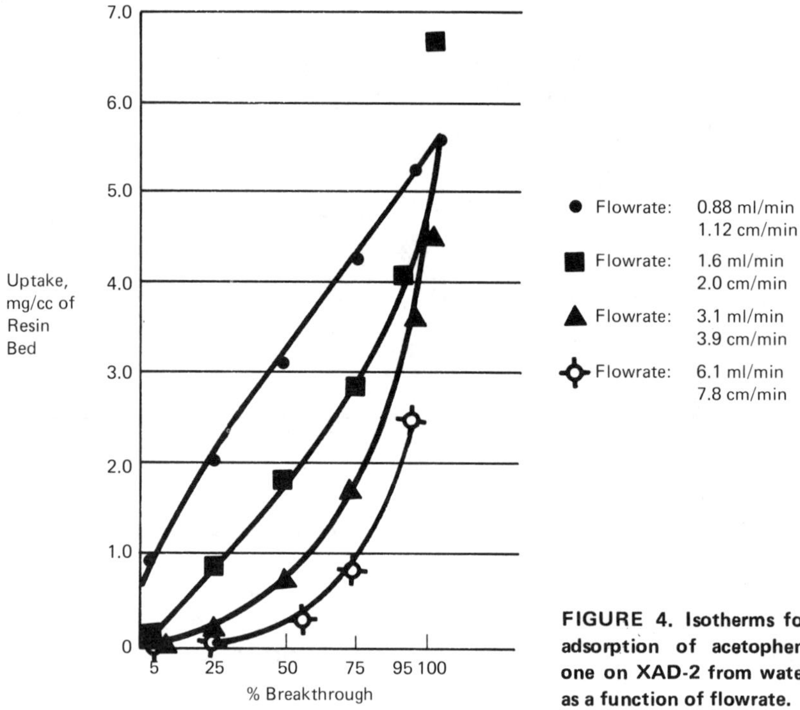

FIGURE 4. Isotherms for adsorption of acetophenone on XAD-2 from water as a function of flowrate.

These pronounced mass transfer effects on adsorption at relatively low flowrates had not been expected based on results previously published. For example, Strup *et al.* (17) had reported good recoveries of PAH using XAD-2 at water flowrates corresponding to $>$ 600 cm/min linear velocity. One hypothesis that reconciles these disparate observations is that adsorption of species for which XAD-2 has limited affinity, such as acetophenone, is limited by diffusion into the pores of the macroreticular resin particles, while more readily sorbed species, such as PAH, are efficiently captured on the exterior surface of the particles. Support for this hypothesis is provided by reports (2,3,12) that PAH can be sampled using relatively low surface-area sorbents such as foam plugs. Also, the present work has shown that very poorly sorbed species, such as phenol, show mass transfer effects even at the lowest velocities tested. The flowrate dependence study illustrates the fact that sampling approaches proved effective for PAH in water are often not directly applicable to other organics.

It should be noted that once an appropriate linear velocity criterion has been determined for a sorbent/analyte combination in a chromatographic experiment, it can be used in design of a large-scale sampling system. For example, assuming a 4 cm/min velocity criterion and a 4-hr, 10 liter water-sampling constraint, one can calculate that a sorbent cartridge of at least 4 cm diameter would be required.

Effect of Compound Type

A variety of organic compounds in aqueous solution were examined to determine their frontal chromatographic behavior on XAD-2. Because of the interest in extrapolating to field sampling of industrial effluents, pH adjustments were not made in these tests. The 50% breakthrough parameters are presented in the first column of Table I. The results confirm the observations of other workers (8,9,17,23) that XAD-2 has a high affinity for nonpolar aromatic species; no measureable breakthrough was obtained for these compounds. The XAD-2 also seems to be an adequate sorbent, for sampling purposes, for species such as ketones and esters. The low affinity of XAD-2 for very polar organics such as phenol and benzoic acid is confirmed.

It has been suggested (20,21) that the efficiency of XAD-2 for removing organics from aqueous streams is correlated to their water solubility. In the present work, no correlation was found between water solubility and the breakthrough parameter, Vg, or chromatographic capacity factor, K′. This may reflect the rather narrow range of aqueous solubilities (3000 to 40,000 ppm) of compounds for which breakthrough curves could be determined.

Effect of Sorbent Type

Several other macroreticular resins (Table II) were tested to explore their potential for use as alternatives or supplements to XAD-2 for collection of polar species. The breakthrough parameters for these resins are summarized in Table I. The XAD-7 was found to have higher affinity for phenol, but lower affinity for benzoic acid, than XAD-2. A similar result had been reported by Thurman *et al.* (20), who interpreted their results in terms of specific π-π interactions between benzoic acid and XAD-2 not possible with polyacrylic XAD-7.

The ion exchange materials IRA-904 and IRA-93 showed $>$ 10 times more capacity for phenol and benzoic acid in distilled water than did XAD-2 or XAD-7. However, breakthrough on these resins was virtually instantaneous if the aqueous solution contained as much as 250 ppm of sodium chloride, an amount commonly found in aqueous effluents. This is not entirely unexpected in that the manufacturer of these resins recommends use of salt solutions as the most effective regeneration method. Figure 5 shows the unusual breakthrough curve obtained in these tests. This curve shape can be explained by competition between the phenol and chloride ions for the hydroxide exchange sites on the resin. When the flow through the resin is started, the

TABLE I. Comparison of Uptake on Various Resins at 50% Breakthrough for 10 ppm Challenge Concentration

Resin	XAD-2		XAD-7		IRA-904	IRA-93	XE-347
Analyte	Weight (mg/cc) Capacity	Volume BVs Capaci	Weight (mg/cc) Capacity	Volume Capacity	Weight Capacity	Weight Capacity	Weight Capacity
Phenol	0.43	74.7	0.60	67.9	[2]*(>14.3)	*[1]	*[1]
Benzoic Acid	0.54	102	0.32	39.2	[2]*(>14.3)	*[1]	*[1]
Benzyl Alcohol	0.53	98.9			0.164		*[1]
Acetophenone	3.2	527	0.89	96.6			[2]*(>712)
Chlorobenzene	*(>3.8)						
Methyl Benzoate	11.4	1,750					
Napthalene	[1]*(>3.8)						
Carbazole					0.58[3]		[2]*(>0.72)
Acridine					0.32[3]		[2]*(>1.42)

[1] No breakthrough in 4 hr, therefore volumetric capacity not determinable. (Number in parenthesis is amount sorbed up to that point.)
[2] No breakthrough in 16 hr, therefore volumetric capacity not determinable. (Number in parenthesis is amount sorbed up to that point.)
[3] Uptake at 40% breakthrough.

TABLE II. Sorbents Studied as Alternatives to XAD-2

Sorbent	Description	Structure
XAD-7	Nonionic Adsorbent Acrylic Ester Polymer	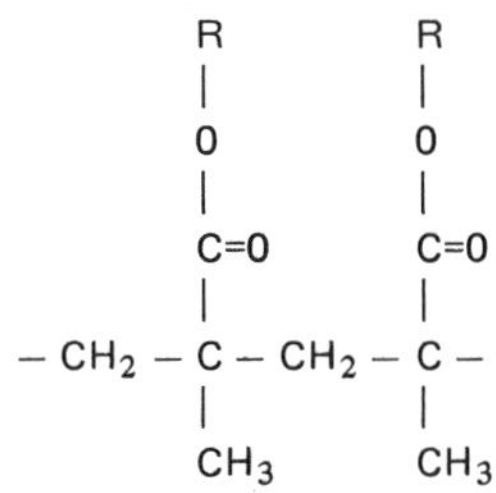
IRA-904	Strong Anion Exchange Resin Styrene Divinyl Benzene Quaternary Ammonium Functionality	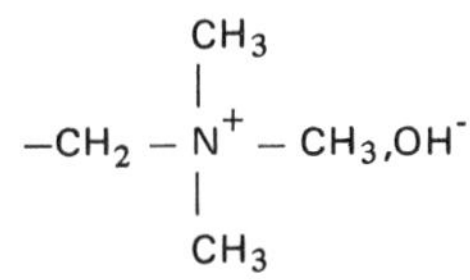 Hydroxide Form
IRA-93	Weak Anion Exchange Resin Styrene Divinyl Benzene Polystyrene Polyamine Functionality	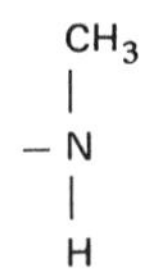 Free Base Form
XE-347	Carbonaceous Adsorbent	

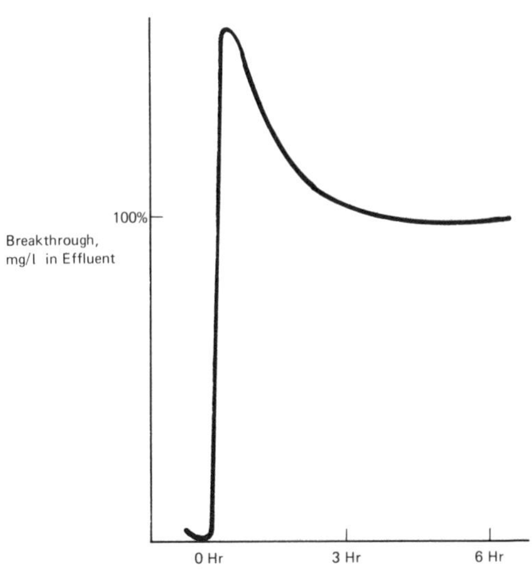

FIGURE 5. Unusual breakthrough curve obtained using ion exchange resins in the presence of 250 ppm sodium chloride.

concentration of the two ions is small and they both exchange onto the resin. As the exchange sites are filled, chloride ions exchange for phenol ions as well as hydroxide ions. Hence the breakthrough exceeds 100% due to an increase in concentration of phenol in the effluent above that of the influent.

The carbonaceous resin, XE-347, showed very high affinity for all of the compounds tested. Except for the very polar analytes, no breakthrough was observed in a 16-hr period.

Scaleup of Breakthrough Experiments

A principal advantage of doing the frontal breakthrough experiments in an equilibrium regime is that one can expect to be able to predict the behavior of the sorbent/analyte system under other experimental conditions. Figure 6 illustrates that this expectation was met in one pair of experiments. A fivefold scaleup at constant linear velocity and challenge concentration gave an isotherm very similar to that of the smaller-scale test and showed $<10\%$ difference in the specific retention volume (mL/BV).

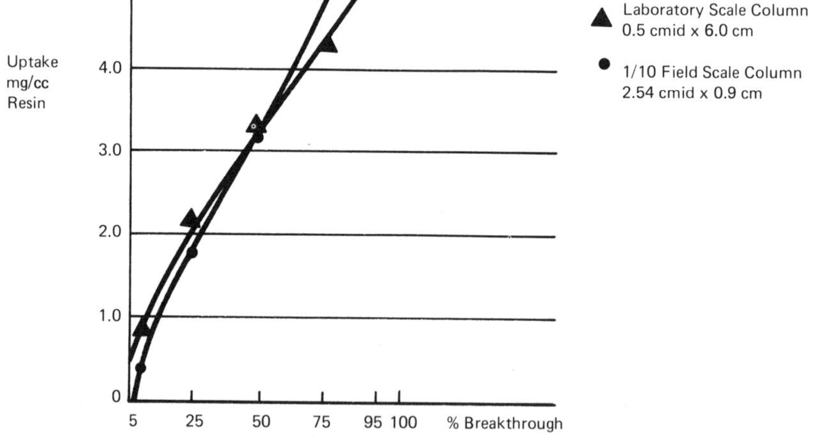

FIGURE 6. Comparison of isotherms obtained for fivefold experimental scaleup.

CONCLUSIONS

The conclusions that can be drawn from this ongoing study include the following:

1. XAD-2 is an effective resin for sampling nonpolar and moderately polar organics; it is not effective for very polar species.
2. The linear velocity of water passed through an XAD-2 bed should be very low for efficient collection of moderately polar organics. PAH collection is not especially sensitive to flowrate.
3. Of the macroreticular resins tested, as possible backups to an XAD-2 sampler, the carbonaceous XE-300 series appears most promising for very polar species.
4. The chromatographic approach used in this work allows rapid simple quantitative evaluation of resins for sampling applications. The results can be extrapolated to larger-scale systems.

ACKNOWLEDGMENTS

This work was performed under EPA Contract No. 68-02-2150 from the Process Measurements Branch, Industrial Environmental Research Laboratory, Research Triangle Park, N.C. The authors wish to thank Ms. Karen C. Weaver for her contribution to the laboratory work.

REFERENCES

1. Anonymous (1977): Sampling and analysis procedures for screening of industrial effluents for priority pollutants. Available from Environmental Monitoring and Support Laboratory, EPA, Cincinnati, OH 45268.
2. Basu, D. K., and Saxena, J. (1978): Monitoring of polynuclear aromatic compounds in water II. Extraction and recovery of six representative compounds with polyurethane foams. *Environ. Sci. Technol.* 12:791-795.
3. Basu, D. K., and Saxena, J. (1978): Polynuclear aromatic hydrocarbons in selected U.S. drinking waters and their raw water sources. *Environ. Sci. Technol.* 12:795-798.
4. Chasanov, M., Kunin, R., and McGarvey, F. (1956): Sorption of phenols by anion exchange resins. *Ind. Eng. Chem.* 48(2):305-309.
5. Coburn, J., Valdmanis, I., and Chau, S. (1977): Evaluation of organochlorine pesticides and polychlorinated biphenyls from natural waters. *J. Assoc. Official Anal. Chem.* 60(1):224-228.
6. Gallant, R. F., King, J. W., Levins, P. L., and Piecewicz, J. F. (1978): Characterization of sorbent resins for use in environmental sampling. EPA-600/7-78-054.

7. Hamersma, J. W., Reynolds, S. L., and Maddalone, R. F. (1976): *IERL-RTP* Procedures manual: Level 1 environmental assessment. EPA-600/2-76-16a.
8. Junk, G. A., Richard, J. J., Fritz, J. S., and Svec, H. J. (1976): Resin sorption methods for monitoring selected contaminants in water. In: *Identification and Analysis of Organic Pollutants in Water,* L. H. Keith, Ed., pp. 135-158, Ann Arbor Science, Ann Arbor, MI.
9. Junk, G., Richard, J., Grieser, M., Witiak, D., Witiak, J., Arguello, M., Vick, R., Svec, H., Fritz, J., and Calder, G. (1974): Use of macroreticular resins in the analysis of water for trace organic contaminants. *J. Chromatog.* 99:745-762.
10. Kim, B.; Snoeyink, V., and Saunders, F. (1976): Adsorption of organic compounds by synthetic resins. *J. Water Poll. Control Fed.* 48(1): 120-133.
11. Mullins, R., Zogorski, J., Hubbs, S., Allgeier, G., Arbuckle, A., and Wilding, D. (1978): Effectiveness of several brands of granulated activated carbon for the removal of trihalomethanes from drinking water. Presented before the Division of Environmental Chemistry, American Chemical Society, Miami, FL (200-203).
12. Navratil, J., Sievers, R., and Walton, H. (1977): Open-pore polyurethane columns for collection and preconcentration of polynuclear aromatic hydrocarbons from water. *Anal. Chem.* 49(14):2260-2263.
13. Neely, J. (1978): A model for the removal of trihalomethanes from water by Ambersorb XE-340. Presented before the Division of Environmental Chemistry, American Chemical Society, Miami Beach, FL (204-206).
14. Pollio, F., and Kunin, R. (1967): Sorption of phenols by anion exchange resins. *Environ. Sci. Technol.* 1(2):160-163.
15. Simpson, R. (1972): The separation of organic chemicals from water. Presented at the Third Symposium of the Institute of Advanced Sanitation Research, International.
16. Stepan, S., and Smith, J. (1977): Some conditions for use of macroreticular resins in the quantitative analysis of organic pollutants in water. *Water Res.* 11(4):339-342.
17. Strup, P. E., Wilkinson, J. E., and Jones, P. W. (1978): Trace analysis of polycyclic aromatic hydrocarbons in aqueous systems using XAD-2 resin and capillary column gas chromatography-mass spectrometry analysis. In: *Polynuclear Aromatic Hydrocarbons: Second International Symposium on Analysis, Chemistry and Biology,* P. W. Jones and R. E. Freudenthal, Eds., pp. 131-138, Raven Press, New York.
18. Taras, M. J., Greenberg, A. E., Hoak, R. D., and Rand, M. C., Eds. (1971): Organic contaminants. In: *Standard Methods for the Examination of Water and Wastewater,* 13th ed., pp. 259-270, American Public Health Association, Washington, D.C.
19. Tateda, A., and Fritz, J. (1978): Mini column procedure for concentrating organic contaminants from water. *J. Chromatog.* 152:329-340.

20. Thurman, E. M., Aiken, G. R., and Malcolm, R. L. (1977): The use of macroreticular nonionic resins to preconcentrate trace organic acids from water. Presented at the 4th Joint Conference on Sensing of Environmental Pollutants, New Orleans.
21. Thurman, E. M., Malcolm, R. L., and Aiken, G. R. (1978): Prediction of capacity factors for aqueous organic solutes adsorbed on a porous acrylic resin. *Anal. Chem.* 50:775-779.
22. Van Rossum, P., and Webb, R. (1977): Isolation of organic water pollutants by XAD resins and carbon. Unpublished draft report.
23. Webb, R. G. (1975): Isolating organic water pollutants: XAD resins, urethane foams, solvent extraction. EPA 660/4-75-003.

Chemistry

Session Chairmen

A. P. D'Silva
K. T. Knapp
L. D. Johnson
R. B. Gammage

Polynuclear Aromatic Hydrocarbons, edited by P. W. Jones and P. Leber. Ann Arbor Science Publishers, Inc. Ann Arbor, MI ©1979.

PAH EMISSIONS FROM A STRATIFIED-CHARGE VEHICLE WITH AND WITHOUT OXIDATION CATALYST: SAMPLING AND ANALYSIS EVALUATION

Frank S.-C. Lee, T. J. Prater and F. Ferris

Scientific Research Staff
Ford Motor Company
Scientific Research Laboratories
20000 Rotunda Drive
Dearborn, Michigan 48121

INTRODUCTION

The presence of complex mixtures of polynuclear aromatic hydrocarbons (PAH) in automobile exhaust has long been recognized and demonstrated (1,8,9). Since some of the PAH are reported carcinogens (14,17), there has been concern about the possible health effects of automobile PAH emissions. Such concern has given rise to a number of studies regarding the measurements of PAH in gasoline (3-6) and, more recently, diesel (18,19,21) engine exhaust. Although a wealth of data has been accumulated from these reports, the need still exists for the development and validation of an analytical and, even more so, a sampling procedure for the quantitative measurement of PAH in the exhaust. It was toward this goal that the present work was initiated. The design of the experiment to be reported here was intended to serve as a preliminary study preceding a more comprehensive investigation of the exhaust PAH emission problem.

Methods for PAH analysis have been reported extensively in the literature (10). Recent advances in high-performance liquid chromatography (HPLC) and gas chromatography/mass spectrometry (GC/MS) techniques have now made it possible to determine ng to pg levels of PAH in complex environmental samples. We have applied both techniques in our PAH measurement procedure, with emphasis placed on the validation of the analytical method and the comparison of HPLC and GC/MS techniques in exhaust PAH analysis.

Since automobile exhaust contains a wide variety of components of which PAH represent only a minor portion, a large volume of exhaust needs to be collected for subsequent PAH analysis. We have used a dilution tube for exhaust sampling. All the exhaust from the test vehicle was injected into the dilution tube. In the tube the exhaust was cooled and diluted by filtered air. The diluted exhaust was then sampled isokinetically by a filter-plus-absorbent collection system placed near the end of the dilution tube. With this sampling design, we were able to obtain reasonably reproducible PAH emission measurements from the test vehicle running under the same operational conditions. However, a number of problems such as sampling losses and PAH degradations were also noticed from a series of validation experiments. This preliminary but potential crucial observation is further discussed in the Sampling Recovery Section. It should also be noted that the utilization of the dilution tube for exhaust sampling is currently under considerable debate centering about the application of results obtained from the method to real-world situations (21). We are continuing our effort to assess this problem. Accordingly, the work discussed herein should be viewed as a status report and the measured values for PAH emissions should be judged on a relative, rather than absolute, basis.

The sampling and analytical procedures developed have been used to measure the PAH emissions from a test truck. The truck was powered by a prototype stratified-charge engine currently being developed at Ford. Of particular interest here is the effect of the oxidation catalytic converter on PAH emissions. It was found that such catalysts effectively oxidize the PAH in the exhaust, resulting in a large reduction of PAH emissions as compared with noncatalytic control systems.

EXPERIMENTAL

Engine Testing and Dilution Tube Facilities

The engine testing and dilution tube facilities used in this work have been reported previously (13) and are, therefore, only briefly described here. The vehicle used was a Ford F-250 light-duty truck. It was powered by a 460 CID, prototype stratified-charged gasoline-powered internal combustion engine under development at Ford. The vehicle was operated on a chassis dynamometer. The catalysts used were commercial 1,557 cm^3 oxidation catalysts. Indolene clear fuel spiked with di-t-butyl-disulfide (0.03 wt % S) was used throughout the work.

All the exhaust from the vehicle was injected into a dilution tube and diluted by prefiltered air to about 10-20 to 1. A diagram showing the design of the dilution tube is given in Figure 1.

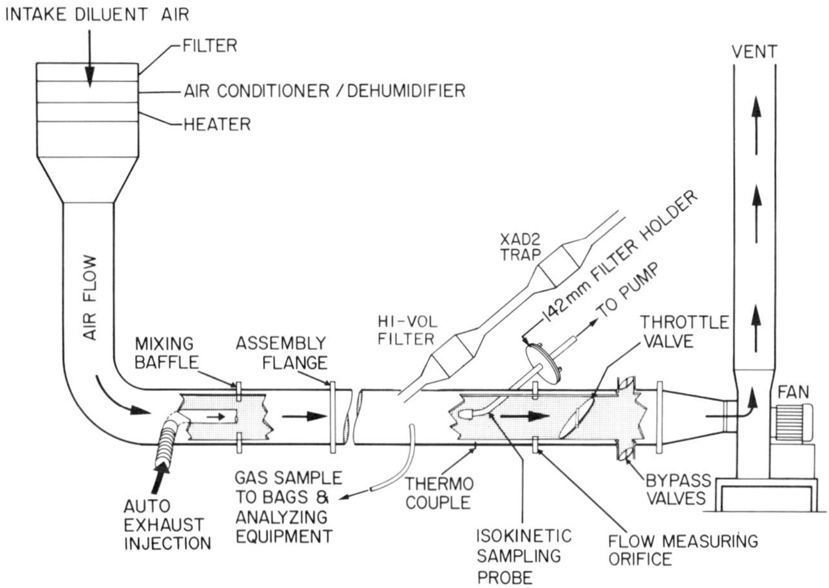

FIGURE 1. Diagram of the exhaust dilution tube facilities.

Before each sample collection, the truck was accelerated several times from 50 to 90 kph. It was then run steadily at 50 kph and allowed to equilibrate for 10-20 min before sampling.

Sample Collection System

A portion of the diluted exhaust in the dilution tube was removed at the sampling site for PAH analysis (Figure 1). The sample collection system consisted of a quartz-fiber filter (20 x 25 cm) for filterable particulate-associated PAH, and an adsorbent trap containing 20/50 mesh Amberlite XAD-2 resin downstream of the filter for the collection of gas-phase components. The total flowrate in the tube was about 350 cfm for noncatalyst runs and about 600 cfm for catalyst runs. Up to 10% of the total flow, and therefore 10% of the vehicle exhaust was sampled. A collection time of 30 min to 2 hr at a sampling rate of 30 cfm generally provided sufficient sample for analysis.

The quartz-fiber filter was used directly without pretreatment. It was placed in a Hi-Vol sampler during sampling. The collecting conditions were 37°C and 20% relative humidity. The filters were equilibrated and weighed conventionally before and after sampling, and the amount of particulate matter was determined by difference. Before use, the XAD-2 resin was precleaned by a procedure reported by Strup *et al.* (20). The resin was packed in a 6 in. diameter stainless-steel adsorbent sampler to approximately 2 in. in thickness. The resin was held inside the sampler by two pieces of stainless-steel screen.

Reagent and Analytical Instrumentation

All the PAH standard material were obtained from Aldrich, Analabs Inc., and Columbia Organics Chemical Co. All the solvents used in this work were UV-grade, distilled-in-glass solvents obtained from Burdick and Jackson. Radioactive-labeled ^{14}C-benzo[a]pyrene and ^{14}C-benzo[a] anthracene were obtained from the Amersham Company. Each was diluted to a specific activity of about one μCi/ml, and stored in benzene solution before use. The scintillation cocktail used for counting was Insta-Gel for water-containing samples and Instra-Fluor for organic-based samples, respectively, both from Packard.

An HPLC system from Waters Associates, Inc. was used, which consists of two Model 6000 pumps, a Model U6K injector, a Model 660 gradient programmer and a Model 440 dual micro UV detector. A Schoeffel model FS-970 LC spectrofluorometer was connected downstream of the UV detector in the HPLC system for fluorescence measurements. HPLC separation was carried out using either Waters micro-Bondapak C_{18} columns or a Hi-Eff C_{18} column obtained from Applied Sciences, Inc.

The gas chromatograph/mass spectrometer (GC/MS) system consisted of a Perkin-Elmer 910 gas chromatograph interfaced to a VG micromass MM16 magnetic sector mass spectrometer and a Finnigan Model 2400 Incos data system. The GC columns used were either packed or capillary Dexsil 300 columns. Oncolumn sample injection was used for packed columns and a falling-needle solid injector and concentrator was used for capillary columns. The MM16 mass spectrometer was an EI, positive-ion CI and negative-ion CI instrument, with an approximate resolving power of 2,000. The Finnigan data system was employed for data acquisition and reduction of the mass spectral data. This system provides quantitative analysis and component identification from a 25,000-compound library. Identifications were also made by the Finnigan data system from a limited PAH library created by the users.

Analytical Procedures and Controls

After sample collection, the particulate-laden filter and XAD-2 traps were analyzed separately by a procedure involving solvent extraction, PAH-enrichment step and, finally, HPLC and GC/MS analysis. Analytical procedures were performed in subdued room light in order to minimize possible photodegradation of PAH.

Solvent extraction was carried out in a Soxhlet extractor. Extraction was generally performed overnight, *i.e.*, 14-16 hr, with approximately 250 ml of methylene chloride at a temperature of 40°C. After extraction, the crude extract was allowed to cool to room temperature and then evaporated to near dryness in a rotary evaporator followed by nitrogen blowdown. About 20 ml of cyclohexane was added to the residual methylene chloride extract and the mixture was further concentrated to a volume of about 3 to 5 ml.

The aromatic components including PAH in the above cyclohexane concentrate were then selectively enriched using the procedure developed by Grimmer, Bohnke and Glaser (6). This method involves the use of two solvent-solvent extraction steps in which the selective partitioning of PAH between dimethyl formamide/water and cyclohexane phases essentially concentrates the PAH in the cyclohexane phase during the last extraction step.

The PAH-enriched cyclohexane extract was further purified by either a Sil-Gel or an Alumina column (10% H_2O deactivated: 15 cm long). Residual aliphatic hydrocarbons were removed for GC/MS analysis with a Sephadex LC separation using a methanol mobile phase. In some cases, the last step was omitted for the HPLC analysis, since the aliphatics generally lack UV and fluorescence responses.

The exhaust samples obtained from the above procedure were generally reduced to a volume of 100-1,000 μl for HPLC or GC/MS analysis. When samples were not analyzed immediately after LC separations, they were stored in the dark in a 0°C freezer.

HPLC analysis was accomplished by reverse-phase column separation and spectrophotometric and spectrofluorometric detection. Initially, only one micro-Bondapak column was used. After preliminary calibration experiments, more columns were added to the system to improve the separations. The results reported here are all obtained with a column system consisting of three micro-Bondapak columns connected in series. The column system gives a minimum efficiency of 10,000 theoretical plates. The columns were operated at room temperature. Typically, a 40- to 50-min linear gradient of the mobile phase from 70/30 CH_3CN/H_2O to 100% CH_3CN was used for separations. For a freshly cleaned column system, the retention times and peak shapes of all the PAH calibrated were found to be reproducible to within ± 1 to 2% in a six-month time period.

Generally, the retention times and the efficiency of the column series were checked with a calibration standard at least once every two or three sample runs. The chromatographic peaks in the sample were identified by: (1) retention time, (2) conjection of the standard mixture, and (3) the ratios of their UV-Vis absorption response (peak heights) at different wavelengths. Their fluorescence emission response provides yet another means of confirmation. The sample size injected into the HPLC was in the range of 2 to 20 μl which provided sufficient quantity for PAH measurements.

The GC/MS and HPLC conditions for exhaust-sample analysis are summarized in Table I. Several technical problems were encountered during the

TABLE I. Instrument Apparatus and Parameters

	Packed GC/MS	Capillary GC/MS	HPLC
Column	3% Dexsil 300 on 100/120 Chromosorb W (AW)	Quadrex Dexsil 300[b]	Micro-Bondapak C_{18}
Length	3.048 m x 6.35 mm x 2 mm i.d. glass	25 m x 0.25 mm i.d. x 0.75 mm o.d.	(3) 3.9 mm x 30 cm
Mobile Phase	Helium	Helium	Gradient elution—70/30 Acetonitrile/H_2O to 100 Acetonitrile
Flowrate	35 ml/min constant flow	20 psi (1.5 ml/min at room temp.) constant pressure.	1 ml/min
Temperature	100° (2 min) to 300° at $8^\circ C$/min	Ambient (2.0 min), 90° (2.0 min) to 280° at $2^\circ C$/min	Ambient
GC/MS Interface	VG micromass jet separator	VG micromass direct inlet jet	—
Interface Temperature	$250^\circ C$	$300^\circ C$	—
Sample Injector	Perkin-Elmer oncolumn injector @ $300^\circ C$	Falling-needle solid[a] injector @ $300^\circ C$	Septumless injector
Sample Sizes	1-3 μl[a]	1 to 20 μl	3-20 μl
Helium Makeup	None	4.0 ml/min	—
Magnet Scan Rate	2.0 sec	2.0 sec	—
Analysis Time	40 min	124 min	30-60
B[a]P Detection Limit	100 ng	10 ng	0.1-1 ng

[a]Obtained from Alltech.
[b]Obtained from Applied Science Laboratories, Inc.

development of the GC/MS method utilizing capillary columns. Because of the dilute concentrations of PAH in the exhaust sample, large sample injections were necessary for MS detection and mass-spectral identifications. Attempts to use Grob injectors improved the capillary separations but limited the size of samples injected to 0.5 to 1.0 μl. Consequently, higher-capacity packed columns with larger-sample-size injections were used during the initial phase of the experiment. All of the quantitative exhaust measurements reported here were obtained with packed columns.

The technical problems found in capillary-column operations were later circumvented by using a commercially available falling-needle solids injector mentioned in the previous section. In this method, sample sizes from 2 to 20 μl could be injected through the septum and were deposited on the tip of the needle. The total injection was carried out by semicontinuous 2 μl injections, allowing time for solvent evaporation from the tip of the needle between each injection. The solvent was swept away by helium carrier passed over the needle and was vented through the restrictor thus leaving only PAH and heavy aliphatics deposited on the needle tip. Some of the more volatile PAH such as phenanthrene were partially lost using this injector. This loss was reduced significantly by coating the tip of the needle with a thin film of SE-54 Silicone. Subsequent to the solvent removal, the needle was inserted into the 300°C injection port at the point of constriction, where the PAH and heavy aliphatics were volatilized onto the Dexsil 300 capillary column. During this sample transfer process the column was held at ambient temperature (2.0 min). The column was then heated to 90°C (held for 2.0 min) and programmed to 280°C at 2°C per min. The column effluent went directly to the source of the mass spectrometer through gold-lined nickel tubing and a glass capillary restrictor. Helium makeup gas was accurately metered at 4.0 ml/min to the column effluent to minimize the effects of postcolumn dead volume and to accurately control the backpressure of the column. The entire GC/MS interface and source were heated to 300°C. Mass spectral data were obtained using an electron impact (EI) ionization source. The computer-controlled magnet scan rate for 10-500 amw was 2.0 sec. Qualitative and quantitative data were obtained from mass chromatograms and reconstructed ion chromatograms (RIC) such as those shown in Figure 2. The RIC is much like an FID trace in which the computer accumulates the total ion signal produced from each scan and plots this signal vs the scan number. The mass chromatogram plots only the signal from a given ion vs scan number. The m/e:252 ion in Figure 2 represents the parent ion of PAH isomers such as benzo[a]pyrene (B[a]P), benzo[e]pyrene (B[e]P), and perylene. The mass spectra of these three PAH are indistinguishable by the computer and one must rely on accurate retention times for their identification.

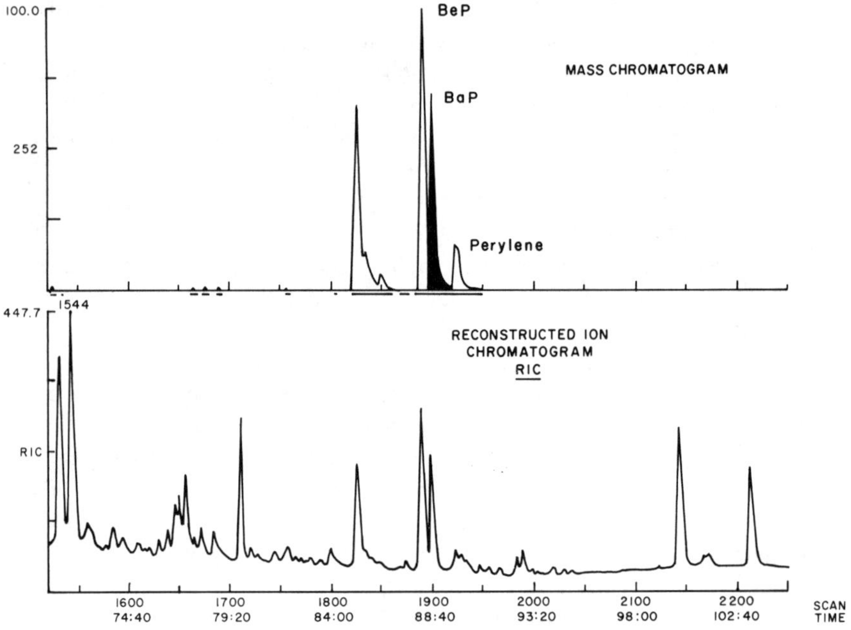

FIGURE 2. RIC and mass chromatogram.

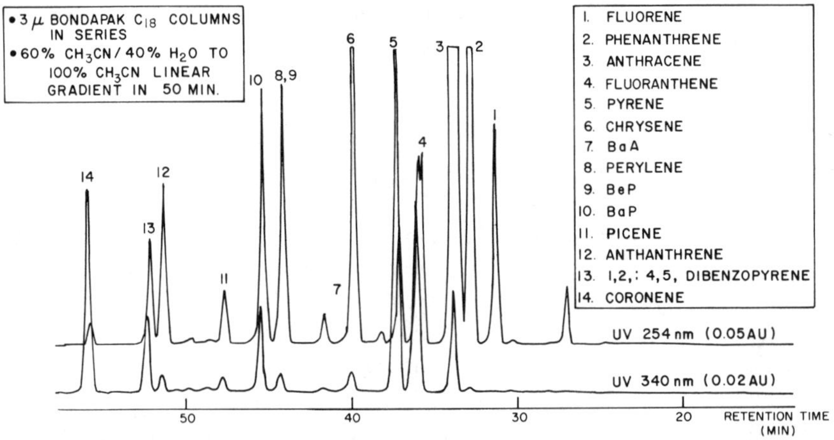

FIGURE 3. High-performance liquid chromatogram of 14 PAH separated by a μ-Bondapak column system.

The PAH components were quantitated by comparing the areas of the parent ion mass chromatograms to either internal or external standards, weighting each PAH with an appropriate response factor. The concentration of B[a]P, for example, in Figure 2 is represented by the shaded area in the m/e: 252 mass chromatogram.

Quantification by HPLC was achieved using peak heights or peak areas compared with standard calibration mixtures. A synthetic mixture containing about 10 ng/μl each of the 14 selected PAH in cyclohexane solution was prepared and used for both HPLC and GC/MS calibrations. This mixture was standardized against several other calibration mixtures prepared separately, and they were found to be in agreement to within ± 10%. The former mixture was therefore used as the working standard for all the runs reported here. This standard mixture was found to be stable with no detectable PAH degradation, when stored in the dark at room temperature, for a time period of six months.

RESULTS AND DISCUSSION

Analytical Recovery

A series of recovery studies was performed to assess the possible loss of PAH during the analytical procedures. Results from these experiments are summarized in Table II. In the first experiment, a known amount of the standard mixture was placed in a PAH-free engine oil sample and the same analytical procedures for real exhaust samples were carried out. Results from duplicate runs show that on the average, over 90% of the PAH were recovered with no apparent difference observed among the 14 individual PAH tested. Similar results were observed for spiked blank-filter samples, indicating that minimum PAH losses occurred during the analytical processes.

As seen from Table II, the recovery from spiked XAD-2 resins is slightly lower than those from blank filters, *i.e.*, a mean of 82% vs 96%. The results seem to be statistically significant, which would then indicate a slight loss of PAH in the former case. It is worth noticing that this recovery is comparable to those generally observed for spiked particulate samples, *i.e.*, about 85%, as seen from the later results of radiotracer runs listed in the same table. The slightly lower recovery in both cases is probably due to the inability of solvent extraction to completely remove PAH adsorbed on high-surface-area soots or XAD-2 resins. In separate experiments, it was observed that benzene is among the most efficient solvents in extracting organics from soots. Insofar as extracting PAH is concerned, however, methylene chloride is at least comparable to benzene, as indicated by the absence of detectable PAH in samples reextracted with benzene after the initial methylene chloride extraction.

TABLE II. Analytical Recovery of PAH

Sample	Procedure	No. of Runs	% Recovery (Compounds)
Engine Oil[a]	Spiked with standard	2	94 (average of 14 PAHs)
Blank Filter	Mixtures of 14 PAH	2	96
Blank XAD-2	PAHs	2	82
Exhaust particulate-laden filter collected from the prototype engine[b]	Spiked with ^{14}C-B[a] A	3	87 (Total ^{14}C-activity) 80 (^{14}C-B[a] A activity)
Exhaust particulate-laden filter collected from the prototype engine[b]	Spiked with ^{14}C-B[a] P	3	88 (Total ^{14}C-activity) 85 (^{14}C-B[a] P activity)
Exhaust particulate-laden filter collected from a diesel engine[b]	Spiked with ^{14}C-B[a] P	2	85 (Total ^{14}C-activity) 61 (^{14}C-B[a] P activity)

[a]The amount of original oil sample contains no detectable PAH of interest.
[b]These spiked samples were extracted and analyzed with the same procedure as those for exhaust sample runs. Spiked engine oil runs were analyzed using the Grimmer separation without Soxhlet extraction. Quenching corrections were made for radiotracer experiments.

In the second experiment, a ^{14}C-radiotracer technique was used to assess the possible loss of PAH during the analysis of real exhaust samples. In addition to testing the efficiency of extracting soot-adsorbed PAH, as stated earlier, the experiment also provides information about the potential degradation of PAH by other reactive species in the exhaust. In these runs, the particulate-laden filter was first spiked with a known amount of ^{14}C-B[a] P or ^{14}C-B[a] A. The radioactive compounds were dispersed on the filter, from solution, using a 25-μl syringe. The doped filter was then extracted and concentrated. A portion of this concentrate was counted directly for its total radioactivity. The remaining portion was first separated by HPLC, the ^{14}C-B[a] A or ^{14}C-B[a] P fraction eluted along with the nonradioactive carriers was then collected and its corresponding radioactivity measured.

Results listed in Table II show that the total activities recovered from these runs were 87% to 88%, comparable to what we observed previously for blank XAD-2 runs. This similarity indicates loss due to adsorption in both cases. More significantly, nearly all the activity was still associated with the original compound B[a] A or B[a] P. This suggests that, at least for B[a] A or B[a] P, little or no degradation occurred during the analysis even though some of these samples had been stored in solution for a time period of 5 to 10 days. It is interesting to compare this with a diesel exhaust run listed at the bottom of the table. With a similar procedure, the diesel run gave comparable recovery in total activity but considerably lower recovery of B[a]P originally spiked, *i.e.,* 61% vs 85%. Thus, in the diesel sample, some reaction loss of B[a]P seems to have occurred. The cause for this difference observed between our prototype engine and the diesel runs is not understood at the present. One possible explanation is that some reactive species responsible for B[a]P degradation in diesel soot is absent, or present in much lower concentrations in the gasoline-powered prototype engine exhaust. We are currently continuing our efforts along this direction.

HPLC Analysis

A typical chromatogram showing the separation of the PAH in the working standard by the three micro-Bondapak column system is given in Figure 3. It can be seen that all the PAH in the mixture are well separated in a 60-min run. The baseline separations of isomeric pairs such as anthracane and phenanthrene, B[a]P and perylene, and anthanthrene and dibenzopyrene are noteworthy.

Through the coinjection of several calibration mixtures, a chromatogram showing the separation of 25 PAH is given in Figure 4. Several unresolved PAH isomers are observed in this figure, *i.e.,* picene and 1,2:5,6-dibenzoanthracene, perylene and B[a]P and the partial separation of chrysene from B[a]A. The presence of alkyl groups on the PAH presents another separation difficulty. From the figure, it is seen that the addition of methyl or dimethyl groups on anthracene or phenanthrene causes a large increase in their retention times as compared with their corresponding unsubstituted parents. Several alkyl-substituted PAH were identified in the exhaust samples by GC/MS analysis, as we will report later. The overlapping of these alkyl-homologues with other still higher member-ring compounds makes peak identification and peak-purity evaluation critical in HPLC analysis.

We have used the combination of fluorescence emission and selective wavelength spectrophotometric absorption detection to approach the problem of peak identification and quantification. Identification is achieved by measuring the UV absorption ratio at several selective wavelengths. The ones most frequently used are 254 nm, 280 nm, 313 nm, 340 nm, and 405 nm. Since a dual fixed wavelength detector was employed in our work, at least two runs per sample were needed to obtain all the information. Generally, three or four member-ring PAH were monitored using the three shorter wavelengths stated above, while the three longer ones were used for PAH with five or more rings. The quantitation was obtained from the observed UV absorption at the optimal wavelengths, or selective wavelength excited fluorescence-emission

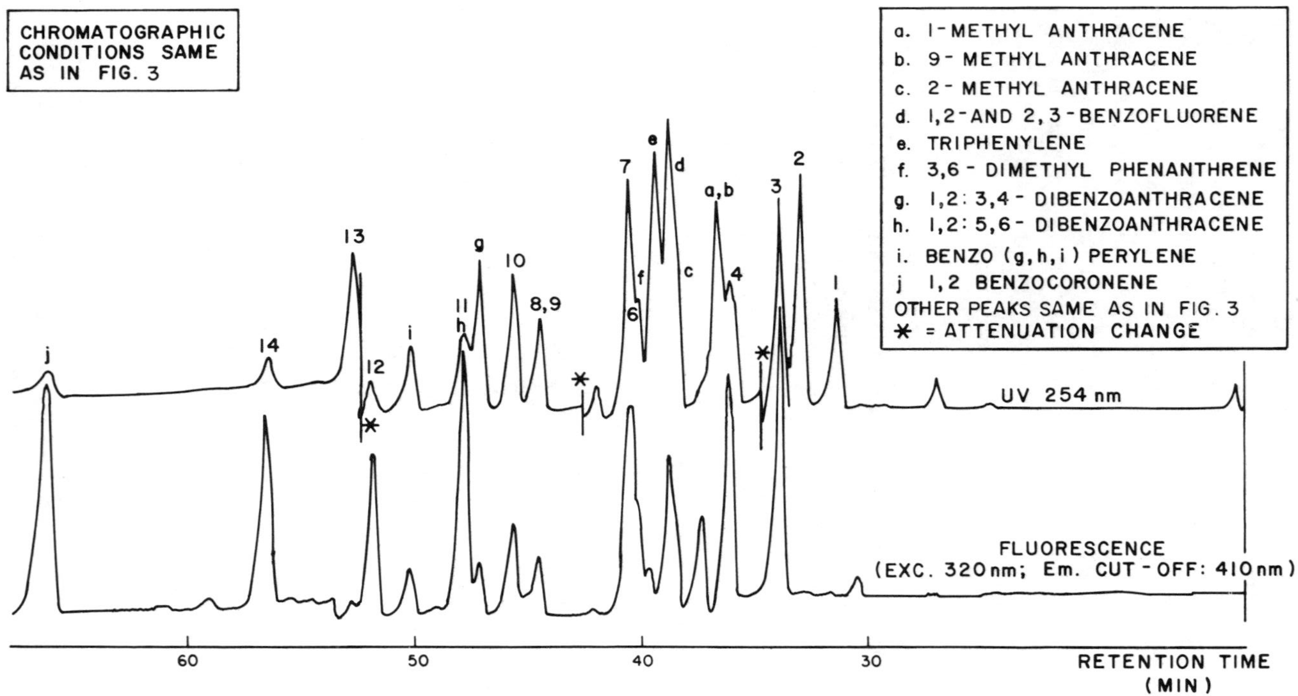

FIGURE 4. High-performance liquid chromatogram of 25 PAH separated by a μ-Bondapak column system.

response, measured from the spectrofluorometer connected downstream of the UV detector.

A typical exhaust sample run is illustrated in Figures 5 and 6. In Figure 5, a chromatogram showing 254-nm detection is given. Phenanthrene and anthracene are clearly separated in this run. Considerable interferences were observed, however, for some of the other PAH, noticeably pyrene, fluoranthene, B[a]A/chrysene and B[a]P/perylene. By using 340 nm or 313 nm pyrene and fluoranthene were measured quite easily without interference, as illustrated in the same figure and in Figure 6. Similarly, interferences for B[a]A and chrysene due to triphenylene or other peaks were largely removed by choosing these wavelengths. The identities of these interference peaks observed on 254 nm were not confirmed. They are probably alkyl-substituted anthracenes or phenanthrenes judging from their retention times, UV-absorption characteristics and comparison with GC/MS analysis.

In Figure 6, the analysis for higher-ring PAH for the same run is illustrated. At 405 nm, most of the five or more ring PAH can be quite selectively detected. Thus, B[a]P and perylene remain unseparated in the column system. Most of the 405-nm absorption, however, is due to perylene since it gives very strong absorption at this wavelength. In addition, by choosing 365-nm excitation, emission from perylene can be monitored by a fluorescence detector free of B[e]P interference. The amount of B[e]P can then be estimated from its absorption at other wavelengths after subtracting contributions due to perylene. The use of the fluorescence detector for HPLC analysis of PAH has been extensively reported in the literature (10,22). We have used a setting of 365-nm excitation and 418-nm filter cutoff for our HPLC-fluorescence measurements. The combination provides quite uniform high sensitivity and selectivity for most of the PAH of interest here.

GC/MS Analysis

The need for chromatographic separation and measurement of PAH in automotive exhaust and atmospheric samples has been largely satisified in the past using HPLC and high-resolution glass capillary columns using FID detectors. GC/MS has been largely confined to low-resolution packed columns or short (10-15m) capillary columns. The need for high-performance glass capillary column GC/MS has been accelerated in the past few years. This is due to the discovery of more complex PAH-type emission samples.

Acceptable GC/MS analysis of PAH has been obtained by using a combination of the solids injector, a 25-m Dexsil 300 column, accurate flow controls and heated (300°C) GC/MS interface lines. The solids injector facilitated the injection of large sample volumes, while reducing solvent overloading of the capillary column to a negligible level. Quantitative losses and

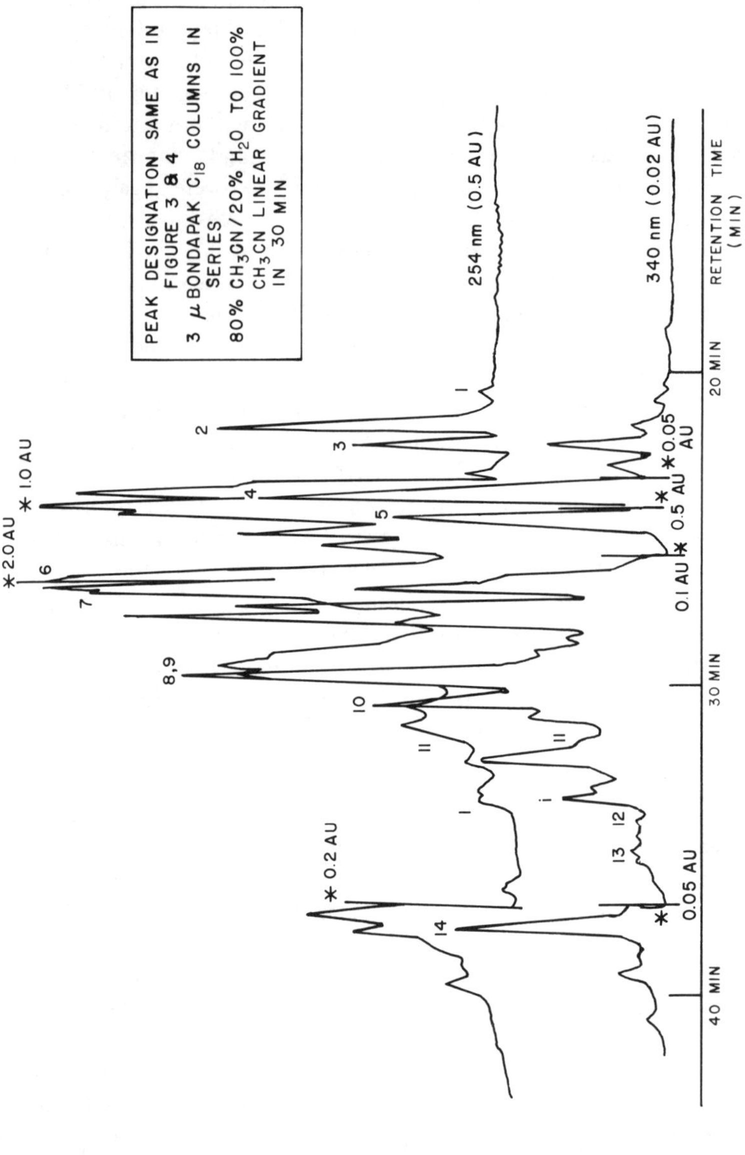

FIGURE 5. High-performance liquid chromatogram of the PAH fraction isolated from the exhaust particulate samples (I).

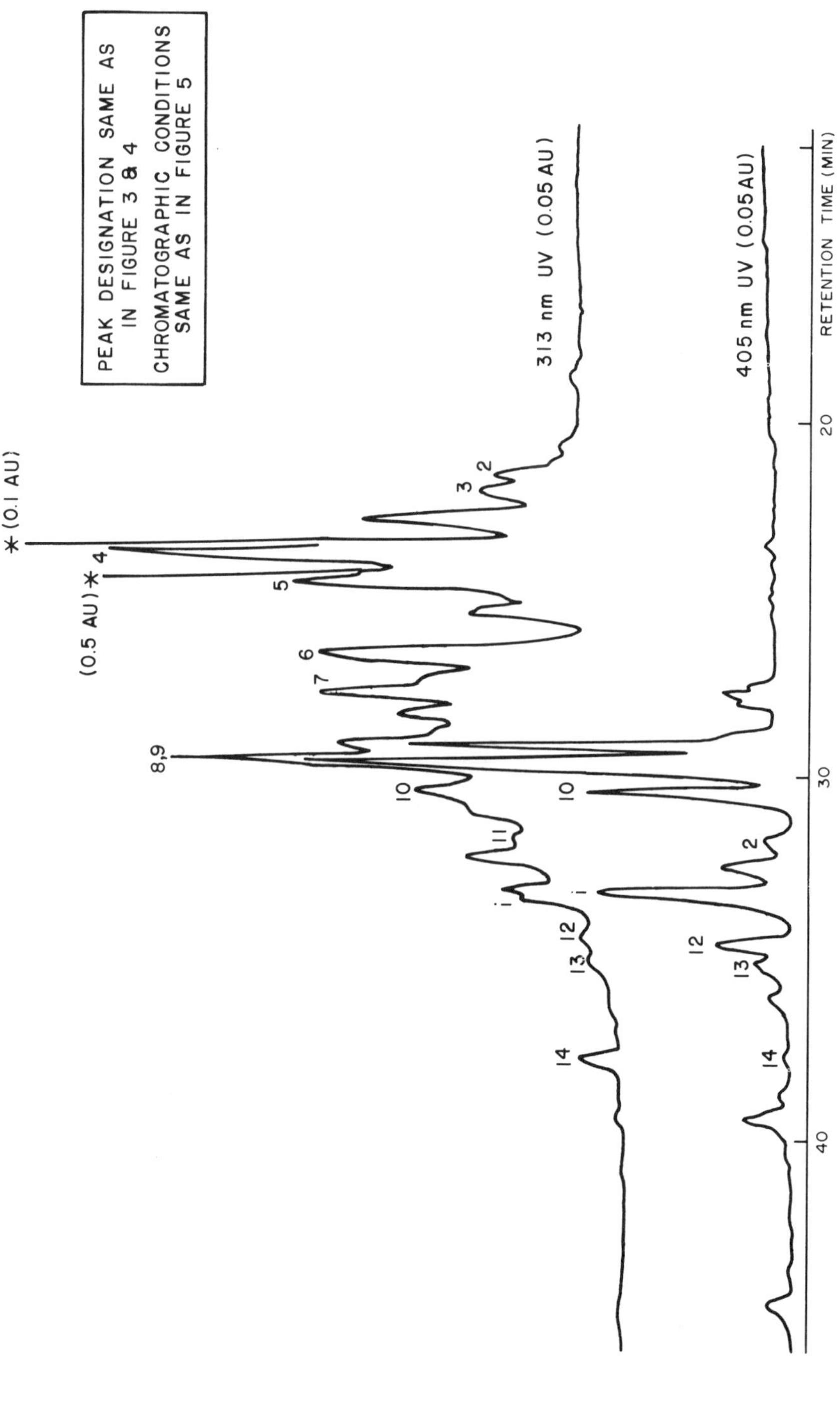

FIGURE 6. High-performance liquid chromatogram of the PAH fraction isolated from the exhaust particulate (II).

peak tailing due to adsorption of PAH in the interface between the GC and MS were explicitly reduced by heating the lines and ion source to 300°C. Irreproducible retention times were a major problem encountered using capillary columns. The variations were traced to helium makeup gas added at the end of the column to reduce postcolumn dead volume. This makeup gas caused fluctuations in backpressure at the column exit. The backpressure fluctuations were eliminated by using a Brooks mass flowmeter to accurately set and control the helium makeup flowrate. After this change was made the retention times were very reproducible, yielding standard deviations in relative retention times from 0.001 to 0.009.

Exhaust samples from the prototype engine analyzed both by packed and capillary column GC/MS are compared in Figures 7 and 8. The packed column separates the several m/e: 252 isomers into only two peaks. Near-baseline separation of B[e]P and B[a]P was achieved with the Dexsil 300 capillary column and perylene was well separated from both. However, it can be seen from the proximity of the B[e]P and B[a]P that accurate retention times are essential to the positive identifications of each isomer. The accurate determination of retention times also facilitates the computerization of a complete qualitative and quantitative PAH analysis.

Sampling Recovery

Three factors may potentially cause PAH losses during the sampling process. First, the filter-plus-absorbent sampler may not be able to completely trap the gaseous or particulate-associated PAH, resulting in losses due to sample penetration. Second, losses may occur due to wall deposition as the exhaust flows through the dilution tube. Third, losses due to reactions between collected PAH and other components in the gaseous exhaust should also be considered. The results presented here provide some information concerning these aspects. However, most of these data are preliminary and the conclusions are basically qualitative in nature because of the experimental complications involved.

Recovery experiments were performed in which known amounts of perylene or pyrene were injected directly into the dilution tube exhaust stream, and recovered in the filter-plus-absorbent sampler subsequently analyzed by HPLC. In these experiments, the solid samples being injected were placed in a small glass tube sealed at both ends but with one hole on the top. The tube was then placed directly in the exhaust tailpipe, which was at a temperature of approximately 250°C. At this temperature, the sample has sufficiently high vapor pressure to be carried by the exhaust into the dilution tube at a constant evaporation rate. Separate bench experiments using thermal gravimetric analysis method showed that the inejctions of perylene or pyrene

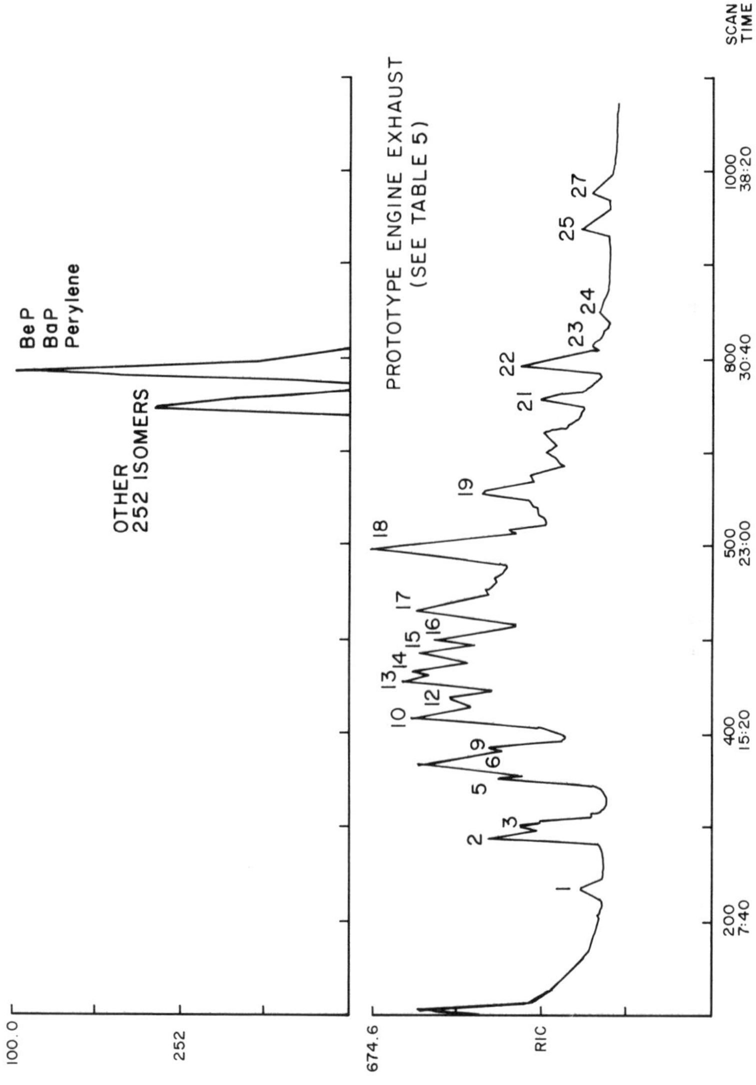

FIGURE 7. PAH analysis using packed Dexsil 300—GC/MS.

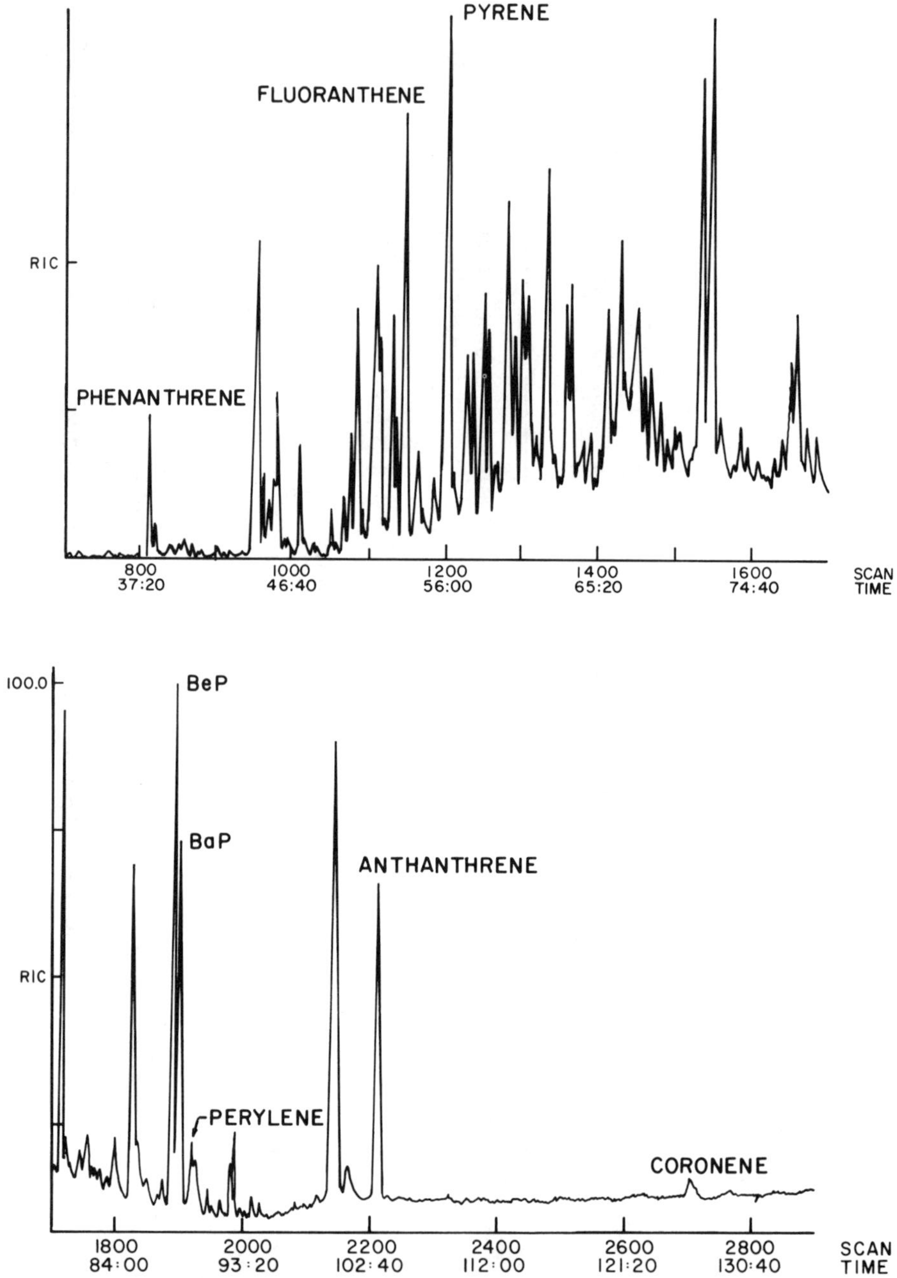

FIGURE 8. PAH analysis using capillary Dexsil 300—GC/MS.

with this design were highly reproducible, and a constant injection rate of several microgram samples per minute was observed.

The recovery results obtained from these experiments were irreproducible. Three separate runs for perylene gave recoveries of 64, 20 and 37%, respectively, and 20% recovery was observed in the pyrene run. In the perylene case, all the recovered perylene was found in the filter and in no case was perylene detectable in the gas phase as determined by analysis of the XAD-2 traps. In the pyrene run, however, the recovery was found to be 5% in the filter and 15% in the XAD-2 trap. In all the above runs, a background exhaust measurement was made immediately preceding the injection experiment in order to provide background subtraction. The quantities being injected were approximately 15 to 20 times those of the background runs. The uncertainty of the recovery measurements was estimated from the recovery studies to be within ± 15%, which cannot explain the low recoveries and large variations found in the injection experiments.

Although a definite conclusion cannot be made from the results of the limited runs stated above, it is believed, however, that degradation of PAH during filter sampling is probably the major factor responsible for the irreproducibility of the recoveries. In an extensive study of diesel PAH emissions, Spindt (18) observed similar irreproducibility of B[a]P and B[e]P recoveries. In his work, the diesel exhaust was precooled first in a short heat-exchange tube, and then sampled with a filter-plus-chromosorb 102 trap sampler. The recoveries for B[a]P he observed were generally in the range of 9-40% and indirect evidence indicating B[a]P degradation was observed. Further experiments showing the ease of B[a]P degradation during filter sampling of atmospheric particulates were recently reported by Pitts *et al.* (15) and Grundel *et al.* (7). In Pitts' experiment, he found that severe degradation of B[a]P occurred as the glass-fiber filter impregnated with B[a]P was subject to exposure of air containing ozone, NO_2 and other pollutants. Suggestions were made by him that similar reactions may potentially occur during exhaust sampling.

We have recently initiated a series of experiments to further assess the above PAH-degradation problem. With the use of an exhaust-gas simulator (2), we have confirmed Pitts' results about gaseous reactions with B[a]P during filter sampling by a procedure similar to his. In addition, we have found that even more severe B[a]P degradations occurred as the B[a]P-impregnated quartz-filter was exposed to sulfuric acid aerosols (11). Our most recent results indicated yet another complication. In these experiments, it was observed that the degradations of PAH are at least partially enhanced, if not mostly caused, by the catalytic effects of the filter media. The extent of such degradation thus varies drastically with the types and surface characteristics of the filters being used (12).

From the above discussion, it seems reasonable to expect also the occurrence of similar degradation of PAH during our dilution tube recovery experiments. Thus, quantitative data regarding dilution tube recoveries cannot be achieved until the degradation problem is clarified. In our perylene (or pyrene recovery experiments stated earlier, the samples were probably injected in the vapor phase. Since the excess amount of sample being injected was probably not completely adsorbed on the particulate matter, they were condensed and collected on the filter. They were then degraded by the continuous sampling of exhaust gases containing NO_x, SO_x and other reactants, and the extent of degradation varied with the time of exposure, the surface conditions of the filter during exposure and other factors.

It is significant to emphasize at this time that the evidence seems to indicate that the PAH are degraded during filter sampling partly, if not mostly, because of the filter-surface catalysis. It was observed, for example, that the percentage recovery for B[a]P varies from less than 10% in the case of tissue-quartz filter to about 60% in the case of Fluoropore or Teflon®[1]-coated filters under otherwise identical sampling conditions (12). Thus, with the proper selection of filters with low surface activities, it is believed that this degradation problem can be minimized to a large extent. It should also be pointed out that in both experiments, by Pitts and ourselves, the degradation of PAH was observed in a model system in which B[a]P (or other PAH) is placed directly on the filter surface and exposed to air or testing gases. Further experiments are needed in order to demonstrate to what extent such reactions may occur for PAH adsorbed on particulate matter or encapsulated in carbonaceous soots as in real-world situations.

The absence of perylene in the XAD-2 trap observed from the recovery experiments indicates that most of the perylene was particulate-associated and filter-collectable. This is consistent with Spindt's finding (18) in which very little B[a]P was found in the Chromosorb 102 backup trap. In Table III, PAH measured in filters and XAD-2 traps were compared for three separate runs. In none of these runs, detectable amounts of five or more member-ring PAH from B[a]P to coronene were observed in the XAD-2 trap. Different results were observed, however, for lighter PAH such as phenanthrene, anthracene, pyrene and probably chrysene/B[a]A. In run 155, it is seen that comparable or larger amounts of these PAH were actually observed in the XAD-2 trap than in the filter. Unfortunately, we were unable to obtain

---

[1]Registered trademark of E. I. duPont de Nemours and Company, Inc., Wilmington, DE.

TABLE III. Comparison of PAH Found in Filter and XAD-2 Samples

Test Run No.	184^{++}		155		198	
Catalyst	No		No		*Yes*	
Collection Time (Min)	30		120		120	
Wt of Filter Collected Particulates (mg)	53.6		331		41.6	
PAH	**μg/sample**					
	Filter	**XAD Trap**	**Filter**	**XAD Trap**	**Filter**	**XAD Trap**
Phenanthrene	2.1	+	40	16	1.6	+
Anthracene	1.2	+	8	30	0.4	+
Fluoranthrene	6.3	+	34	30	1.6	+
Pyrene	15*	38*	36	40	7	+
Chrysene	6	+	10	$\leqslant$60	2	+
Benzo[a]anthracene	$\leqslant$*14*	+	$\leqslant$60			+
Benzo[e]pyrene	8.5	(0.4)	28	(0.1)	1.6	(0.2)
Benzo[a]pyrene	4.2	(0.1)	9	(0.1)	0.8	(0.1)
Benzo[g,h,i]perylene	0.9	(0.1)	31	(0.2)	(0.2)	(0.1)
Picene plus dibenzo-anthracenes	0.4	(0.1)	9	(0.2)	0.4	(0.1)
Anthanthrene	0.9	(0.05)	13	(0.2)	(0.2)	(0.1)
Dibenzopyrene	0.5	(0.05)	3	(0.2)	0.4	(0.1)
Coronene	1.	(0.01)	4	(0.2)	0.4	(0.1)

NOTE: + = Not measured due to background contamination from the adsorbent XAD-2 resin.
* = Including pyrene injected into the dilution tube, see text.
++ = Not observed; run 184 was designed for pyrene injection experiments (see text). The engine operating conditions for this run were not well defined.
( ) = Values listed represent upper limits.

similar information from runs 184 and 198 due to heavy contamination of dirty XAD-2 resin inadvertently used in these runs. In run 184, pyrene was injected as part of our recovery study discussed earlier. The amount of pyrene in the XAD-2 trap in this run was measured by collecting the fraction of the HPLC effluent containing pyrene and subsequently quantified by UV spectrophotometry. The observation of more pyrene in the XAD-2 trap than in the filter from this run is consistent and in qualitative agreement with those observed in run 155.

The vapor pressure for the above lighter PAH at 25°C ranges from 6.8 x 10^{-4} Torr for phenanthrene, 2 x 10^{-4} Torr for anthracene to 10^{-7} Torr for

B[a] A (16). Thus, significant amounts of these PAH may exist in the gas phase and penetrate through the filter during sampling. The observation of less phenanthrene in the XAD-2 than in the filter (run 155) contradicts that observed for anthracene in the same run although the reverse is expected from vapor pressure considerations. This would then suggest the possibility of sampling loss resulting from XAD-2 trap penetration for some light PAH such as phenanthrene or anthracene. More experiments are needed to verify this.

Exhaust PAH Measurement

In Tables IV and V, the PAH concentrations in the exhaust measured by HPLC and GC/MS techniques are listed separately. These samples were collected from the engine running under the same steady speed (50 kph) operating conditions. Steady-state runs were chosen because this mode of operation is much easier to reproduce experimentally, and thus best suited for the method validation experiments reported here. In all these runs, the particulate emission rate and other exhaust emission characteristics such as hydrocarbons, NO_x and CO emissions were monitored separately and were generally found to be quite reproducible (13).

In Table IV the exhaust PAH emissions from four runs are shown; contamination problems in the XAD-2 resin, as stated in the last section, prevented the inclusion of some data for PAH collected in the XAD-2 traps. Results for light PAH from phenanthrene to B[a] A therefore represent probably only a portion of the total exhaust PAH. On the other hand, for all the other PAH from B[a] P to coronene, the filter results also represent the total exhaust PAH since no detectable amount of those were observed in the XAD-2 traps.

Judging from the PAH-emission results in Table IV, the repeatability of the engine testing system is considered satisfactory. The qualitative picture of the overall PAH distribution compared from run to run is very similar, and a quantitative comparison for individual PAH demonstrates reasonable agreement in most cases, with variations generally less than a factor of 2.

The HPLC data obtained from runs 132 and 167 are also in good agreement with those obtained from GC/MS analysis of the same sample, also included in Table IV for comparison. The agreement is within ± 10 to 50% despite the fact that two completely different chromatographic separation and quantification processes were used in these two methods. In addition to the PAH discussed earlier, a number of alkyl-substituted PAH were also detected in the GC/MS runs. Detailed data for these species are seen in Table V.

TABLE IV. Particulate-Associated PAH in the Exhaust Measured by HPLC and GC/MS

Test Run No.	132		167		147	155	Avg.	Std. Dev.
Wt of Particulates (mg)	261		237		314	331		
PAH	μg/sample							
	GC/MS	HPLC	GC/MS	HPLC	HPLC	HPLC		
Phenanthrene	12	16	22	9	14	40	20	6
Anthracene		3		2	12	8	6	5
Fluoranthrene	37	25	25	21	22	34	26	6
Pyrene	29	29	33	25	40	36	33	7
Chrysene	54	35	56	15	50 + 20	10	20	10
Benzo[a] anthracene		⩽110		⩽70		⩽60	⩽80	–
Perylene	36	25	38	8	20	15	17	8
Benzo[a] pyrene		14		10	14	9	12	3
Benzo[e] pyrene		40		26	42	28	34	8
Benzo[g,h,i] perylene	20	12	18	20	7	31	18	10
Anthanthrene		6		2	8	13	7	5
Picene plus dibenzo-anthracene	8	6	–	3	6	9	6	3
Coronene	6	6	4	4	6	4	5	1
Dibenzopyrene	4	3	0.4	0.3	5	3	3	2

TABLE V. Particulate-Associated PAH in the Exhaust Measured by GC/MS Using a Dexsil 300 Packed Column

Peak No.[a]	Compound(s)	Mass	Run 132 (μg/sample)	Run 167 (μg/sample)
1.	Phenanthrene or anthracene	178	12.1	22.4
2.	Methyl (phenanthrene) or (anthracene)	192	23.7	28.9
3.	Methyl (phenanthrene) or (anthracene)	192	13.2	16.1
4.	Dihydro-fluoranthene or phenylnaphthalene	204	4.4	4.4
5.	Dimethyl (phenanthrene) or (anthracenes)	206	22.8	30.8
6.	Dimethyl (phenanthrene) or (anthracenes)	206	43.9	54.4
7.	Benzylnaphthalenes	218	12.9	14.4
8.	Isomers of peaks 5 & 6 or tetrahydrofluoranthene	206	3.1	3.9
9.	Fluoranthene	202	37.2	24.7
10.	Pyrene	202	29.2	32.7
11.	Trimethylphenanthrenes (or anthracenes)	220	14.7	17.3
12.	Trimethylphenanthrenes (or anthracenes)	220	12.8	10.7
13.	Benzofluorenes	216	19.5	25.6
14.	Benzofluorenes	216	27.1	30.0
15.	Methylpyrenes	216	19.1	22.5
16.	Methylpyrenes	216	20.4	21.3
17.	Diphenylbenzene	230	50.3	52.9
18.	B[a] A + chrysene + triphenylene	228	53.9	56.0
19.	Methyl (B[a] A) or (chrysene) or (triphenylene)	242	31.6	33.5
20.	Unknown $C_{20}H_{16}$	256	6.8	11.7
21.	Benzo fluoranthenes	252	26.7	22.8
22.	B[e] P + B[a] P + Perylene	252	35.7	37.6
23.	Methyl benzo fluoranthenes	266	9.9	8.1
24.	Methyl (B[e] P), (B[a] P) or Perylene	266	13.2	11.5
25.	Dibenzanthracene, Picene	278	8.1	—
26.	Indeno (1,2,3-c,d)-(fluoranthene) or (pyrene)	276	7.2	25.1
27.	Benzo[g,h,i] perylene or anthanthrene	276	20.3	17.5
28.	Dibenzopyrenes	302	3.6	0.4
29.	Coronene	300	5.6	3.9

[a]See Figure 7.

Despite the complex sampling problems discussed in the last section, the reasonably reproducible results obtained from the above PAH measurement indicate that a comparison of PAH emissions can still be made on a relative basis. A comparison was made between the exhaust emissions with and without the use of an oxidation catalytic converter. Results are compared in Table VI. It is clearly seen from the data that the emissions of PAH are drastically reduced using a catalytic converter to treat the exhaust. The reduction in B[a]P, for example, is more than fourfold; a factor of two- to tenfold or higher reductions are observed for other PAH. This reduction in PAH is consistent with the finding of three times less organic extractables in the catalyst runs as compared with noncatalyst cases. Thus, it is clearly demonstrated that the catalyst effectively oxidizes the PAH as well as other organic components.

TABLE VI. PAH Emissions from the Ford Prototype Engine

PAH	μg/km without Catalyst[a]	μg/km with Catalyst[b]	Percent Reduction
Phenanthrene	1.9	0.2	89
Anthracene	0.6	0.04	93
Fluoranthrene	2.3	0.2	91
Pyrene	2.9	1.5	49
Chrysene plus benzo[a]anthracene	7	0.4	94
Perylene	1.2	0.4	66
Benzo[a]pyrene	0.9	0.2	78
Benzo[e]pyrene	2.8	0.4	86
Dibenzopyrenes	0.3	0.2	33
Coronene	0.4	0.3	25
Picene plus dibenzoanthracene	0.5	0.3	40
Benzo[g,h,i]perylene	1.8	N.O.[c]	–
Anthanthrene	0.7	N.O.	–
Others identified but not quantified: See Table V			
Fluorene, Benzophenanthrene, Benzofluorenes, Benzofluoranthene, Benzo[a] Anthracene			
Other Related Emission Measurements			
Particulate	27-31 mg/km	5-6 mg/km	
Organic Extractables	90%	30%	
Free Carbon	6%	50%	
Sulfate	4%	20%	

[a]Average of 4 measurements.
[b]Average of 2 measurements.
[c]N.O. = Not observed.

In both chemical and physical nature, the exhaust from the prototype engine more closely resembles that of a diesel than a conventional gasoline engine (13). From this consideration, it is interesting to compare the PAH emissions from these two engines. It has been reported that the emission of B[a]P from a diesel engine without catalyst-treatment is about 3 μg/km (19). This is higher than the B[a]P emission from the prototype engine measured here. Such comparison should obviously be treated with extreme care due to the problems observed in the sampling system, especially because the measurements in the two studies involved completely different engine testing and running conditions. Nevertheless, it seems reasonable to state that the B[a]P emission from the prototype engine is lower than that from the disel after taking into account the beneficial effects of the oxidation catalysts observed here (0.2 vs 3 μg/km). On the other hand, it is known that the use of catalysts for diesel-exhaust treatment is still in the research stage because of the technical difficulties involved.

SUMMARY

Exhaust emissions from a prototype stratified-charge vehicle with and without an oxidation catalyst were measured for polynuclear aromatic hydrocarbon (PAH) content. Samples were collected from a dilution tube, using filter and XAD-2 resin traps in series and subsequently analyzed by high-performance liquid chromatography and gas chromatography/mass spectrometry. The catalyst reduced individual PAH up to 94%, to a level significantly below the PAH emission levels reported in the literature for diesel-powered vehicles.

The analytical accuracies were determined by spiking engine oil, a blank filter and blank XAD-2 resin with a 14-component PAH mixture. Average recoveries for the three experiments were 94%, 96% and 82% respectively. Pyrene and perylene were injected directly into the exhaust system near the end of the tail pipe. Recoveries of only 20 to 60% of the amount injected were achieved. It is postulated that the losses are due to chemical reactions, rather than adsorption losses. PAH analysis of samples analyzed both by HPLC and GC/MS agreed generally within ± 50%.

The collection efficiency of quartz-tissue filters was assessed using a XAD-2 resin backup trap. The filters collected over 90% of PAH less volatile than pyrene, however, only collected 25 to 50% of the PAH more volatile than pyrene.

ACKNOWLEDGMENTS

The authors wish to thank D. E. McKee for his technical assistance relating to sampling and dilution tube facilities. We are also indebted to T. M. Harvey for his technical assistance relating to the GC/MS method development.

REFERENCES

1. Boyer, K. W., and Laitinen, H. A. (1975): In *Environ. Sci. Technol.* 9:457.
2. Butler, J., *et al.* Unpublished results.
3. Candeli, A., Mastrandrea, V., Morozzi, G., and Toccaceli, S. (1974): *Atmos. Environ.* 8:693.
4. Colucci, J., and Begeman, C. (1965). *J. Air Poll. Control Assoc.* 15:113.
5. Grimmer, G., Bohnke, H., and Glaser, A. (1977): *Zbl. Bakt. Hyg.; I. Abt. Orig. B* 164:218.
6. Gross, G. (1970-1973). Annual reports on gasoline composition and vehicle exhaust gas polynuclear aromatic content, The Coordinating Research Council Project Report.
7. Grundel, L., Chang, S. G., and Novakov, T. (1978): Heterogeneous reactions of polynuclear aromatic hydrocarbons and soot extracts with NO_2. Presented before American Chemical Society National Meeting, Anaheim, CA.
8. Hoffman, D., Theiss, E., and Wynder, E. L. (1965): *J. Air Poll. Control Assoc.* 15:162.
9. International Agency for Research on Cancer (1973): Monograph on the evaluation of carcinogenic risk of certain polynuclear aromatic hydrocarbons and heterocyclic compounds. World Health Organization, Geneva, Switzerland.
10. For more recent review, see articles by Lao, *et al.* and others (1978): In *Carcinogenesis, Vol. 3: Second International Symposium on Analysis, Chemistry, and Biology of Polynuclear Aromatic Hydrocarbons* P. W. Jones, and R. I. Freudenthal, Eds., Raven Press, New York.
11. Lee, F. S.-C., Harvey, M. and Prater, T., to be published in 1979.
12. Lee, F. S.-C. and Pierson, W., to be published in 1979.
13. McKee, D. E., Ferris, F. C., and Goeboro, R. E. (1978): Unregulated emissions from a PROCO engine powered vehicle. Paper No. 780592, Society of Automotive Engineers Meeting, Detroit.
14. National Academy of Science Report (1972): Biological effects of atmospheric pollutants: Particulate polycyclic organic matter. U.S. Government Printing Office, Washington, D.C.
15. Pitts, Jr., J. N., Grosjean, D., Van Canwenberyli, K., and Belsen, W. L. (1978): Mutagenic activity of nitrogenous pollutants in ambient and simulated atmosphere. Presented at the American Chemical Society National Meeting, Anaheim, CA.

16. Radding, S. B., *et al.* (1976): The environmental fate of selected polynuclear aromatic hydrocarbons. See data and reference listed in Table 6, p. 22. EPA Report PB-250 948, National Technical Information Service.
17. Searle, C. E., Ed. (1976): Chemical carcinogens, ACS monogram no. 173. American Chemical Society, Washington, D.C.
18. Spindt, R. S. (1974): First and second annual reports on polynuclear aromatic content of heavy duty diesel engine and exhaust gases. The Coordinating Resarch Council Project Report.
19. Springer, K. J., and Baines, T. M. (1977): Emissions from diesel versions of production passenger cars. Paper no. 770818, Society of Automotive Engineers Meeting, Detroit.
20. Strup, P. E., Wilkinson, J. E., and Jones, P. W. (1978): Trace analysis of polynuclear aromatic hydrocarbons in aqueous systems using XAD-2 resin and capillary column gas chromatography–mass spectrometry analysis pp. 131-138 in Ref. 13.
21. Symposium on Diesel Particle Emissions Measurement Characterization. Reports by Baines, T. M., Seizinger, D. E., Springer, K. J., and others, EPA Symposium, Ann Arbor, April 1978.
22. Tehada, S. B. (1977): PNA estimation in diesel exhaust. Presented at the CRC-APRAC Diesel Exhaust Emission Measurement Symposium, Chicago, April 5, 1977.

Polynuclear Aromatic Hydrocarbons,
edited by P. W. Jones and P. Leber.
Ann Arbor Science Publishers, Inc.
Ann Arbor, MI ©1979.

ANALYSIS OF ORGANIC POLLUTANTS BY SYNCHRONOUS LUMINESCENCE SPECTROMETRY

T. Vo-Dinh, R. B. Gammage and A. R. Hawthorne
Health and Safety Research Division
Oak Ridge National Laboratory
Oak Ridge, Tennessee 37830

INTRODUCTION

Because of the health hazards associated with many polynuclear aromatic compounds (PAH) released in our environment by the production of synthetic fuels, it is important to develop analytical techniques that are sufficiently simple and selective for monitoring these species on a routine basis. Our work has dealt with the development and use of the synchronous luminescence technique for the identification and quantification of PAH in mixtures. The synchronous method has been applied to fluorescence as well as phosphorescence analysis of samples of environmental interest.

METHODOLOGY AND INSTRUMENTATION

Of the common spectroscopic methods, fluorescence and phosphorescence have evoked the most interest among analytical chemists. Due to its inherent sensitivity, the luminescence technique is capable of detecting subnanogram quantities of individual compounds. The possibility of utilizing excitation and emission wavelengths is another attribute of the luminescence technique. Both sensitivity and selectivity, however, are limited by the broad-band emission that frequently results from multiple components in complex mixtures.

Synchronous luminescence is a relatively simple technique that can offer significant improvement in sensitivity and selectivity in comparison with conventional fixed-excitation luminescence methods. It is a relatively new technique in which both excitation (λ_{ex}) and emission wavelengths (λ_{em}) are simultaneously, or synchronously, scanned with a constant wavelength interval ($\Delta\lambda$) between them (2,3). As a result, the intensity of the synchronous signal I_s can be considered as a product of two functions (3):

$$I_s(\lambda_{ex}, \lambda_{em}) = k \cdot c \cdot E_X(\lambda_{ex}) \cdot E_M(\lambda_{em})$$

where k = a constant

c = concentration of the analyte

E_X = excitation function

E_M = emission function

Simplification of individual spectra and decrease in spectral interferences from multiple components in complex mixtures are the main advantages of this scanning procedure. The spectral pattern of the system becomes more structured because the spectrum of each component is simplified (often to a single peak) and the bandwidth narrower (3). Correlation of the signal position with the structure of the compounds is also straightforward, generally with an increasing wavelength location with ring size. The effect of limiting each individual spectrum to a definite spectral band provides the most useful feature to locate the presence of a specific compound or group of compounds in complex samples.

Furthermore, the technique of taking the derivative can also be used to enhance the specificity of the analysis. Just as the second-derivative (d^2) technique has improved the selectivity of conventional fluorimetry (1), it should also improve the specificity of synchronous spectroscopy. In the d^2 mode, one measures the curvature of a peak. Resolution of strongly overlapping bands is improved by this technique (7).

The basic spectrometer required for synchronous luminescence measurements is often the same instrument used for conventional fixed-excitation luminescence work. No additional equipment is required, and synchronous measurement can be performed directly using any commercial spectrometer in which excitation and emission monochromators can be interlocked. Whenever derivative signals are needed, a commercial electronic differentiator is used to convert the output from the spectrofluorimeter into signals that are proportional to the first (d^1) and the second derivative (d^2) with respect to time. Scanning the monochromator wavelength at a uniform rate generates, therefore, a signal that is equivalent to a wavelength-derivative spectrum.

It has been demonstrated that for luminescence (1), there is no significant difference between the signal-to-noise values in the d^2 spectra generated by either electronic differentiation or wavelength modulation. The availability of an inexpensive commercial device that can be added to an existing spectrometer makes the choice of electronic differentiation most suitable for luminescence work.

ANALYSIS OF ENVIRONMENTAL SAMPLES

Fluorescence Analysis

The synchronous luminescence technique will have useful applications in many cases of environmental sample analysis. It is most valuable in situations where the variations in the composition of complex samples containing a large number of organic compounds (*e.g.*, oil spill, exhaust soots or by-product water) are unable to cause significant changes in the conventional fluorescence spectra.

A situation where the synchronous technique is also particularly useful is when a mixture contains several isomers or compounds of analog structures that exhibit similar and featureless spectra overlapping one another strongly. In such situations conventional luminescence has only limited use in providing data of sufficient accuracy. The example in Figure 1d shows the fluorescence spectra of a solution of by-product water from the Synthane gasifier. The wastewater was diluted in ethanol, its concentration being 5 ml/l. The two spectra in Figure 1d exhibit a broad-band and featureless structure. Different excitation wavelengths (270 mm and 280 mm) do not induce any significant change in the spectral profile. The spectra in Figures 1a, 1b and 1c are typical of the fluorescence spectra of cresols in ethanol. All spectra of the three cresol isomers exhibit the same featureless structure. Phenol also shows the same fluorescence spectral pattern. It is clear that direct comparison of the conventional fluorescence spectrum from the wastewater with those from phenol and the individual cresol isomers could not provide accurate determination of the content of the aqueous sample. This example underlines the necessity of improved methodologies for analysis of real-life environmental samples.

It has been demonstrated that synchronous spectroscopy enhances the selectivity in multicomponent analysis of complex mixtures (3). The band-narrowing effect resulting from synchronous scanning is illustrated in Figure 2 for *o*-cresol. The remarkable sharpening of the emission band in the synchronous spectrum (Figure 2, lower curve) greatly improves the selectivity of fluorescence analysis. Determination of band position is more accurate and spectral interference is also reduced.

Further enhancement can be achieved with derivation of the synchronous signal. The Synthane aqueous by-product sample has been subjected to analysis by second-derivative synchronous fluorescence and comparison made with gas-chromatographic (GC) data obtained by the Oak Ridge National Laboratory (ORNL) Analytical Chemistry Division. The two most abundant pollutants determined in this wastewater are phenol (2.1 mg/ml) and cresols (2.5 mg/ml). Due to interference effects, direct injection of aqueous material

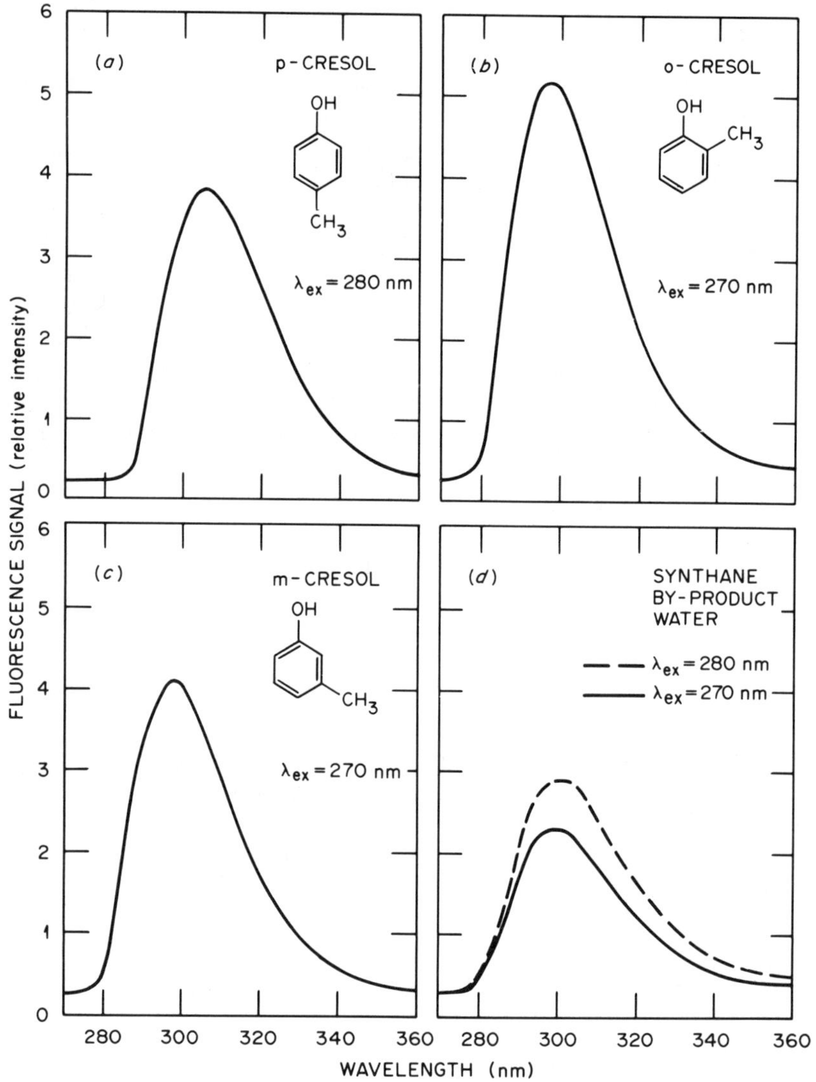

FIGURE 1. **Fluorescence spectra of the three isomers of cresol and the Synthane wastewater.**

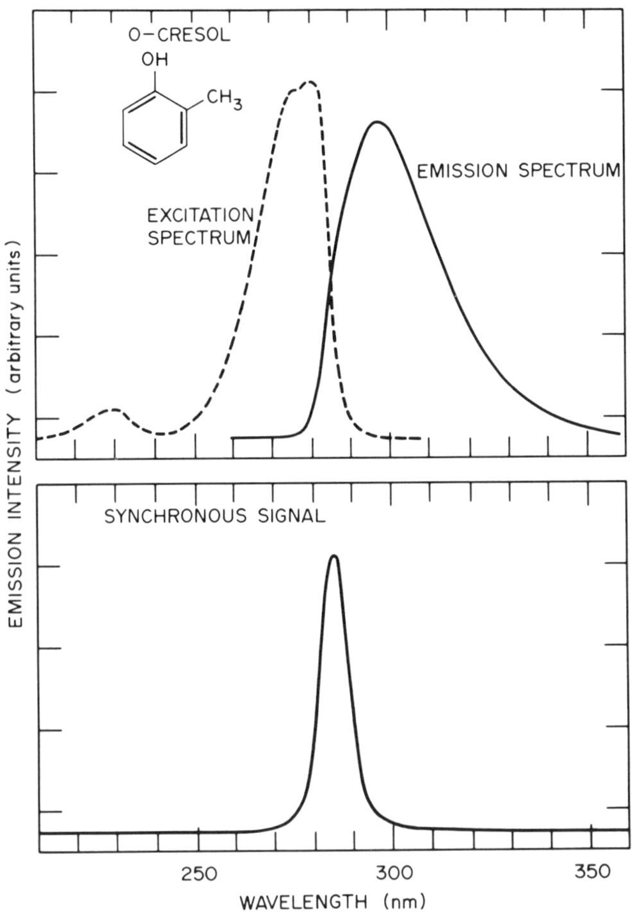

FIGURE 2. The band narrowing effect in the synchronous fluorescence technique.

onto a Tenax column gave unequivocal data only for the ortho isomer of cresol (0.7 mg/ml). Synchronous fluorescence, in the second-derivative mode, provided complementary data on the para isomer (1.1 mg/ml) during direct and rapid analysis of the unfractionated aqueous sample (7). The content of the meta isomer of cresol is derived by simple subtraction (0.7 mg/ml). This dual analysis is an example of synchronous fluorescence being used in a complementary capacity to yield complete quantitative analysis of all of the three cresol isomers in a typical coal-gasifier wastewater.

Room Temperature Phosphorescence (RTP) Analysis

The methodology of synchronous scanning can also be applied to phosphorimetry. Of interest is the use of the synchronous technique in the analysis of PAH by room temperature phosphorimetry (RTP). Room temperature phosphorimetry is a relatively new analytical technique based on phosphorescence at room temperature from organic compounds adsorbed on filter paper (4,5,8). The RTP method, as opposed to conventional low-temperature phosphorimetry, does not require the use of cryogenic equipment. It allows direct characterization of compounds collected on various substrates. Solutions from environmental samples can also be directly spotted on filter paper and the solvent evaporated under an infrared lamp prior to measuring phosphorescence.

The synchronous technique provides a new approach to identify components in mixtures selectively. Use can be made of the singlet-triplet splitting (Δ_{ST}) properties of organic compounds. Our study has revealed that this parameter, Δ_{ST}, can be effectively exploited as an additional factor of selectivity (6). Selection of optimal value $\Delta\lambda$ (matching the parameter Δ_{ST}) should enhance the corresponding synchronous phosphorescence signal, the spectral overlap from other components in the mixture being decreased. The physical parameter Δ_{ST} depends on the size and the chemical nature of the physical compounds. By exploiting the discrete singlet-triplet separation of each compound, this novel approach offers another possibility to analyze selectively specific components in mixtures, which should be useful in those situations where conventional fixed-excitation methods are ineffective. The results of analyses obtained with various PAH are given in Table I. These data indicate the usefulness of the parameter $\Delta\lambda$ and the specificity of the technique.

An example of application of the synchronous technique to RTP analyses is the identification of pyrene in a Synthoil sample, Figure 3(6). Gas chromatographic analysis has identified pyrene as one of the most abundant compounds in the Synthoil sample. Synchronous RTP can offer a rapid and simple method for identifying pyrene in this sample without resorting to separation techniques. With $\Delta\lambda$ = 250 nm (Table I) a sample of pure pyrene exhibits a peak of 597 nm (Figure 3, dashed curve). Under the same experimental conditions, the synchronous RTP signal of Synthoil (Figure 3, solid curve) directly reveals the presence of pyrene by the band at 597. The synchronous method is most suitable for differentiating isomers of PAH compounds, such as benzo[a]pyrene (B[a]P) and benzo[e]pyrene (B[e]P) or 1,2,3,4-dibenzanthracene (1,2,3,4-DBA) and 1,2,5,6-dibenzanthracene (1,2,5,6-DBA) (Figure 4). Accurate determination of these isomers is very important for health effect assessment because the degree of toxicity varies drastically with the structure of the isomers.

TABLE I. RTP excitation and Emission Bands for Various PAH and Experimental $\Delta\lambda$ Values for Several Synchronous RTP Analysis (5)

Compounds	Excitation Peaks (nm)	Emission Peaks (nm)	Experimental Optimal $\Delta\lambda$ Values (nm)
Acridine	360	640	280
Chrysene	325	515	190
Fluorene	305	430	125
Naphthalene	275	472	197
Phenanthrene	250	460	210
Pyrene	350	600	250
Quinoline	315	465	150
Benzo[e]pyrene	335	543	208
Benzo[a]pyrene	392	690	298
1,2,5,6-DBA	301	555	254
1,2,3,4-DBA	296	570	274

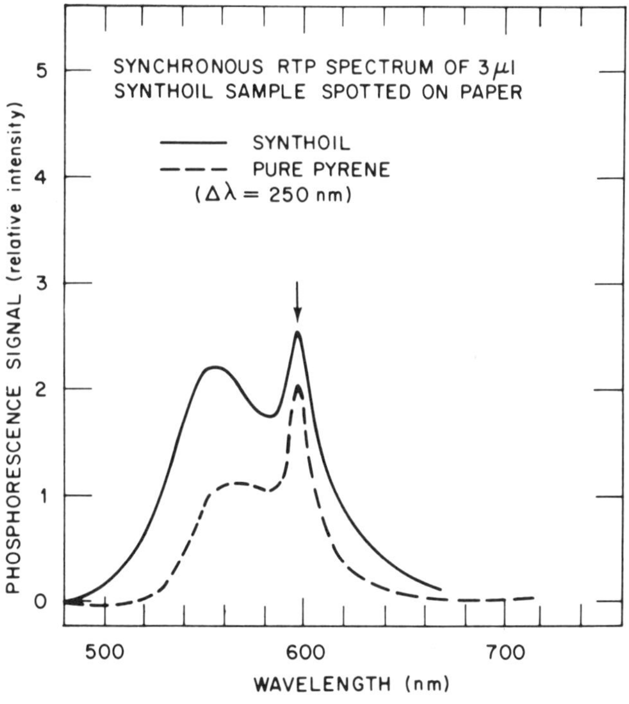

FIGURE 3. Identification of pyrene in a 3 μl sample of Synthoil: synchronous RTP spectrum of pure pyrene (dashed curve); synchronous RTP signal of Synthoil (solid curve). ($\Delta\lambda$ = 250 nm).

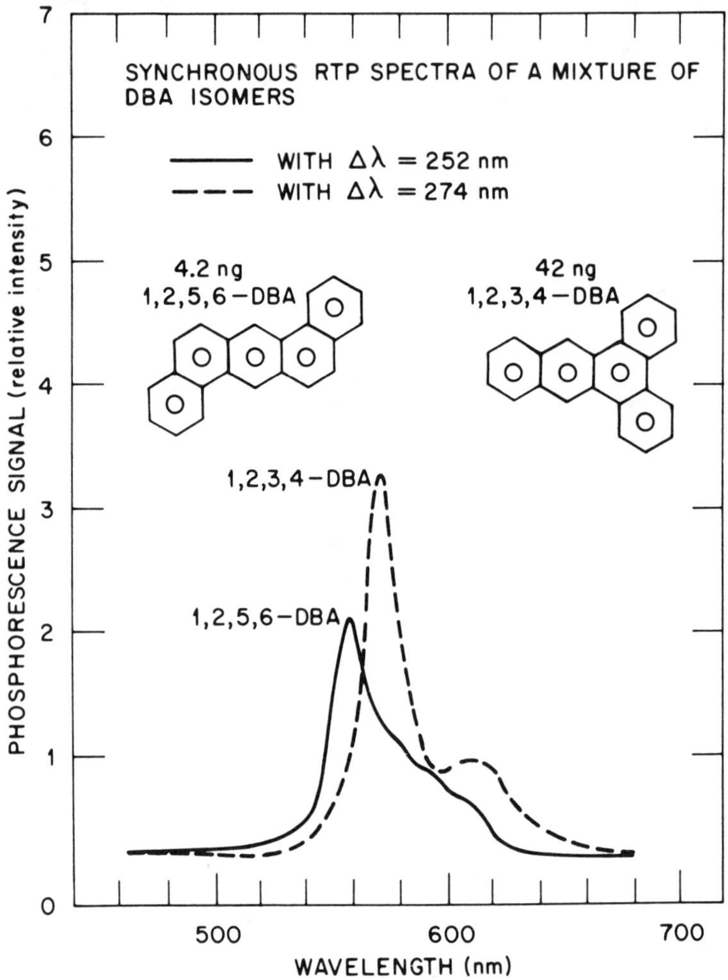

FIGURE 4. Differentiation of isomeric dibenzanthracenes by synchronous phosphorescence analysis at room temperature.

CONCLUSION

Synchronous luminescence spectroscopy has a great potential as a rapid, easy-to-use and cost-effective tool for analyzing large numbers of unfractionated or coarsely fractionated environmental samples. It should improve the effectiveness of analysis in the initial stages of screening organic pollutants in samples originating from coal-energy technologies, such as samples taken from process streams or collected as fugitive emissions.

ACKNOWLEDGMENTS

Research for this paper was sponsored by the Division of Biomedical and Environmental Research, U.S. Department of Energy under Contract W-7405-eng-26 with the Union Carbide Corporation. By acceptance of this article, the publisher or recipient acknowledges the U.S. Government's right to retain a nonexclusive royalty-free license in and to any copyright covering this article.

REFERENCES

1. Green, G. L., and O'Haver, T. C. (1974): *Anal. Chem.* 46:2191-2196.
2. Lloyd, J. B. F. (1971): *Nature* 231:64-65.
3. Vo-Dinh, T. (1978): *Anal. Chem.* 50:396-401.
4. Vo-Dinh, T., Walden, G., and Winefordner, J. D. (1977): *Anal. Chem.* 49:1126-1130.
5. Vo-Dinh, T., and Winefordner, J. D. (1977): *Appl. Spectrosc.* 13(2): 261-294.
6. Vo-Dinh, T., and Gammage, R. B. (1978): *Anal. Chem.* 50:2054-2058.
7. Vo-Dinh, T., and Gammage, R. B.: *Characterization of Cresols in Waste by Second-derivative Synchronous Spectroscopy,* submitted for publication.
8. Wellon, S. L., Paynter, R. A., and Winefordner, J. D. (1974): *Spectrochim. Acta* 30A:2133-2140.

Polynuclear Aromatic Hydrocarbons,
edited by P. W. Jones and P. Leber.
Ann Arbor Science Publishers, Inc.
Ann Arbor, MI ©1979.

THE IDENTIFICATION OF POLYNUCLEAR AROMATIC HYDROCARBON MIXTURES IN HIGH-PERFORMANCE LIQUID CHROMATOGRAPHY FRACTIONS UTILIZING THE SHPOL'SKII EFFECT

Anders Colmsjö[1] and Ulf Stenberg[2]

Department of Analytical Chemistry
University of Stockholm
Arrhenius Laboratory
Stockholm, Sweden

INTRODUCTION

The separation of polynuclear aromatic hydrocarbons (PAH) in samples originating from automobile exhaust and from other combustion processes has been carried out with Reversed-Phase High-Performance Liquid Chromatography (RPHPLC) by a number of investigators (9,13). As yet, complete separation has not been achieved so that the detectability of individual PAH has been dependent on various selective detectors, of which UV and/or fluorescence detectors are the most common (3,8,9,12). This paper examines the continuous monitoring with a UV-detector combined with low-temperature fluorescence analysis of collected fractions.

PAH are compounds known to be suitable for fluorescence analysis due to their rigid structure of condensed unsaturated rings. Four-, five- and six-membered rings usually show fluorescence in the region of 360-480 nm. These spectra are broad-banded at room temperature. If two or more compounds are present in the same solution, these can interfere with one another, making identification difficult or impossible. However, one way to avoid this is by utilizing the so-called Shpol'skii effect (4,14,15) that is, the specificity of fluorescence emission spectra can be greatly increased by irradiating the sample in a suitable frozen *n*-alkane solvent at temperatures below 100°K.

---

[1]Working on a grant from the Swedish Natural Science Research Council.
[2]Working on a grant from the National Swedish Environment Protection Board.

The broad-band emission can thus be converted into so-called quasilinear fluorescence emission, where half-bandwidths of less than 1 nm are frequent. This is not the usual type of band narrowing that appears when the temperature is decreased and can only be explained by means of the quantum theory (17). Quasilinear fluorescence takes place when the emitted photon does not interact with the surrounding solvent matrix (so called phonon free bands). The Shpol'skii spectrum is very specific for a compound in a certain solvent and can be used as a fingerprint for PAH identification. However, certain parameters influence the spectrum, such as temperature (5), concentration (20), choice of solvent (6), impurities, excitation wavelength and freezing rate, and it is important to take these into consideration. In this investigation the applicability of quasilinear fluorescence when identifying PAH in HPLC fractions is discussed.

PROCEDURE

Samples from automobile exhaust gases were taken. The particulate phase was collected on a glass-fiber filter, vacuum-sublimated and dissolved in cyklohexane:acetone (2:1). This was done by an all glass device (7). The solution was treated by a dimethylformamide (DMF) clean-up procedure and injected on a Spectra Physics liquid chromatograph (Model 3500 B). The HPLC system consisted of a reversed-phase column, polarity gradient programmer and a UV-detector. Eluted fractions of interest were collected immediately after detection. The fractions were evaporated to dryness using a stream of nitrogen, and then dissolved in *n*-hexane, *n*-heptane or *n*-octane. Each solution was frozen to 63°K (solid nitrogen) and irradiated with a 450-W Xe-lamp, using the wavelength region below 348 nm by means of a cutoff filter. Spectra were recorded with a Jarrel Ash 0.5-m Ebert scanning monochromator with a photomultiplier (Hammatzu R 928). The signal was amplified, filtered and recorded.

RESULTS AND DISCUSSION

A typical chromatogram from a gasoline automobile exhaust sample is shown in Figure 1. This chromatogram is divided into 15 parts, each representing a collected fraction. The first two fractions were taken from an unresolved peak. The Shpol'skii spectrum of fraction two is shown in Figure 2. This is a good fingerprint of pyrene dissolved in *n*-hexane (reference spectrum Figure 16). About 50 peaks correspond in intensity and wavelength with the reference spectrum (±0.1nm). Pyrene exhibits such strong fluorescence that it completely dominates the spectrum region recorded. The spectrum shows typical concentration dependence of Shpol'skii fluorescence, *i.e.*, the peaks in the shorter-wavelength part of the spectrum are relatively smaller. This is due to the concentration having exceeded the upper limit of linearity. This strongly indicates, apart from the retention data, that pyrene is preferentially present in the second fraction. Fraction three exhibited no Shpol'skii fluorescence. The Shpol'skii spectrum of fraction four dissolved in *n*-hexane is shown in Figure 3. It consists of several spectra superimposed, where 1-methylpyrene, 4-methylpyrene and benzo[a]anthracene are found when compared to the library reference spectra (Figures 17,18 and 19). The identification of 1-methylpyrene and 4-methylpyrene is based primarily on position and intensity of the peaks at 374.8 and 374.0 nm, respectively, but about ten of the smaller peaks correspond to the reference spectra as well. The identification of benzo[a]anthracene is based on position and intensity of two comparably weak peaks at 383.4 and 383.9 nm, while the peaks at 372.4 and 382.7 nm belong to unidentified compounds.

Fraction five (Figure 4) shows a sharp and line-abundant Shpol'skii spectrum of a compound that usually appears in the particulate phase of automobile exhaust. This spectrum is not found among the reference spectra, see Table I, which exclude certain possibilities. The doublet at 385.6 and 385.8 nm does not derive from the compound with major peaks at 377.4 and 377.7 nm.

TABLE I. Compounds with Available Reference Spectra (arranged in order of increasing ring number)

Carbazole	9,10-Dimethylbenzo[a]anthracene
Anthracene	Chrysene
2-Methylanthracene	Perylene
9-Methylanthracene	Dibenzo[a,c]anthracene
9,10-Dimethylanthracene	Dibenzo[a,h]anthracene
Phenanthrene	Benzo [g,h,i] perylene
2-Phenylnaphthalene	Benzo[a]pyrene
Benzo[a]fluorene	Benzo[e]pyrene
Pyrene	Indeno[1,2,3-cd]pyrene
1-Methylpyrene	Pentacene
4-Methylpyrene	Picene
Triphenylene	Benzo[b]chrysene
Dibenzo[a,i]carbazole	Coronene
Dibenzo[b,g]carbazole	Dibenzo[a,h]pyrene
Benzo[b]fluoranthene	Dibenzo[a,i]pyrene
Benzo[k]fluoranthene	Anthanthrene
Benzo[a]anthracene	

The spectrum of fraction six (Figure 5) shows when recorded with *n*-hexane as solvent, the typical line of benzo[e]pyrene at 387.7 nm (reference spectrum Figure 20). The spectral part above 400 nm is reminiscent of the spectrum of benzo[k] fluoranthene (B[k] F) dissolved in *n*-hexane. However, since B[k] F dissolved in *n*-hexane gives a poorly resolved spectrum, the fraction was evaporated and dissolved in *n*-octane, where B[k] F is known to give a sharper spectrum. The result is shown in Figure 6 where more than ten peaks correspond to the reference spectrum (Figure 21) in intensity and position (± 0.2 nm). This is a good illustration of the necessity of selecting the right solvent for a certain compound.

In the seventh fraction (Figure 7) the strongly fluorescing compound benzo[a] pyrene (B[a] P) is detected (reference spectrum Figure 22). B[a] P gives its sharpest (but not most line-abundant) spectrum in *n*-octane, and since B[a] P was expected to be found in this fraction, *n*-octane was chosen as solvent. It is important that B[a] P and B[e] P are collected in two separate fractions because B[a] P quenches the fluorescence spectrum of B[e] P since the emission maximum of B[e] P coincides with a strong absorption maximum of B[a] P. If the sample is concentrated enough, this can be avoided by studying the phosphorescence spectrum of B[e] P.

Fraction eight (Figure 8) exhibits another compound not found among the reference spectra but, as suggested by Grimmer *et al.* (10), 8,9-methylenbenzo[e] pyrene is expected to be found in this region. The UV absorption spectrum of fraction eight corresponds well with that of 8,9-methylenbenzo[e] pyrene.

Fraction nine (Figure 9) shows a very line-abundant and sharp spectrum nearly identical to the reference spectrum of benzo[g,h,i] perylene dissolved in *n*-hexane (Figure 23). About 35 peaks correspond in intensity and position (±0.2 nm) and the spectrum is remarkably free from any emission lines other than those deriving from benzo [g,h,i] -perylene.

The spectrum of fraction ten (Figure 10) can easily be identified as anthanthrene dissolved in *n*-hexane, Figure 24. It is noteworthy that the spectrum of the HPLC fraction seems to be cleaner from fluorescence caused by impurities than is the spectrum of the reference substance.

The spectra of fraction 11 and 12 (Figure 11 and Figure 12) contain unidentified compounds with a large number of sharp and intense peaks. In particular, the spectrum of fraction 12 has been reported by Soviet scientists (1) as a spectrum of a very common compound in curtisite (the peak at 410.7 nm does not belong to this compound). Fraction number 13 (Figure 13) dissolved in *n*-hexane shows a low-intensity spectrum of coronene (reference spectrum Figure 25), but the identification is unambiguous.

Neither fraction 14 (Figure 14) nor fraction 15 (Figure 15) has been identified. We think it is remarkable that we have not found any dibenzopyrenes at all with the molecular weight 302.

It is possible to obtain even better resolved spectra by lowering the temperature to below 20°K (2,11,16,18,19), but that would make the method much more time-consuming and complicated. The step from 77°K (liquid nitrogen) to 63°K (solid nitrogen) is not complicated and worth the effort regarding the better resolution achieved (5).

The remaining identification of the unidentified peaks ought to be done by expanding the library of spectra by a proper number of quasilinear spectra of common PAH. However, complications can arise when using the Shpol'skii fluorescence as a quantitative method due to the possibility of quenching phenomena in unknown mixtures, and due to the restricted region of linearity of the emission. The effect of the latter is that the concentration must be maintained under a certain level to achieve a linear relationship between concentration and emission.

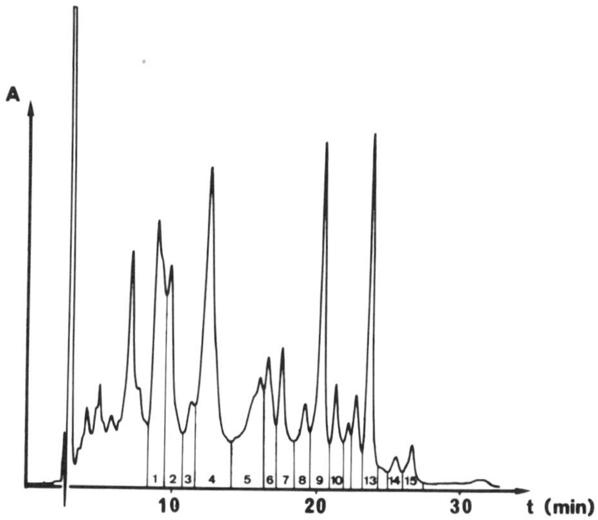

FIGURE 1. HPLC chromatogram of automobile exhaust sample. 10 μl Injected on a ODS 5 μm-column. Inner Diameter: 4 mm; Length: 150 mm; Flowrate: 0.6 ml/min; Gradient Elution: 15%-0% H_2O in MeOH; Delay: 8 min; Sweep Time: 10 min; Absolute Wavelength: 289 nm.

Identified Compounds

Fraction no.	Name	Fraction no.	Name
2	Pyrene	6	Benzo[k]fluoranthene
4	1-Methylpyrene	7	Benzo[a]pyrene
4	4-Methylpyrene	9	Benzo[g,h,i]perylene
4	Benzo[a]anthracene	10	Anthenthrene
6	Benzo[e]pyrene	13	Coronene

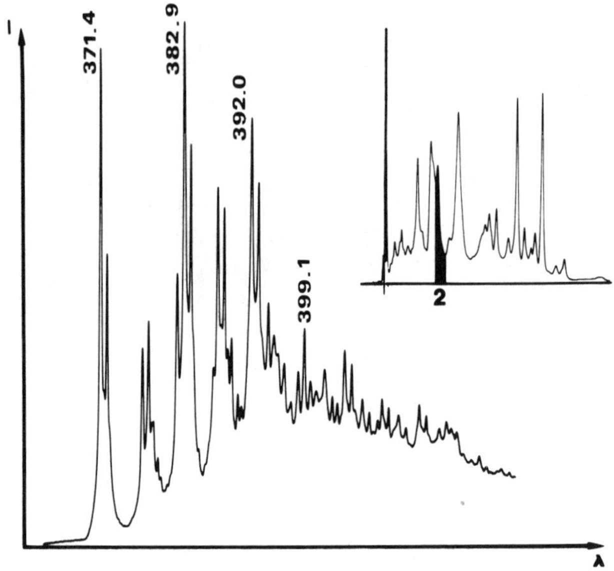

FIGURE 2. Quasilinear fluorescence spectrum of fraction two at 63°K. Solvent: *n*-hexane. Identified compound: pyrene.

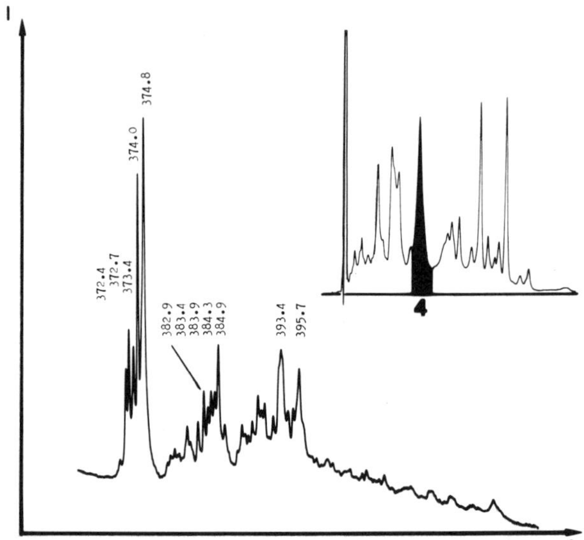

FIGURE 3. Quasilinear fluorescence spectrum of fraction four at 63°K. Solvent: *n*-hexane. Identified compounds: 1-methylpyrene, 4-methylpyrene and benzo[a] anthracene.

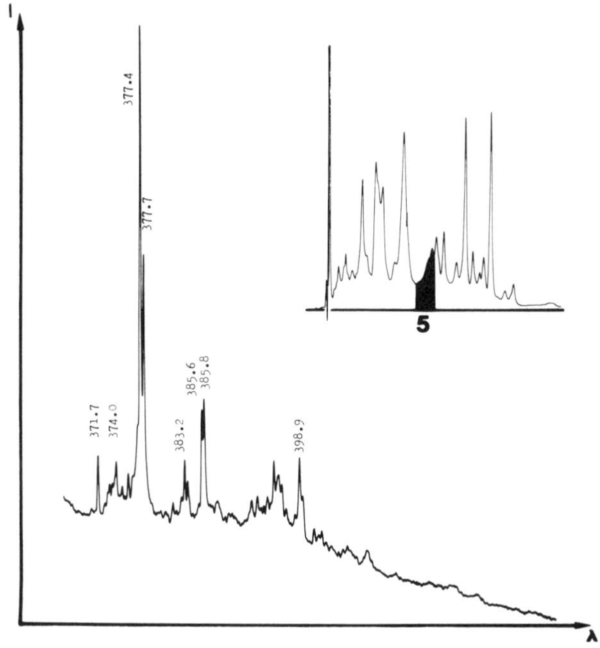

FIGURE 4. Quasilinear fluorescence spectrum of fraction five at 63°K. Solvent: *n*-hexane.

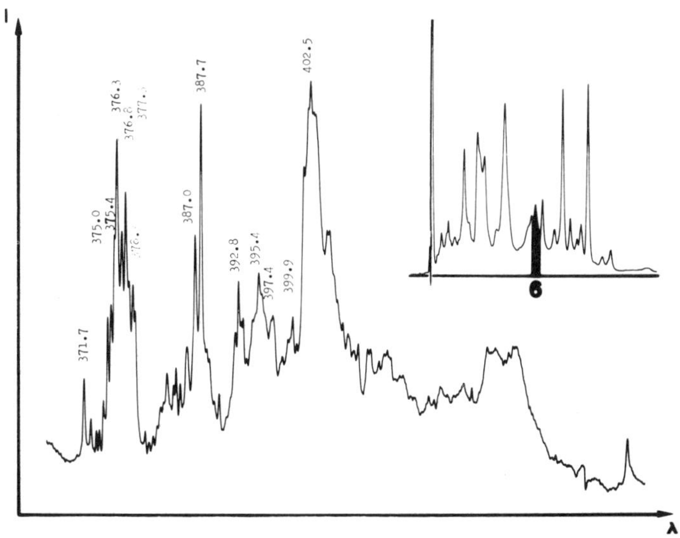

FIGURE 5. Quasilinear fluorescence spectrum of fraction six at 63°K. Solvent: *n*-hexane. Identified compound: benzo[e]pyrene.

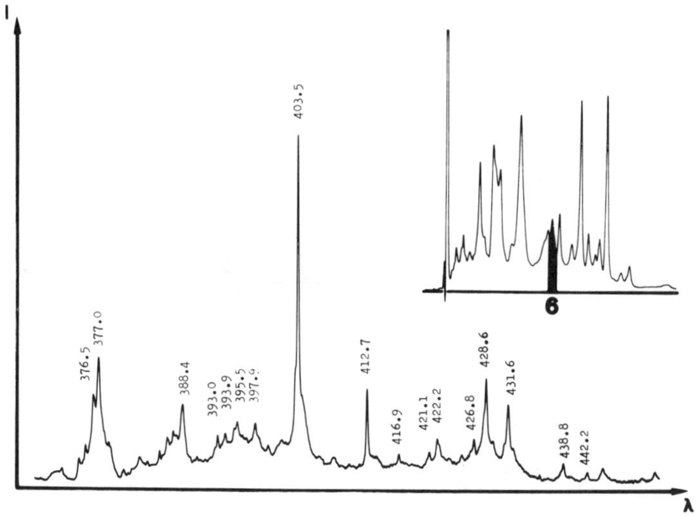

FIGURE 6. Quasilinear fluorescence spectrum of fraction six at 63°K. Solvent: *n*-octane. Identified compound: benzo[k]fluoranthene.

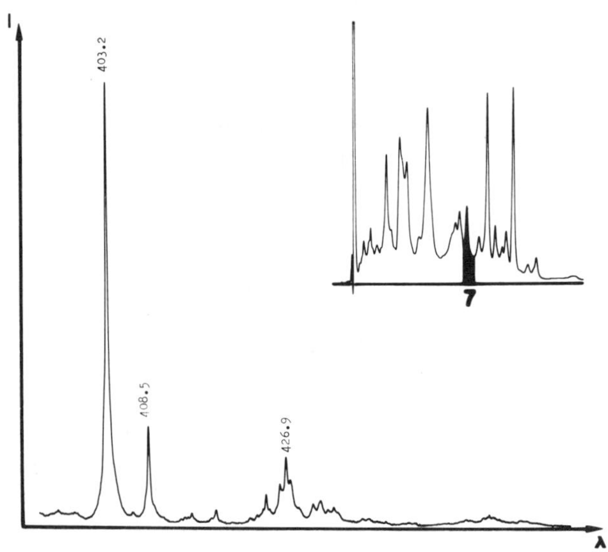

FIGURE 7. Quasilinear fluorescence spectrum of fraction seven at 63°K. Solvent: *n*-octane. Identified compound: benzo[a]pyrene.

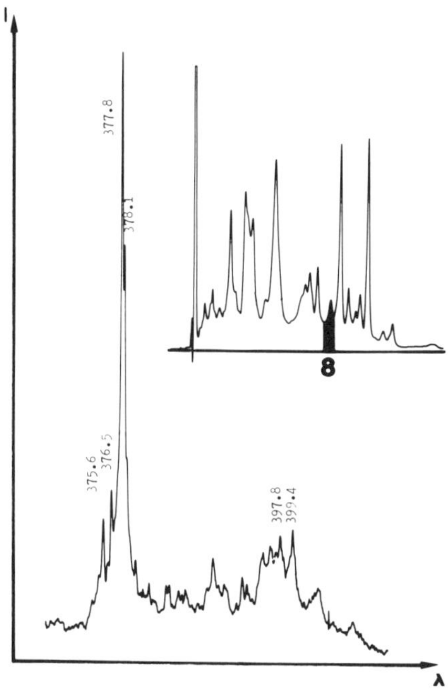

FIGURE 8. Quasilinear fluorescence spectrum of fraction eight at 63°K. Solvent: *n*-hexane.

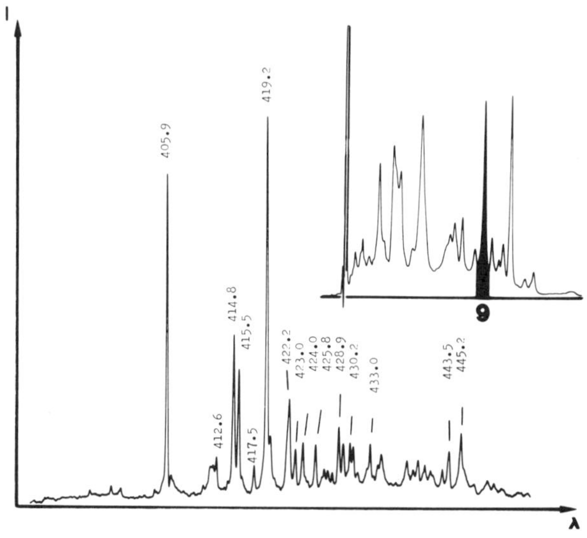

FIGURE 9. Quasilinear fluorescence spectrum of fraction nine at 63°K. Solvent: *n*-hexane. Identified compound: benzo[ghi]perylene.

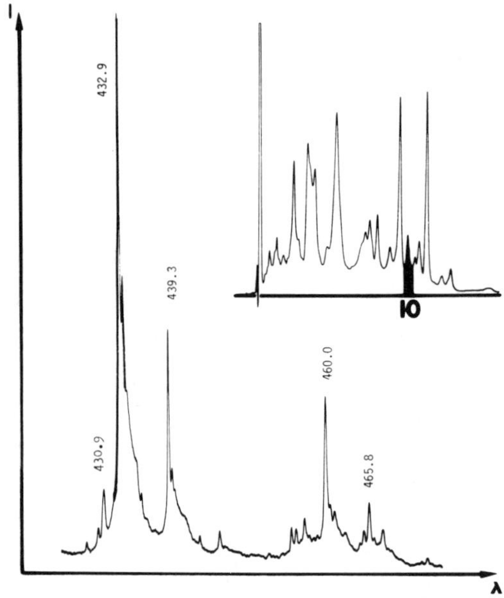

FIGURE 10. Quasilinear fluorescence spectrum of fraction ten at 63°K. Solvent: *n*-hexane. Identified compound: anthanthrene.

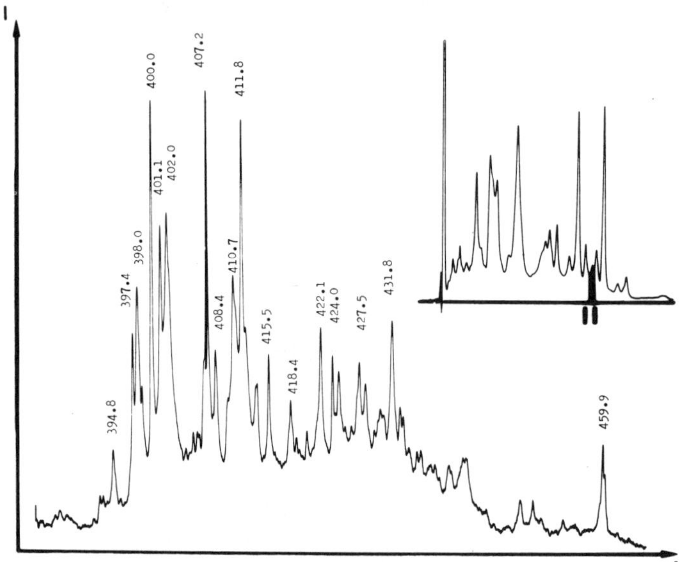

FIGURE 11. Quasilinear fluorescence spectrum of fraction eleven at 63°K. Solvent: *n*-hexane.

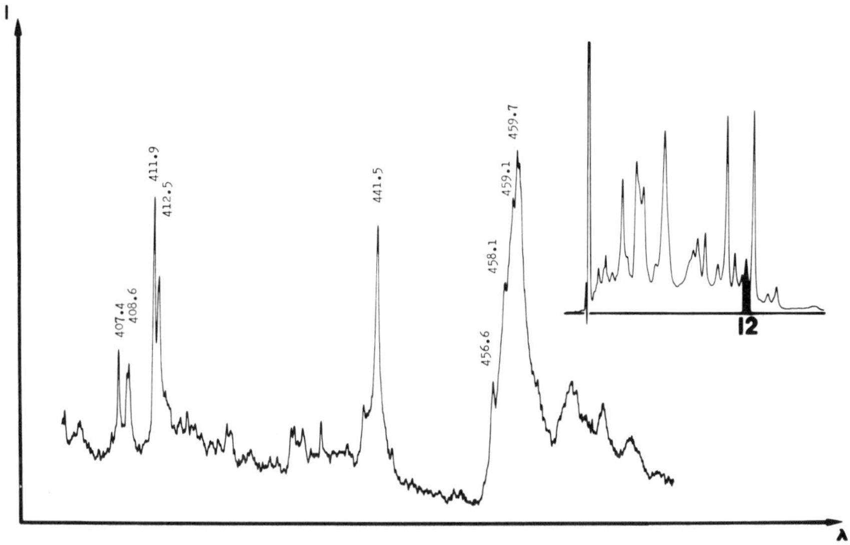

FIGURE 12. Quasilinear fluorescence spectrum of fraction twelve at 63°K. Solvent: *n*-hexane.

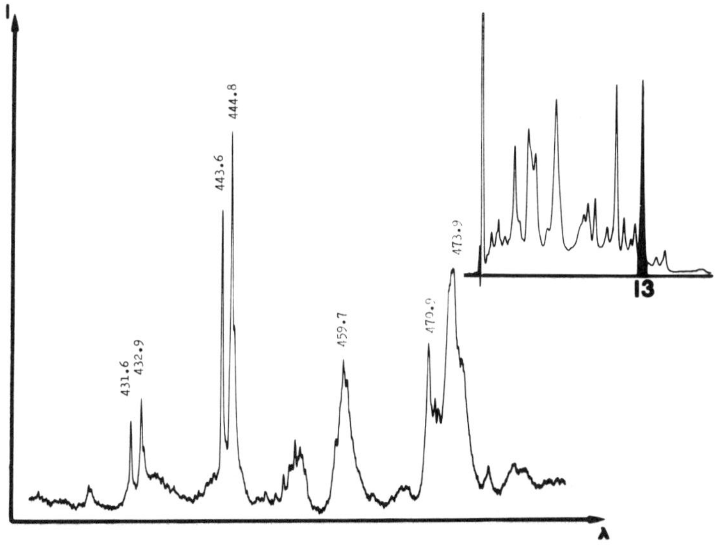

FIGURE 13. Quasilinear fluorescence spectrum of fraction 13 at 63°K. Solvent: *n*-hexane. Identified compound: coronene.

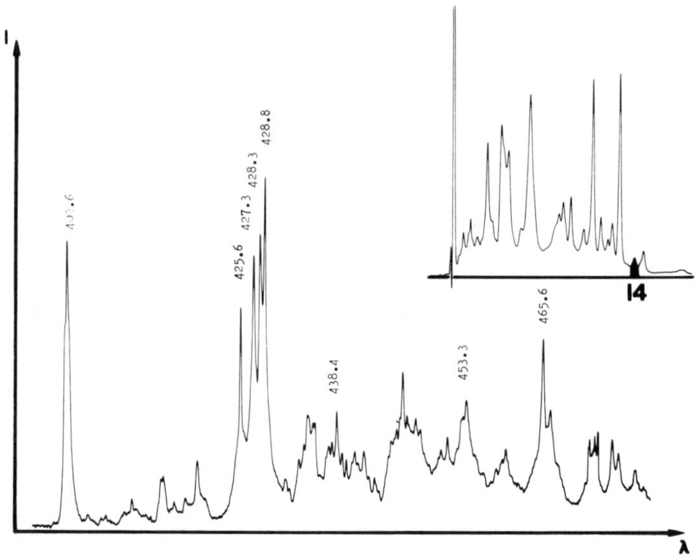

FIGURE 14. Quasilinear fluorescence spectrum of fraction 14 at 63°K. Solvent: *n*-hexane.

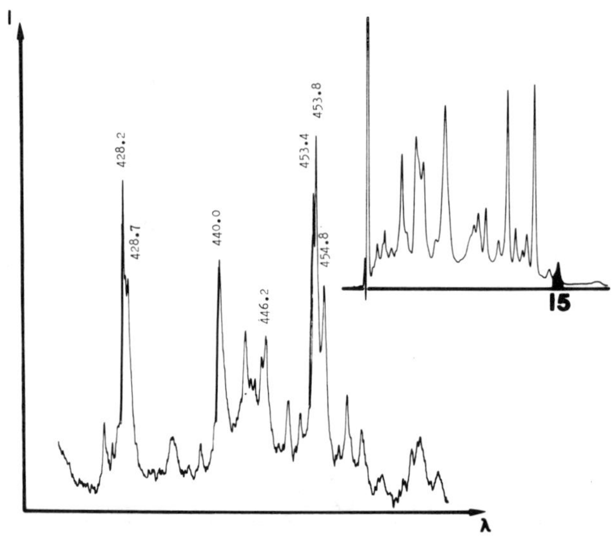

FIGURE 15. Quasilinear fluorescence spectrum of fraction 15 at 63°K. Solvent: *n*-hexane.

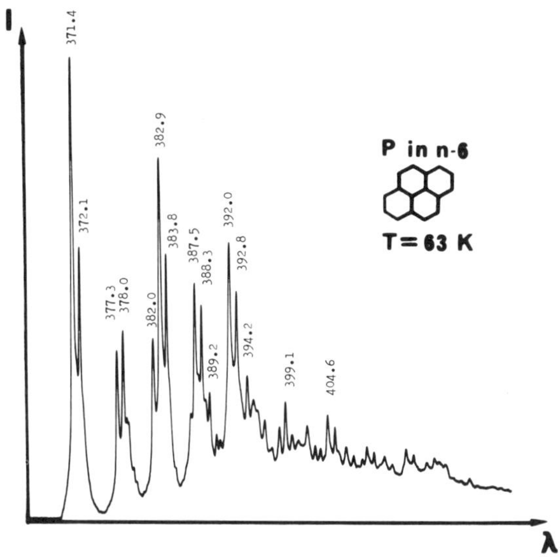

FIGURE 16. Fluorescence spectrum of pyrene, at 63°K, dissolved in *n*-hexane.

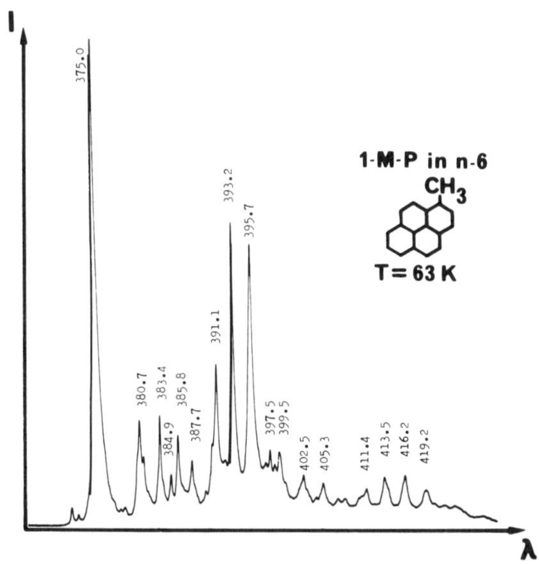

FIGURE 17. Fluorescence spectrum of 1-methylpyrene, at 63°K, dissolved in *n*-hexane.

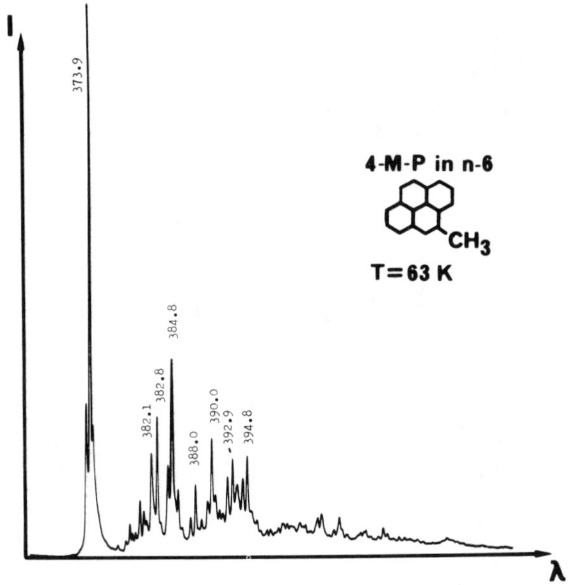

FIGURE 18. Fluorescence spectrum of 4-methylpyrene, at 63°K, dissolved in *n*-hexane.

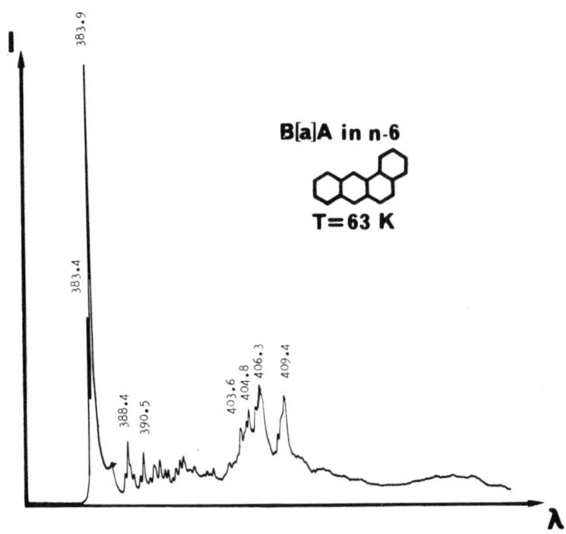

FIGURE 19. Fluorescence spectrum of benzo[a]anthracene, at 63°K, dissolved in *n*-hexane.

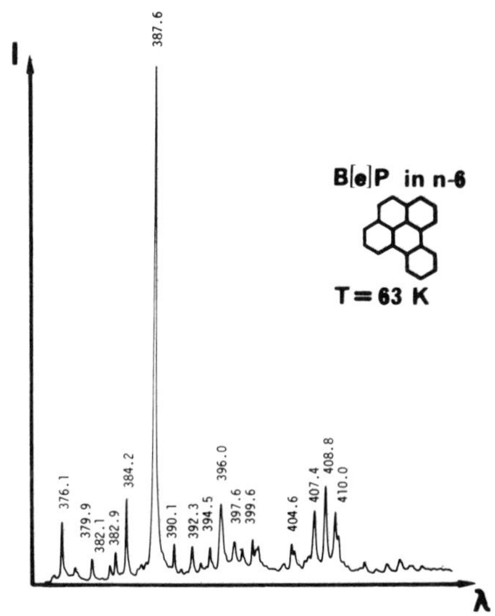

FIGURE 20. Fluorescence spectrum of benzo[e] pyrene, at 63°K, dissolved in *n*-hexane.

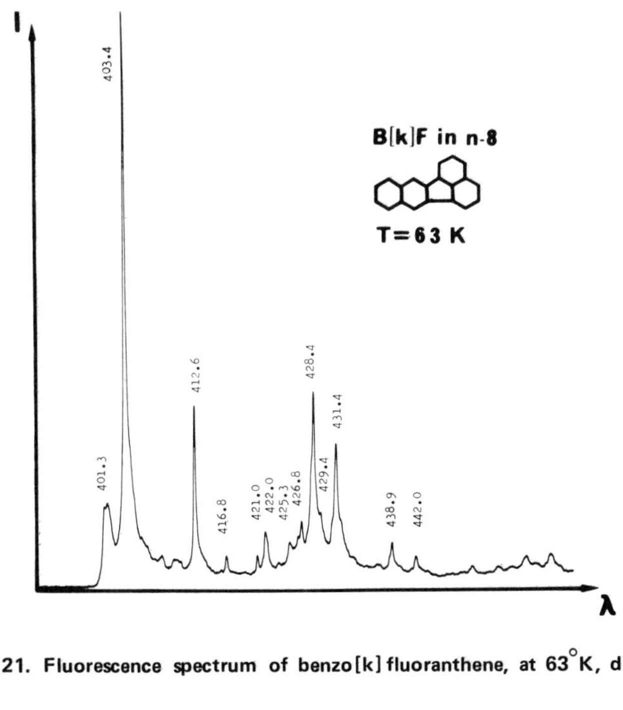

FIGURE 21. Fluorescence spectrum of benzo[k] fluoranthene, at 63°K, dissolved in *n*-octane.

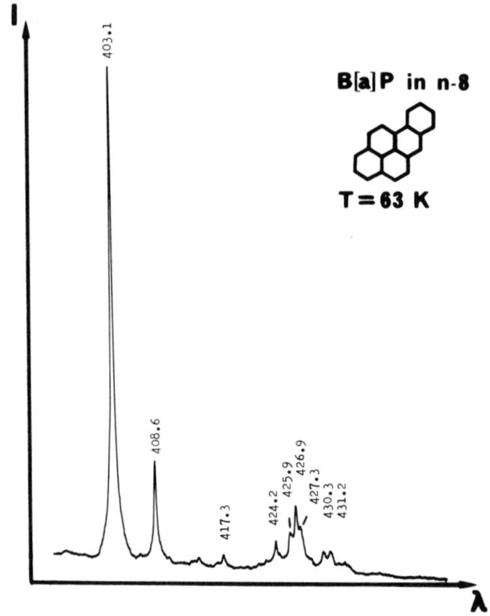

FIGURE 22. Fluorescence spectrum of benzo[a]pyrene, at 63°K, dissolved in *n*-octane.

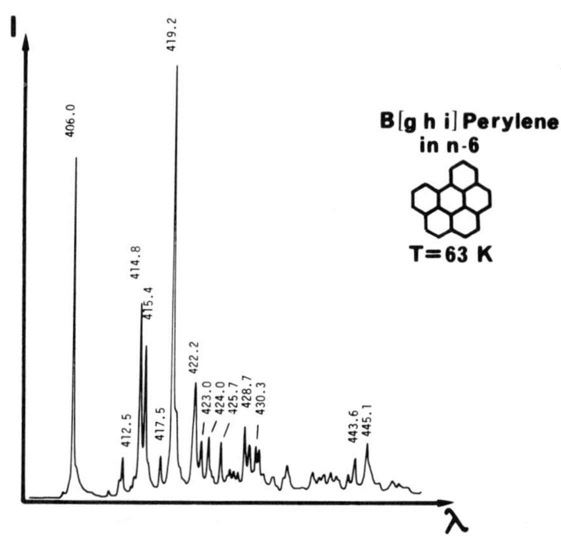

FIGURE 23. Fluorescence spectrum of benzo[g,h,i]perylene, at 63°K, dissolved in *n*-hexane.

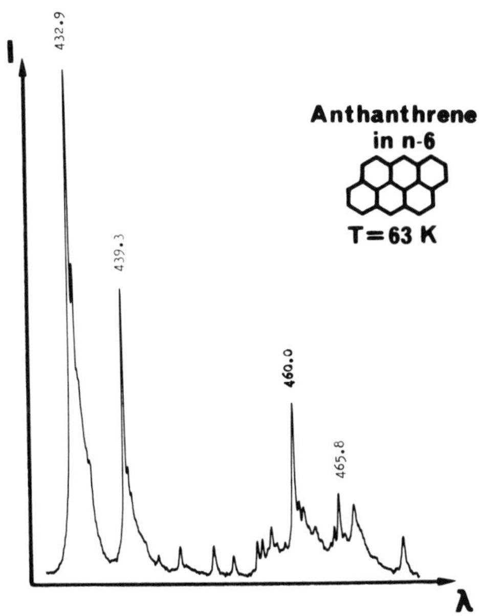

FIGURE 24. Fluorescence spectrum of anthanthrene, at 63°K, dissolved in *n*-hexane.

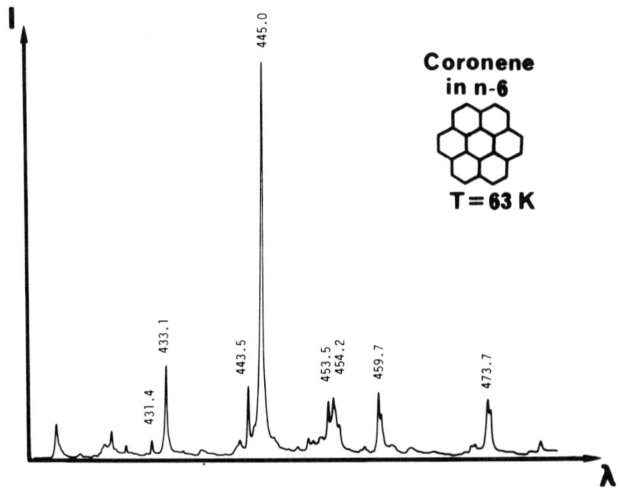

FIGURE 25. Fluorescence spectrum of coronene, at 63°K, dissolved in *n*-hexane.

ACKNOWLEDGMENTS

We thank Mrs. Beryl Holm for reviewing the manuscript. This investigation was kindly supported by the Department of Analytical Chemistry and has been sponsored by the Swedish Natural Science Research Council, Box 23136, 104 05 Stockholm (Contract no. K 0369-015) and the National Swedish Environment Protection Board, Fack, S-171 20 Solna (Contract no. 7-49/77).

REFERENCES

1. Alekseeva, T. A., and Teplitskaya, T. A. (1978): Development of principles for determining the type of molecular structure of unknown compounds of complex mixtures by luminescent spectroscopic methods. *Izv. Akad. Nauk. SSR, Ser Fiz.* 42(3):669.
2. Al'shits, E. I., Godyaev, E. D., and Personov, R. I. (1972): Broadening and shift in the phonon-free lines for trace compounds in n-paraffines below the debye temperature. *Izv. Akad. Nauk. SSR, Ser Fiz.* 36(5): 1117.
3. Boden, H. (1976): The determination of B[a]P in coal tar pitch volatiles using HPLC with selective UV detection. *J. Ch. Science* 14:391.
4. Bowen, E. J., and Brocklehurst, B. J. (1955): The emission spectra of aromatic hydrocarbons in crystalline paraffines at -180°C. *J. Chem. Soc.* 1955:4320.
5. Colmsjo, A., and Stenberg, U. (1977): Temperature effects on resolution of quasilinear fluorescence spectra of some PAH. *Chem. Scr.* 11: 220-23.
6. Colmsjo, A., and Stenberg, U. (1976): The effect of *n*-alkane solvent on low temperature quasilinear fluorescence spectra of some PAH. *Chem. Scr.* 9:227-32.
7. Colmsjo, A., and Stenberg, U. (1979): Vacuum sublimation of Polynuclear Aromatic Hydrocarbons separated by thin layer chromatography for the detection with Shpol'skii low-temperature fluorescence. *J. Chromator.*, 169:205-12.
8. Das, B. S., and Thomas, G. H. (1978); Fluorescence detection in HPLC determination of PAH. *Anal. Chem.* 50(7):967.
9. Fox, M. A., and Staley, S. W. (1976): Determination of PAH in atmospheric particulate matter by HPLC coupled with fluorescence technique. *Anal. Chem.* 48(7):992.
10. Grimmer, G., Bohnke, H., and Glaser, A. (1977): Investigation on the carcinogenic burden by air pollution in man XV. Polycyclic aromatic hydrocarbons in automobile exhaust gas—an inventory. *Zbl. Bakt. Hyg., I. Abt. Orig. B* 164:218.
11. Klimova, L. A. (1963): Absorption and emission spectra of pyrene at 20 and 4 K. *Opt. i Spectr.* 5:344.

12. Krstulovic, A. M., Rosie, D. M., and Brown, P. R. (1976): Selective monitoring of PAH by HPLC with a variable wavelength detector. *Anal. Chem.* 48(9):1383.
13. Schmit, J. A., Henry, R. A., Williams, R. C., and Dieckman, J. F. (1971): Applications of high speed reversed phase liquid chromatography. *J. Ch. Science* 9:647.
14. Shpol'skii, E. V. (1959): Emission spectroscopic analysis of organic compounds. *Usp. Fiz. Nauk.* 68:51-69.
15. Shpol'skii, E. V. (1960): Line fluorescence spectra of organic compounds and their applications. *Usp. Fiz. Nauk.* 71:215.
16. Shpol'skii, E. V. (1963): New data on the nature of the quasilinear spectra of organic compounds. *Usp. Fiz. Nauk.* 80:255.
17. Shpol'skii, E. V. (1962): Problems of the origin and structure of the quasilinear spectra of organic compounds at low temperatures. *Usp. Fiz. Nauk.* 77:321.
18. Shpol'skii, E. V., and Klimova, L. A. (1962): Line spectra of aromatic hydrocarbons in frozen crystalline solutions. *Opt. i Spectr.* 13:174.
19. Shpol'skii, E. V., Klimova, L. A., and Personov, R. I. (1962): Line spectra of aromatic hydrocarbons in frozen crystalline solutions, II. *Opt. i Spectr.* 13:341.
20. Shpol'skii, E. V., Klimova, L. A., Nersesova, G. N., and Glyadkovskii, V. I. (1968): Concentration dependence of the fluorescence and the absorption spectra of frozen paraffine solutions of naphthalene. *Opt. i Spectr.* 24:25.

Polynuclear Aromatic Hydrocarbons, edited by P. W. Jones and P. Leber. Ann Arbor Science Publishers, Inc. Ann Arbor, MI © 1979.

PHOTOREACTIVITY OF SELECTED AROMATIC HYDROCARBONS IN WATER

Richard G. Zepp and Patricia F. Schlotzhauer

Environmental Research Laboratory
U.S. Environmental Protection Agency
College Station Road
Athens, Georgia 30605

INTRODUCTION

Polynuclear aromatic hydrocarbons (PAH) are known to be widespread contaminants of natural water bodies (1). Several hydrophobic PAH such as benzo[a]pyrene (B[a]P) are known carcinogens that have been shown to bioaccumulate in aquatic organisms (23). The more water-soluble bicyclic and tricyclic PAH are not carcinogenic, but recent studies indicate that ultraviolet radiation transforms some of these chemicals into products that are acutely toxic to a variety of aquatic organisms (15).

Concern over the environmental safety and persistence of PAH has prompted studies of processes that transport and transform these chemicals in natural waters. Microbial oxidation is most rapid with benzene and naphthalene and with their alkylated derivatives (7). Tricyclic and especially higher molecular-weight polycyclic aromatics such as B[a]P are considerably more resistant to degradation by microorganisms. Recent studies in an outdoor model ecosystem indicated to Lee and co-workers that photochemical transformation is an important process for several PAH including B[a]P (16).

Although a great deal is known about the photochemistry of aromatic compounds in solution, most past studies have been carried out in organic solvents at high concentrations with oxygen present (9). Under these reaction conditions, direct and indirect (photosensitized) oxygenation are important. Direct photooxygenation is initiated through light absorption by the aromatic compound. Indirect photooxygenation occurs via light absorption by another chemical, called a photosensitizer, that has been added to the system in addition to the aromatic compound. Although a variety of mechanisms have been proposed since these studies began some 50 years ago, it is now generally accepted that singlet oxygen, an excited form of molecular oxygen, is the key intermediate in photooxygenations.

Analytical difficulties associated with the low solubilities of PAH have limited study of their photochemistry in water. Indeed, comprehensive data concerning the aqueous solubilities of the higher-molecular weight PAH have not become available until very recently (17). It has been suggested that the PAH in aquatic environments are completely sorbed to particulate matter (23). Recent studies by Wolfe *et al.* (31), Smith *et al.* (25) and Karickhoff and co-workers (12), however, have demonstrated that sorption of PAH and other chemicals to natural sediments and biogenous material is a reversible equilibrium process and that it is likely that the PAH in the water columns of many inland water bodies and in the ocean exist mainly in the dissolved state. Andelman and Suess (2) and McGinnes (18) established that the PAH, B[a]P and benz[a]anthracene are photoreactive in aqueous media, although interpretation of their results was clouded by the presence of acetone, a known photosensitizer, in their reaction media. More recently, Smith and co-workers (25) have reported kinetic data concerning direct photolysis in water of several aromatic compounds that are found in effluents of fossil-fuel processing plants. In this paper, we present detailed studies concerning the photoreactivity in water of several PAH that are commonly found in environmental samples (8). The effects of light attenuation and sediment-water partitioning on the photolysis of PAH in the aquatic environment are also discussed.

EXPERIMENTAL

Materials

1-Methyl and 2-methylnaphthalenes of greater than 99% purity were obtained from Chemical Samples Co. 9-Methylanthracene, 9,10-dimethylanthracene and chrysene were analytical grade (> 99% pure) obtained from K&K Laboratories. Fluoranthene and naphthacene (> 99%) were obtained from Analabs. Reagent-grade pyrene and phenanthrene were purified by elution chromatography on neutral alumina with subsequent recrystallization from 95% aqueous ethanol. Anthracene and naphthalene (reagent) were each recrystallized from 95% aqueous ethanol prior to use. Laboratory distilled water was redistilled from permanganate and a middle cut was retained for these studies. Spectro-grade isooctane from Burdick and Jackson Co. was used as received. Natural water samples were obtained from the top meter of the water column and stored at 5°C prior to use.

Procedures

Reaction mixtures were analyzed using a Micromeritics Liquid Chromatograph with variable-scan UV-detection or a Farrand MK-1 Spectrofluorometer as detector. An ODS-1 column with methanol-water as the mobile phase was used.

A "merry-go-round" photochemical apparatus that is described by Moses *et al.* (21) was employed for the kinetic studies. Quantum yields for reaction were computed from the slopes of plots of natural log concentration versus time as previously described (32). Light from a 450-W medium-pressure mercury lamp was filtered as follows: 313 nm, a solution of 0.001 *M* potassium chromate and 3% potassium carbonate in conjunction with 4 mm thick pyrex glass; 366 nm, Corning 7-54 and 0-51 filter combination; 436 nm, Corning 3-73 and 5-58 glass filter combination. The light intensity was monitored by the potassium ferrioxalate actinometer (0.15 *M*) at 366 nm and 436 nm (10). The benzophenone-sensitized photoisomerization of *cis*-1,3-pentadiene was employed as an actinometer for 313-nm light (13). All studies were conducted at concentrations lower than the solubility limits (17). The reaction cells were cleaned by aqua regia and rinsed with dilute ammonium hydroxide, then distilled water prior to use; this procedure minimized problems with adsorption of the PAH to the cell walls. Degassing was accomplished by three freeze-pump-thaw cycles with the reaction cells attached to a vacuum rack. The cells were sealed under a vacuum of 10^{-3} mm Hg.

Electronic absorption spectra were obtained using a Perkin-Elmer Model 356 Spectrophotometer employing cells with a calibrated pathlength. Spectra of naphthacene, chrysene and phenanthrene were obtained in water-methanol mixtures.

Rate constants for direct photolysis in the aquatic environment were calculated using a computer program that was previously described (33). The program was modified to permit computation of rate constants at points of various depths in a water body with specified diffuse attenuation coefficients.

RESULTS AND DISCUSSION

Direct Photolysis of PAH

Kinetic data for the direct photochemical reactions in water of all the aromatics studied could be described by a first-order rate equation. Disappearance quantum yields, ϕ_r, were computed from the slopes of the first-order plots (Table I) (32). In general, quantum efficiencies were determined at a wavelength that closely corresponded to that of maximum sunlight absorption by the PAH (*vide infra*). Most of the quantum yields were in the

TABLE I. Kinetic Data for Direct Photochemical Reactions of Polycyclic Aromatic Hydrocarbons in Air-Saturated Water at 25°C

Aromatic Compound	Wavelength (nm)	$10^7 \times [A]$[a] (*M*)	$10^3 \times \epsilon$[b] (1/gmole cm)	$10^3 \times \phi_r$[d]
Naphthalene	313	1.2×10^3	0.11 ± 0.005	15 ± 1
1-Methylnaphthalene	313	1.2×10^3	0.28 ± 0.01	18 ± 1
2-Methylnaphthalene	313	8.8×10^2	0.26 ± 0.01	5.3 ± 0.2
Phenanthrene	313	8.4	0.20 ± 0.01[c]	10 ± 1.6
Anthracene	366	2.3	2.1 ± 0.1	3.0 ± 0.2
9-Methylanthracene	366	3.1	9.9 ± 0.2	7.5 ± 0.5
9,10-Dimethylanthracene	366	1.3	4.0 ± 0.3	4.0 ± 0.4
Pyrene	313	3.9	8.8 ± 0.4	2.0 ± 0.3
Pyrene	366	3.9	0.19 ± 0.01	2.2 ± 0.3
Fluoranthene	313	6.6	3.8 ± 0.2	0.12 ± 0.01
Fluoranthene	366	6.6	3.5 ± 0.2	$0.002 \pm .0002$
Chrysene	313	8.2×10^{-2}	7.4 ± 0.2[c]	2.8 ± 0.7
Naphthacene	436	2.2×10^{-2}	6.1 ± 0.2[c]	13 ± 5

[a]PAH concentration
[b]Molar absorptivity in water.
[c]Obtained in 50% water-methanol.
[d]Disappearance quantum yield; mean of triplicate determination ± standard deviation.

0.001 to 0.01 range. More detailed wavelength studies were conducted with pyrene and fluoranthene because they are both abundant PAH in environmental samples (8). The disappearance quantum efficiency of pyrene exhibited the usual wavelength independence. However, ϕ_r for fluoranthene dropped sharply in going from 313 to 366 nm. Even at 313 nm, fluoranthene photoreacted considerably less efficiently than the other PAH included in this study. Peculiar excited-state behavior has also been observed of fluoranthene in spectroscopic studies (3). The anomolous photochemical behavior is probably related to the fact that fluoranthene, unlike the other PAH studied, is a nonalternant aromatic hydrocarbon (3).

Prior studies (2,9,25,29) suggested that the direct photolysis of PAH in water would involve photooxygenation mediated by singlet oxygen formed by energy transfer from the electronically excited aromatic molecule in its triplet state. Stevens and co-workers (28-30) derived the following equation (Equation 1) for quantum yields of direct photooxygenation, ϕ_{AO_2}.

$$\phi_{AO_2} = \phi_T \{([A] + p\beta)/([A] + \beta\} \quad (1)$$

The term ϕ_T is the intersystem crossing efficiency (4,24), β is a reactivity index that is inversely proportional to the rate constant for reaction of singlet oxygen with the aromatic compound (6,19), and $p\beta$ is a solvent-dependent constant that approximately equals 8×10^{-5} mole/l in water (29). At the low PAH concentrations that were employed in this study, ϕ_{AO_2} is concentration-independent. Values of ϕ_{AO_2} were computed for several PAH and are compared to ϕ_r for the compounds in Table II. The comparison indicates that photooxygenation involving singlet oxygen typically accounts for a negligible traction of the observed photoreactions, even for PAH, naphthacene and 9,10-dimethylanthracene, compounds that have the largest ϕ_{AO_2} values.

Quantum efficiencies for reaction of naphthalene, anthracene, pyrene and naphthacene were observed to be 10 to more than 100 times smaller in air-saturated isooctane than in water, indicating that the observed reaction is strongly dependent on solvent polarity. This solvent effect suggested that the reactions in water may be initiated by electron transfer to oxygen, a "Type I" photooxidation (6) that certainly would be enhanced in polar solvents. However, careful removal of dissolved oxygen by degassing (oxygen concentration less than 10^{-9} *M*) had no effect upon the photolysis rate of 9-methylanthracene (Figure 1). Andelman and Suess (2) and Smith *et al.* (25) have likewise found that photochemical transformations of benzo[a]-pyrene, benz[a]anthracene and some heterocyclic aromatics occur in water in the absence of molecular oxygen. We also found that the 9-methylanthracene reaction was unaffected by the presence of 1.2×10^{-3} *M* mercuric chloride,

TABLE II. Comparison of Observed Quantum Yields for Reaction of PAH in Air-Saturated Water (ϕ_r) with Computed Quantum Efficiencies for Photooxygenation (ϕ_{AO_2})

Aromatic Compound	ϕ_r	Assumed[b] ϕ_T	β_{H_2O} M[e]	ϕ_{AO_2}	ϕ_{AO_2}/ϕ_r
Naphthalene	1.5×10^{-2}	0.86[c]	10^8	7×10^{-13}	5×10^{-11}
Phenanthrene	1.0×10^{-2}	0.87[d]	10^8	7×10^{-13}	7×10^{-11}
Anthracene	3.0×10^{-3}	0.70[c]	3.2	1.8×10^{-5}	6×10^{-3}
9-Methylanthracene	7.5×10^{-3}	0.67[d]	1.6×10^{-1}	3.3×10^{-4}	4×10^{-2}
9,10-Dimethylanthracene	4.0×10^{-3}	0.11[d]	2.4×10^{-2}	3.6×10^{-4}	9×10^{-2}
Chrysene	2.8×10^{-3}	0.83[d]	10^8	7×10^{-13}	2×10^{-10}
Naphthacene	1.3×10^{-2}	0.84[d]	4.1×10^{-2}	1.6×10^{-3}	1×10^{-1}
Benz[a] anthracene	3.3×10^{-3}[a]	0.80[d]	1.0×10	6.4×10^{-6}	2×10^{-3}
Benz[a] pyrene	8.9×10^{-4}[a]	0.7[c]	2.4×10[a]	2.3×10^{-6}	3×10^{-3}

[a] Smith and co-workers (25).
[b] Assumed to equal one minus fluorescence quantum efficiency.
[c] Schwarz and Wasik (24).
[d] Birks (4).
[e] Stevens *et al.* (30) except where noted; corrected for differences in lifetime of singlet oxygen in benzene and water (19).

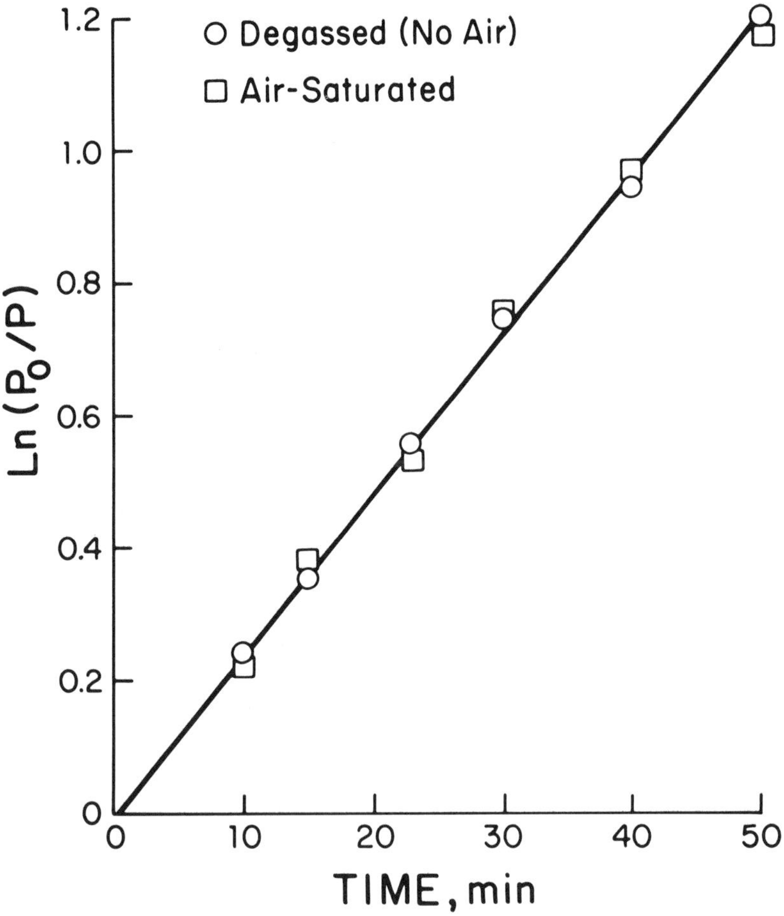

FIGURE 1. Effect of oxygen removal upon direct photolysis (313 nm) of 9-methylanthracene in water.

a known triplet-state quencher (22). Thus, the reactive species is either the excited singlet state or, perhaps, a short-lived upper-triplet state.

Taken together, these kinetic studies suggest that the predominant primary photochemical processes of PAH in water may not involve molecular oxygen. Product studies are being conducted to elucidate the mechanism of these photoreactions. Other studies have demonstrated that aromatic compounds, such as benzene (11), undergo photohydrolysis in water (5). The large change in reaction quantum efficiency in going from nonpolar organic solvents to water suggests that the reactions may involve photoionization to form radical cation intermediates.

Predicted Photolysis Rates of PAH in Natural Water Bodies

The photolysis rate, *i.e.*, conversion per unit time (-d[A]/dt) of an aromatic hydrocarbon in a water body at depth z, can be described by Equation 2:

$$-\left(\frac{d[A]}{dt}\right)_{z,\lambda} = \Sigma\phi_{r,\lambda}\, k_{z,\lambda}\, [A] = k_{p,z}[A] \qquad (2)$$

where $\phi_{r,\lambda}$ is the quantum yield at wavelength λ and $k_{z,\lambda}$ is the specific sunlight absorption rate of the aromatic at wavelength λ and depth z (20). The total rate is obtained by integrating over the effective range of wavelengths (32). The $k_{z,\lambda}$ value is essentially a measure of the spectral overlap between the electronic absorption spectrum of the aromatic compounds and the solar irradiance reaching depth z. We have also found it convenient to define a related term, k_a, that is the *average* specific sunlight absorption rate of a pollutant in a water column of depth z (33). A computer program has been developed that can be used to approximately calculate solar spectral irradiance as a function of time, location and depth in a water body (33). The program uses the simulated irradiance values, quantum yields and the molar absorptivities of chemicals to compute their photolysis rates in natural water bodies. The computed photolysis rate constants, k_p, have been shown to be in close agreement with k_p values obtained from experiments under sunlight (25,33). Some representative half-lives, $0.693/k_p$, for direct photolysis of PAH in water are shown in Table III. The half-lives were computed for a clear day near the surface of a water body where the chemical is fully exposed to sunlight. Effects of light attenuation will be discussed later. Although the photoreactions are very inefficient, with typically less than 1% of the energy absorbed resulting in photoreaction, conversion under sunlight for most of the PAH studied is rapid because sunlight absorption rates are generally very large. Interestingly, the naphthalenes have among the highest disappearance quantum efficiencies but about the lowest photolysis rates. This seeming anomaly is attributable to the much lower sunlight absorption rates of the naphthalenes, which weakly absorb only the less-intense UV-B (280-320 nm) part of solar radiation. Note that there is a ragged trend of increasing photoreactivity with increasing molecular weight; the opposite is the case with microbial oxidation (7,16). This trend is mainly caused by the enhanced spectral overlap of the absorption spectra of the high-weight aromatics with solar radiation. The "bent" PAH, *e.g.*, phenanthrene and chrysene, are less photoreactive than their linear isomers. This effect is best illustrated by comparing half-lives of the isomers, naphthacene (I), benz[a]anthracene (II) and chrysene (III); each "bend" increases the half-life by about an order of magnitude (Table III).

TABLE III. Computed Near-Surface Half-Lives for Direct Photochemical Transformation and Photosensitized Oxygenation of Several Aromatic Hydrocarbons (latitude 40°N, midday, midsummer)

Compound	$10 \times k_a$ (sec^{-1})[a]	λ (nm)[b]	Half-Life (hr)[c]	
			Direct	Photosensitized[d]
Naphthalene	0.0018	310	71	10^9
1-Methylnaphthalene	0.0048	312	22	
2-Methylnaphthalene	0.0069	320	54	
Phenanthrene	0.023	323	8.4	10^9
Anthracene	0.83	360	0.75	2.0×10^2
9-Methylanthracene	2.0	380	0.13	1.0×10
9,10-Dimethylanthracene	1.4	400	0.35	1.5
Pyrene	1.4	330	0.68	
Fluoranthene			21[f]	
Chrysene	0.22	320	4.4	10^9
Naphthacene	4.3	440	0.034	2.6
Benzo[a]pyrene[e]	4.0	380	0.54	1.5×10^3
Benz[a]anthracene[e]	0.99	340	0.59	6.4×10^2

[a] Specific sunlight absorption rate (33).
[b] Approximate wavelength of maximum sunlight absorption.
[c] Period of exposure required for 50% conversion.
[d] Computed by Equation 3 assuming $[O_2{}^*]$ is 6×10^{-12} mole/l.
[e] Reference 25.
[f] Based on photolysis under sunlight.

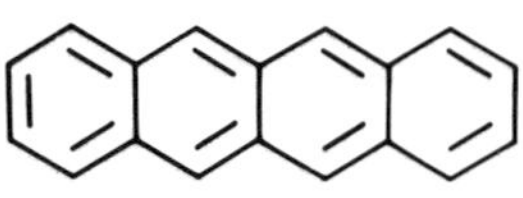

I

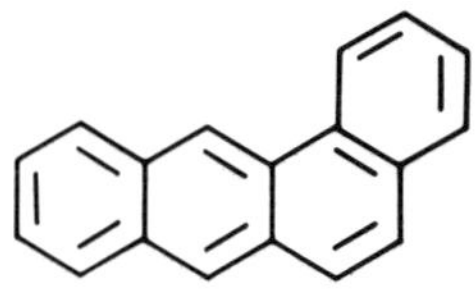

II

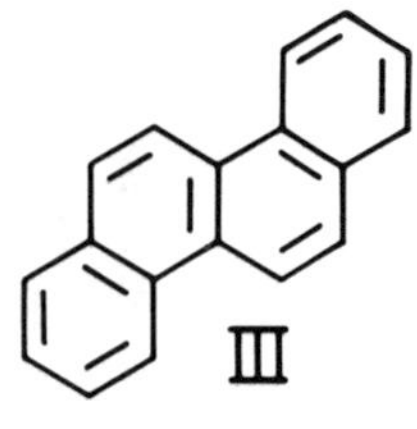

III

Although the data in Table II indicated that direct photooxygenation of the PAH is insignificant, the possibility still existed that photosensitized oxygenation involving singlet oxygen might be important in natural water. Recent studies have indicated that the photosensitized generation of singlet oxygen is a general phenomenon in the aquatic environment (34). To evaluate the potential importance of photosensitized oxygenation of PAH dissolved in natural waters, rate constants and half-lives for the reactions were computed by Equation 3 (34) where the values of k_A, the rate constants

$$t_{1/2} = \frac{0.693}{k_A\ [^1O_2^*]} \tag{3}$$

for reaction of singlet oxygen with the aromatic hydrocarbons, were obtained from studies of Stevens *et al.* (30) and Smith and co-workers (25), and the near-surface steady-state concentration of singlet oxygen $[^1O_2^*]$ was assumed to equal 6 x 10^{-12} M in summer (34). Comparisons of half-lives for direct photolysis and photosensitized oxygenation of the PAH that were studied (Table III) indicate that the former process is considerably more rapid. Thus, photosensitized oxygenation of these PAH is essentially negligible. It should be pointed out that these considerations apply only to PAH dissolved in the water. Larson and Hunt (14) have shown that singlet oxygen is involved in the photooxidation of aromatics dissolved in petroleum films.

To further evaluate the potential importance of photosensitized reactions in natural water, the photolysis of pyrene was studied in water samples obtained from the Gulf of Mexico at Shell Point, Florida, and the Aucilla River at Lamont, Florida. The Aucilla River is a "black water" system that strongly absorbs ultraviolet and visible light; concentrations of singlet oxygen that form when it is exposed to sunlight are among the highest that have been observed in natural waters (34). The observed photolysis rate constant of pyrene was the same at 313 nm in the Gulf of Mexico water as in distilled water (Figure 2) and nearly the same in the Aucilla River water. Similar results were obtained when the solutions were exposed to sunlight. The Aucilla River data, when corrected for the expected retardation caused by light screening by the UV-absorbing substances in the water, indicated that some acceleration,

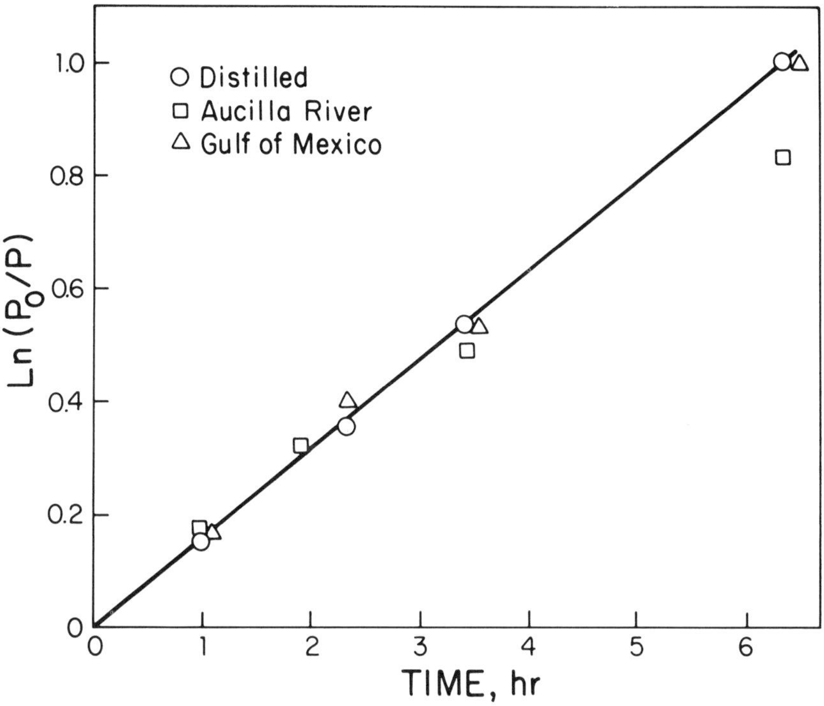

FIGURE 2. Kinetic data for photoreaction (313 nm) of pyrene in distilled water and in two natural water samples.

perhaps photosensitization, was occurring. This effect, however, could also have been caused by a twofold enhancement of the quantum efficiency for direct photolysis. Most significant was the fact that the photolysis rates were nearly the same in both natural and distilled water. Similar observations have been reported for other PAH by Smith and co-workers (25). Dark controls of pyrene in the natural water indicated no microbial degradation over a 10-day period.

Effects of Light Attenuation and Partitioning

Thus far, only factors that determine the photoreactivity of PAH have been discussed. Any assessment of the role of photolysis in a natural water body must also take into account the dose of light that the polycyclic aromatic is likely to receive. Factors influencing the amount of sunlight reaching the earth's surface are discussed in many places (see Reference 33 for lead references). In this section, typical effects of light attenuation and partitioning to biota or sediments are discussed. The procedures used below in making these assessments are discussed more thoroughly elsewhere (20,32,33).

The Sea

The attenuation of solar radiation in a water body is best described in terms of diffuse attenuation coefficients, K_λ, where K_λ is the slope of a logarithmic (Napierian) plot of solar irradiance at wavelength λ versus depth (27). Solar irradiance $I_{z,\lambda}$ at depth z approximately equals $W_\lambda e^{-K_\lambda z}$ where W_λ is the surface irradiance. Diffuse attenuation coefficients for the sea increase with decreasing wavelength in the blue and ultraviolet spectral region where most PAH absorb light. Thus, the decrease in direct photolysis rates of polycyclic aromatics with increasing depth is more pronounced for those PAH that absorb shorter wavelengths of light. This effect is illustrated in Figure 3 for naphthalene, anthracene and naphthacene–PAH that absorb in the UV-B, near-UV, and visible region, respectively (Table III). The depth profiles of Figure 3A were computed for clear ocean water of low productivity in the middle of the Gulf of Mexico employing simulated solar spectral irradiance for cloudless days at the water surface (33) and diffuse attenuation coefficients recently reported by Smith and Baker (26). The profiles computed for the same PAH in moderately productive coastal waters near Tampa,

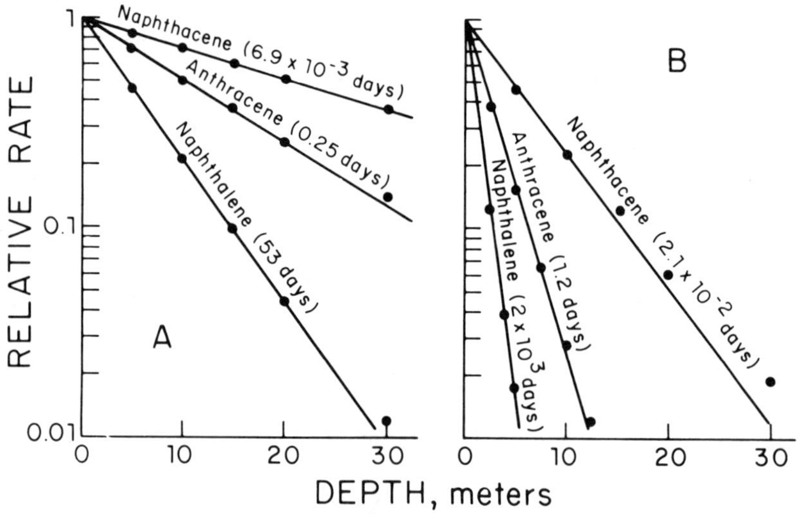

FIGURE 3. Computed depth dependence for direct photolysis of several PAH during summer in the Gulf of Mexico: (A) Mid-Gulf; (B) coastal water near Tampa, FL. Average half-lives in top 35 m are in parentheses.

Florida (26), during October 1977 are illustrated in part B of Figure 3; it was assumed that the algae were uniformly distributed in the water column. Average half-lives inherent to direct photolysis were computed for the PAH in the top 35 meters (Figure 3). The rate constants employed in the half-life calculations were integrated over the full day so that photolysis could be compared to other processes that are not light-dependent. Examination of the data in the figure indicates that direct photolysis of many of the PAH examined in this study is rapid throughout the mixed layer of the ocean. Direct photolysis is generally slower in coastal than in mid-Gulf water, particularly for naphthalene, a UV-B absorbing compound. Smith and Baker (26) attribute the strong light attenuation in the UV-B region mainly to unusually high concentrations of dissolved organics in the coastal water. Several "black water" rivers, *e.g.*, the Suwannee River, drain into the Gulf of Mexico near Tampa. Ultraviolet radiation is not attenuated as strongly in California coastal waters (26).

The above considerations pertain to PAH that are completely dissolved in the sea. Available evidence, discussed in the introduction of the paper, indicates that a large fraction of the PAH that enter the ocean from the atmosphere or a small oil spill become dissolved in the ocean water. A small fraction bioaccumulates in the algae or other biogenous material. With a major oil spill it is likely that the oil-to-water ratio in the mixed layer of the ocean would initially be so high that most of the high-weight PAH would remain in the oil phase. Because sorption to oil is almost certainly a rapidly reversible equilibrium process, however, the PAH would progressively move into the seawater from the oil phase as the oil becomes diluted.

Inland Waters

Light attenuation is much more pronounced in inland water bodies than in the ocean (20,33). Moreover, a portion of the PAH in rivers and lakes partitions into the bottom sediment where no light is present (8).

To approximately assess the effects of this partitioning on the overall rate of loss by photochemical transformation in a river, the following assumptions are made. First, it is assumed that the water body is well mixed and that the PAH dissolved in the water column are equilibrated with the PAH sorbed to suspended sediments and the top layer of bottom sediments. Thus, as the PAH in the water column photodegrade, more PAH are rapidly released to maintain equilibrium. Second, it is assumed that the partitioning can be quantitatively described in terms of a partition coefficient K_p. This coefficient equals $P_s(P_w\rho)^{-1}$ where P_s and P_w are the amounts of dissolved and sorbed PAH per unit volume in the system and ρ is the weight ratio of dry sediment to water. Karickhoff *et al.* (12) have shown that the K_p values for several PAH can be approximately computed using Equation 4.

$$K_p = 0.63\ F_f\ F_{oc} P \tag{4}$$

where F_f = the dry weight fraction of clay and silt-sized particles in the sediment
F_{oc} = the fractional mass of organic carbon in the sediment
P = the octanol-water partition coefficient of the aromatic compound.

It can be shown that partitioning to the bottom sediment slows photolysis (31) by a factor of $(1 + K_p\rho)$ if degradation of the PAH in the sediment is very slow, an assumption that is certainly valid for the higher-molecular weight PAH (7,16,25).

For illustrative purposes, average photolysis rate constants were computed for several PAH in a 5-m-deep well-mixed water body that contained a mean sediment concentration of 20 mg/l (Table IV). This sediment concentration is the average observed in the Ohio River at Cincinnati during the summer (20). The diffuse attenuation coefficient that correspond to the assumed sediment concentration is estimated to be about 11 m^{-1} at 330 nm (20). In computing the rate constants in the table, this attenuation coefficient was assumed for all the PAH, an assumption that probably overestimates light attenuation for most PAH. To permit the rate constants to be better visualized, they have been transformed into half-lives. It is stressed that the term half-life has no meaning in a flowing system with continual inputs as is usually the case with rivers polluted by PAH. To demonstrate the potential "buffering" effect of sediment-water exchange, two sets of rate constants were computed. The first set was computed assuming that the PAH are completely in the water column and the photolysis rates are affected only by light attenuation. It can be readily shown using the K_p values in Table IV, that the PAH in the water column are almost completely dissolved, not sorbed to the suspended sediments; ρ for the suspended sediment is only 2×10^{-5}. The second set of half-lives were computed for photochemical loss of PAH in the same system, but with rapid exchange between the top centimeter of the bottom sediment and the water column. This exchange could involve continual sedimentation and resuspension of the top layer of the bottom sediment. The K_p values for the sediment were computed from octanol-water partition coefficients (Equation 4 assuming that $F_{oc} = 0.015$ and $F_f = 0.5$). The bottom sediments are usually about half water so ρ is assumed to be 0.001.

Results in the table indicate that partitioning to the bottom sediment has the most pronounced effect on photolysis of the higher molecular weight PAH, which have the greatest tendency to sorb. The calculations further suggest that, of the PAH studied, anthracene and naphthacene should be the most susceptible to photochemical transformation in a deep inland water body. Obviously, the above assumptions concerning partitioning are greatly oversimplified. At present, there is little information available on rates of bottom sediment-water column exchange for organic pollutants.

TABLE IV. Illustrative Effects of Light Attenuation and Sediment-Water Partitioning on Direct Photolysis of Several PAH in a 5-m-deep Inland Water Body[a]

Compound	$P \times 10^{-3}$ [b]	K_p [d]	$(1 + K_p \rho)^{-1}$ [e]	Half-Life (days)[f] No Partitioning	Half-Life (days)[f] With Partitioning
Naphthalene	2.3	4.7	0.99	550	550
1-Methylnaphthalene	13	61	0.94	180	190
2-Methylnaphthalene	13	61	0.94	410	440
Phenanthrene	37	180	0.85	59	69
Anthracene	35	160	0.86	4.5	5.2
9-Methylanthracene	117	550	0.64	0.79	1.2
Fluoranthene	60[c]	280	0.78	160	200
Pyrene	84	400	0.71	4.2	5.9
Benz[a]anthracene	310[c]	1,500	0.40	3.7	9.2
Chrysene	880[c]	4,200	0.19	13	68
Naphthacene	800	3,800	0.21	0.20	0.95
Benzo[a]pyrene	650[c]	3,100	0.24	3.2	13

[a] Assumptions: Suspended sediment concentration, 20 mg/l; diffuse attenuation coefficient, 11 m^{-1}

[b] From Karickhoff *et al.* (12) except where noted.

[c] Computed from water solubility (12).

[d] Computed using Equation 13.

[e] Fraction dissolved in water column including partitioning to top cm of bottom sediment.

[f] Integrated over full summer day, latitude 40° N.

SUMMARY

These studies indicate that many of the higher molecular weight polycyclic aromatics are rapidly transformed in water by the action of sunlight. The naphthalenes, fluoranthene and phenanthrene photolyze slowly. Surprisingly, with at least some of the PAH, the disappearance quantum efficiencies are unaffected by molecular oxygen. The photoreactions are considerably more efficient in water than in hydrocarbon solvents. Photosensitized oxygenation of PAH dissolved in natural waters appears to be too slow to compete with direct photolysis. Thus, excluding photoreactions in petroleum slicks, it does not appear that singlet oxygen plays a significant role in the photochemical transformation of PAH in the aquatic environment.

Computations indicate that photoreactions of many PAH are rapid throughout the upper, mixed layer of the ocean (top 35 meters). In turbid rivers, however, photolysis is greatly slowed via light attenuation and partitioning of the PAH to bottom sediments where no light is present. Higher molecular weight PAH are most susceptible to the latter effect.

ACKNOWLEDGMENTS

We thank G. L. Baughman, N. L. Wolfe and S. W. Karickhoff for helpful discussions of portions of this paper. We also express our appreciation to R. A. Larson and R. C. Smith for providing us with prepublication copies of their papers which were discussed above.

DISCLAIMER

Mention of trade names or commercial products does not necessarily constitute endorsement or recommendation for use by the U.S. Environmental Protection Agency.

REFERENCES

1. Andelman, J. B., and Snodgrass, J. E. (1974): Incidence and significance of polynuclear aromatic hydrocarbons in the water environment. *CRC Crit. Rev. Environ. Control* 5:69-83.
2. Andelman, J. B., and Suess, M. J. (1971): The photodecomposition of 3,4-benzpyrene sorbed on calcium carbonate. In: *Organic Compounds in Aquatic Environments,* S. J. Faust and J. V. Hunter, Eds., pp. 439-468, Marcel Dekker, New York.
3. Berlman, I. B. (1971): *Handbook of Fluorescence Spectra of Aromatic Molecules,* pp. 93-94, Academic Press, New York.
4. Birks, J. (1970): *Photophysics of Aromatic Molecules,* pp. 126-132, Wiley, New York.

5. Cornelisse, J., and Havinga, E. (1966): Photosubstitution reactions of aromatic compounds. *Chem. Rev.* 75:353-388.
6. Foote, C. S. (1976): Photosensitized oxidation and singlet oxygen: Consequences in biological systems. In: *Free Radicals in Biology,* Vol. II, W. A. Pryor, Ed., pp. 85-133, Academic Press, New York.
7. Gibson, D. T. (1978): Microbial transformations of aromatic pollutants. In: *Aquatic Pollutants–Transformation and Biological Effects,* O. Hutzinger, I. van Lelyveld and B. C. J. Zoeteman, Eds., Pergamon Press, Fairview Park, NY.
8. Giger, W., and Schaffner, C. (1978): Determination of polycyclic aromatic hydrocarbons in the environment by glass capillary gas chromatography. *Anal. Chem.* 50:243-249.
9. Gollnick, K. (1968): Type II photooxygenation reactions in solution. *Adv. Photochem.* 6:1-122.
10. Hatchard, C. G., and Parker, C. A. (1956): A new sensitive chemical actinometer. II. Potassium ferrioxalate as a standard chemical actinometer. *Proc. Roy. Soc. (London)* A235:518-536.
11. Kaplan, L., Rausch, D. J., and Wilzbach, K. E. (1972): Photosolvation of benzene. *J. Am. Chem. Soc.* 94:8638-8640.
12. Karickhoff, S. W., Brown, D. S., and Scott, T. A. (1979): Sorption of hydrophobic pollutants on natural sediments. *Water Res.* (in press).
13. Lamola, A. A., and Hammond, G. S. (1965): Mechanisms of photochemical reactions in solution. XXXIII. Intersystem crossing efficiencies. *J. Chem. Phys.* 43:2129-2135.
14. Larson, R. A., and Hunt, L. L. (1978): Photooxidation of refined petroleum oil: Inhibition by β-carotene and role of singlet oxygen. *Photochem. Photobiol.* 27 (in press).
15. Larson, R. A., Hunt, L. L., and Blankenship, D. W. (1977): Formation of toxic products from a #2 fuel oil by photooxidation. *Environ. Sci. Technol.* 4:492-496.
16. Lee, R. F., Gardner, W. S., Anderson, J. W., Blaylock, J. W., and Barwell-Clarke, J. (1978): Fate of polycyclic aromatic hydrocarbons in controlled ecosystem enclosures. *Environ. Sci. Technol.* 12:832-838.
17. Mackay, D., and Shiu, W. Y. (1977): Aqueous solubility of polynuclear aromatic hydrocarbons. *J. Chem. Eng. Data* 22:399-402.
18. McGinnes, P. R. (1974): The photodecomposition of polynuclear aromatic hydrocarbons in natural water systems. Ph.D. dissertation, U. of Illinois, Urbana-Champaign, Ill.
19. Merkel, P. B., and Kearns, D. R. (1972): Radiationless decay of singlet molecular oxygen in solution. *J. Am. Chem. Soc.* 94:7244-7253.
20. Miller, G. C., and Zepp, R. G. (1979): Effects of suspended sediments on photolysis rates of dissolved pollutants *Water Res.* 13 (in press).
21. Moses, F. G., Liu, R. S. H., and Monroe, B. M. (1969): The "merry-go-round" quantum yield apparatus. *Mol. Photochem.* 1: 245-249.

22. Nilsson, R., Merkel, P. B., and Kearns, D. R. (1972): Kinetic properties of the triplet states of methylene blue and other photosensitizing dyes. *Photochem. Photobiol.* 16:109-116.
23. Radding, S. B., Mill, T., Gould, C. W., Liu, D. H., Johnson, H. L., Bomberger, D. C., and Fojo, C. V. (1975): The environmental fate of selected polynuclear aromatic hydrocarbons. U.S. Environmental Protection Agency, Office of Toxic Substances, Washington, D.C., Report #560/2-75-009.
24. Schwarz, F. P., and Wasik, S. P. (1976): Fluorescence measurements of benzene, naphthalene, anthracene, pyrene, fluoranthene, and benzo[a]-pyrene in water. *Anal. Chem.* 48:524-528.
25. Smith, J. H., Mabey, W. R., Bohonos, N., Holt, B. R., Lee, S. S., Chou, T.-W., Bomberger, D. C., and Mill, T. (1978): Environmental pathways of selected chemicals in freshwater systems: Part II. Laboratory studies. U.S. Environmental Protection Agency, Athens, GA, Report No. EPA-600/7-78-074.
26. Smith, R. C., and Baker, K. (1979): Penetration of UV-B radiation into natural waters. *Photochem. Photobiol.* 27 (in press).
27. Smith, R. C., and Tyler, J. E. (1976): Transmission of solar radiation into natural waters. *Photochem. Photobiol. Rev.* 1:117-155.
28. Stevens, B. (1973): Kinetics of photoperoxidation in solution. *Acct. Chem. Res.* 6:90-96.
29. Stevens, B., Ors, J. A., and Pinsky, M. L. (1974): The photoperoxidation of unsaturated organic molecules. $O_2\ ^1\Delta_g$-acceptor re-encounter probabilities. *Chem. Phys. Lett.*, 27:157-160.
30. Stevens, B., Perez, S. R., and Ors, J. A. (1974): Photoperoxidation of unsaturated organic molecules. XIV. $O_2\ ^1\Delta_g$ acceptor properties and reactivity. *J. Am. Chem. Soc.* 96:6846-6850.
31. Wolfe, N. L., Zepp, R. G., Paris, D. F., Baughman, G. L., and Hollis, R. C. (1977): Methoxychlor and DDT degradation in water: rates and products. *Environ. Sci. Technol.* 11:1077-1081.
32. Zepp, R. G. (1978): Quantum yields for reaction of pollutants in dilute aqueous solution. *Environ. Sci. Technol.* 12:327-329.
33. Zepp, R. G., and Cline, D. M. (1977): Rates of direct photolysis in aquatic environment. *Environ. Sci. Technol.* 11:359-366.
34. Zepp, R. G., Wolfe, N. L., Baughman, G. L., and Hollis, R. C. (1977): Singlet oxygen in natural waters. *Nature* 267:421-423.

Polynuclear Aromatic Hydrocarbons, edited by P. W. Jones and P. Leber. Ann Arbor Science Publishers, Inc. Ann Arbor, MI ©1979.

PHOTOLYSIS OF TRIPHENYLENE IN RIGID MATRICES AT LOW TEMPERATURE

M. Lamotte, S. Risemberg, J. Pereyre and J. Joussot-Dubien

Laboratoire de Chimie Physique
Photophysique et Photochimie Moleculaire
Universite de Bordeaux I
351 Cours de la Liberation
33405 Talence, France

INTRODUCTION

Light excitation of aromatic hydrocarbons in rigid matrices at room and low temperature has been extensively studied, but except for benzene (6,9), authors have mainly focused their attention on the photoionization and photosensitization processes leading to the formation of various radicals and radical ions (1,2,4,8).

Although photolysis of condensed aromatic solutes has been reported from time to time (1,3), these observations have not attracted much attention.

Irradiating with UV light, coronene or triphenylene included in a single crystal of heptane at 77°K to measure the $\Delta m = 1$ absorption of the triplet species in an ESR cavity, we observed that the solute molecules disappeared very quickly when the full beam of the lamp was used to populate the triplet state. It was found that other condensed aromatic hydrocarbons behaved similarly (*i.e.,* pyrene, benzopyrene, naphthalene, etc.) and that the phenomenon could be reproduced in glassy hydrocarbons and alcoholic matrices at 77°K and, surprisingly, just as well at 4°K.

Intrigued by this reaction we have examined in more detail the case of triphenylene whose spectral properties appear the most favorable for such a study.

RESULTS

Photolysis of triphenylene, C ~ $10^{-4}/10^{-5}$ *M*, in methylcyclohexane (MCH), isopropyl alcohol (iPrOH) glasses and in polyethylene (PE) at 77°K was performed using a nonfiltered medium-pressure mercury arc (HPK 125-W). Under irradiation, the intensity of the triphenylene fluorescence decreases and a new emission appears whose spectrum shares several features with the one from phenanthrene:

1. The first transition is allowed but has a small extinction coefficient.
2. The fluorescence spectrum exhibits a vibronic structure containing the same main fundamental frequencies as phenanthrene (Table I).

Two new absorption band systems appear simultaneously (Figure 1). The first one with maxima at 307, 325 and 340 nm, in isopropanol (iPrOH) corresponds to the fluorescence excitation spectrum of a product (I) responsible for the new fluorescence described above.

The second band system which does not have a counterpart in fluorescence has a maximum at 410 nm (product II). It decreases in the dark upon warming the glass and disappears after melting. Freezing again, the solution leaves the intensity of the first band system almost unchanged while a small increase of the triphenylene absorption intensity is observed compared to the irradiated glass, before warming.

Assignment of the Product II

Our results suggest that the second absorption band system is due to a radical species. Both the monohydrogenated radical ($\dot{\text{Ar}}$H) and the radical cation $\dot{\text{Ar}}^+$ of triphenylene absorb at 410 nm (8), but the latter also absorbs in the red part of the visible range. In our experiment no absorption occurred in this region so we concluded that the photoproduct II is most likely the $\dot{\text{Ar}}$H radical.

This conclusion is confirmed by results obtained with other aromatic hydrocarbons. Indeed irradiation in the same condition as for triphenylene leads to the formation of an $\dot{\text{Ar}}$H-type radical which has been easily identified in the case of naphthalene, 3-4 benzopyrene and pyrene by their known absorption spectra (8).

Identification of Photoproduct I

A general feature of the photolysis of aromatics in conventional solvents is that the reaction occurs only in hydrogenated solvents and at low temperature.

1. The reaction described above is not observed in boric acid glass where aromatics are known to be trapped in hydrogen free zones (7,8).

TABLE I. Band Maxima and Main Frequencies Involved in the Vibronic Structure of the Fluorescence Spectra of Reaction Product Following:
I: Photolyse of Triphenylene in Matrix at 77°K
II: Reduction of Triphenylene in Liquid Solution at Room Temperature

	I			II		
	MCH	iPrOH	P.F.	Li in HMPT–THF as Solvent[a]	Triethylamine in MeCN as Solvent[a]	Phenanthrene
λ_{max}	369.2	369.8	369.8	368.0	369.6	345.9
±0.5 nm	389.2	390.0	389.4	387.8	389.6	363.2
	412.0	413.0	413.3	410.4	412.4	383.0
$\upsilon_{o,o} - \upsilon_{vib}$	1,390	1,401	1,361	1,388	1,389	1,371
$\pm 30\ cm^{-1}$	2,805	2,828	2,794	2,808	2,808	2,793

[a]Spectra taken in MCH at 77°K.

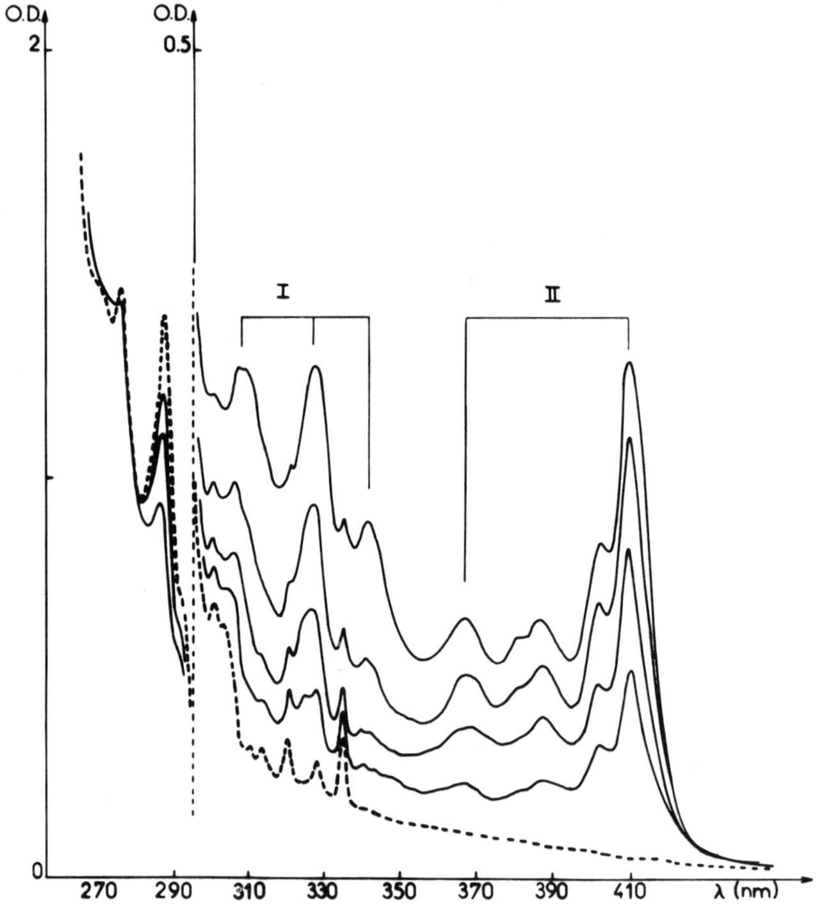

FIGURE 1. Absorption spectra of 1.5 x 10^{-4} *M* solution of triphenylene in an isopropanol glass following irradiation with a 125-W medium-pressure mercury lamp. Irradiation times are 20 s, 45 s, 105 s and 360 s (temp. = 77°K, thickness 1 mm). The broken line corresponds to the absorption of triphenylene before irradiation.

2. In argon and nitrogen matrices, although some photolytic processes do take place they are much less efficient than in hydrogenated solvents and may concern entirely different reactions.
3. No reaction occurs at room temperature and in degassed solution.

Among several possible reactions, isomerization of triphenylene (isomerization of one benzene ring leading to a benzvalene-type structure) could be envisaged. However, the above considerations, added to the fact that such isomerization has not been observed at low temperature with benzene (9) and that coronene (whose structure makes isomerization highly improbable) gives rise to the same photolysis as triphenylene, strongly suggest that the hypothesis of an isomerization must be rejected. We saw rather that photoproduct I as well as Photoproduct II are due to hydrogen abstraction from the solvent by one benzene ring of triphenylene. Unfortunately, the small solubility of triphenylene and the necessity to use low concentrated solutions, because of the large extinction coefficient at 260 nm (~ 180,000) and the instability of the photoproduct I, make it impossible for the moment to isolate product I for definitive identification. However, although we have not been able to determine the structure of the photoproduct I, there is evidence that we observed the reduction of triphenylene leading to a dihydrogenated-type compound.

Such reactions are known to occur in liquid solutions with aromatics in the presence of strong reducing agents such as amines of low ionization potential (10) or alkali metals (5). These reactions lead to the formation of dimers, substituted aromatics and dihydrogenated compounds. We have reproduced two types of these reactions with triphenylene:

1. irradiation in acetonitrile (MeCN) solution with methylamine (TEA),
2. reduction with lithium in a mixed solvent of hexamethylphosphoramide (HMPA) and tetrahydrofurane (THF).

Because of their instability, it was not possible to isolate any reaction products, but we did dissolve some of the reaction solutions in MCH and take the fluorescence spectra at 77°K.

In both cases using a suitable excitation wavelength we have obtained a fluorescence spectrum of a photoproduct which exhibits the same general features as the spectrum of photoproduct I (Table I).

Among the possible photoproducts which could be formed, only the dihydrogenated compounds (with or without the opening of one ring) are expected to have a spectrum red-shifted and similar to that of phenanthrene.

Moreover, the identification of compound I, not only to a dihydrogenated triphenylene but also to a solvent-addition product on one of the ring bonds, is suggested by two other experimental results:

1. The mass spectra performed from fresh irradiated solutions in MCH display two peak systems corresponding respectively to triphenylene and to the solvent molecule; the ratios of the intensity of these two systems are constant whatever the temperature of the inlet probe.
2. Irradiation of triphenylene included in a polyethylene film leads to a photoproduct which cannot be washed from the polymer with chloroform, although aromatic impurities are known to be easily removed.

These results point to the fact that, in all cases, one solvent molecule is attached to triphenylene giving rise to dihydrogenated-type compounds.

In contrast to the case of triphenylene, the formation of stable photoproducts (in O_2-free atmosphere) from other aromatic hydrocarbons has not been observed by absorption or fluorescence, probably because in most cases the loss of the aromaticity on a ring following the addition of the solvent shifts the absorption to shorter wavelengths than the starting molecules.

REFERENCES

1. Astier, R., Bokohza, A., and Meyer, Y. (1969): 20ème réunion de la Société de Chimie Physique, Paris *J. Chem. Phys.* 137.
2. Bernas, A., Gauthiers-Bodard, M., Gand, D. and Leonardi, D. (1969): *Int. J. Rad. Phys. Chem.* 1:229.
3. Gerkin, R. E. and Szenyi, P. (1969): *J. Chem. Phys.* 50:4095.
4. Johnson, P. M. and Rice, S. A. (1968): *Chem. Phys. Lett.* 1:709.
5. Labandibar, P., Lapouyade, R., and Bouas-Laurent, H. (1971): *C. R. Acad. Sci. Paris, 1971* 272:C1257.
6. Leach, S., Migirdicyan, E., and Grajcar, K. (1959): *J. Chem. Phys.* 59:749.
7. Lesclaux, R. and Joussot-Dubien, J. (1968): *J. Chem. Phys.* 65:1313.
8. Lesclaux, R., and Joussot-Dubien, J. (1976): *Organic Molecular Photophysics, VII,* J. Birks, Ed., p. 457.
9. Migirdicyan, E. (1966): *J. Chem. Phys.* 44:520.
10. Weller, A. (1968): *Pure Appl. Chem.* 16:115; Davidson, R. S. (1969): *Chem. Comm.* 1450; Barltrop, J. A. and Owers, Q. F. (1970) (1942): *Chem. Comm.*

Polynuclear Aromatic Hydrocarbons, edited by P. W. Jones and P. Leber. Ann Arbor Science Publishers, Inc. Ann Arbor, MI ©1979.

THERMAL AND PHOTOCHEMICAL DECOMPOSITION OF PARTICULATE PAH

W. A. Korfmacher,[1] D. F. S. Natusch,* D. R. Taylor,*
E. L. Wehry† and G. Mamantov†

Department of Chemistry
University of Illinois
Urbana, Illinois 61801

*Department of Chemistry
Colorado State University
Fort Collins, Colorado 80523

†Department of Chemistry
University of Tennessee
Knoxville, Tennessee 37916

INTRODUCTION

Information about the fate of particulate polycyclic aromatic hydrocarbons (PAH) released to the atmosphere from combustion of carbonaceous fuels is presently fragmentary. It is often assumed, however, that photochemical oxidation processes play an important role since rapid photooxidation is known to occur for many PAH in solution, in pure solid form, or when adsorbed onto certain solid substrates such as alumina (2,3). Consequently, it is widely accepted that the half-lives of particulate PAH in the presence of sunlight may be of the order of only hours or, at most, a few days (2,4,9).

In order to test this supposition and to obtain quantitative data describing the rate and extent of photochemical decomposition of particulate PAH a series of experiments were conducted in which individual PAH were adsorbed onto coal fly ash particles and exposed to both real and simulated sunlight conditions.

[1]Present Address: Chemistry Division, HFT-210, NCTR, Jefferson, Arkansas 72079.

EXPERIMENTAL

To simulate real plume samples, in which PAH are adsorbed onto fly ash particles (8), a model system for adsorption of PAH from the vapor phase onto fly ash was utilized. Samples of fly ash, collected from power-plant electrostatic precipitators, were size-separated and then placed into a vapor deposition apparatus similar to that described by Miguel (6). A simple diffusion cell (7) was used to prepare a known vapor-phase concentration of a particular PAH, and the expanded coal fly ash bed was exposed to this vapor for a known duration (typically 9 hr).

To detect the occurrence of any nonphotochemical decomposition of the vapor-adsorbed PAH which might occur, a portion of fly ash so prepared was immediately subjected to soxhlet extraction (benzene, cyclohexane or methanol). The extract was then analyzed for the PAH and/or degradation products by liquid chromatography, UV-absorption or fluorescence spectrometry. The precision of these analyses was determined to be ± 8% RSD. The remainder of the fly ash sample was subjected to irradiation, using any of several artificial light sources (150-W xenon lamp, 275-W sunlamp or General Electric 500T3/C1 "quartzline" lamp; all unfiltered) or to outdoor sunlight. Extraction followed by analysis, as described above, was performed following illumination.

RESULTS

Rather unexpectedly, only modest (< 20%) photodegradation was observed for the PAH studied (Table I). Furthermore, no significant dependence upon irradiation time or illumination conditions was found. However, under similar illumination conditions, benzo[a]pyrene and anthracene photodecomposed to a much greater extent in solution or when adsorbed from

TABLE I. Typical Changes in Concentration of PAH Adsorbed onto Coal Fly Ash Following Irradiation

PAH Adsorbed on Fly Ash	Light Source	Irradiation Time (hr)	Avg. % Change
Pyrene	sunlamp	24.	-13
Phenanthrene	xenon arc	3.3	+ 7.5
Fluoranthene	xenon arc	14.2	- 9.
Anthracene	xenon arc	24.	-15.
Benzo[a] pyrene	xenon arc	21.	-16.
Benzo[a] pyrene	quartzline	8.7	- 8.

methanolic solution onto alumina. To verify that the observed resistance to photodegradation of benzo[a]pyrene and anthracene was not artifactual, the compound of interest was adsorbed from methanolic solution onto commercial alumina thin-layer chromatography plates and prepared fly ash "thin-layer" plates. In a typical experiment, benzo[a]pyrene adsorbed onto alumina from solution underwent 50% photodecomposition with 80-min exposure to an unfiltered 150-W xenon lamp; under identical conditions, benzo[a]-pyrene adsorbed onto fly ash exhibited only 15% decomposition.

In addition to limited photochemical decomposition it was found that certain PAH (including fluorene, benzo[a]fluorene, benzo[b]fluorene, 9,10-dimethylanthracene, 9,10-dihydroanthracene and 4-azafluorene) underwent rapid, though not immediately complete, decomposition in the absence of light when adsorbed on fly ash. This process was studied most exhaustively for fluorene. The product of nonphotochemical degradation of vapor-adsorbed fluorene is 9-fluorenone. The initial extent of reaction varies from 7% to 90% decomposition, depending upon the type of fly ash used as substrate. The reaction is not observed for solid fluorene, for fluorene vapor-adsorbed on several other solids (alumina, silica gel, Linde Molecular Sieves, Rohm and Haas Ambersorb XE-340), or for fluorene dissolved in cyclohexane or methanol (the solvents used for soxhlet extraction of fluorene from fly ash). However, similar nonphotochemical decomposition is observed for fluorene vapor-adsorbed on activated charcoal.

The nonphotochemical degradation of PAH adsorbed on fly ash is relatively rapid. For example, when fluorene was vapor-adsorbed on fly ash for a duration of only 0.5 hr, fluorene extracted from a portion of the sample 1.0 hr after adsorption had already decomposed to the extent of 35%. The decomposition process continues, albeit much more slowly, if the fly ash samples are "aged" in the dark after adsorption of the PAH has been completed (Figure 1).

DISCUSSION

The foregoing results suggest that two types of chemical processes can operate in the oxidation of PAH adsorbed onto solid substrates. First, it is apparent that PAH adsorbed onto coal fly ash are generally stabilized against photochemical oxidation by comparison with the same compounds present in solution, as the pure solid, or adsorbed onto substrates such as alumina or silica gel. On the other hand, a limited class of compounds is found to undergo spontaneous nonphotochemical oxidation as a result of adsorption onto substrates such as coal fly ash and activated carbon.

The explanation of these findings is not presently clear. Based on the available evidence, however, it seems likely that energetic adsorption of PAH

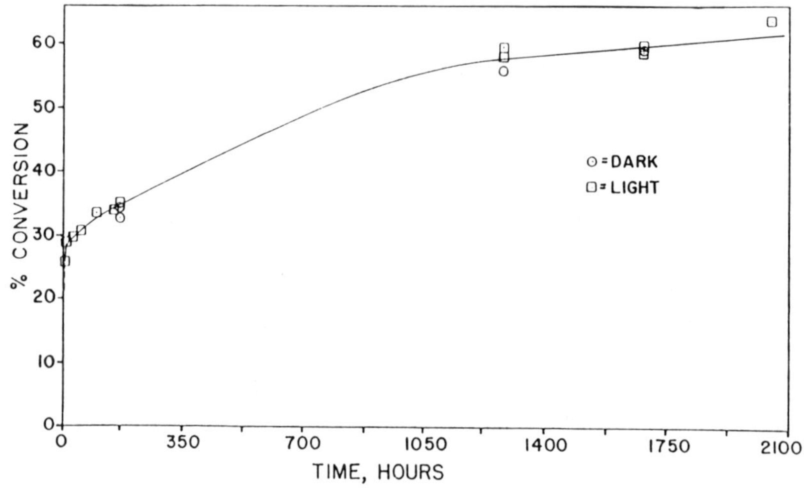

FIGURE 1. Percentage conversion of fluorene adsorbed onto coal fly ash as a function of time following completion of adsorption.

onto a highly active surface, such as that of coal fly ash or activated carbon, effectively stabilizes PAH against photooxidation which either increases the electronic excitation energy or decreases the lifetime of the excited state. The small amount of rapid photooxidation which is observed (Table I) could then be due to a fraction of the PAH which is not directly bonded to the active surface but rather to another PAH molecule (as in a multilayer configuration [4]) or to a weakly adsorbing mineral impurity. The promotion of nonphotochemical degradation could then, in turn, be due either to extensive molecular orbital rearrangement or simply to the ready availability of active oxygen on strongly adsorbing surfaces (1).

It is noteworthy that all of the compounds which undergo spontaneous oxidation as a result of adsorption onto coal fly ash (*i.e.*, fluorene, benzo[a]fluorene, benzo[b]fluorene, 9,10-dimethylanthracene, 9,10-dihydroanthracene and 4-azafluorene) contain a benzylic carbon atom indicating that this structure is especially susceptible to nonphotochemical oxidative attack.

From an environmental standpoint, the foregoing results have several important ramifications.

1. In general PAH which would normally be expected to undergo photochemical decomposition are stabilized when adsorbed onto coal fly ash. This may result in effective retention of the carcinogenic potential of several compounds (*e.g.*, benzo[a]pyrene).
2. Certain compounds (*e.g.*, those containing benzylic carbon atoms) are rapidly oxidized to the corresponding ketones or quinones as a result

of adsorption onto coal fly ash. Whether such oxidation results in a net increase or decrease in the toxic potential of adsorbed PAH is unknown but it does suggest that specific concentrations of particulate PAH in the plume from a coal-fired power plant may differ substantially both between plants and with distance from a given plant stack.

3. Perhaps one of the more interesting ramifications of the nonphotochemical process lies in the fact that certain highly carcinogenic PAH (*e.g.*, dimethylbenzanthracene and 3-methylcholanthrene) contain one or more benzylic carbon atoms and so would be expected to decompose quite rapidly following adsorption onto an active particulate substrate. This process may account for the fact that fluorenone is normally present at much higher levels than fluorene in atmospheric aerosols (2,5).

CONCLUSION

While the results presented herein are not fully definitive, it would appear that particulate association of PAH does not generally promote photochemical decomposition as hitherto supposed (2). Nevertheless, some PAH do undergo rapid oxidation as a result of adsorption onto active substrates such as those likely to be encountered in atmospheric aerosols.

ACKNOWLEDGMENTS

The research described herein was supported in part by Grant No. R803950030 from the U.S. Environmental Protection Agency, Duluth, MN; No. EE-77-S-02-4347 from the U.S. Department of Energy; and by Contract No. RP-332-1 from the Electric Power Research Institute.

REFERENCES

1. Bull, H. I., Hall, M. H., and Garner, W. E. (1931): The reaction between carbon and oxygen at low pressures and room temperature. *J. Chem. Soc.* 1:837-847.
2. Committee on Biological Effects of Atmospheric Pollutants (1972): *Particulate Polycyclic Organic Matter,* National Academy of Sciences, Washington, D.C.
3. Inscoe, N. M. (1964): Photochemical changes in thin layer chromatograms of polycyclic, aromatic hydrocarbons. *Anal. Chem.* 36: 2505-2506.
4. Lane, D. A., and Katz, M. (1977): The photomodification of benzo[a]-pyrene, benzo[b]fluoranthene, and benzo[k]fluoranthene under simulated atmospheric conditions. In: *Fate of Pollutants in the Air and Water Environments,* Part 2, pp. 137-154, I. A. Suffet, Ed., Wiley-Interscience, New York.

5. Lao, R. C., Thomas, R. S., Oja, H., and Dubois, L. (1973): Application of a gas chromatograph-mass spectrometer-data processor combination to the analysis of the polycyclic aromatic hydrocarbon content of airborne pollutants. *Anal. Chem.* 45:908-915.
6. Miguel, A. H., Korfmacher, W. A. Natusch, D. F. S., Wehry, E. L., and Mamantov, G. (1978): Apparatus for vapor-phase adsorption of polycyclic organic matter onto particulate surfaces. *Environ. Sci. Technol.* (submitted for publication).
7. Miguel, A. H., and Natusch, D. F. S. (1975): Diffusion cell for the preparation of dilute vapor concentrations. *Anal. Chem.* 47:1705-1707.
8. Natusch, D. F. S., and Tomkins, B. A. (1978): Theoretical consideration of the adsorption of polynuclear aromatic hydrocarbon vapor onto fly ash in a coal-fired power plant. In: *Carcinogenesis–A Comprehensive Survey, Vol. 3,* P. W. Jones and R. Freudenthal, Eds., pp. 145-153, Raven Press, New York.
9. Thomas, J. F., Mukai, M., and Tebbens, B. D. (1968): Fate of airborne benzo[a] pyrene. *Environ. Sci. Technol.* 2:33-39.

Polynuclear Aromatic Hydrocarbons, edited by P. W. Jones and P. Leber. Ann Arbor Science Publishers, Inc. Ann Arbor, MI ©1979.

RELATIVE RATES OF PHOTOCHEMICAL AND BIOLOGICAL OXIDATION (*in vitro*) OF POLYNUCLEAR AROMATIC HYDROCARBONS

Morris Katz, Cecilia Chan, Helle Tosine and Takeo Sakuma
Centre for Research on Environmental Quality
York University
Downsview (Toronto), Ontario, Canada

INTRODUCTION

Chemical reactions of polynuclear aromatic hydrocarbons (PAH) in the environment appear to represent a major mode of transformation of these compounds. In the case of PAH in the polycyclic organic matter of atmospheric particulates, photochemical oxidation reactions in the presence of sunlight result in a variety of products, including the corresponding phenols and quinones. A number of oxygenated derivatives of PAH have been found in samples of suspended particulate matter of the air collected at this campus by Pierce and Katz (15). A review of the reactions of PAH in the atmosphere by photooxidation, by reaction with atmospheric ozone, oxidants and sulfur oxides has been presented by the U.S. National Academy of Sciences (14).

Recently, the rates of photochemical oxidation of benzo[a]pyrene (B[a]P), benzo[b]fluoranthene (B[b]F) and benzo[k]fluoranthene (B[k]F) have been studied by Lane and Katz (12), under simulated sunlight and atmospheric conditions. Within the large class of PAH, many attempts have been made to relate the electronic structure of these compounds to chemical and potential carcinogenic activity (2,9,17). A review of the chemistry of PAH in relation to carcinogenic activity has been presented by Dipple (5). The mechanisms of *in vitro* chemical carcinogenesis have been discussed by Heidelberger (8).

The principal result of this discussion of the fundamental electronic properties of these compounds and possible correlations with carcinogenic activities has been the evolution of the K- and L-region hypothesis of the Pullmans (17) in the following propositions:

1. The appearance of carcinogenic activity in aromatic hydrocarbons is determined by the existence of an active K-region.
2. If, however, the molecule also contains an L-region, a supplementary condition requires that this region should be rather inactive.

The prototype of this structural element is the 9,10-bond in phenanthrene, called the K-region of the molecule. The L-region is typified by the 9,10-carbon atoms in anthracene. The K,L-region theory postulates that those bonds and atoms where reactions take place *in vitro* are also the region important in *in vivo* carcinogenesis. It is implied that a metabolic reaction initiating the process of carcinogenesis involves the K-region of the carcinogen and a cellular receptor, and that the reaction at the L-region bypasses the development of carcinogenic response. Molecular orbital theory has been useful in calculating reactivity indices to define concurrence with the circumstances of propositions 1 and 2 above by the Pullmans.

Benzo[a]pyrene-4,5-epoxide has been isolated as a metabolite of B[a]P in rat liver by various researchers (6,7,11,20,21) and it has been postulated as the ultimate carcinogen. Weinstein *et al.* (22) showed that a specific isomer of a diol-epoxide derivative of B[a]P is an intermediate in the binding of B[a]P to RNA in cultured bovine bronchial mucosa. Borgen *et al.* (3) reported a metabolic conversion of B[a]P by Syrian hamster liver microsomes and binding of metabolites to deoxyribonucleic acid. They found that a metabolite, 7,8-dihydroxy-7,8-dihydrobenzo[a]pyrene, covalently binds to DNA to a tenfold greater extent than B[a]P itself. Brunette and Katz (4), in a study of the interactions of B[a]P with Chinese hamster ovary cells, reported that the uptake rate of B[a]P was reduced by perylene and by bovine serum.

It is recognized that PAH compounds are not carcinogenic per se, but may be transformed into active intermediate oxidation compounds by microsomal enzyme systems. Therefore, information on rates of photooxidation of PAH under simulated atmospheric conditions and *in vitro* studies of microsomal catalyzed reactions may be useful in providing additional information on their potential carcinogenic properties.

MATERIALS AND METHODS

Photochemical Apparatus

The apparatus consisted of an air source, air purification train, ozone generator, reaction chamber and irradiation source, as shown schematically in Figure 1.

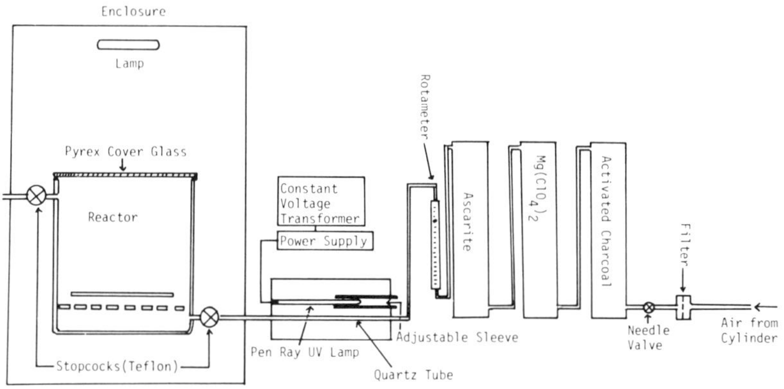

FIGURE 1. Flow diagram of the photooxidation system.

Air from a cylinder (Matheson Zero gas) was passed through a Gelman glass-fiber filter to remove oil droplets and solid particles which might be introduced into the air stream. A needle valve controlled the flow of air to a purification train which consisted of absorbing towers of activated charcoal, magnesium perchlorate and ascarite. A calibrated rotameter measured the airflow to the ozone generator and, hence, to the reaction chamber. All connections downstream of the needle valve were borosilicate glass to glass, or glass to Teflon to glass.

The ozone generator was fabricated in this laboratory after the design of Hodgeson *et al.* (10) and was calibrated by the neutral buffered potassium iodide method with an airflow of 100 ml/min through the generator. The method was checked by a Disibi Model 1003 AH ozone analyzer. A constant voltage transformer regulated the power supply which powered the Pen Ray UV lamp used in the generator.

The reaction flask was a 20-liter glass jar with an air inlet stopcock at the bottom and an air outlet stopcock diametrically opposite but at the top of the jar. Extruded silicone polymer (6.35 mm x 6.35 mm) was sealed to the top of the jar with silicone glue to form a gasket. After 12 months of use, there has been no detectable reaction between the silicone polymer and the ozone used in the reactor. A 3.18 mm thick borosilicate (Pyrex) disc covered the top of the reactor and acted as filter for the irradiation source. Reaction samples were placed on a support provided by glass rods. The glass support was 6.5 cm above the air inlet and presented minimum resistance to air flowing through the reactor. Both the reaction chamber and irradiation source were placed inside an enclosure to exclude laboratory fluorescent illumination from the reaction. Temperature buildup was minimized by installing a muffin fan at the top of the enclosure.

A 500-W General Electric quartzline lamp (Q500T3/CL) mounted in a standard photographic housing, 46 cm above the dishes in the reactor, was used as the irradiation source (12). The spectral output of the lamp was obtained using a Heath 701 spectrophotometer equipped with a 1P-28 photomultiplier tube. Actinic irradiance measurements were made using the standard uranyl nitrate-oxalic acid method and current readings were taken using a number 935 photo diode tube enclosed in a black box with a slit opening of 1.6 mm by 3.18 mm. Both actinic irradiance and current tests were made for the lamp, background irradiation and natural sunlight. When lamp and background irradiation tests were carried out, a 3.18-mm borosilicate glass filter was placed over the uranyl nitrate-oxalic acid absorbing solutions and over the slit on the phototube housing. No filter was used when making sunlight measurements.

Figure 2 shows the spectral output of the sun at solar zenith angles of 40° and 60°, as reported by Leighton (13). The spectral distribution of the quartzline lamp using a 3.18-mm Pyrex disc as a filter is also shown but not to the same absolute intensity scale. The spectral distribution of the quartzline lamp is a continuous spectrum, not a line spectrum, as is the case with fluorescent and sunlamps, and its spectrum very closely resembles that of the sun between 2,950 Å and 5,000 Å.

Sample Preparation and Reaction Under Simulated Atmospheric Conditions

Brinkmann Cellulose MN 300 precoated TLC plates were used as a coating medium for the PAH since these cellulose plates had no effect with the PAH under study. The dimensions (20 x 20 x 0.01 cm) furnished an excellent surface for exposure to irradiation. Stock solutions of 90-100 μg/ml of purified anthracene, benz[a]anthracene, pyrene, benzo[e]pyrene, benzo[a]pyrene, dibenz[a,c]anthracene and dibenz[a,h]anthracene were prepared in glass-distilled chloroform. Each of the prepared solutions was transferred to two 10-ml volumetric flasks and filled up to the mark. These solutions were stored in the dark as standards and analyzed to check the initial concentration of the PAH. A 100-ml portion of each stock solution was transferred to a TLC developing tank. Five precoated plates were put inside the tank at one time until the solution ran up to 1.0 cm from the top of the plates. The plates were then taken out, air-dried in the dark and the upper 2.5 cm of coating material was removed. They were then placed in the reaction chamber, one at a time, and exposed for definite periods of time under the following conditions:

1. simulated sunlight (by irradiation)

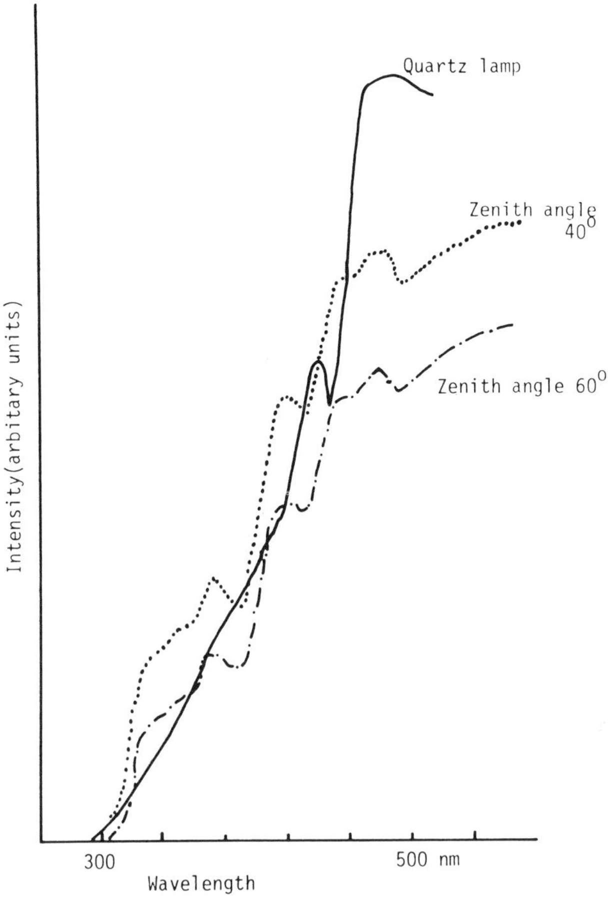

FIGURE 2. Spectral distribution of the quartz lamp and solar radiation in the wavelength range of about 300 to 500 nm.

2. simulated sunlight and 0.2 ppm ozone
3. 0.2 ppm ozone in the dark

The reaction chamber, during irradiation with the quartzline lamp or in the experiments conducted in the dark, was supplied with filtered purified air at a flowrate of 1.3 liter/min, measured by a calibrated rotameter. After irradiation, the remaining coating material on the TLC plates was carefully removed and extracted several times with glass-distilled chloroform. The extract, a pale yellow solution, was then concentrated to about 5 ml in a rotary evaporator and finally evaporated to dryness under a steam of nitrogen. These prepared samples were kept in the dark. Just before analysis by high-pressure liquid chromatography (HPLC), exactly 2 ml of distilled chloroform was added to the extract which was then injected immediately into the HPLC which was equipped with a fixed wavelength (254 nm) UV detector and a Varian Fluorichrome detector. An ODS Spherisorb Column was used to separate the products and determine the rate of photooxidation. The optimum operating conditions for HPLC with the above column were determined to be with a solution of 20% H_2O-80% MeOH and a flow of 0.8 ml/min. Irradiation experiments with ozone were conducted at a concentration of 0.2 ppm ozone in air. Such ozone concentration has been detected in the polluted urban atmosphere.

Photolysis in Aqueous Media

Purified B[a]P was dissolved in methanol and deionized water added to produce stock solutions containing 5-10 ppm (by wt) of B[a]P in 50% methanol/water. To study sensitized photodecomposition of B[a]P, approximately 20 liters of water containing brown, humic matter was collected from Grenadier Pond in West Toronto. The water was filtered through a 0.5-micron filter and subsequently stirred for 72 hr with a small amount of crystalline B[a]P. The solution was allowed to settle for 2-4 days, after which aliquots of solution were carefully siphoned off. This method provided reproducible solutions containing 5-10 ppm of B[a]P.

To increase absorption of light by dissolved organics (specifically fulvic acid) in the pond water, excess fulvic acid was added to the B[a]P/pond water solutions in other irradiation experiments. The fulvic acid was prepared according to the extraction method outlined by Schnitzer and Khan (19), commonly used to remove fulvic acid from soils.

The irradiation experiments were performed in the photochemical reaction apparatus described above, using the General Electric quartzline lamp. A copper cooling coil was fitted around the reaction vessel to prevent any increase in temperature, maintained at about 25°C. The 3-mm-thick borosilicate glass disc was used to eliminate radiation below 300 nm, but removal of the disc provided radiation down to 240 nm. Aliquots containing 250 ml of the prepared stock solutions of B[a]P in deionized water and in pond water were stirred and exposed to UV radiation of (a) wavelengths of 240-400 nm or (b) wavelengths ranging from 300-400 nm (achieved by covering the solutions with the 3-mm-thick glass disc).

The photodecomposition studies involving fulvic acid as the sensitizer were carried out by addition of 0.5 g solid fulvic acid to each 250-ml aliquot, using the pond-water stock solution of B[a]P. The solutions were stirred under the glass disc and irradiated. In all of the above studies, the samples were irradiated for periods of time ranging up to 70 hr.

Following irradiation, the water samples were quantitatively transferred to 500-ml separatory funnels and extracted three times with 100-ml volumes of chloroform. All three chloroform extracts were combined and filtered through 5 g of chloroform-washed sodium sulfate into round-bottomed flasks. After evaporation under vacuum, the combined chloroform extracts were made up to exactly 0.5 ml and stored in the dark at 2°C until analysis by gas chromatography and fluorescence spectrophotometry.

Preparation of Microsomes [After Kinoshita *et al.* (11)]

Adult male Sprague-Dawley rats weighing 160-180 g were used. The control group received intraperitoneal injections of 0.5 ml corn oil 40-48 hr prior to sacrifice, while the rats to be induced were treated with 0.5 ml corn oil containing 5 mg of 3-methylcholanthrene. The two main lobes of the liver and a portion of the lung from each rat were used for the microsome preparation. Two grams from each lobe and lung were homogenized in 10 ml 0.25 *M* sucrose, 0.05 *M* tris buffer, pH 7.5, using a Potter-Elvejhem glass-glass homogenizer (20 strokes). The homogenates were centrifuged at 750 x g for 15 min. The supernatant was transferred to several culture tubes of 10 ml and frozen at -70°C until ready for use. A portion of the frozen supernatant was thawed out in a refrigerator and filtered through a 0.45-μm filter prior to the incubation tests.

Metabolism of PAH

For the microsome-catalyzed reaction, the following assay was performed. Each flask contained, in a total volume of 1 ml, 50 μmoles of tris-chloride buffer (pH 7.5), 0.36 μmole of NADPH, 3 μmoles of magnesium chloride, 0.10 ml of the microsomal preparation (containing 100 to 150 μg of protein), and measured amounts (nmoles) of PAH (added in 0.04 ml methanol or DMSO just prior to incubation). The reaction was started by the addition of substrate. The flasks were gently shaken at 37°C for 30, 60, 120 and 240 min in air under red light. The reaction was stopped by the addition of 1.0 ml cold acetone. The flasks were placed in ice and 2.0 ml of cold ethyl acetate were added. The organic layer was then removed and dried by the addition of anhydrous sodium sulfate. Each extract was divided into two portions, one for HPLC analysis and the other for further tests. These extracts, after preparation, were stored in the dark at 0°C until subsequent tests.

PAH Analysis of Microsomal Extracts

Analyses of the samples were carired out using a Varian 8500 high-performance liquid chromatograph equipped with a "Vydac 201TP" reversed-phase column, a Varian Varichrome variable wavelength UV/VIS absorbance detector and a Schoeffel SF970 variable-wavelength fluorescence detector. Eluent for naphthacene (N), benz[a]anthracene (B[a]A), chrysene (Ch), perylene (Pe) and benz[e]pyrene (B[e]P) was a mixture of acetonitrile (84% v/v) and water (16% v/v). Acetonitrile was the eluent for B[k]F, B[a]P, dibenz[a,c]anthracene (DB[a,c]A) and dibenz[a,h]anthracene (DB[a,h]A). The flowrate was 1 ml/min.

The wavelength of the UV detector was set at 275 nm for analyzing naphthacene. The excitation and emission wavelengths for the fluorescence detector were set at 250 nm and greater than 370 nm, respectively, for B[a]A, Ch, B[e]P and Pe, and 285/389 nm for B[k]F, B[a]P, DB[a,c]A and DB[a,h]A. The quantitation was achieved by comparison of chromatographic peaks with those of standards and the internal standard coronene.

RESULTS AND DISCUSSION

Photolysis of Anthracene and Derivatives

The percentage decomposition with time in hours of anthracene (A), B[a]A, DB[a,c]A, and DB[a,h]A in air under irradiation in simulated sunlight conditions and with similar irradiation in the presence of 0.2 ppm of ozone in air, by volume, is illustrated by the curves in Figures 3 and 4, respectively. Anthracene undergoes rapid photodecomposition both in air and in the presence of ozone under the above conditions. B[a]A undergoes photodecomposition at a rate nearly twice that of DB[a,c]A, DB[a,h]A under simulated sunlight and more than three times that of the latter two PAH compounds when irradiated in the presence of ozone. The rates of photodecomposition of DB[a,c]A and DB[a,h]A are similar, under the above conditions.

Log decomposition plots with time of B[a]A, DB[a,c]A and DB[a,h]A are shown in Figure 5 for irradiation in air and simulated sunlight and in Figure 6 for irradiation in simulated sunlight in the presence of 0.2 ppm ozone. The graphs follow straight lines closely and indicate substantially first-order reactions.

Photolysis of Pyrene and Derivatives

The percentage photodecomposition with irradiation time of pyrene (P), B[a]P and B[e]P is illustrated in Figure 7 for air under simulated sunlight

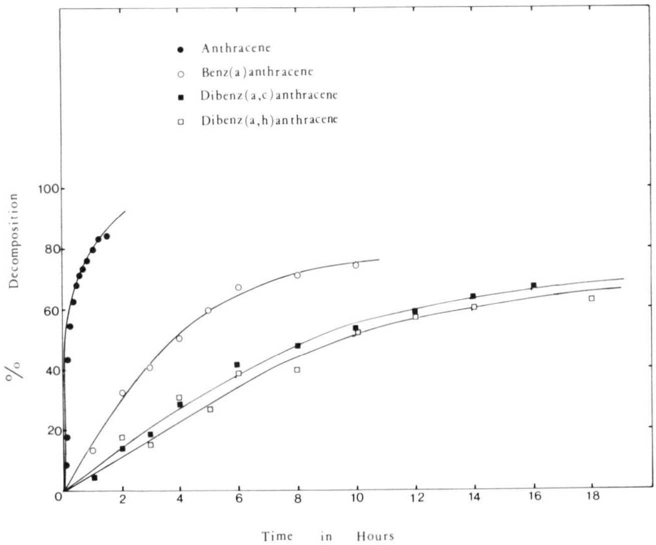

FIGURE 3. Photodecomposition with time of four PAH compounds during exposures to simulated sunlight in air.

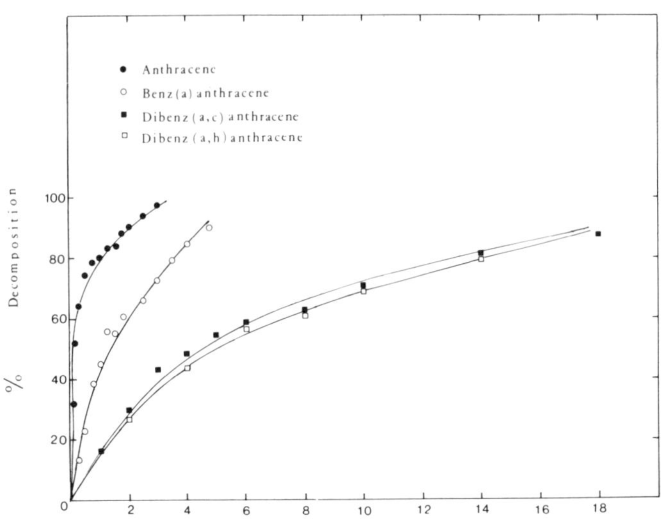

FIGURE 4. Photodecomposition with time of four PAH compounds during exposures to simulated sunlight in the presence of 0.20 ppm (by volume) of ozone in air.

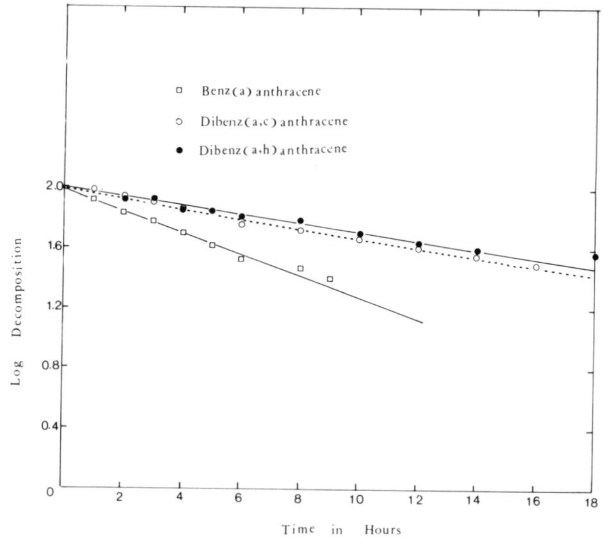

FIGURE 5. Log decomposition with time of benz[a] anthracene, dibenz[a,c] anthracene and dibenz[a,h] anthracene during exposures to simulated sunlight in air.

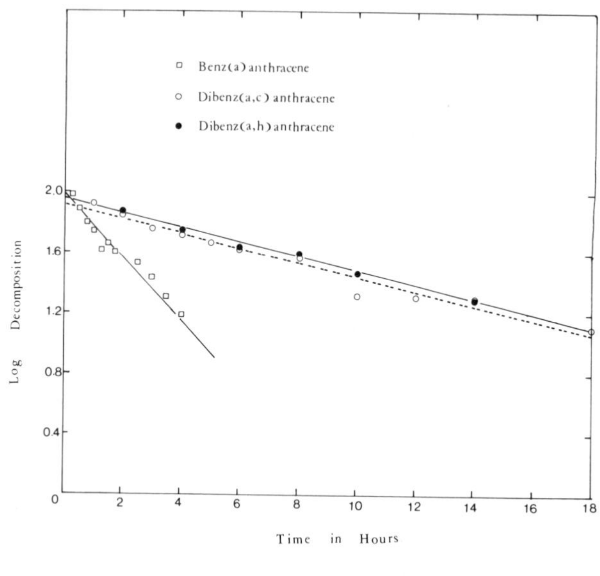

FIGURE 6. Log decomposition with time of benz[a] anthracene, dibenz[a,c] anthracene and dibenz[a,h] anthracene during exposures to simulated sunlight in the presence of 0.20 (by volume) of ozone in air.

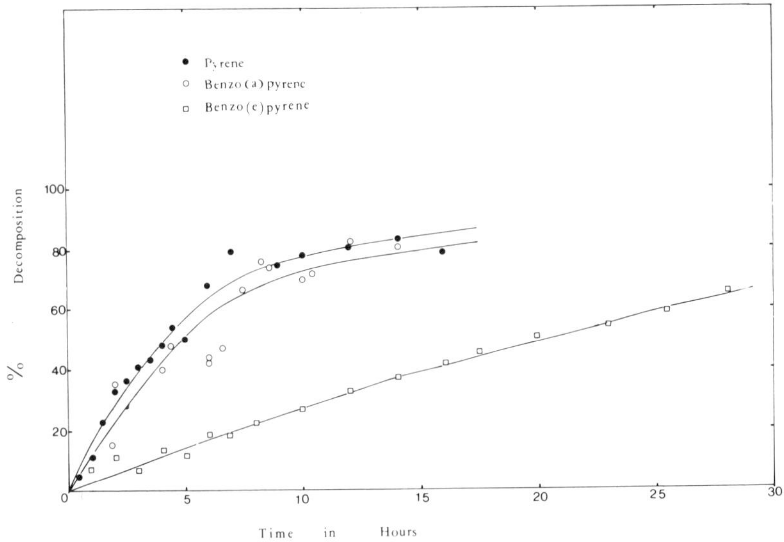

FIGURE 7. Photodecomposition with time of pyrene, benzo[a]pyrene and benzo[e]-pyrene during exposure to simulated sunlight in air.

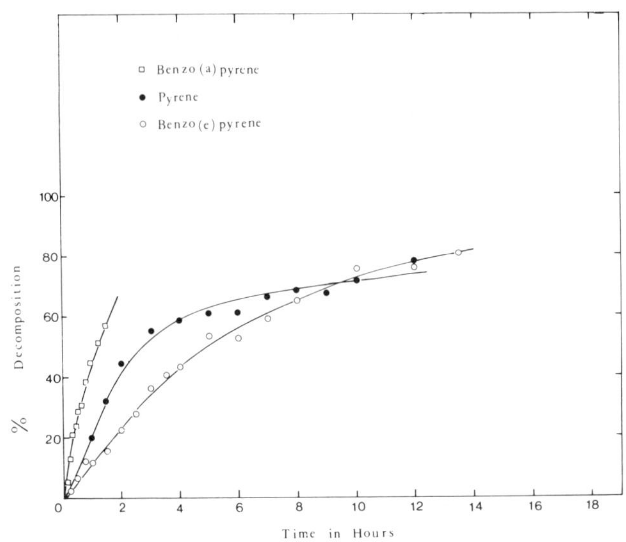

FIGURE 8. Photodecomposition with time of pyrene, benzo[a]pyrene and benzo[e]-pyrene during exposure to simulated sunlight in the presence of 0.20 ppm (by volume) of ozone in air.

and in Figure 8 for comparable irradiation with 0.2 ppm ozone. Photolysis of B[e]P occurs at a much slower rate than that of the other two PAH compounds. After irradiation with ozone, B[a]P photolysis occurs about an order of magnitude more rapidly than B[e]P and about five times as fast at P.

Half-Lives of Irradiated PAH

The relative half-lives in hours of nine PAH compounds, including those discussed above and B[b]F and B[k]F under various irradiation conditions, are presented in Table I. The data for B[a]P, B[b]F and B[k]F have been abstracted from studies by Lane and Katz (12). In the dark, reaction with 0.2 ppm ozone induces rapid oxidation of A, B[a]A, DB[a,h]A and B[a]P. The latter compound undergoes the most rapid rate of oxidation with ozone. DB[a,c]A undergoes dark oxidation at a somewhat slower rate than DB[a,h]A. Pyrene and espeically B[b]F and B[k]F show comparative resistance to dark oxidation.

Exposure to irradiation with simulated sunlight, in the presence of ozone, causes a marked reduction in the half-lives of all of the nine PAH compounds. A, B[a]P and B[a]A, after exposure to these conditions, show the shortest half-lives. After exposure to irradiation with air, B[e]P shows the longest half-life. Anthracene undergoes very rapid photolysis under all reaction conditions.

TABLE I. Half-Lives of Polycyclic Aromatic Hydrocarbons under Simulated Atmospheric Conditions (Expressed in Hours)

	Simulated Sunlight	Simulated Sunlight + Ozone (0.2 ppm)	Dark Reaction Ozone (0.2 ppm)
Anthracene	0.20	0.15	1.23
Benz[a]anthracene	4.20	1.35	2.88
Dibenz[a,h]anthracene	9.60	4.80	2.71
Dibenz[a,c]anthracene	9.20	4.60	3.82
Pyrene	4.20	2.75	15.72
Benzo[a]pyrene	5.30	0.58	0.62
Benzo[e]pyrene	21.10	5.38	7.60
Benzo[b]fluoranthene	8.70	4.20	52.70
Benzo[k]fluoranthene	14.10	3.90	34.90

Photolysis of B[a]P in Aqueous Media

These studies of the photolysis of B[a]P have indicated that the presence of air or oxygen is necessary for reaction since no photolysis or decomposition takes place in the presence of purified nitrogen. In deionized water/methanol solutions, B[a]P irradiated for upwards of 70 hr at wavelengths of 300-400 nm (simulated UV sunlight) showed negligible decomposition. When irradiation was extended to lower wavelength of 240-400 nm, the rate constant for reaction was only k = 0.002/hr.

Irradiation of B[a]P dissolved in pond water, which contains organic acids as the sensitizer, presumably fulvic acid, at wavelengths of 300-400 nm, in the presence of air, caused photodecomposition at a rate of k = 0.004/hr. However, when the photolysis was conducted with B[a]P in pond water containing added fulvic acid, the rate of reaction on irradiation in air at wavelengths of 300-400 nm showed a fourfold increase, with k = 0.016/hr. No reaction was found to occur for either pond water/B[a]P solutions or pond water/B[a]P solutions with added fulvic acid, on irradiation in the absence of oxygen.

B[a]P Photolysis Products

Among the products of B[a]P photolysis in atmospheric simulated-sunlight irradiation experiments and in the irradiation studies in aqueous media, the following three compounds were isolated: 1,6-; 3,6- and 6,12-B[a]P quinones. The structure of these compounds was confirmed by comparison with synthesized standards, using gas chromatography and mass spectrometry. The mass spectra of these compounds are illustrated in Figures 9, 10 and 11. These three quinones of B[a]P were also identified in samples of suspended particulate matter collected from the atmosphere of Toronto by Pierce and Katz (15,16).

Metabolism of PAH by Microsomal Extracts

The metabolism of nine PAH compounds treated *in vitro* with the microsomal preparations from the liver and lung of adult male Sprague-Dawley rats, as described in the above Methods Section, is illustrated by the data in Table II. The results indicate the rate of reaction of each PAH with time in the presence of the microsomal enzyme system. Both B[a]P and DB[a,h]A show a considerable reduction in concentration after reaction for 4 hr. B[e]P, B[k]F and Pe are relatively unaffected, whilst N and DB[a,c]A show a minor reduction in concentration after 4 hr. Both B[a]A and Ch suffer some decrease in concentration on the basis of the liver microsomal test.

The carcinogenic activity of these PAH is also listed in Table II, based on the recent review of this subject by Dipple (5). Although the status as to carcinogenicity of B[a]A, DB[a,c]A and Ch is disputed, Salamone, Heddle and Katz (18) have found these compounds to be mutagenic in the Ames *Salmonella typhimurium* assay (1). The above authors in mutagenic tests of 30 PAH

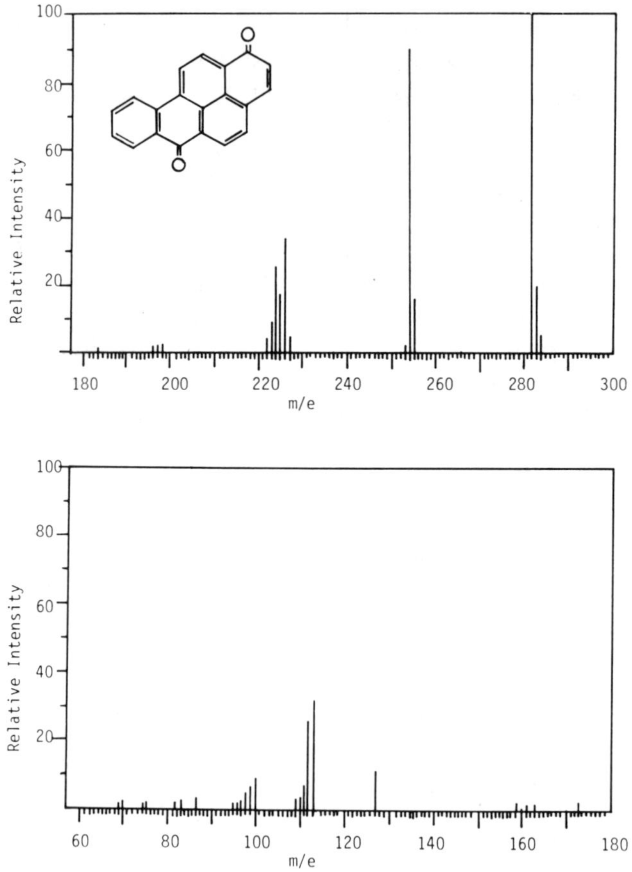

FIGURE 9. Electron impact mass spectrum of benzo[a]pyrene-1, 6-dione.

compounds, found that only the 1,6-, 3,6- and 6,12-benzo[a]pyrene quinones were direct-acting mutagens, whereas the hydrocarbons required activation by induced liver homogenate.

CONCLUSIONS

Photooxidation of the PAH compounds in this study occurs readily in the adsorbed state in a manner similar to reactions which may occur in the atmospheric environment, in sunlight and in the presence of low concentrations of ozone or oxidant. The reactivity of adsorbed PAH is considerably greater than that of corresponding hydrocarbons in aqueous media. In the

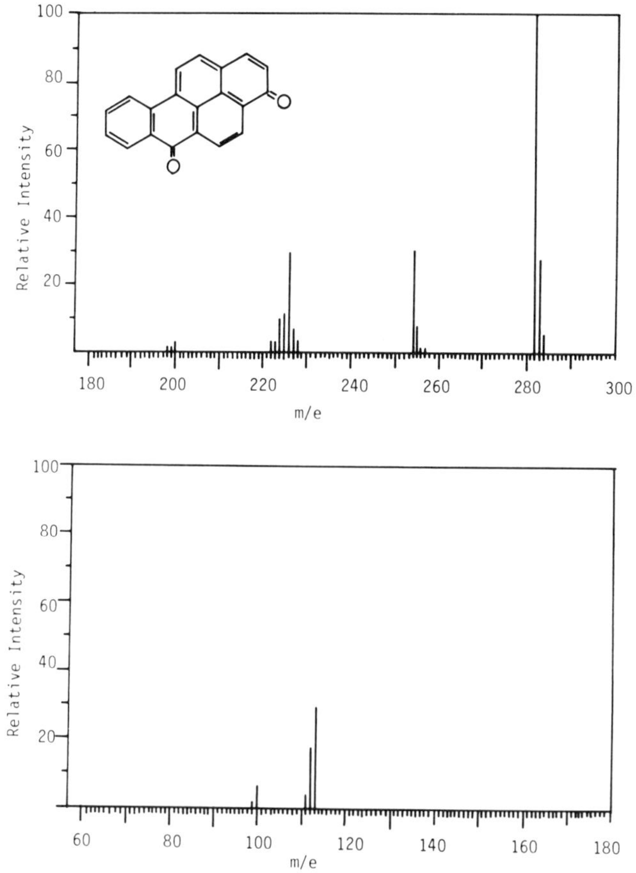

FIGURE 10. Electron impact mass spectrum of benzo[a]pyrene-3, 6-dione.

atmosphere PAH are present as adsorbed products on suspended particulate matter.

B[a]P is oxidized rapidly in simulated sunlight in air and much faster when irradiated in the presence of low concentrations of ozone. Among the resultant products are B[a]P quinones which are direct-acting mutagens.

Photochemical oxidation reactions of PAH appear to represent a major pathway for the removal of such compounds from the atmosphere. However, the products of these reactions may contain substances that are more harmful to health than the original reactants. Studies of such reactions should be conducted under experimental conditions that simulate as closely as possible the actual exposures that prevail in the natural environment.

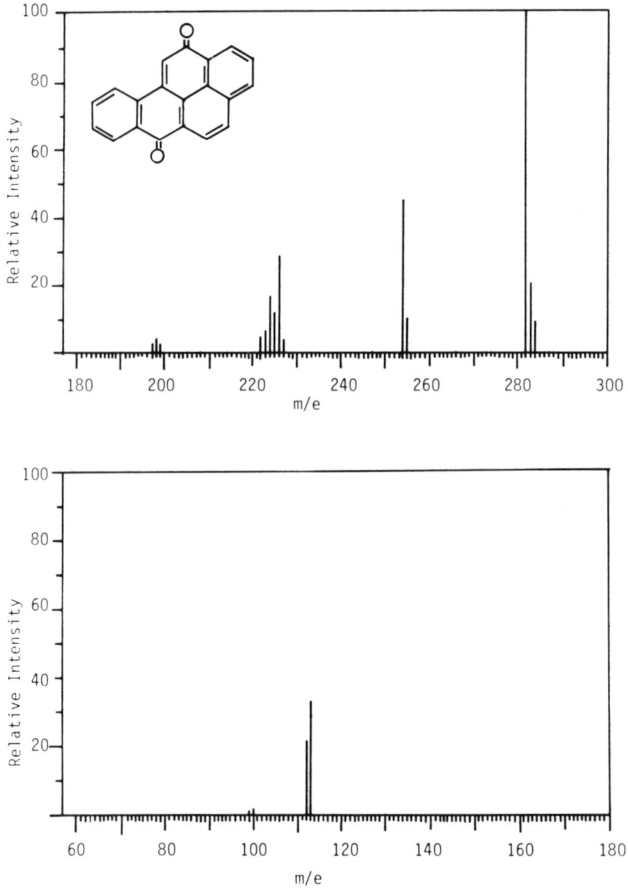

FIGURE 11. Electron impact mass spectrum of benzo[a]pyrene-6,12-dione.

ACKNOWLEDGMENTS

Thanks are due to Dr. David Wang, Organic Trace Contaminants Section, Ontario Ministry of the Environment, for the analysis of the samples of PAH treated with microsomal extracts. This work was partially supported by grants from the National Research Council of Canada and the Ontario Ministry of the Environment.

TABLE II. Metabolism of PAH by Microsomal Extracts of Rat Liver and Lung

Sample and Reaction Time (min)	PAH Concentration in n moles/ml								
	N	B[a] A	DA[a,c] A	DB[a,h] A	B[a] P	B[e] P	B[k] F	CH	Pe
Liver 0	26	20	24	—	38	20	17	25	14
30	25	17	22	40	32	19	17	21	13
60	23	14	22	36	—	18	17	19	14
120	23	14	20	10	33	18	16	19	13
240	20	12	19	19	12	20	15	13	13
Lung 0	26	20	25	—	37	22	26	26	15
30	25	20	24	35	31	21	21	—	12
60	26	22	22	38	—	—	—	25	15
120	24	16	22	20	31	21	21	24	14
240	22	16	19	22	12	18	21	21	13
Carcinogenic Activity	Inactive	Disputed	Disputed	Moderate	High	Inactive	Inactive	Disputed	Inactive

NOTE: N = naphthacene
B[a] A = benz[a] anthracene
DB[a,c] A = dibenz[a,c] anthracene
DB[a,h] A = dibenz[a,h] anthracene
B[a] P = benzo[a] pyrene
B[e] P = benzo[e] pyrene
B[k] F = benzo[k] fluoranthene
Ch = chrysene
Pe = perylene

REFERENCES

1. Ames, B. N., McCann, J., and Yamasaki, E. (1975): Methods for detecting carcinogens and mutagens with the Salmonella/mammalian microsome mutagenicity test. *Mutation Res.* 31:347-364.
2. Bergman, E. D., and Pullman, B., Ed. (1969): *Physico-Chemical Mechanisms of Carcinogenesis,* Israel Academy of Sciences and Humanities, Jerusalem.
3. Borgen, A., Darvey, H., Castagnoli, N., Crocker, T. T., Rasmussen, R. E., and Wang, I. Y. (1973): Metabolic conversion of benzo[a]pyrene by Syrian hamster liver microsomes and binding of metabolites to deoxyribonucleic acid. *J. Med. Chem.* 16:502-506.
4. Brunette, D. M., and Katz, M. (1975): The interactions of benzo[a]-pyrene with cell membranes: Uptake into Chinese hamster ovary cells and fluorescence studies with isolated membranes. *Chem. Biol. Interact.* 11:1-14.
5. Dipple, A. (1976): Polynuclear aromatic carcinogens. In: *Chemical Carcinogens,* C. E. Searle, Ed., pp. 245-314, ACS Monograph 173, American Chemical Society, Washington, D.C.
6. Grover, P. L., Hower, A., and Sims, P. (1972): Formation of K-region epoxides as microsomal metabolites of pyrene and benzo[a]pyrene. *Biochem. Pharm.* 21:2713-2726.
7. Grover, P. L. (1974): K-Region epoxides of polycyclic hydrocarbons: Formation and further metabolism by rat-lung preparations. *Biochem. Pharm.* 23:333-343.
8. Heidelberger, C. (1976): Studies on the mechanisms of carcinogenesis by polycyclic aromatic hydrocarbons. In: *Carcinogenesis–A Comprehensive Survey, Vol. 1,* R. I. Freudenthal and P. W. Jones, Eds., pp. 1-8, Raven Press, New York.
9. Herndon, W. C. (1974): Theory of carcinogenic activity of aromatic hydrocarbons. *Trans. NY Acad. Sci.* 36:200-217.
10. Hodgeson, J. A., Stevens, R. K., and Martin, B. E. (1971): A stable ozone source applicable as a secondary standard for calibration of atmospheric monitors. Presented at the Analysis Instrumentation Symposium, Instrument Society of America, Houston, TX.
11. Kinoshita, N., Shears, B. and Gelboin, H. V. (1973): K-Region and non-K-region metabolism of benzo[a]pyrene. *Cancer Res.* 33:1937-1944.
12. Lane, D. A., and Katz, M. (1977): The photomodification of benzo[a]-pyrene, benzo[b]fluoranthene and benzo[k]fluoranthene under simulated atmospheric conditions. In: *Advances in Environmental Science and Technology, Vol. 9, Fate of Pollutants in the Air and Water Environments,* I. H. Suffet, Ed., pp. 137-154, Wiley-Interscience, New York.
13. Leighton, P. A. (1961): *Photochemistry of Air Pollution,* p. 29, Academic Press, New York.

14. National Academy of Sciences (1972): *Particulate Polycyclic Organic Matter,* pp. 69-73, Washington, D.C.
15. Pierce, R. C., and Katz, M. (1976): Chromatographic isolation and spectral analysis of polycyclic quinones: Application to air pollution analysis. *Environ. Sci. Technol.* 10:45-51.
16. Pierce, R. C., and Katz, M. (1975): Determination of atmospheric isomeric polycyclic arenes by thin-layer chromatography and fluorescence spectrophotometry. *Anal. Chem.* 147:1743-1748.
17. Pullman, A., and Pullman, B. (1955): Electronic structure and carcinogenic activity of aromatic molecules. New developments. *Adv. Cancer Res.* 3:117-169.
18. Salamone, M. F., Heddle, J. A., and Katz, M. (1978): The mutagenic activity of thirty polycyclic aromatic hydrocarbons and oxides in urban airborne particulates. *Int. J. Environ.* (in press).
19. Schnitzer, M., and Khan, S. V. (1972): *Humic Substances in the Environment,* Marcel Dekker, Inc., New York.
20. Selkirk, J. K., Croy, R. G., and Gelboin, H. V. (1975): Isolation by high pressure liquid chromatography and characterization of benzo-[a]pyrene-4,5-epoxide as a metabolite of benzo[a]pyrene. *Arch. Biochem. Biophys.* 168:322-326.
21. Wang, I. Y., Rasmussen, R. E., and Crocker, T. T. (1972): Isolation and characterization of an active DNA-binding metabolite of benzo[a]-pyrene from hamster liver microsomal incubation systems. *Biochem. Biophys. Res. Commun.* 49:1142-1149.
22. Weinstein, I. B., Jeffrey, A. M., Jennette, K. W., Blobstein, S. H., Harvey, R. G., Harris, C., Autrup, H., Kasai, H., and Nakanishi, K. (1976): Benzo[a]pyrene diol epoxides as intermediates in nucleic acid binding *in vitro* and *in vivo*. *Science* 193:592-594.

Polynuclear Aromatic Hydrocarbons, edited by P. W. Jones and P. Leber. Ann Arbor Science Publishers, Inc. Ann Arbor, MI © 1979.

DETERMINATION OF SELECTED POLYNUCLEAR AROMATIC HYDROCARBONS IN SETTLED DUST BY HIGH-PERFORMANCE LIQUID CHROMATOGRAPHY WITH MULTI-WAVELENGTH DETECTION

Detlef Fechner and Bernd Seifert

Institut für Wasser-, Boden- und Lufthygiene des Bundesgesundheitsamtes
Berlin, Federal Republic of Germany

INTRODUCTION

Recent discussions among air pollution specialists in the Federal Republic of Germany have revealed the increasing importance of the determination of dust fallout and its components. It has been recognised in several cases that most daily human intake of hazardous substances is not, to an appreciable extent, caused by breathing dust-polluted air, but rather by the consumption of polluted foodstuffs, thus the substances enter the organism via the gastrointestinal tract. The concentration of pollutants in foodstuffs, however, is greatly influenced by the content of contaminants of the dust fallout which settles on the soil and, thereby, leads to an enrichment of pollutants in plant tissues and—following the food chain—in animal tissues.

In this situation a need arose for the development of a method for determining polynuclear aromatic hydrocarbons (PAH) in dust fallout in order to better understand the importance of PAH in this material. The previously developed thin-layer chromatographic procedure for the routine determination of benzo[a]pyrene in dust (11) had to be replaced by a more detailed analysis of the most important PAH in the sample: fluoranthene (Fl), pyrene (Py), benzo[e]pyrene (B[e]P), benzo[a]pyrene (B[a]P), benzo[b]fluoranthene (B[b]F), benzo[k]fluoranthene (B[k]F), perylene (Per), benzo[g,h,i]-perylene (B[g,h,i]P) and coronene (Cor).

Although a number of gas chromatographic procedures for the separation and determination of PAH in dust do exist (*e.g.*, 1,2,8,10), high-performance liquid chromatography (HPLC) was chosen as the more suitable analytical technique because of its greater speed. This point was of major importance in this case since the method chosen was to be used during a one-year measuring program. Time saving was also a reason for developing a more streamlined procedure for the dust sample cleanup than those described in the literature (3,5,8).

EXPERIMENTAL

Instrumentation

A Bransonic 52 ultrasonic generator (Branson, Soest, The Netherlands) was used for the extraction of PAH from the dust sample. Chromatographic separation was performed on a Waters HPLC chromatograph consisting of a 6000A pump, a U6K injection system and a Model 400 dual-wavelength UV detector. Furthermore, a Model 82.00 fluorescence detector (Dr. Knauer, Oberursel, Federal Republic of Germany) was used.

Columns

A glass column (15 cm x 4 mm i.d.) filled with 1 g of silica gel was used for the cleanup procedure. The silica gel slurried in cyclohexane was deactivated with 10% water.

The separation was carried out using a 25-cm stainless steel column filled with Merck reversed-phase material Lichrosorb 5RP18.

Analytical Procedure

The principal steps of the analytical procedure are as follows; a more detailed description is published elsewhere (6).

The dust sample was extracted ultrasonically with cyclohexane. The extract was evaporated to dryness at 30°C in an evaporator. The residue was dissolved in 500 μl of cyclohexane. This solution was used for cleanup on the silica-gel column. The PAH were eluted with 25 ml of cyclohexane, the eluate being evaporated to dryness and the residue taken up in 100 μl of acetonitrile. The separation was carried out at room temperature after injection of 25 μl of acetonitrile solution onto the Lichrosorb RP18 column. The mobile phase was acetonitrile/water (80:20) with a flowrate of 2 ml/min. For detection, the UV detector was used at 340 nm and 365 nm. The filters chosen for the fluorescence detector transmitted light at 436 nm and $\geqslant$450 nm for excitation and emission, respectively.

RESULTS AND DISCUSSION

Qualitative Aspects

The choice of an appropriate wavelength for detection depends on the type of detector used. The Waters 400 detector is a filter instrument, thus the available wavelengths are restricted to those of the Hg emission spectrum. The most suitable filter has to be chosen according to the wavelength of the absorption maximum of the individual PAH. Table I shows these maxima for the different PAH in acetonitrile/water (80:20). On the basis of these data, the first experiments were carried out at detector wavelengths of 280 nm, 340 nm, 365 nm and 436 nm. Figure 1 shows recordings at these four wavelengths for a standard PAH mixture.

TABLE I. Absorption Wavelengths for PAH and Wavelength used for their Detection Solvent: acetonitrile/water (80:20)

PAH	Absorption Wavelengths[a] (nm)						Detection Wavelength (nm)
Fl	288	277	282	*358*	342	322	365
Py	241	273	*336*	320	263	252	340
B[e] P	290	278	*332*	268	317	303	340
B[a] P	296	265	284	255	280/84	*365*	365
B[b] F	256	302	292	277	350	*367*	365
B[k] F	307	245	296	283	268	402	365
	380	*361*					
B[g,h,i] P	300	289	381	*362*	277	346	365
Cor	303	291	*339*	324	333	346	340
Per	253	247	*436*	410	388		436/≥450[b]

[a] In decreasing order of intensity. The figure nearest to the wavelength used for detection is italicized.
[b] Fluorescence mode: excitation/emission.

From analyses of dust extracts at these four wavelengths, it rapidly became clear that detection at 280 nm was unsatisfactory since, apparently, a number of substances which interfere at this relatively short wavelength were not separated by the cleanup procedure. Thus, there is no additional gain of information in comparison to detection at 340 nm and 365 nm. Additional experiments showed that perylene, which exhibits a strong absorbance maximum at 436 nm, can also be excited to fluorescence at this wavelength, which leads to an easy detection of Per using a coupled fluorescence detector with excitation at 436 nm and measurement of emission at ≥450 nm. Table I indicates the transmission wavelengths of the filters finally used for detection. From this table, it is clear that the wavelength selected for detection

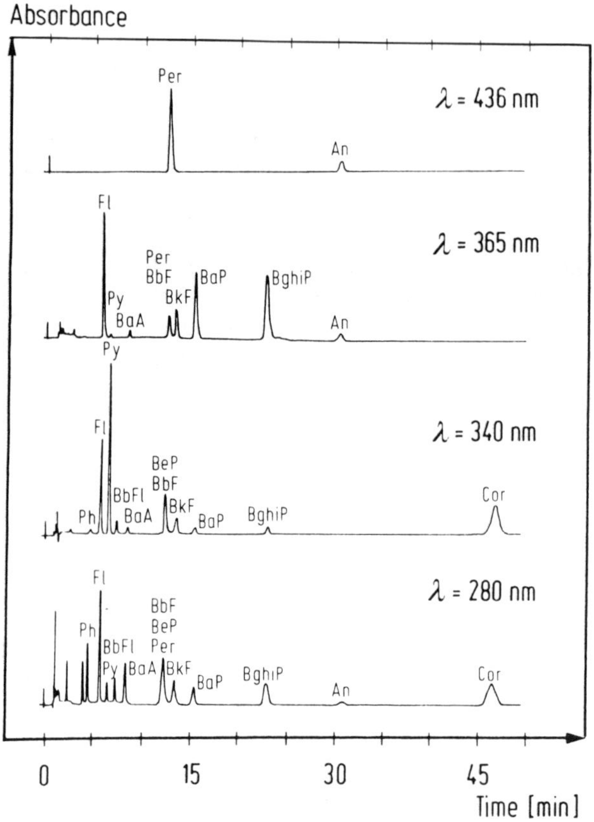

FIGURE 1. Separation of a standard PAH mixture with detection at 280 nm, 340 nm, 365 nm and 436 nm.

does not always represent the optimum. However, a comparison of the intensities of the different absorption maxima revealed that the losses in sensitivity due to an unfavorable wavelength are acceptable in terms of the detection limits for the different PAH. Detection was thus restricted to 340 nm, 365 nm and 436/⩾450 nm for all separations. The advantage of a determination with both types of detectors is clear: all information required is gathered in a single run of less than 45 min. In comparable HPLC methods with multiwavelength detection, published recently by Grant and Meiris (7) and Das and Thomas (4), three to four hours are needed for complete analyses. Figure 2 shows typical recordings obtained with the two detectors on analysis of a dust-sample extract.

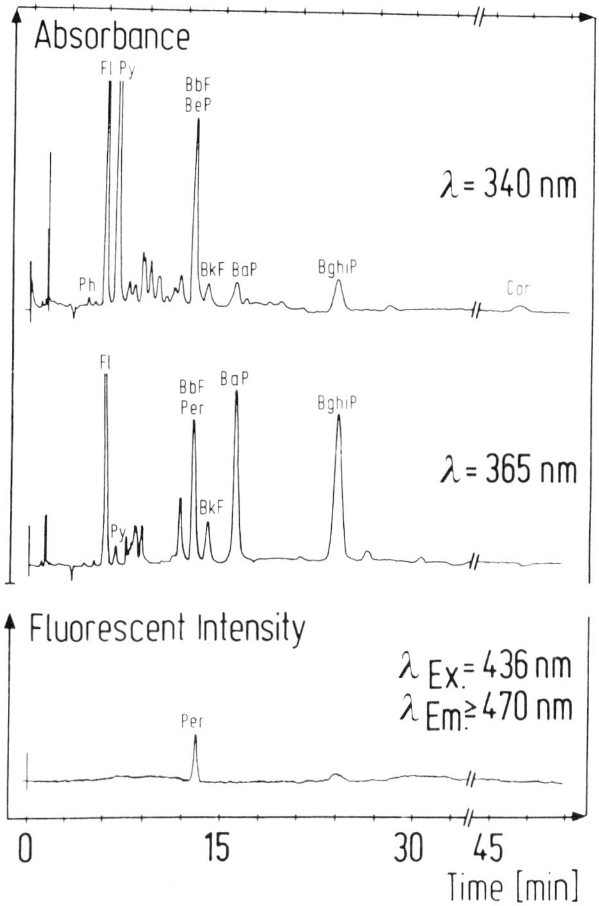

FIGURE 2. Separation of PAH in a dust extract with UV detection at 340 nm and 365 nm, and fluorescence detection at 436 nm/⩾450 nm.

Quantitative Aspects

Figure 2 shows that B[e]P and B[b]F on the one hand, and B[b]F and Per on the other, cannot be determined individually by UV detection at 340 nm or 365 nm. However, their determination is possible with the aid of the fluorescence detection. As the fluorescence detector records only Per, the amount of this compound is easily calculated from the fluorescence track. Knowing the Per content, the concentration of B[b]F can be calculated using the absorption at 365 nm. This concentration, in combination with the detection at 340 nm, finally serves to compute the amount of B[e]P in the sample.

In order to evaluate the results obtained by the HPLC method, four samples were collected in parallel every month at an industrial site in Berlin. The results of a three-month investigation show that the relative standard deviation is highest for the compound with the smallest concentration, *i.e.*, Per, and is then of the order of about 30%. Per concentrations were about 0.02 $\mu g/m^2 \cdot day$. For Fl and Py, the PAH with the highest concentrations in the samples (about 1 $\mu g/m^2 \cdot day$), relative standard deviations of 10% were observed. On considering these figures, it must be taken into account that the Per concentrations were very close to the detection limit (see Table II). These experiments show that the method is well suited for a reproducible determination of PAH in dust fallout.

The accuracy of the method has been checked by comparing the B[a]P value found by HPLC with that obtained by a thin-layer chromatographic procedure (11). Good agreement was found within 10%.

TABLE II. Detection Limits for PAH in Dust Fallout

PAH	Detection Limit[a] ($\mu g/m^2 \cdot day$)
Fl	0.007
Py	0.004
B[e]P	0.021
B[a]P	0.005
B[b]F	0.009
B[k]F	0.021
B[g,h,i]P	0.014
Cor	0.054
Per	0.019

[a]The figures represent the detection limit as obtained with the standard procedure. Lower detection limits are accessible by varying the conditions, *e.g.*, by injecting more sample solution.

Results of a Field Study

During a one-year study started in June 1977, the method was used to determine the amounts of PAH occluded in the dust fallout at 30 sites in the Federal Republic of Germany. The sites were chosen according to their expected PAH burden and were situated in urban and industrial environments. For control purposes, one site was selected in a nonpolluted area. A summary of the results for several typical sites is given in Table III. Comparative concentration data in the literature are scarce. The only results available are from

TABLE III. PAH Fallout at Several Sites in the Federal Republic of Germany from June 1977 to May 1978

	Annual Mean ($\mu g/m^2 \cdot day$)								
Site	Fl	Py	B[b] F	B[e] P	Per	B[k] F	B[a] P	B[g,h,i] P	Cor
Berlin Residential	1.0	0.57	0.23	0.29	0.018	0.13	0.10	0.27	n.d.
Berlin Commerical and Traffic	1.4	0.89	0.36	0.41	0.028	0.19	0.21	0.47	0.093
Berlin Airport	1.0	0.72	0.28	0.40	0.036	0.17	0.21	0.37	n.d.
Frankfurt Airport	0.67	0.40	0.23	0.28	0.016	0.12	0.11	0.29	n.d.
Frankfurt Near Railway Station	2.2	1.4	0.35	0.24	0.026	0.21	0.16	0.33	n.d.
Düsseldorf Urban	1.3	0.81	0.53	0.57	0.037	0.24	0.19	0.62	n.d.
Baltic Summer Resort	0.41	0.22	0.078	0.098	n.d.	0.061	0.035	0.085	n.d.

Note: n.d. = not detected

Japan in 1969/1970 (12) and Hungary in 1965/1970 (9). Urabe *et al.* (12) reported a mean value of 0.82 $\mu g/m^2 \cdot$day for B[a]P at industrial sites in two smaller Japanese cities and indicate that the amount in commercial areas was about one-fifth that in industrial areas. The resulting value thus falls within the range found for B[a]P fallout in this study.

Kertész-Sáringer (9) found an average value of 3.8 μg B[a]P/$m^2 \cdot$day at an urban site in Budapest and of 4.7 μg B[a]P/$m^2 \cdot$day 500 m from a coal-fired power plant. It is not, however, clear whether the differences between these values and those shown in Table III are attributable to different analytical techniques, to a genuinely higher pollution level or to other reasons.

SUMMARY

In a HPLC method for determining the major PAH in settled dust, the sample is filtered and the residue extracted ultrasonically with cyclohexane. The extract is purified using a silica-gel column. The unpolar fraction is carefully evaporated to dryness, and the residue is dissolved in acetonitrile. An aliquot of this solution is injected into the chromatograph equipped with a 25-cm Lichrosorb RP18 column (5 μm). The eluent is acetonitrile/water (80:20). For detection, a dual-wavelength UV instrument is used in combination with 340-nm and 365-nm filters. Additionally, a fluorescence detector records the fluorescence intensity of the column eluate at $\geqslant$450 nm after excitation at 436 nm. This detector configuration affords an elegant but simple possibility for the determination of fluorene, phenanthrene, fluoranthene, pyrene, benzo[b]fluorene, perylene, benzo[b]fluoranthene, benzo[k]fluoranthene, benzo[a]pyrene, benzo[e]pyrene, benzo[g,h,i]perylene, anthanthrene and coronene. The evaluation of the first ten compounds is possible within fifteen minutes of injection.

ACKNOWLEDGMENTS

This work has been granted by the Ministry of the Interior and the Ministry of Research and Technology of the Federal Republic of Germany.

REFERENCES

1. Bjørseth, A. (1977): Analysis of polycyclic aromatic hydrocarbons in particulate matter by glass capillary gas chromatogrpahy. *Anal. Chim. Acta* 94:21-27.
2. Braman, R. S. (1975): Gas chromatographic analysis in air pollution. In: *Chromatographic Analysis of the Environment,* R. L. Grob, Ed., pp. 79-121, Marcel Dekker, New York.

3. Brocco, D., Cantuti, V., and Cartoni, G. P. (1970): Determination of polynuclear hydrocarbons in atmospheric dust by a combination of thin-layer and gas chromatography. *J. Chromatog.* 49:66-69.
4. Das, B. S., and Thomas, G. H. (1978): Fluorescence detection in high performance liquid chromatographic determination of polycyclic aromatic hydrocarbons. *Anal. Chem.* 50:967-973.
5. Dong, M., Locke, D. C., and Ferrand, E. (1976): High pressure liquid chromatographic method for routine analysis of major parent polycyclic aromatic hydrocarbons in suspended particulate matter. *Anal. Chem.* 48:368-372.
6. Fechner, D., and Seifert, B. (1978): Bestimmung von polycyclischen aromatischen Kohlenwasserstoffen in Staubniederschlägen durch Hochleistungs-Flüssigkeits-Chromatographie mit Mehrwellenlängendetektion. 1. Mitt. Qualitative Ergebnisse. *Z. Anal. Chem.* 292:193-198.
7. Grant, D. W., and Meiris, R. B. (1977): Application of thin-layer and high-performance liquid chromatography to the separation of polycyclic aromatic hydrocarbons in bituminous materials. *J. Chromatog.* 142:339-351.
8. Grimmer, G., and Böhnke, H. (1972): Bestimmung des Gesamtgehaltes aller polycyclischen aromatischen Kohlenwassertoffe in Luftstaub und Kraftfahrzeugabgas mit der Capillar-Gas-Chromatographie. *Z. Anal. Chem.* 261:310-314.
9. Kertész-Sáringer, M. (1972): Angaben über die Untersuchung der atmosphärischen 3,4-Benzpyren-Luftverunreinigung in Ungarn. In: *Erfassung und Auswirkungen von Luftverunreinigungen,* Brennstofftechnische Gesellschaft in der DDR, Ed., pp. 107-122, VEB Deutscher Verlag für Grundstoffindustrie, Leipzig.
10. Lao, R. C., Thomas, R. S., Oja, H., and Dubois, L. (1973): Application of a gas chromatograph-mass spectrometer-data processor combination to the analysis of the polycyclic aromatic hydrocarbon content of airborne pollutants. *Anal. Chem.* 45:908-915.
11. Seifert, B., and Steinbach, I. (1977): Dünnschicht-chromatographische Routinebestimmung von Benz[a]pyren im Schwebstaub mit "in situ"-Auswertung. *Z. Anal. Chem.* 287:264-270.
12. Urabe, T., Kodama, Y., and Ishinishi, S. (1973): The quantity of benzo-[a]pyrene in settling and suspended particulates in Kitakyushu City and Fukuoka City. *Nippon Eiseigaku Zasshi* 28:141.

Polynuclear Aromatic Hydrocarbons, edited by P. W. Jones and P. Leber. Ann Arbor Science Publishers, Inc. Ann Arbor, MI ©1979.

SOURCE IDENTIFICATION AND ALLOCATION OF POLYNUCLEAR AROMATIC HYDROCARBON COMPOUNDS IN THE NEW YORK CITY AEROSOL: METHODS AND APPLICATIONS

J. M. Daisey, M. A. Leyko and T. J. Kneip

New York University Medical Center
Institute of Environmental Medicine
New York, New York 10016

INTRODUCTION

The development of methods for source identification and allocation of polynuclear aromatic hydrocarbon (PAH) compounds in the urban aerosol is of importance for a better understanding of the origins, reactions and fate of these compounds in the aerosol. Optimal strategies for the control of these air pollutants depend ultimately on such understanding. The anticipated increase in the use of coal and coal-derived fuels in the near future should give added impetus to such studies.

Most estimates of the contributions of various urban sources to ambient concentrations of total suspended particulate matter (TSP) and other pollutants are based on emission inventories. Although such inventories can provide a rough estimate of source strengths, there are a number of problems associated with this approach. Contributions of various sources are assumed to be proportional to the masses of material released by each source. Because of differences in the size distribution and in the chemistry of particles from various sources this assumption is not generally justified. In addition, there may be significant contributions to TSP and other pollutants from sources outside the inventoried city as well as from physical and chemical processes occurring in the atmosphere.

Several mathematical and statistical methods which utilize measurements of both TSP and trace elements in the TSP have been used in more recent years as another approach to identifying and apportioning the contributions of various sources to the ambient aerosol. It is likely that one or more of these methods may be useful in determining sources of organic constituents of the aerosol.

The methods currently used may be classified in several general categories. Enrichment factor methods (22, 37) examine the concentrations of various aerosol elements relative to their concentrations in natural materials such as soil. Elements originating from anthropogenic activities, such as lead and cadmium, tend to be enriched in the aerosol by several orders of magnitude as compared to soil. Elevated enrichment factors combined with information on characteristic elements or tracers for various sources can be used to identify significant sources of such elements in a given city.

Closely related to this approach is the use of elemental ratios from a source to estimate the contribution of that source to the TSP and various aerosol elements. If, for example, the ratio of lead to particulate matter is known for automobile exhaust and all of the lead in a given urban aerosol can be assumed to originate from automobiles, the aerosol concentration of lead can be used to estimate the contribution of this source to the TSP.

The chemical element balance method, first used by Winchester and Nifong (10) and by Miller and co-workers (31) combines measured emission ratios of certain tracer elements to total emission mass with ambient aerosol measurements to determine the contributions of various sources to the elemental concentrations and to TSP. Friedlander and co-workers (12,31) further supplement these calculated contributions with source emission inventory data for those sources for which characteristic tracers are not available and with theoretical calculations of sedimentation and gas-to-particle conversions. Kowalczyk and co-workers (25) and Gatz (13) have also taken this approach. Ideally applied, the chemical element balance method requires a knowledge of both the fractions of various elements in source emissions and a knowledge of fractionation coefficients between the source and the atmosphere. Relatively little ambient data is required, but the results can be generalized only for periods represented by the sampling data.

Statistical models (2,11,19,23,32) generally begin with a large data base of ambient aerosol measurements of TSP, tracer elements, meteorological data and other variables. Pairwise correlations and factor analysis are then used to identify sources. From a knowledge of tracer elements associated with various sources and such statistical analyses, a set of tracer elements characteristic of a given city can be chosen for the development of a source allocation model.

This approach has been taken in our laboratory for New York City (23) and multiple regression analysis has been used to obtain coefficients relating TSP to various source tracers. The coefficients obtained from the multiple regression analysis can then be used to apportion TSP and/or trace elements among various sources. Although a relatively large data base is required, source emission factors and source inventory data are not.

In view of the success of Kleinman (23) in developing a statistical model for the source allocation of TSP in New York City, and the knowledge of the New York City aerosol developed over ten years of studies at New York University, we have been prompted to investigate the possibility of using such methods for source identification and allocation of organic compounds in New York City. The results of our initial investigations on PAH compounds are presented here.

EXPERIMENTAL

Sampling

TSP samples were collected during the New York Summer Aerosol Study (27) at the sampling station located approximately 50 meters above ground level at NYU Medical Center between First Avenue and the FDR Drive at 30th Street in Manhattan. Twenty-four-hour samples were collected from noon to noon during two alternate weeks in August 1976, on preextracted (1-propanol) Gelman Type A fiberglass filters for organic compound analysis and on Whatman 41 filters for trace metal analysis.

Extraction and Analysis of PAH

The particulate samples and blanks were extracted with cyclohexane (Burdick and Jackson, Inc., "distilled in glass") for 8 hr in a Soxhlet apparatus. Extracts were then filtered and reduced to a 10-ml volume under vacuum using a rotary evaporator with a bath maintained at 35-45°C. Samples were stored in a freezer until analysis. All samples were protected from light from the time of collection through the completion of analysis. Polynuclear aromatic hydrocarbons were determined using a thin-layer chromatography/gas chromatography method which has been described (5). Total PAH in each sample was estimated from the total area of the chromatograms and an average relative response factor.

Trace Metals

Half of each high-volume sample of TSP which had been collected on 8 in. x 10 in. Whatman filters was analyzed for Cd, Fe, Mn, Pb, V and Zn using atomic absorption spectroscopy as described by Eisenbud and Kneip (8).

Other Data

Several other investigators in the New York Summer Aerosol Summer Study (27) provided additional data. Ambient aerosol concentrations of sulfate, nitrate, ammonium and hydrogen ions were provided by R. Tanner of

Brookhaven National Laboratories; S. Bugoff of the New York City Department of Transportation provided traffic count data. Meteorological data were obtained from the National Weather Service (New York City).

PRELIMINARY CONSIDERATIONS

The development and application of mathematical and statistical models for source identification and allocation of PAH in the urban aerosol requires a number of prior considerations.

Tracer Elements

All of the methods for source identification and allocation are based on a knowledge of tracer elements for various sources of particulate matter and on some general knowledge of the types of sources which are significant for a given city.

For New York City, the sources which are probably the most significant contributors to the PAH aerosol burden are anticipated to be the same sources which are significant contributors of TSP. These include transportation, solid waste incineration, combustion of fuel oil for power production and space heating and resuspension of soil (23,24). At present, there is little or no coal burned in New York City.

Kleinman (23) used factor analysis on an extensive aerosol data set to determine the best tracers for these major sources of TSP in New York City. Using lead for automobiles, vanadium for fuel oil burning, copper for incineration and manganese for soil, as well as sulfate as a tracer of secondary aerosol, he could account for approximately 60-75% of the TSP. Although these tracers can and do originate from more than one type of source at various locations in the U.S., they could be used most readily for modeling in New York City where each appears to originate primarily from one type of source (23).

There is no heavy industry located in New York City, although there are a large number of small commerical operations such as printing plants and electroplating facilities, which taken together may contribute significantly to TSP and possibly to PAH. It is not feasible to select an element which is representative of these mixed sources although elevated concentrations of zinc and cadmium have been found in the dust fall in commercial/industrial areas (20). Zinc and cadmium are also emitted by incinerators (16).

Refineries located approximately 20-30 km to the southwest of the city are anticipated to be minor contributors to aerosol PAH in New York City although this may not hold under appropriate meteorological conditions. At present, there are no known elemental tracers for this type of source.

Physical and Chemical Factors

Mathematical relationships between tracer elements and PAH will be affected by both physical and chemical factors. Ideally, the particle-size distributions of the tracer elements and of the PAH compounds from a given source should be similar so that atmospheric lifetimes are the same, and a relatively simple model can be developed. Very little data is available on the particle-size distribution of PAH compounds in source emissions. The work of Natusch (33) indicates that PAH vapors in the stack system of a coal-fired power plant condense onto the surface of coentrained fly ash particles at, or close to, the stack exit. As emission controls tend to be less effective for smaller particles, it is probable that PAH from coal-fired (and fuel oil-fired) plants will be associated with smaller particles. Measurements of PAH compounds in urban atmospheres indicate that approximately 75% of the PAH compounds are found in particles <2.5 μm in diameter (35). Atmospheric particle-size distributions will, of course, reflect the sources which contribute to the aerosol burden of PAH.

Greenberg and co-workers (16) have found 91% of the copper and 90% of the zinc from a municipal incinerator associated with particles $\leqslant 2$ μm diameter. Vanadium, a tracer for fuel oil burning, was found to be enriched in submicron particles from an oil-burning power plant (15). Most of the lead emitted from automobiles which remains suspended following emission has been found to be associated with particles <0.5 μm mass median diameter (17,29). Thus, the tracers for combustion sources are all associated with the fine-mode particles. These tracers have also been found in the fine-mode particles in the New York City aerosol (1,34). Most of the iron (approximately 80%), which is a tracer for resuspension of soil, is found in coarse-mode particles while manganese, which also can be used as a soil tracer, has a bimodal distribution (1,34). The foregoing indicates that the particle-size distribution of the tracers and of the PAH compounds are, in general, very similar.

PAH are known to undergo photodegradation reactions. Thus, the ratios of various PAH compounds to tracer elements from various sources may differ significantly from season to season when measured in ambient aerosol. We anticipate that such seasonal differences will have to be taken into account when yearly data is used in modeling. As a first approach, however, it seems likely that PAH collected within a relatively short sampling period should show approximately the same degree of degradation and relatively constant ratios with respect to tracer elements.

Sampling Artifacts

Sampling artifacts are a very important consideration for aerosol modeling. These are, of course, minimal for the tracer elements. Loss of more volatile

PAH compounds from conventionally collected (fiberglass filter) particulate samples during sampling have been reported by a number of investigators (6,36). Day-to-day variability in sampling losses, particularly in warm weather, could be a significant interference when seeking relationships between these compounds and tracer elements. Data on the distribution of PAH between vapor and particulate phases such as that reported by Cautreels and Van Cauwenberghe (3) as well as improved sampling methods may eventually eliminate this problem. For our initial work, however, we shall investigate relationships for higher-molecular-weight PAH compounds for which such losses are minimal.

All the factors discussed above may introduce complications. In the process of developing a model, we anticipate learning more about the relative significance of each of these aspects.

ENRICHMENT FACTORS AND RATIOS

Enrichment factors for tracer elements in the aerosol are used generally to obtain qualitative information about the nature of the sources which may be significant for a given city. As PAH compounds originate from more than one source, calculation of enrichment factors, relative to soil concentrations, is not very useful. Ratios for individual PAH compounds to other PAH compounds or to tracer elements can, however, be used to obtain clues to the nature of significant sources and sometimes to estimate the contributions of a given source.

Several investigators, Sawicki (38) and Hangebrauck and co-workers (18) in particular, have attempted to develop relations between individual PAH compounds such as the ratio of benzo[a]pyrene (B[a]P) to pyrene and to benzo[g,h,i]perylene (B[g,h,i]P). We have examined ratios of the B[a]P to B[g,h,i]P for samples collected in August 1976. These ratios are presented in Table I for comparison with ratios in Table II which have been measured for

TABLE I. Ratios of Benzo[a]pyrene to Benzo[g,h,i]perylene for New York City (August 1976)

Date	B[a]P/B[g,h,i]P	Date	B[a]P/B[g,h,i]P
8/10	1.38	8/17	1.24
8/11	9.00	8/23	5.77
8/12	8.80	8/24	2.64
8/13	1.85	8/27	0.49
8/14	1.03	8/28	2.39
8/16	1.08	8/29	0.42

a number of combustion sources. The ratios for three days, August 11, August 12 and August 23, stand out in comparison with the remaining values. The average ratio for these three days (7.9 ± 1.8) is strikingly and significantly different ($P \leqslant 0.005$) than the average for the remaining dates (1.4 ± 0.8). The latter is similar to the ratio of 1.3 which can be calculated from the data reported by Dong *et al.* (7) for a composite of six samples collected in New York City in November and December of 1974. The concentrations of B[a]P and B[g,h,i]P which were determined for this August 1976, period were also similar to those measured by Dong *et al.* (7), *i.e.,* 0.3-3.5 ng/m^3. The B[a]P/B[g,h,i]P ratio for the other three days is more in line with the high ratios which have been measured for coal-burning power plants (18) although there are no such sources within the city.

The transport of sulfate aerosol from the Midwest, an area of coal-burning power plants and industries, to the Northeast during the summer months has been demonstrated in recent years by a number of investigators (28,42,43). Of particular interest is the data analysis by Wolff and co-workers (43) of measurements made during the New York Summer Aerosol Study (27). They concluded that as much as 47% of the TSP could have been transported into New York City during the period of August 22 and August 23, 1976. This coincides with one of the dates for which a high B[a]P/B[g,h,i]P ratio was found. Unfortunately, the period of August 11 and August 12 was not covered by the study of Wolff and co-workers (43). An examination of the meteorology of this period, however, is informative. Hurricane Belle was over the New York City area on August 9 and 10th. On August 11, a new air mass, which had been sitting over the Midwest since August 7 arrived in New York City. Ratios of B[a]P/B[g,h,i]P were 9.0 and 8.8 for these dates.

We have examined several other factors which might have accounted for these unusually high B[a]P/B[g,h,i]P ratios, including differences in wind direction, degree of pollution and minutes of sunlight. No significant differences could be found for these factors. Photochemical degradation of PAH

TABLE II. Ratios of Benzo[a]pyrene to Benzo[g,h,i]perylene for Various Emission Sources

Source	B[a]P/B[g,h,i]P	Reference
Baltimore Harbor Tunnel	0.78 ± 0.16	9
Auto Exhaust	0.3-0.44	38
Incinerators	0.14-0.6	18
Oil-Burning Power Plants	>2,3.0	18
Coal-Burning Power Plants	0.9-6.6	18
Petroleum Refineries	0.65-1.7	18

aerosol emitted within the city might be expected to give a reduced rather than an elevated ratio of B[a]P/B[g,h,i]P as the B[a]P seems to be degraded more readily than B[g,h,i]P (21). Further, our ratios increase because of an increase in B[a]P concentrations, precisely the opposite of what could be accounted for by degradation.

Although no firm conclusions can be drawn at this time, the data certainly suggest that PAH compounds may be transported in New York City under certain meteorological conditions. Polynuclear aromatic hydrocarbon compounds have been found to be associated primarily with particles $\leqslant 2.5\ \mu m$ (35) and such particles have a residence time on the order of days (40). Evidence of long-range transport of PAH in aerosols from England and France into Norway has been reported previously by Lunde and Bjorsoth (30). Further investigation into possible transport of PAH into New York City is clearly warranted due to the implications for control strategies.

The average ratio of 1.4 for B[a]P to B[g,h,i]P which was observed for most days of the New York Summer Aerosol Study is much closer to those observed for automobiles, incinerators and oil-burning power plants. An estimate of the contribution of automobiles to the B[a]P in the New York City aerosol can be made from the ratios of B[a]P to lead. A value of 2×10^{-4} can be calculated for B[a]P/Pb from the data reported by Gordon and Bryan (14) for a sampling station location in an area of Los Angeles dominated by heavy traffic. A similar value of 2.9×10^{-4} can be derived from data reported by Fox and Staley (9) for the Baltimore Harbor Tunnel, assuming that lead comprises 10% of the tunnel aerosol by weight (26). Assuming that all of the New York City lead aerosol is derived from automobiles, and using the average aerosol concentrations of lead ($0.81\ \mu g/m^3$) and B[a]P ($0.85\ ng/m^3$) for those days which did not exhibit unusually high B[a]P/B[g,h,i]P ratios, the contribution of automobiles to the B[a]P aerosol burden is estimated to be between 19 and 28% of the total. We note parenthetically that Kleinman (23) has estimated that automobiles contribute 20-25% of the TSP in New York City. As ratios of B[a]P to other tracer elements in source emission are, unfortunately, unavailable it is not possible to estimate the contributions of other sources to B[a]P in the New York City aerosol in this way.

CHEMICAL ELEMENT BALANCE METHOD

The chemical element balance method has been used by a number of investigators (10,12,31,41) to estimate the relative contributions of various sources to the primary aerosol. The ambient aerosol concentration of a given element is assumed to be a linear combination of the contributions of various sources according to the relationship:

$$F_i = \sum_j \alpha_{ij}\, m_j\, f_{ij}$$

where F_i is the fraction of element i in the aerosol, m_j is the fractional mass contribution of source j to the aerosol and f_{ij} is the fraction of element i in emissions of source j. Finally, α_{ij} is a fraction used to make corrections for possible changes in f_{ij} between the source and the ambient aerosol. As the values for α_{ij} are not known they are generally assumed to be 1.

Ambient aerosol concentrations, F_i, of tracer elements chosen for a particular city are combined with the source fractions, f_{ij}, of these elements and the simultaneous equations are solved for the best values of m_j, the source strength coefficients, by the method of least-squares analysis. We have applied this method to the measurements made during the New York Summer Aerosol Study (27). As there was insufficient data on the fraction of total PAH or of most individual PAH compounds in source emissions, we used the values for B[a]P which are more generally available.

Values used for the fraction of B[a]P in various source emissions were those reported by Sawicki (39). A value of 0.51×10^{-6} was reported for the fraction of B[a]P in emissions from power plants (39). The type of fuel is not specified. As no other value was available, this was used for oil-burning power plants.

Four primary sources with their appropriate tracers were used in this model: automobiles (Pb), soil resuspension (Fe), incinerators (Zn) and fuel oil combustion (V). The source fraction values for these tracers are those used by Friedlander (10) for soil, oil burning and automobiles with one exception. Friedlander used a value of 0.40 for f_{Pb} for automobiles. We have used what is probably a more realistic value of 0.1 for this fraction, as measured by Larsen (26) for the Summer Tunnel and found by Kleinman (23) using a statistical model for New York City. Incinerator emission source fractions used were those reported by Greenberg and co-workers (16).

The results of these calculations are reported in Table III as the percent contribution of each of the sources to the aerosol concentrations of the tracers and B[a]P. The results for the tracers for soil resuspension (Fe) and oil-burning (V) are in good agreement with the results of Kleinman (23) for New York City, *i.e.*, each of these tracers is associated with a single type of source. Long-term data indicates that aerosol lead has been derived almost wholly from automotive sources in past years. Summertime reductions in traffic in New York City combined with increased use of low-lead and unleaded fuel may have changed the relative proportions of lead from automobiles and incinerators for this time period. On the whole, the results for the tracer elements are very reasonable.

TABLE III. Contributions of Various Sources to Ambient Aerosol Concentrations of Tracers and Benzo[a]pyrene in New York City Calculated by Chemical Element Balance Method

	Percent of Component From Source				
Aerosol Component	**Auto**	**Incinerator**	**Soil**	**Oil**	**Residual**
Benzo[a]pyrene	61.2	6.1	6.8	0.05	22.9
Lead	76.0	22.7	1.0	0.1	0.04
Zinc	4.0	94.0	1.8	0.1	0.1
Iron	1.8	1.1	92.1	4.9	0.02
Vanadium	0	0	2.9	96.7	3.3

The chemical element balance method used with the source fractions indicated that the greatest proportion of B[a]P was derived from the automobile and that soil resuspension and incinerators made relatively small contributions during this period. Approximately 23% of the B[a]P could not be attributed to the four sources investigated.

There is a large discrepancy between the 64% calculated from this model for automobiles and the 19-28% calculated previously using B[a]P/Pb ratios from tunnel measurements. The ratio of B[a]P/Pb calculated from the source fractions chosen for this model was approximately 12×10^{-4}, which is 4-6 times larger than the ratios measured for a tunnel (9) and for Los Angeles (14). This discrepancy very clearly points up the greatest problem in using the ratio method or the chemical element balance method, *i.e.*, the need for accurate and representative source emission data. Such data are not widely available for PAH compounds and those which are available were, in large part, measured 10-20 years ago and are probably not appropriate today.

STATISTICAL METHODS

There is insufficient data as yet to apply factor analysis to ambient aerosol concentrations of tracer elements, PAH and other aerosol components to determine factors which might be used in the development of a multiple regression model for PAH source allocation. It is possible, however, to take the first step in this direction with the data available from the New York Summer Aerosol Study (27). Parametric (Pearson) and nonparametric (Kendall rank and Spearman) correlations have been examined for B[a]P and total PAH with other aerosol variables as a first approach to determining sources of PAH in New York City. Variables used in this investigation included Pb, V, Zn, Cd, Mn, Fe, $SO_4^=$, NO_3^-, NH_4^+ and TSP. Pearson correlation coefficients for those variables which were significantly correlated are shown in

Table IV. Nonparametric statistics gave essentially the same results although the numerical values of the coefficients differ. Benzo[a]pyrene and total PAH were both significantly correlated with Pb, an automotive tracer, and with traffic. Total PAH showed some correlation to Fe and Zn as well, tracers for soil and incinerators, respectively. These correlations support the choice of Pb, Fe and Zn as source tracers for PAH compounds in New York City. No significiant correlation with V was found for this short period but it is probable that a more extensive data base may be required to observe this relationship. The results to date indicate that automobiles are a significant source of PAH in New York City and that this may also be true for soil resuspension and incinerators.

Colucci and Begeman (4) investigated correlations between B[a]P, benzo-[a]anthracene (B[a]A) and several other aerosol variables including Pb, V, $SO_4^=$ and CO for samples collected from April 1962, through February 1964, in New York City. Significant correlations between B[a]P, B[a]A and Pb, CO and V were found. The lack of correlation with sulfate was attributed to possible sampling anomalies. Our study also showed no correlation between PAH compounds and sulfate although specially treated filters were used to eliminate the conversion of SO_2 to sulfate on the filter. This lack of correlation is, perhaps, not surprising as sulfate is a secondary pollutant in the summer aerosol in the northeastern U.S. (28). Colucci and Begeman (4) did not investigate possible correlations with the tracers iron and zinc.

TABLE IV. Correlations Between Benzo[a]pyrene, Polycyclic Aromatic Hydrocarbons and Other Aerosol Variables During the New York Summer Aerosol Study, 1976

Benzo[a]pyrene with:	r^a $(p)^b$	PAH with:	r^a $(p)^b$
Pb	0.54(0.04)	Pb	0.75(0.004)
Traffic	0.62(0.016)	Traffic	0.77(0.003)
Fe	c	Fe	0.48(0.066)
Zn	c	Zn	0.60(0.026)

[a]r = Pearson correlation coefficient.
[b]p = level of significance.
[c]Not significant at $p < 0.1$ level.

CONCLUSION

Three methods for source identification and allocation of polynuclear aromatic hydrocarbon compounds in the atmosphere have been examined. The results of our preliminary investigations indicate that all three methods can provide useful information. Daily variations in the ratios of B[a]P/B[g,h,i]P in New York City during August 1976, suggested that long-range transport may be a singificant factor in determining aerosol burdens of PAH under certain meteorological conditions.

The chemical element balance method and aerosol ratios of B[a]P to tracer elements provided estimates of the contributions of various sources to aerosol PAH. The lack of source emission data for PAH compounds and uncertainties in the accuracy of such data at this time, however, are severe limitations for this approach.

Statistical modeling does not depend on source emisssion data but does require a large data base of ambient aerosol measurements. Although our data base for New York City is not as large as is desirable, the results of our early investigations of statistical relations between PAH, source tracers and other variables are very promising.

Both the chemical element balance method and statistical correlations indicated that automobiles were the most significant source of PAH in New York City during this period. Soil resuspension and incinerators may also contribute to a lesser extent.

A serious problem with all source allocation modeling is the lack of independent methods for assessing the accuracy of the results. Comparisons of results from various approaches do, however, provide useful information for making such judgments. As greater accuracy develops in each of the independent predictions, agreement should eventually be obtained or the errors in specific methods should become apparent.

ACKNOWLEDGMENTS

The authors wish to thank R. Tanner of Brookhaven National Laboratories and S. Bugoff of the New York City Department of Transportation for data provided for the correlation studies.

This work was supported by Grant No. RP439-1 from the Electric Power Research Institute and by the American Petroleum Institute and is part of a center program supported by the National Institute of Environmental Health Sciences, Grant No. ES00260, and the National Cancer Institute, Grant No. CA 13343.

REFERENCES

1. Bernstein, D. M. (1977): The influence of trace metals in disperse aerosols on the human body burden of trace metals. Doctoral Dissertation, New York University, New York.
2. Blifford, I. H., and Meeker, G. O. (1967): A factor analysis model of large scale pollution. *Atmos. Environ.* 1:147-157.

3. Cautreels, W., and Van Cauwenberghe, K. (1978): Experiments on the distribution of organic pollutants between airborne particulate matter and the corresponding gas phase. *Atmos. Environ.* 12:1133-1141.
4. Colucci, J. M., and Begeman, C. R. (1971): Carcinogenic air pollutants in relation to automotive traffic in New York. *Environ. Sci. Technol.* 5:145-150.
5. Daisey, J. M., and Leyko, M. A. (1979): Thin-layer/gas chromatographic method for the determination of polycyclic aromatic and aliphatic hydrocarbons in airborne particulate matter. *Anal. Chem.* 51:24-26.
6. DeWiest, F., and Rondia, D. (1976): Sur la validite des determinations du benzo[a]pyrene atmospherique pendant les mois d'ete. *Atmos. Environ.* 10:487-489.
7. Dong, M., Locke, D. C., and Ferrand, E. (1976): High pressure liquid chromatographic method for routine analysis of major parent polycyclic aromatic hydrocarbons in suspended particulate matter. *Anal. Chem.* 48:368-372.
8. Eisenbud, M., and Kneip, T. J. (1976): *Trace Metals in Urban Aerosols,* EPRI Report No. RP-117, NTIS # Pb-248-324.
9. Fox, M. A., and Staley, S. W. (1976): Determination of polycyclic aromatic hydrocarbons in atmospheric particulate matter by high pressure liquid chromatography coupled with fluorescence techniques. *Anal. Chem.* 48:992-998.
10. Friedlander, S. K. (1973): Chemical element balances and identification of air pollution sources. *Environ. Sci. Technol.* 7:235-240.
11. Gaarenstroom, P. D., Perone, S. P., and Moyers, J. L. (1977): Application of pattern recognition and factor analysis for characterization of atmospheric particulate composition in Southwest desert atmosphere. *Environ. Sci. Technol.* 11:795-800.
12. Gartrell, Jr., G., and Friedlander, S. K. (1975): Relating particulate pollution to sources: The 1972 California Aerosol Characterization Study. *Atmos. Environ.* 9:279-299.
13. Gatz, D. F. (1975): Relative contributions of different sources of urban aerosols: Application of a new estimation method to multiple sites in Chicago. *Atmos. Environ.* 9:1-18.
14. Gordon, R. J., and Bryan, R. J. (1973): Patterns in airborne polynuclear hydrocarbon concentration in four Los Angeles sites. *Environ. Sci. Technol.* 7:1050-1053.
15. Gordon, G. E., Davis, D. D., Israel, G. W., Landsberg, H. E., O'Haver, T. C., Staley, S. W., and Zoller, W. H. (1974): Study of the emissions from major air pollution sources and their atmospheric interactions. Two-year progress report, Period: November 1, 1972-October 31, 1974, to National Science Foundation, Grant No. GI 35338X, University of Maryland.
16. Greenberg, R. R., Zoller, W. H., and Gordon, G. E. (1978): Composition and size distributions of particles released in refuse incineration. *Environ. Sci. Technol.* 12:566-573.

17. Habibi, K. (1973): Characterization of particulate matter in vehicle exhaust. *Environ. Sci. Technol.* 7:223.
18. Hangebrauck, R. P., Von Lehmden, D. J., and Meeker, J. E. (1967): Sources of polynuclear hydrocarbons in the atmosphere. Public Health Service Publication No. 999-AP-33, Cincinnati, Ohio 45213.
19. Hopke, P. K., Gladney, E. S., Gordon, G. E., Zoller, W. H., and Jones, A. G. (1976): The use of multivariate analysis to identify sources of selected elements in the Boston urban aerosol. *Atmos. Environ.* 10: 1015-1025.
20. Hunt, W. F., Pinkerton, C., McNulty, O., and Creason, J. (1970): A study of trace element pollution of air in 77 midwestern cities. In: *Proceedings of the 4th Annual Conference on Trace Substances in Environmental Health,* University of Missouri, Columbia, MO.
21. Inscoe, M. N. (1964): Photochemical changes in thin layer chromatograms of polycyclic aromatic hydrocarbons. *Anal. Chem.* 36:2505-2506.
22. John, W., Kaifer, R., and Rahn, K., and Wesolowski, J. J. (1973): Trace element concentrations in aerosols from the San Francisco Bay area. *Atmos. Environ.* 7:107-118.
23. Kleinman, M. T. (1977): The apportionment of sources of airborne particulate matter. Doctoral Dissertation, New York University, New York.
24. Kleinman, M. T., Kneip, T. J., and Eisenbud, M. (1976): Seasonal patterns of airborne particulate concentrations in New York City. *Atmos. Environ.* 10:9-11.
25. Kowalczyk, G. S., Choquette, C. E., and Gordon, G. E. (1978): Chemical element balances and identification of air pollution sources in Washington, D.C. *Atmos. Environ.* 12:1143-1153.
26. Larsen, R. I. (1966): Air pollution from motor vehicles. *Ann. NY Acad. Sci.* 136:277-301.
27. Leaderer, B. P., *et al.* (1978): Summary of the New York Summer Aerosol Study (NYSAS). *J. Air Poll. Control Assoc.* 28:321-327.
28. Lioy, P. J., Wolff, G. T., Czachor, J. S., Coffey, P. E., Stasiuk, W. N., and Romano, D. (1977): Evidence of high atmospheric sulfates in the northeast. *J. Environ. Sci. Health* A12:1-14.
29. Little, P., and Wiffen, R. D. (1978): Emission and deposition of lead from motor exhausts-II. Airborne concentration, particle size and deposition of lead near motorways. *Atmos. Environ.* 12:1331-1341.
30. Lunde, G., and Bjorseth, A. (1977): Polycyclic aromatic hydrocarbons in long-range transported aerosols. *Nature* 268:518-519.
31. Miller, M. S., Friedlander, S. K., and Hidy, G. M. (1972): A chemical element balance for the Pasadena aerosol. *J. Colloid Interface Sci.* 39:165-176.
32. Moyers, J. L., Panweiler, L. E., Hopf, S. B., and Korte, N. E. (1977): Evaluation of particulate trace species in southwest desert atmosphere. *Environ. Sci. Technol.* 11:789-794.

33. Natusch, D. F. S. (1976): Characterization of atmospheric pollutants from power plants. In: *Proceedings of the Second Federal Conference on the Great Lakes,* pp. 114-129, Great Lakes Basin Commission, Ann Arbor, MI.
34. Patterson, R. K., and Wagman, J. (1977): Mass and composition of an urban aerosol as a function of particle size for several visibility levels. *J. Aerosol Sci.* 8:269-279.
35. Pierce, R. C., and Katz, M. (1975): Dependency of polynuclear aromatic hydrocarbon content on size distribution of atmospheric aerosols. *Environ. Sci. Technol.* 9:347-353.
36. Pupp, C., Lao, R. C., Murray, J. J., and Pattie, R. F. (1974): Equilibrium vapour concentrations of some polycyclic aromatic hydrocarbons, As_4O6, and ScO_2 and the collection efficiencies of these air pollutants. *Atmos. Environ.* 8:915-925.
37. Rahn, K. A. (1971): Sources of trace elements in aerosols, an approach to clean air. Doctoral Dissertation, University of Michigan, Ann Arbor.
38. Sawicki, E. (1962): Analysis for airborne particulate hydrocarbons: Their relative proportions as affected by different types of pollution. *Nat. Cancer Inst. Monograph* 9:201-220.
39. Sawicki, E. (1976): Analysis of atmospheric carcinogens and their cofactors. In: *INSERM Symposia Series,* 52:297-354, IARC Scientific Publications No. 13.
40. Willeke, K., and Whitby, K. T. (1974): Physical characteristics of Denver area aerosols. Paper No. 74-262, In: *Proceedings of the 67th Meeting Air Pollution Control Association,* Denver, CO, June 9-13, 1974.
41. Winchester, J. W., and Nifong, G. D. (1971): Water pollution in Lake Michigan by trace elements from pollution aerosol fallout. *Water Air Soil Poll.* 1:50-64.
42. Wolff, G. T., and Lioy, P. J. (1977): Transport of suspended particulates into the New York Metropolitan Areas. In: *Proceedings of the 70th Annual Meeting of the Air Pollution Control Association,* No. 77.2.7.
43. Wolff, G. T., Lioy, P. F., Leaderer, B., Kneip, T. J., Bernstein, D. M., and Kleinman, M. T. (1977): Characterization of aerosols upwind of New York City. II. Transport. Paper presented at the American Industrial Hygiene Association 1977 Conference, New Orleans, LA, May 22-27, 1977.

Polynuclear Aromatic Hydrocarbons,
edited by P. W. Jones and P. Leber.
Ann Arbor Science Publishers, Inc.
Ann Arbor, MI ©1979.

QUANTITATIVE ANALYSIS OF SELECTED PAH IN AQUEOUS EFFLUENT BY HIGH-PERFORMANCE LIQUID CHROMATOGRAPHY

JoAnn E. Wilkinson, Paul E. Strup and Peter W. Jones[1]

Battelle-Columbus Laboratories
505 King Avenue
Columbus, Ohio 43201

The EPA has established a list of toxic substances which are either known or likely to be present in selected aqueous effluents. Ultimately, the EPA is bound, under a Consent Decree, to establish a monitoring program with the intent to control the concentration of such substances in the effluents at an environmentally acceptable level.

These toxic substances have been categorized on a chemical basis into thirteen categories. One of these categories contain 16 polynuclear aromatic hydrocarbons (PAH) which are to be monitored in industrial aqueous effluents. They range from two-ring compounds such as naphthalene to as large as six rings (benzo[g,h,i]perylene). The group contains known and suspected carcinogens such as benzo[a]pyrene and benzo[b]fluoranthene as well as noncarcinogens such as the isomer benzo[k]fluoranthene. These PAH compounds which are to be monitored in this consent decree program are: naphthalene, acenaphthalene, acenaphthlene, fluorene, phenanthrene, anthracene, fluoranthene, pyrene, benzo[a]anthracene, chrysene, benzo[b]fluoranthene, benzo[k]fluoranthene, benzo[a]pyrene, dibenzo[a,h]anthracene, benzo[g,h,i]perylene, and indeno[1,2,3-c,d]pyrene. This monitoring program necessitates the use of analytical methodologies within the capability of most industrial laboratories. Numerous methodologies have been reported in the literature for the collection and analysis of these compounds. Most of these methods which use collection techniques involving the use of sorbent resins (1,12,16,18) and analytical procedures using capillary column (2,9,11) combined with gas chromatography/mass spectrometry (GC/MS) were considered too sophisticated and costly for general monitoring. Thus, methods development for this subsequent monitoring program investigated the use of

---

[1]Present address: Electric Power Research Institute, Palo Alto, California 94303.

liquid-liquid extraction for sample recovery and high-performance liquid chromatography (HPLC) for analysis.

Methods development for extraction of PAH from water was conducted in a series of three studies. These were solvent stability studies, extraction studies and preservation studies.

SOLVENT STABILITY STUDIES

Solvent stability studies were conducted using DMSO and acetone as the solvents of choice. These solvents were selected because they are both water-miscible and readily dissolve PAH.

The 16 PAH compounds were initially divided into two groups to alleviate any separation problems. Solutions of PAH in appropriate solvents were prepared at a concentration of 20 ppm, sealed in glass ampules and stored for 30-, 60- and 90-day periods. Some ampules were opened and the solutions analyzed immediately to assure that no problems had occurred during the sealing procedure. Analysis of the freshly opened ampules at the various time periods indicated the PAH to be stable over a minimum of a 90-day period for both DMSO (Table I) and acetone (Table II). Thus, the use of stock PAH standard solutions for calibration and spiking of samples can be drawn from freshly opened ampules over a minimum of a 90-day period.

EXTRACTION STUDIES

Extraction studies were conducted to evaluate the extraction efficiencies of methylene chloride and 15% methylene chloride in hexane for PAH in water at pH 2, 7 and 10.

One milliliter of approximately 20 ppm standard solution in acetone was added to 500 ml of appropriately buffered distilled water in a separatory funnel. The PAH were extracted three times with 0.30-ml portions of solvent. The extract was dried over $MgSO_4$, filtered and the $MgSO_4$ was washed with an additional 30 ml of solvent. The resultant extract was concentrated to 1 ml using a Kuderna-Danish concentrator and then analyzed.

The results of these studies are shown in Tables III through V. Results indicate that at each pH level methylene chloride was somewhat more efficient at extracting the PAH than 15% methylene chloride in hexane. Also, extraction at pH 7 appeared to give slightly higher extraction efficiencies and lower standard deviations than the other pH levels.

PRESERVATION STUDIES

Preservation studies were conducted to evaluate the stability of PAH in water under selected conditions. These conditions include pH 2, 7, 10; 0 and 2 ppm chlorine; and ambient temperature and 4°C.

TABLE I. Solvent Stability Studies

Compound	DMSO 30-, 60-, 90-Day Average
Acenaphthylene	100 ± 3
Fluorene	101 ± 2
Phenanthrene	100 ± 2
Fluoranthene	98 ± 3
B[a] A	100 ± 3
B[a] P	100 ± 4
Benzo[g,h,i] perylene	99 ± 3
Naphthalene	99 ± 4
Acenaphthene	100 ± 5
Anthracene	98 ± 2
Pyrene	100 ± 1
Chrysene	99 ± 1
Indeno [1,2,3-c,d] pyrene	99 ± 2
Dibenzo[a,h] anthracene	100 ± 2

Once again, 1 ml of 20 ppm standard solution was spiked into 500 ml of distilled water that had previously been buffered and/or chlorinated in accordance with the analytical matrix. These solutions were stored in the dark for seven days.

TABLE II. Solvent Stability Studies

Compound	Acetone 30-, 60-, 90-Day Average
Acenaphthylene	102 ± 3
Fluorene	103 ± 4
Phenanthrene	102 ± 3
Fluoranthene	101 ± 2
B[a] A	99 ± 5
B[a] P	97 ± 8
Benzo[g,h,i] perylene	100 ± 2
Naphthylene	100 ± 4
Acenaphthene	99 ± 4
Anthracene	98 ± 4
Pyrene	99 ± 3
Chrysene	101 ± 4
Indeno[1,2,3-c,d] pyrene	98 ± 4
Dibenzo[a,b] anthracene	97 ± 6

TABLE III. Extraction Studies pH 2

Compound	Average Percent Recovery	
	15% CH_2Cl_2/Hexane	CH_2Cl_2
Naphthalene	66 ± 8	102 ± 5
Acenaphthene	81 ± 9	100 ± 5
Anthracene	87 ± 10	100 ± 7
Chrysene	72 ± 31	103 ± 10
Benzo[k] fluoranthene	63 ± 25	104 ± 14
B[a] P	72 ± 6	86 ± 2
Dibenzo[a,h] anthracene	64 ± 20	97 ± 11
Indeno [1,2,3-c,d] pyrene	96 ± 32	106 ± 13

TABLE IV. Extraction Studies pH 7

Compound	Average Percent Recovery	
	15% CH_2Cl_2/Hexane	CH_2Cl_2
Naphthalene	66 ± 11	100 ± 7
Acenapthene	76 ± 13	99 ± 9
Anthracene	77 ± 12	100 ± 7
Chrysene	79 ± 10	100 ± 10
Benzo[k] fluoranthene	64 ± 18	101 ± 5
B[a] P	74 ± 16	101 ± 7
Dibenzo[a,h] anthracene	60 ± 10	99 ± 2
Indeno[1,2,3-c,d] pyrene	88 ± 17	100 ± 4

TABLE V. Extraction Studies pH 10

Compound	Average Percent Recovery	
	15% CH_2Cl_2/Hexane	CH_2Cl_2
Naphthalene	57 ± 9	89 ± 8
Acenaphthene	70 ± 9	101 ± 8
Anthracene	66 ± 9	91 ± 4
Chrysene	67 ± 8	78 ± 1
Benzo[k] fluoranthene	59 ± 7	63 ± 4
B[a] P	91 ± 24	89 ± 7
Dibenzo[a,h] anthracene	41 ± 11	59 ± 6
Indeno[1,2,3-c,d] pyrene	56 ± 19	64 ± 11

Each solution was extracted with methylene chloride. Each portion of solvent was used to rinse the storage bottle prior to extraction of the solutions. This was to minimize PAH adsorbtion onto the surfaces of the glass bottle.

The preservation study data is presented in Tables VI through VIII. All data has been corrected for the extraction efficiencies shown in the previous tables. In spite of the wide variance in data, one can see that there is usually better preservation at 4°C than at room temperature, especially for the smaller molecular weight PAH. There is also a definite effect on some PAH (benzo[a]pyrene, anthracene) when stored in chlorinated water. These effects appear to be in accordance with that reported in the literature (3,10).

The lack of recovery of naphthalene at ambient temperature without the addition of chlorine may be due to an anomalous surface effect. However, these experiments were repeated with similar results.

CONCLUSIONS

The following conclusions can be drawn from these preliminary extraction studies:

- PAH solutions are stable in DMSO and acetone for at least 90 days.
- Methylene chloride is a suitable extraction solvent for PAH in water.
- For best preservation of PAH in water, the pH should be around 7, stored at subambient temperatures, with removal of free chlorine possibly by addition of sodium bisulfate.

ANALYTICAL METHODS DEVELOPMENT

Analytical methods development involved evaluation of water extract cleanup procedures using liquid chromatography (LC) with alumina and silica gel as the stationary phase, and reverse-phase and normal-phase HPLC.

LC CLEANUP

Recoveries of PAH from two LC cleanup procedures were investigated. Each LC separation scheme was performed by applying 0.5 ml of a 20 ppm cyclohexane solution of representative PAH. The PAH in solution were chosen on the basis of representative ring number and relative elution volumes. These PAH were: naphthalene, anthracene, fluoranthene, chrysene and dibenzo[a,h]anthracene.

The alumina LC column was prepared by placing 10 g of alumina (3% H_2O w/w) in a 25 x 1 cm column, and wetted with 15 ml pentane. As the

TABLE VI. Preservation Studies pH 2

Compound	Average Percent Recovery			
	RT 0 ppm Cl	RT 2 ppm Cl	4C 0 ppm Cl	4C 2 ppm Cl
Naphthalene	77 ± 5	89 ± 9	81 ± 9	61 ± 5
Acenaphthene	86 ± 5	79 ± 7	90 ± 9	54 ± 5
Anthracene	83 ± 7	75 ± 13	85 ± 8	39 ± 7
Chrysene	63 ± 12	88 ± 10	79 ± 10	59 ± 10
Benzo[k] fluoranthene	80 ± 14	91 ± 14	80 ± 14	54 ± 14
B[a]P	44 ± 10	26 ± 3	112 ± 5	26 ± 4
Dibenzo[a,h] anthracene	73 ± 11	88 ± 11	75 ± 11	61 ± 11
Indeno[1,2,3-c,d] pyrene	77 ± 13	88 ± 13	75 ± 13	46 ± 13

TABLE VII. Preservation Studies pH 7

Compound	Average Percent Recovery			
	RT 0 ppm Cl	RT 2 ppm Cl	4C 0 ppm Cl	4C 2 ppm Cl
Naphthalene	0	93 ± 7	89 ± 7	97 ± 8
Acenaphthene	69 ± 9	66 ± 9	97 ± 9	100 ± 9
Anthracene	23 ± 7	22 ± 7	88 ± 7	45 ± 7
Chrysene	84 ± 11	53 ± 10	83 ± 10	77 ± 10
Benzo[k] fluoranthene	69 ± 7	82 ± 5	73 ± 5	65 ± 6
B[a]P	84 ± 7	35 ± 7	86 ± 7	46 ± 7
Dibenzo[a,h] anthracene	74 ± 9	78 ± 16	70 ± 2	60 ± 8
Indeno[1,2,3-c,d] pyrene	75 ± 16	85 ± 4	78 ± 4	73 ± 4

TABLE VIII. Preservation Studies pH 10

Compound	Average Percent Recovery			
	RT 0 ppm Cl	RT 2 ppm Cl	4C 0 ppm Cl	4C 2 ppm Cl
Naphthalene	104 ± 8	83 ± 8	111 ± 8	101 ± 8
Acenaphthene	102 ± 8	77 ± 8	102 ± 8	98 ± 11
Anthracene	100 ± 4	25 ± 4	100 ± 4	38 ± 4
Chrysene	109 ± 3	91 ± 1	105 ± 1	95 ± 13
Benzo[k] fluoranthene	125 ± 4	105 ± 4	121 ± 6	103 ± 4
B[a]P	111 ± 7	72 ± 7	89 ± 9	89 ± 7
Dibenzo[a,h] anthracene	124 ± 18	88 ± 6	105 ± 6	90 ± 8
Indeno[1,2,3-c,d] pyrene	142 ± 11	117 ± 11	133 ± 11	102 ± 12

pentane reached the tip of the alumina, 0.75 in. of $NaSO_4$ was added. The column was washed with 20 ml of methylene chloride and then reconditioned with 20 ml of pentane. The PAH extract is then placed on the head of the column and eluted with 25 ml of pentane and 25 ml of methylene chloride. The latter methylene chloride fraction contains the PAH. The LC scheme is presented in Figure 1.

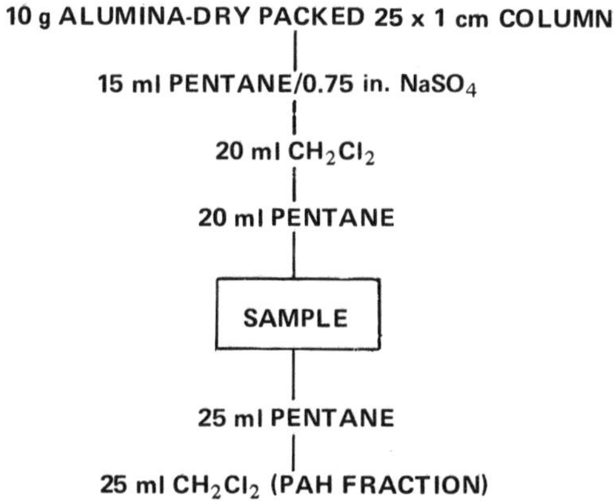

FIGURE 1. Alumina LC clean-up scheme.

The silica gel LC column was prepared by slurry packing 10 g of silica gel with methylene chloride into a 1 x 25 cm LC column. The column is then washed with 40 ml of pentane. The sample extract is then placed on the head of the silica gel and eluted with 25 ml of pentane and 25 ml 40% methylene chloride in pentane. This latter fraction contains the PAH of interest. This LC scheme is presented in Figure 2.

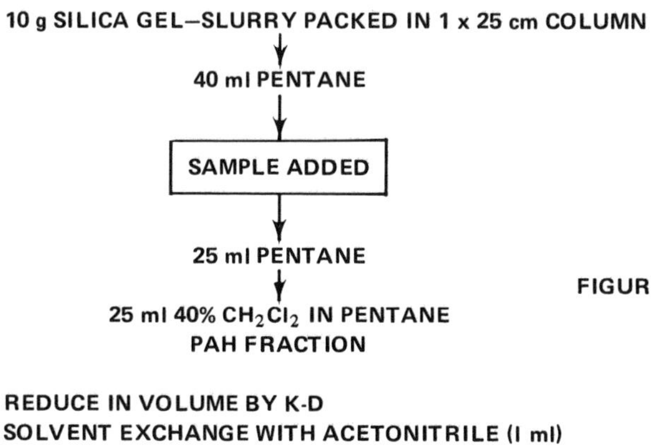

FIGURE 2. LC cleanup scheme.

Each PAH fraction from both LC cleanup methods was reduced in volume and solvent exchanged with acetonitrile to an exact volume of 1 ml and analyzed by HPLC using fluorescence detection. Results of these analyses are given in Table IX.

From the above data, it is apparent that either LC separation scheme would be suitable as a cleanup procedure. However, silica gel gave somewhat higher recoveries for the larger molecular weight components as well as exhibiting a considerably faster flowrate during the separation procedure. Thus, the use of silica gel was favored over alumina as the stationary phase for these cleanup procedures.

TABLE IX. PAH Recovery

Compound	Percent Recovery	
	Alumina	Silica Gel
Naphthalene	100	93
Anthracene	99	105
Fluoranthene	96	97
Chrysene	91	98
Dibenzo[a,h] anthracene	83	92

HIGH-PERFORMANCE LIQUID CHROMATOGRAPHY

Normal and reverse phase HPLC were investigated for the separation of the 16 PAH of interest. There is much in the literature describing separation of a limited number of PAH using reverse-phase HPLC (4,5,7,8,13,14,15,17, 19); however, no references could be located on the separation of all 16 PAH of interest using a single stationary phase.

Moderate success was initially achieved using a 5-μ Spherisorb ODS column. The mobile phase was a gradient of 50% acetonitrile in 50 min. Fluorescence detection was used with excitation at λ_{ex} 280 nm and a cutoff filter of $\lambda_{em} < 389$ nm. A chromatogram of this separation is shown in Figure 3. Note the important isomer pair, benzo[a]anthracene/chrysene as well as dibenzo[a,h]anthracene/benzo[g,h,i]perylene, are completely unresolved. This latter pair could be resolved using an isocratic step program, but these conditions still left benzo[a]anthracene and chrysene unresolved.

Quantitation of benzo[a]anthracene/chrysene involving selective wavelength detection [based on the work of Das and Thomas (6)] was attempted. This method involves the use of the ratio of the response at λ_{ex} 250 and $\lambda_{em} > 370$ to the response at λ_{ex} 280 and $\lambda_{em} > 389$. However, in real-world samples spiked with standard PAH, we found this ratio to fluctuate widely. Consequently this method would be unreliable for use in a monitoring program.

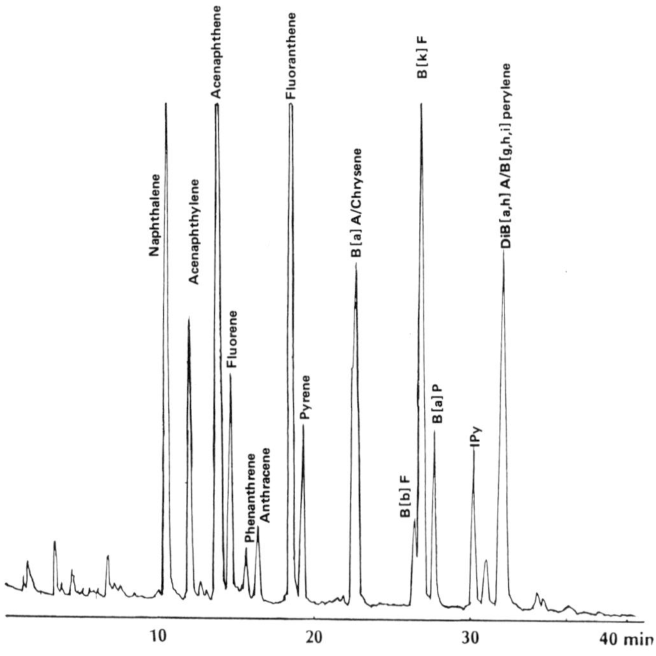

FIGURE 3. Distilled water extract of PAH standard.

Several normal phases for the separation of PAH were investigated. These phases included Lichrosorb Si60, alumina and Lichrosorb NH_2. Hexane was the mobile phase in all cases. The results of these studies indicated that there was not enough efficiency to separate all 16 PAH nor enough selectivity to resolve the above PAH pairs.

Further investigation of various reverse-phase columns indicated that the separation of all 16 PAH could be achieved using a Perkin-Elmer 10 μ, 0.26 x 25 cm HC-ODS column. Figure 4 shows the separation of all 16 consent decree PAH using a single stationary phase. The conditions used for this separation were 40% acetonitrile in water isocratic for 5 min. followed by a gradient elution from 40-100% acetonitrile in 25 min.

Evaluation of the above procedures utilizing this HC-ODS column were undertaken on selected real world samples. Figure 5 shows a chromatogram obtained from an extract of an aqueous effluent from a plastics manufacturing facility. Analysis using retention time and standard addition indicated

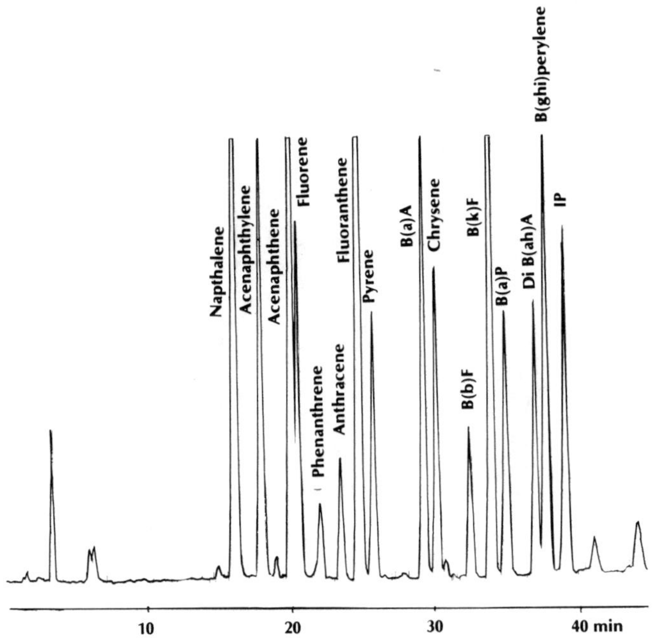

FIGURE 4. PAH mixture.

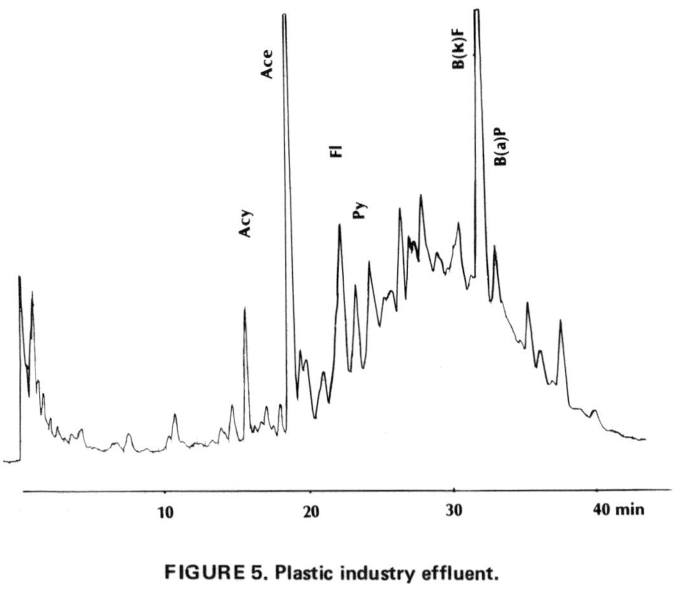

FIGURE 5. Plastic industry effluent.

acenaphthene and benzo[k]fluoranthene to be the major PAH components in this effluent at a flowrate of 0.5 ml/min.

The utilization of selective wavelength detection is demonstrated by comparing Figure 6 with Figure 7. The extract of fly ash wash obtained from a coal-fired power plant was analyzed using fluorescence detection at λ_{ex} 280, λ_{em} >389 (Figure 6) and λ_{ex} 254, λ_{em} >350 (Figure 7). Evidence of fluoranthene can be detected at λ_{ex} 280 nm, λ_{em}>389 nm while phenanthrene and anthracene are the predominate PAH identified at λ_{ex} 254 nm, λ_{em} >350 nm. The PAH in Figure 7 were quantitated to be at a level of 28 ppt (phenanthrene) and 200 ppt (anthracene).

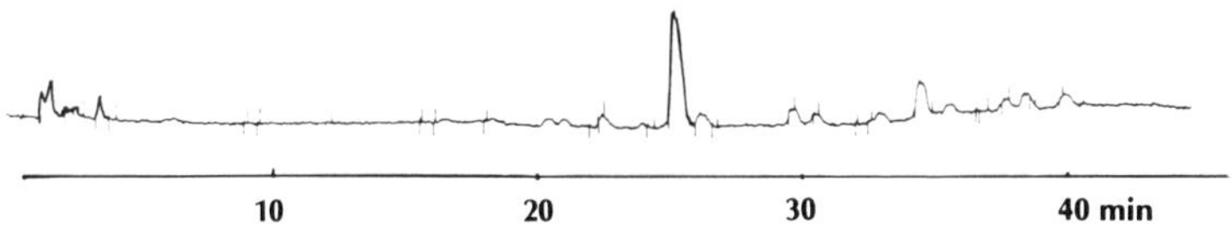

FIGURE 6. Fly ash wash extract (ex 280, em 389).

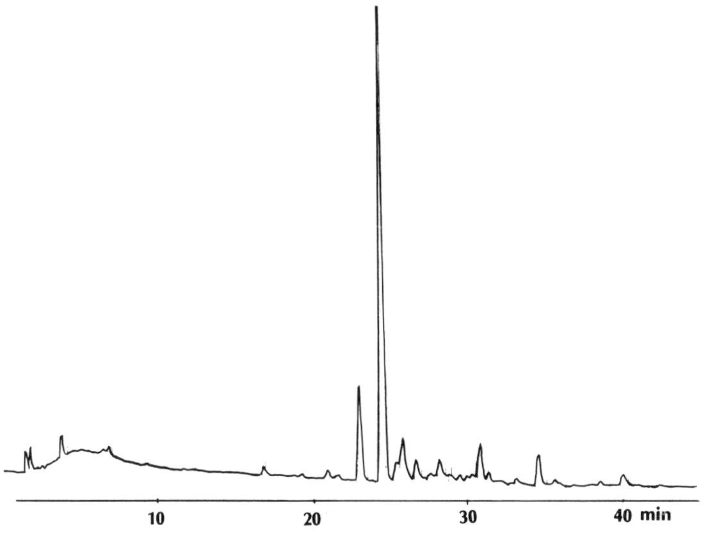

FIGURE 7. Fly ash wash extract (ex 254, em 350).

CONCLUSIONS

A number of conclusions may be drawn from this preliminary analytical methods development study.

1. All 16 consent decree PAH can be separated in a single HPLC run using a HC-ODS column.
2. With fluorescent detection and proper wavelength selection all consent decree PAH may be quantitated at the ppt level.
3. There remains a need for improved cleanup techniques to lower sample background.
4. There is a need for unambiguous peak identification; perhaps involving stop-flow scanning detectors.
5. The above method should prove to be an excellent screening technique for monitoring PAH in aqueous effluent.

ACKNOWLEDGMENTS

The authors wish to thank Dr. Ken Ogan of Perkin-Elmer Corporation for his suggestions and assistance in the use of their HC-ODS column. This study has been sponsored by the Environmental Protection Agency, Physical and Chemical Methods Branch at Cincinnati, Ohio, under Contract No. 68-03-2624.

REFERENCES

1. Basu, D. K., Safena, J. (1978): *Environ. Sci. Technol.* 12:791-794.
2. Bjorseth, A. (1978): *Carcinogenesis: A Comprehensive Survey, Vol. 3,* P. W. Jones and R. I. Freudenthal, Eds., Raven Press, New York.
3. Borneff, J. (1977): Fate of Pollutants in Air and Water Environments. In: *Advances in Environmental Sciences and Technology, Vol. 8, Part 2,* I. H. Suffett, Ed., pp. 393-408, Wiley-Interscience, New York.
4. Dang, M., Locke, D. C., and Ferrand, E. (1976): *Anal. Chem.* 48: 368-372.
5. Dark, W. A., McFadden, W. H., and Bradford, D. L. (1977): *J. Chromatog. Sci.* 15:454-460.
6. Das, B. S., and Thomas, G. H. (1978): *Anal. Chem.* 50(7):967-973.
7. Fox, M. A., and Staley, S. W. (1976): *Anal. Chem.* 48(7):992-998.
8. Grant, D. W., and Meiris, R. B. (1977): *J. Chromatog.* 142:339-351.
9. Grob, K. , and Grob, K., Jr. (1978): *J. Chromatog.* 151:311-320.
10. Harrison, R. M., Perry, R., and Wellings, R. A. (1976): *Environ. Sci. Technol.* 10(12):1151-1156.
11. Jungclaus, G. A., Lopez-Avila, V., and Hites, R. A. (1978): *Environ. Sci. Technol.* 12(1):88-96.
12. Junk, G. A., *et al.* (1976): *Z. Anal. Chem.* 282:331-337.

13. Krstubuvic, A. M., Rosie, D. M., and Brown, P. R. (1976): *Anal. Chem.* 48:1383-1385.
14. Smillie, R. D., Wang, D. T., and Meresz, D. (1978): *J. Envrion. Sci. Health* A1 3(1):4.
15. Sorrell, R. K., Dressman, R. C., and McFadden, E. F. (1977): Paper presented at Water Quality Technology Conference, Kansas City, MO.
16. Strup, P. E., Wilkinson, J. E., and Jones, P. W. (1978): *Carcinogenesis: A Comprehensive Survey, Vol. 3.*, P. W. Jones and R. I. Freudenthal, Eds., Raven Press, New York.
17. Supelco, Inc. (1977): Analysis of Polycyclic aromatic hydrocarbons. Bulletin 773.
18. Van Rossum, P., and Webb, R. G. (1978): *J. Chromatog.* 150:381-392.
19. Waters Associates (1978): Polynuclear aromatic hydrocarbons. Bulletin H90, April.

Polynuclear Aromatic Hydrocarbons, edited by P. W. Jones and P. Leber. Ann Arbor Science Publishers, Inc. Ann Arbor, MI ©1979.

METHODS FOR CHARACTERIZATION OF COMPLEX MIXTURES OF POLYNUCLEAR AROMATIC HYDROCARBONS

M. E. Snook, R. F. Severson, H. C. Higman, R. F. Arrendale and O. T. Chortyk

Tobacco and Health Research Laboratory
Science and Education Administration/Federal Research
U.S. Department of Agriculture
Athens, Georgia 30604

Cigarette smoke condensate (CSC) is one of the most complex mixtures of compounds in the environment. Almost 1,000 different polynuclear aromatic hydrocarbons (PAH) have been identified in CSC (23-25). The unambiguous identification of individual PAH in such a mixture necessitates separation by a combination of advanced chromatographic techniques, including high-pressure liquid chromatography (HPLC) and characterization by gas chromatography/mass spectrometry (GC/MS). Methodology was developed for the isolation and identification of the PAH of CSC that allows not only the determination of ring type but also gives a good estimation of the number of isomers of alkylated PAH present. The methodology involves silicic acid, gas, gel filtration (GF) chromatography and HPLC. A key step in the procedure is the use of gel chromatography to separate the PAH into two fractions: parent PAH and multialkylated PAH. Also, GF chromatography removes interfering substances and yields a relatively pure PAH isolate. The method was successfully applied to the analysis of PAH in other important environmental samples, such as crude oil and pitch. Familiarity with isomeric PAH is required for the determination of structures and the developed tables of PAH molecular weights used in MS data interpretation are also presented.

ISOLATION OF PAH FROM COMPLEX MIXTURES

Polynuclear aromatic hydrocarbon sources in the environment with few exceptions are also sources of heteroatom analogues of PAH. These oxygen and nitrogen PAH (0- and N-PAH) also contribute to air pollution (4,15) and are found along with sulfur PAH (S-PAH) such as dibenzthiophene, in petroleum (17) and coal-derived fuels (21). They often cochromatograph with many of the PAH and thus interfere in PAH analyses. Therefore, in choosing a method for the analysis of PAH from an environmental sample, it is important to consider the source of the PAH and the possible presence of their heteroatom analogues.

The PAH fraction generally obtained by the Rosen method (19) or by other solvent-partitioning methods also contains heteroatom aromatics. Furthermore, many solvent-partitioning systems fail to quantitatively recover PAH. Howard and Haenni (8) showed that the recovery of PAH ranged between 62 and 96% when they were partitioned between dimethylsulfoxide (DMSO) and cyclohexane. It is probable that the use of nitromethane in PAH extractions would also result in similar losses. Consequently, organic solvent-partition separations of PAH must be avoided.

We have found that silicic acid chromatography effectively separated nitrogen PAH from PAH while gel chromatography separated the alkylated PAH from the parent PAH. The isolation procedure is shown in Figure 1. Although the method was developed for CSC, it has been successfully applied to other environmental samples (see below). Experimental details have been reported (20). Briefly, CSC (dissolved in a mixture of benzene/methanol/ether) was extracted with water and the resulting "organic solubles" were chromatographed on silicic acid and eluted with petroleum ether (PE),

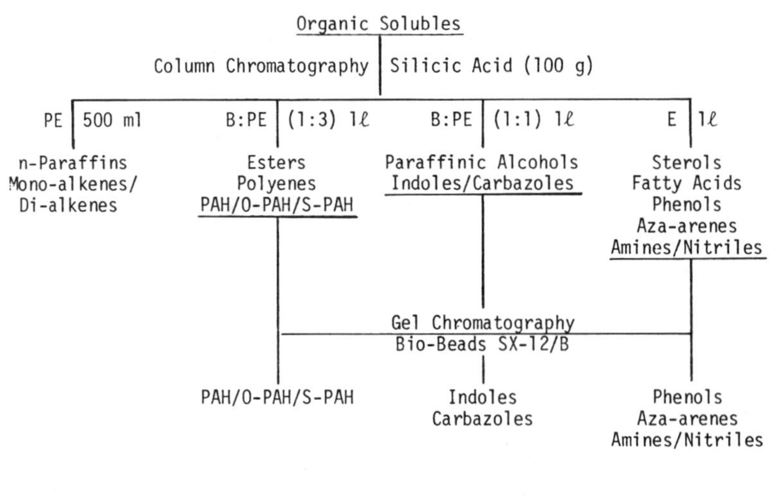

FIGURE 1. Fractionation scheme for CSC (PE = petroleum ether; B = benzene; E = ether).

25% benzene (B)/PE, 50% B/PE and ethyl ether. The PAH were found in the 25% B/PE fraction while the nitrogen PAH eluted in the two succeeding fractions. Standards indicated that the PAH fraction would also contain O-PAH and S-PAH. Thus, we have also found O-PAH in CSC (23) and S-PAH in crude oil (see below). These analogues, however, did not interfere with the PAH analysis.

PAH ISOLATIONS BY GEL CHROMATOGRAPHY

Much research has been reported on the gel chromatography of PAH (1,5-7,9-14,18,22-29,32). Most of this work has involved Sephadex LH-20 (a partially alkylated cross-linked dextran) (10,12,18,27-29,32) and the Bio-Beads SX-series of gels (a styrene-divinylbenzene copolymer) (1,9-12,22,26). In certain solvents, both gels have been shown to separate aromatic compounds from aliphatic compounds by an adsorption-type mechanism. Further, when adsorption occurs individual aromatic compounds are eluted from the gels in order of increasing ring number. This adsorption phenomena has been extensively investigated for LH-20 and Bio-Beads. Apparently, the order of PAH elution from these gels is solvent-dependent. In tetrahydrofuran (11,28) and dimethylamine (10), both gels cause the larger PAH to be eluted first, as expected for gel permeation. LH-20 in dimethylformamide (28) and chloroform (32) exhibits a similar effect. Adsorption of PAH occurs when LH-20 is eluted with methanol (12,27-29), isopropanol (12,18,27-29,32), acetone (12) and acetonitrile (12,28), and when Bio-Beads are eluted with acetone (12,26), methylethylketone (12), methylene chloride (1) and benzene (22, 24,25). No adsorption or permeation of PAH is observed with Bio-Beads in methanol, isopropanol (12,18,27-29) or acetonitrile (30). The adsorption effect of PAH appears to be due to a π-π-type interaction between the PAH and the gels (LH-20 and Bio-Beads) and has been correlated with the resonance energy of the PAH (LH-20 system) (27,29).

In addition to separating PAH according to the number of aromatic rings, both gels exhibit a "methyl effect" in that alkylated PAH are eluted earlier than their parent hydrocarbons: the order of elution is trimethyl, dimethyl, methyl and parent PAH (22). The data in Tables I and II show these two gel properties well for Bio-Beads SX-12 in benzene and for Sephadex LH-20 in the commonly used solvent isopropanol. The data show that PAH are adsorbed more strongly on LH-20 in isopropanol than on Bio-Beads in benzene. However, Bio-Beads in benzene are better for the purification of PAH fractions because:

1. they are more efficient in separating the multialkylated PAH from their parents;

TABLE I. Gel Elution of Standard PAH from Bio-Beads SX-12 in Benzene (four-column system)

PAH	Percent Distribution of PAH														
	Fraction Number[a]														
	35	36	37	38	39	40	41	42	43	44	45	46	47	48	49
2,3,6-Trimethylnaphthalene	3.1	21.9	35.9	31.7	6.4	0.9									
2,6-Dimethylnaphthalene		11.7	45.3	30.0	12.8	0.3									
1,4-Dimethylnaphthalene		0.1	10.1	43.3	29.1	13.3	3.9	0.3							
2-Ethylnaphthalene		26.1	50.0	17.8	4.3	0.1									
1-Methylnaphthalene				23.3	32.7	27.9	15.5	0.6							
Naphthalene					2.3	14.2	35.6	44.5	3.4						
Acenaphthene						4.8	65.3	27.1	2.2	0.7					
Acenaphthylene						0.1	24.2	51.2	14.5	9.9					
2-Methylfluorene			7.2	53.8	31.3	6.8	0.8								
Fluorene				11.8	35.3	35.3	15.6	2.1							
3,6-Dimethylphenanthrene			0.2	4.4	19.5	42.9	21.8	8.3	2.3	0.4					
2-Methylphenanthrene				1.2	17.3	21.9	29.7	17.1	5.3	0.5					
Phenanthrene						0.4	5.7	21.5	35.6	27.0	8.3	1.5			
Fluoranthene						0.8	11.0	40.5	36.5	36.3	19.1	1.3			
1-Methylpyrene						1.6	8.3	23.1	31.5	19.9	9.9	5.0	0.7		
Pyrene							0.5	6.1	20.3	27.4	26.4	14.5	4.9		
3-Methylchrysene						11.5	24.9	31.3	22.9	8.9	0.8				
Chrysene						1.9	12.0	29.8	34.1	16.6	5.1				
Benzo[a]pyrene									1.2	10.7	26.2	35.7	20.2	5.9	
Dibenzo[a,h]pyrene										4.0	10.5	30.0	38.5	14.6	2.4

[a]Number of 8-ml fractions from point of injection; column 109 cm x 1.25 cm (4 in series); flowrate 120 ml/hr.

TABLE II. Gel Elution of Standard PAH from Sephadex LH-20 in Isopropanol (one-column system)

PAH	Percent Distribution of PAH: Fraction Number[a] 14	15	16	17	18	19	20	21	22	23	24	25	26	27	28	29	30	31	32
2,3,6-Trimethylnaphthalene	1.9	24.0	48.9	19.4	4.7	0.8	0.4												
2,6-Dimethylnaphthalene	0.3	6.7	38.0	40.4	13.2	0.9	0.2												
2-Methylnaphthalene	0.2	4.7	25.8	45.0	17.9	5.5	0.8	0.1											
Naphthalene		0.3	17.6	26.0	40.5	10.6	4.3	0.6	0.1										
3,6-Dimethylphenanthrene					2.1	6.3	21.0	29.4	31.5	7.1	1.9	0.4	0.1						
3-Methylphenanthrene					1.1	1.2	7.1	22.2	46.9	15.7	4.7	0.9	0.2	0.1					
Phenanthrene						0.1	1.6	8.1	38.0	29.9	15.9	5.1	1.1	0.2					
Fluoranthene								0.3	1.4	3.6	11.5	24.0	27.3	18.4	8.7	3.2	1.2	0.2	0.1
Pyrene								0.2	1.3	2.3	8.2	18.5	25.3	22.4	13.1	5.9	1.8	0.6	0.2

[a]Number of 8-ml fractions from point of injection; column 109 cm x 1.25 cm; flowrate 120 ml/hr.

2. they result in less zone spreading of the higher PAH;
3. evaporation of the more volatile solvent, benzene, reduces PAH losses; and
4. PAH are more soluble and appear to be more stable in benzene.

The adsorption characteristic of Bio-Beads in benzene was utilized to great advantage in purifying the PAH fraction obtained from silicic acid chromatography. Figure 2 shows the gel elution of the 25% B/PE silicic acid fraction (F-BPE) from four 109 cm x 1.25 cm columns connected in series (120 ml/hr flowrate). Notice that the PAH cut, which includes gel fractions 36 to 47, represents less than one percent of the initial weight of F-BPE, a remarkable purification. The purification obtained by gel chromatography is illustrated in Figure 3. In the crude GC profile of the crude silicic acid fraction (F-BPE) of CSC, the PAH are barely detected in the envelope of other components. However, after refinement by gel chromatography, a clean PAH-GC profile is obtained.

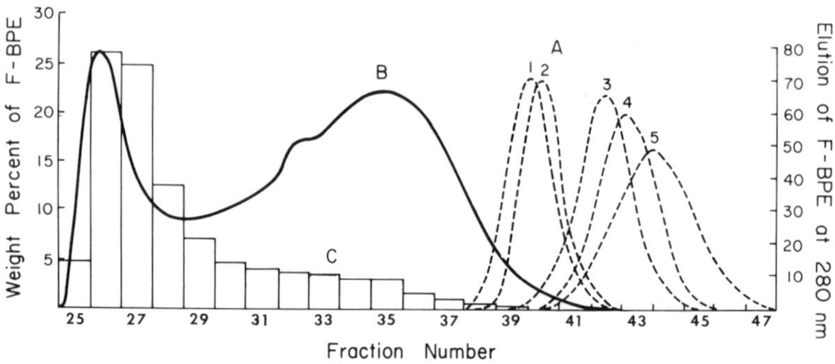

FIGURE 2. Gel elution characteristics of fraction F-BPE on Bio-Beads SX-12 in benzene. A = elution curves of PAH standards; 1 = 2,6-dimethylnaphthalene, 2 = naphthalene, 3 = fluoranthene, 4 = phenanthrene, 5 = pyrene; B = UV-absorbance curve; C = weight percent distribution of 8-ml fractions from point of injection.

Some environmental samples are sufficiently clean and do not require GF chromatography, especially if they are low in alkylated PAH and intefering material. Figure 4 is a gas chromatogram of PAH isolated from a coal tar pitch sample; the compounds are identified in Table III. The spectrum of PAH is relatively simple, consisting mainly of parent PAH with very low levels of alkylated derivatives.

In environmental samples containing large amounts of alkylated PAH, GF chromatography is required to obtain an analyzable GC chromatogram, even those obtained by use of capillary columns. As noted before (Tables I and II),

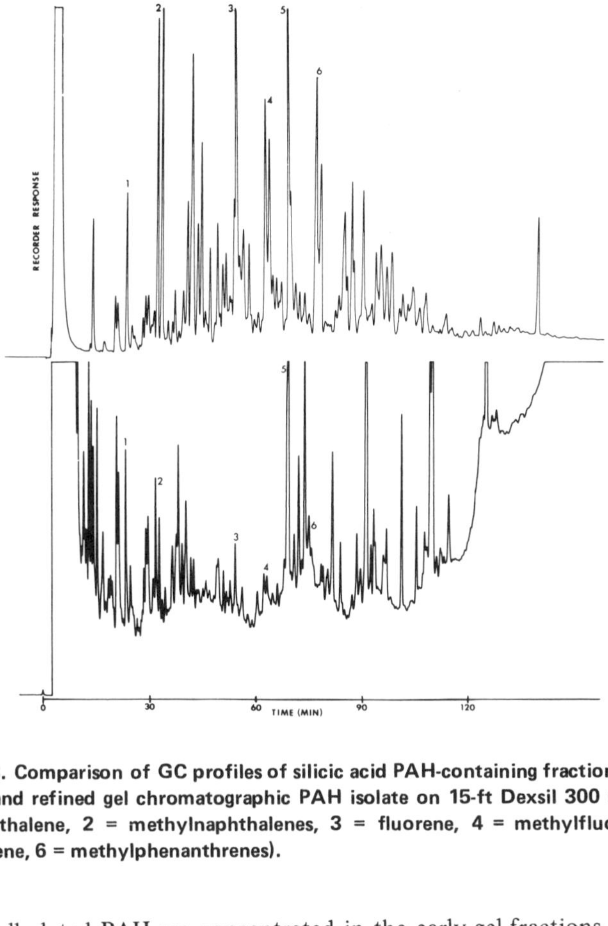

FIGURE 3. Comparison of GC profiles of silicic acid PAH-containing fraction from CSC (bottom) and refined gel chromatographic PAH isolate on 15-ft Dexsil 300 GC column (1 = naphthalene, 2 = methylnaphthalenes, 3 = fluorene, 4 = methylfluorenes, 5 = phenanthrene, 6 = methylphenanthrenes).

the multialkylated PAH are concentrated in the early gel fractions. For complex PAH mixtures, the gel fractions can be combined in such a way as to concentrate the multialkylated PAH into one fraction (GF-B) and the parent and monomethyl PAH in the following fraction (GF-C). The Bio-Beads gel columns were standardized with trimethylnaphthalene, the earliest-eluting commercially available PAH, and with fluoranthene, the earliest-eluting major parent PAH (20). Baseline separation of these two PAH was achieved with a four-column gel system. The elution of trimethylnaphthalene defined the multialkylated PAH fraction (GF-B) while the beginning of the elution of fluoranthene to gel fraction 55 defined the major parent PAH/methyl-PAH fraction (GF-C). Removal of the multialkylated PAH facilitated the identification and quantitation of the major parent PAH. The merits of separating the multialkylated PAH from the parent/methyl-PAH are illustrated in

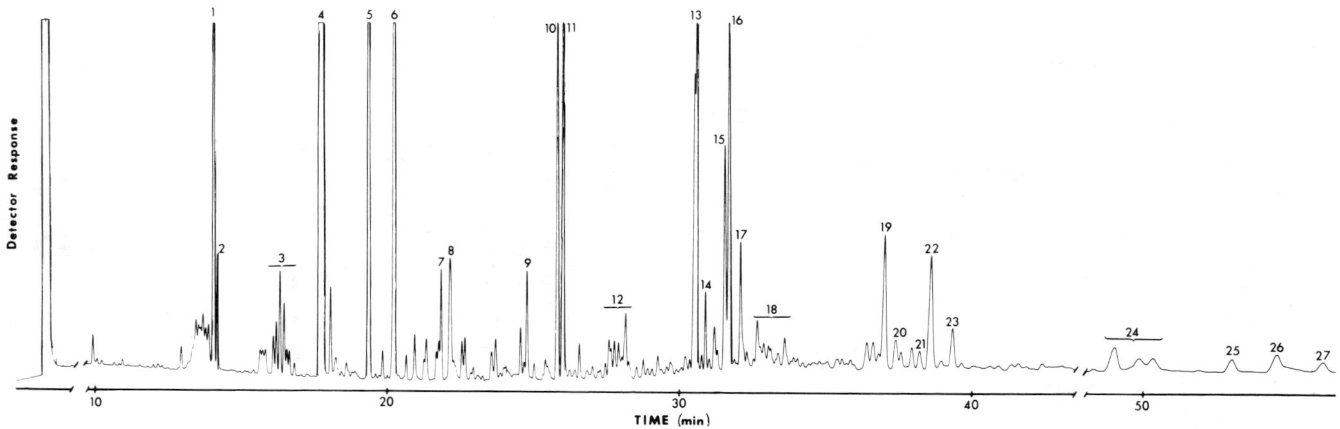

FIGURE 4. Gas chromatogram of PAH isolated from coal tar pitch on a 30-m SE-54 glass capillary column (80-250°C at 5°/min). Peak identities in Table III.

TABLE III. Peak Identities for PAH in Coal Tar Pitch (Figure 4)

1	Phenanthrene
2	Anthracene
3	Methylphenanthrenes/Anthracenes
4	2-Phenylnaphthalene (Internal Standard)
5	Fluoranthene
6	Pyrene
7	Benzofluorenes/Methylfluoranthenes
8	Methylpyrenes
9	Benzo[g,h,i] fluoranthene
10	Benz[a] anthracene
11	Chrysene
12	Methylbenz[a] anthracenes/Chrysenes
13	Benzo[b,j, and k] fluoranthenes
14	Benzo[a] fluoranthene
15	Benzo[e] pyrene
16	Benzo[a] pyrene
17	Perylene
18	Methylbenzo[e and a] pyrenes
19	Indenopyrene
20	Dibenz[a,c and a,h] anthracene
21	Picene
22	Benzo[g,h,i] perylene
23	Anthanthrene
24	Dibenzofluoranthenes
25	Coronene/Dibenzopyrene
26-7	Dibenzopyrenes

Figure 5 for the analysis of PAH of a crude oil sample, Venezuelan asphalt base, on a 30 m SE-54 capillary column. Obviously, an analysis of the total PAH isolate, without prior separation into the two gel cuts, would have given very crude quantitative results for the parent PAH. This sample shows that our procedure also isolated sulfur-containing aromatics. Even though present at high levels in the sample, the parent and alkyl-dibenzothiophenes and the 9-thia-benzofluorenes (Table IV) did not interfere with the analysis of the major parent PAH or their monomethyl derivatives.

The broad spectrum of PAH found in CSC was ideally suited to separation by gel chromatography on Bio-Beads SX-12 in benzene (23-25). The gas chromatograms of three individual gel fractions, are given in Figures 6, 7 and 8 and illustrate the two gel characteristics of early elution of alkylated PAH and late elution of multi-ring PAH.

For cigarette smoke condensate, ^{14}C-naphthalene, anthracene and benzo-[a] pyrene were quantitatively recovered by the above isolation scheme. When the procedure was applied to coal tar pitch and crude oil samples, similar

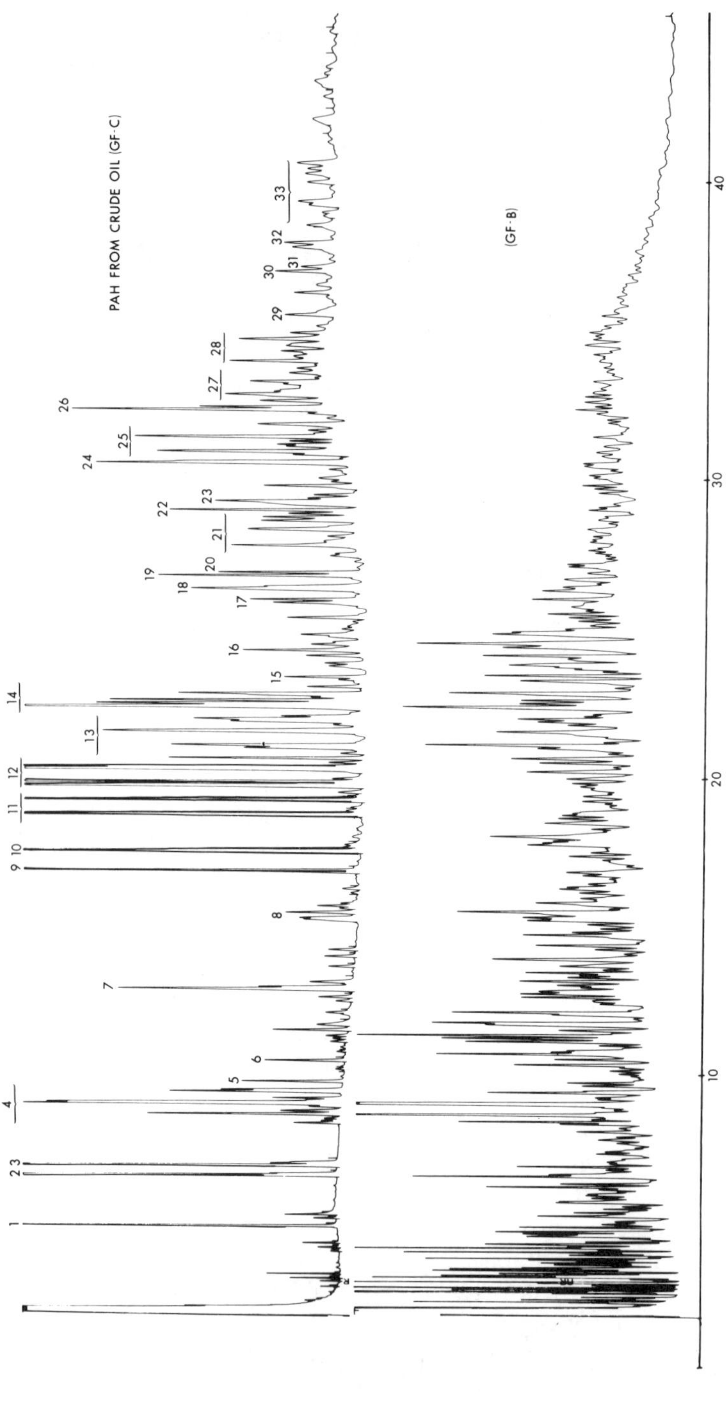

FIGURE 5. PAH isolated from crude oil sample (Venezuelan asphalt base) on 30-m SE-54 glass capillary column (80-250°C at 5°/min). GF-B = multialkylated PAH fraction, GF-C = major parent/methyl parent PAH fraction. Peak identities in Table IV.

TABLE IV. Peak Identifies for PAH in Crude Oil (Figure 5)

1	Naphthalene
2	2-Methylnaphthalene
3	1-Methylnaphthalene
4	Dimethylnaphthalenes
5	Acenaphthylene
6	Acenaphthene
7	Fluorene
8	Methylfluorenes
9	Dibenzothiophene
10	Phenanthrene
11	Methylphenanthrenes/Anthracenes
12	Dimethyldibenzothiophenes
13	Dimethylphenanthrenes/Anthracenes
14	Dimethylphenanthrenes
15	Fluoranthene
16	Pyrene
17	Methylfluoranthenes
18	2-Methylpyrene
19	4-Methylpyrene
20	1-Methylpyrene
21	Dimethylfluoranthenes/Pyrenes
22	9-Thia-1,2-Benzofluorene
23	9-Thia-3,4-Benzofluorene
24	Chrysene/Triphenylene
25	Methyl-9-thia-benzofluorenes
26	Methylchrysenes/Triphenylenes
27	Dimethyl-9-thia-benzofluorenes
28	Dimethylchrysenes/Triphenylenes
29	Benzofluoranthenes
30	Benzo[e]pyrene
31	Benzo[a]pyrene
32	Perylene
33	Methylbenzopyrenes/Perylenes

recoveries of ^{14}C-B[a]P were obtained. However, for true quantitation, the recovery of the environmental aromatics from the trapping system must also be determined. Often standards are added to the trapping medium and the recoveries are used to quantitate the PAH in the environmental sample. We found that (Table V) 500 ml of methylene chloride quantitatively recovered B[a]P from clean glass-fiber filters of the type used to trap air samples. However, recovery of the standards must also be determined in the presence of the trapped environmental sample. When a sample of ^{14}C-B[a]P was placed on a glass-fiber filter previously used for the collection of a polluted air sample, the recovery of the radioactive B[a]P was less than 65%. Thus, when determining extraction recoveries, standards should always be spiked into the environmental sample.

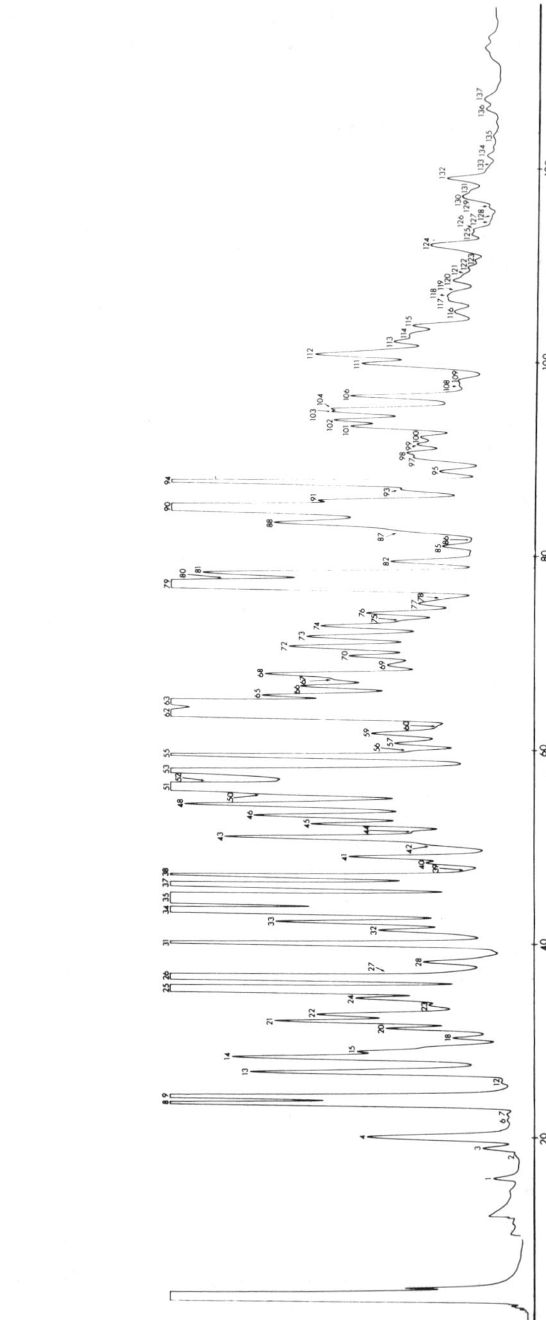

FIGURE 6. Gas chromatogram of gel fraction 40 from CSC on 15-ft Dexsil 300 GC column (100-325°C at 2°/min). (8,9 = methylindenes; 25,26 = methylnaphthalenes; 31 = biphenyl; 34-38 = dimethylnaphthalenes; 43 - dibenzofuran; 51 - fluorene; 62,63 = methylfluorenes; 70-78 = dimethylfluorenes; 79-81 = methylphenanthrenes/anthracenes; 88-94 = dimethylphenanthrenes/anthracenes; 101-106 = trimethylphenanthrenes/anthracenes; 111-115 = dimethylfluoranthenes/pyrenes; 124-126 = methylchrysenes; 130-132 = dimethylchrysenes; see reference 3 for additional peak identities).

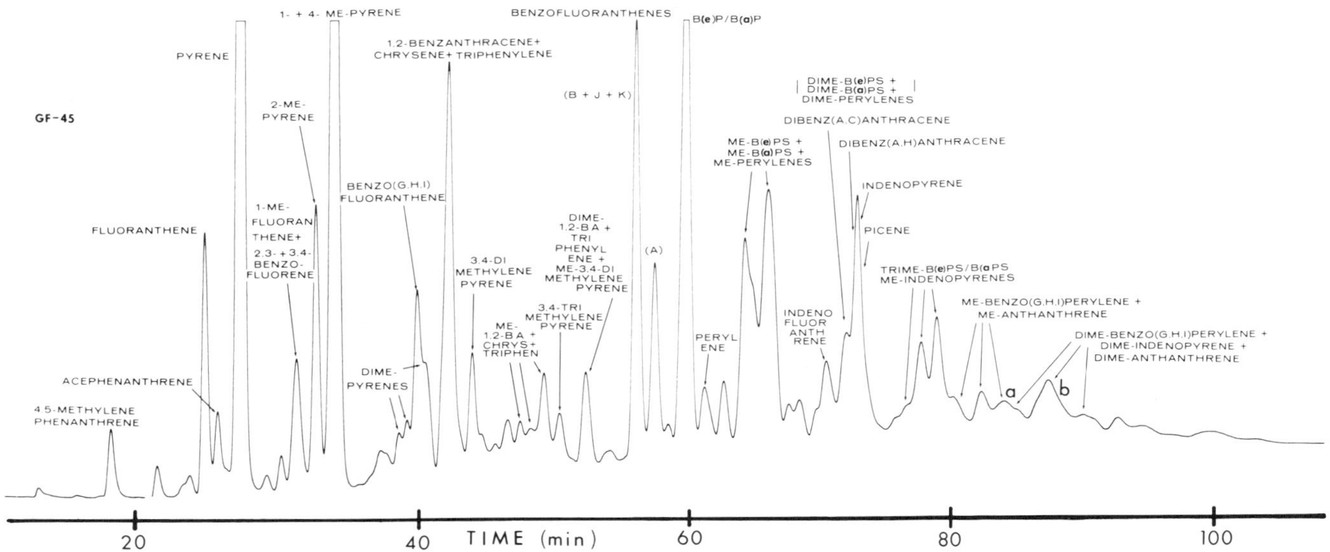

FIGURE 7. Gas chromatogram of gel fraction 45 on 15-ft Dexsil 300 GC column (200-325°C at 2°/min). Collected cuts for "a" and "b" peaks submitted to HPLC (see text and Figures 13 and 14, respectively).

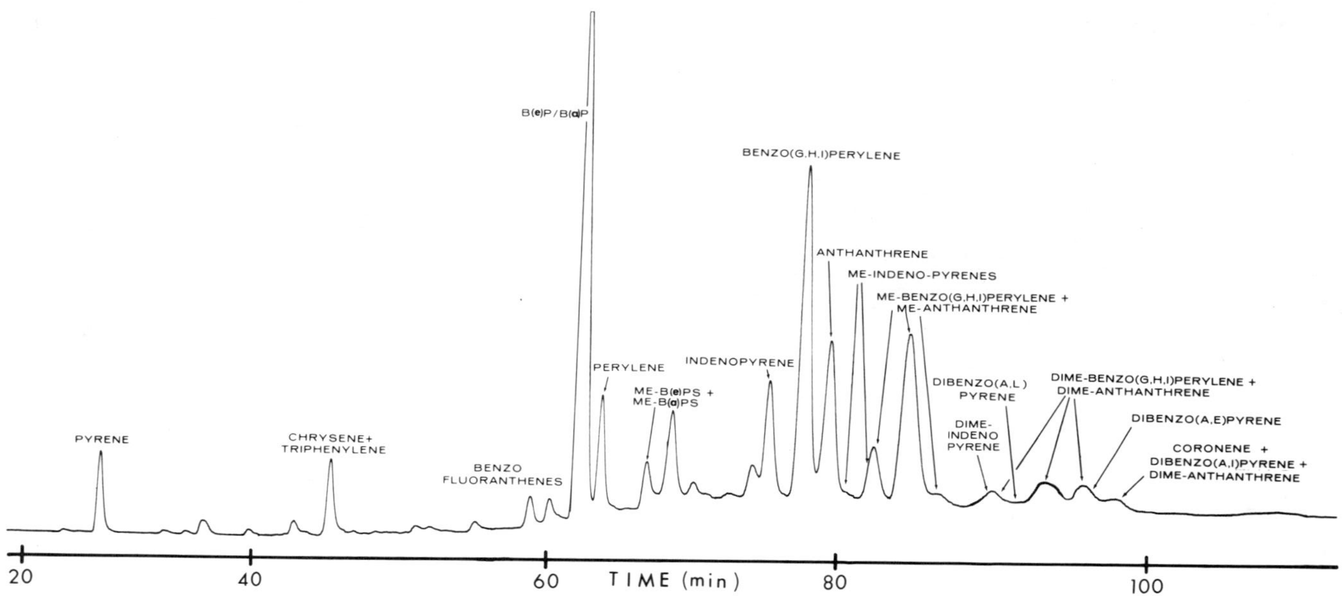

FIGURE 8. Gas chromatogram of gel fraction 49 on 15-ft Dexsil 300 GC column (200-325°C at 2°/min).

TABLE V. Percent Recovery of ^{14}C-B[a]P from Glass-Fiber Filter[a]

CH_2Cl_2 (ml)	Filter Blank	Filter Plus Air Sample
100	52.2	
200	84.1	
300	94.6	58.0
400	97.8	
500	99.0	63.2

[a]Based on ^{14}C-B[a]P placed onto blank filter and filter containing trapped air sample.

IDENTIFICATION OF PAH IN COMPLEX MIXTURES

Very complex PAH mixtures, like those isolated from CSC or petroleum, make the identifications of individual components difficult. The advent of capillary GC/MS has proved that more than ever standard PAH are urgently needed. Because of the almost exclusive use of GC/MS data as identification criteria, researchers must be familiar with all possible isomeric PAH when assigning structures. As an aid to other researchers, Table VI lists the molecular ions of PAH and associated compounds. Heteroatom polycyclics frequently detected in PAH samples are listed in Table VII.

Ultraviolet (UV) spectra in conjunction with MS will continue to be the definitive criteria in PAH identifications. Good references for UV spectra of PAH can be found in *UV Atlas of Organic Compounds* (31), Clar's *Aromatische Kohlenwassertoffe, Polycyclische Systeme* (2) and *Polycyclic Hydrocarbons* (3) and Friedel's *Ultraviolet Spectra of Organic Compounds* (6). We have tried to develop methods of PAH characterization that allow UV spectra to be obtained for isolated PAH in order that more correct and unambiguous structures may be assigned. Reverse-phase HPLC of preparative GC cuts was found to be an excellent method for separating isomeric PAH and allowing UV spectra to be obtained.

The key to the identification of over 900 PAH in CSC was the analysis of each individual gel fraction from GF-36 (Figure 2), the beginning of the elution of the multialkylated PAH to GF-55, the end of the elution of coronene. In this way, changes in GC peak composition could easily be followed and compounds otherwise hidden under large GC peaks were often times resolved. Each of the gel fractions (36-55) was subjected to preparative GC on 1/8-in. x 15-ft stainless-steel Dexsil 300 GC columns, with subsequent UV analysis of the collected cuts corresponding to all GC peaks. In addition, each gel fraction was submitted to GC/MS. Capillary GC showed that virtually every GC

TABLE VI. Molecular Ions Associated with PAH

m/e	Compound(s)[a]
78	Benzene
92	Toluene
102	Phenylacetylene
104	Styrene
106	Xylenes, Ethylbenzenes
116	Indene
118	Indan, Methylstyrenes
120	Trimethylbenzene, Isopropylbenzenes, Ethyltoluenes
128	Naphthalene, Azulene
130	Dihydronaphthalenes, Methylindenes
132	Dimethylstyrenes, Tetrahydronaphthalene (Tetralin), Methylindans
134	Tetramethylbenzenes (*i.e.,* Durene, Methylcumenes, etc.)
142	Methylnaphthalenes
144	Dimethylindenes
146	Trimethylstyrenes
148	Pentamethylbenzenes
152	Acenaphthylene, Diphenylene, Indacene
154	Acenaphthene, Biphenyl, Vinylnaphthalenes
156	Dimethylnaphthalenes, Ethylnaphthalenes
158	Trimethylindenes
160	Tetramethylstyrenes
166	Fluorene, Methylacenaphthylenes/Diphenylenes,[a] Benzindenes, Perinaphthene
168	Methylacenaphthenes/Biphenyls/Vinylnaphthalenes, Allylnaphthalenes, Diphenylmethane
170	Trimethylnaphthalenes, Isopropylnaphthalenes, etc.
172	Tetramethylindenes
174	Pentamethylstyrenes
178	Phenanthrene, Anthracene, Diphenylacetylene
180	Methylfluorenes/Benzindenes, Dimethylacenaphthylenes/Diphenylenes/Vinylbiphenyls, Stilbenes, Dihydrophenanthrene/Anthracene
182	Dimethylbiphenyls (bitolyls), Ethylbiphenyls, Dibenzyl, Dimethylacenaphthenes, Dimethylvinylnaphthalenes
184	Tetramethylnaphthalenes
186	Pentamethylindenes
190	4,5-Methylenephenanthrene
192	Methylphenanthrenes/Anthracenes, Phenylindenes
194	Dimethylfluorenes/Benzindenes, Trimethylacenaphthylenes/Diphenylenes, Phenylindan, Methyl-m/e 180
196	Trimethylacenaphthenes, Methylbibenzyl, Trimethyl-m/e 154
198	Pentamethylnaphthalenes
200	Hexamethylindenes
202	Pyrene, Fluoranthene, Acephenanthrylene, Aceanthrylene
204	Phenylnaphthalenes, Vinylphenanthrenes/Anthracenes, Acephenanthrene, Aceanthrene, Methyl-4,5-methylenephenanthrenes
206	Dimethylphenanthrenes/Anthracenes, Methylphenylindenes
208	Trimethylfluorenes/Benzindenes, Tetramethylacenaphthylenes/Diphenylenes, Dimethyl-m/e 180

TABLE VI, continued

m/e	Compound(s)[a]
210	Tetramethylacenapthenes, Tetramethyl-m/e 154, Dimethylbibenzyl
212	Hexamethylnaphthalenes
216	Methylpyrenes/Fluoranthenes/+ other m/e 202, Benzofluorenes, Benzanthrene
218	Cyclopentenophenanthrenes/Anthracenes (Trimethylenephenanthrenes/Anthracenes), Benzylnaphthalenes, Methylphenylnaphthalenes/+ other m/e 204, Dimethyl-m/e 190
220	Trimethylphenanthrenes/Anthracenes, Dimethylphenylindenes
222	Tetramethylfluorenes/Benzindenes, Pentamethylacenaphthylenes/Diphenylenes, Trimethyl-m/e 180
224	Pentamethyl-m/e 154
226	Benzo[g,h,i] fluoranthene, Cyclopenta[c,d] pyrene
228	Benz[a] anthracene, Chrysene, Triphenylene, Benzo[c] phenanthrene, Tetracene (Naphthacene), 3,4-Dimethylenepyrene
230	Dimethylpyrenes/Fluoranthenes/+ other m/e 202, Methylbenzofluorenes, Terophenyls, Bi-idenyl
232	Methylcyclopentenophenanthrenes/Anthracenes, Methylbenzylnaphthalenes, Dimethylphenylnaphthalenes/+ other m/e 204, Trimethyl-m/e 190
234	Tetramethylphenanthrenes/Anthracenes, Trimethylphenylindenes
236	Pentamethylfluorenes/Benzindenes, Tetramethyl-m/e 180
240	Methylbenzo[g,h,i] fluoranthenes, Methylenechrysene/+ other m/e 228, Naphthanthrene
242	Methylbenz[a] anthracenes/Chrysenes/Triphenylenes/+ other m/e 228, 3,4-Trimethylenepyrene, Phenylfluorenes
244	Trimethylpyrenes/Fluoranthenes/+ other m/e 202, Dimethylbenzofluorenes, Methylterphenyls, Triphenylmethane
246	Dimethylcyclopentenophenanthrenes/Anthracenes, Dimethylbenzylnaphthalenes, Trimethylphenylnaphthalenes/+ other m/e 204, Tetramethyl-m/e 190
248	Pentamethylphenanthrenes/Anthracenes, Tetramethylphenylindenes
250	Dibenzo[g,h,i:m,n,o] fluoranthene, Tetramethyl-m/e 180
252	Benzo[a and e] pyrenes, Benzo[a,b,j and k] fluoranthenes, Perylene, Dibenzo- or Naphthoacenaphthylenes/Diphenylenes
254	Phenylphenanthrenes/Anthracenes, Binaphthyls, Dibenzo- or Naphthoacenaphthenes, Cholanthrene, Dimethylenechrysenes/+ other m/e 228, Dimethylbenzo-[g,h,i] fluoranthenes
256	Dimethylbenz[a] anthracenes/Chrysenes/Triphenylenes/+ other m/e 228, Methylphenylfluorenes, Methyl-3,4-trimethylenepyrenes
258	Tetramethylpyrenes/Fluoranthenes/+ other m/e 202, Trimethylbenzofluorenes, Dimethylterphenyls
260	Trimethylcyclopentenophenanthrenes/Anthracenes, Trimethylbenzylnaphthalenes, Tetramethylphenylnaphthalenes/+ other m/e 204, Pentamethyl-m/e 190
266	Methylbenzopyrenes/Benzofluoranthenes/Perylenes/+ other m/e 252, Dibenzo- or Naphthofluorenes, Coeranthrene
268	Methylcholanthrenes/+ other m/e 254, Trimethylbenzo[g,h,i] fluoranthenes, Cyclopenteno- (Trimethylene) m/e 228
270	Trimethylbenz[a] anthracenes/Chrysenes/Triphenylenes/+ other m/e 228, Dimethylphenylfluorenes, Dimethyl-3,4-trimethylenepyrene

TABLE VI, continued

m/e	Compound(s)[a]
272	Pentamethylpyrenes/Fluoranthenes/+ other m/e 202, Tetramethylbenzofluorenes, Trimethylterphenyls
274	Tetramethylcyclopenetenophenanthrenes/Anthracenes, Tetramethylbenzylnaphthalenes, Pentamethylphenylnaphthalenes/+ other m/e 204
276	Benzo[g,h,i] perylene, Anthanthrene, Indeno[1,2,3-c,d] fluoranthene, Indeno-[1,2,3-c,d] pyrene (*o*-phenylenepyrene), Dibenzo[b:g,h,i] fluoranthene/+ other isomers, Cyclopenta[c,d] perylene, Acenaphth[1,2-a] acenaphthylene
278	Dibenzo[a,c;a,g;a,h;a,i;b,h;b,g;c,g;] phenanthrenes, Dibenzo[a,c;a,h;a,i;a,j;b,i] anthracenes
280	Dimethylbenzopyrenes/Benzofluoranthenes/Perylenes/+ other m/e 252, Methyldibenzo or Naphthofluorenes/+ other m/e 266
282	Dimethylcholanthrenes/+ other m/e 254, Tetramethylbenzo[g,h,i] fluoranthenes
284	Tetramethylbenz[a] anthracenes/Chrysenes/Triphenylenes/+ other m/e 228, Trimethylphenylfluorenes, Trimethyl-3,4-trimethylenepyrenes
286	Pentamethylbenzofluorenes, Tetramethylterphenyls
288	Pentamethylcyclopentenophenanthranthenes/Anthracenes/+ other m/e 218
290	Methylbenzo[g,h,i] perylenes/Anthanthrenes/Idenofluoranthenes or Pyrenes/+ other m/e 276, 2,3-Benzylenepyrene
292	Methyldibenzophenanthrenes/Anthracenes
294	Trimethylbenzopyrenes/Benzofluoranthenes/Perylenes/+ other m/e 252, Dimethylbenzo- or Naphthofluorenes/+ other m/e 266
296	Trimethylcholanthrenes/+ other m/e 254, Pentamethylbenzo[g,h,i] fluoranthenes
298	Pentamethylbenz[a] anthracenes/+ other m/e 228, Tetramethylphenylfluorenes, Tetramethyl-3,4-trimethylenepyrene
300	Coronene, Pentamethylterphenyls
302	Dibenzo[a,e;a,h;a,i;a,l;e,l] pyrenes, Dibenzo[a,e;a,f;a,j;a,k;a,l;b,e;b,j;b,k;b,l;j,l]-fluoranthenes, Naphthopyrenes/Fluoranthenes, Benzo[a and b] perylenes, Zanthrene
304	Dimethylbenzo[g,h,i] perylenes/Anthanthrenes/Idenofluoranthenes or Pyrenes/ + other m/e 276, Methyl-2,3-benzylenepyrenes
306	Dimethyldibenzophenanthrenes/Anthracenes, Triphenylbenzenes (Quaterphenyls)
308	Tetramethylbenzopyrenes/Benzofluoranthenes/Perylenes/+ other m/e 252, Trimethyldibenzo- or Naphthofluorenes/+ other m/e 266
310	Tetramethylcholanthrenes/+ other m/e 254
312	Pentamethyl m/e 242
314	Methylcoronene (only one isomer)
316	Methyldibenzopyrenes/Dibenzofluoranthenes/+ other m/e 302
318	Trimethylbenzo[g,h,i] perylenes/Anthanthrenes/Indenofluoranthenes or Pyrenes/ + other m/e 276
320	Trimethyldibenzophenanthrenes/Anthracenes, Tetraphenylmethane
322	Pentamethyl m/e 252, Tetramethyl m/e 266
324	Pentamethyl m/e 254
328	Dimethylcoronene

[a]The / notation denotes that the same substituents are present on the basic ring structure indicated.

TABLE VII. Molecular Ions Associated with Heteroatom PAH Frequently Found in PAH Fractions

m/e	Compound(s)[a]
O-PAH	
118	Benzo[b] furan
120	Phthalan(*o*-Xylylene Oxide), 2,3-dihydrobenzo[b] furan
132	Methylbenzo[b] furans, Benzopyrans
134	Methyl-2,3-dihydrobenzo[b] furans
146	Dimethylbenzo[b] furans
148	Dimethyl-2,3-dihydrobenzo[b] furans
160	Trimethylbenzo[b] furans
168	Dibenzofuran, Naphthofurans
174	Tetramethylbenzo[b] furans
182	Methyldibenzofurans/Naphthofurans
188	Pentamethylbenzo[b] furans
196	Dimethyldibenzofurans/Naphthofurans
202	Hexamethylbenzo[b] furans
218	Benzo[b] naphthofurans
220	Trimethyldibenzofurans/Naphthofurans
232	Methylbenzo[b] naphthofurans
234	Tetramethyldibenzofurans/Naphthofurans
246	Dimethylbenzo[b] naphthofurans
260	Trimethylbenzo[b] naphthofurans
268	Dinaphthofurans
S-PAH	
134	Benzothiophene
148	Methylbenzothiophenes
162	Dimethylbenzothiophenes
176	Trimethylbenzothiophenes
184	Dibenzothiophene
186	Diphenylsulfide
190	Tetramethylbenzothiophenes
198	Methyldibenzothiophenes
208	4,5-Thia-phenanthrene (phenanthro[d,e,f] thiophene)
210	Phenylbenzothiophene
212	Dimethyldibenzothiophenes
222	Methyl-4,5-thia-phenanthrenes
226	Trimethyldibenzothiophenes
234	9-Thia-benzofluorenes
236	Dimethyl-4,5-thia-phenanthrenes
240	Tetramethyldibenzothiophenes
248	Methyl-9-thia-benzofluorenes
258	Benzo-4,5-thia-phenanthrenes
260	Phenyldibenzothiophenes
262	Dimethyl-9-thia-benzofluorenes
276	Trimethyl-9-thia-benzofluorenes
N-PAH	
105	Vinylpyridines

TABLE VII, continued

m/e	Compound(s)[a]
117	Indole
129	Quinoline, Isoquinoline
131	Methylindoles (3-methyl = skatole)
143	Methylquinolines/Isoquinolines
145	Dimethylindoles
157	Dimethylquinolines/Isoquinolines
159	Trimethylindoles
167	Carbazole, Azafluorenes, Benzindoles
171	Trimethylquinolines/Isoquinolines
173	Tetramethylindoles
179	Benzoquinolines, Isoquinolines
181	Methylcarbazoles
185	Tetramethylquinolines/Isoquinolines
193	4H-Benzo[d,e,f] carbazole, Methylbenzoquinolines/Isoquinolines
195	Dimethylcarbazoles
203	Azapyrenes/Fluoranthenes
207	Dimethylbenzoquinolines/Isoquinolines
209	Trimethylcarbazoles
217	Benzocarbazoles
221	Trimethylbenzoquinolines/Isoquinolines
229	Azabenz [a] anthracenes/Chrysenes/Triphenylenes/Tetracenes/Benzo [c] phenanthrenes
231	Methylbenzocarbazoles
245	Dimethylbenzocarbazoles
253	Azabenzopyrenes/Fluoranthenes/Perylenes
259	Trimethylbenzocarbazoles
279	Azadibenzphenanthrenes/Anthracenes

[a]The / notation denotes that the same substituents are present on the basic ring structure indicated.

peak from our packed columns contained multiple components. Since capillary GC/MS could not tell us which structural PAH isomers were present, we employed reverse-phase HPLC to separate the preparative GC cuts and collected the separated compounds for UV analysis. HPLC of the preparative GC cuts resolved these into multiple components (as well as capillary GC) and allowed determination of their exact structures. In some instances, probe MS data were obtained on some HPLC separated components for further proof. Examples of the excellent separations of PAH obtained by reverse-phase HPLC are given in Figures 9 through 15. The separations were performed on a 25-cm x 2.4-mm DuPont Zorbax ODS column with a linear gradient of 3%/min (except as noted) ranging from 65% CH_3OH/H_2O to 85% CH_3OH/H_2O. These illustrations show that it was also possible to determine the number of isomers of multialkylated PAH present. Thus, HPLC showed that the cut for GC peak no. 101 in the gas chromatogram of gel fraction 40 (Figure 9) contained at least three separate isomers of trimethylphenanthrene, a trimethylanthracene, a methylfluoranthene and a small amount of benzo-[a]fluorene. Figure 10 (HPLC of GC peak no. 129 of gel fraction 40) illustrates the merits of using HPLC in the identification of compounds. It is doubtful that even capillary GC would have separated all of the components in this one GC peak. Further, the first four HPLC-separated compounds all had the same molecular weight and could only be differentiated by UV data.

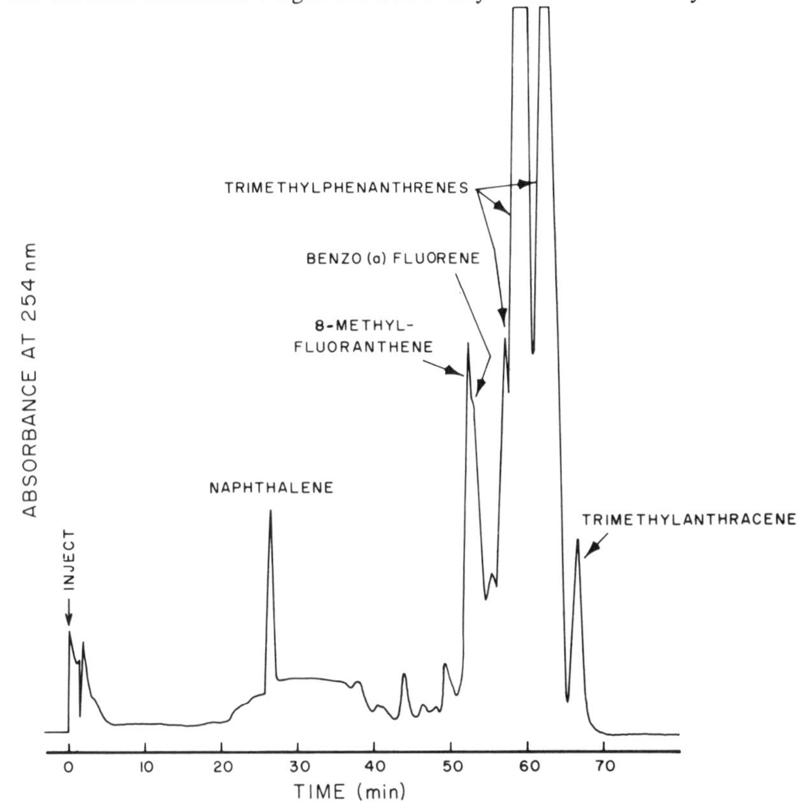

FIGURE 9. HPLC chromatogram of cut for GC peak no. 101 of gel fraction 40 (Figure 6). Gradient elution of 2%/min from 50-85% CH_3OH/H_2O. Naphthalene added as internal standard.

Figure 11 [HPLC of the benzolfuorene/methylfluoranthene peak from gel fraction 45 (GC profile—Figure 7)] shows that HPLC resolved 2, 3- and 3,4-benzofluorene and two methylfluoranthenes. UV data confirmed that the broadness and asymmetry of the methylfluoranthene was due to the presence of the 1- and 2-methyl isomers. The HPLC separation of the cut for the perylene peak in the gas chromatogram of gel fraction 45 (Figure 7) is given in

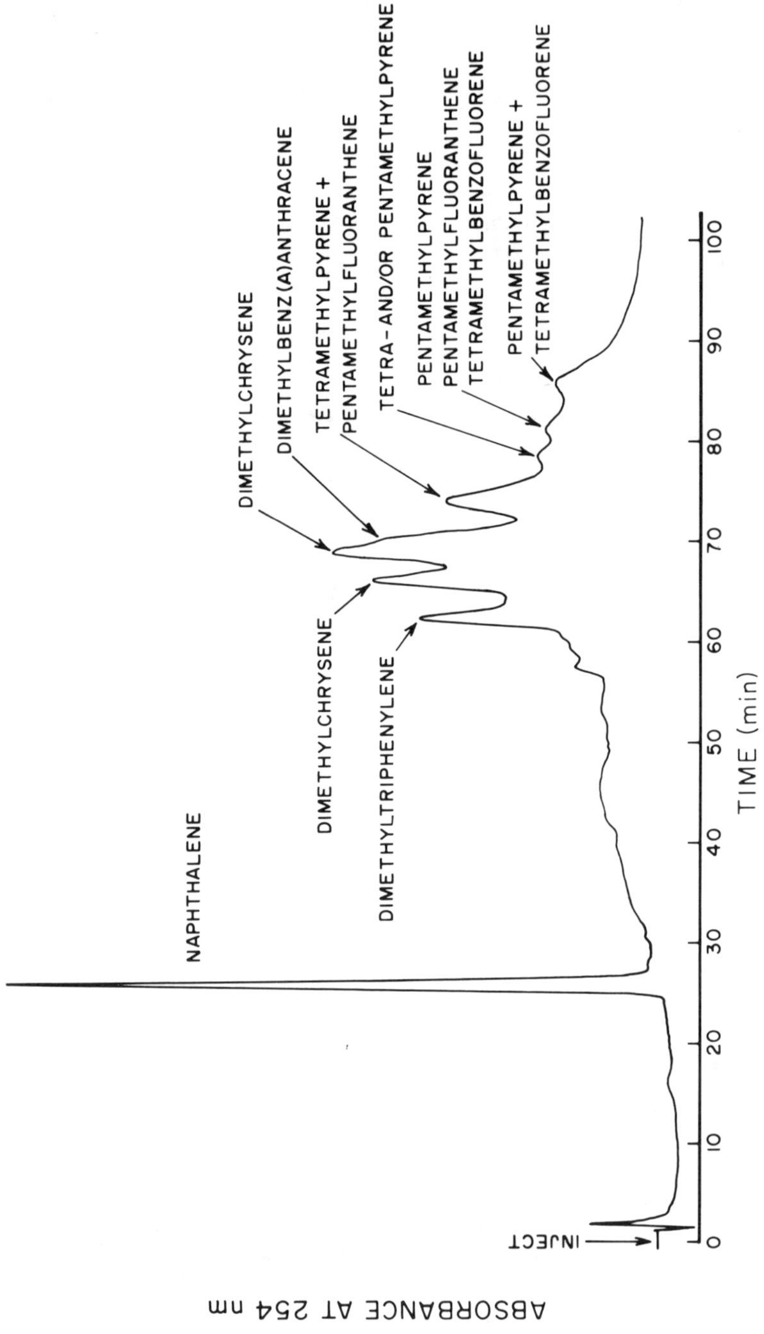

FIGURE 10. HPLC chromatogram of cut for GC peak no. 129 of gel fraction 40 (Figure 6). Gradient elution of 2%/min from 50-85% CH_3OH/H_2O. Naphthalene added as internal standard.

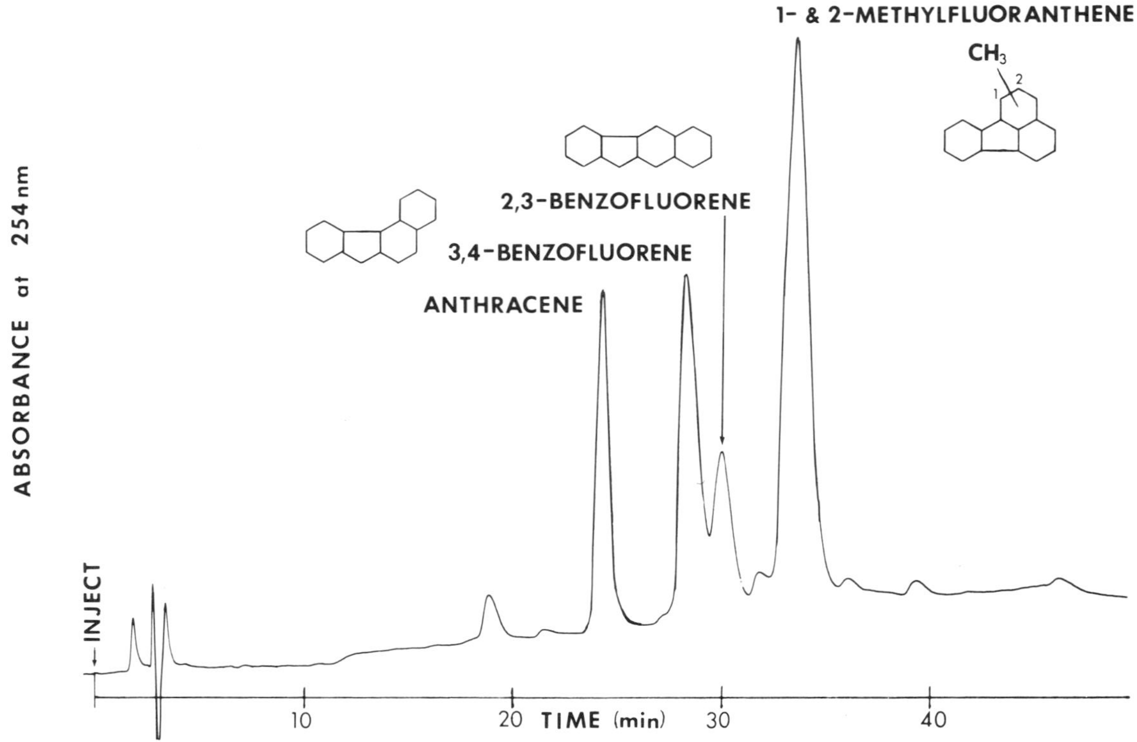

FIGURE 11. HPLC chromatogram of the benzofluorene/methylfluoranthene peak in gas chromatogram of gel fraction 45 (Figure 7). Anthracene added as internal standard.

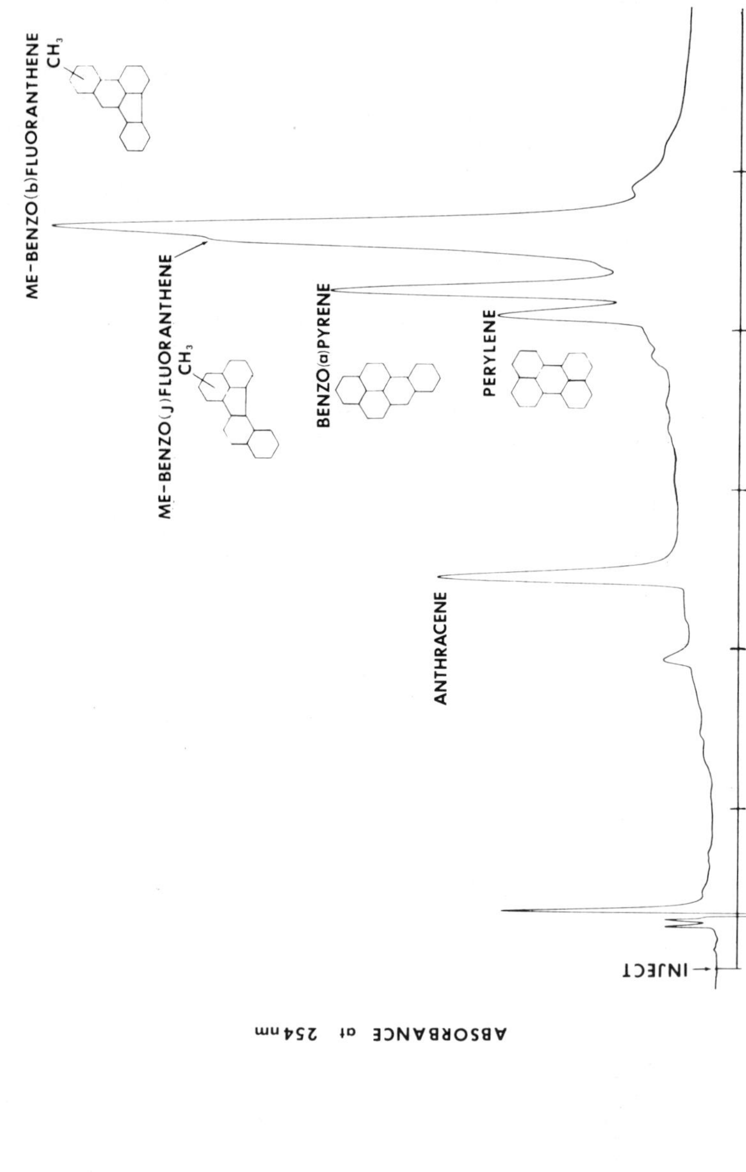

FIGURE 12. HPLC chromatogram of the cut for the perylene peak in gas chromatogram of gel fraction 45 (Figure 7). Anthracene added as internal standard.

Figure 12. Perylene, benzo[a]pyrene and two methylbenzofluoranthenes were separated. The benzo[a]pyrene was a carryover from the previous GC peak. Such carryover was found in many cases with the high-molecular-weight, low-volatility PAH on packed GC columns.

Figure 13 is the HPLC profile of GC peak marked "a" (eluting at about 84 min) in Figure 7. As evident, HPLC was as effective in separating isomeric alkylated high-molecular-weight PAH as it was in separating low-molecular-weight PAH (Figure 9). Ultraviolet data showed the presence of two separate methylanthanthrene isomers.

The compounds, resolved by HPLC in Figure 14 (GC peak "b" in Figure 7, eluting at about 87 min), are of special interest. They represent the first known report of dibenzofluoranthenes in CSC (24). This important group of compounds–of which there are ten isomers–has been overlooked by researchers who have reported the isomeric dibenzopyrenes but not the dibenzofluoranthenes. This oversight is understandable. We have been able to locate UV spectra for only four of the ten isomers in the literature. Further, an important paper by Lavit-Lamy and Buu-Hoi (16) has shown that the reported compound dibenzo[a,1]pyrene (3) is in fact dibenzo[a,e]fluoran-

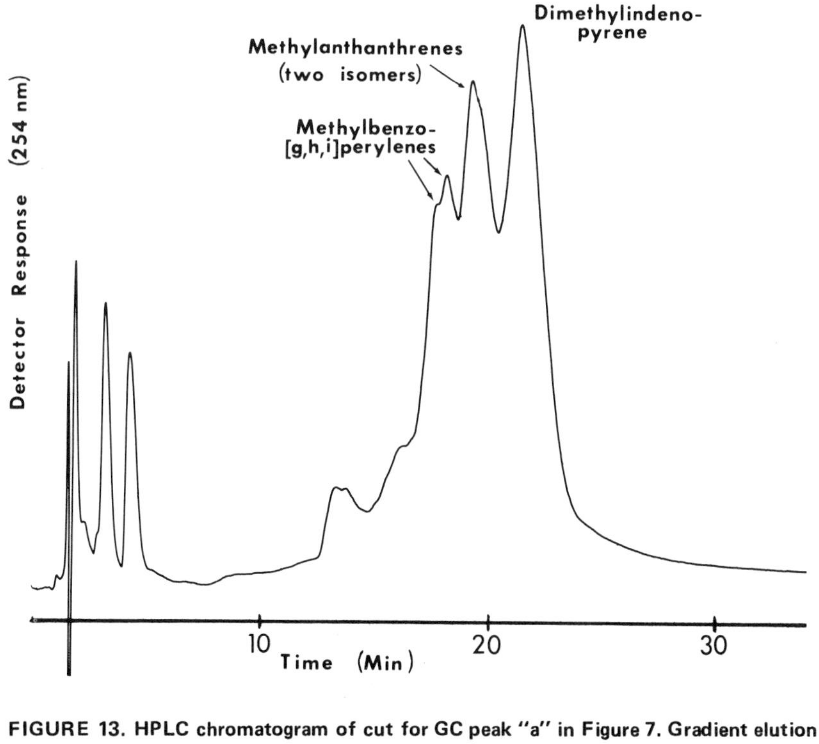

FIGURE 13. HPLC chromatogram of cut for GC peak "a" in Figure 7. Gradient elution of 2%/min from 85-90% CH_3OH/H_2O.

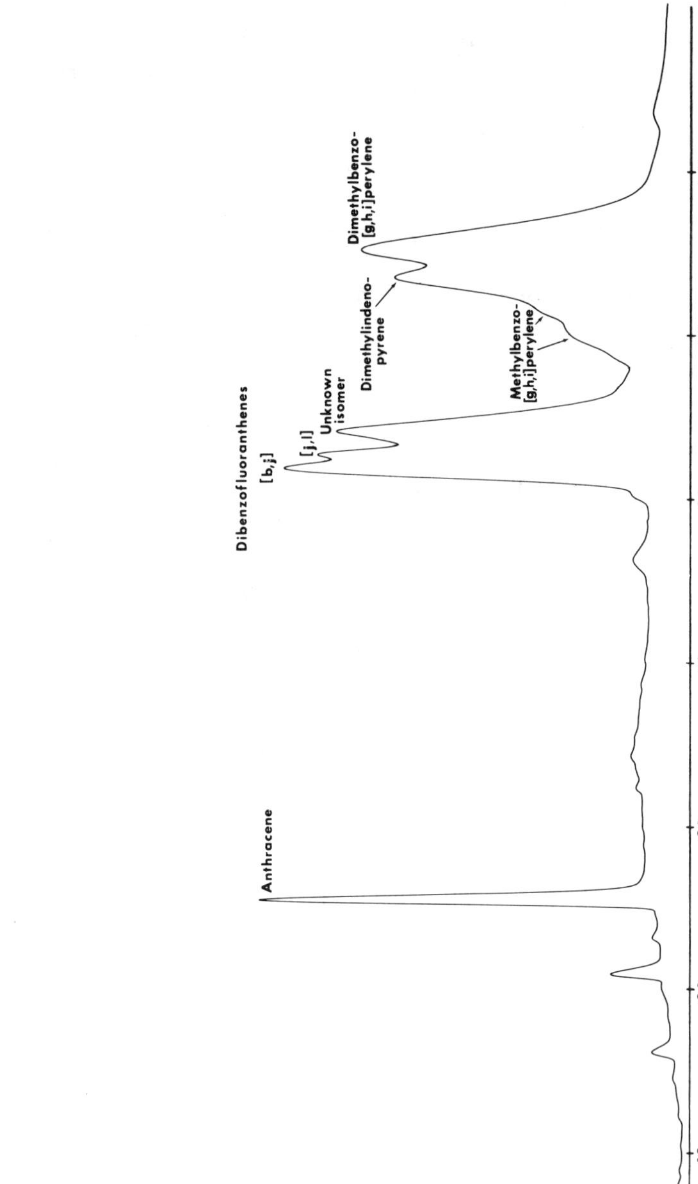

FIGURE 14. HPLC chromatogram of cut for GC peak "b" in Figure 7. Anthracene added as internal standard.

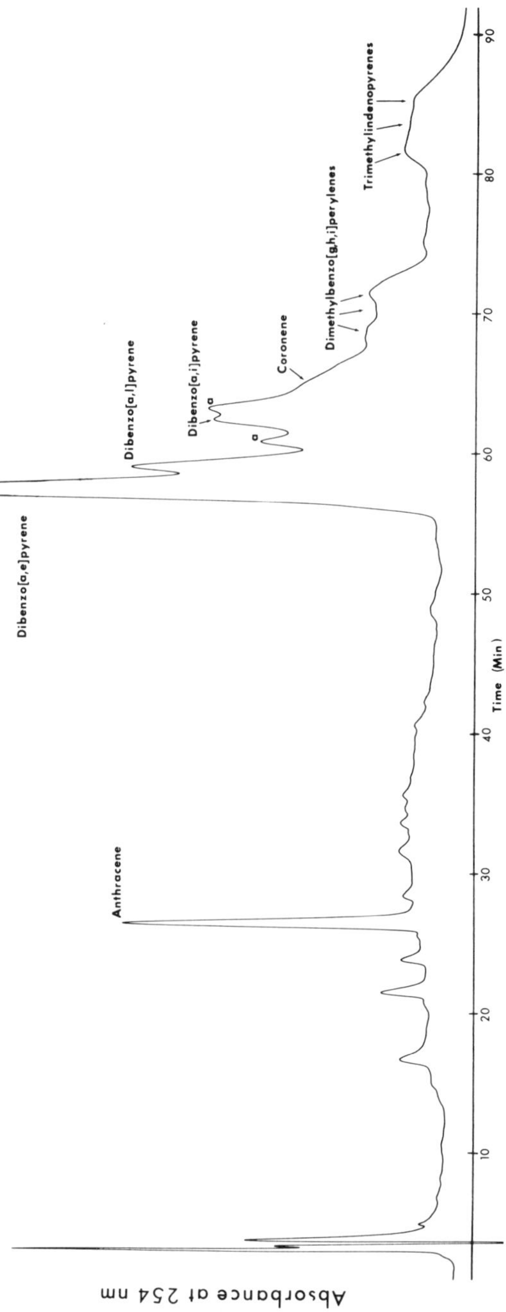

FIGURE 15. HPLC chromatogram of combined dibenzo[a,e] pyrene and coronene peaks in gas chromatogram of gel fraction 49 (Figure 8). a = Unidentified.

thene. Unfortunately, they did not give the UV spectra of authentic samples. Researchers using dibenzo[a,l]pyrene in cancer studies should be aware of this discrepancy. Figure 14 shows the parent PAH dibenzofluoranthenes clearly separated from the alkylated PAH.

The final example of HPLC separations is given in Figure 15. The separated compounds were contained in the cuts for the last two labeled GC peaks in the gas chromatogram of gel fraction 49 (Figure 8). The compound labeled dibenzo[a,l]pyrene had a UV spectrum (λ max 230, 240, 248, 271, 282, 294, 308, 344, 361 and 381 nm). We, therefore, believed the compound to be the true dibenzo[a,l]pyrene. According to Clar (3), three of the five possible dibenzpyrenes–[a,e], [a,h] and [a,i]–have pyrene-type spectra, whereas the other two isomers–[a,l] and [e,l]–do not. However, the [e,l] isomer can be thought of as a naphthacene derivative and, therefore, its spectrum probably would not look like that of pyrene. As noted before, the [a,l] isomer is in reality a dibenzofluoranthene and this fact may explain its peculiar spectrum. The spectrum listed above has a typical pyrene spectrum and the compound had a molecular weight of 302. Therefore, we concluded that the compound was dibenzo[a,l]pyrene. However, the possibility exists that this compound is a naphthopyrene. Confirmation will have to await publication of the spectrum of authentic dibenzo[a,l]pyrene.

SUMMARY

The described methodology, although developed for CSC, has been shown to be applicable to other environmental sources of PAH. The procedures effectively separate nitrogen analogues from the PAH and result in quantitative recovery of the PAH. The PAH isolate is relatively free of contaminants and amenable to analysis by gas chromatography, both packed or capillary, GC/MS or preparative GC-UV. The tremendous separating power of reverse-phase HPLC coupled with UV analysis was utilized to obtain unambiguous identification of components in complex GC peaks and to determine the number of closely isomeric PAH present.

REFERENCES

1. Asche, W., and Oelert, H. H. (1975): Zur systematik der anomalien bei der gelchromatographie von kohlenwasserstoffen an polystyrolgelen. *J. Chromatog.* 106:490-497.
2. Clar, E. J. (1952): *Aromatische Kohlenwasserstoffe, Polycyclische Systeme,* Springer-Verlag, Berlin-Gottingen, Heidelberg.
3. Clar, E. (1964): *Polycyclic Hydrocarbons,* Academic Press, London and New York.
4. Dong, M., Locke, D. C., and Hoffmann, D. (1977): Separation of

Aza-arenes by high-pressure liquid chromatography. *J. Chromatog. Sci.* 15:32-35.

5. Edstrom, T., and Petro, B. A. (1968): Gel permeation chromatographic studies of polynuclear aromatic hydrocarbon materials. *J. Polymer Sci.* 21:171-182.
6. Friedel, R. A., and Orchin, M. O. (1951): *Ultraviolet spectra of organic compounds,* John Wiley & Sons, Inc., New York.
7. Hendrickson, J. G., and Moore, J. C. (1966): Gel permeation chromatography. III. Molecular shape versus elution. *J. Polymer. Sci.* 4:167-188.
8. Howard, J. W., and Haenni, E. O. (1963): The extraction and determination of polynuclear hydrocarbons in paraffin waxes. *J. Assoc. Official Anal. Chem.* 46:933-941.
9. Klimisch, H. J. (1973): Anreicherungs-und trennverfahren zur analyse polycyclischer aromatischer kohlenwasserstoffe. Fraktionierung von zigarettenrauchkon densat. *Z. Anal. Chem.* 264:275-278.
10. Klimisch, H. J., and Ambrosius, D. (1974): Gel chromatographie polyzyklischer aromatischer kohlenwasserstoffe. *J. Chromatog.* 94:311-312.
11. Klimisch, H. J., and Reese, D. (1972): Das gel chromatographische trenverhatten von polystyrolgel fur kohlenwasserstoffe, amine und phenole. Praparative fraktionierung von cigarettenrauch kondensat fur biologische versuche. *J. Chromatog.* 67:299-307.
12. Klimisch, H. J., and Reese, D. (1973): Saulen-chromatographie trennung polyzyklischer aromatischer kohlenwasserstoffe. Trennung an organischen gelen in protischen losungsmitteln. *J. Chromatog.* 80:266-270.
13. Klimisch, H. J., and Stadler, L. (1972): Gel verteilungschromatographischer verfahren zur praparativen abtrennung polarer substanzen von polyzyklischen aromatischen kohlenwasserstoffen fraktionierung von cigarettenrauchkondensat. *J. Chromatog.* 67:175-178.
14. Klimisch, H. J., and Stadler, L. (1972): Trennung komplexer stoffgemische lipophiler und kydrophiler substanzen durch gel-verteilungsschromatographie an Sephadex LH-20. *J. Chromatog.* 67:291-297.
15. Lao, R. C., Thomas, R. S., and Dubois, L. (1973): Application of a gas chromatograph-mass spectrometer-data processor combination to the analysis of the polycyclic aromatic hydrocarbon content of airborne pollutants. *Anal. Chem.* 45:908-915.
16. Lavit-Lamy, D., and N. P. Buu-Hoi (1966): The true nature of dibenzo [a,l]-pyrene and its known derivatives. *Chem. Comm.* 4:92-94.
17. McKay, J. R., Weber, J. H., and Latham, D. R. (1976): Characterization of nitrogen bases in high-boiling petroleum distillates. *Anal. Chem.* 48:891-898.
18. Oelert, H. H. (1969): Atypische gel chromatographie im system Sephadex-isopropanol. *Z. Anal. Chem.* 244:91-101.
19. Rosen, A. A., and Middleton, F. M. (1955): Identification of petroleum refinery wastes in surface waters. *Anal. Chem.* 27:790-794.
20. Severson, R. F., Snook, M. E., Arrendale, R. F., and Chortyk, O. T. (1976): Gas chromatographic quantitation of polynuclear aromatic

hydrocarbons in tobacco smoke. *Anal. Chem.* 48:1866-1872.

21. Sharkey, A. F., Jr. (1976): Mass spectrometric analysis of process streams for coal-derived fuels. In: *Carcinogenesis, Vol. 1,* R. I. Freudenthal and P. W. Jones, Eds., pp. 341-347, Raven Press, New York.
22. Snook, M. E. (1976): Gel filtration of methyl-substituted polynuclear aromatic hydrocarbons. *Anal. Chim. Acta.* 81:423-427.
23. Snook, M. E., Severson, R. F., Higman, H. C., Arrendale, R. F., and Chortyk, O. T. (1976): Polynuclear aromatic hydrocarbons of tobacco smoke: Isolation and identification. *Beitr. Tabakforsch.* 8:250-272.
24. Snook, M. E., Severson, R. F., Higman, H. C., Arrendale, R. F., and Chortyk, O. T. (1977): The identification of high molecular weight, polynuclear aromatic hydrocarbons in a biologically-active fraction of cigarette smoke condensate. *Beitr. Tabakforsch.* 9:79-106.
25. Snook, M. E., Severson, R. F., Higman, H. C., Arrendale, R. F., and Chortyk, O. T. (1978): Multi-alkylated polynuclear aromatic hydrocarbons of tobacco smoke: Separation and identification. *Beitr. Tabakforsch.* 9:222-247.
26. Stedman, R. L., Miller, R. L., Lakritz, L., and Chamberlain, W. J. (1968): Method for concentrating polynuclear aromatic hydrocarbons in cigarette smoke condensate. *Chem. Ind.* 394.
27. Streuli, C. A. (1971): The interaction of aromatic compounds in alcohol solution with Sephadex LH-20 dextran. *J. Chromatog.* 56:219-223.
28. Streuli, C. A. (1971): The effect of solvent change on the separation processes of Sephadex LH-20 modified dextran. *J. Chromatog.* 56: 225-229.
29. Streuli, C. A., and Orloff, M. (1971): Chromatographic adsorption constants and π electron delocalization energies. *J. Chromatog.* 62: 73-77.
30. Unpublished data.
31. *UV Atlas of Organic Compounds* (1966): Plenum Press, New York.
32. Wilk, M., Rochlitz, J., and Bende, H. (1966): Saulenchromatographie von polycyclischen aromatischen kohlenwasserstoffen an lipophilem Sephadex LH-20. *J. Chromatog.* 24:414-416.

Polynuclear Aromatic Hydrocarbons, edited by P. W. Jones and P. Leber. Ann Arbor Science Publishers, Inc. Ann Arbor, MI ©1979.

SOME ANALYTICAL ASPECTS OF THE QUANTITATIVE DETERMINATION OF POLYNUCLEAR AROMATIC HYDROCARBONS IN FUGITIVE EMISSIONS FROM COAL LIQUEFACTION PROCESSES

Curt M. White, A. G. Sharkey, Jr., Milton L. Lee*[1] and Daniel L. Vassilaros*[1]

U.S. Department of Energy
Pittsburgh Energy Technology Center
Pittsburgh, Pennsylvania 15213

*Department of Chemistry
Brigham Young University
Provo, Utah 84602

INTRODUCTION

Coal liquefaction products and by-products are known to contain materials which are carcinogenic (6,8,30); among these are polynuclear aromatic hydrocarbons (PAH). Environmental concerns have led to much effort to determine the PAH content of coals, coal-derived products and fugitive emissions from liquefaction and gasification processes. Fugitive emissions can be defined as emissions from nonpoint sources, in contrast to emissions from stacks or controlled process streams. Fugitive emissions can be released from open operations, leakage around pipes and seals, spills and from inadequate control of point sources. A recent Battelle Laboratory Report (3) of potential hazards associated with coal-derived fuels cautions that "any point in a process where high-boiling coal-derived liquids or their vapors may escape must be considered a point of hazard." Recognition of this problem is not new; nearly 20 years ago researchers at the Union Carbide coal hydrogenation plant at

---

[1]A portion of this work was accomplished at the Pittsburgh Energy Technology Center during the summer of 1978, at which time Milton L. Lee and Daniel L. Vassilaros were Fossil Energy Research Participants under the administration of the Oak Ridge Associated Universities, University Programs.

Institute, West Virginia, described some of the industrial hygiene problems associated with coal liquefaction processes (12,28,29,34). They showed that the incidence of skin lesions among coal hydrogenation workers was significantly higher than among control populations sampled in the United States. Additionally, it is thought that the higher-than-normal incidence of lung cancer in workers in a sister industry, coal coking operations, is due to exposure to PAH in coal tar volatiles (2).

Concern about exposure of workers to PAH at coal conversion plants has prompted several recent investigations. Jackson and Cupps (9) have reported on the sampling and proposed analytical procedures for PAH at the 50 ton/day Solvent Refined Coal (SRC) plant at Tacoma, Washington. NIOSH has funded a project to determine, among other things, the PAH content of air around coal gasification processes (23), while Gammage *et al.* (4) are attempting to develop new monitoring devices to determine airborne PAH from synthetic fuel processes.

The objective of this study is to develop a method for the quantitative collection and analysis of PAH in fugitive emissions in the ambient air of coal liquefaction plants. The investigation is divided into three distinct areas: (1) sample collection, (2) sample separation and purification of a PAH-rich fraction, and (3) identification and quantitation of individual PAH.

This report describes the results of an investigation of the collection efficiency of Tenax-GC[2] (32) for PAH, the recovery of PAH from Tenax-GC via solvent extraction, and the development of a new gas chromatographic retention index system unique for PAH. The applicability of the PAH retention index system is demonstrated by the analysis of a coal liquefaction product.

Collection Efficiency of Tenax-GC for PAH

Organic materials occur in air as either vapors or aerosols, often at sub-part per million (ppm) levels—below the detection limits of currently available detectors; thus, the sample must be concentrated prior to analysis PAH are usually collected from air using a two-stage sampling apparatus consisting of a particulate filter followed by a sorbent trap. It is most desirable that the concentration and subsequent recovery steps be quantitative and the sorbent have a sufficiently high capacity for the retention of PAH in order that an enriched sample may be collected without breakthrough. A cartridge containing Tenax-GC (32) was selected for concentration of PAH because it reportedly met the above criteria (19,21) and it has been widely used in many

[2]Use of brand names facilitates understanding and does not necessarily imply endorsement by the U.S. Department of Energy.

successful investigations for collection of vapors, including PAH (11,23). In this study, experiments were conducted to determine the collection efficiency of Tenax-GC toward selected PAH under specific conditions. Pellizzari (21) previously reported that Tenax-GC had a less-than-quantitative collection efficiency for several carcinogenic compounds of high volatility. Pellizzari defined the collection efficiency of a sorbent as "the fraction of solute vapor in a polluted gas stream retained by the sorbent" before concentration equilibrium is attained.

A dynamic generating and sampling system similar to that described by both Raymond and Guiochon (26) and Rappaport *et al.* (25) was used in the collection-efficiency experiments. The entire apparatus was housed in the oven of a Varian 1700 GC, shown schematically in Figure 1. The diffusion cell was identical to that described by Rappaport *et al.* (25); the glass cartridges used to hold the Tenax-GC were obtained from Nu-Tech Corporation. The filter holder on the Nu-Tech glass cartridge was not fitted with a filter so that the collection efficiency of only Tenax-GC would be observed. Virgin Tenax-GC was preextracted for 24 hr in a Soxhlet using 150 ml of pesticide-grade acetone and purged at 270°C for 1 hr in a flow of helium prior to being charged into the glass cartridge. The collection efficiencies of Tenax-GC with respect to vapors of naphthalene, 1-methylnaphthalene and indene were measured at a temperature of 80°C. The experimental procedure was identical for the three compounds investigated and is described for naphthalene.

Naphthalene was charged into the diffusion cell and approximately 0.5 g of preextracted Tenax-GC was placed in the Nu-Tech glass cartridge on the right (Figure 1) while the cartridge on the left remained empty. The diffusion cell was isolated and a flow of nitrogen (120 ml/min) was established through the empty glass cartridge and to the flame ionization detector (FID) (A-1, closed, B-1, C-1, D-1).[3] The chromatographic oven temperature was raised to 80°C and operated isothermally while the FID signal was monitored. The nitrogen flow was then diverted through the diffusion cell and through the empty glass cartridge (A-2, open, B-1, C-1, D-1) to the FID. The FID responded indicating that naphthalene was being detected. The response continued to increase until the output from the diffusion cell was constant. A graphic record of the FID signal is depicted in Figure 2. The diffusion cell was then isolated and pure nitrogen was directed into the Nu-Tech glass cartridge containing Tenax-GC (A-1, closed, B-2, C-2, D-1). The nitrogen flow through the sorbent bed was maintained at 120 ml/min while the FID signal was monitored (Figure 2). When the diffusion cell was brought back on

[3]This denotes the various valves and their positions in Figure 1. Valve A is in the 1 position, the on/off valve is closed, and valves B, C and D are in the 1 position.

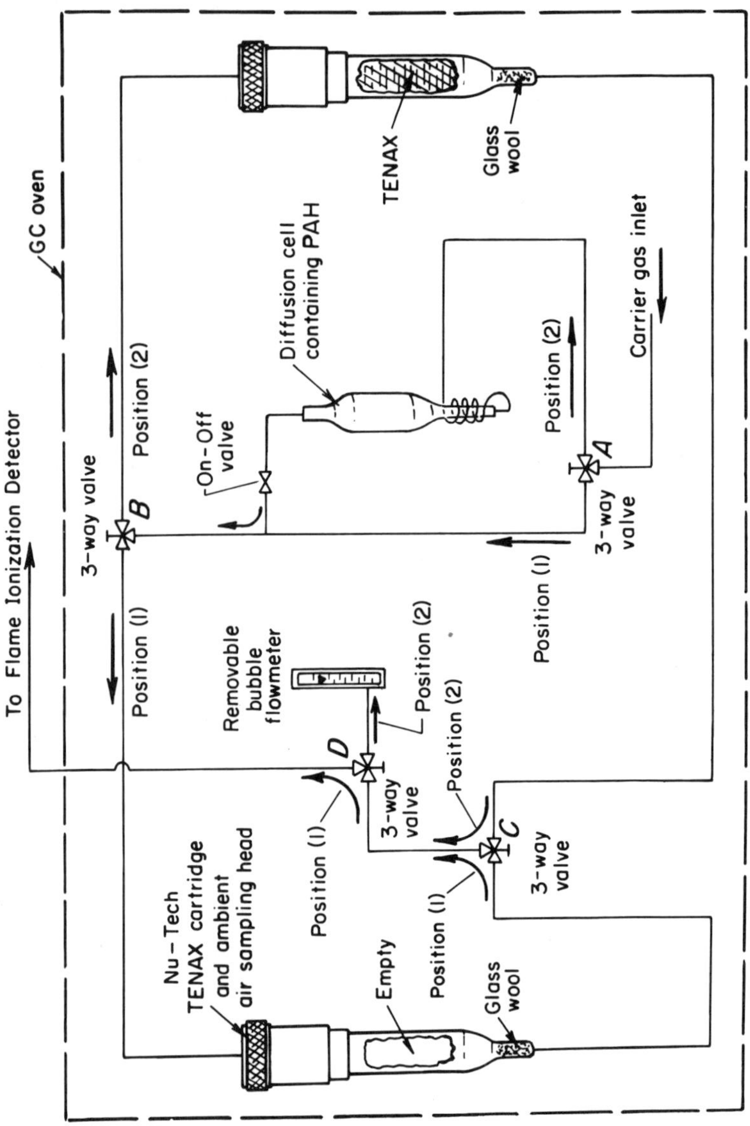

FIGURE 1. Apparatus used to determine PAH collection efficiency on Tenax.

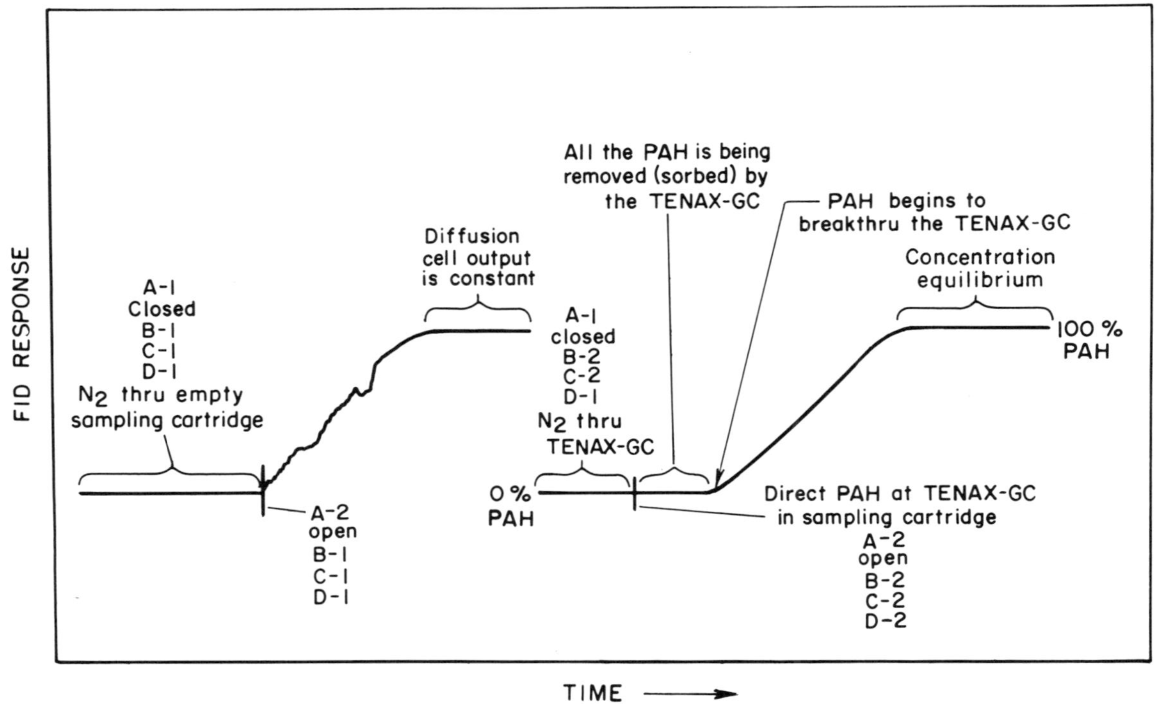

FIGURE 2. Collection-efficiency experimental results for the general case of a PAH. The designations A-1, closed, B-1, C-1 and D-1 refer to valve positions in Figure 1. For example, in the above designation, valve A is in the 1 position, the on/off valve is closed, and valves B, C and D are in the 1 position.

stream, the naphthalene-polluted nitrogen was directed through the glass cartridge containing Tenax-GC while the effluent from the cartridge was monitored continuously by the FID (A-2, open, B-2, C-2, D-1). Initially, the FID response was the same whether the naphthalene-polluted nitrogen or purified nitrogen was directed at the Tenax-GC bed (Figure 2). After approximately 8 hr, naphthalene began to break through the Tenax-GC bed and the FID signal increased until concentration equilibrium was attained. This experiment was performed in triplicate for naphthalene, 1-methylnaphthalene and indene. Results indicated quantitative removal of PAH from the gas streams until initial breakthrough occurred. Under these conditions, the initial breakthrough of 1-methylnaphthalene and indene required approximately 9 hr and 2 hr, respectively. When sampling is performed at ambient temperatures, breakthrough times should be greater. Although the collection efficiency toward only three of the more volatile PAH was determined, these results should also apply to the less volatile higher-molecular-weight PAH.

Recovery of PAH from Tenax-GC

Once a sample has been collected, it is necessary to quantitatively recover the sample from the collection media. Two commonly used methods of sample recovery are solvent extraction (9,11,23) and thermal desorption (22, 26). Although thermal desorption has been shown to be a viable method for recovery of volatile components from Tenax-GC (22) and graphitized carbon black (26), it has not proven effective in the recovery of higher-molecular-weight, relatively nonvolatile species. Thermal desorption requires heating the collected sample to high temperatures; some investigators using this method have reported heating the sample to 300°C. It is important to consider that some components present in coal liquefaction products are thermally labile. For example, dihydroanthracene and dihydrophenanthrene easily undergo hydrogen elimination and aromatize when heated. Another disadvantage of thermal desorption is that the entire sample is exhausted during a single analysis, preventing repetitive analyses to verify experimental results or to obtain statistical data on quantitative measurements. Solvent extraction of the collection media for sample recovery was therefore investigated as the primary means of recovering sorbed PAH.

Results of experiments using Soxhlet extraction methods with *n*-pentane for recovery of PAH spiked onto Tenax-GC are given in Table I. The recovery efficiencies of pyrene, anthracene, perylene, naphthalene and chrysene were determined for the individual compounds as well as for selected mixtures, (see Table I). Brooman and Edgerley (1) had previously reported that the

TABLE I. Percent Recovery of PAH from Tenax-GC using Soxhlet Extraction with *n*-Pentane

	Extraction Time		
	4 hr	8 hr	24 hr
Pyrene	98	98	98
Anthracene	101	94	95
Perylene	96	99	100
Naphthalene	103	98	104
Chrysene	102	105	100
Mixture of			
Pyrene	97		
Anthracene	96		
Phenanthrene	98		

recovery efficiencies of certain compounds on activated carbon were not independent of the presence of other compounds.

Since the experimental procedure used during all the Soxhlet extraction experiments was identical, only one experiment, that using pyrene, will be discussed in detail. Three weighed amounts of Tenax-GC (358 ± 1 mg) were loaded with 0.400 ml of a 600 ppm pyrene in *n*-pentane solution. The *n*-pentane was permitted to evaporate at room temperature and the pyrene-spiked samples were transferred to identical Soxhlet extraction thimbles. These thimbles were placed in microchemical Soxhlet extractors that had been previously cleaned by a 24-hr rinse with pesticide-grade *n*-pentane, which was used throughout this study. These samples were then continuously extracted using *n*-pentane. The first sample was extracted for 4 hr, the second for 8 hr, and the third for 24 hr. The resulting extracts were then quantified for pyrene using UV spectroscopy. The results of these experiments appear in Table I. For the anthracene, perylene, naphthalene and chrysene experiments, solutions of 1200, 10, 2060 and 70 ppm, respectively, were used to spike the Tenax-GC; whereas the three-component solution of PAH used in the mixture experiment consisted of 824 ppm of pyrene, 405 ppm of phenanthrene and 390 ppm of anthracene in *n*-pentane. In all experiments, 0.400 ml of standard solution was loaded onto Tenax-GC. The study indicates that PAH are quantitatively recovered from Tenax-GC after continuous Soxhlet extraction with *n*-pentane for 4 hr. Furthermore, it was shown that even with the presence of more than one PAH on the surface of Tenax-GC, the recovery of each PAH is quantitative after 4 hr of extraction (Table I).

During several steps of the analytical procedure for the determination of PAH, the sample is in a highly diluted form in a volatile solvent. This occurs after the sample has been extracted from the collection media. It is often necessary to concentrate the PAH-rich extract by solvent removal. Care must be exercised or significant PAH losses can occur; Pupp *et al.* (24) warned that vacuum drying should be avoided during analyses.

The Kuderna-Danish evaporative concentrator has found widespread applications for concentrating materials in volatile solvents (7). A 100.0-ml portion of a 6.6 ppm naphthalene in *n*-pentane solution was concentrated to 1.0 ml using this apparatus in order to determine the loss of naphthalene. In four experiments naphthalene had an average recovery of 88.0 ± 3.7%. Naphthalene has a higher vapor pressure than most PAH and therefore should suffer higher losses than less volatile PAH. This is supported by the finding that when 100.0 ml of 2.4 ppm anthracene in *n*-pentane solutions were evaporated recovery was 97.0%. These results suggest that the PAH losses incurred when using a Kuderna-Danish evaporative concentrator to remove *n*-pentane from PAH solutions are minimal for PAH having vapor pressures greater than that of naphthalene.

Development of a New Gas Chromatographic Retention Index Unique for PAH

The resolution and subsequent identification of polynuclear aromatic hydrocarbons in complex mixtures are best accomplished by combined glass capillary gas chromatography/mass spectrometry (GC/MS). Mass spectral data and GC retention data together constitute sufficient and necessary proof for positive identification of many isomeric PAH. In cases where sophisticated GC/MS equipment is not available, it would be useful to have reproducible gas chromatographic retention indices available for routine gas chromatographic screening of fugitive-emission samples. Preliminary identification and estimated quantitative data can be obtained from gas chromatographic measurements alone. Lunde (18) has recently used capillary gas chromatography alone to identify and quantitate the PAH content of long-range transported air pollutants. Compilations of standard mass spectra are available; however, there is insufficient glass capillary GC retention data of PAH in the literature because of the nonreproducibility of capillary column preparation in different laboratories, the unavailability of standard PAH for retention studies, and the difficulties encountered in providing a statistically acceptable retention index for PAH in temperature-programmed glass capillary gas chromatography.

The Kovats' retention indices (13,31,33) for each of 12 PAH ranging from naphthalene to perylene varied widely (average 95% confidence limits for 6 to 10 measurements for each PAH is ± 4.2 index units) with the temperature program rate, film thickness and condition of the column. Therefore a new retention index system based on the four internal standards, Table II–naphthalene, phenanthrene, chrysene and picene–was defined and evaluated. The retention indices were calculated using the equation suggested by Van Den Dool and Kratz (31) for linear-temperature-programmed gas chromatography:

$$I = 100 \left[n + \frac{X - M_{(n)}}{M_{(n+1)} - M_{(n)}} \right] \quad (1)$$

Retention indices were calculated for over 200 PAH standard compounds based on their comparison with these standards using Equation 1. In Equation 1, I is the retention index and n refers to the number of rings in the standard compound that elutes just prior to the compound of interest. X, $M_{(n)}$ and $M_{(n+1)}$ are the measured retention times of the compound of interest, the standard aromatic that elutes just prior to the compound of interest, and the standard aromatic that elutes just after the compound of interest, respectively. The average 95% confidence limits for 3 to 8 measurements on the same 12 compounds measured previously was ± 0.25 index units. A manuscript describing all of the results and experimental details is in preparation (17).

TABLE II. Bracketing Standards used in the PAH Retention Index System

Compound	Defined Index
Naphthalene	200
Phenanthrene	300
Chrysene	400
Picene	500

The analysis of an aromatic fraction of a coal liquefaction product illustrates the application of the new PAH retention index system. The liquefaction product was obtained from the one-half ton per day SYNTHOIL Process Development Unit at Bruceton, Pennsylvania (35). The oils were separated from other materials in the liquefaction product by solvent separation (27). The oils were further separated using the SARA (10) technique to obtain an aromatic fraction. This aromatic fraction was dissolved in methylene chloride and subjected to combined GC/MS analysis. The GC/MS analysis was performed with a Dupont 490 mass spectrometer interfaced to a Varian 3700 GC. A Hewlett-Packard 2100A computer was used for spectrometric data storage and reduction. The mass spectrometer was operated at a resolution of 600, an ionizing voltage of 70 ev, and a scanning speed of 4 sec per decade. The gas chromatographic separation of the aromatic fraction was achieved using a 12 m x 0.30 mm glass capillary column coated with an SE-52 methylphenylsilicone stationary phase and temperature programmed from 50°C to 250°C at 2°C/min. A Varian CDS 101 chromatogaphic data system was used to measure precise retention times (± 0.01 min) for calculation of retention indices. The gas chromatographic profile of the aromatic fraction appears in Figure 3. The numbered chromatographic peaks are identified in Table III. These identifications are based on mass spectral data and were, in most cases, confirmed through a comparison of the measured retention index of the unknown chromatographic peaks with those of standards. This combination of mass spectral and gas chromatographic retention index data led to the positive identification of 24 compounds.

It is expected that the kinds of PAH in fugitive emissions around coal liquefaction plants will closely parallel those in the coal liquefaction product. The aromatic fraction of coal liquefaction products is known to be an extremely complex mixture of compounds consisting of many structural and geometric isomers, representing hundreds of compounds. Some of these compounds are known or suspected carcinogens. When monitoring fugitive emissions from coal liquefaction plants for individual PAH, it will be necessary to differentiate between isomers which may have very similar chemical and physical properties, such as benzo[a]pyrene and benzo[e]pyrene. Evaluation of the presently available chromatographic methods for the separation of complex mixtures of PAH indicated that glass capillary gas chromatography, employing high-resolution wall-coated open tubular (WCOT) columns coated with SE-52, is an excellent, possibly the best, chromatographic method available. This gas chromatographic method has provided the means for extremely detailed studies of complex mixtures of PAH (5,14-16, 20) with baseline separation of benzo[a]pyrene from benzo[e]pyrene.

SUMMARY

From the results of this investigation, it appears that PAH can be both sampled from ambient air and recovered from the collection media quantitatively using Tenax-GC. The solvent can be effectively removed from the recovered sample with minimal sample losses using a Kuderna-Danish evaporative concentrator. The chromatographic resolution of complex mixtures of PAH expected to be present in fugitive emissions from coal liquefaction processes is best attained using glass capillary gas chromatography. The development of a new PAH retention index system provides for the tentative identification of numerous PAH in these complex mixtures. Positive

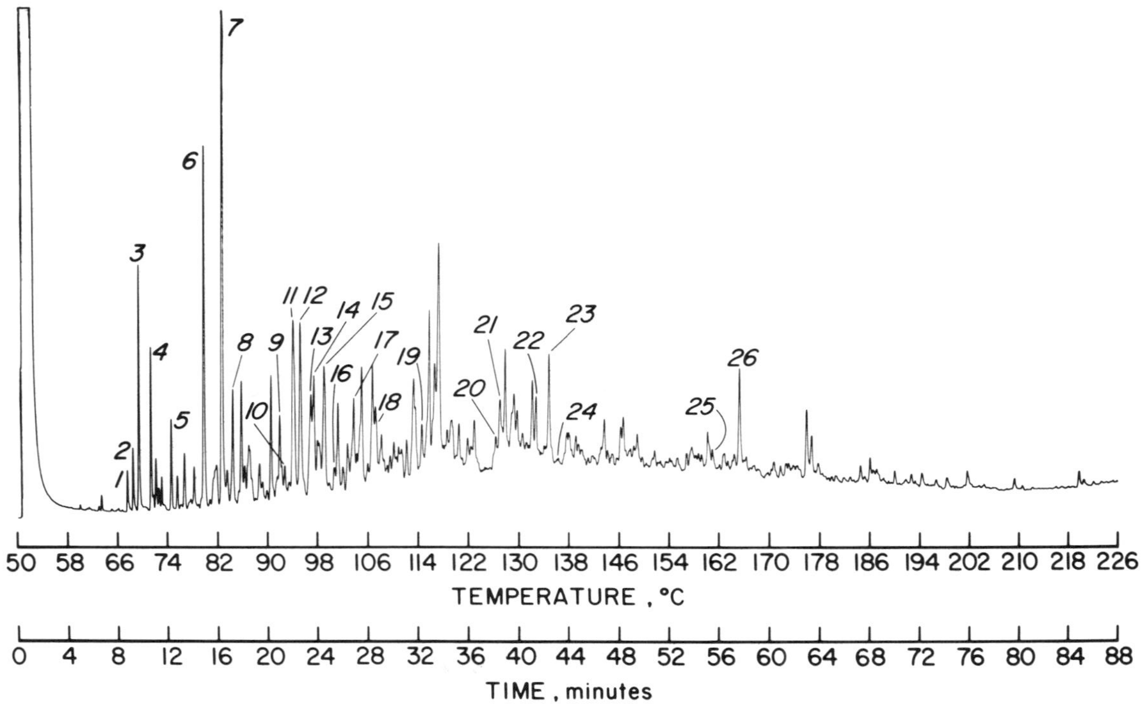

FIGURE 3. Gas chromatographic profile of the aromatic fraction of SYNTHOIL. The numbered chromatographic peaks are identified in Table III.

TABLE III. GC/MS Results from the Analysis of the Aromatic Fraction of SYNTHOIL

GC Peak Number	Compound	Measured Retention Index of Chromatographic Peak from Sample (Average of Three Determinations $\bar{x} \pm S$)	Measured Retention Index Using Standards ($\bar{x} \pm S$)
1	Methylindane	194.04 ± .09	[a]
2	Methylindane	195.48 ± .09	[a]
3	Tetralin	197.17 ± .20	197.04 ± 0.05
4	Naphthalene	200.00	200.00
5	Methyltetralin	205.07 ± .33	[a]
6	Methyltetralin or C_2-Indane	213.61 ± .10	[a]
7	2-Methylnaphthalene	218.31 ± .24	218.14 ± 0.28
8	1-Methylnaphthalene	220.90 ± .04	221.04 ± 0.25
9	1,2,2a,3,4,5-Hexahydroacenaphthylene	232.63 ± .06	232.70[b]
10	Biphenyl	233.90 ± .04	233.96 ± 0.24
11	2-Ethylnaphthalene	236.02 ± .16	236.08 ± 0.16
12	2,6- and 2,7-Dimethylnaphthalenes	237.75 ± .10	237.58 ± 0.17, 237.71 ± 0.07
13	1,3- and 1,7-Dimethylnaphthalenes	240.52 ± .12	240.25 ± 0.16, 240.66 ± 0.25
14	1,6-Dimethylnaphthalene	240.87 ± .19	240.72 ± 0.09
15	1,5-Dimethylnaphthalene	243.83 ± .12	243.98 ± 0.16
16	1,2-Dimethylnaphthalene	246.29 ± .11	246.49 ± 0.30
17	Acenaphthene	251.17 ± .08	251.29 ± 0.14
18	Dibenzofuran	256.94 ± .20	257.17 ± 0.05
19	Fluorene	268.30 ± .10	268.17 ± 0.15
20	9,10-Dihydrophenanthrene	286.86 ± .29	287.09 ± 0.16
21	1,2,3,4,5,6,7,8-Octahydroanthracene	287.76 ± .12	287.69 ± 0.20
22	1,2,3,4-Tetrahydrophenanthrene	296.93 ± .21	297.21[b]
23	Phenanthrene	300.00	300.00
24	Anthracene	301.84 ± .19	301.69 ± 0.08
25	Fluoranthene	344.12 ± .25	344.01 ± 0.16
26	Pyrene	351.38 ± .29	351.22 ± 0.08

[a]The retention index of these compounds were not determined because standards were not available.
[b]Only two determinations were made, therefore no standard deviation is available.

identification is obtained when the retention indices are combined with mass spectral data on the same chromatographic peaks. This was demonstrated by the analysis of the aromatic fraction of a coal liquefaction product.

ACKNOWLEDGMENTS

By acceptance of this article, the publisher or recipient acknowledges the U.S. Government's right to retain a nonexclusive, royalty-free license in and to any copyright covering this article.

REFERENCES

1. Brooman, D. L., and Edgerley, E., Jr. (1966): Concentration and recovery of atmospheric odor pollutants using activated carbon. *J. Air Poll. Control Assoc.* 16:25-29.
2. Fannick, N., Gonshor, L. T., and Shockley, J., Jr. (1972): Exposure to coal tar pitch volatiles at coke ovens. *Am. Ind. Hyg. Assoc. J.* 33: 461-468.
3. Freudenthal, R. I., Lutz, G. A., and Mitchell, R. I. (1975): Carcinogenic potential of coal and coal conversion products. A Battelle Energy Program Report, pp. 1-34.
4. Gammage, R. B., Vo-Dinh, T., Hawthorne, A. R., Thorngate, J. H., and Parkinson, W. W. (1978): A new generation of monitors for PAH from synthetic fuels production. In: *Carcinogenesis, Vol. 3,* P. W. Jones and R. I. Freudenthal, Eds., pp. 155-174, Raven Press, New York.
5. Giger, W., and Schaffner, C. (1978): Determination of polycyclic aromatic hydrocarbons in the environment by glass capillary gas chromatography. *Anal. Chem.* 50:243-249.
6. Guerin, M. R., Epler, J. L., Griest, W. H., Clark, B. R., and Rao, T. K. (1978): Polycyclic aromatic hydrocarbons from fossil fuel conversion processes. In: *Carcinogenesis, Vol. 3,* P. W. Jones and R. I. Freudenthal, Eds., pp. 21-33, Raven Press, New York.
7. Gunther, F. A., Blinn, R. C., Kolbezen, M. J., Barkley, J. H., Harris, W. D., and Simon, H. S. (1951): Microestimation of 2-(*p*-tert-butylphenoxy) isopropyl-2-chloroethyl sulfite residues. *Anal. Chem.* 23: 1835-1842.
8. Hueper, W. C. (1956): Experimental carcinogenic studies on hydrogenated coal oils. Part I. Bergius Oils. *Ind. Med. Surgery* 25:51-55.
9. Jackson, J. O., and Cupps, J. A. (1978): Field evaluation and comparison of sampling matrices for polynuclear aromatic hydrocarbons in occupational atmospheres. In: *Carcinogenesis, Vol. 3,* P. W. Jones and R. I. Freudenthal, Eds., pp. 183-191, Raven Press, New York.
10. Jewell, D. M., Weber, J. H., Bunger, J. W., Plancer, H., and Latham, D. R. (1972): Ion-exchange, coordination, and adsorption chromatographic separation of heavy-end petroleum distillates. *Anal. Chem.* 44:1391-1395.

11. Jones, P. W., Giammar, R. D., Strup, P. E., and Stanford, T. B. (1976): Efficient collection of polycyclic organic compounds from combustion effluents. *Environ. Sci. Technol.* 10:806-810.
12. Ketcham, N. H., and Norton, R. W. (1960): The hazards to health in the hydrogenation of coal. Part III. The industrial hygiene studies. *Arch. Environ. Health* 1:194-201.
13. Kovats, E. (1958): Gas chromatographic characterization of organic compounds. I. Retention indexes of aliphatic halides, alcohols, aldehydes, and ketones. *Helv. Chim. Acta* 41:1915-1932.
14. Lee, M. L., Novotny, M., and Bartle, K. D. (1976): Gas chromatography/mass spectrometry and nuclear magnetic resonance determination of polynuclear aromatic hydrocarbons in airborne particulates. *Anal. Chem.* 48:1566-1572.
15. Lee, M. L., Novotny, M., and Bartle, K. D. (1976): Gas chromatography/mass spectrometric and nuclear magnetic resonance spectrometric studies of carcinogenic polynuclear aromatic hydrocarbons in tobacco and marijuana smoke condensates. *Anal. Chem.* 48:405-46.
16. Lee, M. L., and Hites, R. A. (1976): Characterization of sulfur-containing polycyclic aromatic compounds in carbon blacks. *Anal. Chem.* 48: 1890-1893.
17. Lee, M. L., Vassilaros, D. L., White, C. M., and Novotny, M. V.: A new retention index system for programmed-temperature capillary-column gas chromatography of polycyclic aromatic hydrocarbons *Anal. Chem.* (in press).
18. Lunde, G., and Bjorseth, A. (1977): Polycyclic aromatic hydrocarbons in long-range transported aerosols. *Nature* 268:518-519.
19. Novotny, M., Lee, M. L., and Bartle, K. D. (1974): Some analytical aspects of the chromatographic headspace concentration method using a porous polymer. *Chromatographia* 7:333-338.
20. Novotny, M., Lee, M. L., and Bartle, K. D. (1974): The methods for fractionation, analytical separation and identification of polynuclear aromatic hydrocarbons in complex mixtures. *J. Chromatog. Sci.* 12: 606-612.
21. Pellizzari, E. D., Bunch, J. E., and Carpenter, B. H. (1975): Collection and analysis of trace organic vapor pollutants in ambient atmospheres. Technique for evaluating concentration of vapors by sorbent media. *Environ. Sci. Technol.* 9:552-555.
22. Pellizzari, E. D., Carpenter, B. H., Bunch, J. E., and Sawicki, E. (1975): Collection and analysis of trace organic vapor pollutants in ambient atmospheres. Thermal desorption of organic vapors from sorbent media. *Environ. Sci. Technol.* 9:556-560.
23. Philips, R. D. (1977): Preliminary investigation of airborne pna emissions from the hygas pilot plant. Presented at Workshop on Exposure to Polynuclear Aromatic Hydrocarbons in Coal Conversion Processes, Oak Ridge National Laboratory, Oak Ridge, TN.

24. Pupp, C., Lao, R. C., Murray, J. J., and Pottie, R. F. (1974): Equilibrium vapor concentration of some polycyclic aromatic hydrocarbons, As_4O_6 and SeO_2 and the collection efficiencies of these air pollutants. *Atmos. Environ.* 8:915-925.
25. Rappaport, S. M., Morales, R., Weeks, R. W., Jr., Campbell, E. E., and Ettinger, H. J. (1975): Development of sampling and analytical methods for carcinogens. Los Alamos Scientific Laboratory Progress Report. LA-6387-PR, pp. 1-19.
26. Raymond, A., and Guiochon, G. (1975): The use of graphitized carbon black as a trapping material for organic compounds in light gases before a gas chromatographic analysis. *J. Chromatog. Sci.* 13:173-177.
27. Schweighardt, F. K. (1977): Chromatographic characterization of coal liquefaction products. PERC/RI-77/3, February, Pittsburgh Energy Research Center, Pittsburgh, PA.
28. Sexton, R. D. (1960): The hazards to health in the hydrogenation of coal. Part I. An introductory statement on general information, process description, and a definition of the problem. *Arch. Environ. Health* 1:181-186.
29. Sexton, R. J. (1960): The hazards to health in the hydrogenation of coal. Part IV. The control program and the clinical effects. *Arch. Environ. Health* 1:208-231.
30. Staff (1977): Spectral studies of coal-derived liquids. PERC/RI-77/14, November, Pittsburgh Energy Research Center, Pittsburgh, PA.
31. Van Den Dool, H., and Kratz, P. D. (1963): A generalization of the retention index system including linear temperature programmed gas-liquid partition chromatography. *J. Chromatog.* 11:463-471.
32. Van Wijk, R. (1970): The use of poly-*p*-2,6-diphenyl-phenylene oxide as a porous polymer in gas chromatography. *J. Chromatog. Sci.* 8: 122-124.
33. Watts, R. B., and Kekwick, R. G. O. (1974): A simple relationship between retention parameters in isothermal and programmed-temperature gas chromatography. *J. Chromatog.* 88:165-167.
34. Weil, C. S., and Condra, N. I. (1960): The hazards to health in the hydrogenation of coal. Part II. Carcinogenic effect of materials on the skin of mice. *Arch. Environ. Health* 1:187-193.
35. Yavorsky, P. M., Akhtar, S., Lacey, J. J., Weintraub, M., and Reznik, A. A. (1975): The SYNTHOIL process. *Chem. Eng. Prog.* 71:79-80.

Polynuclear Aromatic Hydrocarbons, edited by P. W. Jones and P. Leber. Ann Arbor Science Publishers, Inc. Ann Arbor, MI ©1979.

PRECURSORS OF POLYNUCLEAR AROMATIC HYDROCARBONS IN TOBACCO SMOKE

Ray F. Severson, William S. Schlotzhauer, Orestes T. Chortyk, Richard F. Arrendale and Maurice E. Snook

Tobacco and Health Research Laboratory
Science and Education Administration/Federal Research
U.S. Department of Agriculture
Athens, Georgia 30604

Several animal bioassays of cigarette smoke condensate (CSC) have shown that condensate fractions containing the polynuclear aromatic hydrocarbons (PAH) possess significant biological activity (1,2,4,6). A recent study in our laboratory indicated that the PAH are the primary tumor initiators in CSC (16). Since the PAH are pyrosynthesized during the smoking process, it is of considerable importance to determine which tobacco-leaf components are the most efficient PAH producers. Reduction of these tobacco-leaf components would produce a safer tobacco.

Numerous pyrolytic studies have shown that the hexane or petroleum ether (PE) extract of tobacco contains the major precursors of smoke PAH (11,13,23). However, many of these studies were made under conditions which produced optimum PAH yields (13); thus, the PAH distributions were not comparable to those in CSC and the data obtained could not be exactly correlated with PAH production during the smoking process.

In this paper, we describe how we determined the pyrolytic conditions that yielded PAH profiles of tobacco pyrolyzates that could be correlated with CSC PAH profiles, and how we applied these conditions in a systematic search for the major precursors of cigarette-smoke PAH.

MATERIALS AND METHODS

Pyrolysis

The pyrolyses were performed in a horizontal 5-ft x 2-in. i.d. Vycor tube inserted into a Lindberg Heavy-Duty furnace. Nitrogen flow through the pyrolysis system was maintained at 150 ml/min. The oven was preheated to the desired temperature (650, 700, 750, 800 or 850°C ± 5°C at the center of the heated zone). The sample (5 g of PE extract or subfraction or standard, 10 g of tobacco or marc) contained in a Vycor boat was placed at the cool end of the pyrolysis tube. The boat was drawn in the direction of the gas flow at a rate of 33.97 cm/min by a wire attached to a kymograph (gear-driven rotating drum) into the center of the heated zone. Products were flushed by the nitrogen into three glass traps, cooled with dry ice/acetone.

Analysis of Phenols, Acids and PAH from Pyrolyzates

The pyrolyzate traps were washed out with 1:1 benzene-diethyl ether solvent and the phenols, acids and PAH were isolated according to the fractionation scheme shown in Figure 1. The phenols and acids were isolated from the basic extract by a series of pH adjustments and diethyl-ether extractions. The PAH were isolated for analyses by gas chromatography (GC) from the other organic solubles by silicic acid chromatography and gel-filtration chromatography on Bio-Beads SX-12 in benzene. The PAH isolate was separated into two portions based on the elution volumes of trimethylnaphthalene (to yield GF-B containing multialkyl PAH) and fluoranthene (to yield GF-C containing parent and monomethylated PAH). The PAH in the pyrolyzates

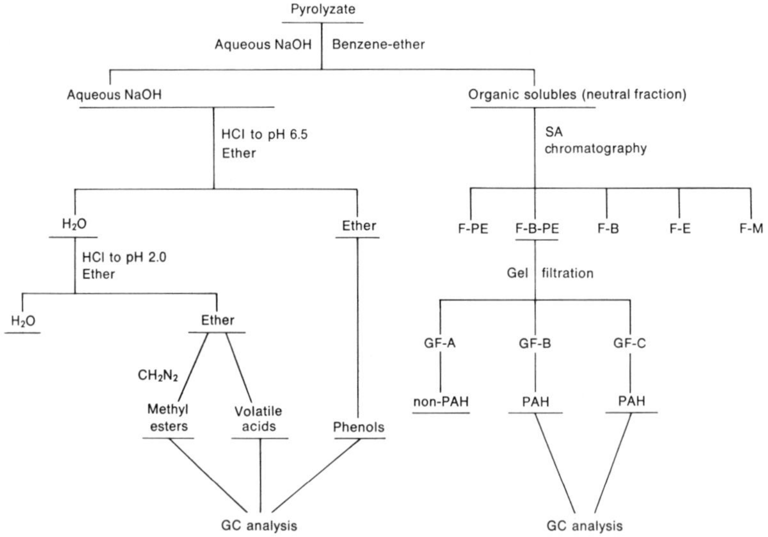

FIGURE 1. Isolation scheme for acids, phenols and PAH of tobacco pyrolyzates.

of tobacco and PE fraction were quantitated using hexatriacontane as an internal standard on a 1/8-in. x 15-ft stainless-steel Dexsil 300 GC column. Gas chromatograms were divided into groups according to relative retention time (RRT; pyrene = 1.00), and each group was named for the major PAH present (Figure 2). The following PAH groups were defined: naphthalene, RRT 0.239-0.568; fluorene, 0.581-0.731; phenanthrene, 0.754-0.951; pyrene, 0.964-1.180; chrysene, 1.203-1.359; benzopyrene, 1.384-1.594. The peak areas of each group in the chromatogram of GF-B and GF-C were summed. Assuming detector responses identical with those of the major parent PAH, weight data based on GC volatiles were calculated for each group.

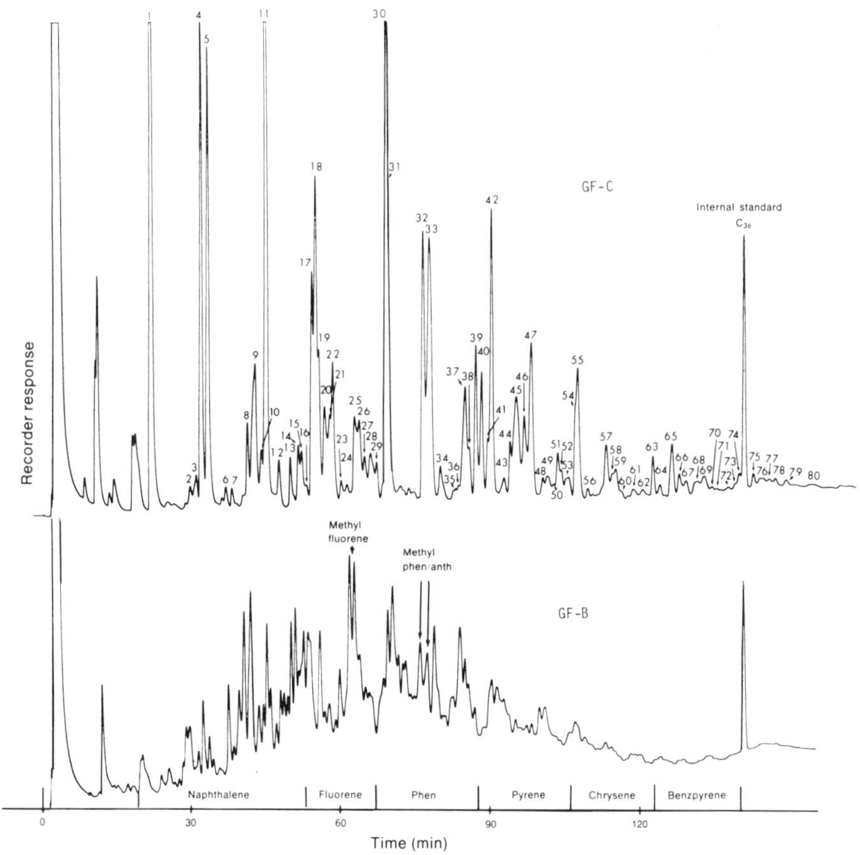

FIGURE 2. Gas chromatograms of parent PAH fraction (GF-C) and multialkyl PAH fraction (GF-B) of 700°C PE extract pyrolyzate.

The GF-C fraction for the standard pyrolyzates was analyzed on a 0.38-mm i.d. x 35-m SE-54 glass capillary column.

Experimental details on the tobacco, its PE fractions and the pyrolyzate collection, fractionation and analysis method have been reported (12,15,17, 18). PAH components in the pyrolyzates of the tobacco, its PE extract and marc were identified by GC and gel-filtration retention data, GC/MS and UV analysis of preparative GC and HPLC cuts according to methods developed in this laboratory for the identification of PAH in CSC (19-21). The major components in the subfractions of the PE extract and in the pyrolyzates of standard compounds were characterized by GC and gel-filtration elution data. The presence of cyclopentenophenanthrenes [general name given to a number of trimethylenephenanthrenes found in CSC (5,19,21)] in the sterol and steryl ester pyrolyzates was confirmed by UV and MS data on preparative GC cuts.

Standards

The standards were obtained from commercial sources and were pyrolyzed without further purification. The sterol mixture used consisted of equal amounts of cholesterol, β-sitosterol and stigmasterol. The fatty acid mixture consisted of the major tobacco acids: palmitic, oleic, linoleic, linolenic and stearic acids, in proportions normally found in flue-cured tobacco (22).

RESULTS AND DISCUSSION

Determination of Pyrolytic Conditions that Produced Correlatable Pyrolyzate-CSC PAH Profiles

Because the majority of the PAH pyrolytic precursor studies on tobacco and its components have been performed at high temperatures, the PE extract of tobacco was initially pyrolyzed at 850°C (13). However, the gas chromatograms of PAH in GF-C (containing parent PAH, most of their methyl derivatives, and some multimethyl compounds of the higher members) from the CSC of Kentucky 1R1 Reference cigarettes and from the 850°C pyrolyzate of the PE extract (Figure 3) differed in the PAH distribution (Table I). A significant reduction of methylated compounds was apparent in the extract pyrolyzate as compared to CSC. As would be expected, pyrolysis at 850°C favored the production of the more thermally stable PAH as well as of larger ring systems. It can, therefore, be concluded that even though the burning cone in a cigarette reaches and exceeds 850°C (13), most of the PAH in cigarette smoke are formed at lower temperatures.

Since the PAH distribution in the 850°C pyrolyzate did not favorably compare with that in CSC, the PE extract of tobacco was also pyrolyzed at

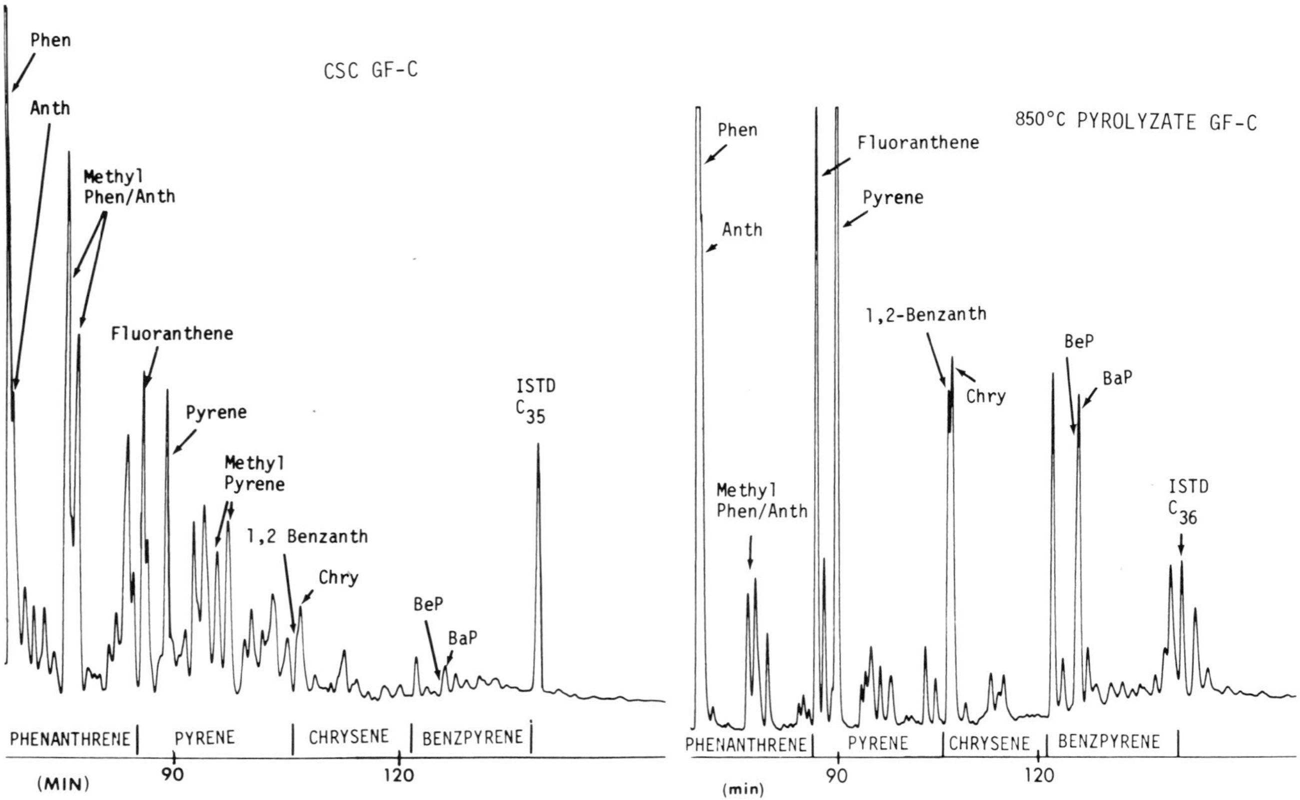

FIGURE 3. Gas chromatograms of parent PAH fraction (GF-C) of cigarette smoke and 850°C tobacco pyrolyzate.

TABLE I. PAH Components in Figures 2, 4 and 5

Compound	Peak Number	
	Figure 2	Figures 4 & 5
Indene		1
Methylindenes		2
Naphthalene	1	3
2-Methylnaphthalene	4	4
1-Methylnaphthalene	5	5
Biphenyl	7	6
2- and 1-Ethylnaphthalene		7
2,6-Dimethylnaphthalene	8	8
1-Vinylnaphthalene	8	9
1,3-Dimethylnaphthalene	9	10
1,6-Dimethylnaphthalene	9	11
2-Vinylnaphthalene	9	12
1,4- 2,3- and 1,5-Dimethylnaphthalene	10	13,14
Acenaphthylene	11	15
1,2-Dimethylnaphthalene	11	16
1,8-Dimethylnaphthalene	12	17
Acenaphthene	12	18
Methylacenaphthylenes	17,18,21	19,21
Fluorene	18	20
9-Methylfluorene	19	21
Benz[f] indene	22	22
2-Methylfluorene	25	23
1-Methylfluorene	26	24
Phenanthrene	30	25
Anthracene	31	26
3-Methylphenanthrene	32	27
2-Methylphenanthrene	32	28
4,5-Methylenephenanthrene	33	29
2-Methylanthracene	33	29
4-Methylphenanthrene	33	30
1-Methylphenanthrene	33	31
9-Methylphenanthrene	34	32
9-Methylanthracene	34	33
Fluoranthene	39	34
Acephenanthrylene	40	35
Cyclopentenophenanthrenes	46,47	36,42,44
Pyrene	42	37
Methylfluoranthenes	44,45	38,39
Benzofluorenes	45	39,40
2-Methylpyrene	46	40
1-Methylpyrene	47	41
4-Methylpyrene	47	43
Dimethylfluoranthenes	48-50	45
Benzo[g,h,i] fluoranthene	53	46

TABLE I, continued

Compound	Peak Number	
	Figure 2	Figures 4 & 5
Dimethylpyrenes	50-53	47
1,2-Benzanthracene	54	49
Chrysene/Triphenylene	55	50
Methyl-1,2-Benzanthracenes/Chrysenes	57-60	51
Benzo[b,j and k] fluoranthene	63	52
Benzo[a] fluoranthene	64	53
Benzo[e] pyrene	65	54
Benzo[a] pyrene	65	55
Perylene	66	56
Methyl-B[e] P, B[a] P	68-70	57
Indenopyrene	74	59
Benzo[g,h,i] perylene	75	60
Coronene		61

650, 700, 750 and 850°C. The following general trends in product distribution were observed (Table II): the level of parent PAH increased with temperature; the level of methyl components peaked at 800°C and then declined; the relative amounts of the large-ring compounds increased with increasing pyrolytic temperatures (for example the phenanthrene/anthracene to benzopyrene ratio was about 7:1 at 650°C, but less than 5:1 at 850°C).

The temperature study clearly showed that pyrolysis of the PE extract in our system at a temperature in the range of 650 to 750°C would yield a PAH profile most similar to that of CSC. However, the PAH composition of CSC reflects the total reactions of all the constituents in the burning tobacco. Therefore, whole tobacco was pyrolyzed at 650, 700 and 750°C, and the resulting PAH profiles were compared with CSC PAH of 1R1 cigarettes (17). The methyl PAH to parent PAH ratios for the 650 and 700°C tobacco pyrolyzates were in good agreement with the ratios for CSC. The relative amounts of selected PAH in condensate and in tobacco pyrolyzates are compared in Table III. At all three temperatures, the fluoranthene level was lower in the pyrolyzates than in CSC. The best overall compromise appeared to be pyrolysis at 700°C. And indeed, peak-by-peak comparison of the corresponding gas chromatograms showed that the GC profile of the 700°C tobacco pyrolyzate was the most similar to that of CSC (15). The gas chromatograms of the small-ring, mono- and multimethyl PAH (contained in GF-B) and of the monomethyl and parent PAH (in GF-C) of the 700°C tobacco pyrolyzate are shown in Figure 2.

TABLE II. Selected PAH Levels in Pyrolyzates of the PE Extract of Tobacco as a Function of Pyrolysis Temperature

PAH	Amount (mg/g)[a] Pyrolysis Temperature (°C) 650	700	750	800	850
Naphthalene	3.48	6.30	8.00	14.60	25.00
Methylnaphthalenes	7.09	12.30	8.64	9.17	4.78
Fluorene	1.88	2.58	2.92	4.92	3.28
Phenanthrene/Anthracene	2.12	3.26	6.64	15.30	17.80
Methylphenanthrenes/ Methylanthracenes	3.89	5.82	7.36	7.60	3.78
Fluoranthene	0.56	0.70	1.16	3.02	3.28
Pyrene	1.02	1.32	2.32	5.34	6.86
Methylpyrenes	1.19	1.55	2.20	2.40	1.36
1,2-Benzanthracene/ Chrysene/Triphenylene	0.82	1.08	2.08	1.94 2.88	2.26 3.36
Benzo[e]pyrene/ Benzo[a]pyrene	0.32	0.38	1.14	3.04	3.73
Methylbenzopyrenes	0.32	0.44	0.74	1.26	1.24

[a]Corrected for differences in detector response; methyl PAH assumed to respond identical to parent; average of two determinations.

TABLE III. Comparison of Relative Abundances of Selected PAH in CSC and Tobacco Pyrolyzates

PAH	Relative Amount[a] CSC (1R1)	Pyrolyzates 650°C	700°C	750°C
Phenanthrene/Anthracene	3.3	2.4	4.2	3.5
Methylphenanthrenes/ Methylanthracenes	9.2	3.6	7.7	3.2
Fluoranthene	0.8	0.5	0.5	0.6
Pyrene	1.0	1.0	1.0	1.0
Methypyrenes	1.3	1.4	1.3	0.9
1,2-Benzanthracene/ Chrysene/Triphenylene	0.5	0.9	0.8	0.5
Benzo[e]pyrene/ Benzo[a]pyrene	0.2	0.1	0.3	0.3

[a]Relative to pyrene; corrected for difference in detector response.

Thus, in our system, a temperature of 700°C produced a pyrolytic PAH profile for both the whole tobacco and its PE extractables similar to that of CSC. However, the PE extract produced somewhat higher yields of alkyl PAH than tobacco. Thus, it appeared that the PE extractable material from tobacco produced not only parent PAH but also their alkyl derivatives and in the proper (CSC) ratio. To test the validity of this observation, we pyrolyzed the PE-extracted tobacco (marc) at 700°C. The levels of selected PAH in the 700°C pyrolyzates of tobacco, PE extract and marc are compared in Table IV.

TABLE IV. Comparison of PAH Fraction Levels in 700°C Pyrolyzates from Tobacco, PE Extract and Marc

PAH Group	Amount[a] Tobacco (mg/100 g)	PE Extract[b] (mg/8 g)	Marc[c] (mg/92 g)
Naphthalene	320	390	130
Fluorene	110	110	86
Phenanthrene	160	150	78
Pyrene	63	82	35
Chrysene	18	26	8
Benzopyrene	19	14	5
Total	690	772	342

[a]Based on GC volatile material, assuming detector response for the group identical to that of major parent PAH.
[b]PE extract represents about 8% of tobacco weight.
[c]Marc represents about 92% of tobacco weight.

The data are based on the relative amount of each fraction present in whole tobacco. Thus, 8 g of PE extract, 92 g of marc and 100 g of tobacco are equivalent. In agreement with previous work (13), the PE extract produced the most PAH. The 8 g of extract yielded 2 to 3 times as much PAH as the 92 g of the marc. The sum of the weights of PAH produced separately by the PE extract and the marc was considerably higher than the amount produced by the whole tobacco. This is readily explained by mechanisms proposed by Badger *et al.* (3) and Jones *et al.* (8). According to the proposal, a series of bimolecular reactions occur in the pyrolytic synthesis of the PAH. Concentrating the major PAH precursors into the PE extract greatly increased the concentration of the PAH-producing intermediates in the gas phase. Since a higher concentration of reactive species greatly increases the rate of a bimolecular reaction, the pyrolysis of the PE extract would be expected to produce substantially higher levels of PAH, as was observed. When tobacco is pyrolyzed, the PAH-producing intermediates, resulting from the decomposition of PE solubles, are diluted with other non-PAH-producing products so that the probability of a PAH-forming bimolecular collision decreases. As a result, the amount of PAH produced during tobacco pyrolysis decreases.

Since the interactions of PAH with phenols and acid constituents have been implicated in explanations of the biological effects of CSC (4,6,16), it was also of interest to determine the patterns of phenol and acid formation at the various pyrolytic temperatures. Table V lists the GC volatile phenols formed by the 700°C pyrolysis of tobacco, PE extract and marc.

TABLE V. Comparison of the Phenolic Yields in 700°C Pyrolyzates of Tobacco, PE Extract and Marc

Phenol	Yield[a] Tobacco (mg/100 g)	PE Extract[b] (mg/8 g)	Marc[c] (mg/92 g)
Phenol	361	5	262
o-Cresol	75	5	63
m/p-Cresol	162	5	118
Ethyl-/Dimethylphenols	91	13	70
1-Naphthol	16	3	11
2-Naphthol	14	2	11

[a]Corrected for differences in GC detector response, based on one determination.
[b]PE extract represents about 8% of tobacco weight.
[c]Marc represents about 92% of tobacco weight.

Although optimum phenol formation may occur at a slightly lower temperature, the phenol distribution at the 700°C pyrolysis temperature should be very similar to that found in smoke. In good agreement with previous work (10), the components of the marc were the major precursors of the volatile phenols. The sum of the phenolic yields from the PE extract and marc roughly equalled the yield from tobacco. Therefore, phenols appear to be formed by thermal decomposition mechanisms different from those proposed for PAH. The marc constituents were the major precursors of the volatile acids (Table VI). Most of the high-molecular-weight fatty acid materials originated in the PE extract. The major identifiable nonvolatile acid in the tobacco and PE-extract pyrolyzates was palmitic acid. The other components in the marc pyrolyzate could not be readily identified by GC retention data.

Evaluation of Petroleum Ether Extractables as PAH Precursors

The PE extract of flue-cured tobacco constitutes from 5 to 10% of dry-leaf weight and consists of paraffinic hydrocarbons, terpenes, aliphatic esters,

TABLE VI. Comparison of the Acid Yields in 700°C Pyrolyzates of Tobacco, PE Extract and Marc

	Yield[a]		
Acid	Tobacco (mg/100 g)	PE Extract[b] (mg/8 g)	Marc[c] (mg/92 g)
Volatile Acids			
C_1-C_7[d]	33	14	23
Nonvolatile Acids			
C_{11}-C_{34}[e]	1.2	1.1	0.4

[a]Based on GC volatile material in fraction (one determination).
[b]PE extract represents about 8% of tobacco weight.
[c]Marc represents about 92% of tobacco weight.
[d]Analyzed on 10% SP-1200/2% H_3PO_4.
[e]Methyl esters analyzed on 10% Apolar 10-C.

phytosterols, steryl esters, carotenoids, chlorophyll and related compounds, fatty acids and other acids and bases. It was apparent that any meaningful evaluation of the contribution of PE-extractable leaf constituents to smoke PAH would require separation of these constituents prior to pyrolytic studies. The PE extract of flue-cured tobacco (7.63% of dry-leaf weight) was fractionated on a silicic acid column using a series of solvents of increasing polarity, ranging from PE to methanol. The 42 eluted fractions were characterized by thin-layer chromatography and combined on the basis of similar physical and chemical characteristics into eight major fractions (F-1 through F-8) for pyrolytic and compositional studies (Table VII). The major components in each fraction were characterized by various techniques including gas chromatography, column chromatography, high-pressure liquid chromatography,

TABLE VII. Silicic Acid Fractionation of Petroleum Ether Extract of Flue-Cured Tobacco

Extract Fraction	Eluting Solvents	% Total Extract
F-1	Petroleum Ether (PE)	5.27
F-2	5% Benzene/PE-20% Benzene/PE	4.35
F-3	30% Benzene/PE-40% Benzene/PE	8.93
F-4	60% Benzene/PE-90% Benzene/PE	29.25
F-5	Benzene	6.58
F-6	20% Diethyl Ether/Benzene-60% Diethyl Ether/Benzene	20.26
F-7	Diethyl Ether	15.78
F-8	Acetone-Methanol	9.58

ultraviolet and infrared spectrophotometry, and mass spectrometry (9,12,14, 20) (Table VIII).

The percentages of PAH-containing F-BPE fractions from pyrolyzate neutrals of F-2 through F-8 were similar to that from the PE extract (Table IX). Extract fraction F-1 yielded a pyrolyzate in which only 39% of the neutrals eluted in F-BPE and 52% in F-PE, which primarily contained the free paraffins and neophytadiene. F-1 and F-8 yielded relatively low amounts of PAH on pyrolysis (Table X), the former because of its thermally stable hydrocarbon content, and the latter because of its polar oxygenated constituent

TABLE VIII. Component Distribution in Fractions from Tobacco PE Extract

PE Extract Components	Relative Distribution[a]							
	F-1	F-2	F-3	F-4	F-5	F-6	F-7	F-8
Free Compounds								
C_{25}-C_{36} Hydrocarbons	+++							
Neophytadiene	+++							
Squalene		+						
Sterols					+	++		
Triterpenols					+	+		
Chlorophyll and Related Compounds						0	0	
Carotenoids						0	0	
α-Tocopherol			+					
Solanesol				+++	+			
Solanesenes				+				
Fatty Acids					+++	+++		
Bound Compounds[b]								
Fatty Acids Esters of:								
Sterols		+++	++					
Triterpenols		++	+					
Phytol		+						
Paraffinic Alcohols		+						
Solanesol			+++					
Polar Esters of:[b]								
Fatty Acids							+++	+++
Sterols							+	+
Triterpenols							+	+
Solanesol							+	+

[a]Based on GC volatiles.
[b]Determined after hydrolysis.
NOTE: +++ = greater than 40% of fraction.
++ = 20 to 40% of fraction.
+ = less than 20% of fraction.
0 = present but quantity not determined.

TABLE IX. Comparison of the Yields of F-PE Through F-M Obtained from the Organic Solubles of Tobacco and its Subfractions at 700°C

	% Yields of Chromatographic Eluates[a]										
Eluates	Marc	Tobacco	PE Extract	F-1	F-2	F-3	F-4	F-5	F-6	F-7	F-8
F-PE	3.1	7.3	10.6	51.6	4.8	4.2	3.5	5.0	2.3	1.6	1.9
F-BPE	6.6	25.9	62.5	39.1	74.3	65.9	74.0	60.4	67.3	61.9	52.3
F-B	10.8	9.9	6.8	12.3	6.1	15.3	10.6	7.5	12.3	7.8	8.8
F-E	4.5	22.9	16.2	13.5	10.9	12.0	10.1	17.6	10.2	19.5	22.4
F-M	28.3	21.6	6.8	3.9	2.1	2.5	0.8	1.9	2.0	3.2	7.3
Total	53.3	87.6	102.3	120.4	98.2	99.9	99.0	92.4	94.1	94.0	92.7

[a]Based on amount of neutral fraction placed on SA column.

TABLE X. Yield of Selected and Total PAH Obtained from Pyrolysis of PE Extract Fractions

Compounds	Yield (mg/g pyrolyzed)[a]							
	Fraction Number							
	F-1	F-2	F-3	F-4	F-5	F-6	F-7	F-8
Naphthalene	2.04	7.78	6.42	4.38	2.12	6.18	4.12	1.23
2-Methylnaphthalene	0.96	3.49	5.78	6.56	3.11	6.08	5.44	2.10
1-Methylnaphthalene	0.82	5.21	5.39	4.35	2.60	3.90	4.09	1.69
Acenaphthylene	0.80	3.00	3.45	2.60	1.39	2.84	2.17	0.92
Acenaphthene	0.12	0.29	0.33	0.27	0.24	0.26	0.31	0.16
Fluorene[b]	0.78	2.28	3.08	2.55	1.91	2.85	2.15	0.72
Phenanthrene	1.55[c]	7.80[c]	6.03[c]	3.38	1.31	3.71	1.94	0.82
Anthracene				1.05	0.37	0.94	0.71	0.29
Methylphenanthrenes/Anthracenes	1.11	8.86	8.07	5.76	2.71	6.24	2.15	2.04
Fluoranthene	0.20	1.40	0.97	0.79	0.32	0.77	0.55	0.20
Acephenanthrylene	0.13	0.87	0.77	0.46	0.20	0.54	0.38	0.14
Pyrene	0.51	0.82	1.87	1.45	0.74	1.50	1.26	0.44
Methylpyrenes & Cyclopentenophenanthrenes	0.40	2.89	2.32	1.50	0.89	1.69	1.58	0.63
1,2-Benzanthracene	0.09	3.86[d]	2.57[d]	0.53	0.47[d]	1.45[d]	0.31	0.11
Chrysene	0.20			0.66			0.42	0.16
Methyl-1,2-benzanthracenes/Chrysenes	0.15	2.31	0.04	1.20	0.46	1.31	0.65	0.24
Benzo[b,j and k] fluoranthene	0.013	0.76	0.46	0.46	0.096	0.41	0.14	0.047
Benzo[a and e] pyrenes	0.029	1.82	0.88	0.93	0.22	0.67	0.37	0.12
Total PAH[e]	21.78	101.8	100.4	78.2	58.3	81.7	96.1	36.3

[a]Corrected for difference in detector response; methyl PAH assumed to respond identically to parent PAH.
[b]Contains considerable amounts of methylacenaphthylene.
[c]Not resolvable because of large amounts of phenanthrene present.
[d]Not resolvable because of larger amounts of chrysene present.
[e]Based on total GC volatile material in PAH groups of GF-B and GF-C, assuming detector response for the group identical to that of major parent PAH.

content. Such oxygenated compounds, having a relatively low carbon content yield low amounts of the alkyl residuals essential for PAH formation (3,13). F-2, F-3 and F-4 produced the greatest amounts of PAH on pyrolysis.

For correlation with cigarette-smoke PAH, the yields of PAH from F-1 to F-8 pyrolyzates should be expressed as percentages of the total PAH formed on pyrolysis of the entire PE extract, reflecting the relative abundance of each fraction in tobacco leaf. Inspection of Table XI, which reflects the relative abundance of each fraction, shows that the components in F-1, F-5 and F-8 contributed very little to the total PAH production. F-4 and F-6 were the major PAH producers. F-2 and F-3, the most potent PAH precursors per gram, produced considerable quantities of chrysene.

TABLE XI. Relative Yields of Selected PAH Produced by Individual Fractions F-1 to F-8[a]

	F-1	F-2	F-3	F-4	F-5	F-6	F-7	F-8
Phenanthrene/Anthracene	2	9	13	32	8	23	10	2
Methylphenanthrenes/ Anthracenes	1	8	15	35	4	26	7	4
Pyrene	2	9	12	32	4	23	14	3
Methylpyrenes/ Cyclopentenophenanthrenes	1	9	12	30	4	23	17	4
1,2-Benzanthracene/Chrysene	2	16	22	16	3	28	11	2
Methyl-1,2-benzanthracenes/ Chrysenes	1	10	18	34	3	26	6	2
Benzo[e and a] pyrenes	1	12	12	42	2	20	9	2
Total PAH Yield	1	6	11	31	6	26	17	2

[a]Based on relative abundance of each fraction in the total tobacco PE extract.

The data obtained here as well as in other studies (13) clearly indicate that the sterols and related triterpenes are important in the pyrosynthesis of tobacco-smoke PAH. However, the PE-extract fractions with high PAH production efficiency, except for the F-4 solanesol fraction, also contained other components such as fatty acids, carotenoids, and chlorophyll and related compounds. Thus, we pyrolyzed representative standards of these materials to determine their relative importance in the pyrosynthesis of smoke PAH (Table X). Solanesol and the constituents of F-4 produced pyrolyzates with essentially identical PAH distributions (Figure 4). GC profiles of PAH from solanesol and F-4 were also very similar to CSC PAH profile.

The sterol mixtures produced PAH levels very comparable to that from solanesol. As observed in other studies, the sterols and sterol esters, due to their phenanthrene-type ring structure, produced high levels of phenanthrenes

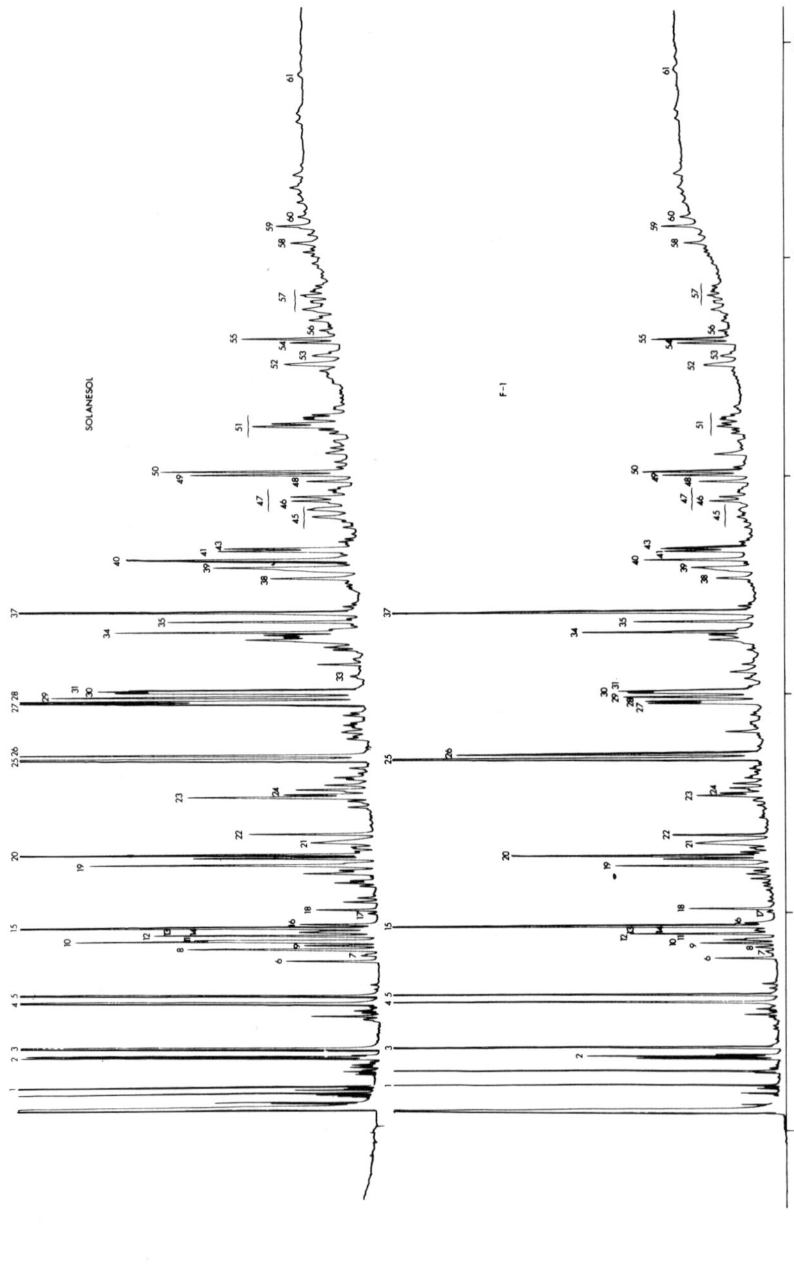

FIGURE 4. Gas chromatograms of PAH fractions from pyrolyzates of solanesol and tobacco hydrocarbons (contained in F-1).

and chrysenes (13). The chrysenes were probably formed by alkyl insertion into the cyclopentane ring of the sterol moiety, analogous to the mechanism proposed by Schmeltz *et al.* (13) for the formation of naphthalenes from alkylated indenes. If the insertion mechanism is important in chrysene production, considerable amounts of 1-methylchrysene should have been formed. Relative to the pyrolyzates of the other standards, the sterol pyrolyzate did contain higher levels of 1-methylchrysene. The same mechanism would also explain the unusually high levels of 1-methylnaphthalene observed in the sterol and steryl ester pyrolyzates. However, relative to solanesol, the sterol-ring moiety is not an efficient alkyl-chrysene precursor. Not unexpectedly, the sterols are also the predominant producers of the cyclopentenophenanthrene peaks found in CSC (5,20,21). Except for the lower levels of the benzopyrenes, the sterols and sterol esters produced essentially identical PAH profiles (Figure 5).

The large C_{40} terpene, β-carotene, also produced high levels of PAH, especially alkylated PAH (Figure 5). Along with solanesol, it may be responsible for the large amounts of the multialkylated PAH found in CSC (5,17,20). The more-saturated C_{30} terpene, squalene, chlorophyll and the fatty acids were relatively poor PAH producers. The squalene pyrolyzate produced a PAH profile high in alkylated PAH and nearly identical to that of β-carotene. The PAH profiles for the pyrolyzates of fatty acids and chlorophyll very closely resembled the profile for the pyrolyzate of the tobacco hydrocarbons (Figure 4).

Using the PAH profiles obtained from the pyrolyzates of the standard compounds, it was then possible to interpret the PAH data obtained in the pyrolysis of the PE extract fractions. The high levels of naphthalene, 1-methylnaphthalene, phenanthrenes, cyclopentenophenanthrene and chyrsene produced by F-2 indicate that its major PAH precursors were the steryl esters and the triterpenol esters. It can also be postulated that the esters of β-amyrin and cycloartenol, the major triterpenols in tobacco (14) are more efficient benzopyrene precursors than the sterol esters. The steryl and solanesyl esters in F-3 probably contributed about equally to the PAH products from F-3. Solanesol is completely responsible for the PAH from F-4. The very high concentration of the fatty acids in F-5 must have far outweighed the effect of the sterols and triterpenols, greatly reducing PAH production by this fraction. The cartenoids and free sterols in F-6 were responsible for its high production of alkylated and parent PAH, respectively. The carotenes and chlorophyll-related compounds were most likely responsible for the PAH distribution obtained from F-7.

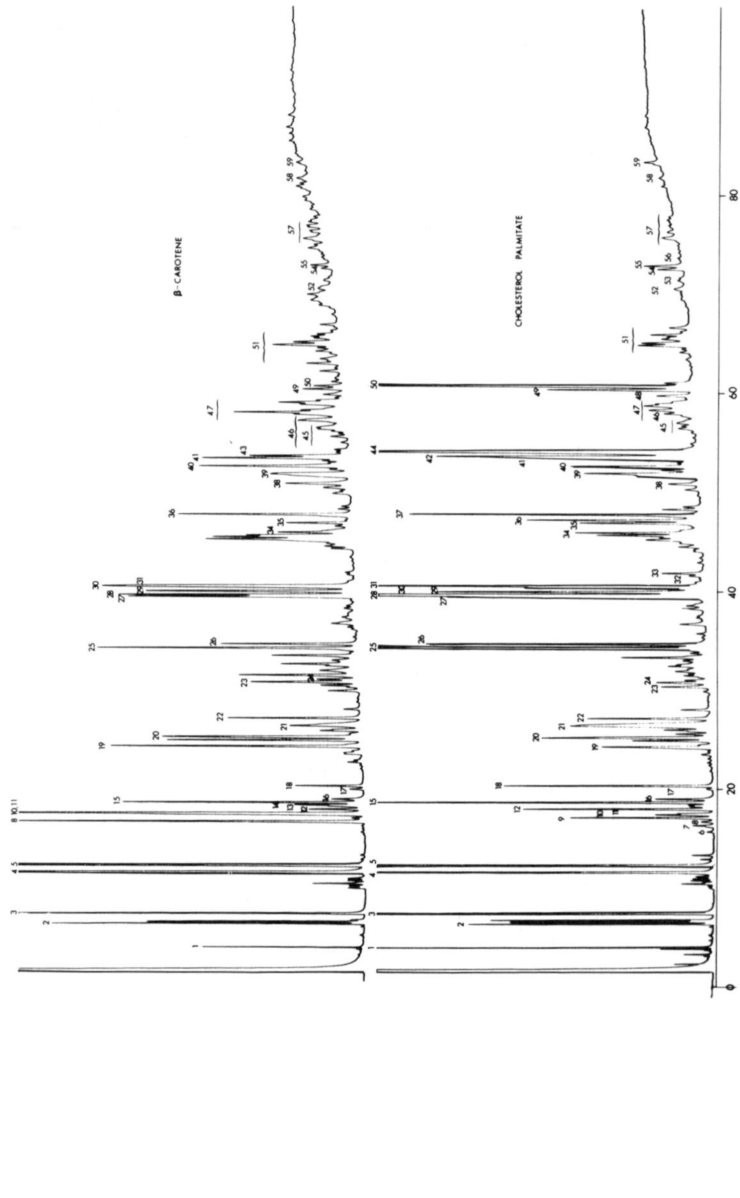

FIGURE 5. Gas chromatograms of PAH fractions from pyrolyzates of β-carotene and cholesterol palmitate.

TABLE XII. Yield of Selected PAH in Standard Pyrolyzates[a]

PAH	Yield (mg/g pyrolyzed)						
	Standards Pyrolyzed						
	Solanesol	Sterols	Cholesterol Palmitate	β-Carotene	Squalene	Chlorophyll	Fatty Acids
Naphthalene	5.54	5.88	2.51	1.50	0.83	0.52	0.37
2-Methylnaphthalene	4.53	1.60	0.848	4.25	0.90	0.23	0.088
1-Methylnaphthalene	3.08	3.05	2.18	2.84	0.62	0.24	0.120
Acenaphthylene	1.64	2.42	0.75	0.91	0.21	0.14	0.092
1-Methylacenaphthylene	0.14	0.81	0.26	0.22	0.056	0.027	0.020
Fluorene	1.29	1.27	0.34	0.43	0.17	0.057	0.079
Phenanthrene	3.44	7.32	2.58	0.59	0.17	0.11	0.14
Anthracene	1.04	0.94	0.39	0.32	0.059	0.032	0.040
Methylphenanthrenes/Anthracenes	4.16	11.9	4.01	2.52	0.42	0.16	0.13
Fluoranthene	0.69	0.64	0.22	0.19	0.037	0.017	0.030
Acephenanthrylene	0.42	0.58	0.25	0.17	0.030	0.013	0.018
Pyrene	1.19	1.74	0.61	0.44	0.076	0.055	0.053
Methylpyrenes/Cyclopentenophenanthrenes	1.21[b]	4.57[c]	8.36[c]	1.04[b]	0.14[b]	0.11[b]	0.059[b]
1,2-Benzanthracene	0.55	0.84	0.24	0.16	0.023	0.013	0.011
Chrysene	0.66	3.07	0.79	0.13	0.025	0.019	0.012
Methylchrysenes and 1,2-Benzanthracenes	1.26	1.98	0.43	0.88	0.046	0.042	0.016
Benzo[b,j and k] fluoranthene	0.32	0.23	0.037	0.062	0.011	0.0074	0.0040
Benzo[e] pyrene	0.22	0.27	0.047	0.035	0.009	0.0053	0.0037
Benzo[a] pyrene	0.40	0.44	0.074	0.077	0.014	0.0092	0.0053

[a]Based on amount of PAH in GF-C only.
[b]Contains only trace amounts of cyclopentophenanthrenes.
[c]Cyclopentophenanthrenes are major components.

SUMMARY

Analyses showed that the pyrolysis of tobacco and its PE extract at 700°C produced PAH profiles comparable to those found in CSC. The data indicated that most of the alkyl PAH and the major PAH in cigarette smoke are derived from the PE extractables of tobacco. The constituents of the marc (tobacco remaining after extraction with PE) were the major precursors of phenols, oxygenated PAH and low-molecular-weight acids; the components of the PE extract were the major producers of high-molecular-weight acids. Comparison of pyrolysis data on subfractions of the PE extract and of standard compounds showed that the high-molecular-weight terpenes, solanesol, the sterols and related compounds, and the carotenoids were the major PAH precursors in the PE-soluble fraction of tobacco. The sterols and related compounds were important precursors of phenanthrenes and chyrsenes and responsible for essentially all of the cyclopentenophenanthrene found in the pyrolyzate of the PE extract. The carotenoids and solanesol are most likely responsible for the high levels of the multialkylated PAH found in the PE-extract pyrolyzate and, by analogy, in CSC. Because of its relative abundance in leaf, solanesol may contribute as much as 40% of the benzopyrenes produced on pyrolysis of the PE extract of tobacco.

REFERENCES

1. Akin, F. J., Chamberlain, W. J., and Chortyk, O. T. (1975): Mouseskin tumorigenesis and induction of aryl hydrocarbon hydroxylas by tobacco smoke fractions. *J. Nat. Cancer Inst.* 54:907-912.
2. Akin, F. J., Snook, M. E., Severson, R. F., Chamberlain, W. J., and Walters, D. B. (1976): Identification of polynuclear aromatic hydrocarbons in cigarette smoke and their importance as tumorigens. *J. Nat. Cancer Inst.* 57:191-195.
3. Badger, G. M., Kimber, R. W. L., and Novotny, J. (1964): The formation of aromatic hydrocarbons at high temperature. XXI. The pyrolysis of *n*-butylbenzene over a range of temperatures from 300-900°C at 50°C intervals. *Aust. J. Chem.* 17:778-786.
4. Bock, F. G., Swain, A. P., and Stedman, R. L. (1970): Composition studies on tobacco. XLI. Carcinogenesis assay of subfractions of the neutral fraction of cigarette smoke condensate. *J. Nat. Cancer Inst.* 44:1305-1310.
5. Brunneman, K. D., and Hoffmann, D. (1976): Analysis of polynuclear aromatic hydrocarbons in the respiratory environment. In: *Carcinogenesis, Vol. 1*, R. I. Freudenthal and P. W. Jones, Eds., pp. 283-297, Raven Press, New York.

6. Hoffman, D., and Wynder, R. L. (1971): A study of tobacco carcinogenesis. XI. Tumor initiators, tumor accelerators, and tumor-promoting activity of condensate fractions. *J. Nat. Cancer Inst.* 52:613-615.
7. Jenkins, R. W., Jr., Newman, R. H., Carpenter, R. D., and Osdene, T. S. (1970): Cigarette smoke formation studies. I. Distribution and mainstream products from added ^{14}C-dotriacontane-16,17. *Beitr. Tabakforsch.* 5:292-298.
8. Jones, T. C., and Schmeltz, I. (1969): Pyrolysis of *t*-cinnamic acid, sodium *t*-cinnamate, styrene, distyryl, and *c*- and *t*-stibbene. Products and implications. *J. Org. Chem.* 34:645-649.
9. Schlotzhauer, W. S. (1978): Liquid chromatographic separation of chlorophyll products in tobacco leaf. *Tob. Sci.* XXII:44-45.
10. Schlotzhauer, W. S., Severson, R. F., Chortyk, O. T., and Snook, M. E. (1975): Pyrolytic precursors of polynuclear aromatic hydrocarbons in the petroleum ether extract of tobacco. Paper presented at 29th Tobacco Chemists' Research Conference, College Park, MD.
11. Shlotzhauer, W. S., and Schmeltz, I. (1968): Pyrogenesis of aromatic hydrocarbons present in cigarette smoke. I. Role of the hexane-soluble fraction of tobacco. *Beitr. Tabakforsch.* 4:176-181.
12. Schlotzhauer, W. S., Severson, R. F., Chortyk, O. T., and Snook, M. E. (1976): Pyrolytic formation of polynuclear aromatic hydrocarbons from petroleum-ether extractable constituents of flue-cured tobacco leaf. *J. Agric. Food Chem.* 24:992-996.
13. Schmeltz, I., and Hoffmann, D. (1976): Formation of polynuclear aromatic hydrocarbons from combustion of organic matter. In: *Carcinogenesis, Vol. 1,* R. I. Freudenthal and P. W. Jones, Eds., pp. 225-239, Raven Press, New York.
14. Severson, R. F., Ellington, J. J., Arrendale, R. F., and Snook, M. E. (1978): A quantitative method for GC-volatile aliphatic hydrocarbons, terpenes, fatty alcohols, fatty acids, and sterols in tobacco. *J. Chromatog.* 106:155-168.
15. Severson, R. F., Schlotzhauer, W. S., Arrendale, R. F., Snook, M. E., and Higman, H. C. (1977): Correlation of polynuclear aromatic hydrocarbon formation between pyrolysis and smoking. *Beitr. Tabakforsch.* 9:23-37.
16. Severson, R. F., Snook, M. E., Akin, F. J., and Chortyk, O. T. (1978): Correlation of biological activity with polynuclear aromatic hydrocarbon content of tobacco smoke fractions. In: *Carcinogenesis, Vol. 3,* P. W. Jones and R. I. Freudenthal, Eds., pp. 115-130, Raven Press, New York.
17. Severson, R. F., Snook, M. E., Arrendale, R. F., and Chortyk, O. T. (1976): Gas chromatographic quantitation of polynuclear aromatic hydrocarbons in tobacco smoke. *Anal. Chem.* 48:1866-1872.
18. Severson, R. F., Snook, M. E., Chortyk, O. T., and Arrendale, R. F. (1976): A chromatographic analysis for polynuclear aromatic hydrocarbons in small quantities of cigarette smoke condensate. *Beitr. Tabakforsch.* 9:273-282.

19. Snook, M. E., Severson, R. F., Arrendale, R. F., Higman, H. C., and Chortyk, O. T. (1977): The identification of high molecular weight polynuclear aromatic hydrocarbons in a biologically-active fraction of cigarette smoke condensate. *Beitr. Tabakforsch.* 9:79-106.
20. Snook, M. E., Severson, R. F., Arrendale, R. F., Higman, H. C., and Chortyk, O. T. (1978): Multi-alkylated polynuclear aromatic hydrocarbons of tobacco smoke: Separation and identification. *Beitr. Tabakforsch.* (in press).
21. Snook, M. E., Severson, R. F., Higman, H. C., Arrendale, R. F., and Chortyk, O. T. (1976): Polynuclear aromatic hydrocarbons of tobacco smoke: Isolation and identification. *Beitr. Tabakforsch.* 8:250-272.
22. Tso, T. C., and Chew, H. (1970): Effect of growth . . . and flue-curing on fatty acid composition of tobacco. *Agron. J.* 62:512-514.
23. Wynder, E. L., Wright, G., and Lam, J. (1958): A study of tobacco carcinogenesis. II. The role of pyrolysis. *Cancer* 11:1140-1148.

Polynuclear Aromatic Hydrocarbons, edited by P. W. Jones and P. Leber. Ann Arbor Science Publishers, Inc. Ann Arbor, MI ©1979.

DEVELOPMENT OF A PROTOTYPE INSTRUMENT FOR FIELD MONITORING OF PAH VAPORS

A. R. Hawthorne, J. H. Thorngate,[1]
R. B. Gammage and T. Vo-Dinh

Health and Safety Research Division
Oak Ridge National Laboratory
Oak Ridge, Tennessee 37830

Many polynuclear aromatic hydrocarbon (PAH) compounds have been shown to be carcinogenic and many others may act as cocarcinogens or tumor promoters. Considerable emphasis is being placed on the mechanisms of formation of PAH and their metabolic products, as well as on the biological interaction of PAH and the separation, detection and characterization of these compounds in environmental and biological samples. The first two meetings of the International Symposium on Polynuclear Aromatic Hydrocarbons (3,11) revealed the level of interest in PAH measurement, chemistry, metabolism and carcinogenesis.

A new potentially large source of human and environmental exposure to PAH is from coal conversion facilities (2,12). With the ever dwindling sources of petroleum fuels and the increased United States dependence on foreign sources of petroleum, the Department of Energy is committed to the development of our extensive coal resources. An important part of this commitment is the production of synthetic fuels as a substitute for both oil and gas, with a concurrent commitment to ensure that minimal adverse effects occur to human health and the environment as a result of new energy technologies. This will include efforts to ensure that adequate industrial hygiene measures are provided to protect occupational workers from exposure to PAH. This paper describes the development of a field-portable monitor that can provide real-time measurement of volatile PAH in various work areas.

[1]Present address: Lawrence Livermore Laboratory, Livermore, California 94550.

MONITORING NEEDS

It is well established that coal conversion facilities provide potential exposure to a wide range of toxic compounds (2,12); among these are the PAH. Exposure to these compounds can occur through multiple pathways, including contaminated surfaces, respirable particulate matter and, for the lower-mass PAH, through the vapor phase. Workers may encounter both chronic and acute exposures to PAH, in contrast to the general public and most of the environment where exposures are more likely to be persistent low-level exposures.

Since there is a potential for worker exposure to compounds that give no immediate warning signs and yet may cause cancer decades later, a real-time continuous portable monitor is needed to warn of elevated concentrations in work areas. This monitoring is needed in addition to measurements of integrated exposures obtained by personnel monitors requiring subsequent laboratory analysis. Portable real-time monitors will also prove valuable in detecting small leaks in process streams and in pinpointing sources of exposures.

It has been proposed that an easily detected proxy compound be monitored at conversion facilities to provide an indication of potential exposure to PAH. The National Institute for Occupational Safety and Health (NIOSH) report, *Recommended Health and Safety Guidelines for Coal Gasification Pilot Plants,* suggests the use of CO as an indicator substance for all the compounds in a process stream (12). There are several positive factors for using this compound:

1. It can be easily monitored in real time by commercially available personal or remote samplers.
2. It can be analyzed in areas where resources and technical skills are limited.
3. It is not present in ambient air at high or widely fluctuating concentrations.
4. It is free from interfering substances in the process stream or ambient air.
5. It is a regulated compound, so that measurements serve a dual purpose.
6. The ratio of CO to PAH is approximately constant at a given location in the process stream under equilibrium conditions.

Under normal operations, where detection of exposures due to small leaks is the object of monitoring, the presence of this gas serves as an adequate indicator that other compounds may be escaping into the work zones. It does not, however, serve to indicate buildup of low-boiling-point PAH vapors from spills, contaminated surfaces or open process equipment, where there is no continuous source of CO to provide an indication of possible exposure. This situation may lead to false assurances that there is no exposure to PAH.

We propose that, in addition to CO as an indicator substance, one of the more volatile PAH be used as an indicator of possible PAH buildup. For example, 2-methylnaphthalene, which occurs as a major PAH component in synthetic fossil fuels (5), not only can be used as an indicator of a leak in a process stream where CO is found, but also can lead industrial hygienists to spills and point sources of PAH exposure where CO is not present.

This paper discusses the development of a portable monitor that can be used to measure volatile PAH as a part of an industrial-hygiene area monitoring and leak-detection program. Results are preliminary, as the prototype instrument only recently has been completed further optimization will be required before field testing.

PROTOTYPE INSTRUMENT

Ultraviolet (UV) absorption spectrometry provides good sensitivity for the detection of most PAH. However, selectivity between various PAH in a mixture poses a problem since these compounds often exhibit considerable spectral overlap. Derivative UV absorption spectrometry is a technique that permits improved selectivity between overlapping spectral bands (9,13,14). In particular, this procedure tends to enhance the signal from compounds that exhibit narrow-band absorption while diminishing the signals from compounds with broad-band absorption.

Work is in progress at Oak Ridge National Laboratory (ORNL) to develop a prototype second-derivative UV absorption spectrometer (DUVAS) for analyzing parts-per-billion levels of many trace gases in air. Figure 1 shows the prototype spectrometer with the 1-m multipass air-sampling cell attached. A particular emphasis is being placed on the detection of volatile PAH compounds. This instrument is designed to be smaller and substantially faster in spectrum scanning compared to a commercially available spectrometer, the Lear-Seigler, Inc., Model SM-400 second-derivative spectrometer (6). It is also microcomputer-controlled whereas the SM-400 is not.

Theory

Detailed treatment of the theory of wavelength-modulated derivative absorption spectrometry may be found elsewhere (1,7,9,14). Briefly, the technique consists of modulating the incident wavelength by a few nanometers and detecting the second harmonic of the modulation frequency. This signal gives a response that is proportional to the concentration of the compound being measured. Since a second-derivative spectrum is a measure of the curvature of the absorption spectrum and not of the transmitted light, the technique of second-derivative absorption has the important advantage of

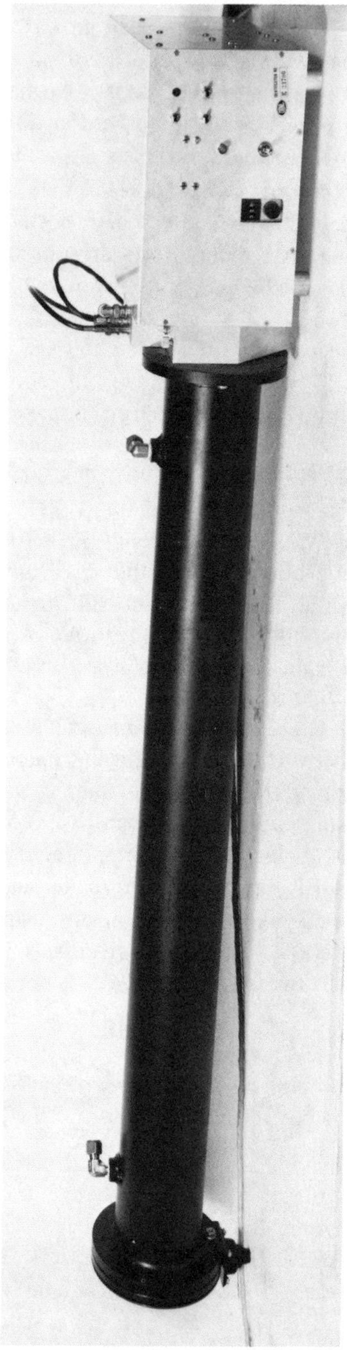

FIGURE 1. Photograph of DUVAS with 1-m multipass air-sampling cell attached.

being independent of sample opacity, light intensity fluctuations and source-energy variations. This advantage is particularly useful when no sample preparation to remove particulate matter is performed prior to analysis.

Optics

The optical system of the prototype DUVAS consists of a deuterium-lamp light source, monochromator, sample compartment and photomultiplier tube (PMT). The small monochromator is a 10-cm holographic grating monochromator having a resolution of 8 nm/mm. Wavelength modulation is achieved by replacing one of the mirrors in the monochromator with a commercially available taut-band vibrator that oscillates with a frequency of 400 Hz. The second harmonic signal at 800 Hz is sufficiently removed from the 120-Hz line voltage noise to allow good filtering, thus improving the noise level.

Another distinct advantage of the higher modulation frequency relative to the commercial SM-400 second-derivative spectrometer is that, when coupled with an integrating technique (discussed in the next section), much faster scanning speeds can be obtained. Whereas the SM-400 has scanning speeds of 2.5 or 10.0 nm/min, the DUVAS scanning speed can be varied between 30 and 240 nm/min, thus allowing a sample scan to be completed in considerably less time.

Monochromator positioning and scanning are accomplished with a four-phase bidirectional stepping motor. When used with a 2:1 gear reduction, this motor gives a stepping resolution of 0.25 nm/step.

Electronics

Figure 2 is a block diagram of the DUVAS showing the relationships of the major electronic components of the prototype instrument. In addition to the previously mentioned deuterium lamp, PMT and preamp, stepping motor, and mirror coils, the electronic components consist of three printed circuit (PC) boards.

One PC board contains an F8 microcomputer that provides the control logic for scanning a spectrum. It positions the stepping motor to a beginning wavelength, scans to the stopping wavelength at a selected speed from 30 nm/min to 240 nm/min, and stores data in its memory at resolutions of 0.25, 0.50, 0.75 or 1.00 nm/data point. The data points are obtained by counting the pulses from a voltage-to-frequency (V/F) converter for a time interval determined by the scanning speed (*e.g.*, a scanning speed of 120 nm/min gives an integration time of 0.125 sec/point). This integration of the second-derivative signal helps to improve signal-to-noise (S/N) ratios without requiring the much slower scanning speeds necessary to a low-pass electronic filter

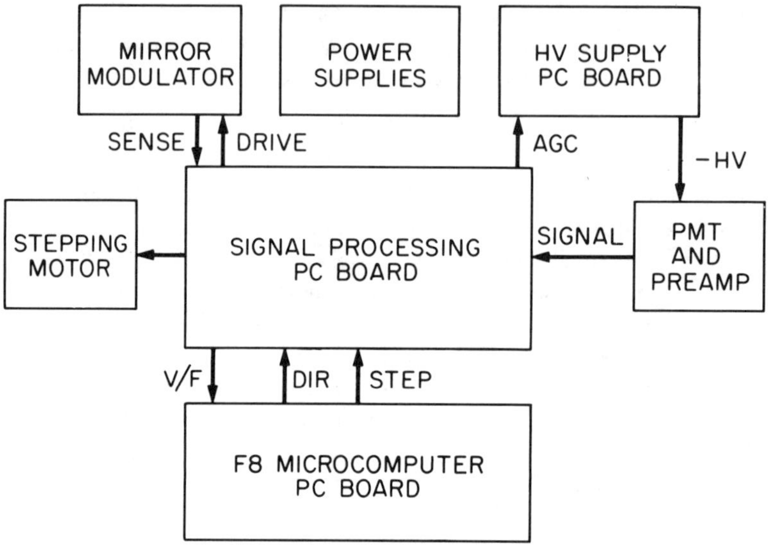

FIGURE 2. Block diagram of DUVAS major electronic components.

having the same S/N ratio. The F8 microcomputer also provides a parallel interface to external computers that can transfer scanning parameters to the F8 and receive a spectrum scan from the F8. A teletype interface is also provided, permitting the use of a teletype as an input/output device when desired.

The second PC board is a commercially available voltage-programmable high-voltage (HV) supply for the photomultiplier tube. A control voltage input of 0 to -5 V gives a 0- to -1,200-V output. The programmable feature allows an automatic-gain-control (AGC) signal to the HV board to control the gain of the PMT by varying the applied voltage. The second-derivative signal is not affected if the time constant of the AGC is slow relative to the 400-Hz modulation frequency because varying the gain is equivalent to changing the light intensity; the second-derivative signal is insensitive to light intensity changes. The AGC feature acts as an automatic iris to ensure that dc signals from the PMT are neither too small nor too large, thus ensuring that signals fall within the optimum range for the signal processing electronics.

The final PC board contains signal processing and control circuits. Figure 3 is the block diagram of this PC board. The AGC circuit can be adjusted for the desired dc signal from the PMT. This AGC circuit then acts as a servo control to adjust the PMT voltage to give the desired output. Also, on this PC board is the mirror-driving and -sensing circuit. This circuit oscillates the

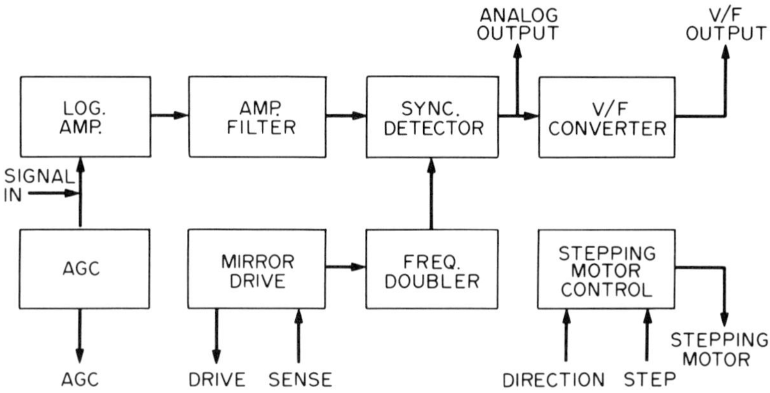

FIGURE 3. Block diagram of DUVAS signal-processing and control circuits.

mirror and provides a 400-Hz reference signal to a frequency-doubling circuit, which gives an 800-Hz reference signal for use in the synchronous detector circuit. A stepping-motor control circuit takes a direction signal and step pulse from the F8 microcomputer and provides the proper sequence of pulses to the four-phase stepping motor.

The signal from the PMT, which has an 800-Hz ac component containing the second-derivative information, is input to a logarithmic amplifier circuit. This circuit eliminates the need for the divider circuit that the SM-400 requires and also eliminates a nonlinearity in the second-derivative signal (9). Following the log circuit is a bandpass filter/amplifier with a central frequency of 800 Hz. This circuit filters out both high and low frequencies, including 120-Hz noise from line voltage and some of the 400-Hz signal due to the first-derivative signal. A synchronous detector circuit then selects only signals with an 800-Hz frequency and, with a subsequent filter/amplifier circuit, converts this ac signal into a dc signal proportional to the second derivative of the absorption curve of the measured compound at the measured wavelength. This analog signal is available at the rear panel of the instrument for use with a chart recorder. The final circuit on the board is a V/F converter; this provides a frequency signal proportional to the second-derivative signal for the F8 microcomputer to record as digital data.

Computer Interface

As stated, the F8 microcomputer has a parallel interface for communication with an external computer. The DUVAS has been operated with a Digital Equipment Corporation (DEC) DECLAB 11/03 minicomputer and a

Commodore PET 2001 microcomputer. Figure 4 shows a spectrum as displayed on the 11/03 system. For laboratory use the PDP 11/03 provides much flexibility, with dual floppy discs for mass data storage and program storage plus a video display for plotting spectra and displaying analyses.

For field use the PET 2001 is an inexpensive microcomputer. It provides a 9-in. video display, a communication keyboard, a magnetic-tape cassette data recorder, and a microcomputer programmable in BASIC for data analysis (Figure 5). With the PET keyboard, the user may define the spectral region to be scanned, as well as the speed and resolution. The spectrum is displayed on the screen and may be permanently saved on magnetic tape. Software development is currently in progress on a series of programs for controlling the spectrometer, plotting spectra in a variety of modes, and analyzing the data by comparison with a cassette of library spectra.

Modes of Operation

The DUVAS can be used to analyze both liquid and gaseous samples. As a trace gas analyzer, a 1-m multipass sampling cell is attached to the spectrometer. The total pathlength of light through the sample is adjustable, with a 12-m pathlength typically used. An air-sampling pump pulls air through the cell with a flowrate of approximately 4 liter/min. A heating blanket surrounds the sample cell and can be used to prevent condensation of low-vapor-pressure compounds within the cell that would otherwise produce anomolies or background signals.

The DUVAS can also be operated in a liquid analysis mode. A 1-cm quartz cuvette is used to hold a 1-ml sample. Water, cyclohexane and ethanol are suitable UV-transparent solvents for use in second-derivative spectrometry. Since DUVAS is insensitive to sample turbidity (10), it may be operated as an online analyzer for monitoring wastewater streams. A small sampling pump is used to pull the water through a flowthrough cell.

ANALYTICAL EXAMPLES

Since the construction of the field version of DUVAS has been completed only recently, analytical examples using this instrument are limited. Optimization of the electronics and operating parameters is still in progress to improve signal-to-noise ratios and to enhance selectivity between compounds. Figure 6 shows, however, two calibration curves for PAH compounds in cyclohexane solvent obtained using the new instrument. Second-derivative spectra of several PAH compounds in ethanol obtained from DUVAS are given in Figure 7.

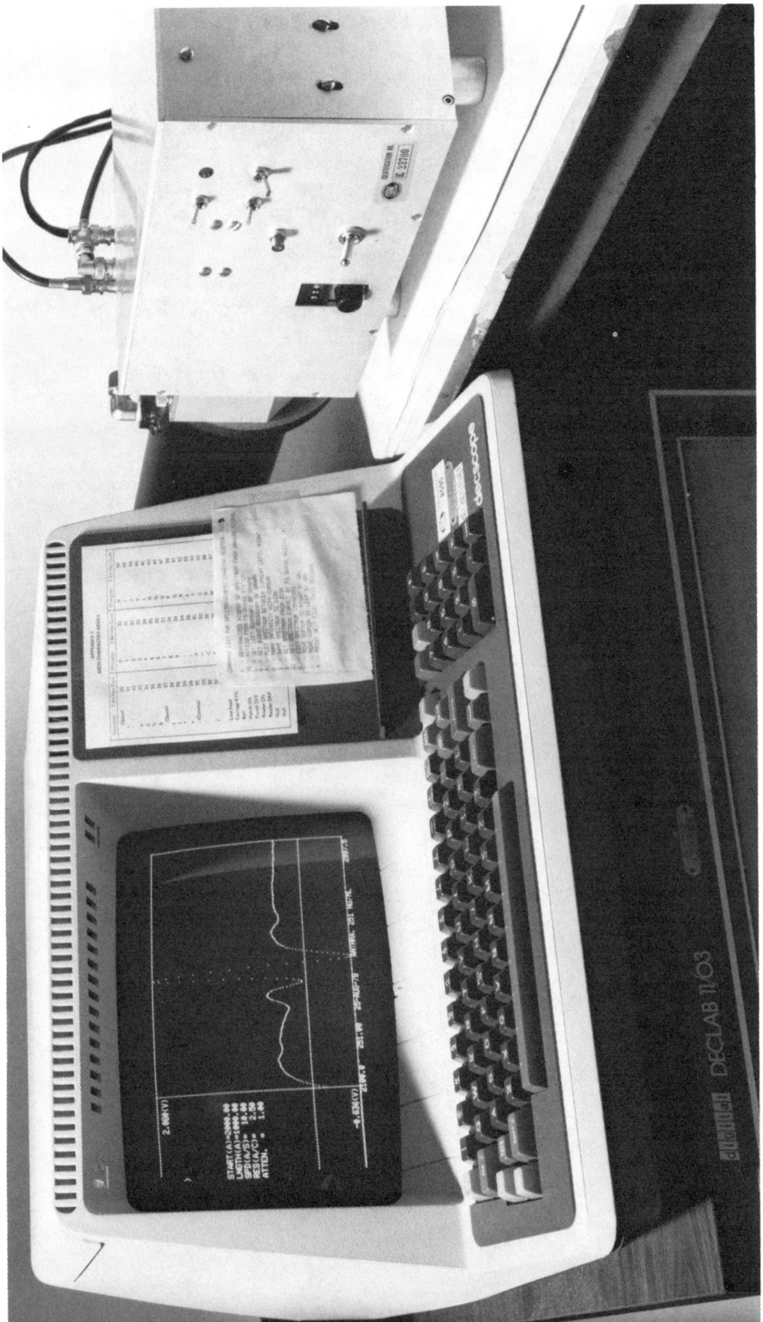

FIGURE 4. DUVAS spectrum displayed on a DECLAB 11/03 system.

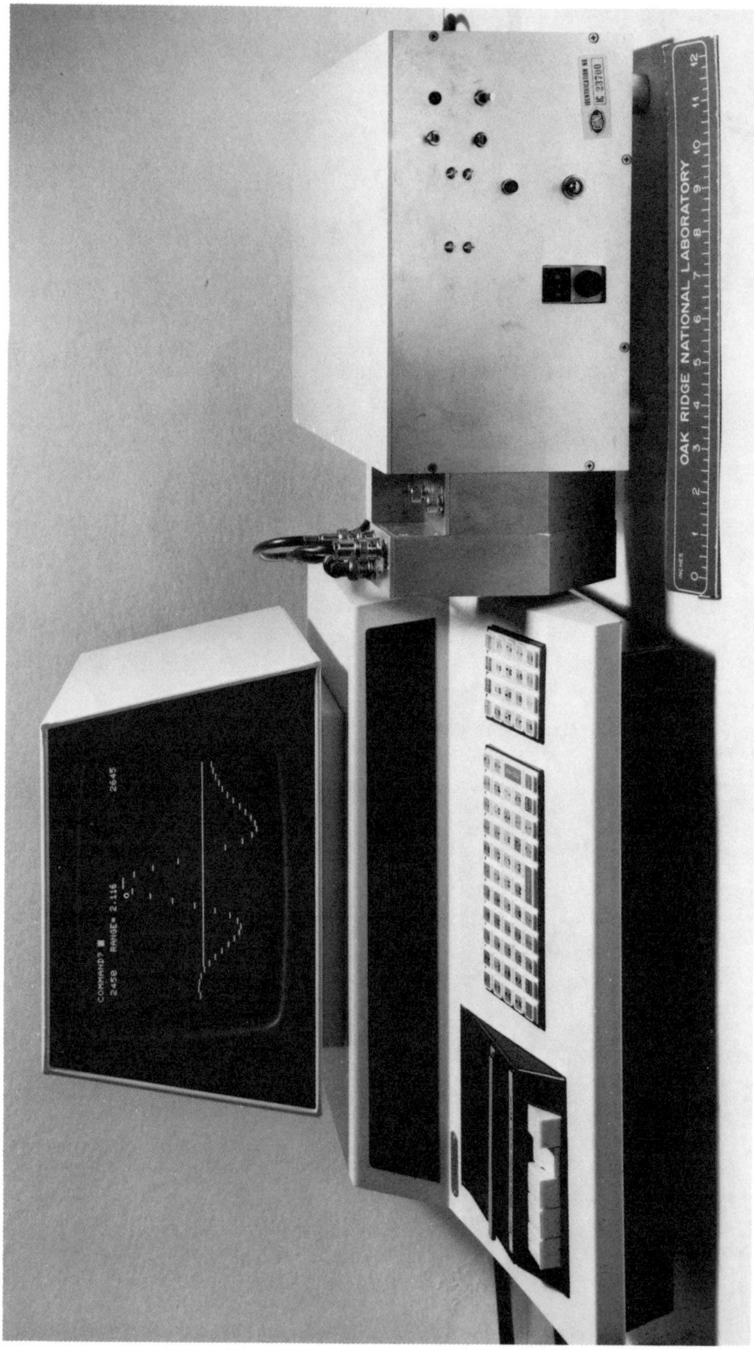

FIGURE 5. DUVAS with liquid sampling compartment attached and PET-2001 microcomputer used for data analysis and I/O purposes.

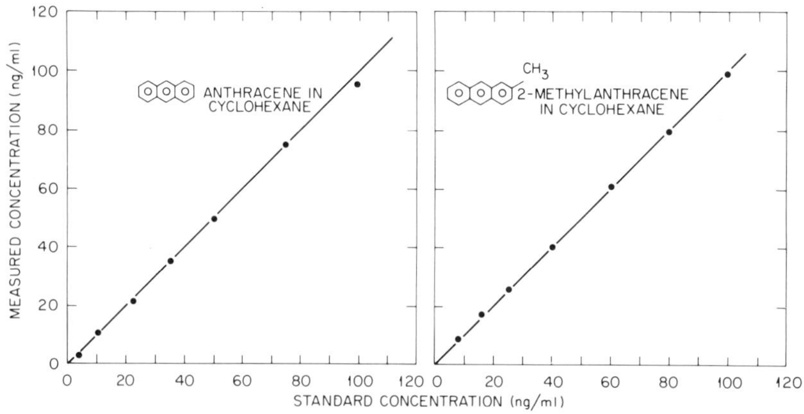

FIGURE 6. Calibration curves for two PAH compounds.

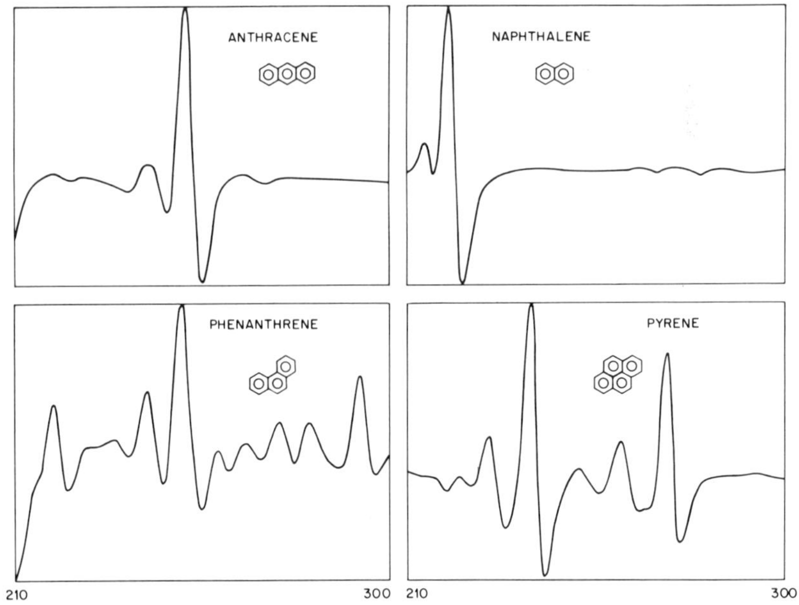

FIGURE 7. Standard second-derivative spectra of some PAH compounds.

Illustration of the advantages of this technique is provided by further examples taken on an earlier second-derivative spectrometer (4,8,10). An example revealing the insensitivity of this technique to water turbidity is given for a wastewater containing phenol (10). The ability to distinguish between naphthalene and 2-methylnaphthalene vapors in air at parts-per-billion concentrations has also been demonstrated (4). Reference 8 gives a final example of the capability of second-derivative spectrometry, coupled with a least-squares data-analysis computer program, to analyze a mixture of PAH.

SUMMARY

The derivative UV absorption spectrometer described in this chapter helps fill the need for multicomponent real-time monitors of the hazardous compounds which may be present in fugitive emissions at coal conversion facilities. It should prove a useful addition to the monitoring instruments available to the industrial hygienist. Optimizations and field evaluations will be performed on this monitor and the results reported in the near future.

ACKNOWLEDGMENTS

Research for this paper sponsored by the Division of Biomedical and Environmental Research, U.S. Department of Energy under Contract W-7405-eng-26 with the Union Carbide Corporation. By acceptance of this article, the publisher or recipient acknowledges the U.S. Government's right to retain a nonexclusive, royalty-free license in and to any copyright covering this article.

REFERENCES

1. Bonfiglioli, G., and Brovetto, P. (1964): Principles of self-modulating derivative optical spectroscopy. *Appl. Opt.* 3:1417-1424.
2. Braunstein, H. M., Copenhaver, E. D., and Pfuderer, H. A., Ed. (1976): *Environmental Health, and Control Aspects of Coal Conversion: An Information Overview,* Oak Ridge National Laboratory, ORNL/EIS-94, Oak Ridge, TN.
3. Freudenthal, R. I., and Jones, P. W., Eds. (1977): *Carcinogenesis, Vol. I.,* Raven Press, New York.
4. Gammage, R. B., Vo-Dinh, T., Hawthorne, A. R., Thorngate, J. H., and Parkinson, W. W. (1978): A new generation of monitors for poly-nuclear aromatic hydrocarbons from synthetic fuel production. In:

Carcinogenesis, Vol. 3, P. W. Jones and R. I. Freudenthal, Eds., pp. 155-174, Raven Press, New York.

5. Guerin, M. R., Epler, J. L., Griest, W. H., Clark, B. R., and Rao, T. K. (1978): Polycyclic aromatic hydrocarbons from fossil fuel conversion processes. In: *Carcinogenesis, Vol. 3,* P. W. Jones and R. I. Freudenthal, Eds., pp. 21-33, Raven Press, New York.
6. Hager, R. N. (1973): Derivative spectroscopy with emphasis on trace gas analysis. *Anal. Chem.* 45:1131A-1138A.
7. Hager, R. N., and Anderson, R. C. (1970): Theory of the derivative spectrometer. *J. Opt. Soc. Am.* 60:1444-1449.
8. Hawthorne, A. R., and Thorngate, J. H. (1978): Application of second-derivative UV-absorption spectrometry to PNA analysis, *Appl. Spectrosc.* (in press).
9. Hawthorne, A. R., and Thorngate, J. H. (1978): Improving analysis from second-derivative UV-absorption spectrometry. *Appl. Opt.* 17: 724-729.
10. Hawthorne, A. R., Thorngate, J. H., Gammage, R. B., and Vo-Dinh, T. (1978): Trace organic analysis using second-derivative UV-absorption spectroscopy. *Proceedings of NBS Conference on Trace Organic Analysis, 1978* (in press).
11. Jones, P. W., and Freudenthal, R. I., Eds. (1978): *Carcinogenesis, Vol. 3,* Raven Press, New York.
12. National Institute for Occupational Safety and Health (1978): *Recommended Health and Safety Guidelines for Coal Gasification Pilot Plants,* DHEW (NIOSH) 78-120.
13. O'Haver, T. C., and Green, G. L. (1975): Numerical error analysis of derivative spectrometry for the quantitative analysis of mixtures. *Anal. Chem.* 48:312-318.
14. Williams, D. T., and Hager, R. N. (1970): The derivative spectrometer. *Appl. Opt.* 9:1597-1605.

Polynuclear Aromatic Hydrocarbons,
edited by P. W. Jones and P. Leber.
Ann Arbor Science Publishers, Inc.
Ann Arbor, MI ©1979.

GAS CHROMATOGRAPHIC SEPARATION OF HIGH-MOLECULAR POLYNUCLEAR AROMATIC HYDROCARBONS IN SAMPLES FROM DIFFERENT SOURCES, USING TEMPERATURE-STABLE GLASS CAPILLARY COLUMNS

Ulf Stenberg, Tomas Alsberg, Lars Blomberg and Thomas Wännman

Department of Analytical Chemistry
University of Stockholm
Arrhenius Laboratory
Stockholm, Sweden

BACKGROUND

The widespread occurrence of polynuclear aromatic hydrocarbons (PAH) (14) in the environment calls for rapid and adequate methods in order to perform qualitative and quantitative measurements.

It should be stressed that for reliable results, the analysis of samples containing PAH puts heavy demands on the separation. First, some PAH are known to be potent carcinogens, while others, occasionally close isomers, have so far been found inactive (11). Further, the identity of many of the components in such samples is unknown and it is an urgent task to identify these. Moreover, good separation would help to clarify the differences in the composition of samples obtained from different sources. Factors complicating the separation are that the PAH samples contain components covering a wide range of boiling points and concentrations.

Glass capillary gas chromatography is probably the only method currently available for making such intricate separations. However, to facilitate the mass spectrometric (MS) interpretation of high boiling compounds, glass capillaries are required which possess low bleeding properties at high temperatures.

In this chapter we will demonstrate the broad range of applications possible in PAH analysis, using a standard cleanup procedure and a glass capillary chromatographic technique developed at this laboratory. For these separations we have used glass capillary columns coated with chemically bonded, nonextractable methyl silicone gums (2,3). In addition, we present some preliminary results from MS interpretation of compounds in ambient air, previously also found in automobile exhaust (5).

SAMPLING PROCEDURES

In spite of the high boiling points of PAH, *e.g.*, benzo[a]pyrene (B[a]P), 495°C, a significant amount is present in the gas phase of samples taken after the particle precipitators at stationary sources (10) or in the exhaust pipe of an automobile (6). Thus, in order to make an accurate estimation of the emission it is necessary to collect the gas phase as well as the particulate phase in the investigated effluent. The gas-phase trapping procedure differs with the source investigated, but gives a condensed water phase and/or a phase on a lipophilic adsorbent.

SAMPLE PREPARATION

Extraction of collected gas phases is performed with cyclohexane. Samples in the water phase are shaken three times in a separating funnel; the adsorbent is Soxhlet-extracted for 6 hr. The obtained solutions are evaporated under reduced pressure.

All particulate samples are collected on glass-fiber filters. Volatiles from the particulate matter are collected by vacuum sublimation (1 hr, 0.1 torr) in an all-glass apparatus (13). The collected sublimate is transferred to a centrifuge tube with 2 ml of cyclohexane which is selectively extracted with dimethylformamide (DMF):water (9:1) (Reference 1) (1 x 2 ml + 1 x 1 ml). The PAH in the DMF phase are back-extracted into cyclohexane, same volumes as above, after adding 3 ml of water to the DMF phase. This semimicro extraction technique is a solution to the problem with emulsions which can be difficult to avoid when using ordinary separation funnels. The semimicro method is also less time-consuming. The final volume of 3 ml cyclohexane is reduced almost to dryness by flushing with nitrogen at 30°C, and the remaining residue is diluted with an appropriate volume of cyclohexane:acetone (2:1). The cyclohexane phases obtained from the adsorbent and from the extraction of the water condensate are treated in the same manner.

For mass spectrometric determination of certain samples, especially those from automobile exhausts and ambient air, further purification is necessary. This cleanup includes treatment with:

1. Silica column deactivated with 10% water, cyclohexane as mobile phase. The silica column removes polar constituents which would otherwise cause contamination of the Sephadex gel.

2. Gel filtration chromatography on Sephadex LH-20 with isopropanol. The gel filtration facilitates a division of the PAH sample into a low- ($<$4 rings) and a high-molecular-weight fraction. This aids the determination of minor constituents with high molecular weight, which would otherwise be less easily separated due to overloading by excessive amounts of other components. This is especially important when using capillary columns which are sensitive for the dynamics of the concentration range in the sample.

SEPARATION AND IDENTIFICATION

Separation and detection is achieved with a gas chromatograph equipped with a glass capillary column and FID. Identification and quantification on a routine basis is performed by electronic integration, response factors and the comparison of retention times.

In these investigations we have used splitless injection according to Grob and Grob (7), and in order to achieve a quantitative transfer of high-boiling PAH into the column, we have found that the injector temperature should be no lower than 275°C. This injection system has proven to be reliable, but the high temperature may involve the possibility of artifact formation when labile components are injected.

The separations have been performed on glass capillary columns coated with chemically bonded methyl silicone gums. The advantage of such columns is that the film of stationary phase deposited on the column walls is very stable due to the relatively rigid structure of the polymer. The columns used here thus show low bleeding and high efficiency at high temperatures (300°C).

In this work we have used nonpolar columns because they give short retention times, which makes feasible the elution of components with boiling points higher than that of coronene. Further, we obtain higher efficiencies with nonpolar columns than with polar. For the separation of some close isomers, such as chrysene/triphenylene, columns coated with selective stationary phases are required (2).

RESULTS AND DISCUSSION

It is evident that the composition of the emitted PAH differs with its origin. The formation of PAH is a consequence of incomplete combustion and it is obvious that some structures are more easily formed or more stable than others. The tangible difference between the sources investigated seems to be the amount of minor components, some of which are unknown both in structure and in biological activity.

Mobile Sources

Figure 1 shows the PAH emission profile from an automobile driven in the L.A. 4-test cycle using methanol:gasoline fuel. The sample consists of both the gas and the particulate phase; the composition of the PAH is very complex and by no means fully separated. It is noticeable that the amounts of benzo-[g,h,i] perylene and coronene are rather high compared with benzo[a] pyrene. From our experience, this is typical for automobile exhaust. It should be stated here that the PAH emission profiles from automobiles running on ordinary gasoline do not differ considerably from the one shown (12).

In Figure 2 the PAH emission of a diesel engine running on diesel oil: methanol is shown. The sample was taken in step 12 in the 13-mode Federal Test Procedure. The profile differs considerably, especially in the low molecular-weight region, from that of the Otto engine (Figure 1) although the ratio between benzo[g,h,i] perylene, coronene and benzo[a] pyrene still has a close resemblance.

Stationary Sources

The combustion of fossil fuels is the dominant source of polycyclics which can be found in the environment (11). However, very little information is given in the literature about the combustion in commercial boilers of wood and peat, which have been considered as partial substitutes for oil and coal in energy production. It is essential that the polluting potential of these fuels is determined before they attain widespread use, in order to avoid increasing the burden on the environment (15). Figures 3 and 4 show the PAH emission of wood and peat combustion, respectively. The sampling procedure resembles that reported by Jones *et al.* (8), and the samples were taken after the particle precipitators. The filter for particulate sampling was in all cases held at 150°C in order to avoid condensation of sulfuric acid. The trapping temperature for the gas phase was 20°C.

It can be seen that the amount of trapped 3- and 4-ring polycyclics on the filter is rather low, due to the filter temperature. These figures also demonstrate the high resolution which can be achieved with these cross-linked stationary phases in capillary columns. Peaks 13, 14 and 15 constitute a "triplet" which is critical. Peak 13 is benzo[e] pyrene and 15 is benzo[a]-pyrene. The component in peak 14 has a molecular weight of 254 compared with 252 for the two others. The retention time together with the mass spectrum (Figure 5) indicates that the compound is polycyclic, but since no reference substance or spectrum was available, further investigations must be performed before a strucure can be assigned. We have detected this component in samples from wood and peat combustion and also in automobile exhaust.

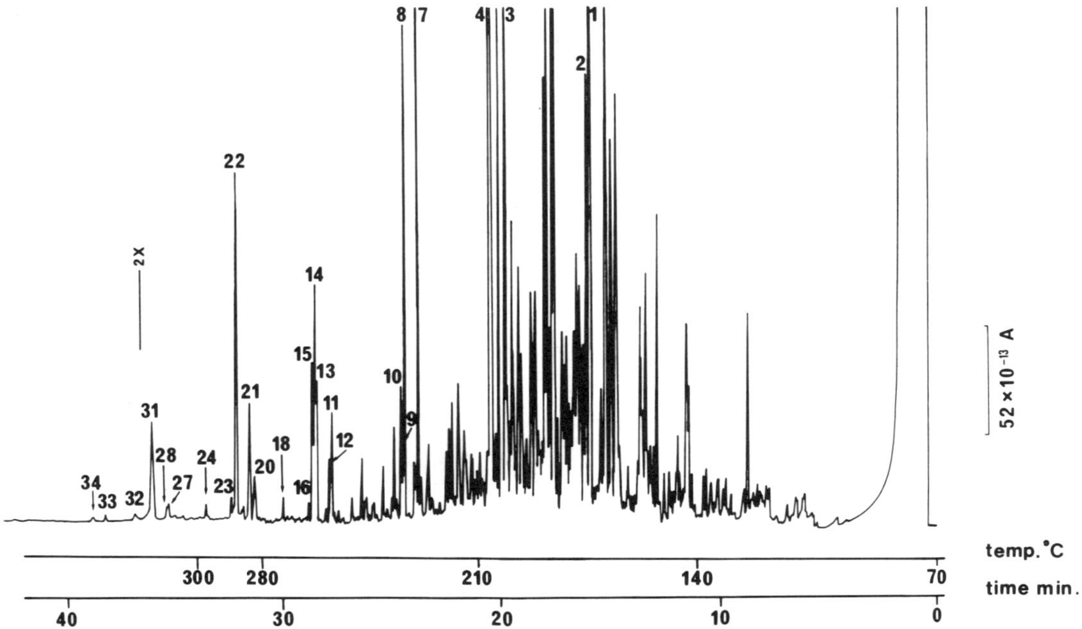

FIGURE 1. Gas chromatogram (FID) of PAH from Otto engine exhaust. Fuel: gasoline-methanol-isobutanol (80-18-2 vol %). Pyrex-glass capillary column (20 m x 0.2 mm i.d.) coated with a cross-linked methyl siloxane polymer, CH_3/Si ratio = 1.6. Chromatographic conditions: initial temperature on injection, 70°C, after 1 min programmed to 300°C at 7°/min, then isothermal for 10 min. Carrier gas (hydrogen) velocity at 70°C, 67 cm/sec. Inlet-splitter opened 1 min after injection. Peak assignment in Table I.

TABLE I. Peak Assignment of Polynuclear Aromatic Hydrocarbons

1.	Phenanthrene	18.	Mol wt 264, Methylene-252[a]
2.	Anthracene	19.	Mol wt 264, Methylene-252[a]
3.	Fluoranthene	20.	Indeno[1,2,3-c,d] fluoranthene
4.	Pyrene	21.	Indeno[1,2,3-c,d]pyrene/Dibenzo[a,h]-anthracene
5.	Benzo[a] fluorene		
6.	Benzo[b] fluorene	22.	Benzo[g,h,i] perylene
7.	Benzo[g,h,i] fluoranthene	23.	Anthanthrene
8.	Cyclopenteno[c,d] pyrene	24.	1,12-Methylene benzo[g,h,i] perylene[a]
9.	Benzo[a] anthracene	25.	Mol wt 302
10.	Chrysene/Triphenylene	26.	Dibenzopyrene
11.	Benzo[b] fluoranthene	27.	Mol wt 300 A
12.	Benzo[k] fluoranthene	28.	Mol wt 300 B
13.	Benzo[e] pyrene	29.	Mol wt 300 C
14.	Mol wt 254	30.	Dibenzopyrene
15.	Benzo[a] pyrene	31.	Coronene
16.	Perylene	32.	Mol wt 314
17.	Mol wt 266, Methyl-252[a]	33.	Mol wt 324
		34.	Mol wt 326

[a]Tentatively determined.

The combustion of peat seems to form more PAH in the high-molecular weight-region (mol wt > 270) than the combustion of wood. In the peat samples we have been able to determine PAH within the 304-326 mol wt range.

PAH emission results not only from combustion processes with fossil fuels, but also from certain industrial activities such as alumina reduction and coke production (4), Figure 6. The PAH profiles from both these processes are very similar and the relatively low amount of coronene emanating from these sources is noticeable.

Ambient Air

The emitted PAH can undergo reactions in the atmosphere and it is presumed that the lifetimes of these pollutants are dependent on several factors, such as UV intensity, inclination for UV absorbance, etc. In spite of this degradation, it is reported that PAH could be transported long range on particles (9). Thus, measurement of ambient air will be the sum of close and distant sources of PAH. Figure 7 shows a PAH profile from the central part of Stockholm. The sample was taken during a period of 8 hr in the beginning of September and consists only of particles. The chromatogram shown here is that of a late fraction from gel filtration on Sephadex and thus containing PAH with four or more rings. Mass spectrometric analysis indicates the presence of compounds previously found in automobile exhausts (5), such as

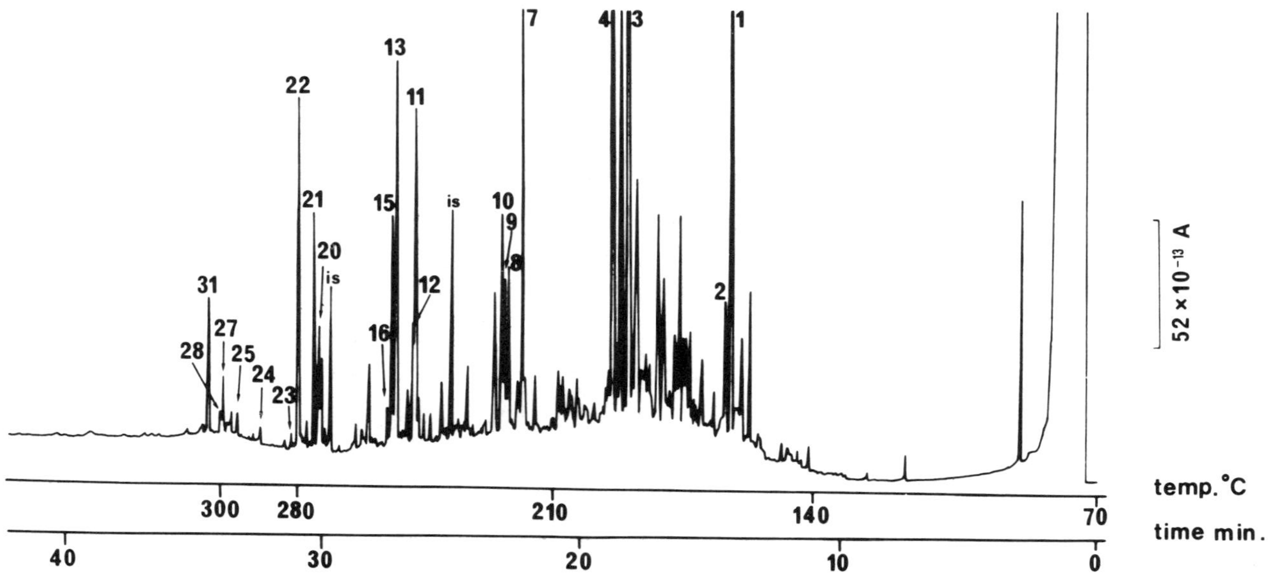

FIGURE 2. Gas chromatogram (FID) of PAH from diesel engine exhaust. Fuel: diesel oil-methanol (51-49 vol %). Pyrex-glass capillary column (18 m x 0.2 mm i.d.) coated with a methyl siloxane polymer, CH_3/Si ratio = 1.8. Chromatographic conditions: See Figure 1. Peak assignment in Table I.

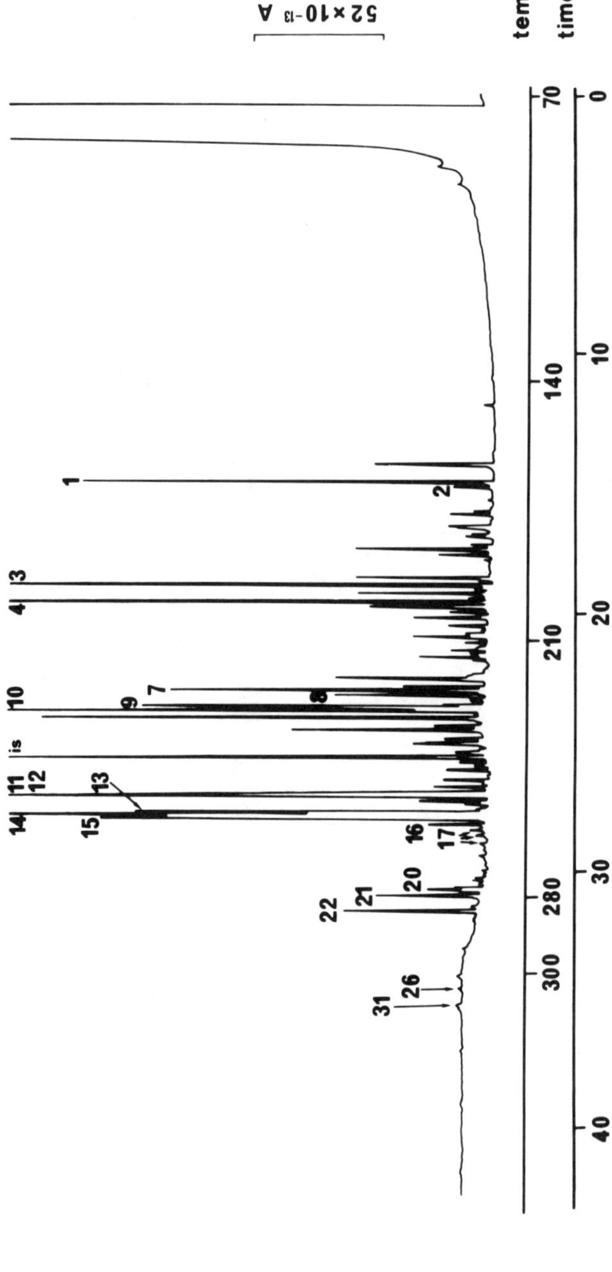

FIGURE 3. Gas chromatogram (FID) of PAH from combustion of wood chips. Column and chromatographic conditions: See Figure 1. Peak assignment in Table I.

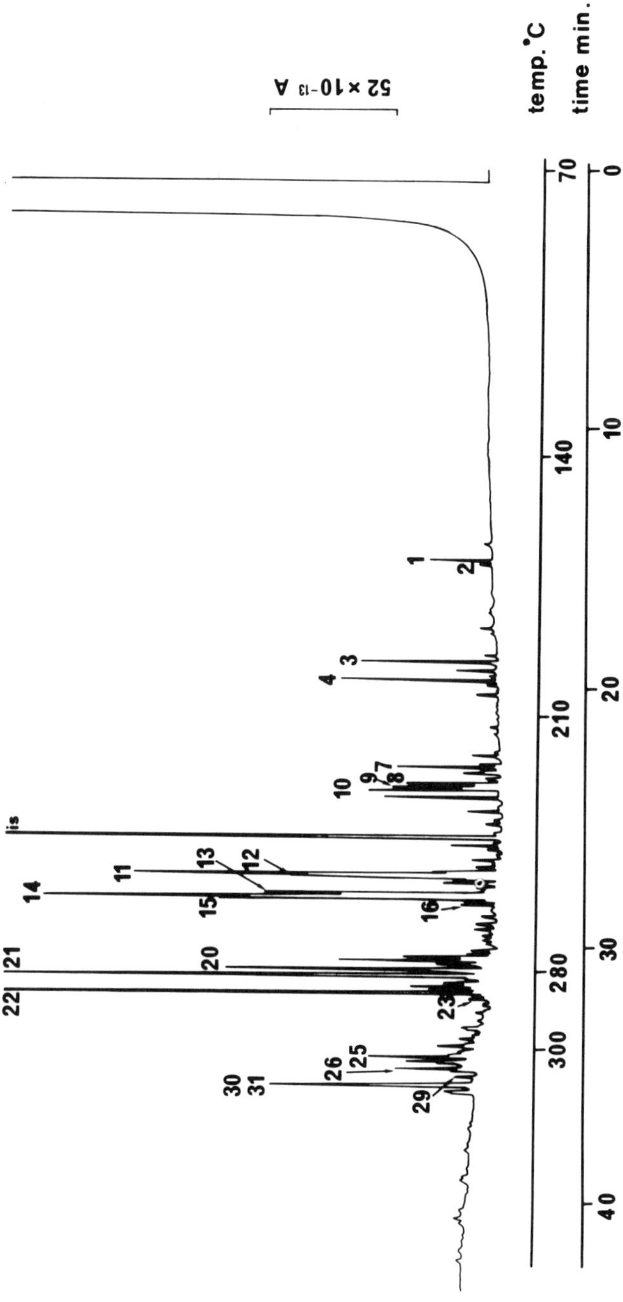

FIGURE 4. Gas chromatogram (FID) of PAH from combustion of peat. Column and chromatographic conditions: See Figure 1. Peak assignment in Table I.

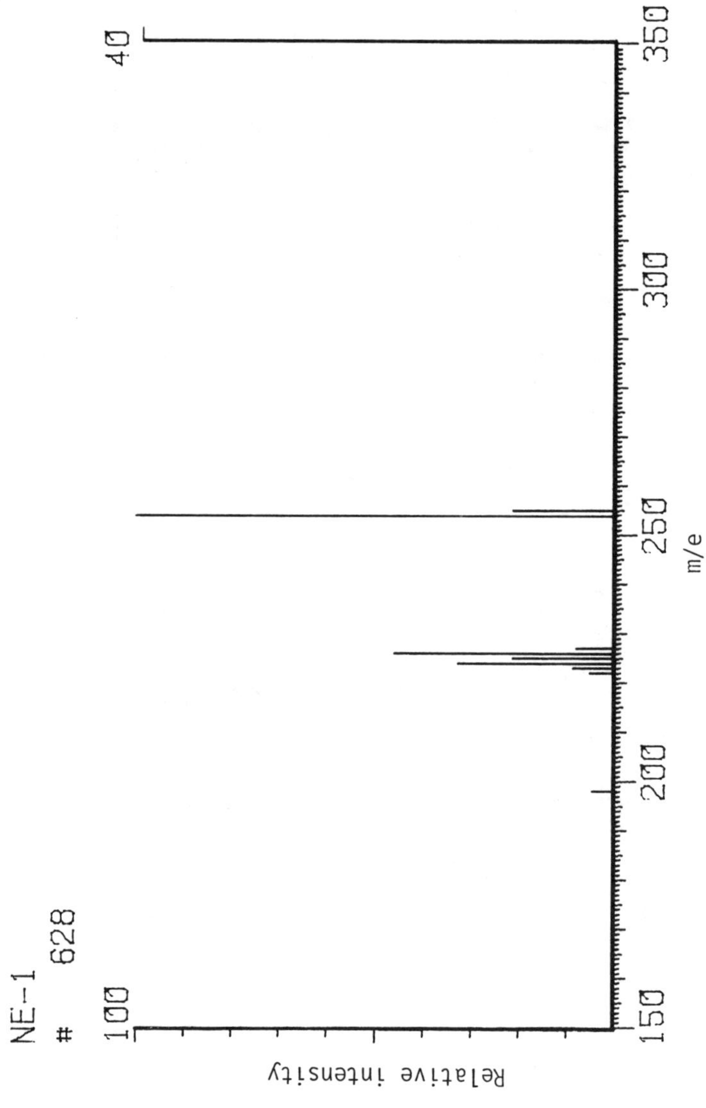

FIGURE 5. Mass spectrum of peak number 14. Electron impact 70 eV.

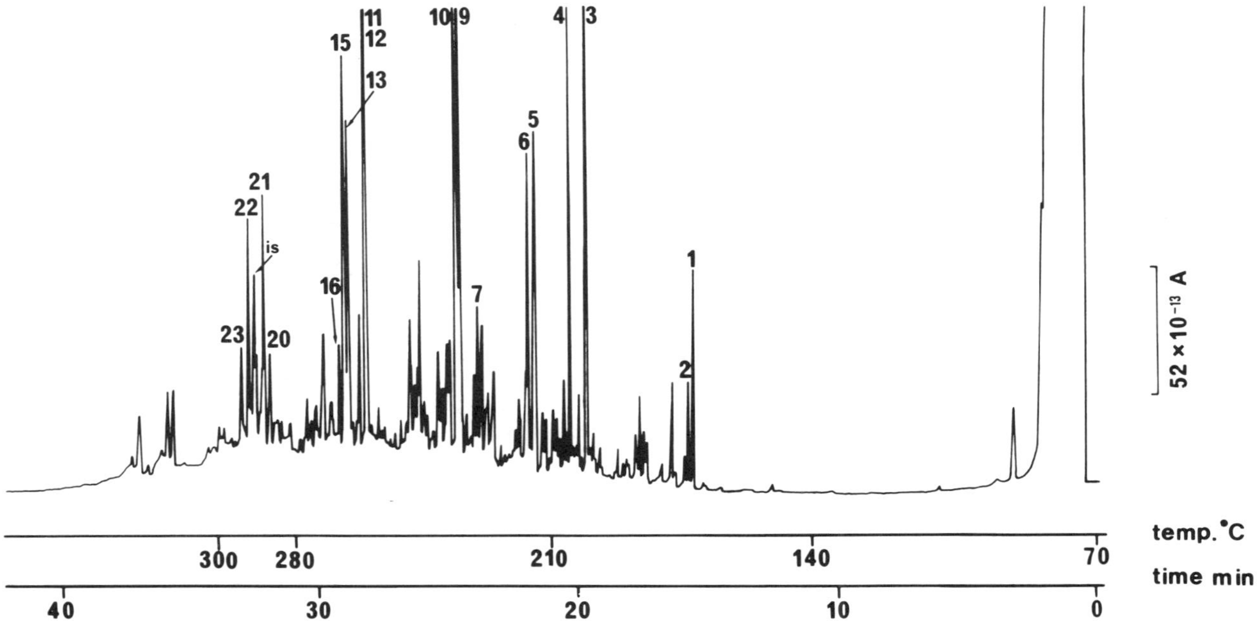

FIGURE 6. Gas chromatogram (FID) of PAH from particulate matter from coke production. Pyrex-glass capillary column (19 m x 0.2 mm i.d.) coated with a methyl siloxane polymer, CH_3/Si, ratio = 2.0. Chromatographic conditions: See Figure 1. Peak assignment in Table I.

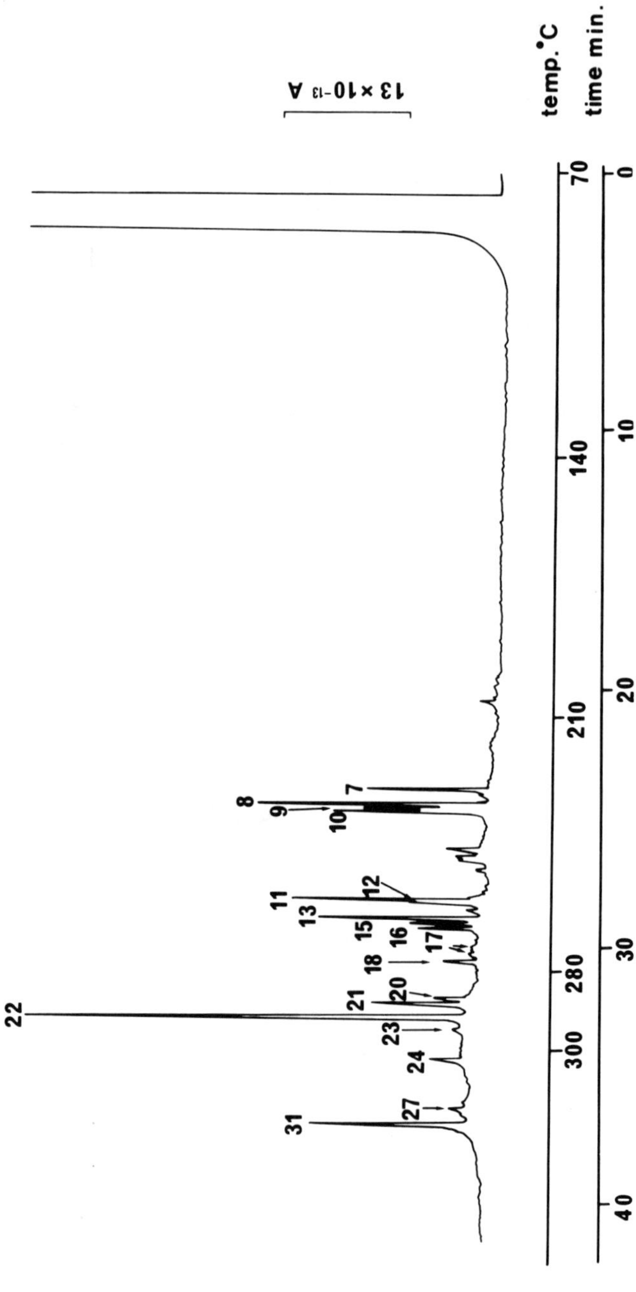

FIGURE 7. Gas chromatogram (FID) of PAH from particulate matter in ambient air. Late fraction from gel filtration chromatography on Sephadex LH-20, (isopropanol 45-120 ml). Column: See Figure 6. Chromatographic conditions: See Figure 1. Peak assignment in Table I.

1,12-methylenebenzo[a]pyrene and methyl derivatives of mol wt 252. It should also be noted that relatively high amounts of benzo[g,h,i]perylene and coronene are present.

The measurement of ambient air with high-resolution chromatography is valuable both in order to obtain an accurate estimation of the amount of PAH found, but also for the facility of obtaining a fingerprint of the air. However, the latter has to be further improved since many sources have PAH profiles which are relatively similar, and atmospheric reactions might degrade certain components.

ACKNOWLEDGMENTS

This work has kindly been supported by the Department of Analytical Chemistry at the University of Stockholm. Financial support was given by the National Swedish Environment Protection Board, Contract No. 7-49/76. We also thank Mr. Jan Bergström at the Swedish Steam Users' Association and Mrs. BrittMarie Bertilsson at the Swedish Methanol Development Company. Thanks are also due Mrs. Beryl Holm for reviewing the manuscript.

REFERENCES

1. Björseth, A. (1977): Analysis of polycyclic aromatic hydrocarbons in particulate matter by glass capillary gas chromatography. *Anal. Chim. Acta.* 94:21-27.
2. Blomberg, L., Buijten, J., Gawdzik, J., and Wännman, T. (1978): Preparation of thermostable, phenyl silicone coated, glass capillary columns for separation of polynuclear aromatic hydrocarbons. *Chromatographia* 11:521-525.
3. Blomberg, L., and Wannman, T. (1979): In situ synthesis of highly thermostable, non-extractable methyl silicone gum phases for glass capillary gas chromatography. *J. Chromatog.* 168:81-88.
4. Burchill, P., Herod, A. A., and James, R. G. (1978): A comparison of some chromatographic methods for estimation of polynuclear aromatic hydrocarbons in pollutants. In: *Carcinogenesis, Vol. 3.* P. W. Jones and R. I. Freudenthal, Eds., pp. 35-46, Raven Press, New York.
5. Grimmer, G., Böhnke, H., and Glaser, A. (1977): Investigation of the carcinogenic burden by air pollution in man. XV. Polycyclic aromatic hydrocarbons in automobile exhaust gas. *Zbl. Bakt. Hyg., I. Abt. Orig.* B164:218-234.
6. Grimmer, G., Hildebrand, A., and Böhnke, H. (1973): Investigation of the carcinogenic burden by air pollution in man. II. Sampling and analytics of polyaromatic aromatic hydrocarbons in automobile exhaust gas. *Zbl. Bakt. Hyg., I. Abt. Orig.* B158:22-34.

7. Grob, K., and Grob, G. (1969): Splitless injections on capillary columns, Part 1. The basic technique, steroid analysis as sample. *J. Chromatog. Sci.,* 7:584-586.
8. Jones, P. W., Giammar, R. D., Strup, P. E., and Stanford, T. B. (1976): Efficient collection of polycyclic aromatic compounds from combustion effluents. *Environ. Sci. Technol.* 10:805-810.
9. Lunde, G., and Björseth, A. (1977): Analysis of polycyclic aromatic hydrocarbons in long-range transported aerosols. *Nature* 268:518-519.
10. Natusch, D. F. S., and Tomkins, B. A. (1978): Theoretical consideration of the adsorption of polynuclear aromatic hydrocarbon vapor onto fly ash in a coal-fired power plant. In: *Carcinogenesis, Vol. 3.* P. W. Jones and R. I. Freudenthal, Eds., pp. 145-154, Raven Press, New York.
11. *Particulate Polycyclic Organic Matter* (1972): National Academy of Science, Washington DC
12. Stenberg, U. (1978): A comparison of the emission of polynuclear aromatic hydrocarbons emitted from automobiles using gasoline or gasoline/methanol as fuel. Report to Swedish Methanol Development Company, Stockholm (in Swedish).
13. Stenberg, U., and Colmsjö, A. (1979): Vacuum sublimation of polynuclear aromatic hydrocarbons separated by thin layer chromatography for the detection with Shpol'skii low temperature fluorescence. *J. Chromatog.* 169:205-212.
14. Suess, M. J. (1976): The environmental load and cycle of polycyclic aromatic hydrocarbons. *Sci. Total Environ.* 6:239-250.
15. Swedish Steam Users' Association (1978): Determination of NO_x and PAH. NE Project 3066 261. Report to the National Swedish Board for Energy Sources Development, Stockholm (in Swedish).

Polynuclear Aromatic Hydrocarbons,
edited by P. W. Jones and P. Leber.
Ann Arbor Science Publishers, Inc.
Ann Arbor, MI ©1979.

SORPTION PROPERTIES OF ENERGY-RELATED POLLUTANTS AND SEDIMENTS

J. C. Means J. J. Hassett, S. G. Wood and W. L. Banwart
Institute for Environmental Studies
and Department of Agronomy
University of Illinois
Urbana, Illinois 61801

The production of energy from fossil fuels, whether it involves the combustion of fuels such as distillate fractions, coal and oil or more advanced processes such as conversion of coal, oil shale and tar sands into acceptable liquid or gaseous fuels, is known to result in a wide variety of organic pollutants which enter the environment as gaseous, liquid or solid wastes (13,32, 35). Polynuclear aromatic hydrocarbons (PAH) have long been recognized as components of these waste streams and therefore have been the subject of concern by environmental researchers.

Numerous papers report the presence of PAH in surface water, soils and sediments (7,16,34). Sorption/desorption equilibria are important parameters in assessing the environmental fate and impact of organic pollutants in aqueous systems. However, little experimental data exist on the sorption properties of PAH on representative sediments (22,33). In this paper, we present research protocol and data on the sorption behavior of selected polynuclear aromatic hydrocarbons on sediments which have been chemically and physically characterized. We also relate this sorption behavior to mathematical models of sorption and the associated chemical properties to the observed sorption properties.

THEORY OF SORPTION

The sorption of organic materials by sediments and soils is dependent on the nature of the adsorber and the adsorbate. The majority of the experimental studies on sorption of organics on soils/sediments have been performed on pesticides. A number of soil/sediment parameters have been related to the sorption of organics. These include: organic matter, clay mineral content, clay mineral texture, pH, and amorphous and crystalline iron and aluminum hydrous oxide content. Organic matter and clay mineral content are the main sources of cation exchange capacity and along with the hydrous oxides are the most active sorptive fractions of soils. The nature of the organic molecule, whether it is a cation, anion or neutral molecule, its water solubility, its polarizability, its effective chain length and molecular configuration are a few of the properties of the adsorbate that interact with the adsorber to influence the amount of adsorption (5).

Neutral molecules are generally not adsorbed to nonswelling-type clays, such as kaolinite, since only external surfaces are involved. Adsorption of neutral organic molecules by swelling-type clays is dependent on the polarizability of the molecule in the clay interlayer. Adsorption of neutral molecules may or may not result in a corresponding release of exchangeable inorganic ions. When an organic molecule is adsorbed to a clay surface, there may be a release of several water molecules resulting in a positive change in entropy. Parfitt (30) using calcium-montmorillonite found that positive entropy changes contributed to the adsorption of neutral molecules with molecular weights greater than 200 and was the dominant force when the molecular weight was greater than 400.

Although clay minerals and to a certain extent hydrous oxides of iron and aluminum may be the most influential in adsorbing organic cations and some polarizable neutral molecules, soil organic matter is undoubtedly of much more general importance in controlling neutral organic adsorption (2). Bailey and White (3) gave the affinity of soil components for pesticide adsorption in the order: organic matter > vermiculite > montmorillonite > illite > chlorite > kaolinite. MacNamara and Toth (28) found that seventy times more of the pesticides linuron and malathion were adsorbed by a humic acid than by a K^+-montmorillonite. Some researchers have considered that for many pesticides adsorption is dependent solely on the relative affinity of the pesticide for the stationary phase (soil organic matter) and the mobile phase (water) (9,15).

Adsorption measurements of organic materials by soil and soil constituents generally have been expressed as adsorption isotherms. One of the most commonly used methods for describing the adsorption isotherms is the Freundlich equation (14).

The Freundlich adsorption isotherm is generally written as:

$$\frac{x}{m} = KC_o^{1/n}$$

where $\frac{x}{m}$ = the amount of material adsorbed per gram of adsorber

C_o = the equilibrium concentration of the adsorbate left in solution

K and n = constants

An alternate form of the Freundlich isotherm can be written as:

$$C_s^n = K^1 C_w$$

or

$$K^1 = C_s^n / C_w$$

where C_s (the concentration of the adsorbate on the solid phase) = x/m

C_w = C_o

K^1 = K^n

The Freundlich isotherm is an empirical equation that has the disadvantage of predicting that adsorption can continue infinitely (12). The adsorption of ureas, triazines (4), lindane (17) and selected hydrophobic compounds (22) has been shown to conform to the Freundlich isotherm. Lambert (24) criticized the use of the Freundlich isomtherm K unless $n \cong 1$.

Karickhoff and co-workers (22) have demonstrated that the adsorption of hydrophobic organics can be related substantially to the organic carbon content of sediments and sediment fractions. They also state that linear partition coefficients (Kp) are relatively independent of sediment concentration, and mixtures of hydrophobic substances are sorbed independently. Further, they determined that partition coefficients (Kp) which have been normalized on an organic carbon basis (Koc) can be estimated from octanol/water partition coefficients. The relationships they suggest are:

$$\text{Koc} = 0.53\ \text{Kow}$$

and

$$\log \text{Koc} = 1.00 \log \text{Kow} - 0.21$$

MATERIALS AND METHODS

Sediment Preparation

The three sediments (Table I) were selected as a subset of 30 soils/sediments and associated watershed soils collected to provide a range of compositional and physical properties. All the soils/sediments were collected from river systems and watersheds geographically targeted as possible coal conversion facility sites (26). The samples were air-dried, crushed and passed through a No. 20 sieve.

The pH, cation exchange capacity (CEC), percent total nitrogen and percent organic carbon, as well as textural information, were determined on each sample (Table II). Soil reaction pH was determined on 1:1 and 1:2

TABLE I. Location and Field Notes on Selected Soil/Sediments

Soil/Sediment	Notes
EPA-6	Sediment sample from Antelope Creek public use area in lake behind Big Bend Dam on Missouri River, southwest of Pierre, SD. Sample is a grayish clayey material with a fair silt content.
EPA-14	Soil taken from a highly eroded red clay hillside southeast of the mouth of Big Sandy River and Ohio River. Point Park, Ceredo, WV.
EPA-23	Sediment taken from Crane Lake north of blind number 63, Sanganois Wildlife Refuge, confluence of Sangamon and Illinois Rivers.

soil-water mixtures by the method of Peech (31). Cation exchange capacity was measured by the ammonium acetate method as modified by Banwart and Hassett (6). Total nitrogen was determined by the method of Bremner (8), and total organic carbon was measured by the Walkley-Black method as described by Allison (1). Particle size analysis was determined by the hydrometer method of Day (11) using hydrogen peroxide to destroy the organic matter (23). Semiquantitative analysis of the clay mineral content of the $< 2\text{-}\mu$ fraction of the clays was performed using a Phillips-Norelco X-ray

TABLE II. Physical-Chemical Properties of Selected Sediments

Soil/Sediment	pH (1:1)	pH (1:2)	CEC (ml/100 g)	Total N (%)	Organic C (%)	Sand (%)	Clay (%)	Silt (%)
EPA-6 (sed.)	7.83	8.23	33.01	0.097	0.72	0.2	68.6	31.2
EPA-14 (soil)	4.54	4.30	18.86	0.064	0.48	2.1	63.6	34.4
EPA-23 (sed.)	6.70	7.10	31.15	0.195	2.38	17.3	69.1	13.6

Semiquantitative Determination of the Clay Minerals in the $<2\text{-}\mu$ Fraction of Sediment and Soil Samples

	Montmorillonite and/or Vermiculite	Illite	Kaolinite
EPA-6	60.8(M)	4.6	3.2
EPA-14	13.8(V&M)	12.8	37
EPA-23	57.6(M)	4.2	7.3

Note: (M) = montmorillonite.
(V) = vermiculite.

diffractometer equipped with a digital computer. The montmorillonite or vermiculite (14-15 Å), illite (~ 10 Å) and kaolinite (~ 7 Å) content of each soil/sediment is presented in Table II.

Compound Purification and Radiolabelling

7 12-[Dimethyl-^{14}C] benz[a] anthracene (DMBA) was obtained commerically (New England Nuclear) with a specific activity of 114.6 mCi/mmol. It was necessary to purify this material by preparative thin-layer chromatography on silica gel G plates (New England Nuclear) in the dark with a nitrogen atmosphere. The solvent system used was hexane:diethyl ether (100:3, v/v). The final purity of the compound used in these studies was 99.2% as determined by thin-layer chromatography followed by scraping of sequential segments from the plate and liquid scintillation counting of each segment.

Pyrene was obtained in 99^{+}% pure form (Aldrich) and tritiated using BF_3-$^3H_3PO_4$ as described by Yavorsky and Gorin (36) and modified by Hilton and O'Brien (20). The resulting generally labelled pyrene was purified by preparative thin-layer chromatography using a solvent system of hexane: diethyl ether (95:5 v/v). The final purity of the pyrene used was determined to be 99.6% using the TLC procedure described previously, and the specific activity was 300 mCi/mmol.

The radioactive content of samples was measured by liquid scintillation counting in a Packard 2425 Tri-Carb Liquid Scintillation Spectrometer using Aquasol (New England Nuclear) as the counting cocktail. Soil and sediment samples processed in a Packard Model 306 Sample Oxidizer were analyzed using Carbo-Sorb (Packard) and Monophase-40 (Packard) as collecting media for $^{14}CO_2$ and ^3H_2O, respectively, and Permafluor V (Packard) as the scintillation counting cocktail. All samples were counted for 10 minutes and all counts were corrected for background and counting efficiency.

Determination of Water Solubilities and Octanol/Water Partition Coefficients

Water solubilities were determined using the procedure outlined in Figure 1 which was developed from portions of the methods of Hague and Schmedding (18) and Mackay and Shiu (27). Once the water solubility of a compound was determined at room temperature, the same procedure was used to prepare aqueous stock solutions of the compounds at ~80% of saturation. These solutions were then used to determine the sorption isotherms.

Octanol/water partition coefficients were determined using the procedure outlined in Figure 2. This procedure was developed in our laboratory using elements of the procedures described by several investigators (10,21,25).

Pure radiolabeled compound
(allow at least tenfold excess/flask over predicted solubility level)

↓

Dissolve in suitable solvent

↓

Shell-evaporate (under N_2) on lower walls of
several (*e.g.*, 6) foil-covered screw-capped Erlenmeyer flasks

↓

Add 100 ml degassed ultrapure (>10 megohm/cm) water

↓

Flood flask with N_2

↓

Shake (wrist-action shaker), moderate speed, room temperature

↓

Remove duplicate flasks at 24, 48 and 72 hr (or later, if saturation level not yet reached)

↓

Filter several successive 5-ml samples from flask
through same 0.2-μ Nuclepore filter; save filtrates

↓

Sample each filtrate 3 x 1.0 ml for liquid scintillation counting

↓

Calculate solute concentration in each filtrate in terms of ppm or ppb;
average the values of successive filtrates containing similar solute levels
(disregard any early filtrates reflecting adsorption of solute onto filter membrane)

FIGURE 1. Protocol for determining the water solubility of hydrophobic organic compounds.

Determination of Sorption Isotherms

The sorption isotherms were determined by adding 4.0 g of soil to 40 ml of an aqueous solution of the compound under study contained in a 50-ml stainless steel centrifuge tube. Aqueous solution concentrations were carried over a range of 5% to 80% of saturation by diluting an appropriate aliquot of the aqueous stock solution to 40 ml. Each point on the isotherm was determined in triplicate. The soil/water mixture was shaken in a 24°C water bath for 20 hr and then centrifuged at 10,000 rpm for 10 min in a temperature-controlled centrifuge. Kinetic studies were performed to ensure that equilibria were obtained in this amount of time for each compound on each sediment. Triplicate 1-ml aliquots of the aqueous phase were pipetted into Aquasol and counted. The remaining aqueous phase was decanted and the soil was removed and subdivided vertically into two samples. One subsample was weighed and oven-dried at 110°C for 48 hr to determine the moisture content of the sample. The other subsample was weighed and combusted in

Pure radiolabeled compound

↓

Dissolve in suitable solvent

↓

Shell-evaporate (under N_2) on lower walls of 50-ml foil-covered screw-capped Erlenmeyer flask

↓

Add 2 ml purified 1-octanol

↓

Shake (wrist-action shaker), moderate speed, several hours or overnight, room temperature

↓

Add 10 ml distilled (ultrapure) water; flood flask with N_2

↓

Shake (wrist-action shaker), moderate speed, 15 min, room temperature

↓

Transfer mixture to glass (Corex) centrifuge tube; centrifuge, 30,000 x g, 20 min, 24°C

↓

Repeat centrifugation if interface is not clear

↓

Remove and sample octanol phase 3 x 10 μl for liquid scintillation counting

↓

Remove and discard *all* residual octanol at interface (including some water phase, if necessary)

↓

Carefully sample water phase 3 x 1.0 ml (making sure that pipette is wiped free of any residual octanol) for liquid scintillation counting

↓

Calculate K_{ow} (= dpm/ml octanol phase ÷ dpm/ml water phase)

↓

Return octanol to Erlenmeyer flask and repeat partitioning procedure with fresh distilled water until K_{ow} value remains constant (indicative of compound purity)

FIGURE 2. Protocol for determining the octanol/water partition coefficient of hydrophobic organic compounds.

the Packard 306 Sample Oxidizer. The evolved $^{14}CO_2$ or ^3H_2O was collected and counted for 10 min. In this way, both solution isotherms and solid-phase isotherms could be calculated and a ^{14}C or ^3H mass balance could be obtained. An additional control procedure was used to determine if degradation of the compound had occurred during the isotherm. Both the aqueous and solid phases were extracted with diethyl ether or other appropriate solvent. The extract was concentrated and subjected to thin-layer chromatography followed by counting of sequential segments scraped from the plate.

RESULTS AND DISCUSSION

Pyrene (MW = 202) and 7,12-dimethylbenz[a]anthracene (MW = 256) were found to have water solubilities of 135 ppb and 25 ppb, respectively, at 24°C. Both of the values reported are averages of at least 12 determinations. Equilibrium values were achieved in 48 hr for both compounds. Care was taken to prevent photooxidation of the compounds during the equilibration period. The variation observed in the reported values was less than 5%. Losses of material by adsorption onto glassware were determined to be less than 2%. Some adsorption onto the Nuclepore membranes was observed; however, after two to three successive aliquots were passed through the filter, stable values for the solubility of both compounds were achieved.

The octanol/water partition coefficients (Kow) for pyrene and dimethylbenz[a]anthracene were determined to be 123,000 and 950,000, respectively. Equilibria were achieved in 1 hr. Both the values reported are averages of at least 12 determinations. The variation between replicates for pyrene was 2.3%, while the variation between replicates for dimethylbenz[a]anthracene was 25%. The primary sources of error in the Kow determinations are:

1. trace contaminants or degradation products which are analytically indistinguishable from the parent compound and which have higher water solubilities than the parent compound,
2. incomplete solution of the compound in either phase,
3. incomplete separation of the aqueous phase from the octanol phase due to the formation of stable microemulsions, and
4. contamination of the aqueous phase with foreign material (*i.e.*, dust, bacteria, etc.) which can retain excess solute in the water phase.

The Kow values were determined at a minimum of two different solute concentrations to ensure that no concentration effects were observed. Precautions were exercised to ensure that degradation of the compounds did not occur during the Kow determinations.

The sorption isotherms for the two compounds on each of the three soil/sediment samples were linear over a range of water-phase concentrations. Figures 3 and 4 show the isotherms for pyrene and dimethylbenz[a]anthracene, respectively, on EPA soils/sediments 6, 14 and 23. The Kp values defined by these lines were calculated by fitting the data to the Freundlich equation:

$$Cs = Kp\,Cw^{1/n}$$

where Cs = the concentration of the compound in the solid phase at equilibrium, calculated on a dry weight basis

Cw = the equilibrium solution concentration of the compound

Kp = the partition coefficient

$1/n$ = a constant related to the surface adsorptive capacity of the solids phase (19)

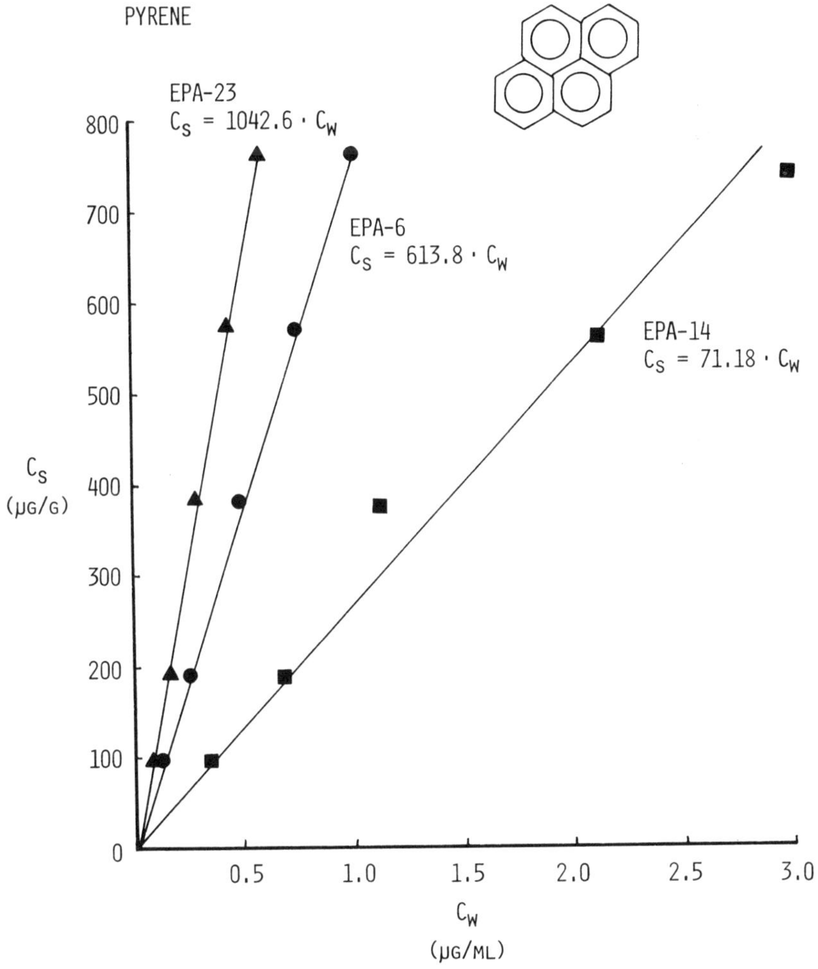

FIGURE 3. Sorption isotherms for pyrene on EPA soils/sediments 6 (●), 14 (■) and 23 (▲).

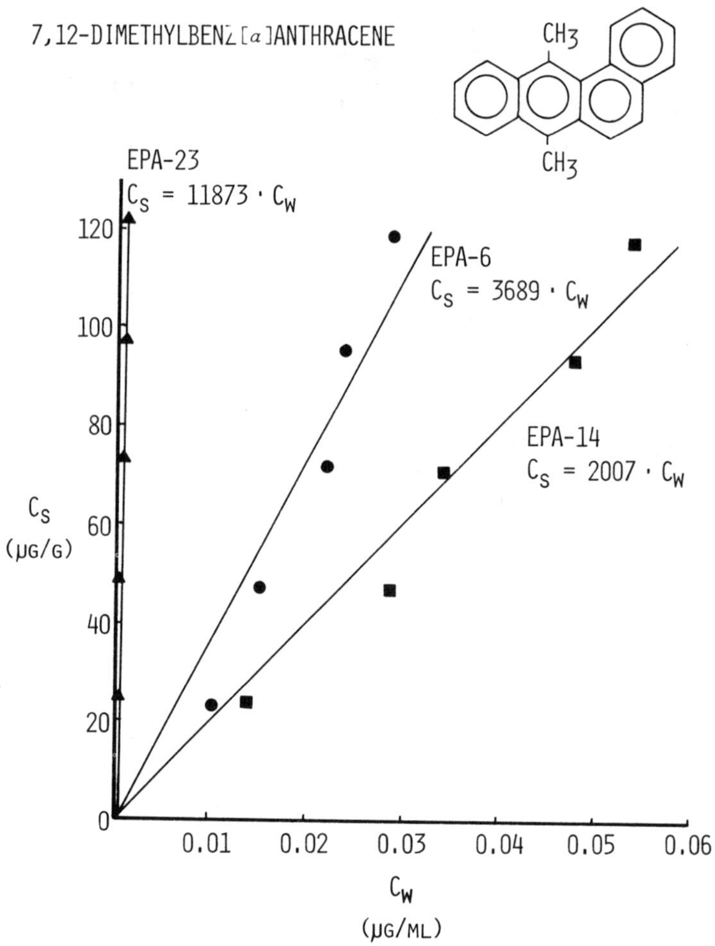

FIGURE 4. Sorption isotherms for 7,12-dimethylbenz[a]anthracene on EPA soils/sediments 6 (●), 14 (■) and 23 (▲).

Table III gives a summary for the sorption coefficients determined in this study. Since organic matter is known to be the source of the major adsorptive capacity of soil/sediments for hydrophobic substances, the Kp values for the individual samples were normalized to the organic carbon content of the samples using the formula:

$$\text{Koc} = \frac{\text{Kp}(100)}{\%\ \text{organic carbon}}$$

TABLE III. Sorption Coefficients for Pyrene and 7,12-Dimethylbenz[a]anthracene

	Pyrene		7,12-Dimethylbenz[a]anthracene	
Soil/Sediment	Kp	Koc	Kp	Koc
EPA-6	614	85256	3690	512500
EPA-14	71	64706	2007	418125
EPA-23	1043	43807	11870	498739
	avg Koc =	64590	avg Koc =	476454

The Koc values for the soil/sediments are also presented in Table III. It should be noted that the normalizations for organic carbon content cause the partition coefficients to converge on an average value. This trend can be verified by the observation that there is a linear relationship between Kp values and increasing organic carbon content of sediments (range = 0.48% to 2.38%). None of the other soil/sediment parameters (Table II) give this type of linear relationship.

When the Kow and Koc values for pyrene and dimethylbenz[a]anthracene (DMBA) were evaluated using the data for the three sediments, the relationships were:

$$\text{(pyrene) Koc (avg)} = 0.525 \text{ Kow}$$

$$\text{(DMBA) Koc (avg)} = 0.502 \text{ Kow}$$

These values compare fairly well with the factor of 0.53 cited by Karickhoff (22). The line defined by these two points and the origin of a log-log plot of Koc vs Kow fits the equation:

$$\log \text{Koc} = 1.00 \log \text{Kow} - 0.30$$

These coefficients agree reasonably well with those determined by Karickhoff (22) and they also are equal to the coefficients we found for several other hydrophobic compounds using these same soil/sediment samples (29).

SUMMARY

Water solubilities and octanol/water partition coefficients are reported for pyrene and 7,12-dimethylbenz[a]anthracene. Sorption isotherms for both hydrophobic compounds were linear over a range of aqueous phase concentrations on soil/sediment samples which exhibit a broad range of physical and chemical properties. The linear partition coefficients (Kp) for both compounds could be related to the organic carbon content of the soil/sediments. Octanol/water partition coefficients for both compounds could be used to estimate the Kp values for each sediment after these values had been normalized for organic carbon content.

ACKNOWLEDGMENTS

This work was supported by U.S. Environmental Protection Agency Contract 68-03-2555. The authors wish to acknowledge the technical assistance of John Ameel, Sandra Dick, Elsa Tong, Carolyn Hanson, David Zierath and Adam Khan.

REFERENCES

1. Allison, L. E. (1965): Organic carbon. In: *Methods of Soil Analysis, Agronomy* 9(2), C. A. Black, Ed., pp. 1367-1378, American Society of Agronomy, Madison, WI.
2. Aomine, S., and Inoue, K. (1967): Retention of mercury by soils. II. Adsorption of phenylmercuric acetate by soil colloids. *Soil Sci. Plant Nutr. (Tokyo)* 13:195-200.
3. Bailey, G. W., and White, J. L. (1964): Review of adsorption and desorption of organic pesticides by soil colloids, with implications concerning pesticide bioactivity. *J. Agric. Food Chem.* 12:324-332.
4. Bailey, G. W., White, J. L., and Rothberg, T. (1968): Adsorption of organic herbicides by montmorillonite: Role of pH and chemical character of adsorbate. *Soil Sci. Soc. Am. Proc.* 32:222-234.
5. Bailey, G. W., and White, J. L. (1970): Factors influencing the adsorption, desorption, and movement of pesticides in soil. *Res. Rev.* 32: 29-92.
6. Banwart, W. L., and Hassett, J. J. (1976): *Laboratory Introduction to Soil Science,* p. 118, Stipes Publishing Co., Champaign, IL.
7. Blumer, M., and Youngblood, W. W. (1976): Polycyclic Aromatic Hydrocarbons in the Environment: Homologous Series in Soils and Recent Marine Sediments. NTIS No. AD A023637, Office of Naval Research.
8. Bremner, J. M. (1965): Total nitrogen. In: *Methods of Soil Analysis, Agronomy* 9(2), C. A. Black, Ed., pp. 1149-1237, American Society of Agronomy, Madison, WI.
9. Briggs, G. G. (1969): Molecular structure of herbicides and their sorption by soils. *Nature (London)* 223:1288.
10. Chiou, C. T., Freed, V. H., Schmedding, D. W., and Kohnert, R. L. (1977): Partition coefficient and bioaccumulation of selected organic chemicals. *Environ. Sci. Technol.* 11:475-478.
11. Day, P. R. (1965): Particle fractionation and particle size analysis. In: *Methods of Soil Analysis, Agronomy* 9(1), C. A. Black, Ed., pp. 545-567, American Society of Agronomy, Madison, WI.

12. de Hann, F. A. M., and Zuerman, P. J. (1976): Pollution of Soil. In: *5A Developments in Soil Science-Soil Chemistry A. Basic Elements,* Bolt and Bruggenwert, Eds, Chapter 10, Elsevier, New York.
13. Forney, A. J., Haynes, W. P., Gasior, S. J., Johnson, G. E., and Strakey, Jr., J. P. (1974): Analysis of Tars, Chars, Gases and Water Found in Effluents from the Synthane Process. Progress Report 76, Bureau of Mines, U.S. Department of the Interior.
14. Freundlich, H. (1922): *Colloid and Capillary Chemistry,* Methuen & Co Ltd., London.
15. Furmidge, C. G. L., and Osgerby, J. M. (1967): Persistence of herbicides in soil. *J. Sci. Food Agric.* 18:269-273.
16. Giger, W., and Blumer, M. (1974): Polycyclic aromatic hydrocarbons in the environment: Isolation and characterization by chromatography, visible-ultraviolet, and mass spectrometry. *Anal. Chem.* 46:1663-1671.
17. Graham-Bryce, I. J. (1967): Adsorption of disulfoton by soil. *J. Sci. Food Agric.* 18:72-77.
18. Hague, R., and Schmedding, D. (1975): A method of measuring the water solubility of hydrophobic chemicals: Solubility of five polychlorinated biphenyls. *Bull. Environ. Contam. Toxicol.* 14:13-18.
19. Henry, D. C. (1922): A kinetic theory of adsorption. *Philos. Mag.* 44:689-705.
20. Hilton, B. D., and O'Brien, R. D. (1964): A simple technique for tritiation of aromatic insecticides. *J. Agric. Food Chem.* 12:236-239.
21. Karickhoff, S. W. (1978): Determination of octanol/water distribution coefficients, water solubilities, and sediment/water partition coefficients for hydrophobic organic pollutants (personal communication).
22. Karickhoff, S. W., Brown, D. S., and Scott, T. A. (1978): Sorption of hydrophobic pollutants on neutral sediments. *Water Res.* (in press).
23. Kilmer, V. J., and Alexander, L. T. (1949): Methods of making mechanical analysis of soils. *Soil Sci.* 68:15-24.
24. Lambert, S. M. (1967): Functional relationship between sorption in soil and chemical structure. *J. Agric. Food Chem.* 15:572-576.
25. Leo, A., Hansch, C., and Elkins, D. (1971): Partition coefficients and their uses. *Chem. Rev.* 71:525-616.
26. Lindquist, A. E. (1977): Siting potential for coal gasification plants in the United States. Bureau of Mines Circular 8735.
27. Mackay, D., and Shiu, W. Y. (1977): Aqueous solubility of polynuclear aromatic hydrocarbons. *J. Chem. Eng. Data* 22:399-402.
28. MacNamara, G. M., and Toth, S. J. (1970): Adsorption of linuron and malathion by soils and clay minerals. *Soil Sci.* 109:234-240.
29. Means, J. C., Hassett, J. J., Banwart, W. L., and Wood, S. G. (1978): Sorption properties of nitrogen and sulfur analogs of polynuclear aromatic hydrocarbons and sediments (manuscript in preparation).
30. Parfitt, R. L. (1969): Mechanisms of adsorption of polymers and polysaccharides by montmorillonite. Ph.D. Thesis, University of Adelaide, Australia.

31. Peech, M. (1965): Hydrogen ion activity. In: *Methods of Soil Analysis, Agronomy* 9(2), C. A. Black, Ed., pp. 914-926, American Society of Agronomy, Madison, WI.
32. Sharkey, A. G., Schultz, J. L., White, C., and Lett, R. (1976): Analysis of organic material in coal, coal ash, fly ash, and other fuel and emission samples. Report No. EPA-600/2-76-075, Office of Research and Development, U.S. Environmental Protection Agency, Washington, D.C.
33. Smith, J. H., Mabey, W. R., Bohovos, N., Holt, B. R., Lee, S. S., Chou, T. W., Bomberger, D. C., and Mill, T. (1977): Environmental pathways of selected chemicals in freshwater systems. Part II: Laboratory studies. Final Report No. EPA-600/7-78-074. Office of Research and Development, U.S. Environemntal Protection Agency, Athens, GA.
34. Suess, M. J. (1976): The environmental load and cycle of polycyclic aromatic hydrocarbons. *Sci. Total Environ.* 6:239-250.
35. TRW Systems and Energy, Inc. (1976): Carcinogens relating to coal conversion processes. U.S. Energy Research and Development Administration, Oak Ridge, TN.
36. Yavorsky, P. M., and Gorin, E. (1962): New reagent for labeling organic compounds with tritium. *J. Am. Chem. Soc.* 84:1071-1072.

Polynuclear Aromatic Hydrocarbons, edited by P. W. Jones and P. Leber. Ann Arbor Science Publishers, Inc. Ann Arbor, MI ©1979.

SEPARATION AND IDENTIFICATION OF SULFUR-CONTAINING POLYCYCLIC AROMATIC HYDROCARBONS (THIOPHENE DERIVATIVES) FROM SOME PAH

W. Karcher, R. Depaus, J. van Eijk and J. Jacob*

Commission of the European Communities
Joint Research Centre
Petten Establishment
The Netherlands

*Biochemisches Institut für Umweltcarcinogene
Ahrensburg, Federal Republic of Germany

In the frame of a research and development program on Standards and Reference Materials, the Commission of the European Communities (EC) has launched a project for the preparation of a set of polynuclear aromatic hydrocarbons (PAH) of certified purity for calibration and reference purposes in collaboration with laboratories of the member states. The first stage of this project aims at the preparation of some pure five-ring compounds (benzopyrenes, benzofluoranthenes, benzo[b]chrysene, indeno[1,2,3-c,d]pyrene, benzo[g,h,i]perylene) (5) and some of the methylchrysenes.

In a screening exercise of commercial materials, it has been found that the major impurity is quite often a thiophene derivative of the parent PAH, where an aromatic benzene ring is replaced by a thiophene ring. This seems to be the case not only for commercial PAH materials derived from coal tar but, in some instances, it has been observed for PAH compounds prepared by synthesis as well. In view of the general interest in five-ring PAH materials and their derivatives, with respect to potential carcinogenic effects and previous reports on the occurrence of thiophene derivatives in coal tar, carbon blacks, lubricating oils, aerosols and automobile exhausts (1-4,7-9), it was thought to be worthwhile separating and identifying some of these thiophene derivatives, where possible, from the accompanying parent PAH compound. As a result, the following compounds have been identified in PAH materials: $C_{14}H_8S$ (I), (phenanthro[4,5-b,c,d]-thiophene), $C_{18}H_{10}S$ (II) (benz[a]anthraceno-[11,11′,12-b,c,d]- and chryseno-[4,4a,4b,5-b,c,d]-thiophene); $C_{20}H_{10}S$ (III) (probably peryleno-[1,12-b,c,d]thiophene) and two isomers of $C_{20}H_{12}S$(IV) (dinaphtho-2,1,1′,2′ and -2,3,2′,3′-thiophene).

EXPERIMENTAL

Materials

For reasons of economy, it was at first hoped possible to rely on commercial sources and to purify them to the required purity levels. To this end, specimens of some candidate materials were obtained from various suppliers. Various analytical techniques were used to assess their purity (see Analysis section). However, it soon became apparent that in most cases the commercial materials contained a number of impurities at significant concentration levels which made purification less attractive. Also, it was found that the purity of commercial PAH materials tended to vary significantly between production batches. Therefore, preference was given to synthesizing the required PAH compounds, aiming at a purity level of $>99\%$ (wt). Synthesis has been carried out by specialist laboratories of the EC member countries under contract to the EC Commission (General Directorate for Research and Education–Community Bureau of Reference).

Analysis

The following analytical techniques were employed and evaluated for the quantitative analysis of the parent PAH, and for the detection of accompanying impurities:

- GC (SCOT and Packed Columns)
- HPLC (adsorption and reversed-phase/analytical and preparative scale)
- GC/MS and direct-inlet MS
- UV and fluorescence spectroscopy
- TLC (thin-layer chromatography)
- DSC (differential scanning calorimetry)

In general, it was found that quantitative GC results tend to slightly overestimate the purity of PAH materials. HPLC is very useful for the qualitative detection and for the separation and enrichment of impurities. Quantification by HPLC using a UV detector, however, is only possible after separation and isolation of impurities, in order to determine their UV-response factors. For the detection of low-level impurities by mass spectrometry, the use of low ionization potentials and temperature programming was especially advantageous so as to limit the mass spectrum, preferably to the appearance of the parent ions. The presence of a peak at M-45 indicated a sulfur-containing impurity. Also, when using a direct-inlet probe, the mass spectrum yielded useful information with respect to the structure of the impurities.

TLC and fluorescence spectroscopy proved useful as complementary techniques but were of limited value for the determination of impurities because of poor sensitivity. DSC was mostly used to assess the overall purity of the respective PAH materials.

Separation and Enrichment of Impurities

The impurities suspected to contain sulfur, because of the presence of a M-45 peak in the mass spectrum resulting from the loss of an S-CH ion, were separated and enriched by preparative liquid chromatography on Partisil 5 columns, using dried hexane as solvent. In most cases, the separation of a sizable quantity of the impurity was a tedious process, since repeated HPLC injections were necessary when sufficient separation could not be achieved in one step. Also, it involved the handling of appreciable quantities of the basic PAH materials and the HPLC solvents. For instance, in order to isolate about 10 mg of the impurity $C_{14}H_8S$ (I-mass 208), which was present at approximately the 1% level in pyrene, 2 g of pyrene had to be injected into the preparative HPLC instrument, thus using more than 40 liters of ultrapure hexane.

Where sufficient material could be obtained (> 1 mg), identification was then attempted by means of direct-inlet mass spectrometry and ^{1}H and ^{13}C nuclear magnetic resonance spectroscopy and UV spectroscopy.

To obtain the NMR spectra for structural identification, at least 5 mg of the sample was dissolved in either carbon disulfide (for ^{1}H-NMR) or in deuterated acetone (for ^{13}C-NMR). The proton-NMR spectra have been taken with a 100-MHz instrument.

IDENTIFICATION OF S-IMPURITIES

Following the above experiments, thiophene derivatives with four different mass numbers could be identified in four PAH compounds (Table I).

Pyrene

In pyrene, which was used for comparing and testing the various techniques for the purity analysis of PAH, the main impurity was found to be a sulfur-containing compound with the mass number 208. The mass spectrum allowed for two possible structures for the composition $C_{14}H_8S$ (Ia and Ib):

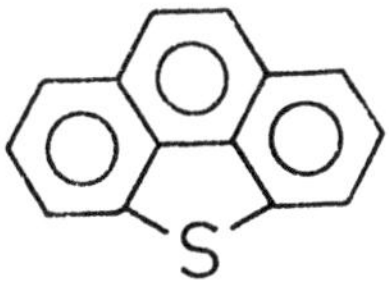

(Ia)

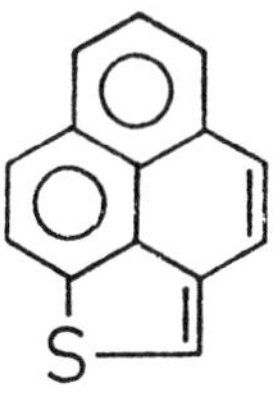

(Ib)

TABLE I. Summary of S-Containing Impurities Found in Some PAH

Parent PAH	Mass	S-Impurities Mass	Composition	Conc. (wt %)		Structure
Pyrene ($C_{16}H_{10}$)	202	208	$C_{14}H_8S$	0.8-1.8		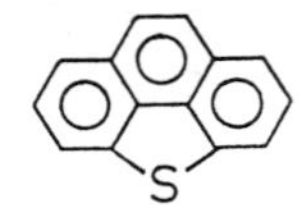 Phenanthro[4,5-b,c,d] thiophene
Benzo[a] pyrene ($C_{20}H_{12}$)	252	258	$C_{18}H_{10}S$	0.5-4		
		284	$C_{20}H_{12}S$	0.1-4		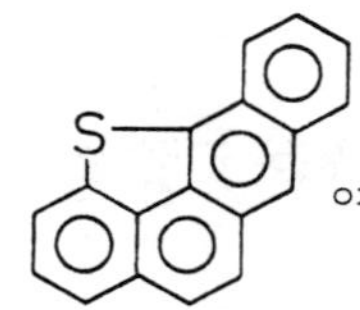 or 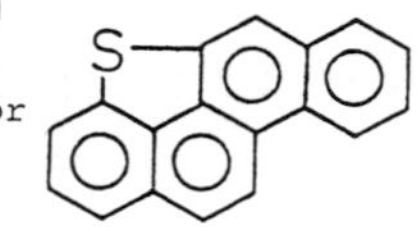
Benzo[g,h,i] perylene ($C_{22}H_{12}$)	276	282	$C_{20}H_{10}S$	2.6-3	probably:	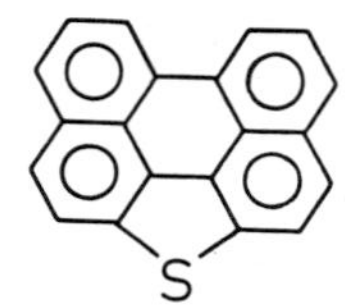 Perylo[1,12-b,c,d] thiophene
Benzo[j] fluoranthene ($C_{20}H_{12}$)	252	284	$C_{20}H_{12}S$	2.0		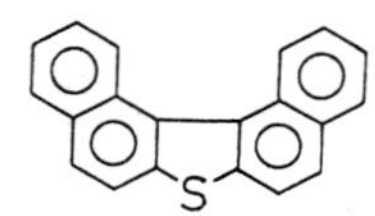 Dinaphtho-2,1,1′,2′-thiophene

The evaluation of the NMR spectra, however, indicated a symmetrical structure (Ia), with two identical sets of four protonated and three quaternary carbons. These findings suggest that the isolated compound $C_{14}H_8S$ is a phenanthro[4,5-b,c,d]thiophene (structure Ia). The mass spectrum and the NMR spectrum of this impurity are given in Figures 1 and 2, respectively. The UV spectrum of the compound $C_{14}H_8S$, which is depicted in Figure 3, is quite similar to the UV spectrum of pyrene. This may be interpreted as a corroboration of the phenanthrothiophene structure (Ia), since in the UV spectrum, the thiophene ring tends to exhibit an absorption behavior comparable to an aromatic ring.

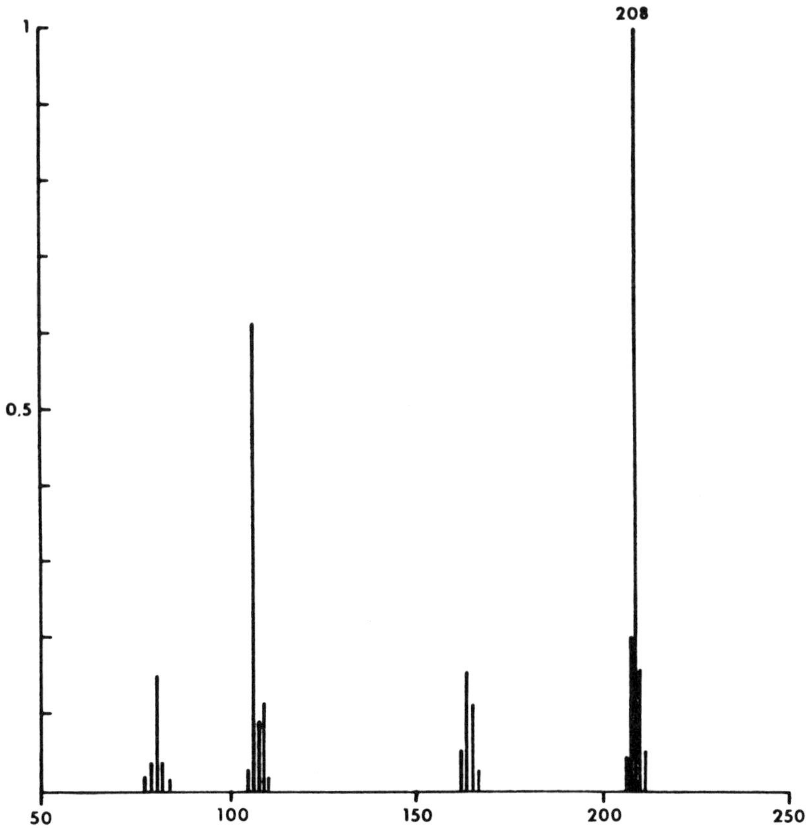

FIGURE 1. Mass spectrum of $C_{14}H_8S$.

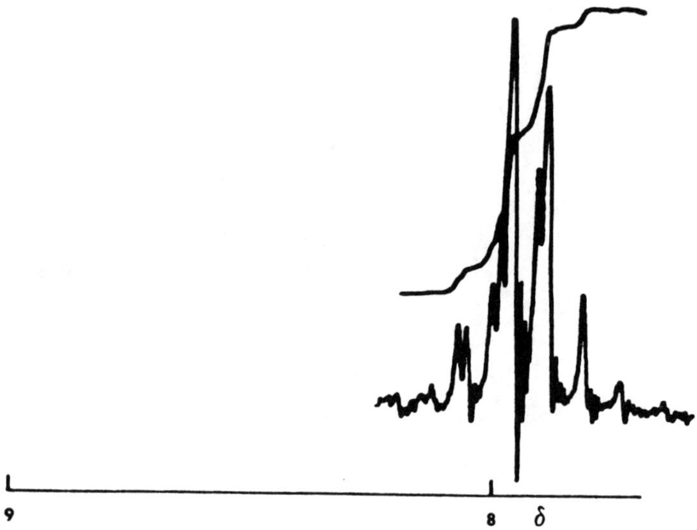

FIGURE 2. NMR spectrum for $C_{14}H_8S$.

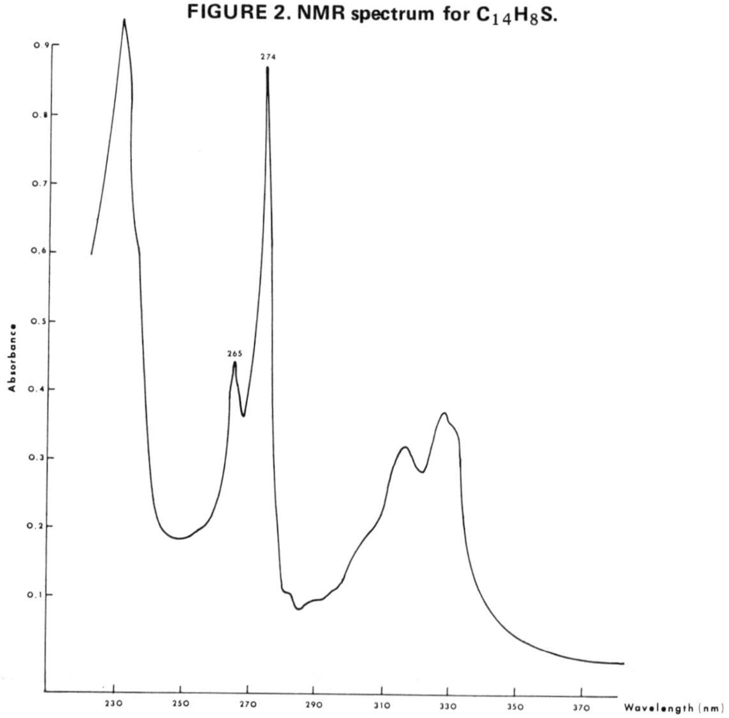

FIGURE 3. UV spectrum of $C_{14}H_8S$.

Benzo[a]pyrene

In commercial samples of benzo[a]pyrene, at least two thiophene derivatives were identified with molecular weights of 258 and 284, respectively. A third impurity, at mass number 314, was also suspected to contain sulfur and, tentatively, the composition $C_{20}H_{10}S_2$ was allocated to this MS peak (see Figure 4). However, the concentration of this impurity was too low to permit separation and identification by a combination of preparative HPLC, direct-inlet MS and NMR.

The concentration of the first two impurities was sufficient to allow separation and enrichment according to the procedure outlined previously.

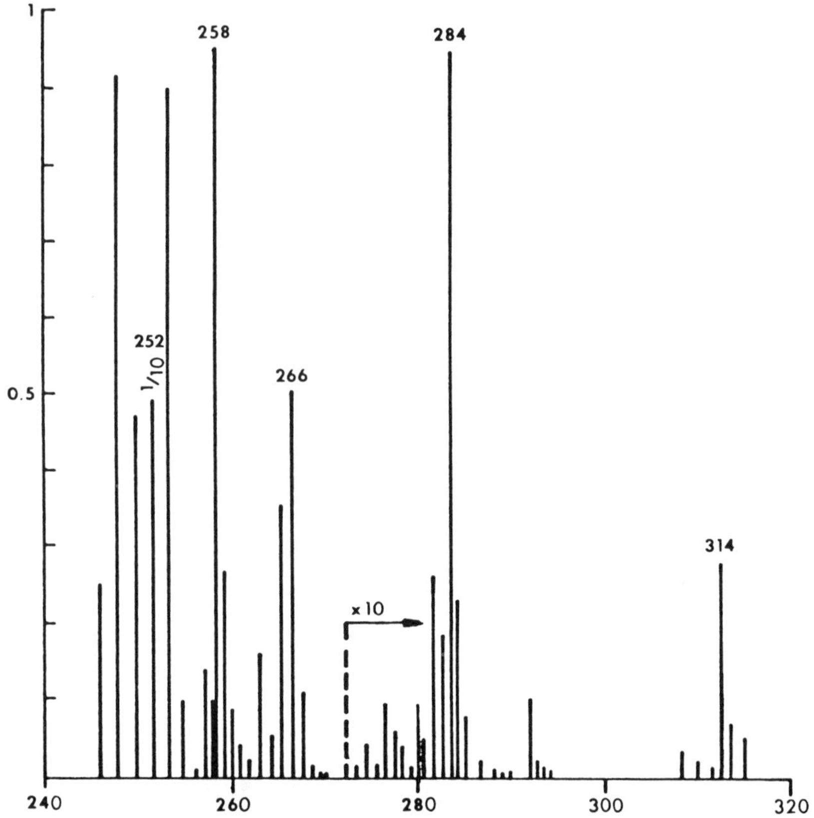

FIGURE 4. Mass spectrum of benzo[a]pyrene.

On the basis of the mass spectrum, the compositions $C_{18}H_{10}S$ and $C_{20}H_{12}S$ were allocated to these impurities. For the major impurity, present at the 1 to 4% level in commercial benzo[a]pyrenes, the mass 258 suggests a structure where one aromatic ring of benzopyrene is replaced by a thiophene ring in a similar way as it was found for pyrene. Indeed, from the interpretation of the mass spectrum (Figure 5) and the ^{1}H-NMR spectrum (Figure 6) two alternative structures can be put forward benz[a]anthraceno-11,11′,12-b,c,d 9-thiophene (IIa) and chryseno[4,4a,4b,5-b,c,d] thiophene (IIb).

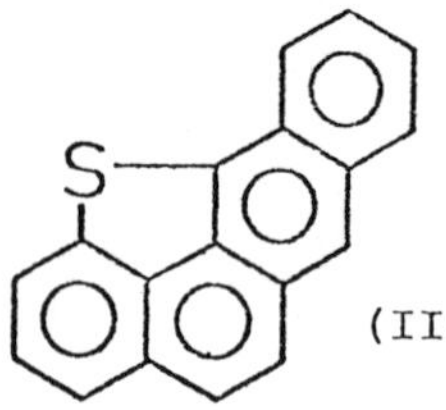

(IIa)

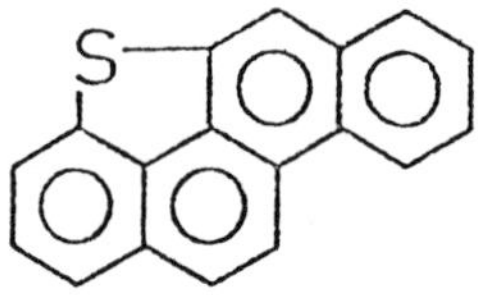

(IIb)

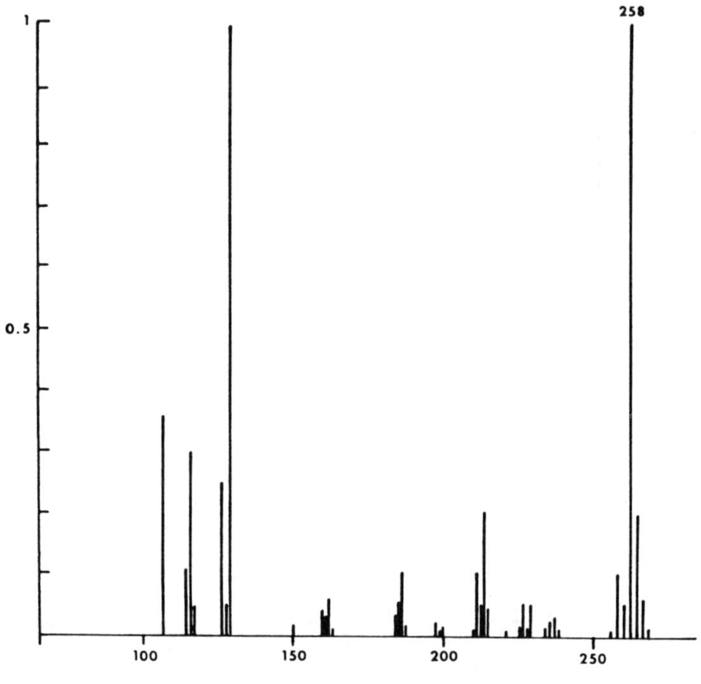

FIGURE 5. Mass spectrum of $C_{18}H_{10}S$.

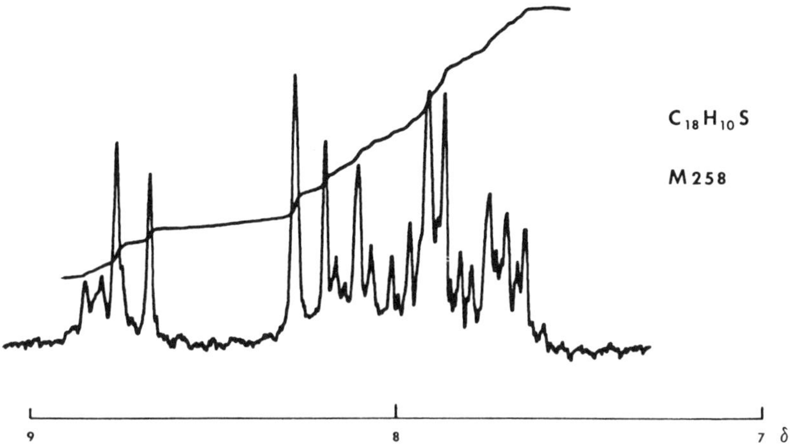

FIGURE 6. NMR spectrum for $C_{18}H_{10}S$.

A third possibility, which is derived from a triphenylene configuration, can be excluded definitively because of the presence of a strong singlet signal (one proton) with a chemical shift of δ = 8.27 (ppm or Hz/MHz) (see Figure 6). Though the evaluation of the NMR spectrum tends to support the alternative (IIb), an unambiguous choice cannot be made on the basis of the information presently available.[1] The UV spectrum of $C_{18}H_{10}S$ dissolved in hexane is shown in Figure 7.

The second S-impurity, which has been found at concentrations between 0.1 and 1% in commercial benzo[a] pyrenes, has been identified as an isomer of dinaphthothiophene. The most likely structure for the isomer is that of a dinaphtho-2,3,2′,3′-thiophene (number 5 in Figure 8), based on the evaluation of the respective mass and NMR spectrum. From the shape of the mass spectrum (see Figure 9), as well as the UV spectrum (see Figure 10), it is clear that this thiophene compound must have a different structure than the impurity of the same mass number which has been found in benzo[j] fluoranthene. Especially significant is the absence of the mass peaks M-1 and M-2 in the mass spectrum of $C_{20}H_{12}S$, isolated from benzo[a] pyrene. This indicates that in contrast to the thiophene isomer observed in benzo[j]-fluoranthene, no ring condensation is possible, thereby excluding structures 1, 2 and 3 of Figure 8.

Benzo[g,h,i] perylene

The principal impurity of commercial benzo[g,h,i] perylene materials again was found to contain sulfur, and by means of mass spectrometry the

[1] Both isomers have now positively been identified in the 258 mass fraction.

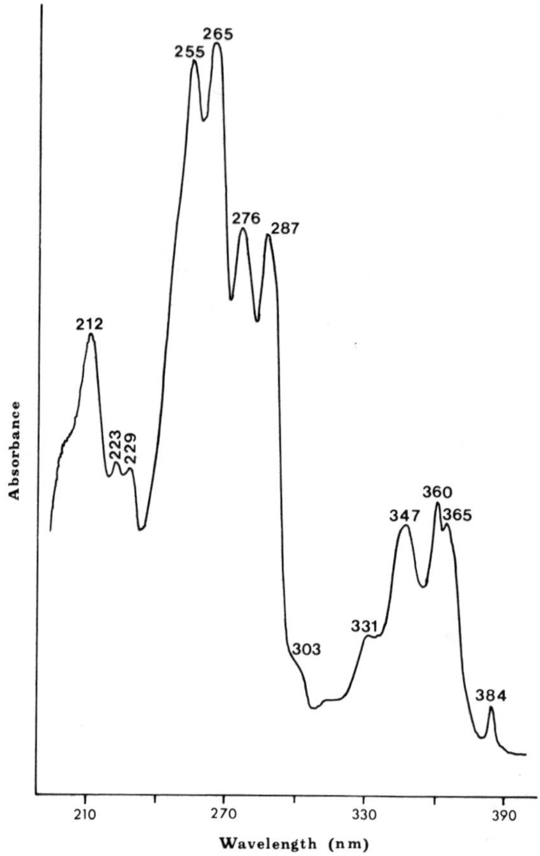

FIGURE 7. UV spectrum of $C_{18}H_{10}S$ from B[a] P.

composition $C_{20}H_{10}S$ could be allocated. As in the case of pyrene and benzo[a]pyrene, the difference in molecular weight between the impurity and the parent PAH (202/208; 252/258; 276/282) suggested the replacement of an aromatic ring by a thoiphene ring. However, it has not been possible thus far to separate sufficient quantities of the material to permit unambiguous identification of this impurity. The most probable structure is that of a peryleno[1,12-b,c,d] thiophene(III)(1).

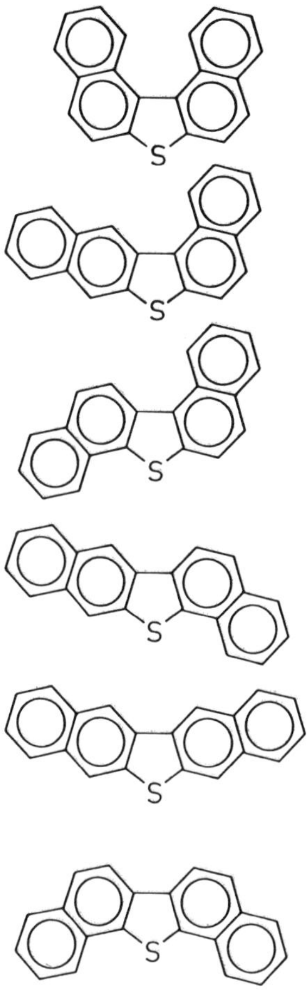

1. Dinaphtho – 2, 1, 1′, 2′– thiophene
2. 〃 – 2, 3, 1′, 2′– 〃
3. 〃 – 1, 2, 1′, 2′– 〃
4. 〃 – 2, 3, 2′, 1′ – 〃
5. 〃 – 2, 3, 2′, 3′– 〃
6. 〃 – 1, 2, 2′, 1′ – 〃

FIGURE 8. Alternative structures for $C_{20}H_{12}S$.

Benzo[j] fluoranthene

So far, the detection of these thiophene derivatives had been restricted to commercial materials which presumably had been derived from coal tar or petroleum fractions. However, a thiophene derivative has been observed, also, in a benzo[j] fluoranthene sample which had been obtained by synthesis, involving a ring condensation step by reduction in the presence of sulfur. Evidently, in dehydrogenation reactions of PAH materials in the presence of sulfur, there is a strong tendency to incorporate a sulfur atom into the PAH molecule, resulting in a PAH-thiophene derivative. Here, the main impurity had the mass number 284 and the composition $C_{20}H_{12}S$, which had already been observed in benzo[a] pyrene. Since a sufficient amount of this impurity

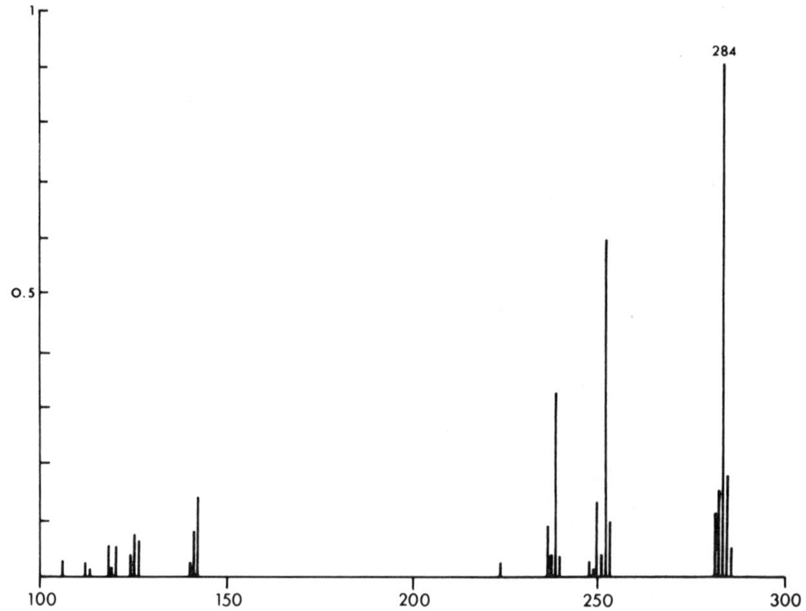

FIGURE 9. Mass spectrum of $C_{20}H_{12}S$, isolated from benzo[a]pyrene.

could be separated, the mass and NMR spectra have been determined. From the NMR spectrum it could be deduced that two naphtho groups were joined to the thiophene ring. Of the six possible structures (1 through 5 of Figure 8) 4, 5 and 6 are excluded on the basis of the mass spectrum (Figure 11). In this way, only structure 1 (Figure 8) remains. As can be seen in the mass spectrum, relatively high intensities are observed for mass fractions M-1, M-2 and M-2/2. In the structure below, this can be explained by the loss of two protons in positions 8 and 8′, respectively, and subsequent formation of another aromatic ring with a double charge for mass M-2, leading to the observed peak at M-2/2 (mass 141).

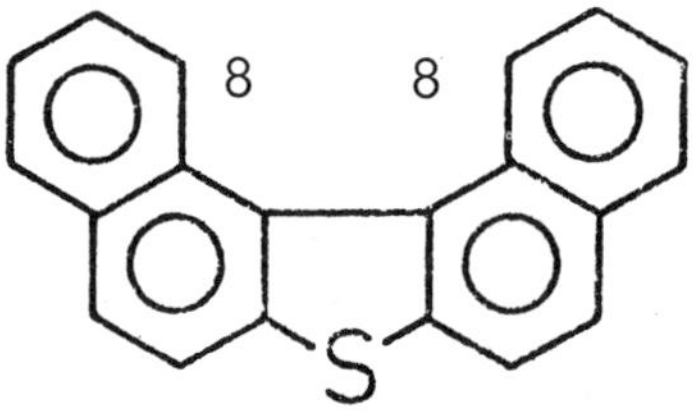

(IV)

The UV spectrum of this dinaphthothiophene isomer, taken in a solution of hexane, is presented in Figure 13.

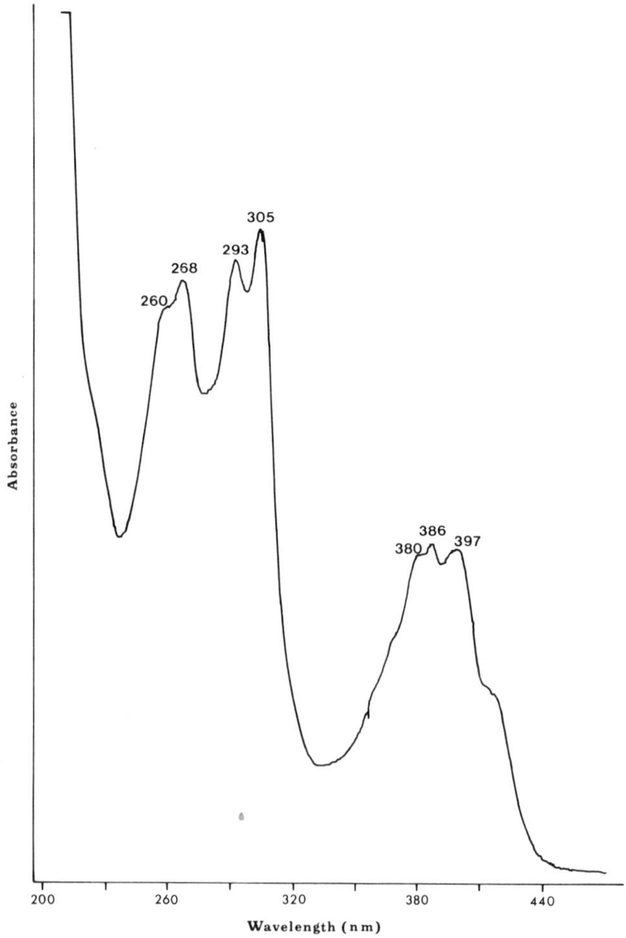

FIGURE 10. UV spectrum of $C_{20}H_{12}S$ from B[a] P.

CONCLUSIONS

Although the work described in this paper is not complete, and is to be continued in an attempt to identify further thiophene derivatives in other PAH materials of general interest, the following conclusions may be drawn at this stage:

1. PAH materials which have been obtained from fossil sources (coal, crude oil/low-temperature genesis) seem to be accompanied consistently by the corresponding thiophene derivatives.

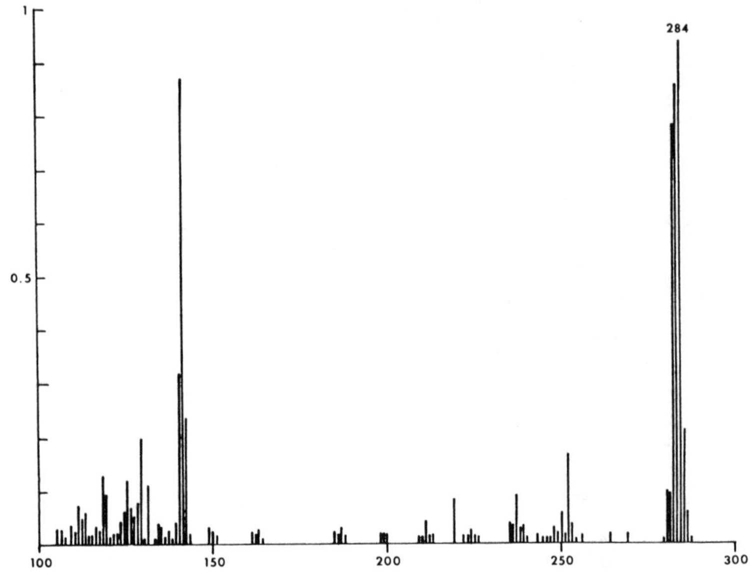

FIGURE 11. Mass spectrum of $C_{20}H_{12}S$ (from benzo[j] fluoranthene).

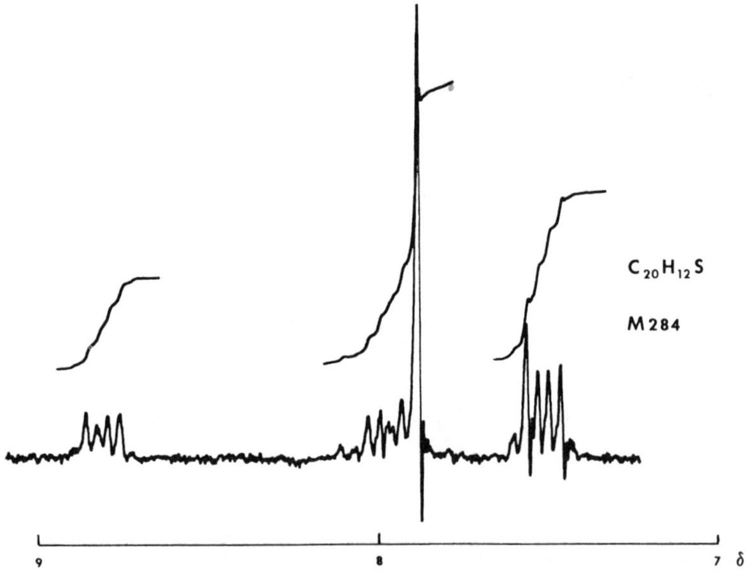

FIGURE 12. NMR spectrum of $C_{20}H_{12}S$.

Fig.12

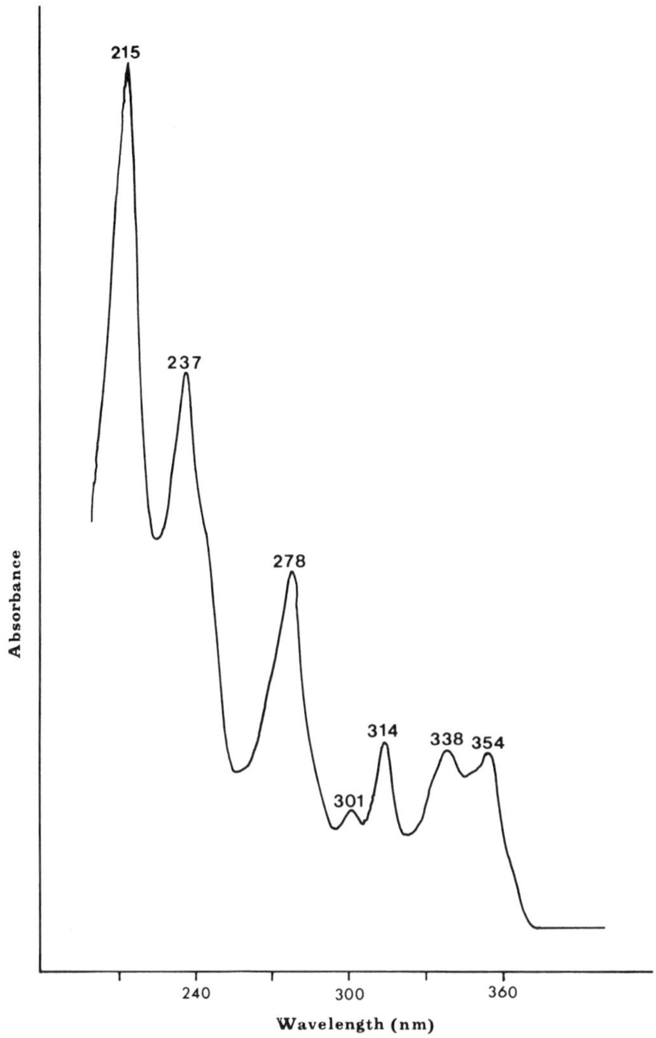

FIGURE 13. UV spectrum of $C_{20}H_{12}S$ from B[j] F.

2. The major impurity in PAH materials prepared from coal or petroleum derivatives is often a thiophene compound resulting schematically from the replacement of one aromatic ring by a thiophene ring ($C_mH_n \rightarrow C_{m-2}H_{n-2}S$). The formation of these thiophene derivatives is probably due to the reaction of sulfur into a bay-region containing PAH (*e.g.*, phenanthro-, chryseno- or benzanthraceno thiophene; general composition $C_{4z+2}H_{2z+2}S$, z indicating the number of benzene rings).

3. In some other cases, thiophene derivatives have been observed which have only two sides of the thiophene ring in common with adjoining benzene rings (*e.g.*, dinaphthothiophene isomers).
4. With increasing molecular weight, the number of thiophene derivatives which appear in the parent PAH tends to become greater.
5. Sulfur-containing impurities may be found, too, in PAH materials which have been prepared by synthesis. This is especially likely in those cases when ring condensation in the course of synthesis is achieved by a reduction step in presence of sulfur.
6. To permit a comparison of the carcinogenic effects of the parent PAH and the accompanying thiophene impurity, the separation and enrichment of larger quantities of sulfur-containing PAH might be desirable.

ACKNOWLEDGMENTS

We wish to thank Messrs R. Fordham, Ph. Glaude and T. Driscoll for their very valuable technical assistance. Also, we thank Dr. T. Wirthlin of Varian, Zurich, for the ^{13}C-NMR Spectrum of $C_{14}H_8S$.

REFERENCES

1. Borwitzky, H., Henneberg, D., Schomburg, G., Sauerland, H. D., and Zander, M. (1977): *Erdöl und Kohle-Erdgas-Petrochemie* 30:370.
2. Burchill, P., Herod, A. A., and James, R. G. (1978): In: *Carcinogenesis, Vol. 3,* R. Freudenthal and P. W. Jones, Eds. pp. 35-45.
3. Grimmer, G., and Böhnke, H. (1976): *Chromatog.* 9:30.
4. Grimmer, A., and Glaser, A. (1975): *Erdöl und Kohle-Erdgas-Petrochemie* 28:570.
5. Karcher, W., Jacob, J., and Haemers, L. (1979): The certification of six PAH-materials (benzo[b] fluoranthene, benzo[k] fluoranthene, benzo [j] fluoranthene, benzo[e] pyrene, indeno[1,2,3-c,d] pyrene, benzo[b] chrysene). EUR-report (in press).
6. Kruber, O., and Grigoleit, G. (1954): *Chem. Ber.* 87:1895.
7. Kruber, O., and Raeithel, A. (1940): *Chem. Ber.* 75:1184.
8. Lee, M. L., and Hites, R. A. (1976): *Anal. Chem.* 48:1890.
9. Ave, W. A., and Flinn, C. G. (1978): *J. Chromatog.* 153:305.

Polynuclear Aromatic Hydrocarbons,
edited by P. W. Jones and P. Leber.
Ann Arbor Science Publishers, Inc.
Ann Arbor, MI ©1979.

POLYCYCLIC AROMATIC HYDROCARBONS IN THE LOWER ATMOSPHERE OF KARLSRUHE

Hans Güsten and Günther Heinrich

Kernforschungszentrum Karlsruhe
Institut für Radiochemie
Karlsruhe, Federal Republic of Germany

INTRODUCTION

The simultaneous measurement of several polynuclear aromatic hydrocarbons (PAH) allows better assessment of the air quality in towns, but it could also reflect the contributions to atmospheric pollution by the individual groups of emitters, such as residential heating systems, automobile traffic and industry. By the simultaneous analysis of PAH and lead, the contribution of automobile traffic can be estimated since the lead emission in towns free from industry is caused solely by tetraethyl lead added to the gasoline as an anti-knock agent.

In this work, field measurements of the concentrations of six PAH are reported, namely the carcinogenic PAH benzo[a]pyrene (B[a]P), benzo-[b]fluoranthene (B[b]F), benzo[a]anthracene (B[a]A) and indeno[1,2,3-c,d]pyrene (IP), as well as the noncarcinogenic PAH fluoranthene (F) and benzo[k]fluoranthene (B[k]F), together with lead. The measurements were carried out in the Karlsruhe region over an extended period of two winter seasons (1974/75 and 1975/76), and during the months of May and June 1975 for comparison, in order to determine the relative amount of PAH particulate matter from residential heating systems and automobile traffic in the Karlsruhe urban area. The first three carcinogenic PAH were proposed by the International Association for Research on Cancer as "reference components" for air monitoring. Besides the four carcinogenic PAH, the noncarcinogenic fluoranthene was selected because among the less-volatile PAH, it is formed at the relatively highest concentration both in the combustion of liquid and solid fuels of domestic heating systems (4,15), as well as in the automobile engine (5). The noncarcinogenic B[k]F, a structural isomeric $C_{20}H_{12}$ aromatic hydrocarbon to B[a]P, is always associated with B[a]P in the environment. Its measurement was chosen for analytical reasons discussed later.

EXPERIMENTAL

The ecotoxicological assay of PAH present in the environment in the ppt range calls for a highly sensitive analytical technique which allows short collection times for the airborne PAH absorbed on aerosols and dusts. Details of the fluorescence analysis used, as well as the analytical technique, were published earlier (3,16).

By means of the corrected absolute fluorescence spectra, a simultaneous quantitative determination can be made of polycyclic aromatic hydrocarbons with similar emission behavior. The sensitivities of given PAH can be enhanced by elimination of the oxygen from the solution prior to measurement by fluorescence spectroscopy (16). The sensitivity of the fluorescence spectrometer with respect to benzo[a]pyrene in oxygen-free *n*-heptane is better than 0.05 ng/ml. Therefore, the sensitivity of the fluorescence spectroscopic analysis of the PAH in the aerosol samples collected allows an air sampling time of only one hour.

The two air samplers were located at a distance of about 550 m, one in the Municipal Garden away from traffic, and the other in the covered railway underpass (length 100 m) near the Albtal Railway Station in Karlsruhe. The collection point in the covered railway underpass was situated at a 3 m height. At these two locations, air samples were taken with the air collector (EM 100, Sartorius) every Monday, Wednesday and Friday from 2 to 7 pm. When glass-fiber filters (type no. 8, Schleicher and Schull) were used, the volume of air throughput reached was 100 m^3/hr. The mean filter removal efficiency is 99.99% according to the supplier as compared with a test aerosol having a maximum particle-size abundancy of 0.3-0.5 μm. The majority of airborne aerosols and particles with the PAH adsorbed had greater diameters (1,20). The filters exclusively contained inorganic binders which did not fluoresce, and thus did not adversely affect PAH determination by fluorescence spectroscopy. The volume of air collected was 500 m^3 per filter. The filters were extracted in benzene and the samples separated by means of two-dimensional thin-layer chromatography at a mixture made from acetylated cellulose, silica gel G and alumina G type E in *n*-heptane-benzene (9:1) and methanol-diethylether-water (4:4:1).

In thin-layer chromatography of PAH, unavoidable losses occur which may range from 10 to 15% depending on the compound (21). This always has to be taken into account when considering the noncorrected absolute concentration values.

RESULTS

The concentration of benzo[a]pyrene and benzo[k]fluoranthene found in the Municipal Garden from October 1975 to March 1976 has been plotted in

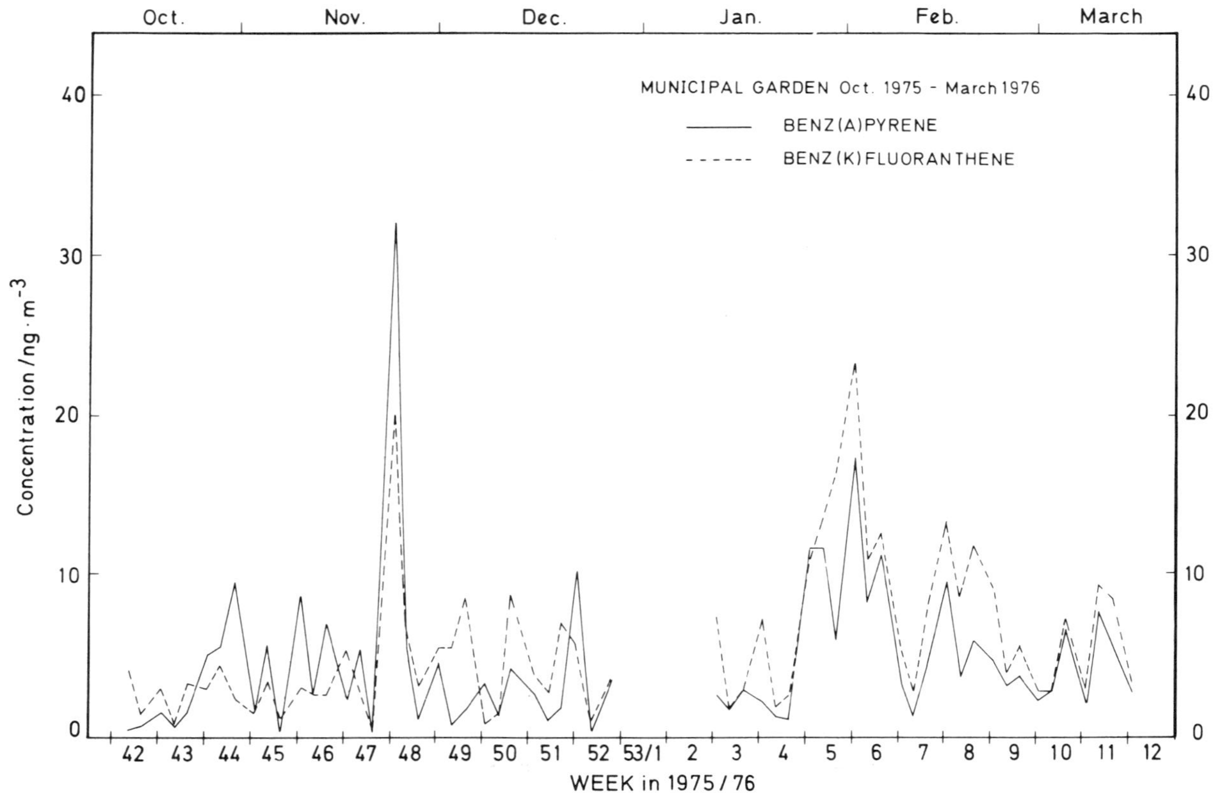

FIGURE 1. Concentration of airborne benzo[a] pyrene and benzo[k] fluoranthene in the Municipal Garden in Karlsruhe from October 1975 to March 1976.

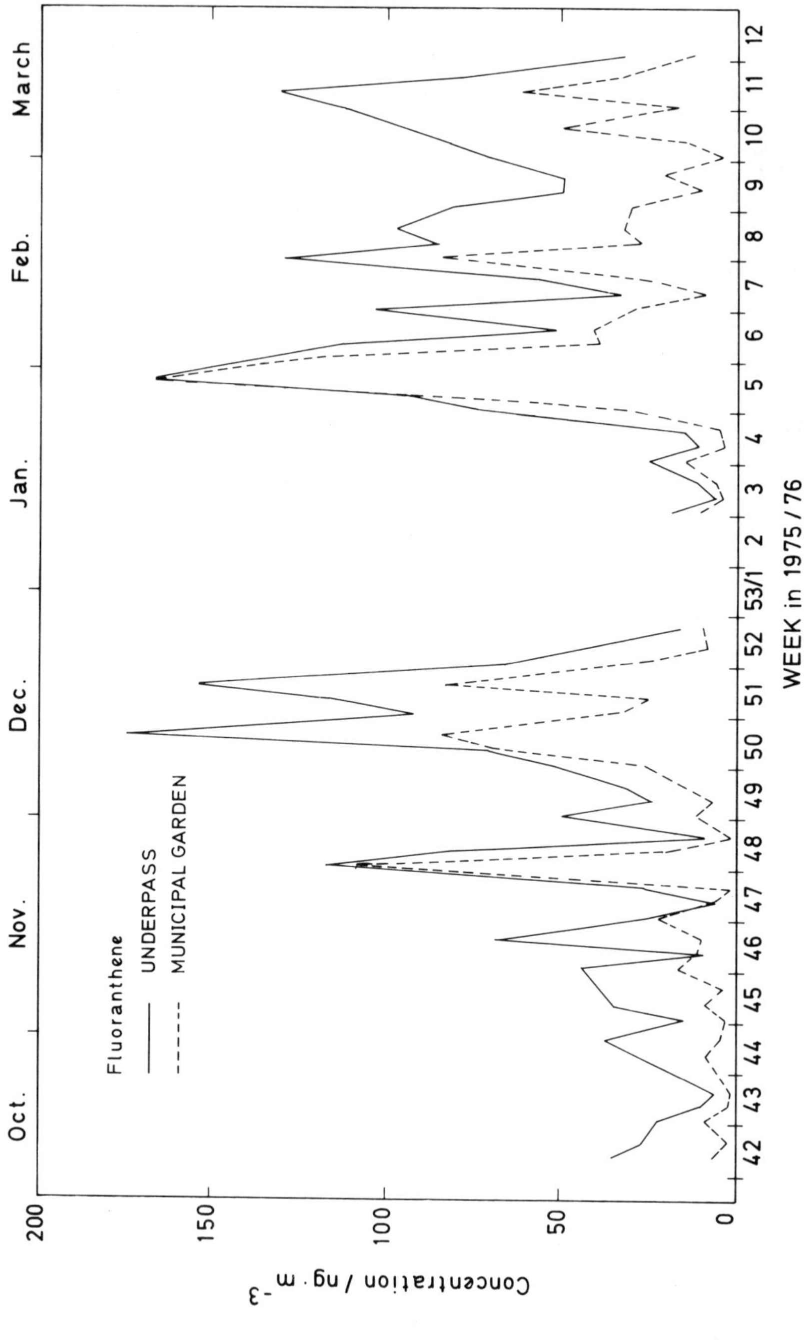

FIGURE 2. Concentration of airborne fluoranthene in the Municipal Garden and in a nearby railroad underpass in Karlsruhe from October 1975 to March 1976.

Figure 1. The same fluctuations in other PAH concentrations are found in the railroad underpass as is shown for example in Figure 2 for fluoranthene.

The striking phenomena are the large temporal variations of the PAH content. For instance, the values of benzo[a]pyrene in the Municipal Garden vary between 0.2 and 32 ng/m^3. The maximum value of 42.5 ng/m^3 was measured in the underpass. The other PAH yield a similar time curve, *i.e.*, they show their highest concentrations in the 44th, 48th and 50th/51st weeks of the year 1975 and in the 5th and 8th weeks of 1976 (Figures 1 and 2). A summary of the distribution of the benzo[a]pyrene content in the atmo-

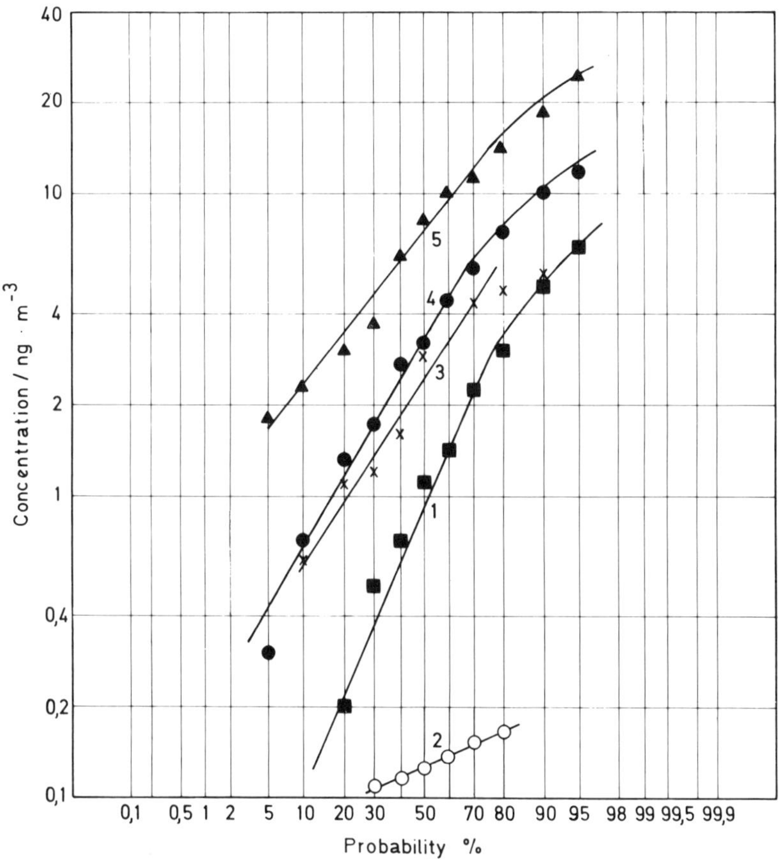

FIGURE 3. Frequency distribution of the airborne benzo[a]pyrene concentration at different measuring sites in the area of Karlsruhe. (1) Nuclear Research Center, Karlsruhe, November 1974-March 1975; (2) Municipal Garden, May to June 1975; (3) railroad underpass, May to June 1975; (4) Municipal Garden, October 1975 to March 1976; (5) Railroad underpass, October 1975 to March 1976.

sphere during the individual seasons, as found at the air sampling sites, is shown in Figure 3. The results of investigations made earlier (November 1974 to March 1975) at the Karlsruhe Nuclear Research Center, located in a woods about 11 km air distance north of the city, are shown in addition to the data measured in the urban area. The benzo[a]pyrene concentrations are always lowest in the summer season at the Municipal Garden. The concentration distribution is shifted to higher values in the underpass during summer and winter. At least 50% of all the values measured at one air sampling site exceed 1 ng/m^3. The lead concentrations simultaneously measured in the particulate matter of the same air samples are about three orders of magnitude above the PAH concentrations (1-4 μg/m^3). Among the PAH, benzo[b]fluoranthene (B[b]F) and fluoranthene (F) reach the highest values at different locations in every season. In Figure 4, the mean values of all the PAH concentrations measured are summarized. Only benzo[a]anthracene (B[a]A) measured exclusively in the winter of 1975/76 occurs at a similarly high average concentration in the underpass. The concentrations of benzo[a]pyrene and benzo[k]fluoranthene are always lowest at the two air sampling sites in the urban area and at the Nuclear Research Center. In the summer months the values of PAH are about 10 to 20 times higher in the underpass than in the Municipal Garden. The comparison between the summer and winter months shows significant rise in atmospheric PAH concentrations in the Municipal Garden during the winter. There, the average values in the winter are higher than summer values by up to 45 times. In the underpass the increase in PAH concentrations in the winter is higher by only 2 to 4 times, on an average, as compared to the summer months. The comparison between the Municipal Garden and the underpass changes the picture inasmuch as the mean concentration values in the underpass in winter are greater than in the Municipal Garden only by a factor of two.

DISCUSSION

Benzo[a]pyrene and Benzo[k]fluoranthene

For the ecotoxicological evaluation of the PAH and their dose/effect relationship, numerous authors have measured benzo[a]pyrene as a reference substance for about 20 years. B[a]P has proved strongly carcinogenic in animal tests and can be detected in a relatively simple way by fluorescence analysis. Moreover, this established a relationship with the carcinogenic effect of cigarette smoke condensates, so that in animal tests a dose/effect analysis can be performed on the condensates of both airborne dust and cigarette smoke. Thus, it is quite possible that on days with high benzo[a]-pyrene concentrations (more than 30 ng/m^3; see Figure 1) and assuming a daily volume of respired air of about 12 m^3, people living in Karlsruhe inhale

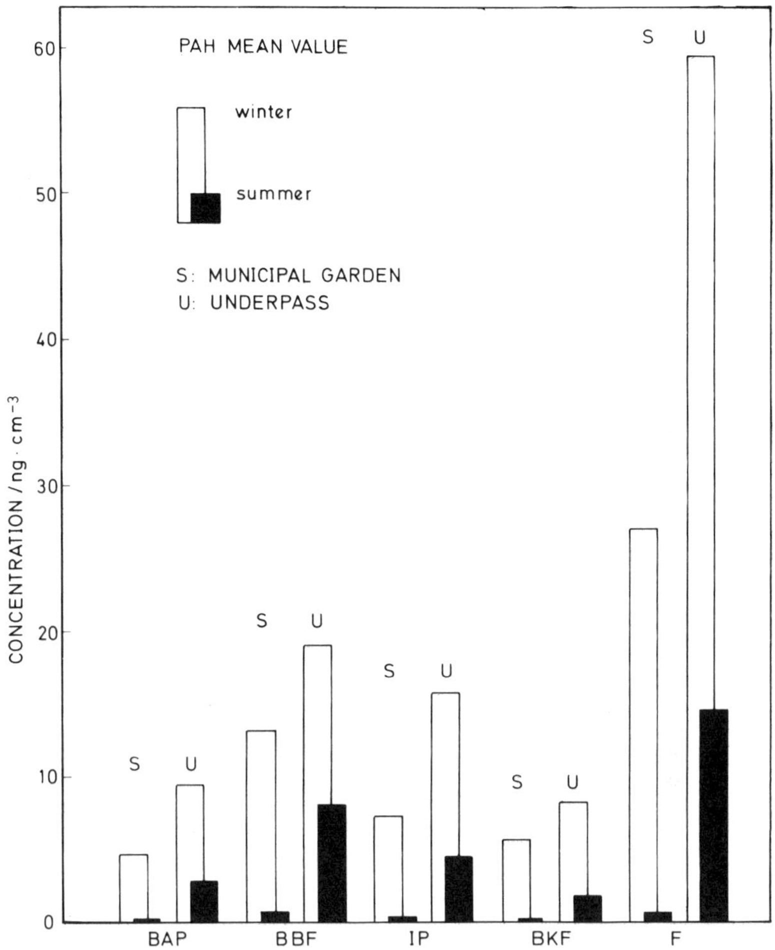

FIGURE 4. Mean concentration of airborne PAH in the Municipal Garden and a nearby railroad underpass in Karlsruhe during May-June 1975 and from October 1975 to March 1976.

approximately the same amount of benzo[a]pyrene as contained in the smoke condensate of 10 cigarettes.

A comparison of concentration values for benzo[a]pyrene in a number of large European cities (London, Copenhagen, Hamburg, Prague, Budapest and Milan) as compiled by Colucci and Begeman (9), and the data measured at 13 air sampling sites located in the Ruhr area (11) with our values measured in Karlsruhe, shows that our values of B[a]P are lower by a factor of 10 to 50. In Europe, the benzo[a]pyrene concentrations we measured were comparable only with the values found in Madrid (2), Lyon (6) and Stockholm (18). However, our values compare in magnitude with many values measured in large U.S. cities (9,13) and in Toronto (20).

We think there are two significant aspects explaining the great differences in benzo[a]pyrene concentrations found at the different locations. A number of measurements have been made using very short periods of air collection, *e.g.*, one day. However, as can be seen from Figures 1 and 2 for Karlsruhe in the winter 1975/76 period, the PAH concentrations vary by a factor of 10 so that "snapshot" measurements of B[a]P cannot give representative concentration values as frequently encountered in the literature (2,6,18). A second cause might be the different analytical techniques employed by different authors. B[a]P and B[k]F are obviously always associated in airborne particulate matter. This can be shown by comparing two sets of concentration data of B[a]P and B[k]F. The plot of the B[a]P concentration and the B[k]F concentration from the Municipal Garden and the underpass shows that the distribution of their concentrations during the winter 1975/76 lie along the straight line Y = X (Figure 5). As assumed by Colucci and Begeman (9), benzo[k]fluoranthene, which is difficult to separate from benzo[a]pyrene by chromatography, was probably measured erroneously as benzo[a]pyrene. Column chromatography and even one-dimensional thin-layer chromatography do not guarantee that these two $C_{20}H_{12}$ aromatic compounds are adequately separated from each other. According to our measurements, benzo[k]fluoranthene occurs at concentrations corresponding to that of benzo[a]pyrene (Figures 1 and 4). However, the determination of benzo[a]pyrene by fluorescence spectroscopy in the presence of benzo[k]-fluoranthene does not result in an error by a factor of 2, as assumed by their concentration ratios, but to a substantially higher factor depending on the measurement conditions selected for fluorescence excitation. The reason is that benzo[k]fluoranthene shows sensitivity by fluorescence up to six times higher than benzo[a]pyrene (16). Our investigation of the atmospheric pollution by PAH in Karlsruhe shows that stating the benzo[a]pyrene content alone as a measure of the total PAH concentration and for air quality evaluation is not sufficient, since substantially higher concentrations of other carcinogenic PAH are in the urban air (see Figure 4).

Lead

The major source of lead emissions is the automobile. As expected the lead concentration downtown was highest at locations with high traffic such as the railway underpass. There the mean values for lead were 3.1 $\mu g/m^3$ in the summer and 3.6 $\mu g/m^3$ in the winter. By contrast, in the Municipal Garden the mean value for lead in the winter reached 0.5 $\mu g/m^3$. Thus, our values lie within the concentration range of many measurements performed in Europe

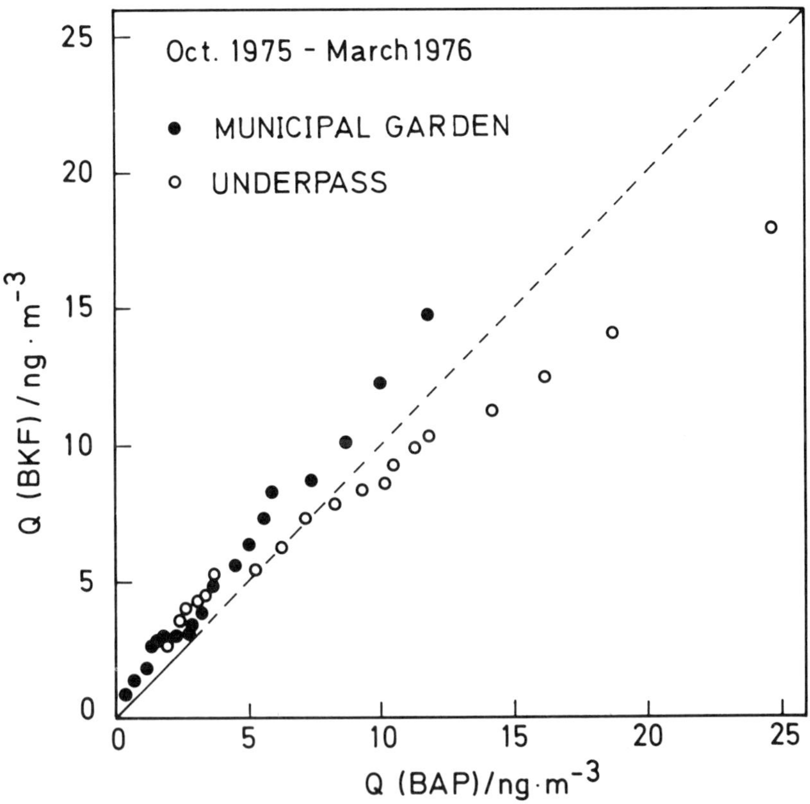

FIGURE 5. Quantiles of airborne benzo[a]pyrene and benzo[k]fluoranthene concentrations from the Municipal Garden and the underpass in Karlsruhe from October 1975 to March 1976.

and America as compiled by Colacino and Lavagnini (7). Even at the Nuclear Research Center, Karlsruhe, situated 11 km distant from the town and away from traffic, a relatively high average lead concentration of 0.3 $\mu g/m^3$ was still measured. However, in comparing the Municipal Garden with the distant Nuclear Research Center, the lead concentrations were similar. It is known that the variation of the lead content in the atmosphere depends little on the meteorological conditions (7,8). By contrast with PAH, the size distribution of airborne lead aerosols is smaller than 0.5 μm (12,17) so it can be assumed that only the small lead aerosols are transported by wind to remote areas, *e.g.*, at the Nuclear Research Center.

The monthly mean values of lead concentrations decreased in the railway underpass from 4.4 $\mu g/m^3$ in October and November 1975 to 4.0 $\mu g/m^3$

in December 1975; 2.9 $\mu g/m^3$ in January 1976; 3.1 $\mu g/m^3$ in February 1976; and 2.6 $\mu g/m^3$ in March 1976. This is probably a consequence of the new German act restricting the organo-lead compounds in gasoline, effective January 1, 1976.

Polycyclic Aromatic Hydrocarbons

The highest concentration of air pollutants was caused by automobile traffic in the railway underpass. Even some 100 m distant in the Municipal Garden, the PAH concentration decreased by at least one order of magnitude. One reason for the strong decrease in PAH concentration in the Municipal Garden is certainly the deposition of the airborne particulate matter on the leaves of the trees and in bushes. The high increase in the mean PAH concentrations in the city area during the winter months was generally attributed to residential heating. For example, the PAH concentrations in the Municipal Garden were up to 40 times higher in winter than in summer. The increase in PAH concentration in the railway underpass was less pronounced. The assumption that the traffic density, and hence the PAH emissions by automobiles, is the same summer and winter is justified due to the near constant lead concentrations found. Even outside the town at the Nuclear Research Center, the PAH content in the winter attained values such as observed in the city area in the underpass only during the summer season. These results clearly show that during the winter months residential heating ranks as a major source of PAH. Comparison of the summer and winter PAH concentrations shows that in the winter the PAH emissions originates by more than 90% from residential heating systems. The considerable rise in benzo[a]pyrene, benzo[k]fluoranthene and fluoranthene shown in Figures 1 and 2 for the end of November (48th week) and the end of January (5th week) was attributable to a ground inversion layer whose influence was also found in the underpass (see Figure 2). In the 48th week the temperature rose from -7 to +13°C and, subsequently, a strong rainfall accompanied by little wind was responsible for the washout of the airborne PAH by the end of that week (see Figures 1 and 2).

Supposing that the PAH concentrations in the atmosphere of Karlsruhe stem essentially from one emission source in the summer, namely automobile traffic, it should be possible to establish a relationship between the respective PAH and the lead concentrations. Table I is a compilatioon of the correlation coefficients between the measured PAH and the lead both in the Municipal Garden and in the nearby railroad underpass. Table I shows that in the summer a good relationship exists for PAH at both air sampling sites with values between 0.7 and 0.9, this means that PAH arise primarily from automobiles. By contrast, the correlation of PAH with the lead concentrations

TABLE I. Correlation coefficients of the PAH and Lead Concentrations in the Municipal Garden (lower left half) and in the Railroad Underpass (upper right half) from May to June 1975

Municipal Garden	Underpass					
	B[a]P	B[k]F	B[b]F	IP	F	Lead
B[a]P		0.90	0.81	0.87	0.50	0.63
B[k]F	0.80		0.77	0.86	0.68	0.56
B[b]F	0.80	0.90		0.67	0.56	0.58
IP	0.71	0.76	0.86		0.50	0.50
F	0.85	0.66	0.78	0.87		0.47
Lead	-0.27	-0.14	-0.33	-0.38	-0.16	

were negative in the Municipal Garden, and low even in the underpass, contrary to expectations.

Table 1 proves that the previous attempts (8,9,22) in deriving the contribution of automobile traffic to the PAH total emission from the comparison of B[a]P and lead concentrations in the air as well as in gasoline cannot provide a significant statement. The cause of the missing correlation should be the differing transport mechanisms and ranges, respectively, of the two types of aerosols from one emission source. While the mass of lead aerosols gets deposited in the immediate vicinity of their origin so that a very strong reduction in lead concentration can be found over the short distance between the underpass and the Municipal Garden, the PAH associated with the particulates are transported over much greater distances (10,20) so that they can still be measured at relatively high concentrations, *e.g.*, at the Nuclear Research Center 11 km from Karlsruhe.

Comparison of PAH concentrations in the underpass and in the nearby Municipal Garden during both seasons yields the correlation coefficients

TABLE II. Correlation Coefficients of PAH in the Municipal Garden and in the Underpass in Karlsruhe for the Periods May to June 1975 and October 1975 to March 1976

PAH	Winter 75/76	Summer 75
B[a]P/B[a]P	0.78	-0.04
B[k]F/B[k]F	0.76	-0.06
B[b]F/B[b]F	0.65	0.93
IP/IP	0.65	0.43
F/F	0.80	0.27
B[a]A/B[a]A	0.76	––

compiled in Table II. Whilst during the winter season the statistically significant correlations between the PAH concentrations in the Municipal Garden and in the underpass show that PAH stem from the same emission sources, residential heating and automobile traffic, the missing correlations—except for benzo[b]fluoranthene—during the summer season are surprising. As a matter of fact, only automobile traffic is a significant emitter in the Karlsruhe area during the summer. We speculate that the different correlation coefficients of PAH concentrations between the underpass, protected from light, and the Municipal Garden during the summer reflect the different photochemical rates of degradation of PAH in the atmosphere.

It is known that the photochemical reactions of atmospheric pollutants, such as SO_2 in the winter, decrease by approximately the factor 10 in Central Europe (9). According to this interpreation, the two PAH with negative correlation coefficients shown in Table II, namely benzo[a]pyrene and benzo[k]fluoranthene, should have the highest photochemical degradation in the summer. In addition to a direct photoreaction of PAH through absorption of the sunlight, they might also react with the oxidizing radicals of the photochemical smog (14,20).

SUMMARY

The concentrations of six polycyclic hydrocarbons (PAH) (fluoranthene, benzo[a]anthracene, benzo[a]pyrene, benzo[b]fluoranthene, benzo[k]-fluoranthene, indeno[1,2,3-c,d]pyrene and lead were simultaneously measured in a railroad underpass with heavy traffic and in the Municipal Garden away from traffic in the city of Karlsruhe. While the average PAH concentrations measured in summer were less than 1 ng/m^3 in the Municipal Garden, the corresponding mean values measured in the underpass were 10 to 20 times higher, depending on the compound. However, while in the winter season the PAH mean values measured in the Municipal Garden were 20 to 45 times higher than in the summer months, they increased by only the factor 2 in the underpass and, hence, their absolute concentrations in winter were only twice as high as in the Municipal Garden. By contrast, the summer/winter mean values for lead varied by a factor of 2 in the Municipal Garden and increased only slightly in the underpass. Correlation analysis of the benzo-[a]pyrene and lead concentrations did not provide conclusive evidence for the PAH fraction from traffic. The correlation analysis of the six PAH concentrations measured in the summer of 1975 and from October 1975 until March 1976, shows that the impact on the atmosphere by PAH during the winter months must be due mainly to the emissions from residential heating systems and that the low summer concentrations of some PAH might have their origin in photochemical degradation or oxidation.

REFERENCES

1. Albagli, A., Oja, H., and Dubois, L. (1974): Size-distribution pattern of polycyclic aromatic hydrocarbons in airborne particulates. *Environ. Lett.* 6:241-251.
2. Anechina, P., Romero, J. M., Sanchez, B., Murias, F., and de la Serna, J. (1970): Evaluation of the atmospheric contamination in Madrid in 1969. *Rev. San. Hig., Spain* 44:885-907.
3. Blume, H., and Güsten, H. (1977): Spektroskopische Verfahren, in *Ultraviolette Strahlen,* J. Kiefer, Ed., pp. 311-348, Walter de Gruyter Verlag, Berlin.
4. Brockhaus, A., and Tomingas, R. (1976): Emission Polyzyklischer Kohlenwasserstoffe bei Verbrennungsprozessen in kleinen Heizungsanlagen und ihre Konzentration in der Atmosphäre. *Staub-Reinh. Luft* 36:96-101.
5. Candeli, A., Morozzi, G., Paolacci, A., and Zoccolillo, L. (1975): Analysis using thin layer and gas-liquid chromatography of polycyclic aromatic hydrocarbons in the exhaust products from a European car running on fuels containing a range of concentrations of these hydrocarbons. *Atmos. Environ.* 9:843-849.
6. Chatot, G., Dangy-Caye, R., and Fontanges, R. (1973): Edude de la pollution atmosphérique par les arènes polynucleaires dans la région lyonnaise à l'aide de duex collecteurs de principe different. *Atmos. Environ.* 7:819.
7. Colacino, M., and Lavagnini, A. (1974): Particulate lead pollution in Rome's atmosphere. *Water Air Soil Poll.* 3:209-216.
8. Colucci, J. M., and Begeman, C. R. (1969): Lead concentrations in Detroit, New York, and Los Angeles air. *J. Air Poll. Control Assoc.* 19:255-260.
9. Colucci, J. M., and Begeman, C. R. (1971): Carcinogenic air pollutants in relation to automotive traffic in New York. *Environ. Sci. Technol.* 5:145-150.
10. De Maio, L., and Corn, M. (1966): Polynuclear aromatic hydrocarbons associated with particulates in Pittsburgh air. *J. Air Poll. Control Assoc.* 16:67-71.
11. Friedrichs, K.-H., Stuke, J., Brockhaus, A., and Steiger, H. (1971): Luftverunreinigungen im Ruhrgebiet-Beurteilung eines Meßprogramms unter lufthygienischen Gesichtspunkten. *Staub-Reinhalf. Luft* 31: 323-326.
12. Gilette, D. A., and Winchester, J. W. (1972): A study of aging of lead aerosols-I. Observations. *Atmos. Environ.* 6:443-450.
13. Gordon, R. J. (1976): Distribution of airborne polycyclic aromatic hydrocarbons throughout Los Angeles. *Environ. Sci. Technol.* 10: 370-373.
14. Güsten, H., and Penzhron, R.-D. (1974): Photochemische Reaktionen Atmosphärischer Schadstoffe. *Naturwiss. Rundschau* 27:56-68.

15. Hangebrauck, R. P., von Lehmden, D. J., and Meeker, J. E. (1964): Emission of polynuclear hydrocarbons and other pollutants from heat-generation and incineration processes. *J. Air Poll. Control Assoc.* 14:267-278.
16. Heinrich, G., and Güsten, H. (1976): Die fluoreszenzspektroskopische Bestimmung in der Atmosphäre vorkommender polycyclischer aromatischer Kohlenwasserstoffe. *Z. Anal. Chem.* 278:257-262.
17. Lee, R. E., Jr., Goranson, S. S., Enrione, R. E., and Morgan, G. B. (1972): National air surveillance cascade impactor network II. Size distribution measurements of trace metal components. *Environ. Sci. Technol.* 6:1025-1030.
18. Linstedt, G. (1968): Routine determination of 3,4-benzopyrene in air samples. *Atmos. Environ.* 2:1-11.
19. Penzhorn, R.-D., Filby, W. G., and Gusten, H. (1974): Die photochemische Abbaurate des Schwefeldioxids in der unteren Atmosphäre Mitteleuropas. *Z. Naturforsch.* 29a:1449-1453.
20. Pierce, R. C., and Katz, M. (1975): Dependency of polynuclear aromatic hydrocarbon content on size distribution of atmospheric aerosols. *Environ. Sci. Technol.* 9:347-353.
21. Pierce, R. C., and Katz, M. (1975): Determination of atmospheric isomeric polycyclic arenes by thin-layer chromatography and fluorescence spectrophotometry. *Anal. Chem.* 47:1743-1748.
22. Pott, F., and Brockhaus, A. (1973): *Kanzerogene Stoffe,* Jahresbericht des Med. Inst. fur Lufthyg. und Silikoseforschung 1972, Dinslaken, pp. 41-53.

Polynuclear Aromatic Hydrocarbons, edited by P. W. Jones and P. Leber. Ann Arbor Science Publishers, Inc. Ann Arbor, MI © 1979.

DETERMINATION OF POLYNUCLEAR AROMATIC HYDROCARBONS IN THE WORKING ENVIRONMENT

Alf Bjørseth[1]

Central Institute for Industrial Research
Blindern, Oslo, Norway

INTRODUCTION

Polynuclear aromatic hydrocarbons (PAH) are formed by a large number of industrial processes involving combustion or high-temperature treatment of coal-derivatized materials (7). Consequently, PAH may appear in working atmospheres in many industries. Due to the biological effects of PAH which have been clearly elucidated here and elsewhere (7,8,11,13), there has been an increasing concern about the presence of PAH in working atmospheres.

There are surprisingly few detailed studies on PAH in working atmospheres. Most work has focused on the gravimetric determination of total benzene soluble matter (16). However, this method is inaccurate (20) and does not correlate very well with determinations of PAH (19). Other studies have been limited to one compound (usually benzo[a]pyrene, B[a]P) or a very limited number of compounds. A characterization of PAH in working atmospheres is possible using gas chromatography, preferably with glass capillary columns, or gas chromatography/mass spectrometry (GC/MS). Recently, studies of PAH in working atmospheres have been reported, giving qualitative and quantitative data from industries such as aluminum plants (3), coke plants (4,6), foundries (18) and a Soderberg paste plant (2).

The effect of high exposure to PAH has to some extent been revealed; epidemiological studies of the cancer mortality rates among steel workers, roofers and gas workers indicate that high PAH exposure is associated with increased mortality from lung cancer (9,12,14,17).

---

[1]Present address: Battelle-Columbus Laboratories, 505 King Avenue, Columbus, Ohio 43201.

This paper reports on a study of PAH in the working atmospheres of an aluminum and a coke plant. The samples were collected by area and personal sampling methods, and analyzed by glass capillary gas chromatography/mass spectrometry. PAH have also been analyzed on different particle size fractions.

In addition to the chemical analysis of air samples, 11 urine samples from coke oven workers have also been tested for mutagenic activity by the Ames Salmonella test.

EXPERIMENTAL

Sampling

The area samples were collected on glass-fiber filters with two subsequent bottles of ethanol cooled in dry ice; about 1 m^3 of air was pulled through the sampling device (3). Personal sampling was performed with a Casella personal monitoring pump and a 38-mm-i.d. glass-fiber filter.

The particle size fractionation was achieved by a Lundgren impactor (10) with the following size fractions: >15 μm, 15-7 μm, 7-3 μm, 3-0.9 μm and <0.9 μm. The smallest particles were collected on a membrane filter while the other particle sizes were collected on drums.

Analysis

The samples were worked up and analyzed as described previously (1). The use of glass capillary gas chromatography has been proven to yield high precision, good accuracy and excellent separation efficiency (1).

The glass capillary gas chromatography/mass spectrometry (GC^2/MS) system used for identifying PAH in the samples is a Carlo Erba gas chromatograph Model 2150 equipped with splitless injector and a 50-m glass capillary column coated with SE-54. The chromatograph is interfaced to a Varian-MAT 112 mass spectrometer with a Spectro system 100 MS computer. Details concerning the GC^2/MS system have been reported elsewhere (10).

Testing of Urine Samples

Samples of urine from occupationally exposed persons were worked up by XAD-2 column chromatography and the extracts tested by Salmonella strains 98, 100 and 1015 according to Yamasaki and Ames (22). Urine from unexposed persons was collected at the Central Institute, to provide control samples.

RESULTS

A typical chromatogram of the PAH fraction of particulates from the working atmosphere in an aluminum plant is given in Figure 1. The multitude of PAH compounds is clearly visualized, indicating the complex nature of these analyses. The main compounds, marked by the numbers in Figure 1, are identified on the basis of retention times only. Further identification has been achieved by analyzing these samples by GC^2/MS (6).

Typical values for occupational exposure of PAH on particulates in an aluminum plant are given in Tables I and II for stationary and personal sampling, respectively. The results are given separately for the anode plant, vertical Soderberg plant and prebaked Soderberg plant. The complete data from this study is published elsewhere (6).

As revealed by Table I, the largest occupational exposure is connected to the vertical-pin Soderberg plant. This technology is at present undergoing much research to reduce emissions, and improvements are expected soon.

Figure 2 gives a total ion chromatogram (TIC) from a GC^2/MS investigation of PAH in a coke plant. As indicated by the numbers in the figure, more than 100 compounds have been identified (6). All the main PAH compounds usually found in environmental samples are present. In addition, more exotic compounds such as naphtylnitrile (peak 25) and sulfur-containing compounds such as benzothionaphthene (peak 87) and several isomers of benzodibenzothiophene (peaks 90 and 91) are present.

Table III shows similar data for occupational exposure of PAH on particulates in a coke plant. In this case, the exposure measured by personal sampling is of the same order of magnitude as the vertical-pin Soderberg plant. The stationary sampling gives a higher value in this case since the sampling unit was placed closer to the source.

In one case, the particles collected at the top of a coke oven battery were fractionated according to particle size. For each of the five size fractions, the PAH content was determined; the results are given in Table IV. The largest amounts of PAH are found for particles in the diameter ranges of 7-3 μm and 3-0.9 μm. Only about 1% of the total PAH identified is connected to particles classified as not respirable, *i.e.,* particle diameter larger than 7 μm.

In much of the work published on PAH in the working environment, B[a]P has been used as a surrogate. The use of B[a]P as a measure of the total amount of PAH in a sample is necessarily based on the assumption that the relative distribution of PAH compounds is the same in all environments. This is not true. We have used the parent PAH profile (PPP) to illustrate this. Figure 3 shows the relative distribution of 10 parent PAH in particulates from two different working atmospheres. The actual process temperatures are quite different in the two plants, resulting in two

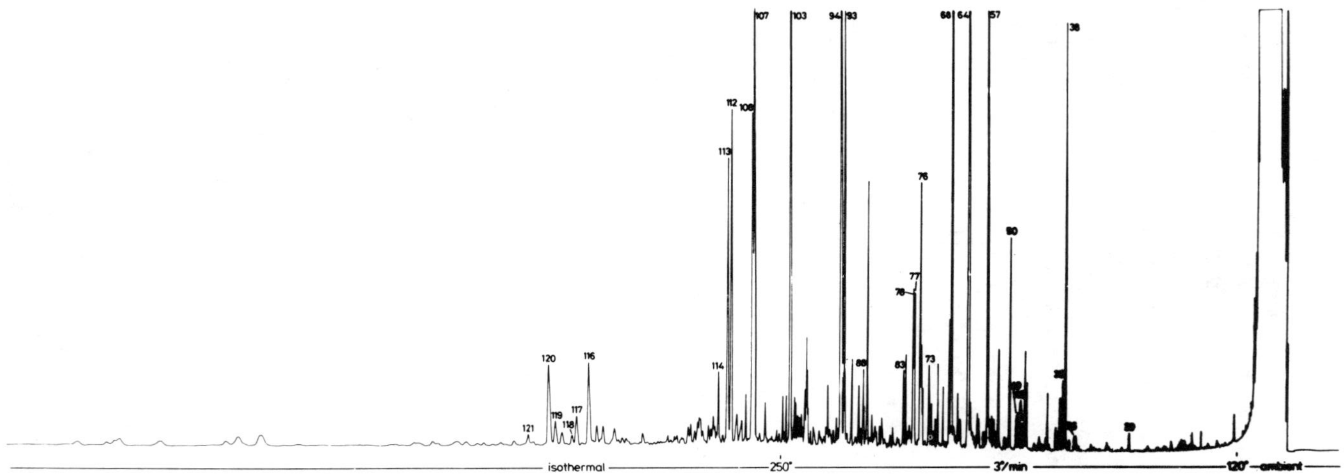

FIGURE 1. Gas chromatogram of PAH in particulate matter from an aluminum plant. The main peaks are (38) phenanthrene; (39) anthracene; (64) fluoranthene; (68) pyrene; (93) benz[a]anthracene; (94) chrysene/triphenylene; (107) benzo[b]fluoranthene; (108) benzo[j,k]fluoranthene; (112) benzo[e]pyrene; (113) benzo[a]pyrene; (114) perylene; (116) *o*-phenylene pyrene; (117 and 118) dibenzanthracenes; (120) benzo[g,h,i]perylene; and (121) anthanthrene.

TABLE I. Total Particulate PAH and Benzo[a]pyrene in Working Atmospheres in an Aluminum Plant, Stationary Sampling

	Total PAH ($\mu g/m^3$)		Benzo[a]pyrene ($\mu g/m^3$)	
	Range	Arithmetic Mean Value	Range	Arithmetic Mean Value
Anode Plant[a]	3.4-73	17.9	0.03-0.3	0.08
Prebaked Soderberg Plant[b]	0.6-1.7	1.2	0.0-0.05	0.01
Vertical-Pin Soderberg Plant[c]	20-202	100	0.7-9.0	4.3

[a]Average of 5 samples.
[b]Average of 6 samples.
[c]Average of 10 samples.

different profiles. It is therefore recommended strongly that the PPP be established for the actual PAH source before using B[a]P or other com pounds as surrogates. On the other hand, it is our experience that when the profile is first established, it does not change significantly, as long as the source conditions remain relatively stable.

TABLE II. Total Particulate PAH and Benzo[a]pyrene in Working Atmospheres in an Aluminum Plant, Personal Sampling

	Total PAH ($\mu g/m^3$)		Benzo[a]pyrene ($\mu g/m^3$)	
	Range	Arithmetic Mean Value	Range	Arithmetic Mean Value
Anode Plant[a]	10.5-374	108	0.8-27.9	8.1
Prebaked Soderberg Plant	0.5-2.0	1.1[a]	b	––
Vertical-Pin Soderberg Plant[c]	38-2,790	1,252	3.4-116	58

[a]Average of 4 determinations.
[b]Not determined.
[c]Average of 3 determinations.

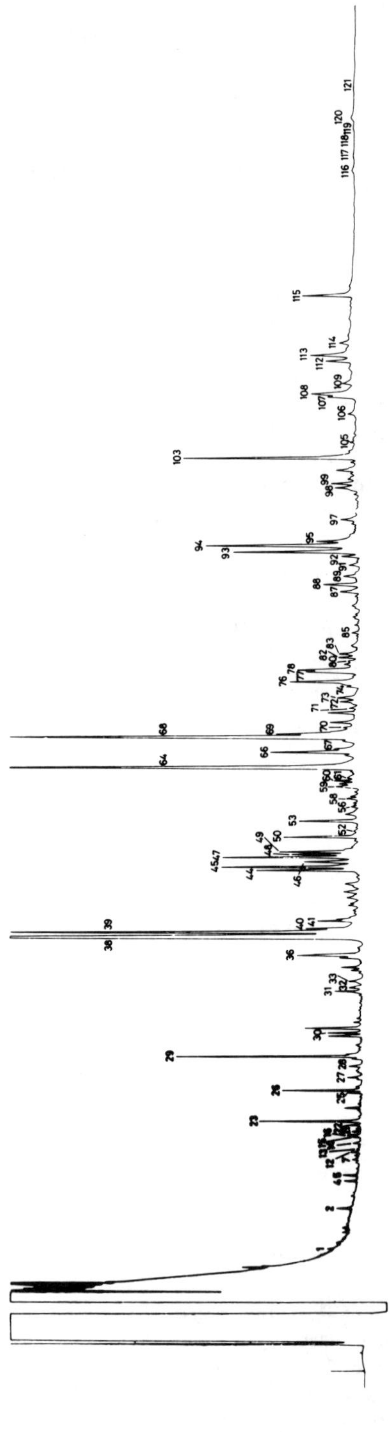

FIGURE 2. Total ion chromatogram of PAH in particulate matter from a coke plant. For peak identities, see Figure 1.

TABLE III. Total Particulate PAH and Benzo[a]pyrene in Working Atmospheres of a Coke Plant

	Total PAH ($\mu g/m^3$)		Benzo[a]pyrene ($\mu g/m^3$)	
	Range	Arithmetic Mean Value	Range	Arithmetic Mean Value
Stationary Sampling				
Battery Top[a]	270-2,423	1,002	14-69	37
Personal Sampling				
Larry Car Operator[b]	168-1,044	370	12-43	22
Coke Car Operator[c]	4.8-73	37	0.5-5.8	3.1
Push Car Operator[d]	9-62	33	0.9-4.4	2.9

[a]Average of 25 samples.
[b]Average of 7 samples.
[c]Average of 3 samples.
[d]Average of 4 samples.

TABLE IV. PAH on Particles as a Function of Particle Size

Size	>15 μm	7-15 μm	3-7 μm	0.9-3 μm	<0.9 μm
PAH	μg	μg	μg	μg	μg
Phenanthrene	0.5	0.5	0.8	2.0	*
Anthracene	0.3	0.2	0.6	0.7	*
Methylphenanthrene/ Methylanthracene[a]	––	––	0.4	1.3	––
Fluoranthene	––	––	4.2	10.7	2.0
Dihydrobenzo[a,b] fluorene	––	––	0.6	2.1	––
Pyrene	––	––	4.3	12.3	1.3
Benzo[a] fluorene	––	––	3.5	9.1	0.5
Benzo[b] fluorene	––	––	2.6	7.2	0.3
Benzo[c] phenanthrene	––	––	2.0	5.2	0.8
Benz[a] anthracene	0.03	0.08	3.3	9.0	4.0
Chrysene/Triphenylene	0.06	0.10	5.8	13.2	1.8
Benzo[b,j,k] fluoranthene	0.05	0.40	4.8	10.6	7.2
Benzo[e] pyrene	0.03	0.14	3.4	8.9	0.5
Benzo[a] pyrene	0.02	0.12	5.2	12.1	1.4
Perylene	––	––	1.3	2.9	––
o-Phenylenepyrene	––	––	3.8	7.2	0.5
Benzo[g,h,i] perylene	––	––	5.2	7.8	1.3
Anthanthrene	––	––	1.2	3.2	0.3
Coronene	––	––	5.9	6.3	0.5
Dibenzopyrene	––	––	0.9	1.5	0.3
TOTAL PAH	0.69	1.54	59.8	133.3	22.7

Note: * Detectable.
–– Not detected.
[a]The isomer not determined.

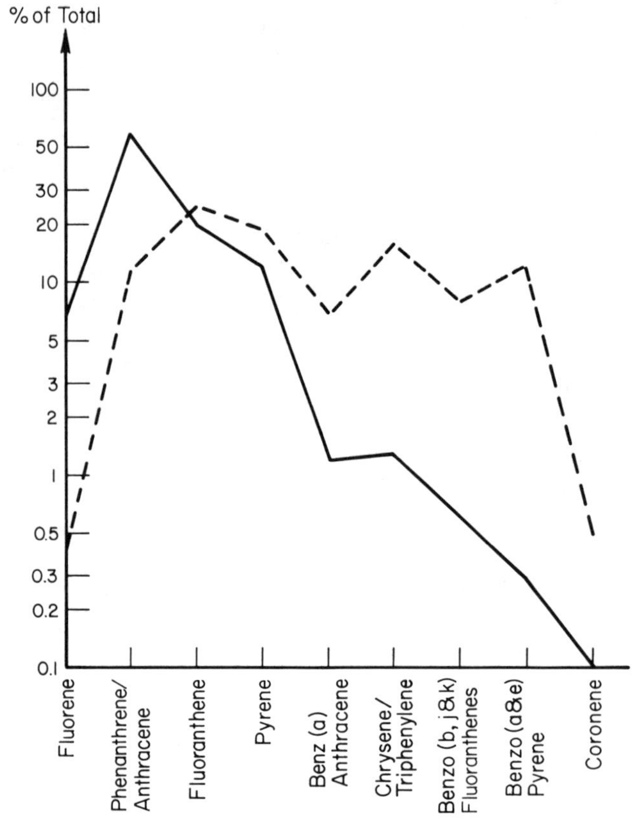

FIGURE 3. Parent PAH profile of PAH in particulate matter from an aluminum plant (– – –) and a Soderberg paste plant (———).

Table V gives a speculative comparison of the occupational exposure to PAH with cigarette smoking. The calculations are made assuming inhalation of 10 m^3 of air during an 8-hr daily exposure period and that smoke from one cigarette contains 20 ng B[a]P. The table shows that the working environments studied cause a substantial exposure to B[a]P (and PAH). Epidemiological studies have, however, never revealed effects comparable to smoking such a large number of cigarettes. It is therefore important to point out that the table does not establish an order of hazard from these sources, since the toxicity of PAH depends on a large number of factors in addition to B[a]P. In fact, it has been shown that although PAH are the primary tumor initiators in cigarette smoke condensates, they must interact with other condensate compounds to exert their tumorigenic effect (21).

TABLE V. Estimated Occupational Exposure and Calculated Cigarette Equivalents

Occupation	Benzo[a]pyrene Intake[a] (μg/day)	Cigarette Equivalents[b]
Aluminum Plant		
Worker—Anode Plant	81	4,050
Worker—Vertical-Pin Soderberg Plant	580	29,000
Coke Plant		
Larry Car Operator	220	11,000
Coke Car Operator	31	1,550
Push Car Operator	29	1,450

[a]Assuming inhalation of 10 m^3 air during an 8-hr daily exposure.
[b]Assuming 20 ng benzo[a]pyrene per cigarette.

To continue our somewhat adventurous comparison of PAH in working atmospheres with cigarette smokers, we decided to use microbiological methods instead of pure mathematics.

Recently, it was shown that urine from cigarette smokers is mutagenic in the Ames Salmonella test system (22). Since PAH are the primary carcinogenic compounds in cigarette smoke, we initiated a study of mutagenic compounds in urine from coke oven workers. All samples were collected from battery-top workers immediately after the shift ended. The preliminary results are given in Table VI. The mutagenic effect of urine from the control group shows a significant difference between smokers and nonsmokers. There are, however, no differences between the smokers in the control group and the exposed workers. This indicates that PAH metabolites are not responsible for the major part of the mutagenic activity found in urine from cigarette smokers.

TABLE VI. Mutagenic Effect in Selected Urine Samples (revertants/25 ml urine)

Test	Number of Samples	Result[a]
Control		
Smokers	5	84 ± 57
Nonsmokers	8	0
Coke Oven Workers		
Smokers	11	126 ± 106

[a]Corrected for spontane mutations.

REFERENCES

1. Bjorseth, A. (1977): Analysis of polycyclic aromatic hydrocarbons in particulate matter by glass capillary gas chromatography. *Anal. Chim. Acta* 94:21-27.
2. Bjorseth, A., and Lunde G. (1977): Analysis of the polycyclic aromatic hydrocarbon content of airborne particulate pollution in a Soderberg paste plant. *J. Am. Ind. Hyg. Assoc.* 38:224-227.
3. Bjorseth, A., Bjorseth, O., and Fjeldstad, P. E. (1978): Polycyclic aromatic hydrocarbons (PAH) in working atmospheres I. Analysis of the PAH-content in an aluminum reduction plant. *Scand. J. Work Environ. Health* 4:212-223.
4. Bjorseth, A., Bjorseth, O., and Fjeldstad, P. E. (1978): Polycyclic aromatic hydrocarbons (PAH) in working atmospheres II. Determination of the PAH-content in a coke plant. *Scand. J. Work Environ. Health* 4:224-236
5. Bjorseth, A., Dybing, E., and Moller, M. Unpublished work.
6. Bjorseth, A., and Eklund, G. (1979): Analysis of polynuclear aromatic hydrocarbons in working atmospheres by computerized gas chromatography/mass spectrometry. *Anal. Chim. Acta* (in press).
7. Committee on Biological Effects (1972): *Particulate Polycyclic Matter,* National Academy of Science, Washington, D.C.
8. Dipple, A. (1976): Polynuclear aromatic carcinogens. In: *Chemical Carcinogens,* C. E. Searle, Ed., ACS Monograph 173, American Chemical Society, Washington, D.C.
9. Doll, R., Fisher, R. E., Gammon, E. J., Gunn, W., Hughes, G. O., Tyrer, F. H., and Wilson, W. (1965): Mortality of gas workers with special reference to cancer of the lung and bladder, chronic bronchitis, and pneumo comiosis. *Brit. J. Ind. Med.* 22:1-12.
10. Eklund, G., Josefsson, B., and Bjorseth, A. (1978): Determination of chlorinated and brominated lipophilic compounds in spent bleach liquors from a sulphite pulp mill. Glass capillary column gas chromatography-mass spectrometry-computer analysis and identification. *J. Chromatograp.* 150:161-169.
11. Freudenthal, R., and Jones, R. W., Eds. (1976): *Carcinogenesis, Vol. 1,* Raven Press, New York.
12. Hammond, E. C., Selikoff, I. Y., Lawther, P. L., and Seidman, H. (1976): Inhalation of benzpyrene and cancer in man. *Ann. N.Y. Acad. Sci.* 271:116-124.
13. Jones, P. W., and Freudenthal, R. I. (1978): *Carcinogenesis, Vol. 3,* Raven Press, New York.
14. Lloyd, J. W. (1971): Long-term mortality study of steel workers V. Respiratory cancer in coke plant workers. *J. Occup. Med.* 13:53-68.
15. Lundgren, D. A. (1967): An aerosol sampler for determination of particle concentration as a function of size and time. *J. Air Poll. Control Assoc.* 17:225-229.

16. National Institute for Occupational Safety and Health (1972): *An Evaluation of Sampling Procedures and Analytical Methods for Coal Tar Pitch Volatiles* PSM-1013.
17. Redmond, C. K., Strobino, B. R., and Cypess, R. H. (1976): Long term mortality of steel workers. VI. Mortality from malignant neoplasms among coke by product workers. *Ann. N.Y. Acad. Sci.* 271:102-115.
18. Schirmberg, R. W., Pfaffli, P., and Tossavainen, T. (1978): Profilanalyse von polycyclischen aromatischen kohlenwasser-stoffen in Eisengiessereien. *Staub-Reinhaltung Luft* 38:273-278.
19. Schulte, K. A., Larsen, D. J., Hornung, R. W., and Crable, J. V. (1974): Report on analytical methods used in coke oven effluent study. The five oven study. *HEW Publication No. (NIOSH) 74-105,* NIOSH, Cincinnati, OH.
20. Seim, H. J., Hanneman, W. W., Barsotti, L. R., and Walker, T. J. (1974): Determination of pitch volatiles in airborne particulates, I. Problems with the benzene-soxhlet extraction method. *J. Am. Ind. Hyg. Assoc.* 35:718-723.
21. Severson, R. F., Snook, M. E., Akin, F. J., and Chortyk, O. T. (1978): Correlation of biological activity with polynuclear aromatic hydrocarbon content of tobacco smoke fractions. In: *Carcinogenesis, Vol. 3,* P. W. Jones and R. I. Freudenthal, Eds., pp. 115-130, Raven Press, New York.
22. Yamasaki, E., and Ames, B. A. (1977): *Proc. Nat. Acad. Sci.* 74:3555-3560.

Polynuclear Aromatic Hydrocarbons, edited by P. W. Jones and P. Leber. Ann Arbor Science Publishers, Inc. Ann Arbor, MI ©1979.

DISTRIBUTION OF POLYCYCLIC AROMATIC HYDROCARBONS WITH RESPECT TO PARTICLE SIZE IN PASADENA AEROSOLS IN THE SUBMICROMETER RANGE

Antonio H. Miguel
Instituto de Química
Universidade Federal do Rio de Janeiro
21941 Ilha do Fundão/Rio de Janeiro, Brasil

INTRODUCTION

Theoretical and experimental studies (2,17) have shown that inhaled particles are deposited in different regions of the respiratory system depending on their effective aerodynamic diameter. In addition, the rate of loss of particles from the atmosphere also depends on their diameter. Thus, the concentration of pollutant aerosols with respect to particle size is important in the evaluation of their atmospheric transport and penetration into the respiratory system.

Reports on atmospheric PAH concentrations indicate that they are associated with particulate matter and, in many instances, with soot. Although the mechanism of this PAH particle association is not well known, it is clear, however, that the PAH are concentrated on the surface of the host particles. Most of the mass of PAH is associated with aerosols of aerodynamic diameters in the respirable size range. Thus, a health hazard exists when carcinogenic PAH are deposited either at bifurcations in the bronchial tubes or in the pulmonary region of the respiratory system.

Benzo[a]pyrene (B[a]P) is a prototype carcinogenic PAH which may cause cancer in humans. Coronene (Cor) is not considered carcinogenic but its concentration in atmospheric aerosols correlates well with traffic density, which allows the calculation of nonautomobile contribution of other PAH (5). This study is focused on B[a]P and Cor.

To date only a limited number of studies have been reported relating the distribution of atmospheric PAH with respect to particle size (1,3,8,12,16). These studies show that, in general, a large percentage of the mass of PAH is associated with particles smaller than the smallest cutoff diameter (~0.3 μm) of the impactor used.

In the following sections we report studies on the distribution of B[a]P and Cor in Pasadena aerosols, with an eight-stage low-pressure impactor with 50% efficiency cutoffs of 4.0, 2.0, 1.0, 0.5, 0.26, 0.12, 0.075 and 0.05-μm aerodynamic diameter. The size distribution data were used to estimate B[a]P and Cor deposition on the respiratory system.

EXPERIMENTAL

Reagents

Benzo[a]pyrene (Eastman Organic Chemicals), coronene (Aldrich Chemical Co.) and spectrograde solvents were used as supplied.

Sampling

Details of the design, calibration and operation of the eight-stage low-pressure impactor used in this study are reported elsewhere (6,7).

All samples were collected 15 m above street level, on the roof of W.M. Keck Labs. at Caltech, in Pasadena, Cal. Each stage of the low-pressure impactor contained a quartz collection disc 25.4 mm in diameter by 3 mm thick. To each disc, 0.5 (μl) of a 2% solution of Vaseline® in benzene was applied at the center. The benzene evaporates leaving a thin film of Vaseline (~5 mm i.d.) on the disc. A Sargent Welch 1403 vacuum pump was used. The impactor samples at a rate of 1 liter/min, verified before and after sampling. Sampling was carried out for 72 hr, representing a volume of 4.32 m^3 air. To prevent overloading of the impaction discs, a new set of Vaseline-coated discs was substituted every 24 hr.

Parallel samples were collected on 8 x 10 in. glass-fiber filters (Gelman type A spectrograde) using a high-volume sampler (General Metal Works, Inc., Cleveland, OH) equipped with an automatic choke which maintained a sampling rate of 1.13 m^3/min. These parallel samples were used to determine the concentration of total suspended particulate matter (TSP) and to compare the total B[a]P and Cor masses by the two sampling methods.

Sample Extraction and Analysis

Details of the extraction and analysis of B[a]P and Cor are reported elsewhere (11). Briefly, the impactor discs and the parallel-fiber glass filter were extracted with benzene in an ultrasonic apparatus for 30 min. The extracts were reduced to appropriate volumes by flowing a stream of high-grade nitrogen gas over their surface at room temperature.

Benzene extracts from the impaction discs and appropriate aliquots of the filter extracts were spotted on precoated thin-layer chromatographic (TLC) plates, (21% acetylated cellulose, Schleicher & Sshuell, Cat. No. G 1800/21 ac). Aliquots of B[a]P (0.2-50 ng) and Cor (1-20 ng) standards were also spotted on the plate. The plate was placed in an equilibrated chromatographic chamber containing 50 ml of developer consisting of a mixture of ethanol: toluene:water (17:4:4 V/V) [Sawicki *et al.* (13)] and allowed to develop until the solvent front had travelled 15 cm (~1 hr 20 min).

The separated PAH were analyzed using an Aminco-Bowman Spectrophotofluorometer (SPF) equipped with a motorized thin-film scanner with reverse optics, shown schematically in Figure 1. The SPF-TLC system permits sampling of the fluorescence directly on the surface of the plate. Fluorescence excitation and emission spectra were obtained from each separated B[a]P and Cor spot. The spectra were traced by the X-Y recorder. Having ascertained that the spot was indeed the sought-after PAH, it was mechanically scanned in the direction of last solvent migration, under optimized excitation and emission wavelengths. The integrated area of the fluorescence intensity was a linear function of the amount of PAH present. From these data and the volume of air sampled, the concentration of B[a]P and Cor as a function of particle size was calculated.

RESULTS AND DISCUSSION

Data on chemical composition as a function of particle size can be reported in a number of ways. A common method is to plot the cumulative mass (percent) of chemical species less than or equal to a given size on log-probability paper. If the species is log-normally distributed as represented by a straight line, it is possible to determine the chemical species mass median diameter (MMD). When Pasadena B[a]P and Cor data are plotted in this way, such log-normal distributions are not followed (Figure 2). As the figure shows, there are two distinct ranges of linearity; one below and one above particles with 0.12-μm aerodynamic diameter. Thus, the MMD for Pasadena aerosols associated with B[a]P and Cor does not coincide with the median diameter of the PAH mode.

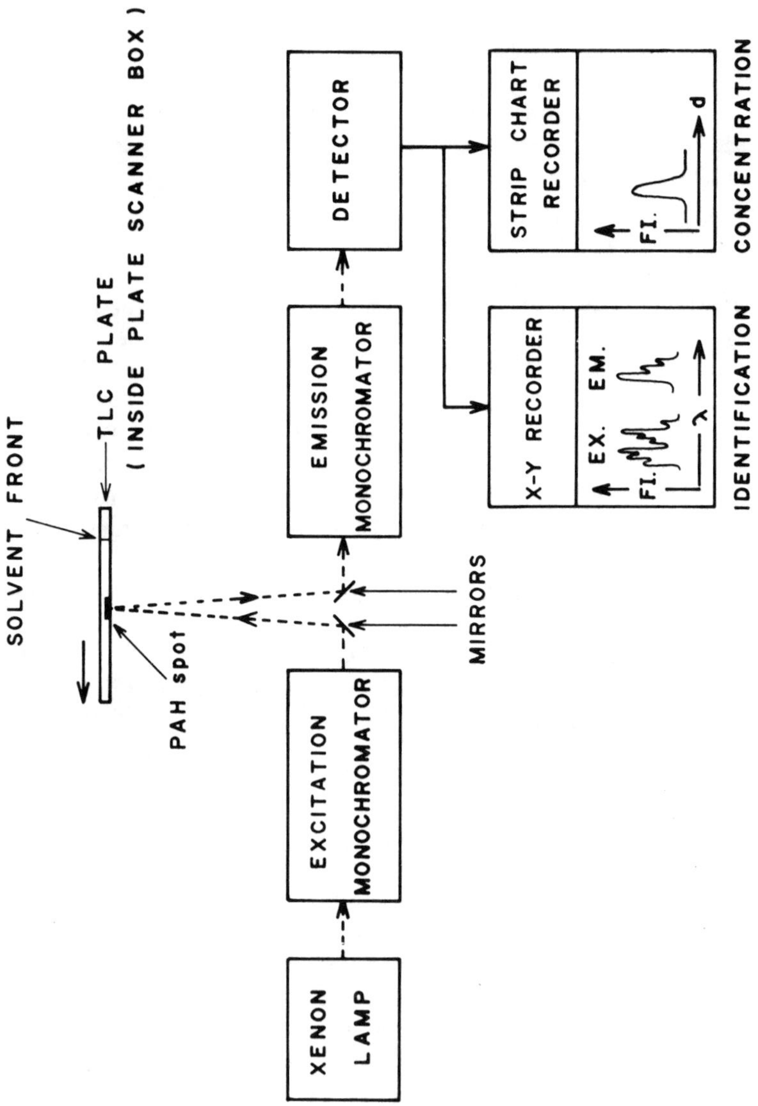

FIGURE 1. Schematic diagram of motorized thin-film scanner mounted on the Aminco-Bowman Spectrophotofluorometer.

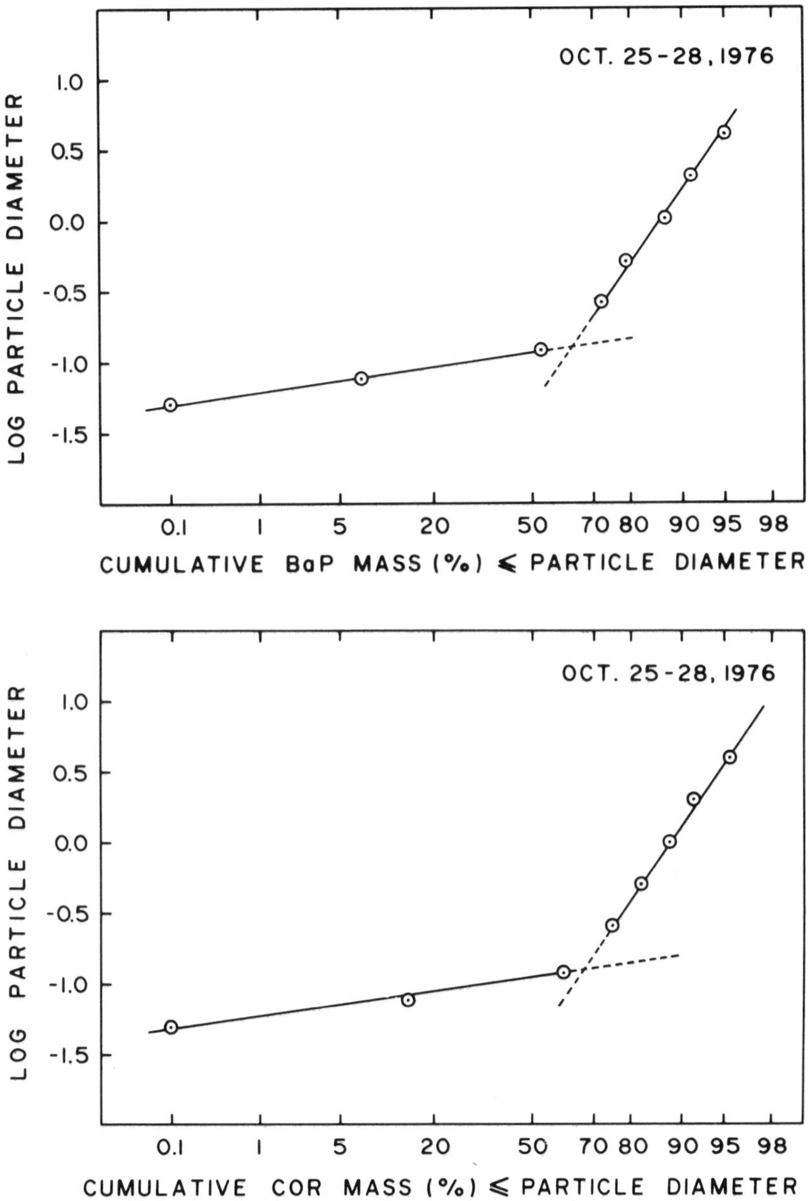

FIGURE 2. Cumulative mass particle size distributions for B[a]P and Cor.

A convenient way of representing mass distribution data is to plot $\Delta M/\Delta \log d_p$ versus $\log d_p$, where ΔM is the weight of PAH in each stage and $\Delta \log d_p$ represents the size interval specified for each impactor stage. The area under the histogram between two values of d_p is proportional to the mass of the PAH in that size range. Pasadena B[a]P and Cor mass distribution data plotted in this way are shown in Figure 3. The upper cutoff for the first stage was taken at 8 μm because particles larger than this are not efficiently collected since sampling was not done isokinetically. This was confirmed by the parallel filter measurements. The average collection efficiency of both PAH in the impactor was about 96 ± 13%.

The histograms of the mass distribution of B[a]P and Cor as a function of particle size (Figure 3) show some very important characteristics. Both are unimodal and show similar shapes within each sampling period. The PAH distributions with respect to particle size always peaked in the size range 0.075 to 0.12 μm.

The total B[a]P and Cor concentrations sharpen in the December period and flatten toward March (legend of Figure 3). Similar results were reported by Gordon (4). This behavior is consistent with the lower inversion heights commonly observed in the winter, which has the effect of concentrating the pollutants in a smaller volume of air. In addition to these meteorological effects, increased temperature, sunlight and oxidant levels in the summer months may contribute to the reduction of the B[a]P and Cor loadings due to their photochemical degradation while airborne.

The effect of the afterfilter (see legend of Figure 3) on the March samples was to increase, slightly, the operating pressure in the last stage, significantly altering the size distribution in this stage. Therefore, use of the afterfilter with this impactor is not recommended.

The B[a]P/Cor ratios (legend of Figure 3) were similar to those reported in other studies (4,5) of samples collected nearby. The relatively small variation in the B[a]P/Cor ratio (0.14-0.22) and the direct relationship between total Cor concentration and vehicle traffic density (5) suggest that B[a]P and Cor in Pasadena aerosols result mainly from motor vehicles. Using carbon monoxide as a tracer for B[a]P emissions, a recent Los Angeles basin study (15) revealed that approximately 90% of the total B[a]P emissions inventoried were due to motor vehicles (mobile sources), and that 96% of the total emissions resulted from area source (mobile plus domestic sources) emissions. This conclusion qualitatively agrees with the results of the study reported here.

The present study revealed that 75% of the B[a]P and 85% of the Cor mass are associated with particles of aerodynamic diameter less than 0.26 μm. Almost one-half the total mass of both PAH is associated with aerosols in a very narrow size range (0.075 to 0.12 μm).

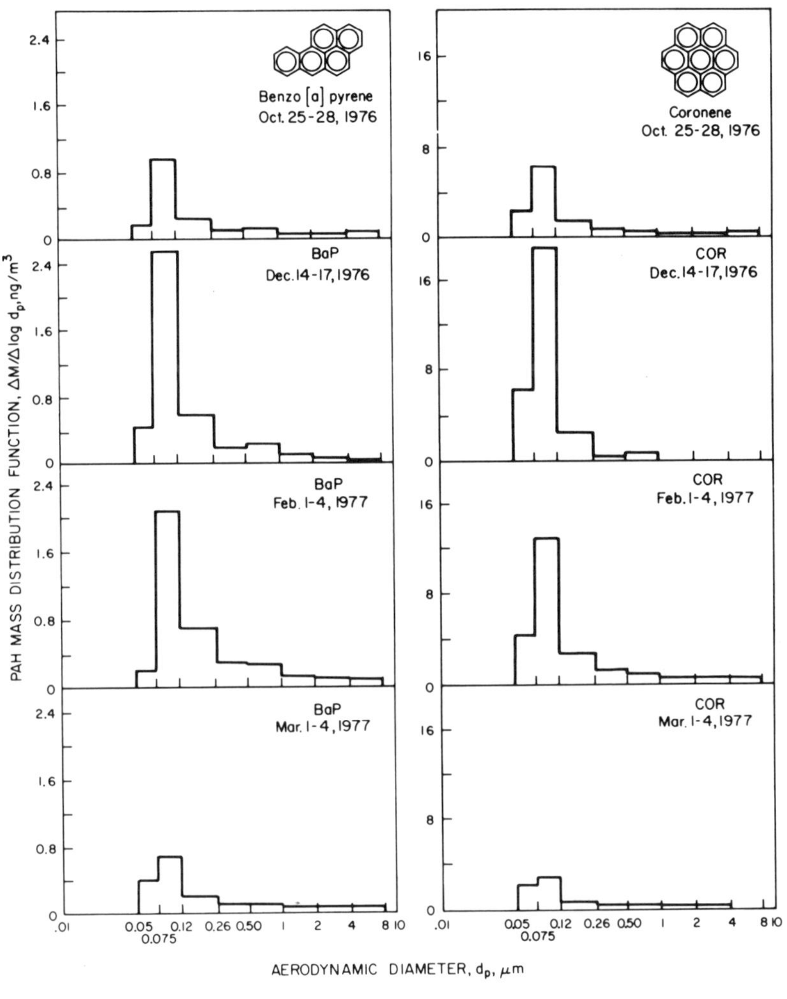

FIGURE 3. Distributions of B[a] P and Cor as a function of aerodynamic diameter. The aerodynamic diameter refers to particles of unit density (for other particle densities see References 6 and 7). For the March samples, a 47-mm afterfilter was placed between the last impactor stage (0.05-0.075 μm) and the vacuum pump. Total suspended particulate matter (TSP) in μg/m^3 and PAH concentration (in parenthesis) in ng/m^3 were: 95 (0.42 B[a] P: 3.01 Cor); 85 (1.06 B[a] P: 6.66 Cor); 91 (1.01 B[a] P: 5.37 Cor); 114 (0.39 B[a] P: 1.81 Cor), respectively, for the sampling periods of October 25-28, 1976 through March 1-4, 1977. The B[a] P/Cor ratio for the same periods were 0.14; 0.16; 0.19 and 0.22, respectively.

From a standpoint of evaluating particulate PAH as a health hazard, the most important physical characteristic lies in the distribution of PAH with respect to particle size. According to a 1966 report by the Task Group on Lung Dynamics (17), particles of aerodynamic diameters less than about 0.6 μm generally pass through the nasopharynx and tracheo-bronchial regions of the respiratory system without significant entrapment. These particles are able to penetrate deep into the pulmonary region of the respiratory system and deposit with high efficiency in the functional area (exchange space) of the lungs. As the histograms show (Figure 3) most of the B[a]P and Cor mass are associated with particles less than 0.6 μm. Therefore, carcinogenic compounds such as B[a]P will find their way into the deep regions of the lung where, presumably, they can cause localized cancer. In addition, the elimination of B[a]P from the lungs appears related to the size of the carbon particle with which it is associated. The elimination progresses more slowly from smaller particles, thus prolonging the residence time of the carcinogen at the target site (15).

The average of the four B[a]P size distribution measurements reported in this study were used to estimate the B[a]P lung deposition in the nasopharynx (NP), tracheo-bronchial (TB) and pulmonary (P) regions of the respiratory tract model of the Task Group on Lung Dynamics. The calculated percentages of airborne B[a]P deposited in each region were 3%, 4% and 35%, respectively, for the NP, TB and P regions (15). These calculations show that the largest fraction of the deposited B[a]P occurs in pulmonary regions where residence times are longest and carcinogenic compounds such as B[a]P may readily enter the bloodstream.

It has been speculated (18) that atmospheric B[a]P and other PAH are adsorbed primarily on the surface of soot by hydrogen bonding. Optical microscopic observation of the aerosols deposited on each impaction disc (prior to PAH extraction) revealed that the degree of darkness of the spot increased with decreasing particle size, peaking in the 0.075 to 0.12 μm diameter range (which corresponds to the peak on the PAH distributions shown in Figure 3) and again decreasing toward smaller particles. This observation was true for all samples collected with the low-pressure impactor. If it is assumed that the degree of darkness of the spot on the impactor disc is a measure of the amount of soot present, these observations support the suggestion that B[a]P is adsorbed by soot (18). It is not possible, however, to definitely establish whether the fractioned PAH are associated with soot as a result of surface adsorption or because PAH may be the precursors of carbon particles, as suggested by Lahaye and Prado (9).

Figure 4 shows that the peak of mass distribution of B[a]P with respect to particle size nearly coincides with the peak in the surface-area distribution of

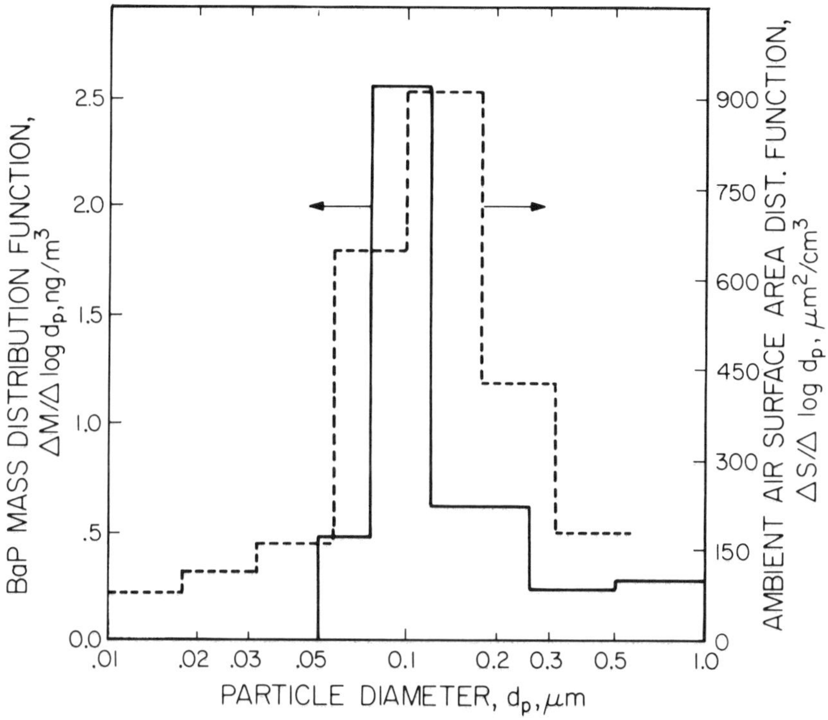

FIGURE 4. Comparison of the B[a]P mass distribution for the October 25-28, 1976, sampling period with typical ambient air aerosol surface area distribution obtained in the same period. The latter distribution was obtained with a Model 3030 Electrical Aerosol Analyser (Thermo-Systems, Inc., St. Paul, MN).

ambient air aerosols. This observation, typical in this study, suggests a surface-area PAH mass relationship, which also indicates that adsorption may be a major factor controlling the mechanism of association. Laboratory experiments showed that the amount of pyrene (a PAH) adsorbed from the vapor phase (at a fixed temperature) on the surface of fly ash particles was directly proportional to the particle surface area (10). Since coal is not burned in the Los Angeles area, soot-like particles in automobile exhaust may serve instead as the adsorbent.

CONCLUSIONS

The present study shows that a considerable amount of the B[a]P and Cor mass in Pasadena aerosols is associated with particles in the respirable size range.

B[a]P as well as other carcinogenic PAH may enter the bloodstream via the respiratory tract.

The consistency in the mass distributions of B[a]P and Cor as a function of particle size with the form of the surface-area distributions of ambient aerosol suggest that PAH particle association may proceed via surface adsorption.

In addition to the evaluation of possible health effects, data on the distribution of atmospheric PAH with respect to particle size allow estimation of their transport, deposition and lifetime in the atmosphere. The atmospheric lifetime of the carrier aerosols may affect the rate of chemical transformation of PAH associated with them.

ACKNOWLEDGMENTS

This research was supported by grants from the Pasadena Lung Association, NIEHS Training Grant No. PSH ES00080-10 and NSF/RANN Grant No. ENV76-04179 to California Institute of Technology, Pasadena, CA. The secretarial assistance of Mr. William Kalaf, of the Institute of Chemistry of the Universidade Estadual de Campinas, S.P., Brazil is appreciated.

REFERENCES

1. Albagli, A., Oja, H, and Dubois, L. (1974): Size-distribution pattern of polycyclic aromatic hydrocarbons in airborne particulates. *Environ. Lett.* 6(4):241-251.
2. Bell, K.A. (1974): Aerosol deposition models of a human lung bifurcation. Ph.D. Thesis, California Institute of Technology, Pasadena, CA.
3. DeMaio, L., and Corn, M. (1966): Polynuclear aromatic hydrocarbons associated with particulates in Pittsburg air. *J. Air Poll. Control Assoc.* 16:67-71.
4. Gordon, R.J. (1976): Distribution of airborne polycyclic aromatic hydrocarbons throughout Los Angeles. *Envion. Sci. Technol.* 10:370-373.
5. Gordon, R.J., and Bryan, R.J. (1973): Patterns in airborne polynuclear hydrocarbon concentrations at four Los Angeles sites. *Environ. Sci. Technol.* 7(11):1050-1053.
6. Hering, S.V., Flagan, R.C., and Friedlander, S.K. (1978): Design and evaluation of a new low pressure impactor, I. *Environ. Sci. Technol.* 12:667-673.
7. Hering, S.V., Friedlander, S.K., Collins, J.G., and Richards, L.W. (1978): Design and evaluation of a new low pressure impactor, II. Submitted to *Environ. Sci. Technol.*
8. Kertész-Saringer, M., Mészaros, E., and Varkonyi T. (1971): On the size distribution of benzo[a]pyrene containing particles in urban air. *Atmos. Environ.* 5:429-431.

9. Lahaye, J., and Prado, G. (1974): Formation of carbon particles from a gas phase: nucleation phenomenon. *Water air soil poll.* 3-473-481.
10. Miguel, A.H. (1976): Studies of gas-solid reactions of environmental significance: I. pyrene adsorption onto fly ash II. thiol oxidation and adsorption by activated carbon. *Ph.D. Thesis*, University of Illinois, Urbana.
11. Miguel, A. H., and Friedlander, S. K. (1978): Distribution of benzo[a]pyrene and coronene with respect to particle size in Pasadena aerosols in the submicrometer range. *Atmos. Environ.* 12:2407-2413.
12. Pierce, R.C., and Katz M. (1975): Dependency of polynuclear aromatic hydrocarbon content on size distribution of atmospheric aerosols. *Environ. Sci. Technol.* 9(4):347-353.
13. Sawicki, E., Stanley, T.W., Elbert, W.C. and Pfaff, J.D. (1964): Application of thin layer chromatography to the analysis of atmospheric pollutants and determination of benzo[a]pyrene. *Anal. Chem.* 36: 497-502.
14. Shabad, L.M., Pylev, L.N., and Kolesnichenko, T.S. (1964): Importance of the deposition of carcinogens for cancer induction in lung tissue. *J. Nat. Cancer Inst.* 33:135-141.
15. Special Report to the U.S. Environmental Protection Agency on Contract No. 68-03-0434 (1978), White, W.H., Friedlander, S.K., Eds. (1978): California Institute of Technology.
16. Starkey, R., and Warpinski J. (1974): Size distribution of particulate benzo[a]pyrene. *J. Environ. Health* 36(5):503-505.
17. Task Group on Lung Dynamics (1966): Deposition and retention models for internal dosimetry of the human respiratory tract. *Health Physics* 12:173-207.
18. Went, F.W. (1960): Organic matter in the atmosphere and its possible relation to petroleum formation. *Proc. Nat. Acad. Sci., U.S.* 46: 212-221.

Polynuclear Aromatic Hydrocarbons,
edited by P. W. Jones and P. Leber.
Ann Arbor Science Publishers, Inc.
Ann Arbor, MI © 1979.

CHARACTERIZATION OF MULTIALKYLATED POLYCYCLIC AROMATIC HYDROCARBONS IN ENERGY-RELATED MATERIALS

W. H. Griest, B. A. Tomkins, J. L. Epler * and T. K. Rao*

Analytical Chemistry Division
*Biology Division
Oak Ridge National Laboratory
Oak Ridge, Tennessee 37830

Multicomponent polycyclic aromatic hydrocarbon (PAH) analysis of many coal-derived materials (5,9), and tobacco or marijuana smokes (10,18) reveals the complex nature of their PAH isolates. This complexity usually is conferred by the multialkylated PAH content. Recent advances in PAH separation methodology (16) have focused on separating the parent and simple alkylated PAH from the interfering multialkylated PAH, to allow quantitation of the former. Most known carcinogenic PAH are found in the parent fraction. In contrast, relatively little attention has been focused specifically on the multialkylated PAH. Bioassays of alkylated naphthalenes (15) have suggested a rough correlation between degree of alkylation and cocarcinogenic activity. Studies (reviewed in references 2 and 8) have shown enhanced carcinogenicity for some isomers of tri- and tetramethyl benz[a]anthracenes and would suggest that multialkylated PAH contribute significantly to the carcinogenicity of PAH as a class. This paper presents the results of some initial studies characterizing the chemical and biological nature of multialkylated PAH from a coal-derived crude oil.

FRACTIONATION AND ANALYSIS OF THE GROSS PAH ISOLATE

The complexity of PAH isolates from many energy-related materials prevents effective quantitative analysis, even by high-resolution gas chromatography (GC). Alumina-column chromatography of the Swain-Stedman-type neutral fraction of a coal-derived crude oil was used to generate the gross PAH isolate of syncrude D, as reported (5) previously. GC analysis was performed under the conditions given in Table I. Figure 1A illustrates that even

TABLE I. GC Conditions

	Capillary Column	Packed Column
Instrument:	HP 5840 A	PE 3920
Column:	Glass capillary, 30 m x 0.2 mm i.d., silanized	1/8 in. o.d. x 10 ft packed glass
Stationary Phase:	Dexsil 400	3% Dexsil 400 on 100/120 Supelcoport
Detector:	FID	FID
Injection Temperature, °C:	250	290
Detector Temperature, °C:	300	270
Gas Flows, Initial Pressure, psi		
Air:	20	40
H_2:	20	20
Carrier (He):	17	80
Temperature Program:	Isothermal 5 min at 100°C, then 3°C/min to 320°C, hold 30 min.	Isothermal 4 min at 100°C, then 2°C/min to 320°C, hold 32 min

capillary-column GC is insufficient to separate completely the constituents of the gross PAH isolate. Gas chromatography/mass spectroscopy (GC/MS) indicated the presence of many overlapping PAH, some of which were multialkylated derivatives. Successful analysis of the PAH isolate obviously required further division of the isolate into smaller, and hopefully simpler, fractions before GC. One aim would be to concentrate the carcinogenic PAH into a single readily analyzed fraction.

The gross PAH isolate was next separated into its multialkylated PAH and simple alkylated plus parent PAH fractions using a gel filtration method (16). The PAH isolate was eluted in two fractions from four 0.5 in. o.d. x 43 in. Biobeads SX-12 columns with redistilled benzene, as shown in Figure 2. The same 2,3,5-trimethylnaphthalene/fluoranthene cut-point (16) was used to separate the multialkylated PAH from the parent plus simple alkylated PAH, but collection of the multialkylated PAH fraction commenced with the rise in 280 nm absorbance of the column eluate as indicated in Figure 3. The collection of the parent plus simple alkylated PAH fraction was extended through the elution of coronene. The solvent was removed from the two collected fractions by evaporation with dry flowing nitrogen under reduced temperature and pressure. The residues were weighed and are reported in Table II. The multialkylated PAH contributed three times more mass to the gross PAH isolate of this coal-derived crude oil than did the parent and simple

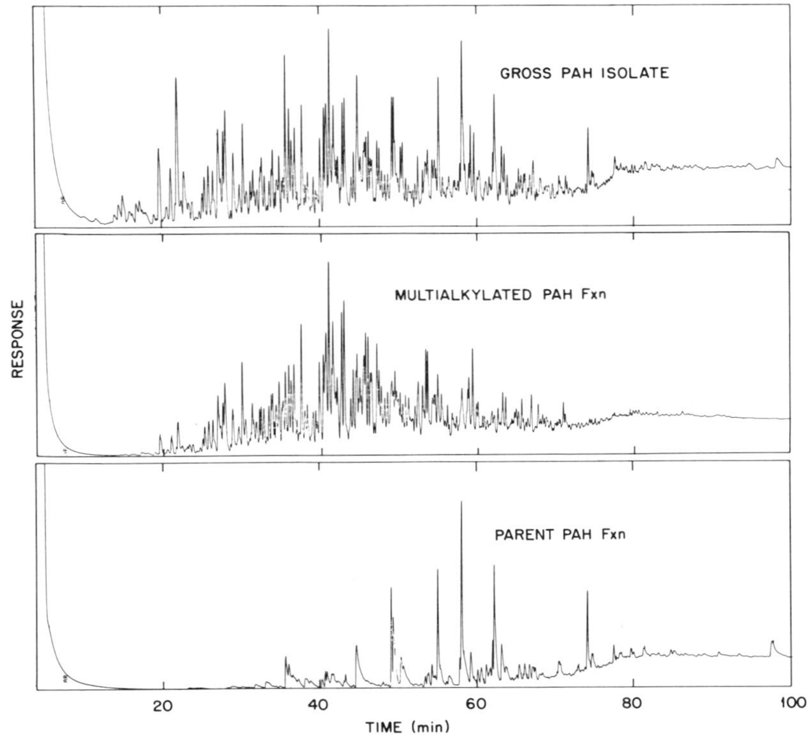

FIGURE 1. Capillary column GC of (A) the gross PAH isolate, (B) the multialkylated PAH fraction, and (C) the parent plus simple alkylated PAH fraction.

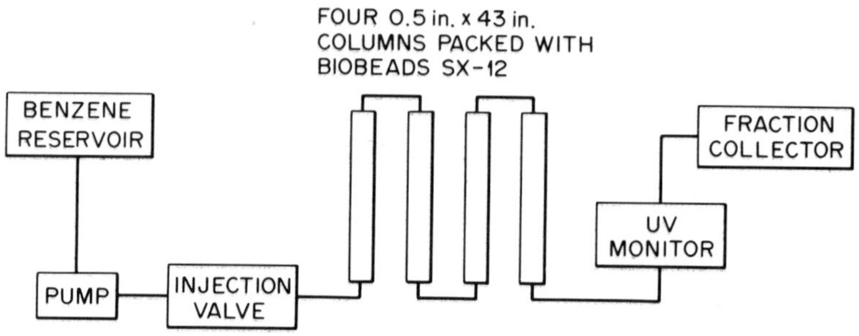

FIGURE 2. Biobeads SX-12 gel filtration apparatus.

alkylated PAH. The fate of the unaccounted material is unknown, but a portion of the more volatile PAH could have been lost during concentration of the fractions.

The GC analysis of the parent plus simple aklylated PAH fraction by capillary-column GC and packed-column GC shown in Figures 1C and 4, indicated a comparatively simple mixture which could be analyzed with reasonable success even by packed-column GC. Nineteen components tentatively identified by GC retention time and confirmed by GC/MS are listed in Table III, along with data previously reported (5) for direct analysis of the gross PAH isolate by packed-column GC. Several carcinogenic PAH such as benzo[a]pyrene and benzo[b,j or k] fluoranthene were recovered in this fraction. The mass of the nineteen constituents reported in Table III accounts for slightly less than 10% by weight of the gross PAH isolate, but some 66%

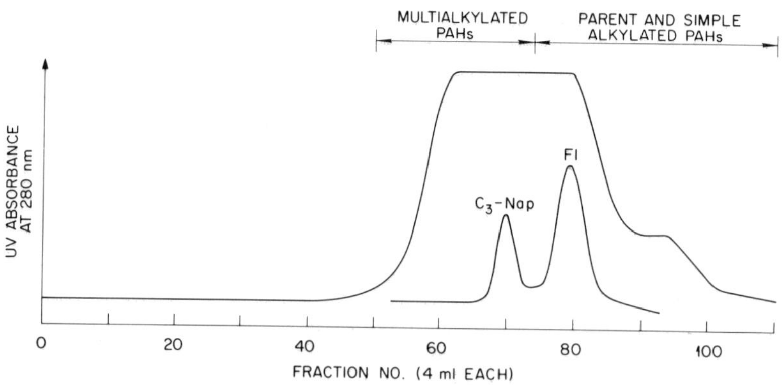

FIGURE 3. UV absorbance (280 nm) of SX-12 gel filtration eluate.

TABLE II. Distribution of Weight and Mutagenic Activity[a] in PAH Fractions of Syncrude

Fraction Name	Weight (mg)	Composition (%)	Specific Activity (rev/mg[b])	Weighted Activity
Original Gross Isolate	54.10		T[c]	0
Parent plus Simple Alkylated PAH	7.13	13.2	3,833	517
Multialkylated PAH	25.38	46.91	2,200	1,034
Unaccounted	21.59	39.91	––	––
Total		100.0%		

[a]All assays were carried out in the presence of rat liver S-9 (Ar-induced).
[b]Rev/mg = determined from the linear portion of a dose response curve (5 μg - 100 μg/plate).
[c]Toxic at higher concentrations and inactive at lower concentrations.

by weight of the parent plus simple alkylated PAH fraction, indicating considerable concentration from fractionation.

Comparison of chromatograms A and B in Figure 1 clearly demonstrates that the complexity of the gross PAH isolate resulted from the presence of numerous multialkylated PAH, and that the multialkylated PAH fraction still was too complicated for GC analysis. Further subfractionation was necessary before the fraction could be analyzed confidently.

BIOASSAY OF PAH ISOLATE FRACTIONS

The gross PAH isolate, its multialkylated PAH fraction and its parent plus simple alkylated PAH fraction were next subjected to a microbial mutagenic bioassay. This provided the first assessment of the relative contributions of these two "classes" to PAH mutagenicity of a coal-derived oil and, hence, would indicate the importance of developing more refined analytical methods for the multialkylated PAH fraction. Carcinogenic compounds generally produce a positive mutagenic response in the Ames test system (11).

Bioassay of the gross PAH isolate and its fractions was accomplished using the *Salmonella* histidine reversion assay developed by Ames (1). Briefly, the test isolates were dissolved in dimethylsulfoxide (DMSO) at concentrations in the range of approximately 10 mg of total material per ml. After preliminary assays to determine toxicity and strain specificity for mutagenicity, a concentration range of 0-0.1 mg of isolate was established and determinations of mutagenicity (reversion to wild-type) were made using the frame-shift strain TA98. All assays included metabolic activation with rat liver

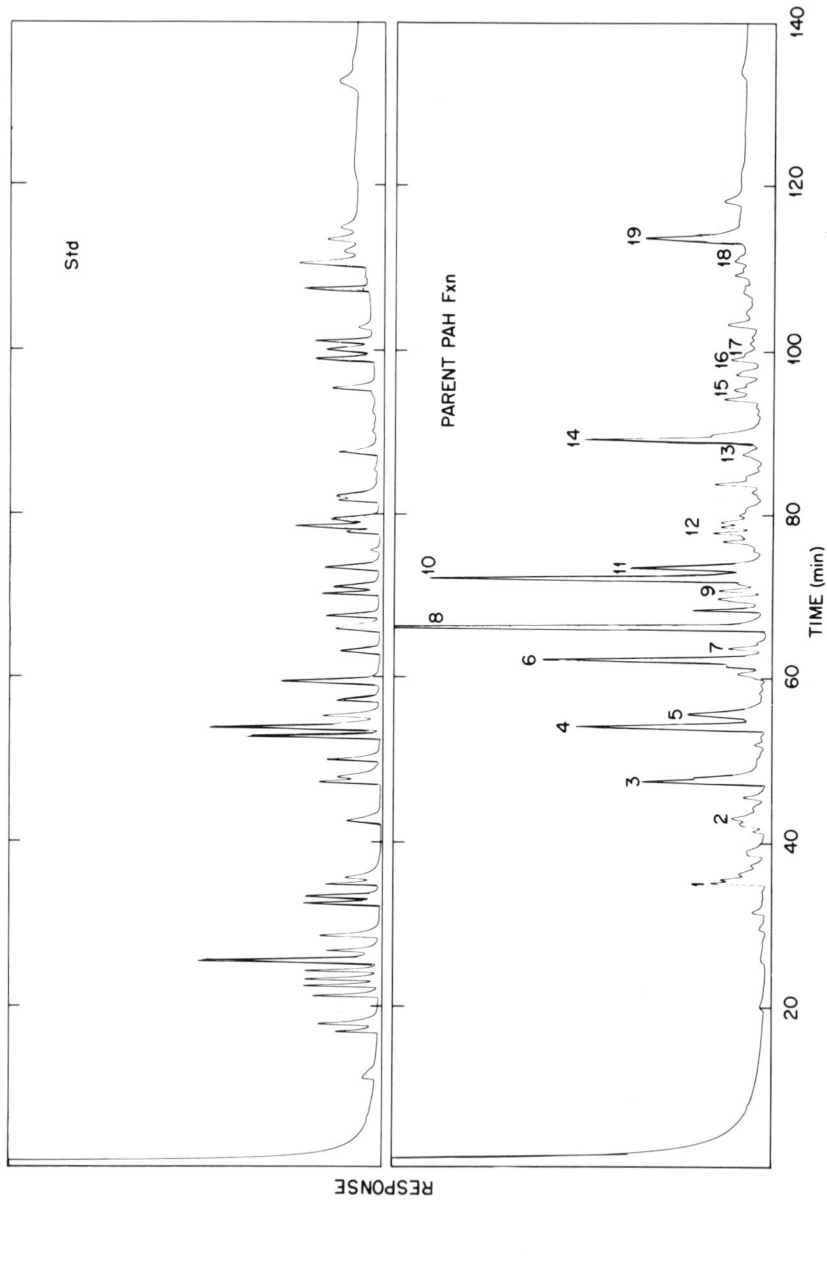

FIGURE 4. Packed-column GC of (A) the PAH standard and (B) the parent plus simple alkylated PAH fraction.

TABLE III. PAH Identifications and Concentrations

Peak No.[a]	Component	Concentration (mg/g)		
		Gross PAH Isolates		Parent PAH Fraction
		Present Work[b]	Past Work (5)	
1	Fluorene	1.8	9.9	13.5
2	1-Methylfluorene	1.8	<4.7	13.9
3	Phenanthrene	5.1	12	38.3
4	2-Methylphenanthrene	7.9		59.7
5	1-Methylphenanthrene	5.7	<5.1	43.0
6	Phenylnaphthalene	7.4		56.2
7	Fluoranthene	1.4	<3.7	11.0
8	Pyrene	20.1	14.2	152.7
9	Benzo[a] fluorene	1.4	2.1	10.8
10	Benzo[b] fluorene	11.1		84.4
11	3-Methylpyrene	7.7		58.5
12	C_2-Fluoranthene/Pyrene	1.6	<2.2	12.4
13	2-Methylchrysene	0.8		6.0
14	Methylchrysene	5.9		44.6
15	Benzo[b] fluoranthene	1.0	<0.5	7.5
16	Benzo[a/e] pyrene	0.9	<0.5	6.8
17	Perylene	0.6	<0.6	4.8
18	Dibenzanthracene	0.7		5.5
19	Benzo[g,h,i] perylene	5.3		40.0
	Accounted	88	55	669

[a]Figure 4.
[b]Calculated from analysis of parent PAH fraction.

crude-enzyme homogenates (S-9), prepared from untreated (uninduced) and Aroclor-induced animals. Results were compared with tests carried out without added activation preparations. The slope of the dose-response curve was used as the indicator of mutagenic potential. Complete results on the primary Swain-Stedman-type acid-base-neutral fractions of syncrude have been reported (4).

The PAH fractions prepared for the study of multialkylated PAH were initially tested as the multialkylated PAH fraction and the parent plus simple alkylated PAH fraction. Both fractions induced revertants in the histidine-dependent strain. The best general indicator was the highly sensitive strain TA98 accompanied by treatment with an Aroclor-induced enzyme preparation. As indicated in Table II, the parent PAH fraction showed a slightly higher specific mutagenic activity than did the multialkylated PAH fraction. The toxicity of the gross PAH isolate to the Ames test system prevented a proper assessment of the additivity of the mutagenic activity of the two fractions or any synergistic/antagonistic effects when present together in the gross PAH isolate. Removal of toxicity or inhibition by fractionation can lead to increased resolution of mutagenic activity.

Although the mutagenic potential (specific activity in rev/mg of a fraction) of the multialkylated PAH appeared to be slightly less than that of the parent plus simple alkylated PAH under these conditions, it should be noted that the overall contribution of the multialkylated PAH class was somewhat larger, based on its greater weight proportion in the total sample of gross PAH isolate. Thus, it appears that multialkylated PAH may contribute significantly to the mutagenicity of PAH in a coal-derived crude oil.

Caution has to be expressed in the execution and interpretation of results from the microbial mutagenesis. Careful evaluation of a crude-test isolate over a wide concentration range using appropriate levels of crude enzyme or S-9 must be considered (13). Not only must the potential for inhibitory effects on the activation process be considered but also the prospect of synergistic effects from the complex array of compounds (14). However, in a *qualitative* sense, it appears that results from this simple, but sophisticated microbial bioassay can be validated by other genetic assays (3) and extrapolated onto the potential for whole-animal carcinogenesis (6,11).

SUBFRACTIONATION AND ANALYSIS OF THE MULTIALKYLATED PAH FRACTION

The significant mutagenic activity which the multialkylated PAH fraction contributed to the gross PAH isolate justified further studies which sought to isolate and concentrate the biologically active components. Normally, PAH with four or more rings show bioactivity (2,8,15). Thus, a ring size separation of the multialkylated PAH fraction seemed most appropriate for a first subdivision.

The multialkylated PAH fraction was subdivided using an apparatus similar to that shown in Figure 2, except that two 6 mm i.d. x 180 cm columns packed with Sephadex LH-20 and an isopropanol eluant at 0.3 ml/min were employed (7,12). Cut-points for three fractions were established with a standard PAH mixture, subdividing the multialkylated PAH into naphthalene, "phenanthrene," and "benz[a]anthracene" fractions bearing two, three, and four or more rings, respectively, as illustrated in Figure 5. Experiments with mono- di- and trimethylnaphthalenes suggested that methylation did not substantially affect the elution volumes; however, the presence of large chain (greater than C_4) alkyl substituents markedly reduces the elution volume compared to the parent species (12).

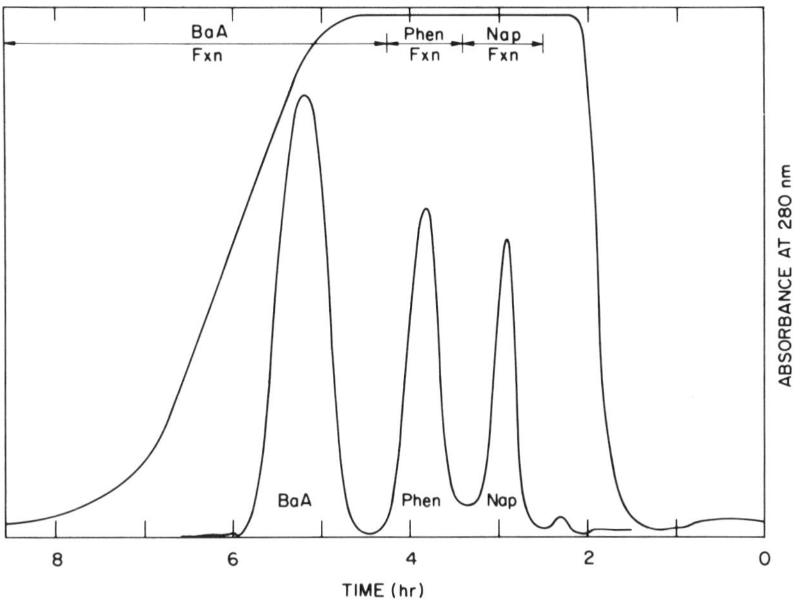

FIGURE 5. UV absorbance (280 nm) of Sephadex LH-20 gel filtration eluate.

To reduce the loss of volatile multialkylated PAH during the concentration step, liquid-liquid extraction was used to transfer the PAH into pentane. Three volumes of water and 0.5 g of ammonium sulfate were added to the isopropanol, and the mixture was partitioned with two 100-ml aliquots of "distilled-in-glass"-grade pentane. (Experiments with carbon-14 labeled naphthalene and phenanthrene indicated quantitative recovery under these conditions.) The solvent was subsequently removed under reduced pressure with a stream of dry nitrogen. The residues were weighed and redissolved in a minimum of redistilled acetone for further characterization.

The mass composition of each subfraction is presented in Table IV. The smaller ring systems accounted for the bulk of the recovered material, and fraction weight dropped off sharply with increasing ring size. Again, on this small scale, significant amounts of material were not recovered. Capillary-column GC analysis of the subfractions is presented in Figure 6. In spite of the subdivision of the multialkylated PAH into three subfractions, the components of each subfraction were still too complex to be properly resolved by GC. Just as the multialkylated PAH subfraction contributed the complexity of the gross PAH isolate, the "naphthalene" subfraction of the multialkylated PAH fraction similarly contributed most of the complexity of the latter. Somewhat lesser, but still significant, complexity existed in the two other

TABLE IV. Distribution of Weight and Mutagenic Activity in Subfractions of Multialkylated PAH Fraction

Fraction Name	Weight (mg)	Composition (%)	Specific Activity (rev/mg)	Weighted Activity
Naphthalene	9.96	39.24	150	59
Phenanthrene	4.84	19.07	5,600	1,070
Benz[a]anthracene	1.37	5.40	6,600	356
Unaccounted	9.31	36.29	––	––
Total		100.0%		

Note: Predicted activity of the multialkylated fraction (by summing weighted activities) 1,485.
Measured activity of the multialkylated fraction: 2,200.

subfractions, indicating the intricate nature of the multialkylated PAH content of this coal-derived oil. Considerable overlap in the GC of the subfractions was evident, indicating that either (1) the elution of long-chain (greater than C_4) alkylated PAH overlapped that of PAH with a smaller parent (12), or (2) the Sephadex column resolution of the fractions was insufficient.

GC/MS analysis of the three multialkylated PAH subfractions indicated the presence of many multiply alkylated derivatives of two-, three- and four-ring PAH. In contrast to the parent PAH fraction, in which the degree of alkylation never exceeded two methyl (or an ethyl) groups, the multialkylated subfractions commonly included derivatives with C_2 and as high as C_6 substitution, similar to those found (17) in cigarette smoke. Only a small number of parent PAH and C_1 derivatives were present. Generic identifications of the PAH constituents of each subfraction are listed in Tables V through VII. The overlap of the fractions observed in the GC analyses appeared to result from substituted fluorenes/acenaphthylenes in the "naphthalene" and "phenanthrene" subfractions, and substituted fluoranthenes/pyrenes in the "phenanthrene" and "benz[a]anthracene" subfractions, indicating insufficient subfraction resolution on the Sephadex gel column. However, it is not easily possible to distinguish with GC/MS between long-chain and multimethyl derivatives, which may behave differently on Sephadex.

Bioassay of Multialkylated PAH Subfractions

Each of the three subfractions of the multialkylated PAH fraction was subjected to strain specificity and inducer specificity assays with the *Salmonella* histidine reversion system. Again as with the gross fractions, strain TA98 and metabolic activation with an Aroclor-induced rat liver enzyme preparation produced the highest levels of mutagenesis. Mutagenic potential, as shown in

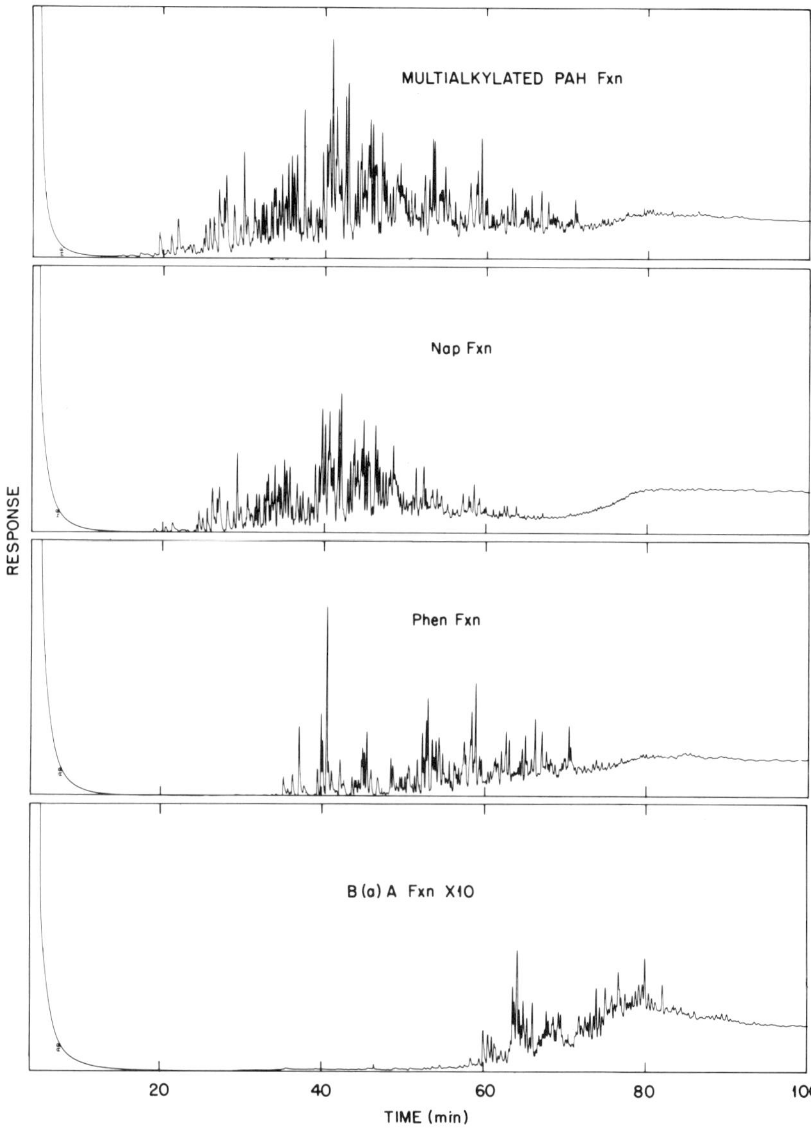

FIGURE 6. Capillary-column GC of (A) the multialkylated PAH fraction, (B) the "naphthalene" subfraction, (C) the "phenanthrene" subfraction, and (D) the "benz[a] anthracene" subfraction.

Table IV, appeared to increase with ring size. The activity of the "phenanthrene" fraction was intermediate between that of the "naphthalene" fraction (lowest activity) and the "benz[a]anthracene" fraction (highest activity).

TABLE V. Composition of the "Naphthalene" Fraction by GC/MS

Parent	C_n Alkylation						Total Number of Species
	1	2	3	4	5	6	
Indene		1		1			2
Indane/Benzofuran[a]		3	3				6
Naphthalene		4	8	5	1		18
C_n-Fluorene/C_{n+1}-Acenaphthylene[a]	1	12	5	8	3	2	31
Acenaphthene/Biphenyl[a]	6	9	14	8	3	2	42
Anthracene/Phenanthrene[a]	2	3	2				7

[a]Indistinguishable by MS data.

TABLE VI. Composition of the "Phenanthrene" Fraction by GC/MS

Parent	C_n Alkylation							Total Number of Species
	0	1	2	3	4	5	6	
C_n Fluorene/C_{n+1} Acenaphthylene[a]	3	7	6	5	3	1		25
Acenaphthene/Biphenyl[a]			2	3	2			7
Phenanthrene/Anthracene[a]		1	10	6	4			21
Cyclopentenophenanthrene	5	9	4					18
Fluoranthene/Pyrene[a]		4		3	1			8

[a]Indistinguishable by MS data.

TABLE VII. Composition of the "B[a] A" Fraction by GC/MS

Parent	C_n Alkylation					Total Number of Species
	1	2	3	4	5	
Phenanthrene/Anthracene[a]		1				1
Fluoranthene/Pyrene[a]	4	2	2	2	2	12
C_n-Fluorene/C_{n+1} Acenaphthylene[a]	1	1				2
Benzo[g,h,i] fluoranthene		2		1		3
B[a] A/Chrysene/Triphenylene[a]	2	2	1			5
Cyclopentenophenanthrene	2			2		4

[a]Indistinguishable by MS data.

These results seem to parallel skin-painting studies (15) which reported that some alkylated naphthalenes act as tumor inhibitors. The cocarcinogenic activity of tobacco-smoke PAH fractions increased with ring size.

When the specific activities of the subfractions were weighted by the actual distribution of the materials in the multialkylated PAH fraction, it was seen that *both* the "phenanthrene" and "benz[a]anthracene" subfractions were significant contributors to the total activity. This effect may have resulted from the overlap in constituents of the latter two fractions. Metabolic activation appeared to be obligatory for these activities. An indication of the additivity of the activity of the subfractions was observable but the overall activity (*i.e.*, that of the multialkylated PAH fraction) was actually greater, suggestive of a synergistic effect. Loss of the more volatile PAH during isolation also may have contributed to this lower summed activity. However, the results do indicate the potential importance of the larger multialkylated PAH.

CONCLUSIONS

These initial studies of the composition and nature of multialkylated PAH indicate that in some complex matrices, the multialkylated PAH constitute a major portion of the PAH isolate. Their overall contribution to the mutagenicity of the PAH isolate appears to be substantial. Studies seeking to identify the bioactive constituents of complex energy-related materials should not overlook the potential importance of this class of PAH.

ACKNOWLEDGMENTS

This work was supported by the Electric Power Research Institute under Union Carbide Corporation contract W-7405-eng-26 with the U.S. Department of Energy. The authors thank M. V. Buchanan, M. P. Maskarinec and C. A. Pritchard for their assistance in obtaining and calculating the MS data.

REFERENCES

1. Ames, B. N., McCann, J., and Yamasaki, E. (1975): Methods for detecting carcinogens and mutagens with the *Salmonella*/mammalian-microsome mutagenicity test. *Mutat. Res.* 31:347-364.
2. Dipple, A. (1976): Polynuclear aromatic carcinogens. In: *Chemical Carcinogens,* C. E. Searle, Ed., pp. 245-314, American Chemical Society, Washington, D.C.

3. Epler, J. L., Larimer, F. W., Rao, T. K., Nix, C. E., and Ho, T. (1978): Energy-related pollutants in the environment: the use of short-term tests for mutagenicity in the isolation and identification of biohazards. Presented at Higher Plant Systems as Monitors of Environmental Mutagens, a workshop co-sponsored by the National Institute of Environmental Health Sciences and the Department of Energy, Marineland, FL, January 16-18, 1978. *Environ. Health Pers.* (in press).
4. Epler, J. L., Young, J. A., Hardigree, A. A., Rao, T. K., Guerin, M. R., Rubin, I. B., Ho, C.-h., and Clark, B. R. (1978): Analytical and biological assessment of test materials from the synthetic fuel technologies. i. Mutagenicity of crude oils determined by the *Salmonella/typhimurium/* microsomal activation system. *Mutat. Res.* 57:265-276.
5. Guerin, M. R., Epler, J. L., Griest, W. H., Clark, B. R., and Rao, T. K. (1978): Polycyclic aromatic hydrocarbons from fossil fuel conversion processes. In: *Carcinogenesis, Vol. 3* P. W. Jones and R. I. Freudenthal, Eds., pp. 21-33, Raven Press, New York.
6. Holland, J. M. (1978): Integrated chemical and biological testing of synthetic oils and effluents, part iii: Carcinogenicity Testing. Presented at Symposium on Potential Health and Environmental Effects of Synthetic Fossil Fuel Technologies, Gatlinburg, TN, September 25-28, 1978. *Proceedings* (in press).
7. Jones, A. R., Guerin, M. R., and Clark, B. R. (1977); Preparative-scale liquid chromatographic fractionation of crude oils derived from coal and shale. *Anal. Chem.* 49:1766-1771.
8. Jones, D. W. (1974): Carcinogenicity and structure in polycyclic hydrocarbons. *Prog. Med. Chem.* 10:159-203.
9. Kubota, H., Griest, W. H., and Guerin, M. R. (1975): Determination of carcinogens in tobacco smoke and coal-derived samples—trace polynuclear aromatic hydrocarbons. In: *Trace Substances in Environmental Health-IX.* D. D. Hemphill, Ed., pp. 281-289, University of Missouri, Columbia.
10. Lee, M. L., Novotny, M., and Bartle, K. D. (1976): Gas chromatography/mass spectrometric and nuclear magnetic resonance studies of carcinogenic polynuclear aromatic hydrocarbons in tobacco and marijuana smoke condensates. *Anal. Chem.* 48:405-416.
11. McCann, J., Choi, E., Yamasaki, E., and Ames, B. N. (1975): Detection of carcinogens as mutagens in the *Salmonella*/microsome test: Assay of 300 chemicals, part i. *Proc. Nat. Acad. Sci., USA* 72:5135-5139.
12. Oelert, H. H. (1969): Atypical gel chromatography in the system Sephadex-isopropanol. *Z. Anal. Chem.* 244:91-101.
13. Rao, T. K., Young, J. A., Hardigree, A. A., Winton, W., and Epler, J. L.: Analytical and biological analyses of test material from the synthetic fuel technologies. ii. Extended genetic and biological studies with mutagenic fractions. *Mutat. Res.* (in press).
14. Rao, T. K., Young, J. A., Weeks, C. E., Slaga, T. J., and Epler, J. L.: Effect of the co-carcinogen benzo[e]pyrene on microsome mediated

chemical mutagenesis in *Salmonella typhimurium. Cancer Res.* (in preparation).

15. Schmeltz, I., Tosk, J., Hilfrich, J., Hirota, N., Hoffman, D., and Wynder, E. L. (1978): Bioassays of naphthalene and alkyl naphthalenes for co-carcinogenic activity. Relation to Tobacco Carcinogenesis. In: *Carcinogenesis, Vol. 3*, P. W. Jones and R. I. Freudenthal, Eds., pp. 47-60, Raven Press, New York.
16. Severson, R. F., Snook, M. E., Arrendale, R. F., and Chortyk, O. T. (1976): Gas chromatographic quantitation of polynuclear aromatic hydrocarbons in tobacco smoke. *Anal. Chem.* 48:1866-1872.
17. Snook, M. E., Severson, R. F., Arrendale, R. F., Higman, H. C., and Chortyk, O. T. (1978): Multialkylated polynuclear aromatic hydrocarbons of tobacco smoke: separation and identification. *Beitr. zur. Tabak.* 9:222-247.
18. Snook, M. E., Severson, R. F., Higman, H. C., Arrendale, R. F., and Chortyk, O. T. (1976): Polynuclear aromatic hydrocarbons of tobacco smoke: Isolation and identification. *Beitr. zur Tabak.* 8:250-272.

Polynuclear Aromatic Hydrocarbons, edited by P.W. Jones and P. Leber. Ann Arbor Science Publishers, Inc. Ann Arbor, MI ©1979.

DEVELOPMENT OF AN AQUEOUS POLYNUCLEAR AROMATIC HYDROCARBON STANDARD REFERENCE MATERIAL

W.E. May, J.M. Brown, S.N. Chesler, F. Guenther, L.R. Hilpert, H.S. Hertz and S.A. Wise

National Bureau of Standards
Washington, D.C. 20234

INTRODUCTION

In the fall of 1975, the National Bureau of Standards (NBS) and the Environmental Protection Agency (EPA) jointly sponsored a series of workshops entitled "Standards and Reference Materials for Environmental Analysis Associated with Energy Development." The objective of these workshops was to obtain input from experts in the field on the methodology and certified standards needed for the accurate analysis of environmental samples associated with the production of alternate fuels.

At the conclusion of these workshops, a number of Standard Reference Materials (SRMs) were recommended by the participants for NBS consideration. One of the SRMs recommended was a polynuclear aromatic hydrocarbon (PAH) in a water matrix. Although many PAH have demonstrated mutagenic properties, this SRM was given a low priority because of the presumed difficulties associated with the preparation and stabilization of such a material.

PROBLEMS ASSOCIATED WITH THE PREPARATION AND STABILIZATION OF A PAH/WATER SRM

Preparation of aqueous PAH solutions of known concentration by gravimetric procedures is difficult because of the extremely low aqueous solubilities of PAH. As shown in Table I, many PAH have aqueous solubilities of less than 500 μg/kg (ppb). Preparation of aqueous solutions of known concentration by serial dilutions of a more concentrated organic solution is both hazardous and wasteful. After small aliquots are taken, large volumes of organic solvent containing toxic and expensive chemicals remain to be disposed of.

TABLE I. The Aqueous Solubilities of Some Aromatic Hydrocarbons as Determined by Several Investigators

Compound	MacKay and Shiu[a]	May *et al.*[b]	Schwarz[c]
Anthracene	73	45	41
Phenanthrene	1,290	1,002	1,151
2-Methylanthracene	39	21	
1-Methylphenanthrene		269	
Fluoranthene	260	206	
Pyrene	135	132	129
1,2-Benzanthracene	14	9.4	
Chrysene	2	1.8	
Naphthacene	0.6		
Triphenylene	43	6.6	
3,4-Benzpyrene	0.4		

[a]Solubilities determined by method described in reference 1.
[b]Solubilities determined by DCCLC method described in reference 3.
[c]Solubilities determined by fluorimetric method described in reference 4.

Preservation of stable aqueous solutions of PAH is hampered by adsorptive losses of the PAH to the surfaces of containers and transfer tools. The magnitude of the adsorptive effect is variable and is a function of the manner in which the solutions are handled. The adsorptive properties of three PAH on four different surfaces are shown in Table II. These results show that losses of PAH from static solutions to surfaces occur in short periods of time. Stirring such solutions only slightly reduces such losses.

PREPARATION AND QUANTITATION OF DILUTE PAH SOLUTIONS BY DYNAMIC COUPLED-COLUMN LIQUID CHROMATOGRAPHY

Recently, we have developed a dynamic coupled-column liquid chromatographic (DCCLC) method (3) for investigating the aqueous solubility behavior of PAH. In the DCCLC method, saturated aqueous solutions of PAH are generated by pumping water through a column packed with glass beads that have been coated with the compound of interest (generator column). The concentration of the desired compound in the effluent of the generator column is measured by a modification of the coupled-column liquid chromatographic process that has been previously described by May *et al.* (2). A flow diagram of this system is shown in Figure 1.

TABLE II. Surface Adsorption Characteristics of PAH in Aqueous Solutions[a]

	Percent Loss from Solution					
	Phenanthrene		Chrysene		Benzo[a] Pyrene	
	1 hr	13 hr	1 hr	13 hr	1 hr	13 hr
Glass	8		46	71	53	82
Silanized Glass	5	73	64	81	73	93
Platinum	35	87	66	85	57	93
Aluminum	5	76	50	89	67	95

[a]The concentration of each PAH was 1μg/kg. surface-to-volume rations are approximately 1 cm^2/cm^3 for all materials studied.

Preparation of Standard Aqueous PAH Solutions Using Generator Columns

The aqueous solubility of a compound is a well-defined thermodynamic quantity. Therefore, saturated aqueous solutions produced by generator columns may also be defined as standard solutions at a given termperature. Generator columns are prepared by packing 60 x 6 cm stainless-steel tubes with 60- to 80-mesh glass beads coated with 1% (wt/wt) of the compound of interest. The beads are coated by adding 20 g of the beads to 200 ml of 0.1% methylene chloride solution of the PAH of interest and removing the solvent with a rotary evaporator.

Stable saturated aqueous solutions are eluted from generator columns after an initial aqueous purge volume of between 200 ml and 100 ml. After this initial conditioning, equilibrium is obtained and the PAH concentration at constant temperature has been shown to be independent of flowrate between 0.1 and 5 ml/min (see Table III). Equilibrium can be reestablished after a change in temperature by passing 10 ml of water through the column under the new conditions. A purge volume of approximately 100 ml is necessary to reequilibrate a generator column after a shelf storage period of more than three months.

Quantitative Analysis of Generated Aqueous PAH Solutions by DCCLC

Extraction of the generated PAH solutions is accomplished by passing a measured volume of that solution through an extractor column (see Figure 1). This 60 x 0.6 cm column is packed with a superficially porous bonded C_{18} stationary phase (Bondapak C_{18}, Water Associates, Milford, MA) and provides better than 99% extraction efficiency for less than 25-ml volumes of aqueous PAH solutions.

After extraction, a water-acetonitrile solvent blend is passed through the extractor column to elute the adsorbed PAH. This eluate is then passed

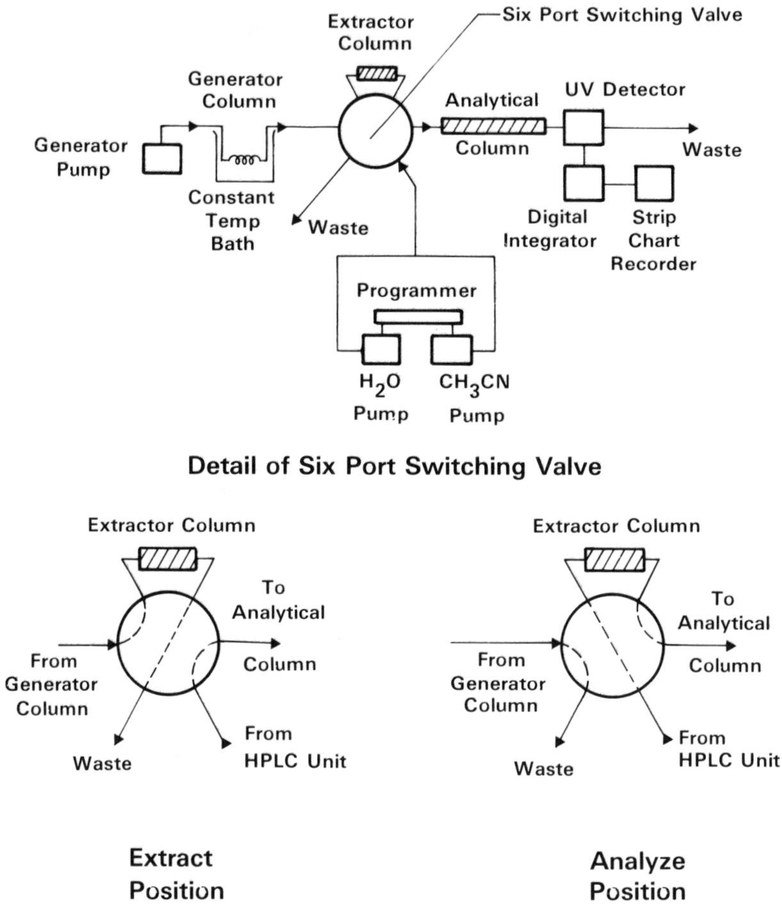

FIGURE 1. Flow diagram of dynamic coupled-column liquid chromatographic system.

through a microparticulate analytical column (μBondapak C_{18}) for separation of the PAH from nonanalyte interferences. Individual response factors are calculated by replacing the extractor column with a calibrated sample loop, and injecting known amounts of the PAH of interest dissolved in acetonitrile.

PREPARATION OF DILUTE ORGANIC PAH SOLUTIONS USING GENERATOR COLUMNS

Small volumes of dilute organic solutions may also be accurately prepared by extracting the generated aqueous solutions with the desired volume of an

TABLE III. Phenanthrene Solubility Dependence on the Aqueous Flowrate Through a Generator Column

Flowrate (ml/min)	Concentration (ppb)
5.0	865 ± 7
0.4	868 ± 4
0.1	866 ± 3
Temperature	22.0°C

immiscible organic solvent, such as hexane. Since PAH hexane/water distribution coefficients are very large, the concentration of these dilute solutions can be calculated if the volume ratios are known. For example, 5 ml of a 10 μg/kg (ppb) solutions of chrysene in hexane cannot be prepared gravimetrically. Such a solution would have to be prepared by serial dilutions of a more concentrated solution. This process is hazardous and wasteful and can be imprecise. However, such a solution can be prepared by extracting 25 ml of a saturated aqueous solution (~2 ppb) of chrysene with 5 ml of hexane. This is only a hypothetical example, but it does demonstrate the utility of using generator columns to indirectly prepare organic PAH solutions of known concentrations.

EVALUATION OF DCCLC AS A METHOD FOR THE PREPARATION AND CERTIFICATION OF AN AQUEOUS PAH SRM

There are several factors that make the DCCLC approach ideal for the preparation and analysis of very dilute aqueous solutions of individual PAH:

1. Saturated solutions are prepared by an equilibrium process. The reversible nature of this process was demonstrated by the following experiment. Distilled water was pumped through two independently thermostated generator columns that were connected in series. The temperature of the first column, A, was maintained at 24.3°C. Concentration measurements were made on the effluent from the second column (B) in the series, at temperatures of 25.3, 12.8 and 6.6°C. The concentration of the solution that eluted from column B at each temperature was identical, within experimental error, to the concentration that had been obtained from column B alone. The data from this experiment are presented in Table IV.

Normally, saturated PAH solutions were generated by pumping distilled water through generator columns. Solutions of the same concentration (saturated solutions) were also obtained when the PAH concentration in the feed solution was greater than that in the effluent. This could only happen through

TABLE IV. A Demonstration of the Reversible Nature of the Generator Column Method for Producing Saturated Solutions

Generator Column A		Generator Column B		Generator Column A + B in Series	
Temp (°C)	Conc (μ/kg)	Temp (°C)	Conc (μg/kg)	Temp (°C)	Conc (μg/kg)
24.3	42.7	25.3	45.5 ± 0.2	25.3	45.4 ± 0.1
		12.8	21.3 ± 0.2	12.8	20.7 ± 0.1
		6.6	14.0 ± 0.1	6.6	14.2 ± 0.2

Note: Flowrate through generator columns was 5.0 ml/min.

an equilibrium process. It has also been shown that PAH concentration is indpendent of the rate of flow through a generator column between 0.1 and 5.0 ml/min.

2. The use of generator columns to produce aqueous PAH solutions circumvents the problems that are usually associated with storing such solutions, since the solutions need not be generated until they are needed.

3. The concentration of the generated aqueous solutions are a function of temperature and may be expressed in terms of least-squares fits of the concentration to temperature. Such equations can be used to interpolate the concentration as a function of temperature to within ±2% of the experimentally determined values between 5 and 30°C (see Table V).

4. Both the short- and long-term precision with which aqueous PAH solutions can be generated appear to be ⩽2% (see Table VI).

TABLE V. Anthracene Solubility = $0.0013\, t^3 \ominus 0.0097\, t^2 \oplus 8.888\, t \oplus 8.21$ (μg/kg)[a]

(°C)	Concentration Measured (mol/l x 10^{-7}	Concentration Measured (μg/kg)	Concentration Calculated (μg/kg)
28.7	3.13	55.7 ± 0.7	55.8
24.6	2.44	43.4 ± 0.1	43.1
22.4	2.09	37.2 ± 1.1	37.5
18.3	1.63	29.1 ± 0.6	29.0
14.1	1.25	22.2 ± 0.1	22.3
10.0	0.98	17.5 ± 0.3	17.4
5.2	0.71	12.7 ± 0.4	12.7

[a] t = temperature in °C.

TABLE VI. Precision with which Anthracene Solutions may be Generated and Measured at 25.4°C

Volume of Distilled Water Eluted Through Column (ml)	Concentration Measured (μg/kg)
1	45.2
190	47.0
230	46.6
500	46.1
775	46.9
910	44.6
1,175	45.5
Average Measured Concentration- (6/21/77)	46.0 ± 0.9
Concentration Calculated from Calibration Curve (12/20/76)	45.7 ± 0.2
Concentration Measured by Solvent Extraction Followed by GC Analysis (6/21/77)	45.8 ± 1.0

5. Shelf storage of generator columns does not seem to present a problem. They have been shown to be stable for longer than 1.5 years and through more than 100 liters of aqueous purge.

6. DCCLC incorporates a rapid and accurate method for analyzing the generated dilute aqueous PAH solutions. Analytical errors due to adsorption are minimized in DCCLC because the solution is extracted and concentrated, online, in less than 500 msec after generation. It has been estimated that this analytical method has an uncertainty of less than 2% (3).

CONCLUSION

The DCCLC technique has been shown to be an ideal method for the preparation and certification of an aqueous PAH SRM. Tentative plans are to issue individually certified anthracene, 1,2 benzanthracene, and 3,4 benzpyrene generator columns by early 1979. Investigations are in progress to determine the feasibility of preparing multi-PAH generator columns.

REFERENCES

1. MacKay, D., and Shiu, W. (1977): *J. Chem. Eng. Data* 22:4.
2. May, W. E., Chesler, S. N., Cram, S. P., Gump, B. H., Hertz, H. S., Enagonio, D. P., and Dyszel, S. M. (1975): *J. Chromatog. Sci.* 13:535.
3. May, W. E., Wasik, S. P., and Freeman, D. H. (1978): *Anal. Chem.* 50:175.
4. Schwarz, F. (1977): *J. Chem. Eng. Data* 22:273.

*Polynuclear Aromatic Hydrocarbons,
edited by P. W. Jones and P. Leber.
Ann Arbor Science Publishers, Inc.
Ann Arbor, MI © 1979.*

MEASUREMENT OF POLYNUCLEAR AROMATIC HYDROCARBONS AND OTHER HAZARDOUS ORGANIC COMPOUNDS IN STACK GASES

Roy L. Bennett, Kenneth T. Knapp, Peter W. Jones,[1] *
JoAnn E. Wilkerson* and Paul E. Strup*

Environmental Sciences Research Laboratory
U.S. Environmental Protection Agency
Research Triangle Park, North Carolina 27711

*Battelle-Columbus Laboratories
505 King Avenue
Columbus, Ohio 43201

INTRODUCTION

An increasing awareness of the hazardous properties of many organic compounds released into our environment has resulted in greater emphasis on the measurement of industrial stacks emissions. Polynuclear aromatic hydrocarbons (PAH), present in some stack emissions, are especially important since many PAH compounds have exhibited mutagenic and carcinogenic activity in bioassay tests. The development of reliable, practical measurement methods for these and other organic pollutants has been a primary objective of an ongoing program to characterize emissions from potentially significant stationary sources.

The amount of quantitative emission data acquired on PAH emissions is limited due to the complexities and attendant cost of the collection and analytical procedures. The sampling strategy must consider numerous compounds, each of which may exist in vapor or particulate form or both. Early sampling designs contained many trapping components (filters, liquid impinger solutions, solid sorbents) to insure effective PAH sample collection. The resulting sampling trains were unwieldy to operate and the distribution

[1]Present address: Electric Power Research Institute, 3412 Hillview Avenue, Palo Alto, California 94304.

of the sample over the large number of train components complicated the sample recovery and analysis. The requirement for advanced techniques for separation, identification and measurement of the individual compounds adds to the expense for collecting emission data. Analysis must be made for individual compounds since the carcinogenic potency and other deleterious environmental effects vary greatly with different compounds.

In the 1960s the National Air Pollution Control Administration of the U.S. Public Health Service, the predecessor in air pollution work to the Environmental Protection Agency (EPA), conducted a study on source-emitted PAH. Hangebrauck *et al.* (5) developed a method and measured the emissions from coal-fired power plants and other sources. Langston Laboratories, Inc. (6) under an EPA contract compared several sampling trains, including a modification of the one used by Hangebrauck, for the measurement of PAH and polychlorinated biphenyls (PCB). All the trains had an organic solvent impinger system which made the sampling train difficult to handle in the field and increased the risk of sample loss and contamination. These organic solvent systems have now been replaced in most instances with solid sorbent systems. The remainder of this chapter will address the use of these solid sorbent systems and the types of organic air pollutants found in stationary source emissions.

MEASUREMENT METHODS

A research group at Battelle–Columbus Laboratories has developed a source sampling system for PAH that uses a macroreticular resin adsorption bed. This system has been described in several publications (7,8,12). A detailed report (9) on the state-of-the-art of the measurement of PAH and other hazardous organic compounds in stack gases has also been prepared for EPA by the Battelle group. The Battelle sampling system consists of a standard EPA Method 5 train (3) with the resin adsorption bed located between the filter and the impingers. The adsorption device consists of a section of coiled glass tubing for cooling followed by an enlarged section with a glass fritted disc. The adsorbing resin bed is held in with a glass-wool plug, and the whole assembly is enclosed in a glass envelope. In sampling hot stack gases for PAH, the adsorption bed assembly is usually water- or air-cooled to about 55°C, a temperature which has provided good collection efficiency.

The Particulate Emissions Research Section of the Environmental Sciences Research Laboratory at EPA and the Battelle group have both been collecting samples at various stationary sources for PAH. The PAH concentrations of the collected samples were determined by GC/MS analysis on solvent extracts of the exposed resin beds. This work was done by Battelle using established analytical procedures described by Jones *et al.* (8,9).

Laboratory studies (11) have demonstrated that the Battelle sampling train is an effective PCB collection system, and field studies are planned to use this train for PCB measurements of sludge incinerator emissions and other possible sources.

Since several investigators have reported detecting hazardous organic compounds other than PAH and PCB in stationary source emissions, EPA has contracted with Southern Research Institute (SoRI) to develop sampling systems for measuring condensable organic compounds emitted from stationary sources. Several systems have been studied: a single-resin sorbent bed after the impingers of a Method 5 train; a two-resin sorbent bed, one bed on each side of the impingers; and the same system with a charcoal bed after the last resin bed. Both laboratory and field studies have been conducted, the latter at a large coal-fired power plant. The system and the results obtained are described in an EPA report (2).

TEST SITES

Site A

This site is a modern oil-fired power plant with a normal load of 525 megawatts (MW) and a maximum load of 560 MW. During the testing, the load levels varied from 400 to 560 MW. However, the load was held constant for each sample collection. Three sampling trips were made to this plant which was burning different fuels each time. All fuels were high-sulfur (about 2.5%) residual oils. The vanadium content was about 200 ppm for the first test, 590 ppm for the second and 290 ppm for the third. A magnesium oxide fuel additive was used as a fire-side corrosion inhibitor during all tests. The Battelle sampling system was used at this site.

Site B

This site is a coal-fired power plant with an electrostatic precipitator (ESP) for particulate emission control. The sampled unit has a normal load of 160 MW and was burning high-sulfur (3.5%) eastern interior coal. The Battelle sampling system was used at this site.

Site G

This site is a modern large coal-fired power plant with an ESP for particulate emission control, and a 700-MW normal load for the unit tested. At the time of testing, the unit was burning Appalachian coal with approximately 1.7% sulfur. The SoRI sampling systems were used at this site by the SoRI staff.

Site M

This site is a modern medium-sized coal-fired power plant with an ESP control system and a 330-MW normal unit load. The unit was burning high-sulfur (4%) eastern interior coal. The Battelle sampling system was used at this site.

Site P

This site is an old small coal-fired power plant, the unit tested having a normal load of 100 MW. The unit was burning eastern interior high-sulfur (3.5%) coal, and an ESP was used for particulate emission control. The Battelle sampling system was used at this site.

Site N

This site is a secondary lead smelter that processes old lead storage batteries. The polymeric organic battery casings are thrown into the furnaces with the lead plates, thus yielding organic emissions. The Battelle sampling system was used at this site.

SORBENT BED RESINS

Macroreticular resins are used in the sorbent beds of the sampling trains. The most widely used resins have been Tenax, XAD resins and polyurethane foam (9). The XAD resins are often preferred to Tenax, still the most commonly used resin in solid sorbent beds, because Tenax presents several drawbacks: its cost, its availability, and its degradation on exposure to hot emission gases having high concentrations of oxides. This decomposition, which results in the contamination of the samples with several diphenyl quinones, has been described by Neher and Jones (10,11) and the SoRI group (2). A high background blank with XAD-2 slowed its acceptance; however, with multisolvent cleaning, a good low-background sample can be obtained.

Extensive testing has been conducted on both Tenax and XAD-2 (1,4,9). The EPA report by Jones *et al.* (9) describes much of this work and has an extensive list of references for additional data. These studies demonstrated that both Tenax and XAD-2 have very good retention for PAH collection.

RESULTS AND DISCUSSION

Oil-Fired Power Plant

Five samples were collected at Site A to measure polycyclic organic matter (POM) emissions. The first three samples were collected on the first visit to this plant, the other two samples were taken on two separate visits. The results of the five runs are given in Table I. The number of asterisks listed under the NAS Notation column indicates the degree of carcinogenicity of the compounds listed by the National Academy of Science. A relatively high level of the lower-molecular-weight PAH was found in the first sample. Sample 4 was collected when the plant was burning a Venezuelan residual fuel with 2.4% sulfur and 590 ppm vanadium; it had about 20 times the carcinogenic PAH of the first three samples and nearly double those of the fifth sample. The total particulate emissions were also the highest during this sampling period. These data indicate that oil-fired power plants, under certain conditioned and burning residual fuels, can emit significant amounts of carcinogenic compounds.

TABLE I. Emissions of PAH from Oil-Fired Power Plants

Compound	NAS Notation	Emission Concentrations (ng/Nm^3)				
		PAH 1	PAH 2	PAH 3	PAH 4	PAH 5
Anthracene/Phenanthrene		2,130	11.0	8.3	89	212
Methyl Anthracenes		210	<0.2	<0.1	14	33
Fluoranthene		98	10.0	6.5	18	52
Pyrene		51	9.9	6.3	22	26
Methyl Pyrene/Fluoranthene		59	4.3	<0.1	<0.1	<1.2[a]
Benzo[c] Phenanthrene	***	<0.1[a]	<0.1	<0.1	3.0	<1.2
Chrysene/Benz[a] Anthracene	*	<0.1	<0.1	<0.1	3.9	<1.2
Methyl Chrysenes	*	<0.1	<0.1	<0.1	4.4	<1.2
7,12-Dimethylbenz[a] Anthracene	****	<0.1	<0.1	<0.1	<0.1	<1.2
Benzo Fluoranthenes	**	<0.1	<0.1	<0.1	9.3	<1.2
Benz[a] Pyrene	***	<0.1	<0.1	<0.1	2.4	<1.2
Benz[e] Pyrene		<0.1	<0.1	<0.1	5.7	<1.2
Perylene		<0.1	<0.1	<0.1	0.8	<1.2
Methylbenzopyrenes		<0.1	<0.1	<0.1	3.1	<1.2
3-Methylcholanthrene	****	<0.1	<0.1	<0.1	<0.1	<1.2
Indeno[1,2,3-c,d] Pyrene	*	<0.1	<0.1	<0.1	0.21	<1.2
Benzo[g,h,i] Perlene		<0.1	<0.1	<0.1	0.47	<1.2
Dibenzo[a,h] Anthracene	***	<0.1	<0.1	<0.1	0.59	<1.2
Dibenzo[c,g] Carbazole	***	<0.1	<0.1	<0.1	<0.1	<1.2
Dibenz[a,i and a,h] Pyrenes	***	<0.1	<0.1	<0.1	<0.1	<1.2
Coronene		<0.1	<0.1	<0.1	0.30	<1.3
Total PAH		2,550	37	23	178	343
Total Carcinogenic Compounds		<1.1	<1.1	<1.1	<24	<13
Total Particulate Emissions, mg/Nm^3		294	161	180	471	202

[a]Minimum detection limit which depends on sample size.

Coal-Fired Power Plants

Samples have been collected at the coal-fired power plants described above, with both the Battelle and the SoRI sampling trains. All plants use ESPs for control of the particulate emission and the ESPs were in use at the time of sampling. The Battelle train was used at three of the plants with the results of the samples collected are given in Table II. Site P was the smallest and oldest unit tested and gave the highest results with a total PAH concentration of about 378 ng/Nm^3. Site M was a new unit burning coal very similar to that used at Site P and had emissions of about 288 ng/Nm^3. Site B was a unit slightly larger and newer than Site P and appeared to be very clean; the PAH emission from it was only 24 ng/Nm^3. Samples from both Sites M and B had amounts of benz[a]pyrene (B[a]P) above the lower detection limit, and several other carcinogenic PAH were detected in amounts too small to be accurately quantified.

TABLE II. PAH Emissions from Coal-Fired Power Plants

Compound	NAS Notation	Emission Concentrations (ng/Nm^3) Sites B[+]	M	P
Anthracene/Phenanthrene		17.5	237	216
Methyl Anthracenes		2.2	14	46
Fluoranthene		1.9	15	60
Pyrene		0.4	11	32
Methyl Pyrene/Fluoranthene		<0.1[++]	7.2	13
Benzo [c] phenanthrene	***	<0.1	<0.2[++]	<0.2
Chrysene/Benz[a] Anthracene	*	<0.1	<0.2	<6.4
Methyl Chrysenes	*	<0.1	<0.2	<0.2
7,12-Dimethylbenz[a] Anthracene	****	<0.1	<0.2	<0.2
Benzo Fluoranthenes	**	<0.1	<0.2	0.27
Benz[a] Pyrene	***	<0.1	0.2	0.27
Benz[e] Pyrene		<0.1	--	--
Perylene		<0.1	0.8	2.2
Methylbenzopyrenes		<0.1	<0.2	<0.2
3-Methylcholanthrene	****	<0.1	<0.2	<0.2
Indeno[1,2,3-c,d] Pyrene	*	<0.1	<0.2	<0.2
Benzo [g,h,i] Perylene		<0.1	<0.2	<0.2
Dibenzo[a,h] Anthracene	***	<0.1	<0.2	<0.2
Dibenzo[c,g] Carbazole	***	<0.1	<0.2	<0.2
Dibenz[a,i and a,h] Pyrenes	***	<0.1	<0.2	<0.2
Coronene		<0.1	<0.2	<0.2

[+] Average of four (4) runs.
[++] Minimum detection limit which depends on sample size.

The organic compounds collected with the two-resin bed SoRI sampling system at Site G are listed in Table III. A red amorphous material was also collected which has been identified as elemental selenium. XAD-2 was the resin used for these collections. Several unexpected organic compounds were found including phthalate esters. A careful check was made of all solvents, but none of the reported phthalate esters was found as a contaminate. In this analysis, the isomers of benzopyrene were not separated from perylene and benzofluoranthene. The sample collection runs were made to demonstrate the system, not for quantitative results.

TABLE III. Compounds Identified in Coal-Fired Power Plant Emissions at Site G

Compound	Formula	Molecular Weight
p-Ethyl Acetophenone	$C_{10}H_{12}O$	148
2-*t*-Butyl Phenol	$C_{10}H_{14}O$	150
Benzoic Acid	$C_7H_6O_2$	122
2-Ethyl-5-Isoamylthiophene	$C_{11}H_{18}S$	182
Fluorene	$C_{13}H_{10}$	166
Diethylphthalate	$C_{12}H_{14}O_4$	222
Benzil	$C_{14}H_{10}O_2$	210
Dichlorobenzoic Acid	$C_7H_4OCl_2$	174
Diphenyl Acetylene or Anthracene or Phenanthrene	$C_{14}H_{10}$	178
Diisobutyl Phthalate	$C_{16}H_{22}O_4$	278
Methylphenanthrene	$C_{15}H_{12}$	192
Di-*n*-Butyl Phthalate	$C_{16}H_{22}O_4$	278
Dibenzanthracene	$C_{22}H_{14}$	278
Diphenyl Diacetylene or Fluoranthene or Pyrene	$C_{16}H_{10}$	202
1-Methyl-7-Isopropyl Phenanthrene or 3,4,5,6-Tetramethyl Phenanthrene	$C_{18}H_{18}$	234
Benzopyrene or Perylene or Benzofluoranthene	$C_{20}H_{12}$	252
Methyl Dihydroabietate	$C_{21}H_{30}O_2$	314
Benzo[g,h,i] Fluoranthene	$C_{18}H_{10}$	226
Dicyclohexyl Phthalate	$C_{20}H_{26}O_4$	330
Diisooctyl Phthalate	$C_{24}H_{38}O_4$	390
A Series of High-Molecular-Weight Hydrocarbons (paraffins)	C_nH_{2n+2}	

Secondary Lead Smelter

In earlier work at a secondary lead smelter for the collection of total particulate emissions, immiscible organic material was unexpectedly collected in the water impingers. The polymeric organic battery casings are thrown in with the lead plates. In this process casings are presumably the source of the organic emissions. A return visit to the smelter, Site N, to collect samples with the Battelle PAH train was made; the results from the tests are given in Table IV. Four samples were taken, two at a time, such that in each of the two sampling runs the gas stream was split and two simultaneous samples were collected. The same amount of stack gas was sampled in each experiment. While the amounts of PAH collected were surprisingly high, the results from the two sets of samples had good agreement, thus demonstrating that the entire sampling and analysis procedures have good reproducibility. Additional secondary lead smelters will be sampled to determine whether these results are typical.

TABLE IV. PAH Emissions from a Secondary Lead Smelter

	Emission Concentrations (ng/Nm^3)			
Compound	1A	1B	2A	2B
Anthracene/Phenanthrene	600	740	770	940
Methyl Anthracenes	25	34	41	33
Fluoranthene	160	170	330	310
Pyrene	31	22	28	30
Methyl Pyrenes/Fluoranthenes	2	2	3	2
Benzo [c] Phenanthrene	10	11	17	13
Chrysene/Benz [a] Anthracene	25	23	28	25
Benz [a] Pyrene	1	1	1	1

SUMMARY

Because of the growing concern of the dangers from hazardous organic air pollutants, EPA has continued to sponsor projects for the development of measurement methods and for the characterization of organic emissions from stationary sources. Several sampling systems which use macroreticular resin adsorption beds to collect the emitted organic compounds are described. Concentrations of PAH collected at an oil-fired power plant varied from over 2,000 to only 23 ng/Nm^3. The sample with the high concentration contained only the lower-molecular-weight PAH, none were found above the detection limit with a molecular weight greater than 220. Several coal-fired power plants were sampled, and the range of concentrations found was from 24 to

378 ng/Nm^3. The concentration found in the samples from a secondary lead smelter was greater than 1,000 ng/Nm^3, generally about four times that of the coal-fired power plants.

ACKNOWLEDGMENTS

The authors wish to acknowledge the work of the group at Southern Research Institute under the direction of Dr. William J. Barrett; the assistance given in collecting the field samples by Robert Griffin and Raymond Steward, U.S. EPA; and the sampling by the Battelle field team.

REFERENCES

1. Adams, J., Menzies, K., and Levins, P. (1977): Section and evaluation of sorbent resins for the collection of organic compounds. EPA-600/7-77-044, U.S. Environmental Protection Agency, Research Triangle Park, NC.
2. Barrett, W. J., and Miller, H. C.: Identification and measurement of high molecular weight organic compounds in emissions from power plants and oil refineries (Interim Report). U.S. Environmental Protection Agency. (In press).
3. EPA Standards of Performance for New Sources. (December 2, 1971): *Federal Register* 36(247):24873.
4. Gallent, R. F., King, J. W., Levins, P. L., and Piecewicz, J. F. (1978): Characterization of sorbent resins for use in environmental sampling. EPA-600/7-78-054, U.S. Environmental Protection Agency, Research Triangle Park, NC.
5. Hangebrauck, R. P., von Lehmden, D. J., and Meeker, J. E. (1967): Sources of polynuclear hydrocarbons in the atmosphere. U.S. Public Health Service Publication No. 999-AP-33, Cincinnati, OH.
6. Hermann, T. S. (1974): Development of sampling procedures for polycyclic organic matter and polychlorinated biphenyl. EPA-650/2-75-007, U.S. Environmental Protection Agency, Washington, D.C.
7. Jones, P. W., Giammar, R. D., Strup, P. E., and Stanford T. B. (1976): *Environ. Sci. Technol.* 10:806.
8. Jones, P. W., Graffeo, A. P., Detrick, R., Clarke, P. A., and Jakobsen, R. J. (1976): Technical manual for analysis of organic materials in process streams. EPA-600/2-76-072, U.S. Environmental Protection Agency, Research Triangle Park, NC.
9. Jones, P. W., Wilkinson, J. E., and Strup, P. E. (1977): Measurement of polycyclic organic materials and other hazardous organic compounds in stack gases–state-of-the-art. EPA-600/2-77-202, U.S. Environmental Protection Agency, Research Triangle Park, NC.
10. Neher, M. B., and Jones, P. W. (1977): *Anal. Chem.* 49:512.

11. Neher, M. B., Jones, P. W., and Perry, P. J. (1978): Validation of the solid sorbent sampler. Final report to EPRI on Contract No. RP383-2, Palo Alto, CA.

12. Strup, P. E., Giammar, R. G., Stanford, T. B., and Jones, P. W. (1976): Improved measurement techniques for polycyclic aromatic hydrocarbons in combustion effluents. In: *Carcinogenesis Vol. I* R. I. Freudenthal and P. W. Jones, Eds., pp. 241-51, Raven Press, New York.

Polynuclear Aromatic Hydrocarbons,
edited by P. W. Jones and P. Leber.
Ann Arbor Science Publishers, Inc.
Ann Arbor, MI ©1979.

THE GAS CHROMATOGRAPHIC SEPARATION AND DETERMINATION OF PAH FROM INDUSTRIAL PROCESSES USING GLASS CAPILLARY AND PACKED COLUMNS

R. C. Lao and R. S. Thomas

Chemistry Division
Air Pollution Control Directorate
Air Pollution Technology Centre
Environment Canada
Ottawa, Canada

A wide variety of analytical schemes have been developed in response to a growing concern over polycyclic aromatic hydrocarbons (PAH) emitted into the environment by many sources. Most of these methods are highly refined and require sophisticated analytical technology and techniques. This precludes frequent monitorings or routine PAH analyses for many industries and regulatory agencies because of high cost per analysis and laboratory facility constraints. Identification and determination of PAH components in complex organic mixtures found in industrial emissions or environmental samples can generally be provided by a computerized gas chromatograph/mass spectrometer (GC/MS) system (6,16), which represents a substantial capital investment or contract expenditure. Automated high-pressure liquid chromatography (HPLC) offers an alternate method, but rising cost of column packings and elution solvents have not allowed this technique to gain the broad appeal of GC. Procedures based on paper chromatography have found little practical application other than academic interest (7). The manual operation has limited the usefulness of ultraviolet absorption or optical fluorescence spectrophotometric methods. New designs for PAH monitors have been developed and are slowly becoming commercially available. However, extensive evaluation will be necessary to assess the feasibility for routine analysis (5).

In the absence of a coupled MS, GC has not been widely accepted for PAH analysis in any complex organic mixture (6) because the identities of all unknown components cannot be unambiguously established. The technique, however, presents great potential for determining PAH in a broad range of samples.

Developments in chromatographic packings and columns have enhanced its applicability and removed many of the disadvantages of GC invariably suffered under the operating conditions previously employed. This study is intended to provide a GC method which is rapid, reproducible and reliable for the routine analysis of a cross section of typical samples. The performances of commercially available GC columns were examined under a series of nominal operational parameters, and the applications and limits for column packings/coating material.

EXPERIMENTAL PROCEDURE

Reagents

The procedures for determining the GC parameters of reference PAH compounds which are available from this laboratory have been reported previously (13,15). Standards were weighed by microbalance and made up volumetrically in spectrograde cyclohexane. Fluoranthene has been selected as the internal reference standard for both GC relative retention times (RRT) and the calculated response factors (RF).

Instrumentation

Two gas chromatographic systems were used in this study. A Perkin-Elmer 990 GC, equipped with heated flash vaporization injectors for 1/8-in. packed columns and stainless-steel capillary columns, was run in parallel to a Perkin-Elmer Sigma I GC with heated injectors for all glass capillary columns. Both systems had dual flame ionization detectors (FID). The Dexsil-300 glass capillary column was supplied by Applied Science Ltd and the SE-52 column was obtained from Dr. T. Sukuma of SCIEX, Thornhill, Ontario, Canada. The Dexsil-300 stainless-steel capillary column and the packed column were purchased from Perkin-Elmer Corp. Operating conditions of both GC and column specifications are summarized in Table I. Qualitative confirmation of known peaks appearing at nonspecified time windows was provided by a Finnigan 1015D GC-MS/Model 6000 system (13,14,16).

SAMPLES

Three samples were selected as representative of the majority of samples submitted related to concerns over environmental contamination or industrial hygiene. Sample No. 1 was drawn from a "pot room" exhaust of a nonferrous metal production unit and collected as particulate matter using a

TABLE I. Chromatographic Operating Conditions

	P-E 990	Sigma I
Detector	*FID*	*FID*
Detector Temperature	300°C	350°C
Column	packed, WCOT, Dexsil-300	glass capillary
Sample Volume	0.2-3 μl	1-5 μl
Sample Injector Temperature	325°C	325°C
Column Specifications	Packed: 12 ft long, 0.125 in. i.d. 6% Dexsil-300 on 80/100 mesh Chromosorb W (HP)	Dexsil-300 50 m long, 0.25 mm i.d.
	WCOT: 150 ft long 0.02 in. i.d. stainless steel preheated and stabilized at 200°C	SE-52: 30 m long, 0.25 mm i.d.
Carrier Gas Flow	packed: N_2 40 ml/min WCOT: N 3.2 ml/min	N_2: 0.8 ml/min
Split Ratio[a]	15:1 for WCOT	150:1
Initial Temperature	175°C	160°C
Programmed Temperature	4°C/min	4°C/min
Final Temperature	280°C	270°C
Recorder Attenuation	from 16 to 128	4

[a]Split ratio: Ratio of flow-of-gas-vented to flow-of-gas-through column.

fiberglass filter. The tared filter was reweighed after sampling and extracted with cyclohexane (15). Instead of Rosen separation, the PAH were isolated by extraction with dimethyl sulfoxide (DMSO) similar to the process developed by Natusch *et al.* (17). Sample No. 2 is a typical process sludge blend from several industries and destined for incineration or other disposal. Sample No. 3 is a sample of extracted coal tar pitch volatiles regularly used as the binder in the production of carbon paste electrodes.

Each of the extracted PAH fractions were concentrated to provide adequate concentrations for the various GC detection limits, and aliquot samples were injected onto the various columns for evaluation. Samples were also run on the GC/MS system for qualitative identifications.

All apparatus and reagents were rinsed with cyclohexane and the final rinsing concentrated to confirm the absence of contamination by GC/FID analysis. All syringes, 1 and 10 μl, were manufactured by the Hamilton Co. and pretested with triple-distilled solvents to verify freedom from contamination. Sample volumes injected varied from 0.2 to 5 μl depending on column used and sample concentration.

RESULTS AND DISCUSSION

The chromatograms for Sample No. 1, the exhaust flue extract, are shown in Figures 1, 2, 3 and 4. Figures 5, 6, 7 and 8 illustrate the chromatograms of the extracted sludge sample on the same four columns. The results for the coal tar pitch volatiles are shown in Figures 9, 10, 11 and 12. Tabulated data from these runs are summarized in Tables II, III and IV.

Prepared standard blends of isomeric pairs were used to evaluate column resolution of the chrysene and benzpyrene "groups." The 50-m glass and stainless-steel WCOT columns were coated with Dexsil-300 and resolved chrysene and benzo[a]anthracene and benzo[e]pyrene from benzo[a]-pryene. Triphenylene and chrysene however remain unresolved.

The concentration ratio of B[e]P:B[a]P is low in all samples examined and the B[a]P/B[e]P separation is apparently incomplete on SE-52 column. By careful optimization of the GC parameters, the SE-52 column will partially resolve these two compounds (18,19). For most field monitor samples, the concentrations of PAH compounds is below 0.1 μg/1,000 m^3 which, after extraction, provides nanogram concentrations. When split by injection onto the capillary columns, the resultant concentration of compound delivered to the detector makes deconvolution difficult.

The chromatograms from the 0.020-in., 150-ft stainless-steel WCOT column, illustrated by Figures 2, 6 and 10 show no improvement over the corresponding runs of the samples on the packed column. Nominal advantages of lower operational temperatures and moderate sample capacity are negated by price and availability compared with packed columns.

The glass capillary columns displayed resolution supporting a theoretical plate number of about 100,000 and provided the highest resolution of all columns evaluated (4,8). Unfortunately, a restrictive sample capacity of about 5 x 10^{-3} microliter requires a split ratio of 1:150 on injection and caused severe detection limits for a number of PAH sampled at ambient concentrations. The solubilities of a number of PAH set serious restrictions on the reduction of sample extract volumes. In contrast, the packed column, because of its higher load capacity, will extend the sensitivity of the method with minor compromises in column resolutions and are preferred by some investigators in ambient PAH determinations (3).

One of the major disadvantages to using GC for evaluation of environmental PAH samples has been the nonselective FID which measures all potentially interfering aliphatics and heterocyclic compounds together with PAH. The cleanup procedure required to separate aliphatic, aromatic and heterocyclic moities of an environmental sample, the Rosen Separation, is effective but time-consuming. HPLC may be used to quickly perform PAH separations by automating this process and using UV detection techniques which have no

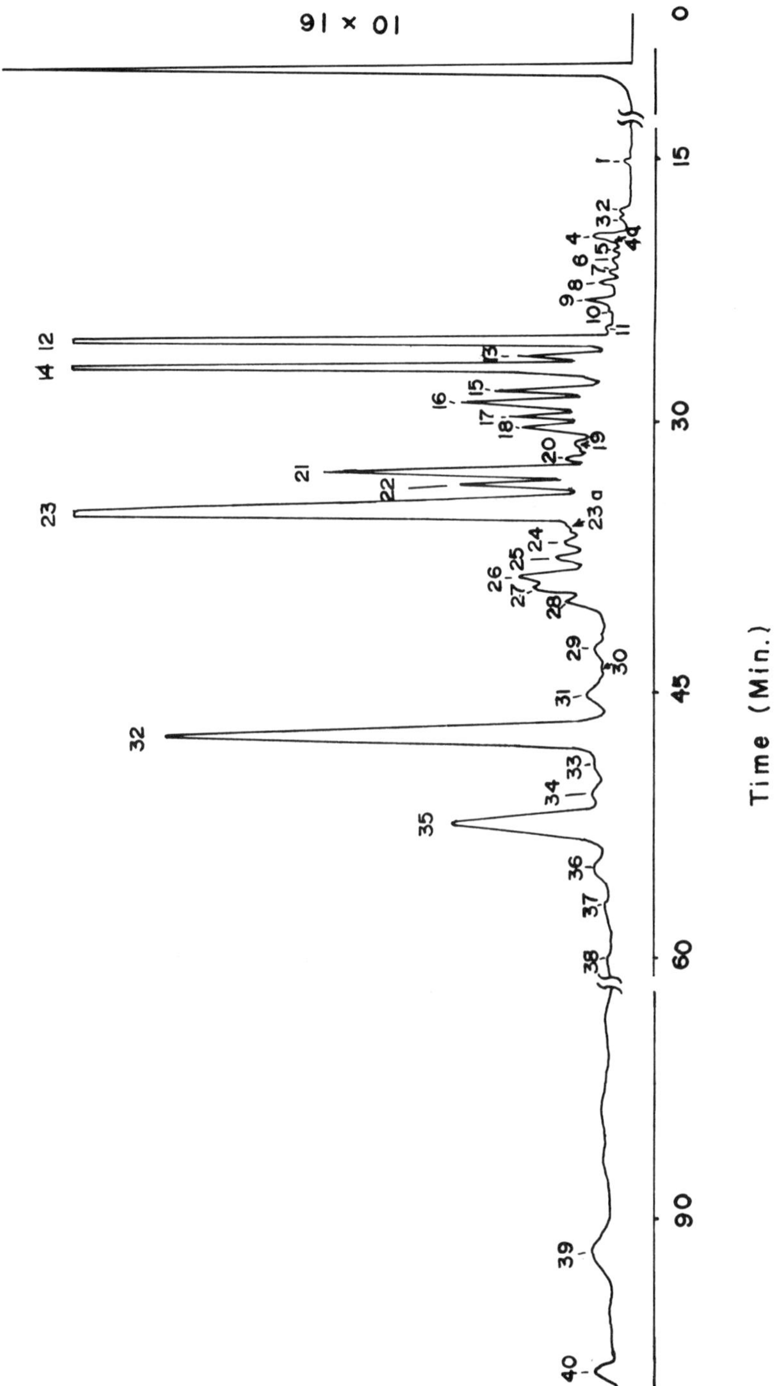

FIGURE 1. Gas chromatogram of airborne particulate on Dexsil-300 packed column.

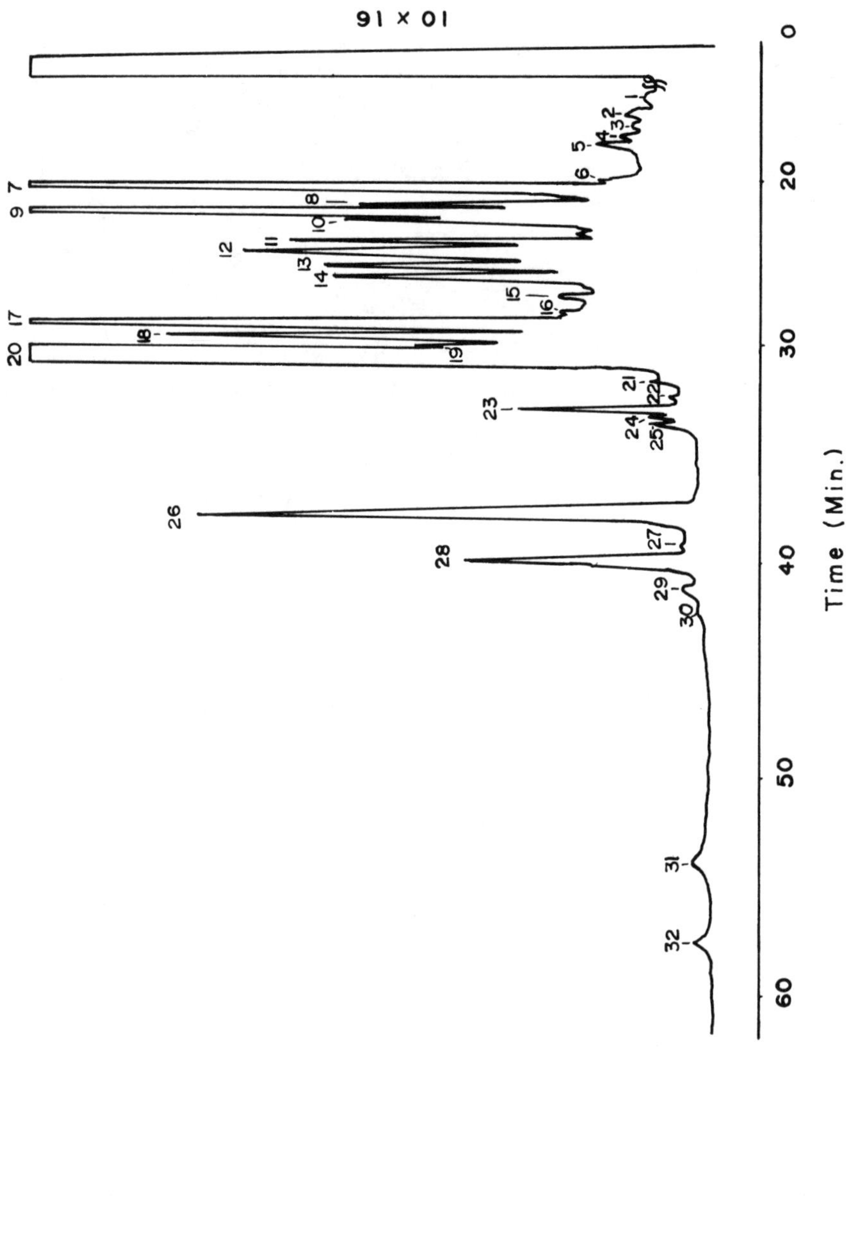

FIGURE 2. Gas chromatogram of airborne particulate on Dexsil-300 stainless-steel WCOT column.

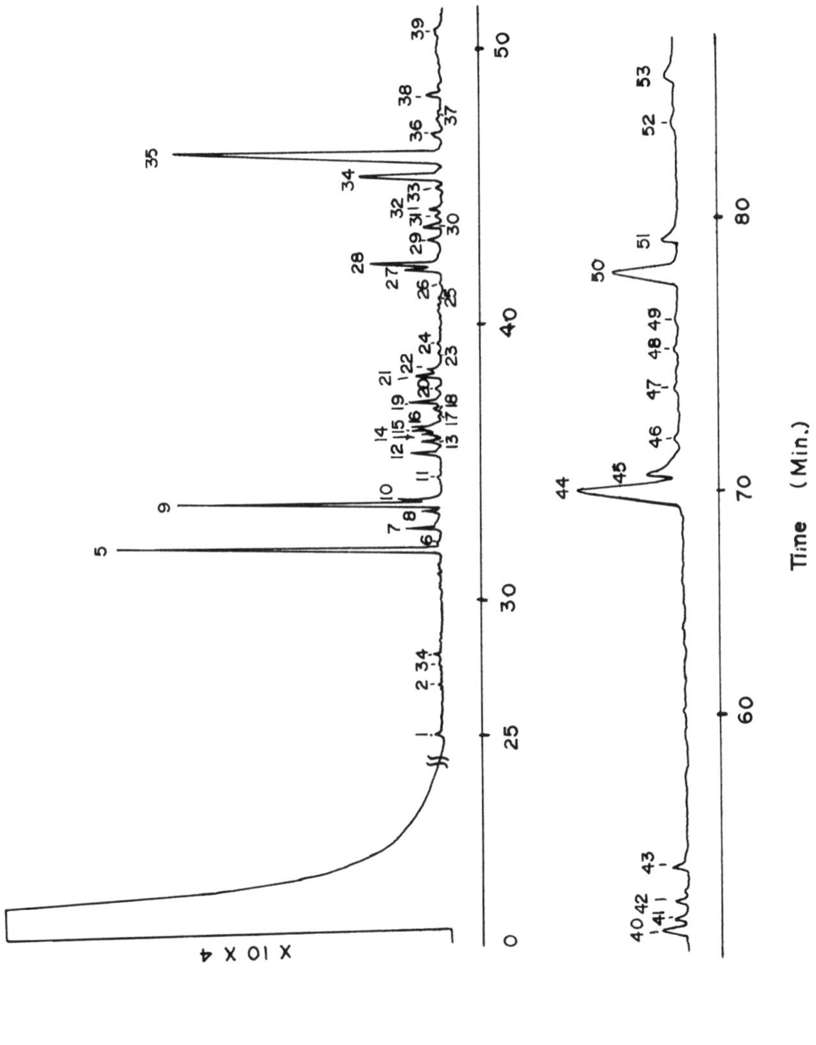

FIGURE 3. Gas chroamtogram of airborne particulate on Dexsil-300 glass capillary column.

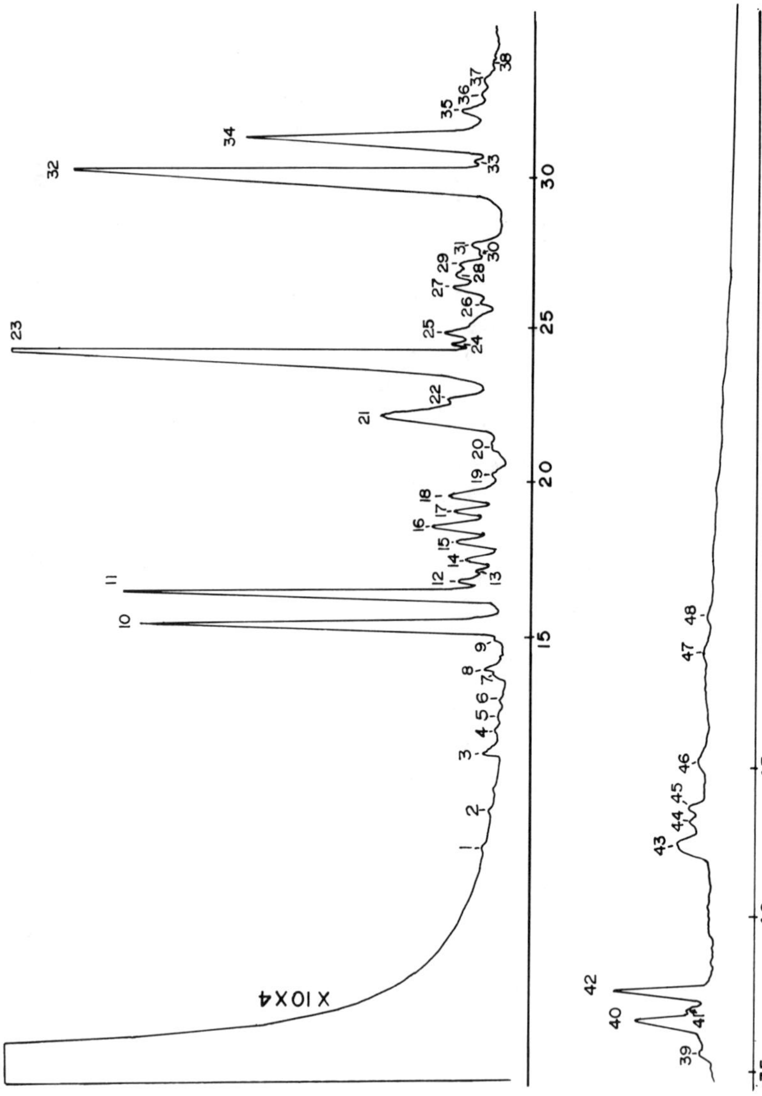

FIGURE 4. Gas chromatogram of airborne particulate on SE-52 glass capillary column.

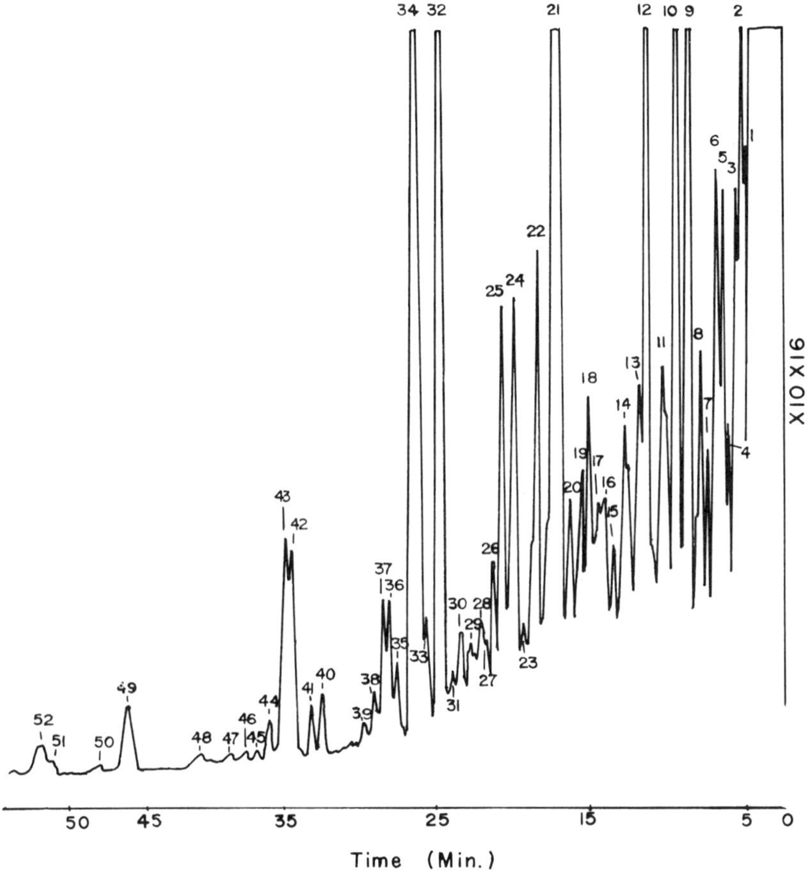

FIGURE 5. Gas chromatogram of process sludge on Dexsil-300 packed column.

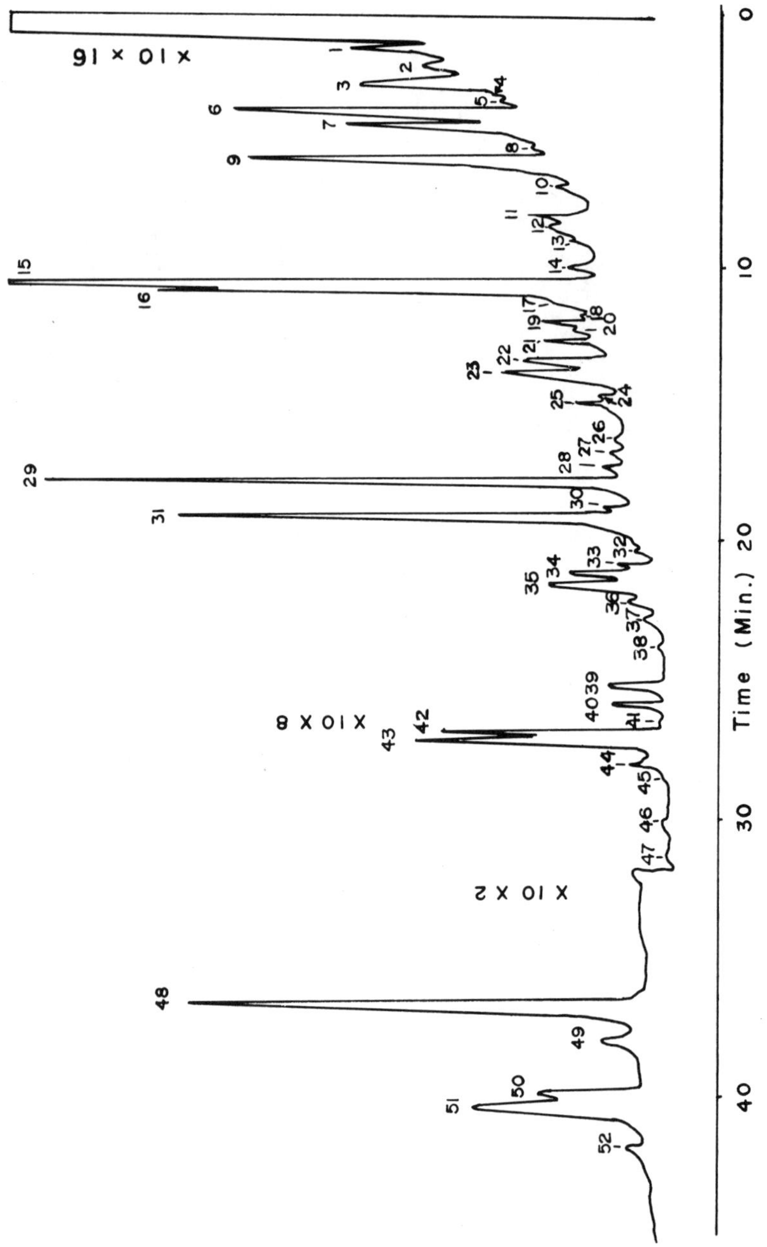

FIGURE 6. Gas chromatogram of process sludge on Dexsil-300 stainless-steel WCOT column.

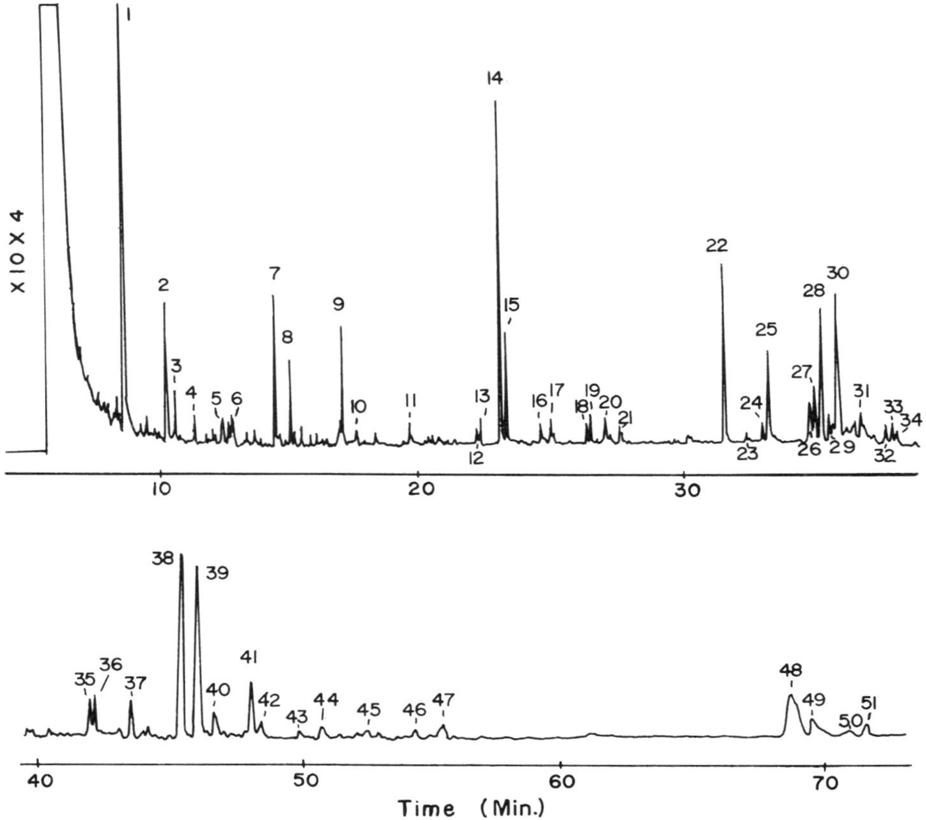

FIGURE 7. Gas chromatogram of process sludge on Dexsil-300 glass capillary column.

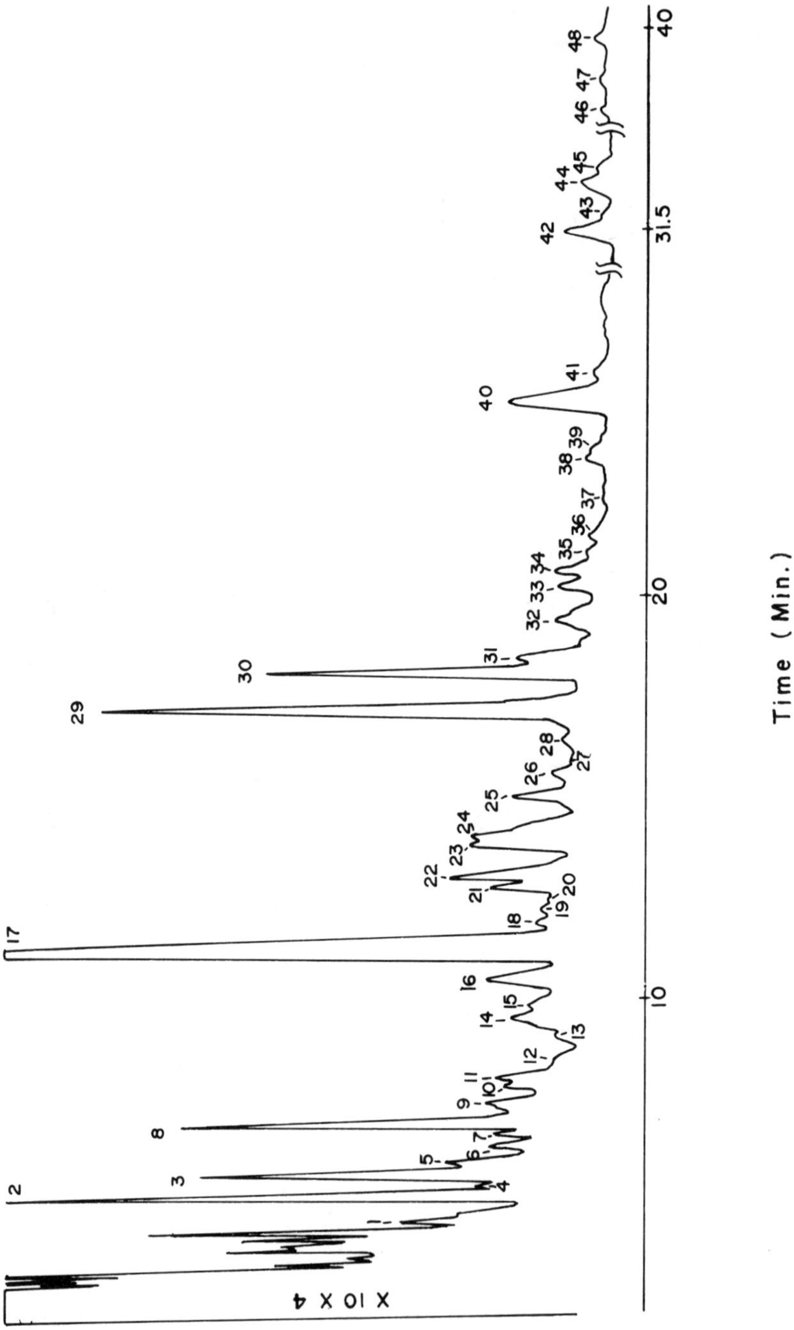

FIGURE 8. Gas chromatogram of process sludge on SE-52 glass capillary column.

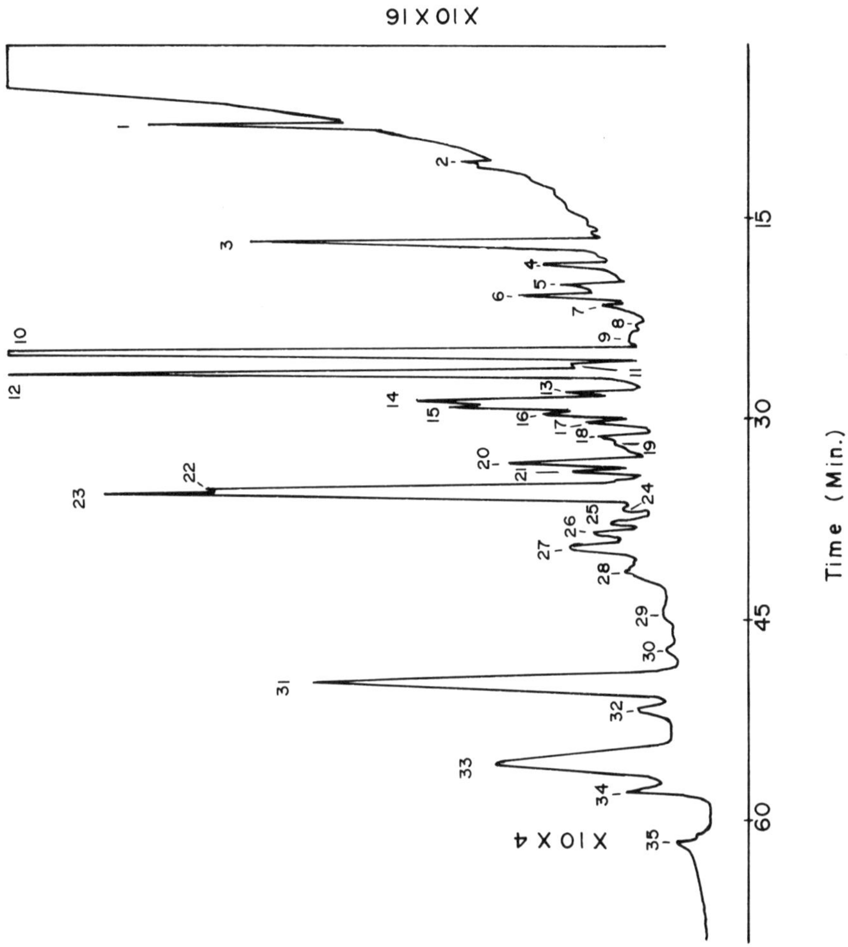

FIGURE 9. Gas chromatogram of coal tar volatile on Dexsil-300 packed column.

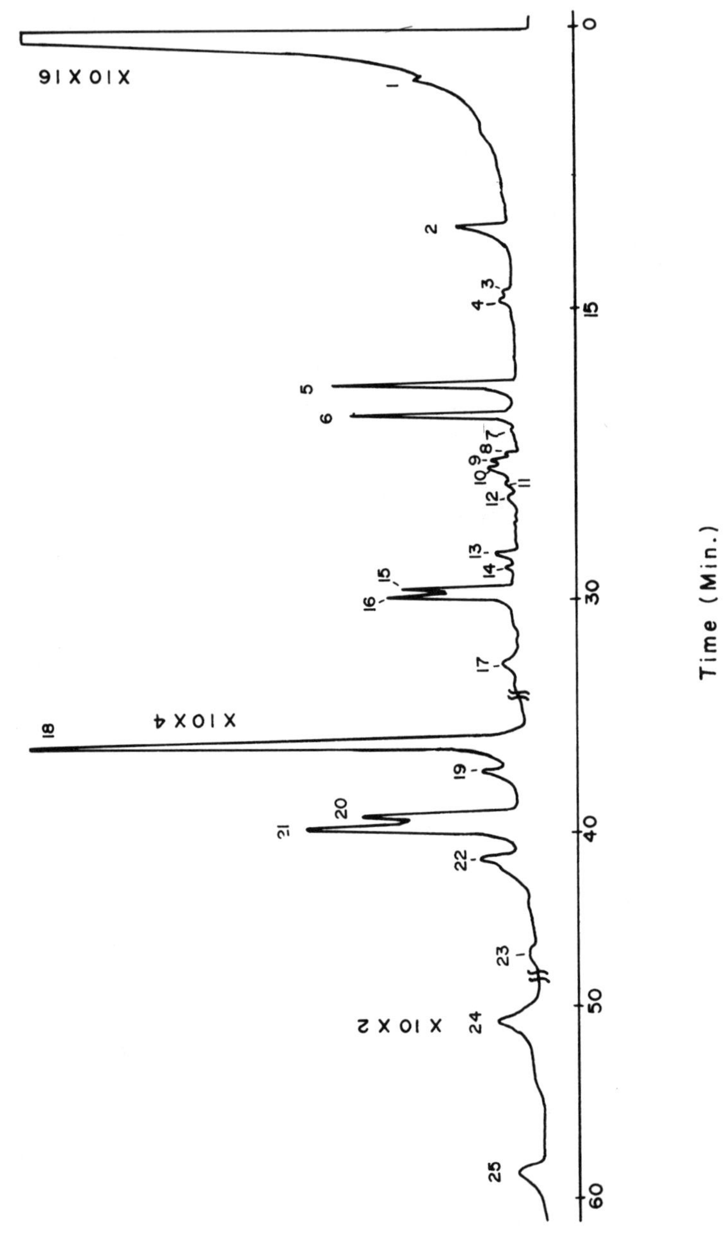

FIGURE 10. Gas chromatogram of coal tar volatile on Dexsil-300 stainless-steel WCOT column.

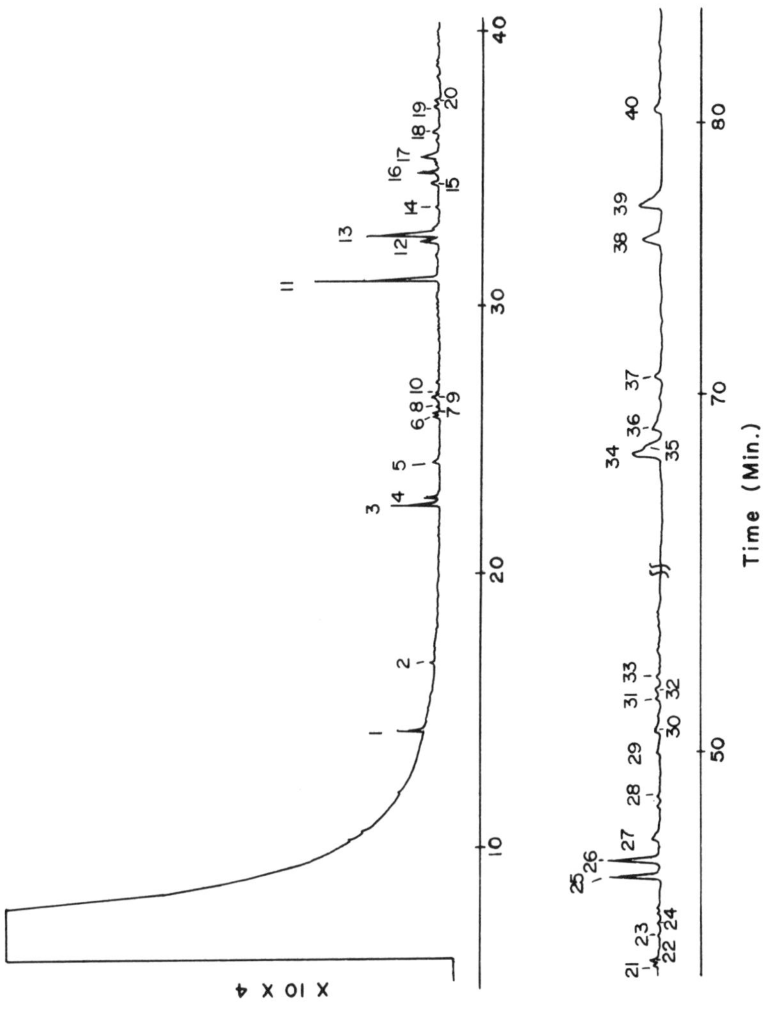

FIGURE 11. Gas chromatogram of coal tar volatile on Dexsil-300 glass capillary column.

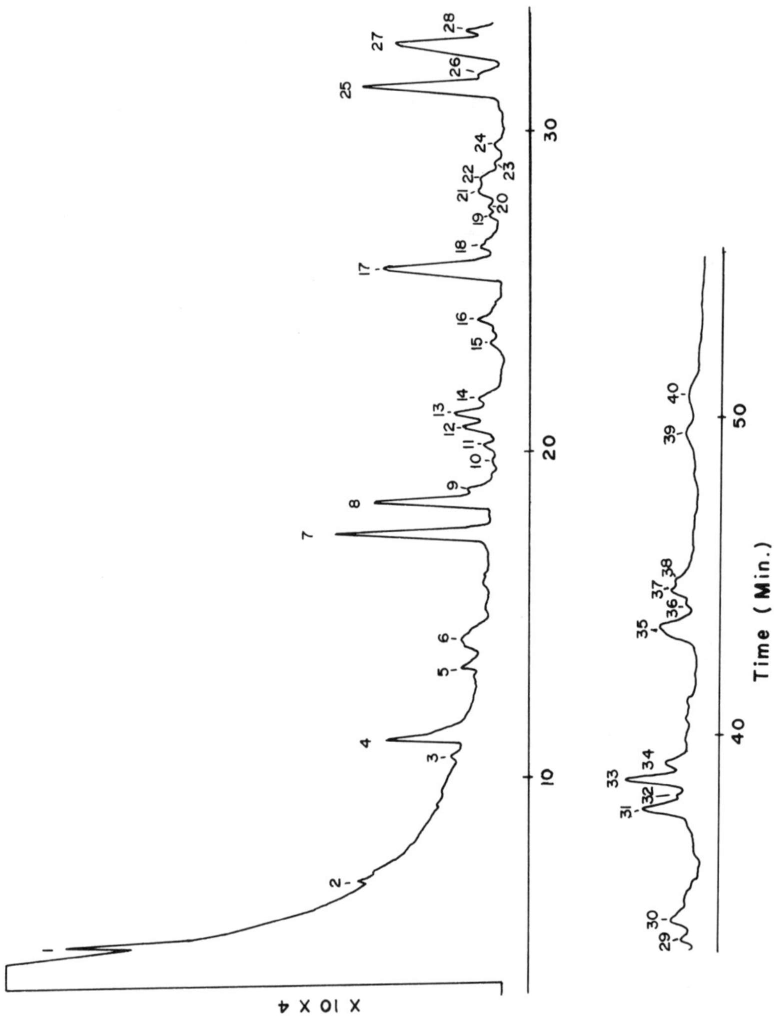

FIGURE 12. Gas chromatogram of coal tar volatile on SE-52 glass capillary column.

TABLE II. PAH in Airborne Particulate

Compounds	Peak No.				Concentration[a] (μg)
	Dexsil-300			SE-52 Glass Capillary	
	Packed	WCOT	Glass Capillary		
Fluorene	1	––	1	1	0.0486
Dihydrophenanthrene	2	1	––	––	0.0173
Dihydroanthracene	2	1	––	––	0.0116[b]
Methylfluorenes	3	––	––	––	0.0265
Phenanthrene/Anthracene	4	2	4	3	0.0879
Benzoquinoline	5	3	––	4	0.0336
Methylphenanthrene	6	––	––	5	0.0228
Methylanthracene	7	––	––	6	0.0735
Ethylphenanthrene/ Dimethylphenanthrene	8	4	––	7	0.2104
Ethylanthracene/ Dimethylanthracene	9	5	––	8	0.2831
Dihydrofluoranthene	10	––	––	––	0.0723
Dihydropyrene	11	6	––	9	0.0654
Fluoranthene	12	7	5	10	7.3621
Dihydrofluorenes	13	8	7,8	––	0.9452
Pyrene	14	9	9	11	8.0477
Benzo[a] fluorene	15	11	12	14	0.6543
Benzo[b&c] fluorene	16	12	13,14	15,16	1.0872
Methylfluoranthene	17	13	15,19	17	1.7322
Methylpyrene	18	14	20,24	18	1.8579
Trimethylfluoranthene/Trimethylpyrene	19	15	25,26	19	0.2461
Dihydrobenzo[c] phenanthrene	20	16	––	20	0.1874
Benzo[c] phenanthrene	21	17	27	21	3.2653
Benzo[g,h,i] fluoranthene	22	18	28	22	2.4781

TABLE II, continued

Compounds	Peak No.				Concentration[a] (μg)
	Dexsil-300			SE-52	
	Packed	WCOT	Glass Capillary	Glass Capillary	
Dihydrobenz[a]anthracene/ Dihydrochrysene	––	––	29,33	––	0.2738
Benz[a]anthracene	23	19	34	23	2.1642[c]
Chrysene/Triphenylene	23	20	35	23	6.5831
Methylbenz[c]anthracene	23[b]	21	36	24	0.0472
Methyltriphenylene	24	22	37	25	0.2658
Methylchrysene	25-27	23,24	38	26-29	2.0411
Dimethylbenz[a]anthracene/ Chrysene/Triphenylene	29,30	––	40-43	––	0.3062
Benzo[j]fluoranthene	31	––	––	––	0.4147
Benzo[k]fluoranthene/ Benzo[b]fluoranthene	32	26	44	32	3.8528
Methylbenzo[k]fluoranthene/ Benzo[b]fluoranthene	34	––	46	––	0.1873
Benzo[e]pyrene	35	27	49	33	0.1352[c]
Benzo[a]pyrene	35	28	50	34	2.2631
Perylene	36	29	51	35	0.2714
Coronene	39	31	52	40	0.6067
Dibenzpyrene	40	32	53	42	0.5384

[a]Calculation based on the response on packed column with fluoranthene response to be 0.2736 μg/unit area.
[b]Calculated from WCOT column. R_F of fluoranthene is 0.5873 μg/unit area.
[c]Calculated from D-300 glass capillary column. R_f of fluoranthene is 0.7462 μg/unit area.

TABLE III. PAH in Process Sludge

Compounds	Peak No.				Concentration[a] (g/l of raw sludge)
	Dexsil-300			SE-52 Glass Capillary	
	Packed	WCOT	Glass Capillary		
Biphenyl	6	1	2	——	1.86
Octahydrophenanthrene/ Octahydroanthracene	7	2	3	1	0.43
Dihydrofluorene	8	3	5,6	——	0.95
Methylbiphenyls	10,11	6,7	7,8	2,3	4.75
Benzidene	12,13	9	9	8	7.31
Fluorene	14,15	10	10	9	2.68
Dihydrophenanthrene	16	11	11	10	0.62
Dihydroanthracene	17	12	12	11	0.58
2-Methylfluorene	18	13	13	13	1.67
1-Methylfluorene	19	——	——	14	1.14
9-Methylfluorene	20	——	——	15	1.08
Phenanthrene/Anthracene	21	15,16	14,15	16,17	11.57
Benzoquinoline	22	17	16,17	18,19	4.69
Fluorene Carbonitrile	23	19,21	18,19	21,22	0.18
Methylphenanthrene	24	22	22	23	3.17
Methylanthracene	25	23	25	24	2.98
Ethylphenanthrene/ Dimethylphenanthrene	26,27	24,25	——	25	0.88
Ethylanthracene/ Dimethylanthracene	28,29	26	——	26	0.64
Dihydrofluoranthene	30	27	26	27	0.72
Dihydropyrene	31	28	27	28	0.21
Fluoranthene	32	29	28	29	9.65
Dihydrofluorenes	33	30	29	30	0.89

TABLE III, continued

Compounds	Peak No.				Concentration[a] (g/l of raw sludge)
	Dexsil-300			SE-52 Glass Capillary	
	Packed	WCOT	Glass Capillary		
Pyrene	34	31	30	31	9.17
Dihydrobenzofluorenes	35	33	31	33	0.55
Methylfluoranthene	36	34	35	34	1.24
Methylpyrene	37,38	35,36	36	35	1.88
Trimethylfluoranthene/ Trimethylpyrene	39	37,38	––	36,37	0.24
Benzo[c] phenanthrene	40	39	––	38	0.62
Benzo[g,h,i] fluoranthene	41	40	37	39	0.61
Benz[a] anthracene	42	42	38	40	1.86
Chrysene/Triphenylene	43	43	39	40,41	1.91
Methylbenz[c] anthracene	44	44	40	––	0.39
Methyltriphenylene	45	45	41,42	––	0.11
Methylchrysene	46	46	43,44	––	0.11
Dimethylbenz[a] anthracene/ Chrysene/Triphenylene	47,48	47	47	––	0.23
Benzo[k] fluoranthene/ Benzo[b] fluoranthene	49	48	48	42	1.27
Methylbenzo[k] fluoranthene/ Benzo[b] fluoranthene	50	49	49	43	0.18
Benzo[e] pyrene	51	50	50	44	0.22
Benzo[a] pyrene	52	51	51	44,45	0.40

[a]See Table II.

TABLE IV. PAH in Coal Tar Volatile

Compounds	Peak No. Dexsil-300 Packed	Dexsil-300 WCOT	Dexsil-300 Glass Capillary	SE-52 Glass Capillary	Concentration[a] (μg/ml prepared sample)
Fluorene	1	1	1	1	1.73
Methylfluorenes	2	––	2	2	0.26
Phenanthrene/Anthracene	3	2	3,4	4	4.85
Fluorene Carbonitrile	4	––	5	––	0.86
Methylphenanthrene	5	3	6	5	0.72
Methylanthracene	6	4	7	6	1.13
Ethylphenanthrene/ Dimethylphenanthrene	7	––	8	––	0.15
Ethylanthracene/ Dimethylanthracene	7	––	8	––	0.15
Dihydrofluoranthene	8	––	9	––	0.08
Dihydropyrene	9	––	10	––	0.16
Fluoranthene	10	5	11	7	11.74
Dihydrofluorenes	11	––	12	––	0.88
Pyrene	12	6	13	8	6.53
Dihydrobenzofluorenes	13	8	14	10,11	0.92
Methylfluoranthene	14	9	15	12	2.74
Methylpyrene	15	10	16	13	2.65
Trimethylfluoranthene/ Trimethylpyrene	16,17	11	17,18	14	1.28
Dihydrobenzo[c] phenanthrene	18	––	19	––	0.44
Benzo[c] phenanthrene	19	12	21	––	0.12
Benzo[g,h,i] fluoranthene	20	13	22	15	1.65
Dihydrobenz[a] anthracene/ Dihydrochrysene	21	14	––	––	1.03
Benz[a] anthracene	22	15	25	16	1.76

TABLE IV, continued

Compounds	Peak No.				Concentration[a] (μg/ml prepared sample)
	Dexsil-300			SE-52 Glass Capillary	
	Packed	WCOT	Glass Capillary		
Chrysene/Triphenylene	23	16	26	17	4.53
Dihydromethylbenz[c] anthracene/ Chrysene/Triphenylene	24	––	––	18,19	0.21
Methylbenz[c] anthracene	25	––	27	20	0.40
Methyltriphenylene	26	––	28,29	21	0.48
Methylchrysene	27	––	31,32	22	0.92
Dimethylbenz[a] anthracene/ Chrysene/Triphenylene	28,29	17	33	23	0.30
Benzo[j] fluoranthene	30	––	––	24	0.18
Benzo[k] fluoranthene/ Benzo[b] fluoranthene	31	18	34,35	25,26	5.83
Methylbenzo[k] fluoranthene/ Benzo[b] fluoranthene	32	19	36,37	––	0.77[b]
Benzo[e] pyrene	33	20	38	27	1.86
Benzo[a] pyrene	33	21	39	27,8	1.97[b]
Perylene	34	22	40	29	0.85
Coronene	––	24	––	31	––
Dibenzpyrene	––	25	––	33	––

[a]See Table II.
[b]See Table II.

response to aliphatic compounds. The development of a scheme based on extraction with DMSO provides a straightforward and high-efficiency isolation of PAH compounds (17). Table V shows the typical PAH recoveries using the DMSO method.

TABLE V. Percentage Recovery of PAH Using DMSO

Compound	Natusch *et al.* (17)	Present Work
Fluoranthene	––	98
Pyrene	99.9	99
Chrysene	99.9	97
B[a]P	––	98
Perylene	––	97

CONCLUSIONS

The trace analysis of PAH in environmental samples is not an easy task and it has been stated (2) that "no existing method separates and resolves adequately the entire PAH fraction on a mg to μg scale. Even in petroleum analysis, where large samples are available, a complete analytical resolution of the PAH fraction exceeds the capability of any existing combination of analytical techniques."

In spite of sensitivity limitations, vapor pressure and thermal stability (9), GC remains the primary method for the separation of PAH from environmental samples. A continuing series of refinements to column design, packings and coatings coupled with advances in preconcentration and UV detection technologies will redefine current resolution and sensitivity limitations. Investigations are continuing into the correlation of connectivity constants and retention indices (10,11) for a variety of column packings to develop a prediction technique for unknown PAH identifications (12). Also a "fingerprint" screening method which will allow onsite PAH analyses at concentrations in the subnanogram range using portable sampling/GC analytical systems (1) is under investigation.

REFERENCES

1. Blihar, R. W., and Amey, G. C. (1978): Evaluation of a new method to determine adsorption/Desorption efficiencies of collector tubes utilizing programmed thermal desorption. Paper No. 44, *American Industrial Hygiene Conference Abstract,* May, Los Angeles.
2. Blumer, M., and Giger, W. (1974): Polycyclic aromatic hydrocarbons in the environment. *Anal. Chem.* 46:1633-1671.

3. Brocco D., Zoccolillo, L., and Liberti, A. (1972): Determination of PAH in air by GC with high efficiency packed columns. *Atmos. Environ.* 6:715-720.
4. Ettre, L. S., (1973): *Open Tubular Columns.* Perkin-Elmer Corp., Norwalk, CT.
5. Gammage, R. B., Vo-Dinh, T., Hawthorne, A. R., Thorngate, J. H., and Parkinson, W. (1978): A new generation of monitors for PAH from synthetic fuel production. In: *Carcinogenesis, Vol. 3,* P. W. Jones and R. I. Freudenthal, Eds., pp. 155-173, Raven Press, New York.
6. Hertz, H. S., May, W. E., Wise, S. A., and Chesler, S. N. (1978): Trace organic analysis. *Anal. Chem.* 50:428A-436A.
7. Hluchan, E., Jenik, H., and Maly, E. (1974): Determination of airborne PAH by paper chromatography. *J. Chromatog.* 91:531-538.
8. Jennings, W. (1978): *Gas Chromatography with Glass Capillary Columns,* Academic Press, New York.
9. Kaiser, R. E. (1974): Gas chromatography in environmental analysis. *J. Chromatog. Sci.* 12:36-39.
10. Kaliszan, R., and Lamparczyk, H. (1978): A relationship between the connectivity indices and retention indices of PAH. *J. Chromatog.* 16:246-248.
11. Kier, L. B., and Hall, L. H., (1976): *Molecular Connectivity in Chemistry and Drug Research.* Academic Press, New York.
12. Lao, R. C., and Thomas, R. S. (1979): Determination of particulate organic matter in environmental samples by GC/MS and HPLC. In: *Proceedings International Congress on Analytical Techniques in Environmental Chemistry,* Pergamon Press, New York.
13. Lao, R. C., Thomas, R. S., and Monkman, J. L. (1975): Computerized GC-MS analysis of PAH in environmental samples. *J. Chromatog.* 112:681-700.
14. Lao, R. C., Thomas, R. S., and Monkman, J. L. (1976): Application of GC-MS to the analysis of PAH in environmental samples. In: *Carcinogenesis, Vol. 1,* R. F. Freudenthal and P. W. Jones, Eds., pp. 271-281, Raven Press, New York.
15. Lao, R. C., Thomas, R. S., Oja, H., and Dubois, L. (1973): Application of GC-MS-computer combination to the analysis of PAH. *Anal. Chem.* 45:908-915.
16. Lao, R. C., Thomas, R. S., Wang, D. T., Robinson, D., and Sakuma, T. (1977): Determination of PAH in atmospheric particulate matter by GC/MS and HPLC. In: *Carcinogenesis, Vol. 3,* P. W. Jones and R. F. Freudenthal, Eds., pp. 9-19. Raven Press, New York.
17. Natusch, D. F. S., and Tomkins, B. A. (1978): Isolation of polycyclic organic compounds by solvent extraction with dimethyl sulfoxide. *Anal. Chem.* 50:1429-1434.
18. Novotny, M., Lee, M. L., and Bartle, K. D. (1976): GC/MS and NMR determination of PAH in airborne particulates. *Anal. Chem.* 48:1566-1572.
19. Onuska, F. I., Comba, M. E., Bistricki, T., and Wilkinson, R. J. (1977): Preparation of surface-modified wide-bore WCOT columns. *J. Chromatog.* 142:117-125.

Biology

Session Chairmen

P. Leber
D. L. Berry
A. W. Wood
E. Cavalieri

Polynuclear Aromatic Hydrocarbons, edited by P.W. Jones and P. Leber. Ann Arbor Science Publishers, Inc. Ann Arbor, MI ©1979.

COMPARATIVE METABOLISM OF A SERIES OF POLYCYCLIC AROMATIC HYDROCARBONS BY RAT LIVER MICROSOMES AND PURIFIED CYTOCHROME P-450

D. R. Thakker, M. Nordqvist, H. Yagi, W. Levin,*
D. Ryan,* P. Thomas,* A. H. Conney* and D. M. Jerina

Section on Oxidation Mechanism
Laboratory of Bioorganic Chemistry
National Institute of Arthritis,
Metabolism and Digestive Diseases
National Institutes of Health, Bethesda, Maryland 20205

*Department of Biochemistry and Drug Metabolism
Hoffmann-La Roche Inc.
Nutley, New Jersey 07110

INTRODUCTION

Polycyclic aromatic hydrocarbons have been known to be carcinogens for almost fifty years. The first evidence for their metabolic activation to chemically reactive species was provided in 1951 by E. C. Miller who showed that metabolites of benzo[a]pyrene (B[a]P)[1] bind covalently to skin proteins after application of the hydrocarbon on mouse skin (22). B[a]P, a

[1]Abbreviations used are: B[a]P, benzo[a]pyrene; B[e]P, benzo[e]pyrene; BA, Benzo [a]anthracene; DBA, dibenzo[a,h]anthracene; MC, 3-methylcholanthrene, B[a]P 7,8-dihydrodiol, *trans*-7,8-dihydroxy-7,8-dihydro B[a]P; B[a]P 4,5- and 9,10-dihydrodiols, other dihydrodiols of B[a]P; dihydrodiols of B[e]P, BA and DBA have been abbreviated in a similar manner; B[a]P 7,8-diol-9,10-epoxide-1, (±)-7β,8α-dihydroxy-9β,10β-epoxy-7,8,9,10-tetrahydro B[a]P; 7,8-diol-9,10-epoxide-2, (±)-7β,8α-dihydroxy-9α,10α-epoxy-7,8,9,10-tetrahydro B[a]P; B[a]P 7,8-diol-9,10-epoxides, both of the diastereomeric diol epoxides; B[a]P 9,10-diol-7,8-epoxides, B[e]P 9,10-diol-11,12-epoxides, BA 8,9-diol-10,11-epoxides, BA 3,4-diol-1,2-epoxides and DBA 3,4-diol-1,2-epoxides are abbreviated in the same manner; 1-HOBA, 1-hydroxy BA; 3-, 4-, 5-, 6-, 8-, 9- and 11-HOBA, other phenols of BA; B[e]P H_4-9,10-epoxide,9,10-epoxy-9,10,11,12-tetrahydro B[e]P.

ubiquitous environmental contaminant, is the hydrocarbon about which the most is known of this metabolic activation process. Hepatic microsomal monooxygenases and epoxide hydrase convert B[a]P into (+)-7β,8α-dihydroxy-9α,10α-epoxy-7,8,9,10-tetrahydro B[a]P (7,13,16,29,41) which is the only known ultimate carcinogenic metabolite of the hydrocarbon (2).

Based on the high biological activity of the diastereomeric B[a]P 7,8-diol-9,10-epoxides and the lack or decrease of activity observed on alkylation or halogenation of the critical benzo ring in selected hydrocarbons, the "bay-region theory" was proposed in order to correlate the physico-chemical properties of the polycyclic hydrocarbons with their biological activity (8,11, 12). The theory predicts that an epoxide on a saturated, angular benzo ring that forms a part of the bay-region (cf. Figure 1) of the hydrocarbon should be highly chemically reactive and presumably highly biologically active. Thus, the bay-region diol epoxides of polycyclic hydrocarbons are prime candidates for ultimate carcinogenic metabolites. In addition to the intrinsic reactivity of bay-region diol epoxides for a given hydrocarbon, the carcinogenicity of the parent hydrocarbon also depends on the extent to which these diol epoxides are metabolically formed. The bay-region theory makes no attempt to take metabolic factors into account. In this chapter, metabolism of several selected carcinogenic polycyclic hydrocarbons (Figure 1) and their derivatives will be discussed with a special emphasis on the role of metabolism in the carcinogenicity of these hydrocarbons.

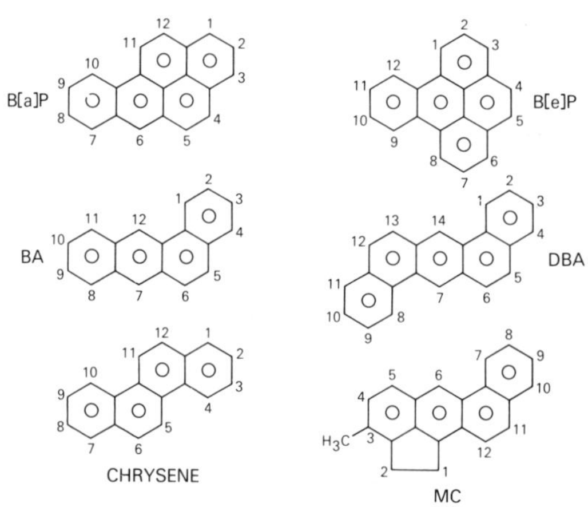

FIGURE 1. Structures of selected polycyclic aromatic hydrocarbons. The bay region in B[a]P, for example, consists of the sterically hindered area between the 10- and 11- positions of the hydrocarbon.

BENZOPYRENES

B[a]P is metabolized by rat liver monooxygenases to the 4,5-, 7,8- and 9,10-oxides. These oxides are further metabolized by epoxide hydrase to 4,5- 7,8- and 9,10-dihydrodiols. In addition, several phenols and quinones are also formed as metabolites of B[a]P (6). At low conversion of B[a]P, only the dihydrodiols, phenols and quinones are detected as metabolites. However, at high conversion of the substrate (77%), small but significant amounts of tetraols, formed by spontaneous solvolysis of B[a]P 7,8-diol-9,10-epoxides, are also detected (Figure 2) (29). The presence of the tetraols is indicative of the secondary metabolism of B[a]P to its 7,8-diol-9,10-epoxides via B[a]P 7,8-dihydrodiol.

Metabolism of B[a]P to its 7,8-dihydrodiol is highly stereoselective. Liver microsomes from control, phenobarbital-treated and MC-treated rats produce 92-96% of the (-)-7R,8R enantiomer (13,29,31,41). The regiospecificity of liver microsomes in the metabolism of B[a]P 7,8-dihydrodiol varies with pretreatment. Microsomes from control and phenobarbital-treated rats form approximately 40% of a phenolic metabolite thought to be hydroxylated at the 6-position in addition to the 7,8-diol-9,10-epoxides, whereas 85-95% of the metabolites formed by microsomes from MC-treated rats are derived from B[a]P 7,8-diol-9,10-epoxides (29). Microsomes from MC-treated rats also show high stereospecificity in the conversion of the (+)- and (-)-enantiomers of B[a]P 7,8-dihydrodiol into the diastereomeric 7,8-diol-9,10-epoxides (29). Metabolism of the (-)-enantiomer produces mainly B[a]P 7,8-diol-9,10-epoxide-2 relative to isomer-1 (6 to 1), whereas the (+)-enantiomer is almost exclusively metabolized to B[a]P 7,8-diol-9,10-epoxide-1 (19 to 1). Since the B[a]P 7,8-dihydrodiol formed by microsomal metabolism of B[a]P consists primarily of the (-)-enantiomer (96%), further metabolism by microsomes produces mainly (+)-B[a]P 7,8-diol-9,10-epoxide-2, the most tumorigenic of the four possible diol-epoxide stereoisomers in newborn mouse (2).

In analogy to B[a]P 7,8-dihydrodiol, the 9,10-dihydrodiol was anticipated to be metabolized to its B[a]P 9,10-diol-7,8-epoxides. However, metabolism of B[a]P 9,10-dihydrodiol by rat liver microsomes resulted in almost exclusive formation of a phenolic product with the new hydroxyl groups at the 1- and/or 3-position. Only trace amounts of metabolites which had pyrene-like UV chromophores were detected (Scheme 1) (30). Thus, the lack of formation of 9,10-diol-7,8-epoxides along with the low intrinsic mutagenic activity of these diol epoxides explains why B[a]P 9,10-dihydrodiol is inactive as a carcinogen (30). These results indicate that the hydroxyl groups in the bay region of B[a]P 9,10-dihydrodiol may direct further metabolism of this dihydrodiol toward the distant end of the molecule. Presumably, the predominant quasi-diaxial conformation of the hydroxyl groups in B[a]P 9,

10-dihydrodiol, as opposed to the predominant quasi-diequatorial conformation of the hydroxyl groups in B[a]P 7,8-dihydrodiol, is responsible for directing the metabolism away from the double bond of the benzo ring.

A second example of the directing effect of bay-region hydroxyl groups is found in the metabolism of B[e]P 9,10-dihydrodiol (38). Microsomes from untreated or Aroclor-treated rats metabolized B[e]P 9,10-dihydrodiol to three phenolic derivatives and a tetraol, 4,5,9,10-tetrahydroxy-4,5,9,10-tetrahydro B[e]P, instead of the 9,10-diol-11,12-epoxides (Scheme 2) (15,

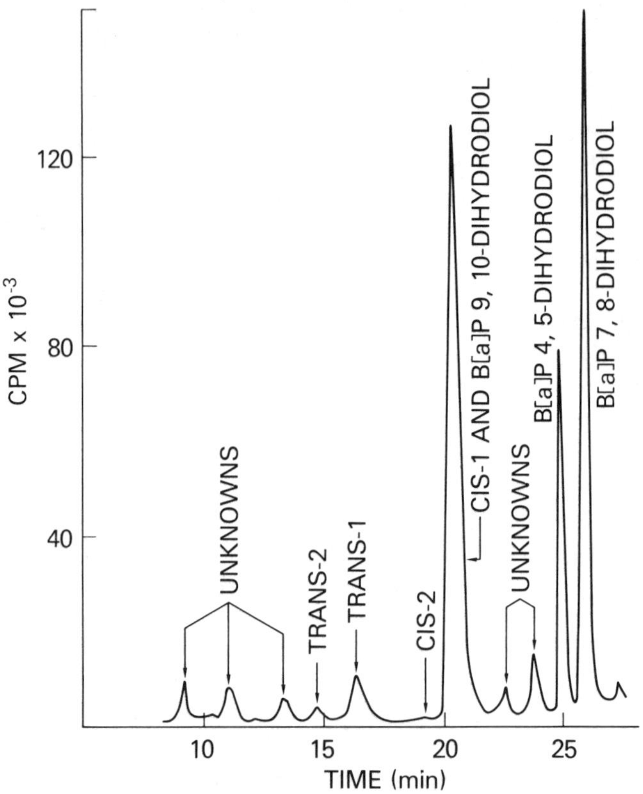

FIGURE 2. HPLC profile (dihydrodiol region only) of the metabolites formed on extensive metabolism of [^{14}C] B[a]P (77%) with microsomes from MC-treated rats (29). The metabolites were eluted on a Du Pont Zorbax ODS column (25 cm x 6.2 mm) with a linear gradient of 60% to 98% methanol in water at a rate of gradient change of 1%/min after 1 min delay and at a flow rate of 1.2 ml/min. *Trans*-1 and *cis*-1 are the tetraols formed by hydrolysis of B[a]P 7,8-diol-9,10-epoxide-1 upon a *trans* and *cis* attack of water at C-10, respectively. Similarly, *trans*-2 and *cis*-2 are the tetraols derived from B[a]P 7,8-diol-9,10-epoxide-2. Relative stereochemistry for the diol epoxides in the isomer -1 and -2 series is indicated in Scheme 2.

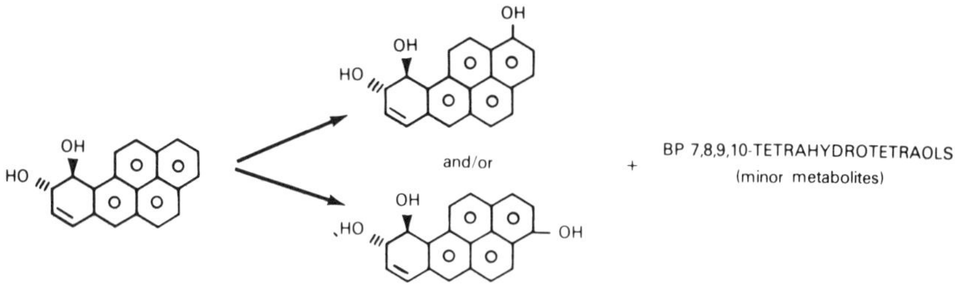

SCHEME 1. Metabolism of B[a]P 9,10-dihydrodiol by rat liver microsomes.

38). Based on these two examples (B[a]P 9,10-dihydrodiol and B[e]P 9,10-dihydrodiol), it is tempting to speculate that further metabolism of all benzo-ring dihydrodiols in which the hydroxyl groups are quasi-diaxial due to a proximate bay region will form little if any diol epoxides on that ring.

Molecular orbital calculations have predicted that B[e]P should have carcinogenic activity somewhere between that of the strong carcinogen B[a]P and the weak carcinogen chrysene based on moderate reactivity anticipated for the B[e]P 9,10-diol-11,12-epoxides (10). Although a B[e]P 9,10-diol-11,12-epoxide has not yet been synthesized in a pure state, the prediction is supported by the fact that B[e]P H_4-9,10-epoxide has strong inherent mutagenic activity (15,38). Nonetheless, B[e]P is inactive as a carcinogen (4), and

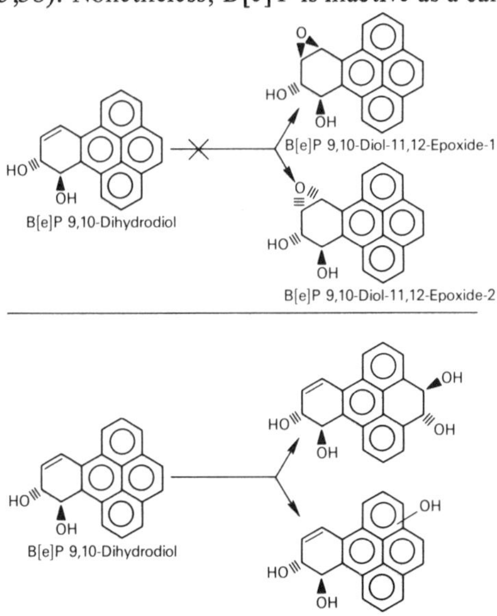

SCHEME 2. Metabolism of B[e]P 9,10-dihydrodiol by rat liver microsomes.

B[e]P 9,10-dihydrodiol undergoes little or no metabolic activation to bacterial mutagens by rat liver microsomes (15,38). Only part of the explanation stems from the failure of liver microsomes to form B[e]P 9,10-diol-11,12-epoxides from B[e]P 9,10-dihydrodiol. In addition to this, preliminary studies of the metabolism of B[e]P with a highly purified and reconstituted cytochrome P-450 system in the presence and absence of epoxide hydrase have indicated that at best only a trace amount of B[e]P 9,10-dihydrodiol is formed if, in fact, it is formed at all (21).

BENZO[a]ANTHRACENE

BA is considered to be a weak carcinogen (4). Biological studies with BA and its synthetic dihydrodiols and diol epoxides have indicated

(1) that BA 3,4-dihydrodiol with a bay-region 1,2-double bond is the most tumorigenic of the metabolically possible dihydrodiols of BA on mouse skin (36) and in the newborn mouse (33)

(2) that the bay-region BA 3,4-diol-1,2-epoxides are 15-125 times more mutagenic to *S. typhimurium* strain TA 100 and Chinese hamster V79 cells than are the other isomeric nonbay-region diol epoxides of BA (34), and

(3) that these bay-region diol-epoxides are highly tumorigenic on mouse skin (17,25).

Previous qualitative studies of the metabolism of BA by rat liver preparations with analysis by thin-layer chromatography have shown that BA 5,6- and 8,9-dihydrodiols are major metabolites (24). *In vivo* studies suggested the formation of small amounts of phenols as well as the 3,4- and 10,11-dihydrodiols (1). However, due to the lack of appropriate synthetic standards, the identification of minor metabolites was only tentative. Since it is essential to know the extent to which BA is converted to its 3,4-dihydrodiol, the metabolism of BA by rat liver microsomes and by a highly purified and reconstituted monooxygenase system in the presence and absence of epoxide hydrase has been reinvestigated with the aid of HPLC (28).

The chromatographic mobilities of the metabolites formed from [^{14}C] BA by liver microsomes from MC-treated rats are shown in Figure 3. Almost all the metabolites emerge from the column in two peaks; one which cochromatographs with BA 5,6-dihydrodiol and a second peak which cochromatographs with BA 3,4-,8,9- and 10,11-dihydrodiols. Only trace amounts of phenols are formed. BA 1,2-dihydrodiol was not detected at low conversion of BA (arrow a in Figure 3). At high conversion ($\sim$77%) of BA, three polar metabolite peaks appeared which cochromatographed with BA 1,2-dihydrodiol (arrow a), BA 8,9-diol-10,11-epoxide-1 (arrow b) and BA 8,9-diol-10,11-

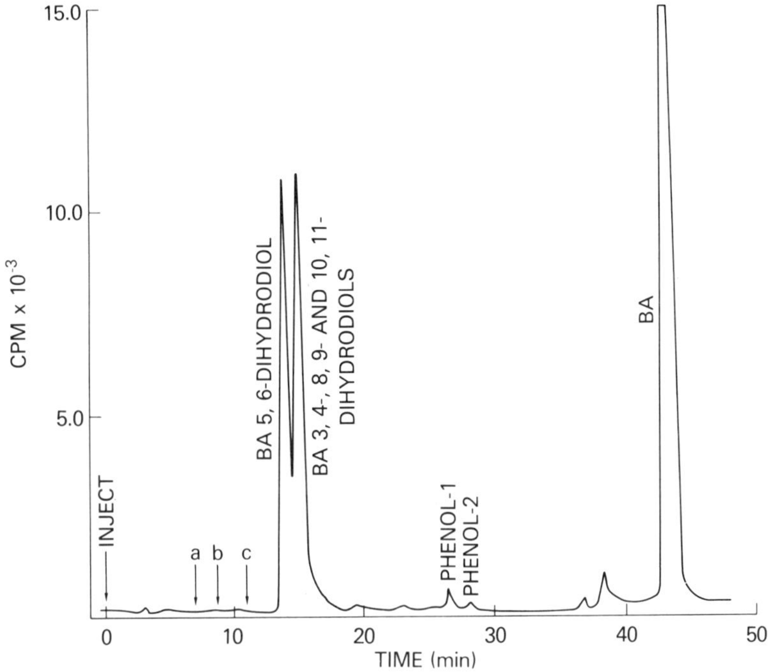

FIGURE 3. HPLC profile of metabolites formed from [^{14}C] BA on metabolism by microsomes from MC-treated rats (28). The metabolites were eluted from a Du Pont Zorbax ODS column (25 cm x 6.2 mm) with a linear gradient of 70% to 85% methanol in water at a rate of gradient change of 1%/min after an initial delay of 1 min at a flow-rate of 1.2 ml/min. The arrows (a, b and c) indicate the retention times of BA 1,2-dihydrodiol and the BA 8,9-diol-10,11-epoxides.

epoxide-2 (arrow c). Thus, even at a very high conversion of BA, only trace amounts of diol epoxides appear to be formed. The chromatographic peak which contains the BA 3,4-, 8,9- and 10,11-dihydrodiol standards was further analyzed (28). BA 10,11-dihydrodiol could be separated from BA 3,4- and 8,9-dihydrodiols on a Du Pont Zorbax Sil column. Separation of the BA 3,4- and 8,9-dihydrodiols was achieved on a Du Pont Zorbax ODS column after acetylation. BA 3,4-dihydrodiol could also be quantitated in the mixture of BA 3,4-, 8,9- and 10,11-dihydrodiols by fluorescence spectrometry. Thus, the conclusion of Yang *et al.* (39,40) that the 3,4-dihydrodiol of BA is not formed is untenable. The metabolites of BA formed by microsomes from MC-treated rats and by purified cytochrome P-448 in the presence and absence of epoxide hydrase are shown in Table I. With microsomes, BA 5,6- and 8,9-dihydrodiols are the two major metabolites with small amounts of BA 3,4-dihydrodiol and BA 10,11-dihydrodiols also detectable. Phenols are formed in extremely small amounts ($<$5% of total metabolites). This is in

TABLE I. Metabolites of BA Obtained with Liver Microsomes from MC-Treated Rats and with Purified Cytochrome P-448 from these Microsomes

Metabolites[a]	Microsomes	Cytochrome P-448 + Epoxide Hydrase	Cytochrome[c] P-448
BA 5,6-dihydrodiol	41.6	48.7	–
BA 3,4-dihydrodiol[b]	1.6	2.5	–
BA 8,9-dihydrodiol[b]	48.8	43.9	–
BA 10,11-dihydrodiol[b]	3.2	3.5	–
Phenol-1	3.9	1.4	28.8
Phenol-2	0.9	0.0	38.6
Total conversion	18.3 (6.94)	8.7 (86.80)	9.4 (94.43)
Recovery	91	81	78

[a]Metabolites are expressed as percent of the total metabolism. The numbers in the parentheses indicate the conversion expressed as nmol products formed/nmol hemeprotein/min. Total conversion was calculated from the total radioactivity above blank which emerges from the column prior to BA. Recovery is the percentage of radioactivity emerging before BA in discrete metabolite peaks. The microsomal incubation contained 0.1 mg protein/ml. Incubations with the reconstituted system used 0.005 nmol hemeprotein/ml (28).

[b]BA 3,4-dihydrodiol was determined in the mixture of BA 3,4-, 8,9- and 10,11-dihydrodiols by fluorescence spectrometry. BA 10,11-dihydrodiol was determined by HPLC on a Zorbax Sil column.

[c]On chromatography BA 5,6-oxide is partially decomposed into three broad peaks, as well as to 5- and 6-HOBA which chromatograph in phenol-1 peak (cf. Figure 3). The three broad peaks constitute 32.6% of the total metabolites.

contrast to microsomal metabolite profiles from B[a]P which contain substantial amounts (>25%) of phenols. As expected, the purified cytochrome P-448 system (0.005 nmol/ml) produced only phenols whereas it gave predominantly dihydrodiols in the presence of epoxide hydrase (Table I).

Unlike the metabolism of B[a]P, metabolism of BA with more than 0.005 nmol/ml of purified cytochrome P-448 occurs with substantial inhibition in the absence of epoxide hydrase (Figure 4). BA (50 nmol/ml) could not be more than 30% metabolized in the absence of epoxide hydrase, even at a cytochrome P-448 concentration of 0.1 nmol/ml; whereas with epoxide hydrase present, 77% of the substrate was metabolized. When metabolism of BA was measured in the presence of bovine serum albumin or epoxide hydrase which had been rendered catalytically inactive by heat denaturation or by the active-site-directed inhibitor *p*-nitrophenacyl bromide (5), "stimulation" of BA metabolism was not observed (Table II). Since catalytically active epoxide hydrase is required to enhance the rate of metabolism of BA,

TABLE II. Effect of Epoxide Hydrase on the Total Metabolism of BA by Purified Cytochrome P-448

Protein[a]	Percent of Maximum Metabolism
Cytochrome P-448 + epoxide hydrase	100
Cytochrome P-448	49
Cytochrome P-448 + heat denatured epoxide hydrase	52
Cytochrome P-448 + epoxide hydrase inactivated with *p*-nitrophenacyl bromide	50
Cytochrome P-448 + bovine serum albumin	46

[a]A cytochrome P-448 concentration of 0.05 nmol/ml was used in these experiments. For incubations with epoxide hydrase, 32 units of enzyme were used. Bovine serum albumin was used at a concentration of 25 μg/ml. Incubation conditions were as described elsewhere (28).

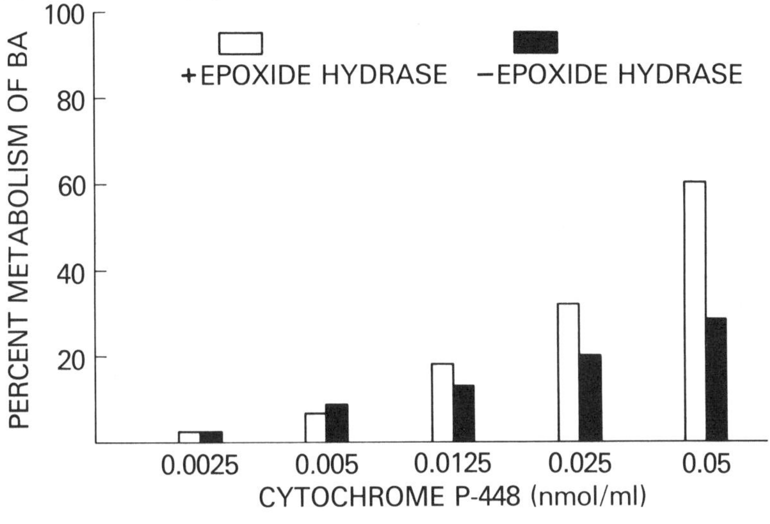

FIGURE 4. Extent of metabolism of [^{14}C] BA by varying amounts of purified cytochrome P-448 in the presence and absence of epoxide hydrase (28).

one or more phenolic metabolites or arene oxides must be inhibitory toward cytochrome P-448. Hence, BA 5,6-oxide and several BA phenols were tested as potential inhibitors of the cytochrome P-448 system (Table III). BA 5,6-oxide was found to be a poor inhibitor, whereas all of the phenols tested were potent inhibitors of the metabolism of BA. The most potent inhibitors among

TABLE III. Inhibition of Monooxygenases by BA Phenols and BA 5,6-Oxide

Inhibitor	Concentration (μM)	Percent Inhibition[a]
BA 5,6-oxide	5	10
1-HOBA	1	20
	5	48
3-HOBA	1	22
	5	50
4-HOBA	1	5
	5	31
5-HOBA[b]	1	58 (71)
	5	95 (89)
5-HOBA + 6-HOBA	5	88
8-HOBA	1	18
	5	40 (26)
9-HOBA	1	15
	5	35
11-HOBA	1	21
	5	68

[a]Percent inhibition was calculated with respect to the conversion of BA (50 μM) by cytochrome P-448 (0.05 μM) in the absence of epoxide hydrase (Figure 4). Numbers in parentheses represent second determinations.

[b]The sample contained predominantly 5-HOBA along with a trace of 6-HOBA.

the phenols tested were 5- and 6-HOBA. Since BA 5,6-oxide survives the incubation conditions, 5- and 6-HOBA could not be the inhibitory metabolites of BA unless they are formed by direct hydroxylation (32). The strong inhibition of hydrocarbon metabolism by phenolic products is thus far unique to BA. Presently it is not known whether the phenols of BA also act as nonspecific inhibitors of cytochrome P-450.

DIBENZO[a,h]ANTHRACENE

The HPLC profile of the metabolites formed from [^{14}C] DBA by microsomes from MC-treated rats is shown in Figure 5. DBA 3,4-dihydrodiol with the bay-region 1,2-double bond is the major metabolite formed. Small amounts of DBA 1,2-dihydrodiol were also detected. The K-region 5,6-dihydrodiol is formed at best in trace amounts (< 2%). The phenolic metabolites of DBA chromatograph as two peaks labeled as phenol-1 (18%) and phenol-2 (15%) in Figure 5. DBA 5,6-oxide also chromatographs with the phenol-1 peak; however, its absence in this peak was established by experiments with

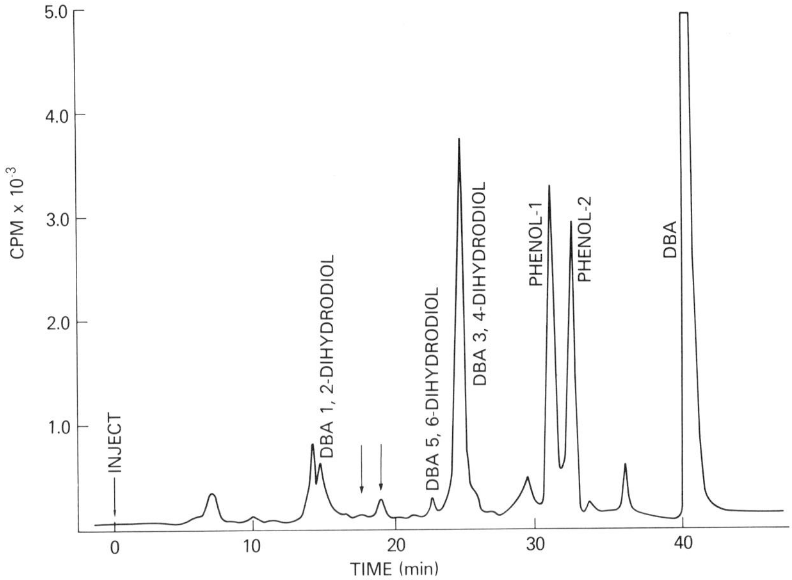

FIGURE 5. HPLC profile of metabolites formed from [^{14}C] DBA on metabolism by microsomes from MC-treated rats (23). The metabolites were eluted on a Du Pont Zorbax ODS column (25 cm x 6.2 mm) with a linear gradient of 70% methanol in water to 100% methanol at a rate of gradient change of 1%/min after an initial delay of 1 min and a flow rate of 1.2 ml/min.

a highly purified and reconstituted monooxygenase system in the presence and absence of epoxide hydrase (23). At high conversion of the substrate, at least two peaks were detected (arrows in Figure 5) which had BA-like chomophores and which cochromatographed with metabolites of DBA 3,4-dihydrodiol that also had BA-like chromophores. These metabolites are suspected to be tetraols derived from DBA 3,4-diol-1,2-epoxides.

When the metabolism of DBA by highly purified cytochrome P-448 was studied in the absence of epoxide hydrase, only the two phenol peaks and an unknown peak were detected in the HPLC profile. In the presence of epoxide hydrase, DBA 1,2-and 3,4-dihydrodiols were formed (11 and 17% of the total metabolites, respectively) at the expense of phenols in the phenol-1 and -2 fractions. Formation of DBA 5,6-dihydrodiol could not be clearly demonstrated. When increasing amounts of epoxide hydrase (21-167 units/ml)[2] were

[2]One unit of epoxide hydrase is defined as that amount of enzyme which forms 1 nmol of styrene glycol from styrene oxide per min.

added to the incubation medium, the formation of DBA 1,2-dihydrodiol continued to increase while the amount of DBA 3,4-dihydrodiol remained constant throughout this range. DBA 5,6-dihydrodiol was present in no more than a constant trace amount at all concentrations of purified epoxide hydrase. These results indicate that the low amount of the K-region diol detected in these experiments as well as with microsomes is not because of the refractory nature of DBA 5,6-oxide toward epoxide hydrase (9,20). The phenol-2 fraction remained constant after an initial decrease with the addition of 21 units of epoxide hydrase per ml, whereas the phenol-1 fraction continued to decrease with addition of increasing amounts of epoxide hydrase. Phenol-1 and phenol-2 fractions were reduced by 53% and 42%, respectively, at 167 units/ml of epoxide hydrase. Similar to the metabolism of B[a]P and unlike that of BA, the total metabolism of DBA increased only slightly upon addition of purified epoxide hydrase.

DBA thus differs from B[a]P (6), B[e]P (21) and BA (28) in its extremely low metabolism at the K-region. Like the metabolism of B[a]P, significant amounts of phenols are formed with microsomes or with purified cytochrome P-448 in the presence of even high concentrations of epoxide hydrase, whereas BA is almost exclusively converted to dihydrodiols under these conditions.

3-METHYLCHOLANTHRENE

MC is essentially an alkylated derivative of BA. Studies of the metabolism of MC with the aid of the purified and reconstituted system in the presence and absence of epoxide hydrase failed to provide proof for the formation of significant amounts of the 9,10-dihydrodiol with the bay-region 7,8-double bond (26). Since the two major metabolites of MC are 1- and 2-hydroxy MC, further metabolism of these alcohols was studied to determine if they are converted into 9,10-dihydrodiols (26,27). (±) 1-Hydroxy MC was further metabolized by rat liver microsomes and by purified cytochrome P-448 in the presence of epoxide hydrase to a pair of diastereomeric *trans* 9,10-dihydrodiols in which the 10-hydroxyl group is either *cis* or *trans* to the 1-hydroxyl group (27). Upon metabolic activation by monooxygenases, one of these dihydrodiols was found to be among the most mutagenic MC derivatives tested toward both bacterial and mammalian cells (35). Interestingly, 2-hydroxy MC, in contrast to 1-hydroxy MC, was found to be a very poor substrate for rat liver monooxygenases.

TABLE IV. Values of $\Delta E_{deloc}/\beta$ for Bay-Region Diol Epoxides and Relative Carcinogenicity of the Corresponding Polycyclic Hydrocarbons

Hydrocarbon[a]	$\Delta E_{deloc}/\beta$ for Bay-Region Diol Epoxide	Carcinogenic Activity[b]
B[a]P	0.794	+++++
BA	0.766	+
DBA	0.738	++++
B[e]P	0.714	0
Chrysene	0.640	++

[a]Arranged in order of predicted decreasing carcinogenic activity.
[b]The relative carcinogenic activity of these hydrocarbons was determined from comparable initiation-promotion experiments on mouse skin (3, 13, 15, 17, 19, 36 and references therein).

COMPARATIVE METABOLISM OF POLYCYCLIC HYDROCARBONS

The bay-region theory attempts to correlate the carcinogenicity of polycyclic aromatic hydrocarbons with the predicted reactivity of their diol epoxides. One method of predicting reactivity is through the use of perturbational molecular orbital calculations which generate the parameter $\Delta E_{deloc}/\beta$. The bay-region theory predicts that the higher the value of $\Delta E_{deloc}/\beta$ (greater ease of carbonium ion formation) for a particular bay-region diol epoxide, the higher will be the biological activity of that diol epoxide (10-12). Indeed, the mutagenicity of several benzo-ring tetrahydro epoxides toward *S. typhimurium* strain TA 100 and Chinese hamster V79 cells show a high correlation with their corresponding $\Delta E_{deloc}/\beta$ values (37). Although a large number of polycyclic hydrocarbons show good correlation between the calculated values of $\Delta E_{deloc}/\beta$ for bay-region diol epoxides and the relative carcinogenicity of the parent hydrocarbon, certain exceptions do exist. For example, although the 3,4-diol-1,2-epoxide of BA has a higher value of $\Delta E_{deloc}/\beta$ than does the 3,4-diol-1,2-epoxide of DBA, BA is a much weaker carcinogen than DBA (Table IV). Similarly, B[e]P is expected to be more active than chrysene, yet the reverse is true (Table IV). These exceptions are thought to occur, at least in part, because of differences in the amount of metabolism of various hydrocarbons to their corresponding bay-region diol epoxides. Thus, metabolism studies complement the bay-region theory in predictions of the relative carcinogenicity of different polycyclic aromatic hydrocarbons.

Metabolism of the polycyclic hydrocarbons described in this study by liver microsomes from MC-treated rats is compared in Table V both in terms of rate of oxidation and percent of total metabolism leading to the dihydrodiol with the bay-region double bond. In the cases of B[a]P 7,8- (14,18),

TABLE V. Comparative Metabolism of a Series of Polycyclic Hydrocarbons

Polycyclic Hydrocarbon	Total Metabolism[a] (nmol/nmol hemeprotein/min)	Benzo-Ring Dihydrodiol with the Bay-Region Double Bond (% of Total Metabolites)
B[a]P	5.2	12
B[e]P	3.7	–[b]
BA	6.9	2
DBA	1.5	24
MC	2.4	–[b]
1-hydroxy MC	2.8	2,5[c]

[a]The conversion was measured under linear conditions. For specific experimental conditions, see references 21, 23, 26, 27 and 28.
[b]Clear evidence for the formation of this dihydrodiol was not obtained.
[c](±)-1-Hydroxy MC gives two diastereomerically related 9,10-dihydrodiols (27).

BA 3,4- (17,36), DBA 3,4- (3), and chrysene 1,2-dihydrodiols (19), tumor studies have shown these dihydrodiols with bay-region double bonds to be the most tumorigenic dihydrodiol formed from each of the respective hydrocarbons. Among these hydrocarbons, BA and B[a]P are the best substrates for the microsomal monooxygenases. Both B[a]P and DBA form higher percentages of the dihydrodiols with bay-region double bonds and are more potent carcinogens than BA and chrysene. Although BA is predicted to be a more potent carcinogen than DBA from the molecular orbital calculations, it is a weaker carcinogen at least in part because of the lower relative amounts of the carcinogenic 3,4-dihydrodiol formed. Similarly, the lack of carcinogenicity of B[e]P in rodents appears to reside in its low metabolic conversion to B[e]P 9,10-dihydrodiol (21) and to the lack of conversion of this dihydrodiol to the bay-region 9,10-diol-11,12-epoxides (15,38). In contrast, chrysene is converted to its 1,2-dihydrodiol with the bay-region 3,4-double bond in high amounts (24). Thus, the higher carcinogenicity of chrysene compared to B[e]P is in accord with bay-region theory after consideration of relative rates and pathways of metabolism. Although MC was not converted in detected amounts to the 9,10-dihydrodiol with a bay-region double bond, a predominant metabolite, 1-hydroxy MC, was metabolized to a 9,10-dihydrodiol (7%)(27). This may be the basis for the high carcinogenic activity of MC although other pathways for metabolic activation of MC cannot be excluded (35). Interestingly, the two most tumorigenic hydrocarbons in Table IV are the same two for which evidence for bay-region diol epoxide metabolites can be detected in high conversion metabolism profiles. The present chapter provides examples of how the comparative metabolism for several hydrocarbons coupled with the bay-region theory provides important insights into why such hydrocarbons differ in their carcinogenic activity.

REFERENCES

1. Boyland, E., and Sims, P. (1964): Metabolism of polycyclic compounds 24. The metabolism of benz[a]anthracene. *Biochem J.* 91:493-506.
2. Buening, M. K., Wislocki, P. G., Levin, W., Yagi, H., Thakker, D. R., Akagi, H., Koreeda, M., Jerina, D. M., and Conney, A. H. (1978): Tumorigenicity of the optical enantiomers of the diastereomeric benzo[a] pyrene 7,8-diol-9,10-epoxides in newborn mice: Exceptional activity of (+)-7β,8α-dihydroxy-9α,10α-epoxy-7,8,9,10-tetrahydrobenzo[a] pyrene. *Proc. Nat. Acad. Sci., U.S.* 75:5358-5361.
3. Conney, A. H., Levin, W. Wood, A. W., Yagi, H., Lehr, R. E., and Jerina, D. M. (1978): Biological activity of polycyclic hydrocarbon metabolites and the bay region theory. In: *Advances in Pharmacology and Therapeutics*, Vol. 9, Y. Cohen, ed., pp. 43-52, Pergamon Press, Oxford.
4. Dipple, A. (1976): Polynuclear aromatic carcinogens. In: *Chemical Carcinogens, ACS Monograph Series, No. 173,* C. E. Searle, ed., pp. 245-314, American Chemical Society, Washington, D.C.
5. DuBois, G. C., Apella, E., Levin, W., Lu, A. Y. H., and Jerina, D. M. (1978): Hepatic microsomal epoxide hydrase: Involvement of histidine at the active site suggests a nucleophilic mechanism. *J. Biol. Chem.* 253:2932-2939.
6. Holder, G., Yagi, H., Dansette, P., Jerina, D. M., Levin, W., Lu, A. Y.,H., and Conney, A. H. (1974): Effects of inducers and epoxide hydrase on the metabolism of benzo[a]pyrene by liver microsomes and a reconstituted system: Analysis by high pressure liquid chromatography. *Proc. Nat. Acad. Sci., U.S.* 71:4356-4360.
7. Huberman, E., Sachs, L., Yang, S. K., and Gelboin, H. V. (1976): Identification of mutagenic benzo[a]pyrene in mammalian cells. *Proc. Nat. Acad. Sci., U.S.* 73:607-611.
8. Jerina, D. M., and Daly, J. W. (1976): Oxidation at carbon. In: *Drug Metabolism–From Microbe to Man*, D. V. Parke and R. L. Smith, eds., pp. 13-32, Taylor and Francis Ltd., London.
9. Jerina, D. M., Dansette, P. M., Lu, A. Y. H., and Levin, W. (1977): Hepatic microsomal epoxide hydrase: A sensitive radiometric assay for hydration of arene oxides of carcinogenic aromatic hydrcarbons. *Mol. Pharmacol.* 13:342-351.
10. Jerina, D. M., and Lehr, R. E. (1977): The bay region theory: A quantum mechanical approach to aromatic hydrocarbon-induced carcinogenicity. In: *Microsomes and Drug Oxidations,* V. Ullrich, I. Roots, A. G. Hildebrant, R. W. Estabrook and A. H. Conney, eds., pp. 709-720, Pergamon Press, Oxford.
11. Jerina, D. M., Lehr, R., Schaefer-Ridder, M., Yagi, H., Karle, J. M., Thakker, D. R., Wood, A. W., Lu, A. Y. H., Ryan, D., West, S., Levin, W., and Conney, A. H. (1977): Bay region epoxides of dihydrodiols- A concept which explains the mutagenic and carcinogenic activity of benzo [a]pyrene and benzo[a]anthracene. In: *Origins of Human*

Cancer, H. Hiatt, J. D. Watson and I. Winsten, eds., pp. 639-658, Cold Spring Harbor Laboratory, Cold Spring Harbor, NY.

12. Jerina, D. M., Lehr, R. E., Yagi, H., Hernandez, O., Dansette, P. M., Wislocki, P. G., Wood, A. W., Chang, R. L., Levin, W., and Conney, A. H. (1976): Mutagenicity of benzo[a]pyrene derivatives and the description of a quantum mechanical model which predicts the ease of carbonium ion formation from diol epoxides. In: *In Vitro Metabolic Activation in Mutagenesis Testing,* F. J. de Serres, J. R. Fouts, J. R. Bend and R. M. Philpot, eds., pp. 159-177, Elsevier, Amsterdam.
13. Jerina, D. M., Yagi, H., Thakker, D. R., Karle, J. M., Mah, H. D., Boyd, D. R., Gadaginamath, G., Wood, A.W., Buening, M., Chang, R. L., Levin, W., and Conney, A. H. (1978): Stereoselective metabolic activation of polycyclic aromatic hydrocarbons. In: *Advances in Pharmacology and Therapeutics*, Vol. 9, Y. Cohen, ed., pp. 53-62, Pergamon Press, Oxford.
14. Kapitulnik, J., Levin, W., Yagi, H., Jerina, D. M., and Conney, A. H. (1977): Benzo[a]pyrene 7,8-dihydrodiol is more carcinogenic than benzo[a]pyrene in newborn mice. *Nature* 266:378-379.
15. Lehr, R. E., Taylor, C. W., Kumar, S., Levin, W., Chang, R. L., Wood, A. W., Conney, A. H., Thakker, D. R., Yagi, H., Mah, H. D., and Jerina, D. M. (1979): Differences in metabolism provides a basis for the low mutagenicity and carcinogenicity of benzo[e]pyrene compared to benzo[a]pyrene. Chapter 3, this volume.
16. Levin, W., Lu, A. Y. H., Ryan, D., Wood, A. W., Kapitulnik, J., West S., Huang, M. -T., Conney, A. H., Thakker, D. R., Holder, G., Yagi, H., and Jerina, D. M. (1977): Properties of the liver microsomal monooxygenase system and epoxide hydrase: Factors influencing the metabolism and mutagenicity of benzo[a]pyrene. In: *Origins of Human Cancer,* H. Hiatt, J. D. Watson and I. Winsten, eds, pp. 659-682, Cold Spring Harbor Laboratory, Cold Spring Harbor, NY.
17. Levin, W., Thakker, D. R., Wood, A. W., Chang, R. L., Lehr, R. E., Jerina, D. M., and Conney, A. H. (1978): Evidence that benzo[a]anthracene 3,4-diol-1,2-epoxide is an ultimate carcinogen on mouse skin. *Cancer Res.* 38:1705-1710.
18. Levin, W., Wood, A. W., Yagi, H., Jerina, D. M., and Conney, A. H. (1976): Benzo[a]pyrene 7,8-dihydrodiol: A potent skin carcinogen when applied topically to mice. *Proc. Nat. Acad. Sci., U.S.* 73:3867-3871.
19. Levin, W., Wood, A. W., Yagi, H., Mah, H. D., Jerina, D. M., and Conney, A. H. (1978): Evidence for bay region activation of chrysene 1,2-dihydrodiol to an ultimate carcinogen. *Cancer Res.* 38:1831-1834.
20. Lu, A. Y. H., Jerina, D. M. and Levin, W. (1977): Liver microsomal epoxide hydrase: Hydration of alkene and arene oxides by the membrane-bound and purified enzymes. *J. Biol. Chem.* 252:3715-3723.
21. MacLeod, M. C., Levin, W., Conney, A. H., Lehr, R. E., Jerina, D. M., and Selkirk, J. K. (In preparation).

22. Miller, E. C. (1951): Studies on the formation of protein bound derivatives of 3,4-benzpyrene in epidermal reaction of mouse skin. *Cancer Res.* 11:100-108.
23. Nordqvist, M., Thakker, D. R., Yagi, H., Ryan, D. E., Thomas, P. E., Levin, W., Conney, A. H., and Jerina, D. M.: The highly tumorigenic 3,4-dihydrodiol is the principal metabolite formed from dibenzo[a,h] anthracene by liver enzymes (submitted).
24. Sims, P. (1970): Qualitative and quantitative studies on the metabolism of a series of aromatic hydrocarbons by rat liver preparations. *Biochem. Pharmacol.* 19:795-818.
25. Slaga, T. J., Huberman, E., Selkirk, J. K., Harvey, R. G., and Bracken, W. M. (1978): Carcinogenicity and mutagenicity of benz[a]anthracene diols and diol-epoxides. *Cancer Res.* 38:1699-1704.
26. Thakker, D. R., Levin, W., Stoming, T. A., Conney, A. H., and Jerina, D. M. (1978): Metabolism of 3-methylcholanthrene by rat liver microsomes and a highly purified monooxygenase system with and without epoxide hydrase. In: *Carcinogenesis, Vol. 3,* P. W. Jones and R. I. Freudenthal, eds, pp. 253-264, Raven Press, New York.
27. Thakker, D. R., Levin, W., Stoming, T. A., Conney, A. H., and Jerina, D. M. (1978): Metabolic formation of 1,9,10-trihydroxy-9,10-dihydro-3-methylcholanthrene: A potential proximate carcinogen from 3-methylcholanthrene. *J. Am. Chem. Soc.* 100:645-647.
28. Thakker, D. R., Levin, W., Yagi, H., Karle, J. M., Lehr, R. E., Ryan, D., Thomas, P. E., Conney, A. H., and Jerina, D. M. (1979): Metabolism of benzo[a]anthracene to its tumorigenic 3,4-dihydrodiol. *Mol. Pharmacol.* 15:138-153.
29. Thakker, D. R., Yagi, H., Akagi, H., Koreeda, M., Lu, A. Y. H., Levin, W., Wood, A. W., Conney, A. H., and Jerina, D. M. (1977): Metabolism of benzo[a]pyrene. VI. Stereoselective metabolism of benzo[a]pyrene and benzo[a]pyrene 7,8-dihydrodiol to diol epoxides. *Chem. -Biol. Interact.* 16:281-300.
30. Thakker, D. R., Yagi, H., Lehr, R. E., Levin, W., Buening, M., Lu, A. Y. H., Chang, R. L., Wood, A. W., Conney, A. H., and Jerina, D. M. (1978): Metabolism of trans-9,10-dihydroxy-9,10-dihydrobenzo[a] pyrene occurs primarily by arylhydroxylation rather than formation of a diol epoxide. *Mol. Pharmacol.* 14:502-513.
31. Thakker, D. R., Yagi, H., Levin, W., Lu, A. Y. H., Conney, A. H., and Jerina, D. M. (1977): Sterospecificity of microsomal and purified epoxide hydrase from rat liver. *J. Biol. Chem.* 252:6328-6334.
32. Thomaszewski, J. E., Jerina, D. M., and Daly, J. W. (1975): Deuterium isotope effects during formation of phenols by hepatic monooxygenases. Evidence for an alternative to the arene oxide pathway. *Biochemistry* 14:2024-2031.
33. Wislocki, P. G., Kapitulnik, J., Levin, W., Lehr, R. E., Schaefer-Ridder, M., Karle, J. M., Jerina, D. M., and Conney, A. H. (1978): Exceptional carcinogenic activity of benzo[a]anthracene 3,4-dihydrodiol in the

newborn mouse supports the bay region theory. *Cancer Res.* 38:693-696.

34. Wood, A. W., Chang, R. L., Levin, W., Lehr, R. E., Schaefer-Ridder, M., Karle, J. M., Jerina, D. M., and Conney, A. H. (1977): Mutagenicity and cytotoxicity of benzo[a]anthracene diol epoxides and tetrahydro epoxides: Exceptional activity of the bay region 1,2-epoxides. *Proc. Nat. Acad. Sci., U.S.* 74:2746-2750.
35. Wood, A. W., Chang, R. L., Levin, W., Thomas, P. E., Ryan, D., Stoming, T. A., Thakker, D. R., Jerina, D. M., and Conney, A. H. (1978): Metabolic activation of 3-methylcholanthrene and its metabolites to mutagens toward bacteria and mammalian cells. *Cancer Res.* 38:3398-3404.
36. Wood, A. W., Levin, W., Chang, R. L., Lehr, R. E., Schaefer-Ridder, M., Karle, J. M, Jerina, D. M., and Conney, A. H. (1977): Tumorigenicity of five dihydrodiols of benzo[a]anthracene on mouse skin: Exceptional activity of benzo[a]anthracene 3,4-dihydrodiol. *Proc. Nat. Acad. Sci., U.S.* 74:3176-3179.
37. Wood, A. W., Levin, W., Chang, R. L., Yagi, H., Thakker, D. R., Lehr, R. E., Jerina, D. M., and Conney, A. H. (1979): Bay region activation of carcinogenic polycyclic hydrocarbons. Chapter 35, this volume.
38. Wood, A. W., Levin, W., Thakker, D. R., Yagi, H., Chang, R. L., Ryan, D. E., Thomas, P. E., Dansette, P. M., Wittaker, N., Turujam, S., Lehr, R. E., Kumar, S., Jerina, D. M. and Conney, A. H.: Biological activity of benzo[e]pyrene: An assessment based on mutagenic activities and metabolic profiles of the polycyclic hydrocarbon and its derivatives, in press.
39. Yang, S. K. (1978): Metabolic formation of 3,4-epoxide, K-region 5,6-oxide and the anti 8,9-diol-10,11-epoxide of benz[a]anthracene. In: *7th International Congress of Pharmacology,* p. 652, Abstract no. 2084, Paris.
40. Yang, S. K., Fu, P. P., Roller, P. P., Harvey, R. G., and Gelboin, H. V. (1978): Evidence for an unstable 3,4-epoxide as a metabolic intermediate of benz[a]anthracene. *Fed. Proc.* 37:597.
41. Yang, S. K., McCourt, D. W., Roller, P. P., and Gelboin, H. V. (1976): Enzymatic conversion of benzo[a]pyrene leading predominantly to the diol-epoxide r-7,t-8-dihydroxy-t-9,10-oxy-7,8,9,10-tetrahydrobenzo[a]-pyrene through a single enantiomer of r-7,t-8-dihydroxy-7,8-dihydrobenzo[a]pyrene. *Proc. Nat. Acad. Sci., U.S.* 73:2594-2598.

Polynuclear Aromatic Hydrocarbons, edited by P. W. Jones and P. Leber. Ann Arbor Science Publishers, Inc. Ann Arbor, MI © 1979.

EFFECTS OF SO_2 ON THE METABOLISM OF BENZO[a]PYRENE IN THE ISOLATED PERFUSED LUNG

D. Warshawsky, R. Niemeier,[1] C. Warren and E. Bingham[2]

University of Cincinnati
College of Medicine
Department of Environmental Health
Keterring Laboratory
Cincinnati, Ohio 45267

We are exposed by inhalation to a complex mixture of potentially hazardous materials, including polynuclear aromatic hydrocarbons (PAH), many of which are carcinogenic, and a variety of gases and particulates which may modify the manner in which the lung disposes of inhaled materials (2,4,24, 30,36). Epidemiological and experimental evidence (5,16) indicate that the interplay of multiple factors is responsible for the induction of respiratory diseases and, in particular, lung cancer (25,34).

Benzo[a]pyrene (B[a]P), a well-known carcinogenic PAH, is formed during the combustion of organic materials (3,17). Other materials, such as particulate matter, which carry a multitude of chemicals–including B[a]P and sulfur dioxide–are also formed during the combustion processes of fossil fuels (7,24). It is well established that the lungs are capable of binding and metabolizing B[a]P (2,4,23,37,40). However, an important requirement in understanding the mechanism of action of B[a]P is a detailed knowledge of the rate and pattern of metabolites formation and the factors controlling formation.

It has been shown experimentally that B[a]P in combination with SO_2 produced squamous cell carcinomas in the rat lung (18,36). However, based on just this one study and the fact that the control group was small, it is difficult to draw any firm conclusions about the cocarcinogenic effects of SO_2. In another study, exposure of rats to high levels of SO_2 did not produce an

---

[1]Present Address: National Institute for Occupational Safety and Health, Cincinnati, Ohio 45226.

[2]Department of Labor, Occupational Safety and Health Administration, Washington, D.C. 20210.

increase in aryl hydrocarbon hydroxylase (AHH) activity of rat lung microsomes, while exposure to SO_2 followed by 3-methyl chlolanthrene treatment, caused an inhibition in the AHH activity when compared to appropriate controls (14).

Besides the aforementioned studies, information on the interaction between carcinogenic PAH, such as B[a]P, and pollutant gases, such as SO_2, is still generally lacking. Even though it is well recognized that 95-99% of SO_2 is adsorbed by the upper respiratory tract (1,26), some small unknown percentage of SO_2 will inevitably get into the lower respiratory tract. Therefore, this study attempts to determine the effects SO_2 and particulates will have on the metabolism of B[a]P in the lung at levels of 1-2 ppm SO_2. This is below the industrial threshold limit value (TLV) of 5 ppm and at levels within exposure possibilities.

The isolated perfused lung (IPL) is well suited for study of these pulmonary interactions without the influence of other organs (4,9,20,28,37).

MATERIALS AND METHODS

Male New Zealand rabbits weighing 4 to 5 kg are fed single-batch standard laboratory chow diets and water *ad libitum* up to the time of lung excision. All animals are allowed to equilibrate with their environment for at least three weeks prior to treatment or use on the system (no pesticides used in housing area). No anesthesia is given and rabbits are heparinized (1,000 I.U./kg) 5 min before cardiac puncture; animals are sacrificed by a blow to the back of the head or CO_2 inhalation before lung removal. Following cannulation, the lung and heparinized autologous whole blood are placed inline and a 15-min stabilization period is allowed before intratracheal administration of the compound. Autologous heparinized whole blood at approximately 160 ml/min is perfused through the lungs at a constant hydrostatic pressure of 23 cm Hg), maintained by electronic level sensors. The lungs are made to respire by alternating subatmospheric pressures created by a commercial respirator and vacuum source: the ventilating gas is humidified, warmed, filtered room air, with CO_2 added to continuously maintain a blood pH of 7.40. The equipment, startup of the perfusion system and preparation of the lungs are described elsewhere (4,27).

One microcurie of pure (7,10-^{14}C)-B[a]P (21 mCi/mole, Amersham/Searle, Arlington Heights, IL) is diluted with unlabeled pure B[a]P (Aldrich Chemical Company, Milwaukee, WI) and evaporated gently to dryness under nitrogen. It is then taken up to a final amount of 1.24 μ moles of B[a]P (0.8 μCi/mmole) or 312 μg of B[a]P in a 1 ml ethanolic saline (1:1) solution and is intratracheally injected into the IPL. When B[a]P and particulate are administered together, the mixed solutions of labeled and unlabeled B[a]P are slowly evaporated under a stream of nitrogen so that B[a]P is adsorbed onto particulate and then taken up in 1 ml of saline and intratracheally injected on the IPL. In each case, the syringe is rinsed once with an additional milliliter of saline and injected on the IPL.

All ^{14}C scintillation counting is performed on Tri-Carb Packard Liquid Scintillation Spectrometer (3200 and 2002). IPL samples are chromatographed on a Varian 8500 HPLC with variable UV-wavelength detector using a Varian reverse-phase ODS, 10-μm particle size, 25-cm x 2-mm columns, and a methanol-water gradient from 62% to 100% methanol, room temperature, monitored at 268 nm with a flow of 1 ml/min.

In order to generate SO_2, a stream of dry compressed air (Figures 1 and 2) is passed over a 2-in. 3/16 in. i.d. x 0.30 in. wall thickness FEP Teflon permeation tube with FEP Teflon plugs maintained at 37°C, containing condensed SO_2 (32). This SO_2-air stream is then mixed with either humidified air containing 5% CO_2 from the ventilating system of the IPL before entering the IPL preparation via the tracheal valve (Figure 1) or dry filtered room air before entering the tracheotomized rabbit (Figure 2). The exhaled and excess SO_2-air is then bubbled through two sodium hydroxide scrubbers and a dessicating column before it is vented out of the system by a house vacuum. The concentration of SO_2 is adjusted by carefully controlling the following systems: (1) the stream of compressed air over the permeation tube, (2) the dry room air (*in vivo*) or humidifed CO_2-air (*in vitro*) and (3) the vacuum source. A critical resistance consisting of a 3-in. piece of P.E. 60 tubing is used to measure flowrate of the compressed air (Figures 1 and 2). The West Gaeke method, a colorimetric determination of sulfur dioxide concentration (ppm) is used (15,38). Sampling of air for specified periods of time throughout the experiment is through a midget impinger containing 10 ml of 0.04 *M* potassium tetrachloromercurate-absorbing reagent located on the exhaust side of the system. Two ml of a 0.016% paraosaniline reagent (Eastman Kodak Co., Rochester, NY) and 1 ml of a 0.2% formaldehyde solution are added to the absorbing reagent, transferred to 25-ml glass-stoppered graduated cylinder and mixed. The purple color is allowed to develop for 20 min at which time the absorbance is determined at 575 nm on a Beckman UV spectrophotometer. The microliters SO_2 of the samples are determined from a standard calibration curve of absorbance (range 0 to 0.6) vs microliters or micrograms of SO_2. The standard curve is prepared daily. Concentration (ppm) of SO_2 is determined by dividing microliters SO_2 by the flow (liter/min) that the vacuum draws through the impinger. The entire system is equilibrated with the SO_2 inline to obtain 1.5-2 ppm before placing the rabbit or the lung in line; the SO_2 is then monitored for the duration of the experiment. The SO_2 pretreatment is performed by restraining the animal on the rabbit board with the level of the head slightly lower than the level of the

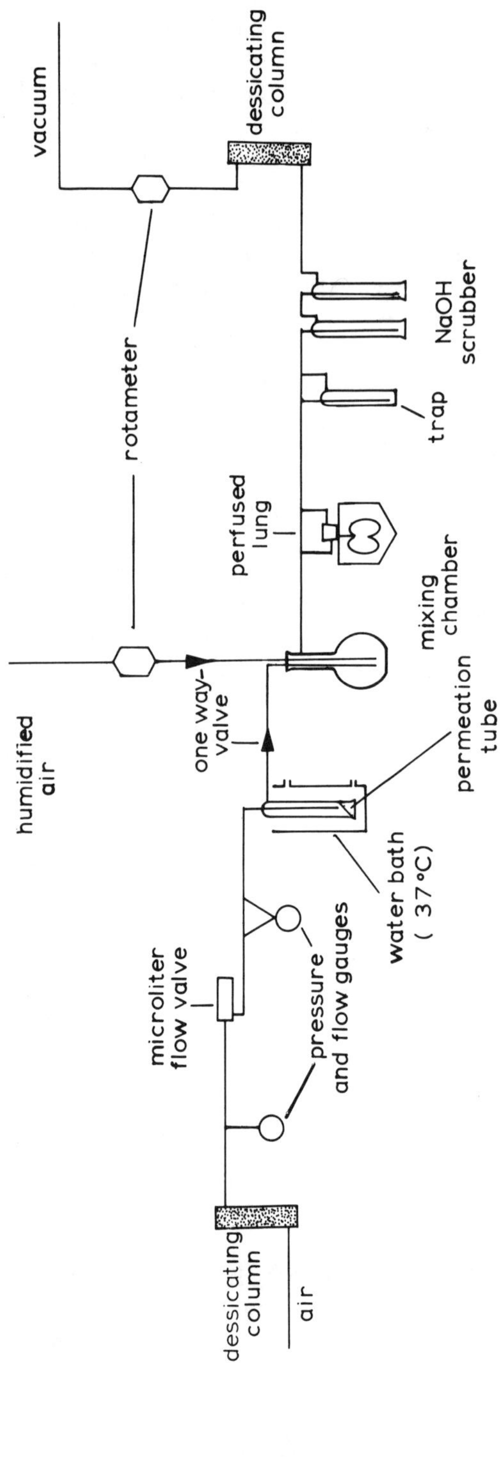

FIGURE 1. Modification of the IPL for *in vitro* SO_2 experiments.

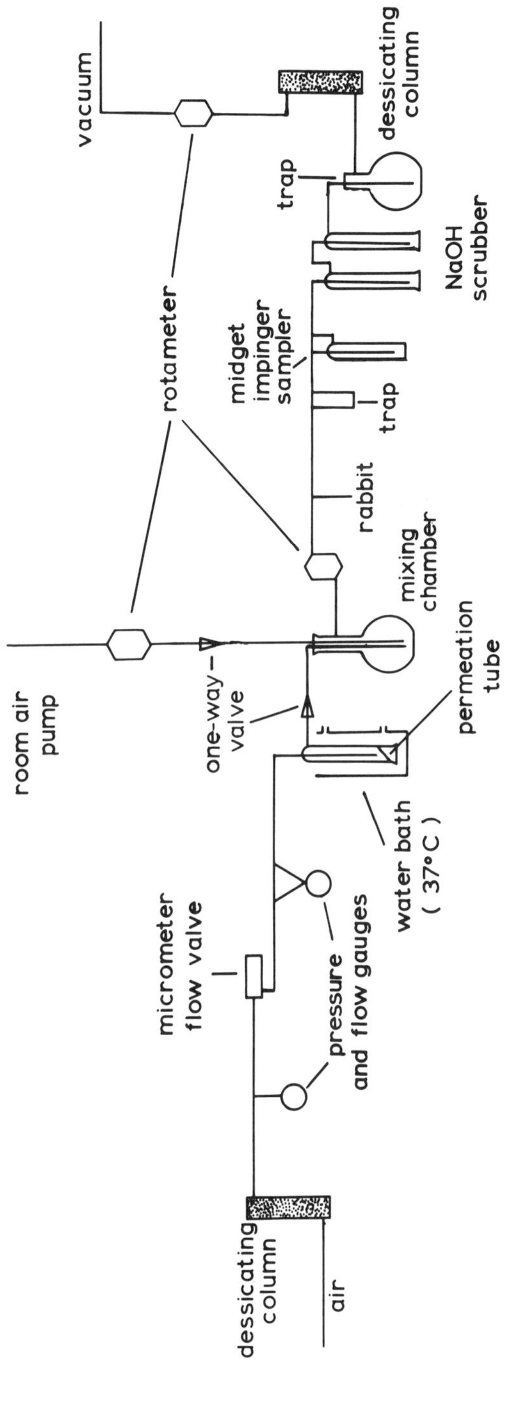

FIGURE 2. Modification of the IPL for *in vivo* SO_2 experiments.

heart and the neck regions shaved with a small animal clipper. One cm^3 of local anesthetic, lidocain 2% in an epinephrine solution (W. A. Butler Co., Columbus, OH) is administered subcutaneously to several sites, injecting a total of 3 to 4 cm^3. After 5 min, a 1-in. midline incision is performed, the trachea isolated, a small hemostat passed behind the trachea (dorsal side) and surgical thread drawn back, a small horizontal cut is made in the trachea and the cannula inserted and ligated. A tee connection inline with the smallest possible distance to the cannula is attached to the cannula through which the rabbit breathes. An additional air reservoir is added to aid the rabbit in case of a sudden loss of positive pressure. A slight positive pressure is maintained by monitoring a respiratory bag placed inline immediately after the rabbit and before the air trap. A rotameter placed on the inhalation side of the rabbit is used to monitor the rabbit's respiration. In these experiments the average respirations prior to tracheotomy are 167 respirations/min and after tracheotomy 153 respirations/min while the pulse rate remains unchanged at 80 pulses/min.

Dr. Harry Gelboin of NCI supplied metabolite standards, 3-hydroxy, 3,6-,6,12- and 1,6-quinones, 9,10-dihydrodiol (9,10-diol), 7,8-dihydrodiol (7,8-diol), 9-hydroxy, 7-hydroxy; Dr. Ranold Harvey of Ben May Cancer Institute supplied the *cis* and *trans* 4,5-dihydrodiol (4,5-diol), 4,5-epoxide and 4,5-quinone; and Drs. Tom Meehan, Ken Straub and Joe Landolph of the Chemical Biodynamics Laboratory, University of California, Berkeley, provided the (±) 7α,8B-dihydroxy-9B,10B-epoxy-7,8,9,10-tetrahydrobenzo-[a]pyrene (7,8-diol-9,10-epoxide), and 6- and 7-hydroxy. The crude air particulate (supplied by EPA) from the Pittsburgh vicinity, analyzed by the analytical division of Kettering Laboratory for size and metal content, contains particulate size 79.0% $\leqslant$ 10 μm, 64.5% $\leqslant$ 6 μm, 47.5% $\leqslant$ 4 μm, and metals (ppm): Cd, 40; Cr, 542; Fe, 47,976; Mg, 6,482; Mn, 2,219; Ni, 1,417; Pb, 6,818; Zn, 5,496. The CAP has been partially characterized for polynuclear aromatic hydrocarbon (PAH) content. Chrysene, pyrene and benzo-[k]fluoranthene have been detected.

Blood samples of 5.5 ml are taken from the perfusion system a minimum of six times over a period of 3 hr, then 0.5 ml is digested in NaOH, heated overnight at 60°C bleached with tert-butyl hydroperoxide, and counted in Scintiverse (Fisher) Scintillation Media with internal standard. The remaining 5 ml of blood are extracted twice with 30 ml acetone:benzene:isoamyl alcohol (10:13:0.1) by shaking for 30 min and centrifuging at 1,500 rpm, 5°C for 30 min. The organic extracted phase is removed and evaporated to dryness in a water bath (50°C) under nitrogen. The remaining blood tissue containing nonextractable material is digested, bleached and counted. The organic portion is reconstituted in benzene for TLC or in chloroform for HPLC and is stored in the cold until analysis. (Metabolite standards are dissolved in benzene or methanol for TLC and in methanol for HPLC.) A portion of the reconstituted benzene sample is chromatographed on TLC (0.1 mm silica gel plates, Eastman Kodak 6061) using benzene:ethanol (19:1). Spots are located in the UV light, the entire chromatogram is cut into strips, and the spots are quantitated by counting using a cocktail of POPOP (25 g) and PPO (4 g) per liter of toluene. Identification is made using R_f values of standard metabolites and published information.

An HPLC chromatogram is recorded using the standard mixture (37) once each morning. The chloroform sample is then chromatographed and the fractions collected with time, such that each of the peaks and spaces between peaks are collected individually. The fractions are then quantitated by counting using a cocktail of toluene:Triton X100:ethanol (8:4:3) plus PPO (6 g/l) and POPOP (125 mg/l). For both TLC and HPLC, amounts of each metabolite are expressed for each blood sample in nanograms per gram lung wet weight used for each perfusion. The total rate of appearance of metabolites (ng/hr/g lung) in the blood is based on a linear regression of a time-count study from 0 through at least 90 min. All the data collection, data reduction and statistics are handled by computer programs. The HPLC chromatograph, at present, is more efficient in resolving the 9-hydroxy, 3-hydroxy, 7-hydroxy, 4,5-epoxide and quinones standards then indicated previously (37).

All samples are processed under nitrogen and subdued yellow lighting to minimize photooxidation. Ninhydrin spray is obtained from Sigma (St. Louis, MO) and all solvents used are redistilled except the methanol (Fisher Scientific, Fairlawn, NJ, spectranalyzed).

RESULTS

Influence of SO_2 Pretreatment on B[a]P Metabolism

SO_2 pretreatment significantly increases the rate of metabolism of B[a]P when compared to its control (Table I); but this increase is significantly smaller than that caused by pretreatment with B[a]P, a well-characterized MFO (mixed function oxidase) enzyme inducer. SO_2, therefore, acts as a biochemical agent which causes an increase in enzyme activity or as a physical agent which causes biochemical changes in the lung due to irritation of SO_2 (10,14,26).

The distribution data do not show any marked changes when the effects of SO_2 pretreatment are compared to B[a]P control. There is a slight increase in the nonextractables and a slight decrease in monohydroxylated compounds and a larger decrease in dione formation. B[a]P pretreatment, on the other hand, shows a marked increase in the 9,10-dihydrodiol and a smaller decrease in monohydroxylated and diones when compared to SO_2

TABLE I. Influence of SO_2 Pretreatment on B[a]P Metabolism

Pretreatment IPL No. of Animals	None B[a]P 9	$(SO_2)_{IT}^a$ B[a]P 5	$B[a]P_{IT}^b$ B[a]P 5
Total rate of appearance of metabolites in blood (ng/hr/g lung ± S.E.)	256 ± 38	514 ± 95[d]	1290 ± 114[d,e]
Metabolic pattern in blood (% ± (S.E.)[c]			
7,8-dihydrodiol	6.6 ± 0.9	6.0 ± 1.0	5.4 ± 1.8
9,10-dihydrodiol	15.4 ± 4.0	15.3 ± 6.6	32.8 ± 3.8[f]
4,5-dihyrodiol	3.3 ± 0.6	1.2 ± 0.4[f]	1.8 ± 0.5[f]
Monohydroxylated	9.7 ± 1.1	7.9 ± 1.7	5.9 ± 1.0
Diones	10.6 ± 1.8	3.8 ± 0.8[e]	2.8 ± 0.9[e]
Nonextractables	54.4 ± 5.4	65.7 ± 0.4	51.8 ± 2.8

Note: All metabolites separated by TLC. All columns compared to each other.
[a] 1.6 ± 0.4 ppm, 45 min.
[b] 10 mg/kg once/wk x 5.
[c] Metabolite pattern values expressed as percent of total rate of appearance of metabolite in blood ± S.E.
[d] $P = 0.05$.
[e] $P = 0.01$.
[f] $P = 0.1$.
Statistics performed by Student-Newman-Kuels test.

pretreatment or its control. This indicates that the SO_2 pretreatment does not have a marked effect on the metabolic pathways, *i.e.*, epoxidation, hydroxylation and/or isomerization and conjugation (21,23,33,39,40).

In order to validate the TLC data, the HPLC data for SO_2 pretreatment is compared to the TLC data (Table II) which not only supports but complements the data in Table I. The rates of formation of the 4,5-dihydrodiols, the nonextractables and the hydroxylated metabolites are very comparable. There are some differences, however, which are not totally unexpected considering the difference in the techniques. There is an increase in the quinones and a small decrease in 9,10-dihydrodiol for the HPLC data when compared to the TLC data. The 7,8-dihydrodiol and the S metabolite (ninhydrin positive and fluorescent which appear after B[a]P) on HPLC, chromatograph together on TLC and show some differences.

Influence of SO_2 Administered to IPL on B[a]P Metabolism

SO_2 administered to the IPL significantly increases the rate of metabolism of B[a]P compared to its control (Table III). This suggests, in the same manner described previously, that SO_2 acts as a biochemical agent which

TABLE II. Comparison HPLC and TLC Data

Pretreatment IPL No. of Animals	SO_2 B[a]P 5 HPLC	 TLC
Total rate of appearance of metabolites in blood (ng/hr/g lung ± S.E.)	547 ± 69	514 ± 95
Metabolic pattern in blood (% ± S.E.)[a]		
9,10-dihydrodiol	10.4 ± 4.3	15.3 ± 6.6
4,5-dihydrodiol	2.5 ± 0.6	1.2 ± 0.3
7,8-dihydrodiol	4.0 ± 0.6	6.0 ± 1.0
4,5-quinone	2.3 ± 0.3	
1,6-,6,12-,3,6-quinones	5.0 ± 1.8	3.8 ± 0.8
4,5 epoxide[b]	2.5 ± 0.5	
9-,6-OH	2.6 ± 0.7	
7-OH	2.2 ± 0.6	7.9 ± 1.6
3-OH	1.6 ± 0.5	
S Metabolite	7.7 ± 2.4	
Nonextractable	59.0 ± 4.7	65.7 ± 0.4

[a] Metabolite pattern values expressed as percent of total rate of appearance of metabolites in blood ± S.E.
[b] ^{14}C counts appear under peak.

causes an increase in the enzyme activity or as a physical agent which causes biochemical changes in the lung (14,26). However, CAP administered concurrently with B[a]P and SO_2 to the IPL, appears to significantly decrease the rate of metabolism of B[a]P when compared to its appropriate control. A factor to consider is that B[a]P may not be leached readily from the CAP and, therefore, is not available to be metabolized.

The metabolite distribution does not show any marked changes when the SO_2 treatment is compared to its control. There are small increases in the 9,10-dihydrodiol and nonextractables and small decreases in monohydroxylated and dione formation. When CAP is administered concurrently with B[a]P and SO_2 to the IPL there is a decrease in the nonextractables and small increases in the 9,10- and 7,8-dihydrodiols. This indicates that SO_2 administered *in vitro* on the IPL does not have major effects on the metabolic pattern.

In order to validate the TLC data, the HPLC data for SO_2 and B[a]P on the IPL, as well as SO_2, CAP and B[a]P on the IPL are compared to TLC (Tables IV and V). The rates, nonextractables, the monohydroxylated and

TABLE III. Influence of SO_2 Administered to IPL on B[a]P Metabolism

Pretreatment IPL No. of Animals	None B[a]P 9	None B[a]P + SO_2[a] 5	None B[a]P + CAP + SO_2[b] 7
Total rate of appearance of metabolites in blood (ng/hr/g lung ± S.E.)	256 ± 38[d]	551 ± 77	177 ± 48[d]
Metabolic pattern in blood (% ± S.E.)[c]			
7,8-dihydrodiol	6.6 ± 0.9	6.0 ± 0.7	11.0 ± 1.6[f]
9,10-dihydrodiol	15.4 ± 4.0	20.0 ± 2.1	26.7 ± 2.9[e]
4,5-dihydrodiol	3.3 ± 0.6	2.0 ± 0.6	1.7 ± 0.7
Monohydroxylated	9.7 ± 1.1[d]	4.2 ± 0.6	6.3 ± 1.1
Diones	10.6 ± 1.8	7.9 ± 2.6	6.9 ± 2.0
Nonextractables	54.4 ± 5.4[e]	59.9 ± 3.4	42.4 ± 2.6[d]

Note: All metabolites separated by TLC. Columns 1 and 3 compared to column 2.
[a]2.2 ± 0.1 ppm SO_2.
[b]10 mg/kg CAP, 2.1 ± 0.1 ppm SO_2.
[c]Metabolite pattern values expressed as percent of total rate of appearance of metabolites in blood ± S.E.
[d]P = 0.01.
[e]P = 0.1.
[f]P = 0.05.
Statistics performed by Student-Newman-Kuels test.

4,5- and 9,10-dihydrodiols are very comparable. The diones in Table IV are comparable, while the diones in Table V show an increase for HPLC data when compared to the TLC data. In both Tables IV and V, the 7,8-dihydrodiol from the TLC data is very comparable to the 7,8-dihydrodiol plus the S metabolite (appears after B[a]P on chromatogram and is fluorescent and ninhydrin positive) for HPLC data.

Influence of SO_2 and CAP Administration to IPL on B[a]P Metabolism

CAP in the IPL slightly inhibits the rate of metabolism of B[a]P (Table VI). SO_2 added concurrently with B[a]P and CAP to the IPL, does not change the rate of metabolism when compared to CAP and B[a]P on the IPL. This may be due to either the slow leaching of B[a]P from CAP or an increase in phagocytic action of the macrophage which might serve to decrease the amount of B[a]P available for metabolism. Additionally, at these low concentrations of SO_2, the gas could very possibly be adsorbed by the particulate and is not available to affect the metabolism of B[a]P.

TABLE IV. Comparison HPLC and TLC Data

Pretreatment IPL No. of Animals	None B[a]P + SO_2 5 HPLC	 TLC
Total rate of appearance of metabolites in blood (ng/hr/g lung ± S.E.)	588 ± 60	551 ± 77
Metabolic pattern in blood (% ± S.E.)[a]		
9,10-dihydrodiol	18.9 ± 2.1	20.0 ± 2.1
4,5-dihydrodiol	2.5 ± 0.6	2.0 ± 0.6
7,8-dihydrodiol	2.0 ± 0.4	6.0 ± 0.7
4,5-quinone	2.0 ± 0.3 }	
1,6-,6,12-,3,6-quinones	3.3 ± 1.1 }	7.9 ± 2.6
4,5 epoxide[b]	4.4 ± 0.5 }	
9-,6-OH	2.7 ± 0.4 }	
7-OH	1.7 ± 0.5 }	4.2 ± 0.6
3-OH	1.4 ± 0.4 }	
S Metabolite	6.0 ± 0.4	
Nonextractable	55.1 ± 4.5	59.9 ± 3.4

[a]Metabolite pattern values expressed as percent of total rate of appearance of metabolites in bood ± S.E.
[b]^{14}C counts appear under peak.

The distribution data due to the particulate indicates an increase in the 7,8- and 9,10-dihydrodiols and a decrease in the nonextractables, monohydroxylated and dione metabolites. It should be noted that the 7,8-dihydrodiol spot contains some (possibly glutathione) conjugation. This suggests that particulate affects the epoxide-hydrase pathway (21,23,33,39,40) more readily than the hydroxylation and/or isomerization and some conjugation pathways. However, when the distribution data for SO_2, CAP and B[a]P on the IPL is compared to B[a]P and CAP alone, there does not seem to be any significant differences except for the 7,8-dihydrodiol, which contains some unknown metabolite (possibly glutathione conjugate), and the 4,5-dihydrodiol. This suggests that SO_2 at these concentrations does not cause any marked changes in the metabolic pathways in the presence of B[a]P and CAP.

DISCUSSION

Our attempt is to simulate environmental conditions. The 312 μg of B[a]P, 1 mg/kg of CAP and 1-2 ppm of SO_2 used in the experiments are realistic human exposure values; however, there are some limitations in our

TABLE V. Comparison HPLC and TLC Data

Pretreatment IPL No. of Animals	None B[a]P + SO_2 + CAP 7 HPLC	 TLC
Total rate of appearance of metabolites in blood (ng/hr/g lung ± S.E.)	191 ± 53	177 ± 47
Metabolic pattern in blood (% ± S.E.)[a]		
9,10-dihydrodiol	17.4 ± 3.5	36.7 ± 2.9
4,5-dihydrodiol	3.5 ± 0.6	1.7 ± 0.7
7,8-dihydrodiol	5.7 ± 0.7	11.0 ± 1.6
4,5-quinone	4.3 ± 1.9	
1,6-,6,12-,3,6-quinones	5.0 ± 0.9	6.9 ± 2.0
4,5 epoxide[b]	5.9 ± 0.6	
9-,6-OH	2.5 ± 0.4	
7-OH	3.0 ± 0.6	6.3 ± 1.1
3-OH	2.8 ± 0.8	
S Metabolite	5.9 ± 1.2	
Nonextractable	43.9 ± 4.0	42.4 ± 2.6

[a]Metabolite pattern values expressed as percent of total rate of appearance of metabolites in blood ± S.E.
[b]^{14}C counts appear under peak.

system. The short time period will show only an immediate effect; with SO_2, CAP and B[a]P together, it may be necessary to run the experiment for longer periods of time, perhaps at higher concentrations of SO_2, or to use larger amounts of B[a]P and CAP to obtain dose-response relationships, or to pretreat with SO_2 by inhalation (6). In the presence of CAP, the SO_2 can be adsorbed on particulates and some of the SO_2 may be catalytically converted to bisulfites, sulfites or sulfates under the right conditions of temperature and humidity: in the presence of sunlight, the SO_2 can be converted photochemically (3,7,13,19,22,35). We are more interested at this point, however, in the effects of SO_2 on B[a]P metabolism under our stricter environmental conditions.

SO_2 increases the metabolism of B[a]P by the IPL, but does not affect the metabolic pattern. It acts as a biochemical agent which causes an increase in enzyme activity or as a physical agent which causes biochemical changes in the lung due to irritation (7,10,14,26). More work in this area needs to be done in order to answer this question. The data indicate, however, that SO_2 can produce changes in the rate of metabolism of a well-defined carcinogen.

TABLE VI. Influence of SO_2 and CAP Administered to IPL on B[a]P Metabolism

Pretreatment IPL No. of Animals	None B[a]P 9	None B[a]P + CAP[a] 5	None B[a]P + CAP + SO_2[b] 7
Total rate of appearance of metabolites in blood (ng/hr/g lung ± S.E.)	256 ± 38	156 ± 42	177 ± 48
Metabolic pattern in blood (% ± S.E.)[c]			
7,8-dihydrodiol	6.6 ± 0.9[d]	19.1 ± 4.4	11.0 ± 1.6
9,10-dihydrodiol	15.4 ± 4.0	28.3 ± 7.9	26.7 ± 2.9
4,5-dihydrodiol	3.3 ± 0.6	3.0 ± 1.3	1.7 ± 0.7
Monohydroxylated	9.7 ± 1.1[e]	5.1 ± 1.4	6.3 ± 1.1
Diones	10.6 ± 1.8	5.2 ± 2.6	6.9 ± 2.0
Nonextractables	54.4 ± 5.4	39.3 ± 13.8	42.4 ± 2.6

Note: All metabolites separated by TLC. Columns 1 and 3 compared to column 2.
[a]1 mg/kg.
[b]1 mg/kg CAP, 2.1 ± 0.1 ppm SO_2.
[c]Metabolite pattern values expressed as percent of total rate of appearance of metabolite in blood ± S.E.
[d]$P = 0.01$.
[e]$P = 0.05$.
Statistics performed by Student-Newman-Kuels test.

The presence of SO_2 at 1-2 ppm with CAP and B[a]P together indicates that SO_2 is either adsorbed by the particulate (10,19,35), or that the B[a]P is not available for metabolism due to either (1) that B[a]P is not leached readily from the particulate, or (2) an increase in phagocytic action of the macrophage which might serve to decrease the amount of B[a]P available for metabolism. The particulates, it should be mentioned, can maintain the environment of the B[a]P in the lung for long periods of time which may be responsible for increased tumorigenic responses (8,12,31) and altered metabolic pathways.

These studies are a first attempt to determine what type of interactions take place when SO_2 and/or CAP are added with B[a]P to a lung model system. This system appears to be a good model system for studying the metabolism of B[a]P. In a recent study (29), the incubation of B[a]P with human and rat lung microsomes produced 30% and 19% total dihydrodiols, respectively, which is consistent with our results for the rabbit IPL (37). There are still many areas related to B[a]P metabolism by SO_2/or CAP which need to be explored in the future due to the increased use of fossil fuels as alternate energy sources (10,24).

The distribution of metabolites in the macrophage (11), lung, blood, wash-out and trachea bronchi, as well as the kinetics of B[a]P in the IPL, are being analyzed to determine the importance of these initial studies.

ACKNOWLEDGMENTS

The authors wish to express their gratitude to Connie Woolums, Janet Mollmann and Connie Menefee for their expert work; to Dr. Martha Radike and William Barkley for helpful discussions; and to Diane Dotson for her secretarial assistance in preparing this manuscript. This research was supported in part by EPA Contract 68-02-1678, NCI Grant CA-15344-03 and USPHS ES-00159 Center for the Study of the Human Environment.

REFERENCES

1. Andersen, I., Lundguist, G. R., Jensen, P. L., and Proctor, D. F. (1974): Human response to controlled levels of sulfur dioxide. *Arch. Environ. Health* 28:31-39.
2. Anderson, M. W., and Eling, T. E. (1976): Studies on the uptake, metabolism and release of endogenous and exogenous chemicals by use of the isolated perfused lung. *Environ. Health Pers.* 16:77-81.
3. Anonymous (1972): *Particulate Polycyclic Organic Matter,* National Academy of Sciences, Washington, D.C.
4. Bingham, E., Niemeier, R. W., and Dalbey, W. (1976): Metabolism of environmental pollutants by isolated perfused lung. *Fed. Proc.* 35: 81-84.
5. Bingham, E., Niemeier, R. W., and Reid, J. B. (1976): Multiple factors in carcinogenesis. *Ann. N.Y. Acad. Sci.* 271:14-21.
6. Bingham, E., Warshawsky, D., and Niemeier, R. W. (1978): Metabolism of benzo[a]pyrene in the isolated perfused rabbit lung following n-dodecane inhalation. In: *Mechanisms of Tumor Promotion, Vol. 2: Carcinogenesis,* T. J. Slaga, A. Sivak and R. K. Boutwell, Eds., pp. 509-516, Raven Press, New York.
7. Coffin, D. L., and Knelson, J. H. (1976): Effects of sulfur dioxide and sulfate aerosol particles on human health. *Ambio* 5:239-242.
8. Creasia, D. A., Poggenburg, J. K., Jr., and Nettesheim, P. (1976): Elution of benzo[a]pyrene from carbon particles in the respiratory tract of mice. *J. Toxicol. Environ. Health* 1:967-975.
9. Eling, T. E., Pickett, R. D., Orton, T. C., and Anderson, M. W. (1975): A study of the dynamics of imipramine accumulation in the isolated perfused rabbit lung. *Drug Metab. Dispos.* 3(5):389-399.
10. Ellison, J. Mck., and Waller, R. E. (1978): A review of sulphur oxides and particulate matter as air pollutants with particular reference to effects on health in the United Kingdom. *Environ. Res.* 16:302-325.

11. Harris, C. C., Hsu, I. C., Storer, G. D., Trump, B. F., and Selkirk, J. (1978): Human pulmonary alveolar macrophages metabolize benzo-[a]pyrene to proximate and ultimate mutagens. *Nature* 272:633-634.
12. Henry, M. C., and Kaufman, D. G. (1973): Clearance of benzo[a]pyrene from hamster lungs after administration on coated particles. *J. Nat. Cancer Inst.* 51:1961-1964.
13. Hidy, G. M., Katz, J. L., and Mirabel, P. (1978): Sulfate aerosol formation and growth in the stratosphere *Atmos. Environ.* 12:887-892.
14. Husain, M. M., and Dehnen, W. (1978): Effect of NO_2 and SO_2 inhalation on benzo[a]pyrene metabolism in rat lung. *Arch. Toxicol.* 40: 207-210.
15. Intersociety Committee (1972): *Methods of Air Sampling and Analysis, Tentative Method of Analysis for Sulfur Dioxide Content of the Atmosphere,* pp. 447-455, American Public Health Association Inc., Washington, D.C.
16. Kotin, P., and Falk, H. L. (1959): The role and action of environmental agents in pathogenesis of lung cancer. I. Air Pollutants, *Cancer* 12: 147-163.
17. Lang, K. F., Buffleb, H., and Zander, M. (1963): Pyrolysis of polynuclear aromatic hydrocarbons. *Erdoel Kohle* 16:944-946.
18. Laskin, S., Kuschner, M., and Drew, R. T. (1970): Studies in pulmonary carcinogenesis. In: *Inhalation Carcinogenesis,* M. G. Hanna, P. Nettesheim and J. R. Gilbert, Eds., pp. 321-351, USAEC Division of Technical Information, CONF-691001.
19. Liberti, A., Brocco, D., and Possanzini, M. (1978): Absorption and oxidation of sulfur dioxide on particles. *Atmos. Environ.* 12:255-261.
20. McGovern, J. P., Lubaway, W. C., and Kostanbauder, W. B. (1976): Uptake and metabolism of nicotine by the isolated perfused rabbit lung. *J. Pharmacol. Exp. Ther.* 199(1):198-207.
21. Meehan, T., Straub, K., and Calvin, M. (1976): Elucidation of hydrocarbon structure in enzyme-catalyzed benzo[a]pyrene-poly G covalent complex. *Proc. Nat. Acad. Sci.* 73(5):1437-1441.
22. Miller, D. F. (1978): Precursor effects on SO_2 oxidation. *Atmos. Environ.* 12:273-280.
23. Moore, B. P., and Cohen, G. M. (1978): Metabolism of benzo[a]pyrene and its major metabolites to ethyl acetate-soluble and water-soluble metabolites by cultured rodent trachea. *Cancer Res.* 38:3066-3075.
24. Natusch, D. F. S. (1978): Potentially carcinogenic species emitted to the atmosphere by fossil fueled power plants. *Environ. Health Pers.* 22:79-90.
25. Nelson, N. (1970): Inhalation carcinogenesis in man: environmental and occupational hazards. In: *Inhalation Carcinogenesis,* M. G. Hanna, P. Nettesheim and J. R. Gilbert, Eds., pp. 3-13. USAEC, Division of Technical Information, CONF-691001.
26. Newhouse, M. T., Dolovich, M., and Obminski, G. (1978): Effects of TLV levels of SO_2 and H_2SO_4 on bronchial clearance in exercising man. *Arch. Environ. Health* 24-32.

27. Niemeier, R. W., and Bingham, E. (1972): An isolated perfused lung preparation for metabolic studies. *Life Sci.* 11(2):807-820.
28. Niemeier, R. W., Warshawsky, D., and Bingham, E. (1977): Influence of pretreatment on B[a]P metabolism in isolated perfused lung. *Toxicol. Appl. Pharmacol.* 41:185.
29. Prough, R. A., Sipal, Z., and Jacobsson, S. W. (1977): Metabolism of benzo[a]pyrene by human lung microsomal fractions. *Life Sci.* 21: 1629-1636.
30. Rhoades, R. A. (1976): Perfused lung preparation for studying altered gaseous environments. *Environ. Health Pers.* 16:73-75.
31. Saffiotti, U. (1970): Experimental respiratory tract carcinogenesis and its relation to inhalation exposures. In: *Inhalation Carcinogenesis,* M. G. Hanna, P. Nettesheim and J. R. Gilbert, Eds., pp. 27-54. USAEC, Division of Technical Information, CONF-691001.
32. Saltzman, B. E., Burg, W. R., and Ramaswamy, G. (1971): Performance of permeation tubes as standard gas sources. *Environ. Sci. Tech.* 5(11): 1121-1128.
33. Selkirk, J. K. (1977): Benzo[a]pyrene carcinogenesis: a biochemical selection mechanism. *J. Toxicol. Environ. Health* 2:1245-1258.
34. Silverberg, E. (1978): Cancer statistics 1978. Ca–*A Cancer Journal for Clinicians* 28(1):17-34.
35. Tartarelli, R., Davini, P., Morelli, F., and Corsi, P. (1978): Interactions between SO_2 and carbonaceous particulates. *Atmos. Environ.* 12: 289-293.
36. Von Nieding, G.: (1978): Possible mutagenic properties and carcinogenic action of the irritant gaseous pollutants NO_2, O_3 and SO_2. *Environ. Health Pers.* 22:91-92.
37. Warshawsky, D., Niemeier, R. W., and Bingham, E. (1978): Influence of particulates on metabolism of benzo[a]pyrene in the isolated perfused lung. In: *Carcinogenesis, Vol. 3,* R. W. Jones and R. I. Freudenthal, Eds., pp. 347-360, Raven Press, New York.
38. West, P. W., and Gaeke, G. C. (1956): Fixation of sulfur dioxide as disulfitoomercurate (11) and subsequent colorimetric estimation. *Anal. Chem.* 28(12):1816-1819.
39. Wood, A. W., Goode, R. L., Chang, R. L., Levin, W., Conney, A. H., Yagi, H., Dansette, P. M., and Jerina, D. M. (1975): Mutagenic and cytotoxic activity of benzo[a]pyrene 4,5,7,8- and 9,10-oxides and six corresponding phenols. *Proc. Nat. Acad. Sci.* 72(8):3176-3180.
40. Yang, S. K., Gelboin, H. V., Trump, B. F., Autrup, H., and Harris, C. C. (1977): Metabolic activation of benzo[a]pyrene and binding to DNA in cultured human bronchus. *Cancer Res.* 37:1210-1215.

Polynuclear Aromatic Hydrocarbons, edited by P.W. Jones and P. Leber. Ann Arbor Science Publishers, Inc. Ann Arbor MI ©1979.

THE GENETIC REGULATION OF ARYL HYDROCARBON HYDROXYLASE ACTIVITY AND METABOLISM OF POLYCYCLIC AROMATIC HYDROCARBONS IN CELL CULTURE

V. Hitchins, B. Laine and J.J. Hutton

Department of Medicine
University of Kentucky Medical Center
Lexington, Kentucky 40506

Medical Service
Veterans Administration Medical Center
Lexington, Kentucky 40507

INTRODUCTION

A most important class of chemical carcinogens found in the environment and in certain foods is polycyclic aromatic hydrocarbons (PAH). The microsomal mixed-function oxidases metabolize these hydrocarbons into a variety of compounds including electrophilic alkylating intermediates and epoxides. These reactive compounds then interact with DNA, RNA and protein in the cell (2,16,18,20,32,33) and are responsible for the mutagenic, toxic and carcinogenic effects of the parent hydrocarbon.

One of the best studied microsomal mixed-function oxidases is the aryl hydrocarbon hydroxylase (AHH) complex. AHH is named more specifically benzo [a] pyrene hydroxylase, since AHH activity is customarily assayed by incubating tissue homogenates or cells with benzo [a] pyrene and measuring the rate of formation of a complex mixture of fluorescent alkali-soluble metabolities (34). AHH is found in most animal tissues and cells *in vitro* and *in vivo* (8,18,22,36,39,45) and is substrate-inducible (34,35). However, the levels of AHH activity vary from animal species to animal species, and from tissue to tissue within the same animal as well as in various cells in culture (4, 24,36,46).

AHH inducibility in mice is under genetic control (3,15,27,28,38,40, 42-44). Inducibility is defined as the ratio of AHH activity in cells after treatment with inducer to the basal activity before treatment. No particular mechanism of induction of AHH activity is assumed. Inbred strains of mice are genetically homogeneous and can be readily studied for the pattern of inheritance of basal and induced levels of AHH activity. When mice are injected parenterally with the inducer 3-methylcholanthrene (3-MC), the activity of AHH increases in livers from some inbred strains of mice, but not in livers from other strains (15,38,40,42,43). Induction is relatively rapid, and AHH activity generally reaches a maximal level 24 to 36 hr after injection. If AHH activity in liver increases twofold or more over basal or control activity within 24 hr of injection of 3-MC, the mice are classified as "AHH-responsive" or "AHH-inducible." If AHH activity in liver increases less than twofold, the mice are classified as "AHH-nonresponsive" or "AHH-noninducible." Strains C3H/HeJ, C57BL/6J and A/J are classified as AHH-inducible and strains AKR/J, DBA/2J and SWR/J are classified as AHH-noninducible (42). Classification does not take into account changes in AHH activity in other organs, or in liver from animals treated with compounds other than 3-MC. Breeding experiments with mice from different strains provide a way of studying the patterns of inheritance of basal and induced levels of AHH. Early genetic analysis with crosses between C57BL/6J and DBA/2J showed that inducibility of AHH in mouse liver by 3-MC segregated as a single autosomal dominant trait (15,37,43). When the 3-MC-induced livers from the offspring of a cross between C3H/HeJ and DBA/2J were assayed for AHH activity, an additive pattern of inheritance of genes at a single locus was demonstrated (41). When the genetic regulation of AHH activity in mouse liver was studied in intercrosses and backcrosses of numerous additional inbred strains (15,40, 42,43), both dominant and additive patterns of inheritance of induction by 3MC were commonly observed. However, genes at only a small number of loci appear to control AHH activity. It is not know whether the same genes control both basal activities and inducibility.

Classification of a mouse strain as AHH-inducible (high inducibility ratio) does not necessarily indicate that basal (as opposed to induced) levels of AHH activity are higher in that strain than in an AHH-noninducible strain (1,3,4). The basal activities of AHH vary from one mouse strain to the next and must be considered when studying metabolism of polycyclic compounds. For example, AHH activity in lung after treatment of SWR/J mice with 3-MC showed a high inducibility ratio (1). This high induced-to-basal activities ratio was associated with very low basal activities rather than unusually high induced activities (1). It is not clear whether carcinogenicity and mutagenicity of a polycyclic compound metabolized by the AHH enzyme complex are associated with the absolute activity or with the inducibility of AHH in target tissue.

Good correlations between high levels of AHH inducibility in livers and susceptibility of animals to 3-MC-induced tumors have been observed in some situations. Kouri and co-workers (23,25,26,28) showed that inbred strains of mice with high AHH inducibility were more susceptible to both 3-MC-and benzo [a] pyrene-induced fibrosarcomas than mice with low AHH inducibility. Thomas and co-workers (42) showed that there was close strain correlation between AHH inducibility and skin ulceration from topically applied 7,12-dimethylbenzanthracene. Presumably, genes affecting AHH activity in liver also affect AHH activity in skin. In contrast to these results with skin, there is no relationship between AHH inducibility and susceptibility to alveolar adenomas in lung (23).

AHH research has been complicated by the fact that the enzyme system responds to a great variety of stimuli which are difficult to control including the hormonal state of the animal (13,14) and chemicals in the environment. Cell culture is a model system to study the metabolic activation of carcinogens and to investigate quantitatively cell toxicity and malignant transformation, since it is free from the influence of the host. Transformation of cells in culture may occur spontaneously or be induced by chemicals and/or viruses (19,31). Transformation of cells in culture is often used as an *in vitro* measure of carcinogenicity, since transformed cells can produce tumors when injected into an isologous host (6,9,10,12). A great variety of animal tissues (5,10,18, 20) have been used to develop primary and transformed cell cultures for study of chemical carcinogenesis. Some of these cell cultures have AHH activities which are similar to levels observed in mammalian liver microsomes (34) and can metabolize a large number of polycyclic hydrocarbons (35). High levels of AHH activity appear to parallel the sensitivity of cells to polycyclic hydrocarbon-induced cytotoxicity and transformation (5,22). In addition, inhibition of AHH activity by 7,8-benzoflavone prevents 3-MC-induced transformation (5,30).

We have chosen to evaluate the inbred mouse genetic system in tissue culture where control of experimental variables is theoretically much better than *in vivo*. Our purpose is to test whether the genotype at the locus controlling AHH inducibility in livers of adult mice also affects inducibility of cells derived from fetuses and from fetal livers in primary tissue culture. If a basic genetic difference is reflected in tissue culture cells, this system is particularly valuable for establishing the effect of genetically controlled metabolic patterns of polycyclic hydrocarbon metabolism on the efficiency of transformation of cells to malignancy. The results are of relevance to understanding the importance of genetic factors in determining individual susceptibility to carcinogens.

MATERIALS AND METHODS

Cell Cultures

Primary cultures of fetal cells were established from six inbred strains of mice: A/J, C3H/HeJ, C57BL/6J, AKR/J, DBA/2J and SWR/J from the Jackson Laboratory, Bar Harbor, ME. The first three strains of mice have hepatic AHH activity inducible by 3-MC; the latter three strains are hepatic-noninducible *in vivo*. All mice were maintained on Purina Mouse Chow and kept on Ab-Sorb-Dry (Garfield, NJ) bedding in a room with a 12-hr light/dark cycle.

Pregnant mothers carrying 19-day-old fetuses were decapitated and the fetuses were removed. The fetuses (minus head, tail, feet and viscera) were minced and treated with an enzyme solution of 0.1% trypsin (Sigma Chemical Co.), 0.1% collagenase (Worthington Biochemicals), and 0.2% hyaluronidase (Sigma) in Hank's buffer (Gibco) to obtain suspensions of single cells. We refer to these cells as "fetal cells." The fetal livers were minced and treated with 0.1% collagenase in Gibco RPMI–1640 tissue culture medium containing 25 m*M* N-2-hydroxyethylpiperazine-N′-2-ethanesulfonic acid (HEPES; Sigma) buffer, pH 7.4, to obtain single-cell suspensions. We refer to these cells as "fetal liver cells." All cells were grown in RPMI-1640 and 25 m*M* HEPES buffer containing 50 μg gentamycin sulfate/ml (Sigma), 100 units penicillin/ml (Gibco) and 100 μg streptomycin/ml (Gibco) in tissue culture flasks (Corning) at 37°C with 5% CO_2-95% air. Cell viability was determined by Trypan Blue dye exclusion and total cell counts were determined with a hemocytometer. Fetal cells were initially plated at a concentration of 5×10^6 cells per 75 cm^2 tissue culture flask while liver cells were plated at a concentration of 1×10^7 cells per 25 cm^2 flask.

Twelve to fifteen hours after establishing a primary culture, dead and unattached cells were poured off and fresh medium was added to the flasks. DNA synthesis (as measured by the incorporation of ^{3}H-thymidine into DNA) and growth of cells (as measured by the amount of DNA in cells harvested from each tissue culture flask) were determined during induction of AHH from 12-15 hr after cells were initially plated until confluency was reached. Cells were pulsed for 1 hr with methyl-^{3}H-thymidine (50 C/mmole; New England Nuclear) and then washed with phosphate-buffered saline (PBS). Cells were harvested by scraping the flask with a glass rod. The entire contents from each flask were collected, washed with PBS, and aliquoted into individual test tubes. The cells were pelleted at 300 x g at room temperature and then stored at -80°C until used for assay of AHH activity, DNA content and incorporation of ^{3}H-thymidine.

Induction with 3-MC and AHH Assay

Stock solutions of 3-methylcholanthrene (3-MC; Sigma) were made in dimethylsulfoxide (DMSO; Sigma) and added to culture media. AHH activity in cells was not affected by the addition of as much as 0.02% (v/v) DMSO in the media, although no more than 0.01% was added in our experiments. Fetal cells were exposed to 3-MC for 12 hr while fetal liver cells were exposed for 24 hr. Control cells contained the solvent DMSO but no 3-MC. One hour prior to harvesting, ^{3}H-thymidine was added.

Frozen cell pellets were thawed and resuspended in 0.05 *M tris*-HCl buffer containing 3 m*M* $MgCl_2$ and 0.2 *M* sucrose at pH 8.0 for assay of AHH activity using benzo [a] pyrene as substrate. The hydroxylated water-soluble products formed from benzo [a] pyrene were measured fluorometrically after hexane and NaOH extractions (1). AHH activity is expressed as pmoles of 3-OH-benzo [a] pyrene (3-OH-BP) formed in 10 min at 37°C per µg DNA in the sample (all the cells from one tissue culture flask). The standard 3-OH-BP is from the NCI Carcinogenesis Research Program, Bethesda, MD. As little as 3-5 x 10^6 cells can be assayed for AHH activity. The reaction is linear with time up to 30-40 min at 37°C.

DNA Determinations

DNA determinations were made following a modified method of Leyva and Kelley (29). After cells had been assayed for AHH activity, they were sonicated for 5 sec. Bovine serum albumin (BSA; 0.5 mg/tube) was added to each tube to enhance precipitation of DNA after addition of cold 2% potassium acetate in 95% ethanol. DNA was pelleted by centrifugation at 500 x g and dried before hydrolysis with 1 *N* $HClO_4$ at 70°C for 30 min. Diphenylamine reagent in acetic acid and H_2SO_4 was added to each sample and incubated overnight at 37°C. The samples were read at 600 nm. Calf thymus DNA in *tris*-HCl buffer was treated with hexane and NaOH in the same manner as cells, before addition of BSA and potassium acetate in ethanol. This DNA was used to standardize the determination. DNA determinations provided better reliability than cell counts when calculating the specific activity of AHH (17). Under our assay procedure, approximately 5 x 10^6 cells contain 30 µg DNA. Aliquots of the completed DNA assay were removed and put into scintillation vials with Aquasol-2 scintillation fluid (New England Nuclear) to measure the amount of ^{3}H-thymidine incorporated into cellular DNA.

Statistical Analysis

Fetuses from at least two mice per inbred strain were tested at each time point. Each of these included control and induced samples in duplicate or triplicate. Thus, each experimental point represents at least four separate assays. Statistical analysis was provided by the University of Kentucky Tobacco and Health Research Institute.

RESULTS

Primary cultures of cells from eviscerated 19-day-old fetuses and from their livers were established from six inbred strains of mice. Cultures produced from eviscerated fetuses contain primarily fibroblasts while cultures produced from fetal livers contain largely hepatocytes. Our references to fetal cell cultures therefore reflect responses of cultured fibroblasts; references to fetal liver cultures reflect responses to hepatic cells in culture. Inbred mouse strains from which cells were cultured include A/J, C3H/HeJ and C57BL/6J, which are defined as inducible, and AKR/J, SWR/J, and DBA/2J, which are defined as noninducible on the basis of the AHH response of their livers to intraperitoneal injection of 3-MC *in vivo* (15,38,40,42,43). AHH activity, incorporation of ^{3}H-thymidine into DNA, and total DNA present in cells were measured at regular intervals from the times cells were initially plated until the stationary phase.

When 3-MC was added to cultures, fetal cells attained maximum fold induction of AHH activity over the basal level after 9-12 hr of exposure to 3-MC and then decreased with longer induction times. Fetal liver cells reached maximum fold induction of AHH activity after 24 hr of exposure to 3-MC. For convenience, fetal cells were induced for 12 hr with 3-MC at intervals of 12 hr; fetal liver cells were induced for 24 hr with 3-MC at intervals of 24 hr. The inducer 3-MC was tested over a 2,000-fold concentration range on cells from each strain to determine optimal concentrations for producing maximum fold induction of AHH activity over the basal level. Figure 1 illustrates typical profiles of 3-MC concentrations vs fold induction of AHH activity in fetal cells from inducible and noninducible strains of inbred mouse. As can be seen in Figure 1 there is a concentration effect on AHH inducibility with both the inducible and the noninducible strains reaching their maximum fold inductions at 3-MC concentrations of 0.05 to 0.10 μg 3-MC/ml. This was true of all six strains. Therefore, 0.05 and 0.10 μg 3-MC/ml were chosen as concentrations for tests of inducibility of AHH in mouse fibroblasts, with cells from at least two animals of each strain being induced at 0.10 μg 3-MC/ml. Statistical analysis of the final data showed no significant concentration effect ($p = 0.31$) on all six strains when concentration was restricted to 0.05 or 0.10 μg 3-MC/ml, firmly supporting the combining of data from these concentrations of the inducer 3-MC.

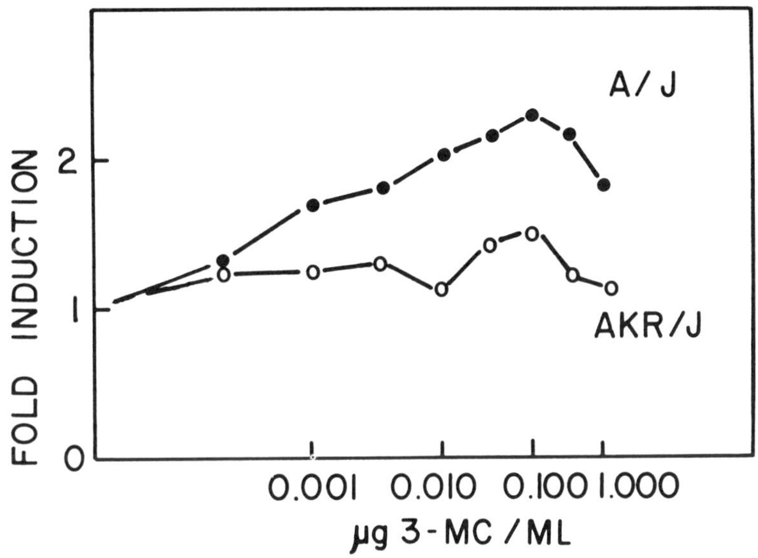

FIGURE 1. Fold induction of AHH activity vs concentration of 3-MC. This figure illustrates typical profiles of fold induction of AHH activity over basal level *vs* concentration of the inducer 3-MC. Fetal cells from an AHH-inducible mouse strain A/J and fetal cells from an AHH noninducible strain SWR/J were treated for 12 hr with various concentrations of 3-MC dissolved in DMSO. Each point represents the mean of duplicate determinations.

Figure 2 illustrates typical data from time-course profiles of cultures from an inducible and a noninducible strain. In all fetal cells both the control and 3-MC-induced levels of AHH activity increased during exponential phase and then reached a plateau during the late exponential and stationary phases of cell growth. At most time points the amount of ^{3}H-thymidine incorporated into 3-MC-induced cells was slightly less than in control cells. However, there were no significant differences in the amount of total DNA in flasks of control and 3-MC-treated cells so that the growth of cells appeared to be the same in the two.

In Figure 3 the data are plotted for AHH activity of control and 3-MC-induced cells over time fetal cells are in culture. Both basal and induced AHH activities increased during exponential growth phase. Various unsuccessful attempts were made to reduce the increase in the basal levels of AHH activity during the cell growth period. Serum was tested for the presence of inducers by varying the serum concentration from 5 to 15% in the media and by pretreating the serum with ether to extract ether-soluble material. In no case was stabilization of the basal AHH activity achieved.

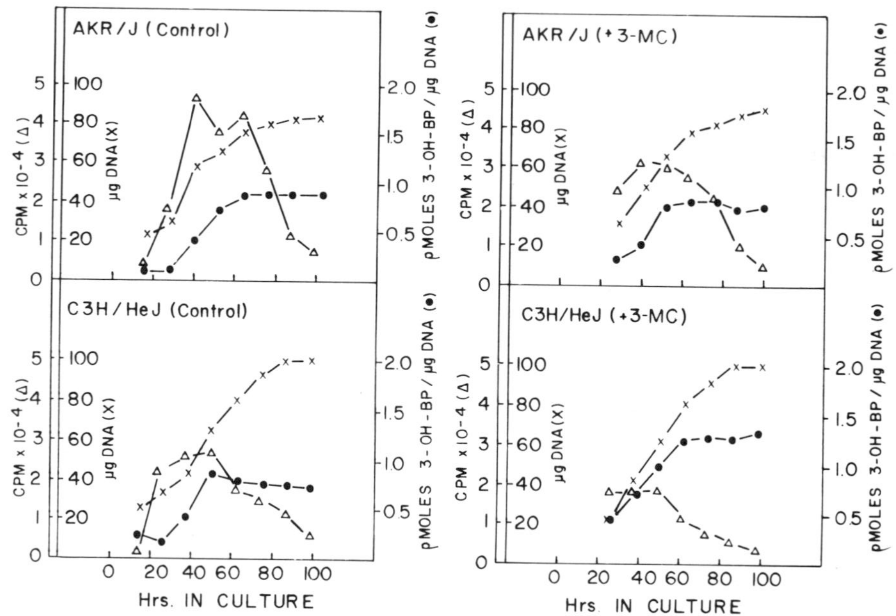

FIGURE 2. Response of cells to 3-MC at various times after cells were placed in culture. Profiles of growth of fetal cells maintained from expo nential to stationary phase are presented. Cells from AKR/J, an AHH-noninducible and C3H/HeJ, an AHH-inducible strain, were placed in culture at zero time. At intervals of approximately 12 hr cells were induced with 0.10 μg 3-MC/ml media for 12 hr. One hour prior to harvesting duplicate control and duplicate 3-MC-induced cells, methyl-^{3}H-thymidine was added to each flask. Cultures were harvested and assayed for AHH activity, DNA and ^{3}H-thymidine content as described in Materials and Methods. Data on the ordinate record experimental values corresponding to the time cells were harvested, *i.e* 12 hr after 3-MC or control solvent was added. Symbols: ●, AHH activity (ρmoles 3-OH-BP/μg DNA); X, μg DNA per flask; △, ^{3}H-thymidine incorporation (cpm per flask).

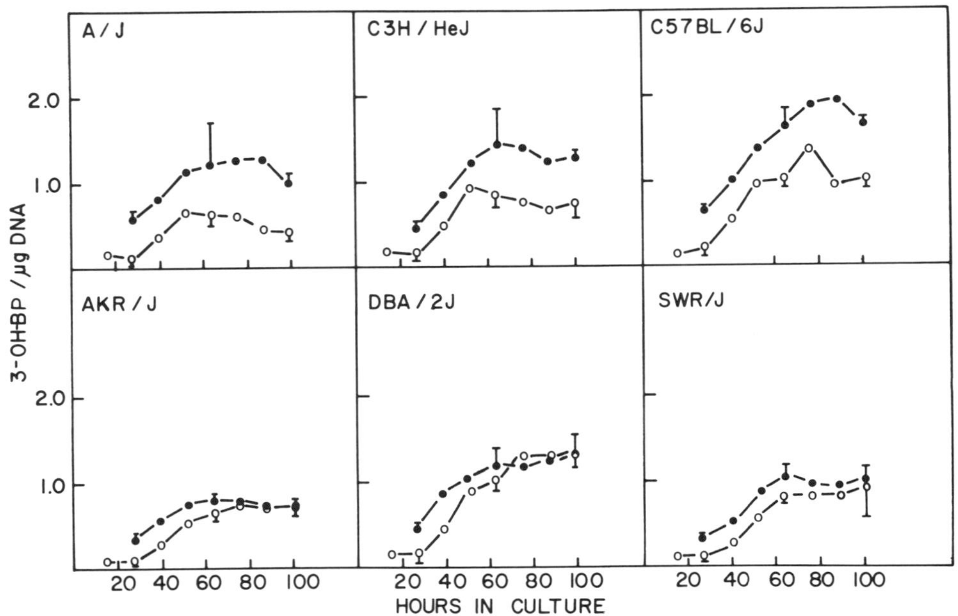

FIGURE 3. AHH activity vs hours in culture for fetal cells from six inbred strains of mice. Fetal cells from three hepatic AHH-inducible and three hepatic AHH-noninducible inbred strains of mice were assayed for 3-MC-induced AHH activity from 16 hr after cells were initially plated to 100 hr when cells were in stationary growth phase. Cultures from two or more mice were used for each strain, and each time point was assayed in duplicate or triplicate for each mouse. Fetal cells were induced with 0.05 or 0.10 µg 3-MC/ml media and harvested 12 hr later. At each time point the average specific activity of AHH is plotted. The standard deviation of the mean is indicated at several points so that the variability of the measurement can be assessed. Symbols: ●, 3-MC-induced cells; 0, control cells.

Data in Figure 4 are taken from Figure 3 and replotted as the fold induction of AHH activity vs time in culture for each of the six strains. A mean plateau value for each experimental set of points (fetal cell cultures from one mother treated identically and sampled at 12-hr intervals from plating until confluency) was calculated as the mean AHH activity of the terminal series of points whose values from one point to the next differed by less than 20%. A ratio of the induced plateau level to the control plateau level was then calculated for each culture. In Figure 4 the final plateau value is illustrated by a horizontal line across the plot of each strain and is the mean induced:control plateau ratio for all individuals tested from a given strain. These plateau values fell into two groups, one with values near 1.0 and the other with values near 2.0. Strains with low values were AKR/J (1.05), DBA/2J (0.96) and SWR/J (1.25). Strains with high values were A/J (2.21), C3H/HeJ (1.74) and C57BL/6J (1.86). Clearly the six strains, on the basis of mean ratio of induced to control plateau level of AHH activity, clustered into the same two groups as they do on the basis of *in vivo* induction of AHH in liver tissue with 3-MC. Statistical analysis (linear model) of the data supports the division of these six strains into two groups on the basis of only one factor–inducibility. Among all six strains of mice, there was a significant inducibility effect ($p < 0.0001$). The concentrations of 3-MC over the range 0 to 1.0 μg/m1 media does affect inducibility. However, over the range of 0.05 to 0.10 μg 3-MC/m1, there was no concentration effect ($p = 0.31$). With the inducer concentration restricted to 0.05–0.10 μg 3-MC/m1, there was no interaction between strain and concentration of inducer ($p = 0.53$). Ninety-one percent of the variance in our data accounted for in the linear model is due to the differences of plateau values when the cell strains are classified as either AHH-inducible or AHH-noninducible, and the remaining 9% of the variance is due to differences of plateau values among the three cell strains within either the inducible or the noninducible group.

AHH activity in fetal liver cells followed the same basic pattern shown by fetal fibroblasts. Both basal and induced values of AHH activity increase during exponential growth and reach a plateau as the cells reach confluency. Typical data of liver cultures representing an inducible strain, C3H/HeJ, and a noninducible strain, SWR/J, are shown in Figure 5. Plateau levels of "fold induction" are 1.6 for SWR/J and 3.2 for C3H/HeJ. Thus, the noninducible strain shows some induction in cultured fetal hepatocytes. However, inducibility of strain C3H/HeJ was twice that of strain SWR/J. In general, fetal liver cells showed higher AHH values than did fetal cells, reaching induced values of AHH activity as high as 4.0 ρmoles 3-OH-BP/μg DNA while the maximum reached by fetal cells was slightly below 2.0 ρmoles 3-OH-BP/μg DNA. It should be noted that incorporation of ^{3}H-thymidine by fetal liver cells in the presence of the inducer, 3-MC , was markedly lower than incorporation into

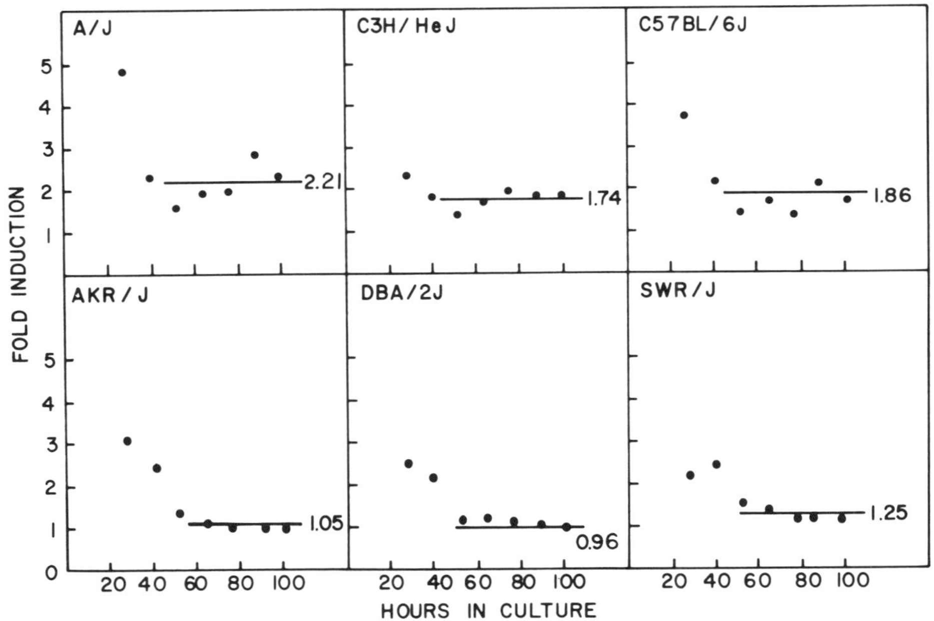

FIGURE 4. Fold induction of AHH activity vs hours in culture. Data from Figure 3 were replotted as fold induction of AHH activity vs hours in culture from early exponential to stationary growth phase. Plateau values were calculated as follows: individual plateau levels of AHH activity for induced and control sets of points for a given mouse were determined by calculating the mean AHH activity of the terminal series of points whose values from one point to the next differed by less than 20%. A ratio of the induced plateau level to the control plateau level was then calculated for each culture. The final plateau value for cells from each mouse strain is the mean of induced:control plateau ratios for all individuals tested from one strain.

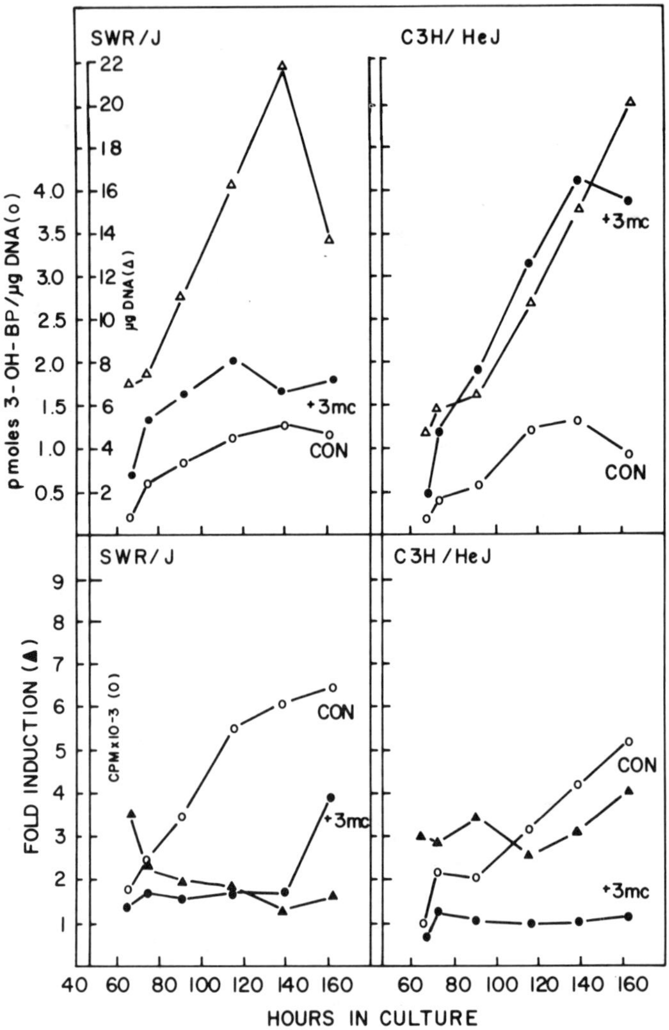

FIGURE 5. Time course of AHH activity during fetal liver cell growth period. Two typical profiles of fetal liver cells maintained during the growth period are illustrated in this figure. SWR/J, an AHH-noninducible, and C3H/HeJ, an AHH-inducible strain *in vivo*, were induced with 0.10 µg 3-MC/ml media for 24 hr. Both control and induced fetal liver cells were pulsed with ^{3}H-thymidine for 1 hr and then harvested. AHH activity, DNA content and ^{3}H-thymidine incorporation were assayed as described in Materials and Methods. Data on the ordinate record the experimental values corresponding to the time cells were harvested. Top two panels show control (0) and 3-MC-induced (●) AHH activities (ρmoles 3-OH-BP/µg DNA) and µg DNA per flask (△) over time. Bottom two panels illustrate the fold induction of AHH activity (▲), and amount of ^{3}H-thymidine incorporated per flask by control (0) and 3-MC induced (●) cells at different times during the cell growth period.

the controls, while amounts of DNA in cultures of 3-MC treated and untreated fetal liver cells did not differ significantly from each other. This change in the rate of incorporation of ^{3}H-thymidine, which is more marked in fetal liver cultures than in other fetal cultures, may reflect changes in the rate of synthesis of DNA. However, this is not necessarily so since changes in the rate of transport or pool size of ^{3}H-thymidine could have occurred.

DISCUSSION

Fetal cells and fetal liver cells were cultured from six inbred strains of mice. Livers from three of the strains are AHH-inducible and three are AHH-non-inducible. We studied the cell cultures for the pattern of AHH induction during the cell growth phase and determined whether there is a correlation between control of AHH inducibility *in vitro* and *in vivo*. Since inducibility of AHH in liver is regulated by a small number of genes *in vivo*, a strong correlation would be taken to imply regulation by the same genes *in vitro*.

We observed that AHH activity could be induced by 3-MC within 24 hr after cells were started in culture. Both basal and induced AHH levels increased during exponential growth phase. The ratio of induced to basal levels of AHH activity was very high in the early exponential growth period followed by a rapid decrease, about 40-50 hr after plating, to a plateau value which was maintained to confluency of cells. Bittner and Ruddon (7) also observed in their mouse 3T3 cells induced with 3-MC an increase in AHH activity during exponential phase with the lowest activity of AHH in confluent nonfed cells. Nebert and Gelboin (35) reported that in hamster fetal cells 3-MC induction of AHH activity was greatest in cells that were beginning exponential growth phase. A similar pattern of increase in AHH activity during early exponential phase followed by a slow decrease in late exponential and early stationary phase was observed in benzanthracene or 7,12-dimethylbenzanthracene-induced hamster fetal cells (34).

A difficulty with cultured cell systems is the dependence on the addition of fetal calf serum to the defined medium and the presence of potential stimuli in serum for AHH induction. We considered the possibility that substances present in fetal calf serum might be causing the rise in basal levels of AHH activity with time during exponential growth. Gielen and Nebert (14) reported significant changes in basal levels of AHH activity in fetal rat liver cultures grown in different concentrations of fetal calf serum. However, our experiments using 5%, 10%, 15% or 20% ether-extracted serum showed no significant reduction in the rising basal level of AHH activity during exponential phase. Thus all our data reflect cultures grown in medium containing 10% heat-inactivated fetal calf serum from the same lot throughout the course of this project so that all cultures were treated identically.

Our observations suggest that high basal activity of AHH as shown with DBA/2J fetal cells is not necessarily associated with high AHH inducibility (ratio of 3-MC induced to basal activity), a view which has been reported by previous investigators (1,11,24,43). For example, for our six inbred strains of mice, the statistical correlations for basal plateau level:induced plateau level, basal plateau level:fold induction, and induced plateau level:fold induction are $r = 0.53$ ($p = 0.01$, $n = 23$), $r = -0.47$ ($p = 0.02$, $n = 23$), and $r = 0.45$ ($p = 0.03$, $n = 23$), respectively.

We examined the mean plateau values of fold induction of AHH activity for our six fetal cell cultures to see whether our results obtained in the *in vitro* system reflect the same genetic control of AHH inducibility observed *in vivo*. Statistical analyses with the linear model showed that our data did distinguish between AHH-inducible and AHH-noninducible fetal cell cultures ($p < 0.0001$) with either 0.05 or 0.10 μg 3-MC/m1 media as inducer. We then compared our plateau values for fetal cells with fold induction of AHH activity in adult livers from mice injected parenterally with 3-MC. Fetal cells derived from A/J, AKR/J, DBA/2J and SWR/J mice had plateau values which are similar to those reported for A/J, AKR/J, DBA/2J and SWR/J livers whose AHH-induction ratios are 3.1, 0.94, 0.85 and 1.3, respectively (42). The plateau values of inducibility of AHH activity for C57BL/6J and C3H/HeJ fetal cells are lower than the AHH-inducibility ratios in livers *in vivo*, which are 4.8 and 4.9 for each respective mouse strain (42). Nebert and Gelboin (36) reported lower AHH-inducibility ratios in nonhepatic tissues than in liver and found that the extent of AHH induction was greater in adult than in fetal tissues.

Cultured fetal liver cells follow the same basic patterns which have been discussed for cultured fetal cells except that absolute activities were notably higher, reaching values twice as high as those attained in nonhepatic cultures. As in fetal tissue, the inducible strain *in vivo*, C3H/HeJ, showed a considerably higher fold induction in culture than the noninducible strain, SWR/J.

In conclusion, our experimental observations support the idea that induction of AHH activity by 3-MC in fetal cell cultures developed from inbred mice are under the same genetic control as induction of AHH activity by 3-MC in liver *in vivo*. Fetal cell cultures can be classified as either AHH-inducible or AHH noninducible by their AHH-induction ratios obtained during the late exponential to stationary phases of cell growth. Fetal cells and fetal liver cells that are AHH-inducible were derived from inbred strains of mice whose livers are AHH-inducible *in vivo*; fetal cells and fetal liver cells with low AHH induction ratios *in vitro* were derived from inbred strains of mice whose livers are AHH noninducible *in vivo*. The strain-dependent metabolic differences of these six fetal cell cultures also point out the probable importance of strain of origin of cell cultures when testing substances for carcinogenicity and mutagenicity.

ACKNOWLEDGMENTS

This research was supported by the University of Kentucky Tobacco and Health Research Institute Project 22043 and by Veterans Administration Research Project 596-3843. We thank Drs. John Turbek, Dennis Haack and Harley McKean for their invaluable help with the statistical analyses.

REFERENCES

1. Abramson, R.K., and Hutton, J.J. (1975): Effects of cigarette smoking on aryl hydrocarbon hydroxylase activity in lungs and tissues of inbred mice. *Cancer Res.* 35:23-29.
2. Ames, B.N., Durston, W.E., Yamasaki, E., and Lee, F.D. (1973): Carcinogens are mutagens: A simple test system combining liver homogenates for activation and bacteria for detection. *Proc. Nat. Acad Sci., U.S.* 70:2281-2285.
3. Atlas, S.A., Taylor, B.A., Diwan, B.A., and Nebert, D.W. (1976): Inducible monooxygenase activities and 3-methylcholanthrene-initiated tumorigenesis in mouse recombinant inbred sublines. *Genetics* 83:537-550.
4. Benedict, W.F., Considine, N., and Nebert, D.W. (1973): Genetic differences in aryl hydrocarbon hydroxylase induction and benzo [a] pyrene-produced tumorigenesis in the mouse. *Mol. Pharmacol.* 9:266-277.
5. Benedict, W.F., Gielen, J.E., and Nebert, D.W. (1972): Polycyclic hydrocarbon-producing toxicity, transformation, and chromosomal aberrations is a function of aryl hydrocarbon hydroxylase activity in cell culture. *Int. J. Cancer* 9:435-451.
6. Benedict, W.F., Rucker, N., Faust, J., and Kouri, R.E. (1975): Malignant transformation of mouse cells by cigarette smoke condensate. *Cancer Res.* 35:857-860.
7. Bittner, M.A., and Ruddon, R.W. (1976): Induction and decay of aryl hydrocarbon hydroxylase activity in mouse 3T3 cells. *Mol. Pharmacol.* 12:966-976.
8. Burki, K., Liebelt, A.G., and Bresnick, E. (1973): Expression of aryl hydrocarbon hydroxylase induction in mouse tissues *in vivo* and in organ culutres. *Arch. Biochem. Biophys.* 158:641-649.
9. Casto, B.C., and DiPaulo, J.A. (1973): Viruses, chemicals and cancer. *Prog. Med. Virol.* 16:1-47.
10. Chen, T.T., and Heidelberger, C. (1969): Quantitative studies on the malignant transformation of mouse prostate cells by carcinogenic hydrocarbons *in vitro. Int. J. Cancer* 4:166-178.
11. Diamond, L., and Baird, W.M. (1977): Relationship of aryl hydrocarbon hydroxylase activity to benzo [a] pyrene-metabolizing activity of cells in culture. *Cancer Lett* 3:189-195.

12. Evans, C.H., and DiPaolo, J.A. (1975): Neoplastic transformation of guinea pig fetal cells in culture induced by chemical carcinogens. *Cancer Res.* 35:1035-1044.
13. Gelboin, H.V., and Wiebel, F.J. (1971): Studies on the mechanism of aryl hydrocarbon hydroxylase induction and its role in cytotoxicity and tumorigenicity. *Ann. N.Y. Acad. Sci.* 179:529-547.
14. Gielen, J.E., and Nebert, D.W. (1972): Aryl hydrocarbon hydroxylase induction in mammalian liver cell culture. III. Effects of various sera, hormones, biogenic amines, and other endogenous compounds on the enzyme activity. *J. Biol. Chem.* 247:7591-7602.
15. Gielen, J.E., Goujon, F.M., and Nebert, D.W. (1972): Genetic regulation of aryl hydrocarbon hydroxylase induction. II. Simple Mendelian expression in mouse tissues *in vivo. J. Biol. Chem.* 247:1125-1137.
16. Grover, P.L., Hewer, A., Pal, K., and Sims, P. (1976): The involvement of a diol-epoxide in the metabolic activation of benzo [a] pyrene in human bronchial mucosa and in mouse skin. *Int. J. Cancer* 18:1-6.
17. Gurtoo, H.L., Minowada, J., Paigen, B., Parker, N.B., and Hayner, N.T. (1977): Factors influencing the measurement and the reproducibility of aryl hydrocarbon hydroxylase activity in cultured human lymphocytes. *J. Nat. Cancer Inst.* 59:787-798.
18. Harris, C.C., Autrup, H., Stoner, G., Yang, S.K., Leutz, J.C., Gelboin, H.V., and Trump, B.F. (1977): Metabolism of benzo [a] pyrene and 7,12-dimethylbenzo [a] pyrene in cultured human bronchus and pancreatic duct. *Cancer Res.* 37:3349-3355.
19. Heidelberger, C. (1973): Chemical oncogenesis in culture. *Adv. Cancer Res.* 18:317-366.
20. Heidelberger, C. (1975): Chemical carcinogenesis. *Ann. Rev. Biochem.* 44:79-121.
21. Huberman, E., Aspiras, L., Heidelberger, C., Grover, P.L., and Sims, P. (1971): Mutagenicity to mammalian cells of epoxides and other derivatives of polycyclic hydrocarbons. *Proc. Nat. Acad. Sci., U.S.* 68:3195-3199.
22. Huberman, E., Selkirk, J.K., and Heidelberger, C. (1971): Metabolism of polycyclic aromatic hydrocarbons in cell cultures. *Cancer Res* 31: 2161-2167.
23. Kouri, R.E. (1976): Relationship between levels of aryl hydrocarbon hydroxylase activity and susceptibility to 3-methylcholanthrene and benzo [a] pyrene-induced cancers in inbred strains of mice. In: *Carcinogenesis, Vol. I*, R. I. Freudenthal and P.W. Jones, eds., pp. 139-151, Raven Press, New York.
24. Kouri, R.E., Kiefer, R., and Zimmerman, E.M. (1974): Hydrocarbon metabolizing activity of various mammalian cells in culture. *In Vitro* 10:18-25.
25. Kouri, R.E., Ratrie, H., and Whitmire, C.E. (1973): Evidence of a genetic relationship between methylcholanthrene-induced subcutaneous

tumors and inducibility of aryl hydrocarbon hydroxylase. *J. Nat Cancer Inst.* 51:197-200.

26. Kouri, R.E., Ratrie, H., and Whitmire, C.E. (1974): Genetic control of susceptibility to 3-methylcholanthrene-induced subcutaneous sarcomas. *Int. J. Cancer* 13:714-720.
27. Kouri, R.E., Rude, T., Thomas, P.E., and Whitmire, C.E. (1976): Studies on pulmonary aryl hydrocarbon hydroxylase activity in inbred strains of mice. *Chem-Biol. Interact.* 13:317-331.
28. Kouri, R.E., Salerno, R.A., and Whitmire, C.E. (1973): Relationships between aryl hydrocarbon hydroxylase inducibility and sensitivity to chemically induced subcutaneous sarcomas in various strains of mice. *J. Nat. Cancer Inst.* 50:363-368.
29. Leyva, A., Jr., and Kelley, W.N. (1974): Measurement of DNA in cultured human cells. *Anal. Biochem.* 62:173-179.
30. Marquardt, H., and Heidelberger, C. (1972): Influence of "feeder cells" and inducers and inhibitors of microsomal mixed-function oxidases on hydrocarbon-induced malignant transformation of cells derived from C3H mouse prostrate. *Cancer Res.* 32:721-725.
31. Marquardt, H., Kuroki, T., Huberman, E., Selkirk, J.K., Heidelberger, C., Grover, P.L., and Sims, P. (1972): Malignant transformation of cells derived from mouse prostate by epoxides and other derivatives of polycyclic hydrocarbons. *Cancer Res.* 32:716-720.
32. Miller, E.C. (1978): Some current perspectives on chemical carcinogenesis in humans and experimental animals: Presidential Address. *Cancer Res.* 38:1479-1496.
33. Miller, E.C., and Miller, J.A. (1976): The metabolism of chemical carcinogens to reactive electrophiles and their possible mechanisms of action in carcinogenesis. In: *Chemical Carcinogenesis,* American Chemical Society Monograph, Number 173, C.E. Searle, ed., pp. 737-762.
34. Nebert, D.W., and Gelboin, H.V. (1968): Substrate-inducible microsomal aryl hydrocarbon hydroxylase in mammalian cell culture. I. Assay and properties of induced enzyme. *J. Biol. Chem.* 243:6242-6249.
35. Nebert, D.W., and Gelboin, H.V. (1968): Substrate-inducible microsomal aryl hydroxylase in mammalian cell culture. II. Cellular responses during enzyme induction. *J. Biol. Chem.* 243:6250-6261.
36. Nebert, D.W., and Gelboin, H.V. (1969): The *in vivo* and *in vitro* induction of aryl hydrocarbon hydroxylase in mammalian cells of different species, tissues, strains, and developmental and hormonal states. *Arch. Biochem. Biophys.* 134:76-89.
37. Nebert, D.W., Goujon, F.M., and Gielen, J.E. (1972): Aryl hydrocarbon hydroxylase induction by polycyclic hydrocarbons: simple autosomal dominant trait in the mouse. *Nature New Biol.* 236:107-110.
38. Nebert, D.W., Robinson, J.R., Niwa, A., Kumaki, K., and Poland, A.P.

(1975): Genetic expression of aryl hydrocarbon hydroxylase activity in the mouse. *J. Cell Physiol.* 85:393-414.

39. Pelkonen, O. (1976): Metabolism of benzo [a] pyrene in human adult and fetal tissues. In: *Carcinogenesis, Vol. I,* R.I. Freudenthal and P.W. Jones, eds., pp. 9-21, Raven Press, New York.
40. Robinson, J.R., Considine, N., and Nebert, D.W. (1974): Genetic expression of aryl hydrocarbon hydroxylase induction. Evidence for the involvement of other genetic loci. *J. Biol. Chem.* 249:5851-5859.
41. Thomas, P.E., and Hutton, J.J. (1973): Genetics of aryl hydrocarbon hydroxylase induction in mice: additive inheritance in crosses between C3H/HeJ and DBA/2J. *Biochem. Genetics* 8:249-257.
42. Thomas, P.E., Hutton, J.J., and Taylor, B.A. (1973): Genetic relationship between aryl hydrocarbon hydroxylase inducibility and chemical carcinogen induced skin ulceration in mice. *Genetics* 74:655-659.
43. Thomas, P.E., Kouri, R.E., and Hutton, J.J. (1972): The genetics of aryl hydrocarbon hydroxylase induction in mice: a single gene difference between C57BL/6J and DBA/2J. *Biochem. Genetics* 6:157-168.
44. Thorgeirsson, S.S., and Nebert, D.W. (1977): The *Ah* locus and the metabolism of chemical carcinogens and other foreign compounds. *Adv. Cancer Res.* 25:149-193.
45. Whitlock, J.P., Jr., and Gelboin, H.V. (1974): Aryl hydrocarbon (benzo [a] pyrene) hydroxylase induction in rat liver cells in culture. *J. Biol. Chem.* 249:2616-2623.
46. Wiebel, F.J., Gelboin, H.V., and Coon, H.G. (1972): Regulation of aryl hydrocarbon hydroxylase in intraspecific hybrids of human, mouse, and hamster cells. *Proc. Nat. Acad. Sci. U.S.* 69:3580-3585.

Polynuclear Aromatic Hydrocarbons.
edited by P. W. Jones and P. Leber.
Ann Arbor Science Publishers, Inc.
Ann Arbor, MI ©1979.

DIFFERENCES IN PATHWAYS OF POLYCYCLIC AROMATIC HYDROCARBON METABOLISM AS DETECTED BY ANALYSIS OF THE CONJUGATES FORMED

William M. Baird, Ruth Chemerys, Andrew A. Erickson
Ching Jer Chern and Leila Diamond

The Wistar Institute of Anatomy and Biology
36th Street at Spruce
Philadelphia, Pennsylvania 19104

The metabolites of polycyclic aromatic hydrocarbons that are formed by cells in culture are retained in the culture vessle, thus permitting quantitative studies of hydrocarbon metabolism. Studies on hydrocarbon-induced transformation, mutation and cytotoxicity can be carried out in the same cell cultures so that it is possible to determine, in these systems, the role of various hydrocarbon metabolic pathways in the induction of biological effects (see, *e.g.*, 11,17,18,20,24).

Analyses of the hydrocarbon metabolites present in the medium of the cell cultures are complicated by the fact that the medium contains salts and proteins that interfere with direct chromatography. Therefore, analyses are routinely carried out by extracting the hydrocarbon metabolites into an organic solvent and chromatographing this organic extract, using either thin-layer chromatography (TLC) or high-pressure liquid chromatography (HPLC) (10, 13-15,19,22,23). With many types of cells, however, a major portion of the metabolites formed remains in the water phase. For example, in early-passage hamster embryo cell cultures, more than 80% of the metabolites formed from a low concentration of benzo[a]pyrene (BP) after 24 hr are water-soluble (2).

To understand the pathways of hydrocarbon oxidation in cells, it is necessary to be able to identify the water-soluble metabolites. We have found that a major portion of the water-soluble hydrocarbon metabolites formed in hamster embryo cells are glucuronic acid conjugates (1,2). In the case of BP, these are mainly glucuronides of 9-hydroxy-BP and 3-hydroxy-BP (2). With the methyl-substituted hydrocarbon, 7,12-dimethylbenz[a]anthracene (DMBA), the major water-soluble metabolites are glucuronides of at least two DMBA phenols, but the identity of these phenols has not yet been established (1). Cohen and coworkers (5-8) have found that short-term organ cultures from rodent tissues can form glucuronic acid and sulfate conjugates of BP metabolites.

The studies described here were done to determine whether metabolism of BP to phenols and conjugation of these phenols with glucuronic acid is a major pathway of BP metabolism in cells from various species. We also describe a new extraction procedure which permits the isolation of intact hydrocarbon-glucuronide conjugates from cell culture medium.

METHODS

Hamster embryo cell cultures were prepared from 14-day-old Syrian hamster embryos as described previously (3). For studies with late-passage hamster embryo cells, cells were passaged at a 1:10 ratio once weekly. The sources of the other cell lines studied are as follows: ME1109, mouse embryo cell line derived from C57B1/10J mice by Dr. T. Shimada of the Wistar Institute; IT22, cell line derived from a clone of Swiss mouse 3T3 cells (9); WIRL-3C, cell line derived from liver of a weanling rat (12); WI-38, human diploid fibroblast line derived from embryonic lung (16); human embryonic kidney (third passage); bovine endothelial cell line derived from vascular tissue (21).

Hydrocarbon metabolism was measured by treating the cells with 0.5 nmole/m1 of [G-^{3}H]BP (3 to 21 Ci/mmole) for 24 hr or [G-^{3}H]DMBA (12 Ci/mmole) for 48 hr (1,2). Medium samples for β-glucuronidase treatment were diluted 1:1 with 0.2 *M* sodium acetate buffer (pH 4.5) and incubated with 2,000 Fishman units of β-glucuronidase from bovine liver at 37°C for 2 hr prior to organic solvent extraction; untreated samples were incubated without enzyme (1,2). Metabolites were extracted from medium samples (0.2 to 0.4 m1) either with the chloroform-methanol-water procedure described previously (3) or by extraction twice with two volumes of ethyl acetate after diluting each sample to a volume of 1 m1 with water. The amount of radioactivity in each phase was determined by liquid scintillation counting of 0.1-m1 aliquots. The metabolites in the organic solvent phases were analyzed by TLC on silica gel chromatogram sheets as described previously (2,3) or by HPLC (22). Media samples from BP-treated cells were prepared for HPLC as described (1) and chromatographed on a 4.6-mm x 25-cm Spherisorb 5-μm ODS reverse-phase column at 30°C on a Model 312 Altex chromatograph. Samples were eluted with a 60-min concave gradient (Altex exponent 3) from 3:2 to 4:1 methanol-water followed by 10 min at 4:1 with a flowrate of 1 m1/min. One hundred forty fractions (0.5 min) were collected in scintillation vials and counted (1). Elution positions of markers of BP metabolites were determined by UV absorbance.

RESULTS AND DISCUSSION

Extraction of Hydrocarbon-Glucuronides

In earlier studies, we showed that the BP-phenol-glucuronides present in the medium can be cleaved to release primary oxidation products of BP by direct treatment of medium samples with β-glucuronidase (2). By comparing the chromatograms of organic solvent extracts of enzyme-treated samples with those from control incubations without enzyme, both the primary oxidation products that were free and those that were originally conjugated to glucuronic acid can be determined (1,2). HPLC analysis of ethyl acetate extracts of two such samples are shown in Figure 1. The extract of the control sample contained mainly BP-9,10-dihydrodiol and BP whereas the extract of the β-glucuronidase-treated sample contained similar amounts of those compounds as well as a small peak of quinones and a large peak of 9-hydroxy-BP. The control sample, however, also contained a large peak of polar material in fractions 5-8 that was absent from the β-glucuronidase-treated sample; this polar material was not present in control samples extracted with chloroform.

This suggested to us that BP-glucuronides could be selectively extracted. Therefore, we investigated the effects of solvent and pH on the extraction of BP metabolites from medium (Table I). When medium or medium diluted with the pH 4.5 acetate buffer used for enzyme treatment was extracted with chloroform-methanol-water, the amount of radioactivity recovered in the chloroform phases was the same. β-Glucuronidase treatment of the medium, however, converted 43% of the radioactivity from water-soluble to chloroform-soluble metabolites (Table I).

When medium was extracted with ethyl acetate, the proportion of organic solvent-soluble metabolites was the same as in the chloroform extract and β-glucuronidase treatment increased the percentage of organic solvent-extractable radioactivity by the same amount. However, if the medium was diluted with 0.2 *M* acetate buffer (pH 4.5) prior to extraction, the same percentage of radioactivity was extracted into the ethyl acetate as in the enzyme-treated sample (Table 1). The effect was due to the acidic pH and not to an increase in salt concentration: dilution of medium with 0.2 *M* NaCl, 0.01 *M* tris (pH 7.0) did not increase the amount of radioactivity extracted into ethyl acetate above that from the medium alone whereas the addition of either hydrochloric or acetic acid to a pH of 4.5 resulted in extraction of the same amount of radioactivity as dilution with 0.2 *M* acetate buffer (pH 4.5). The acidic pH was not sufficient to cleave BP-phenol-glucuronides to phenols, for no free BP-phenols were detected in the HPLC of the control sample (Figure 1). We have also shown that BP-dihydrodiols are not converted to phenols by incubation in pH 4.5 buffer at 37°C for 2 hr (2). Thus, extraction of

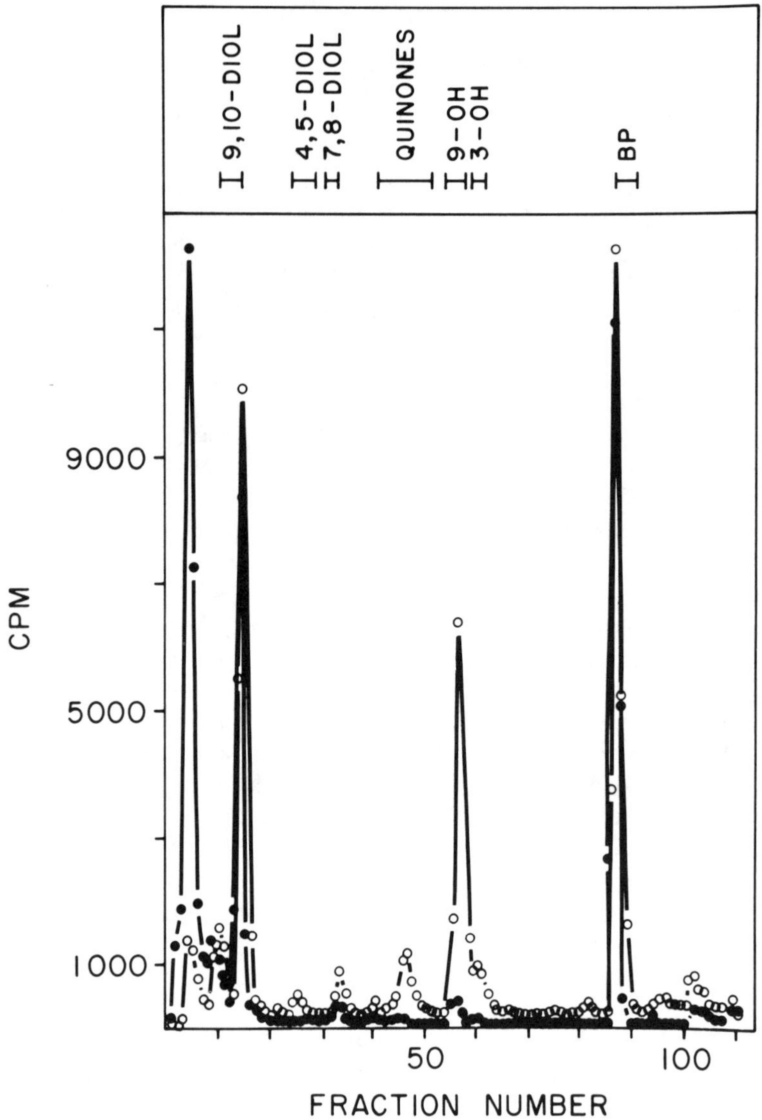

FIGURE 1. HPLC of BP metabolites in ethyl acetate extracts of medium samples from a third-passage culture of hamster embryo cells exposed to 0.5 nmole [^{3}H] BP/ml medium for 24 hr. ○- - - -○, β-Glucuronidase-treated medium; ● ——●, control medium incubated without enzyme. The samples were prepared as described in Methods, but chromatography was carried out on a 3.2-mm x 25-cm Lichrosorb RP18 10-μ column with a 40-min linear gradient from 1:1 to 17:3 methanol:water at 50°C. The elution positions of BP metabolite markers are at the top.

TABLE I. Effect of pH on BP Metabolite Extraction

Extraction Conditions	Percent Radioactivity in Organic Solvent Phase
Chloroform-Methanol Procedure	
Medium	42
β-Glucuronidase-treated medium	75
Medium + 0.2 *M* acetate buffer (pH 4.5)	41
Ethyl Acetate Procedure	
Medium	42
β-Glucuronidase-treated medium	75
Medium + 0.2 *M* acetate buffer (pH 4.5)	74
Medium + 0.2 *M* NaCl (pH 7.0)	44
Medium + hydrochloric acid to pH 4.5	77
Medium + acetic acid to pH 4.5	78

medium samples with ethyl acetate, followed by acidification to pH 4.5 and reextraction with ethyl acetate is a rapid procedure for isolating BP-phenol-glucuronides. These may then be cleaved with β-glucuronidase to identify the primary oxidation products or chromatographed directly for further analyses.

The glucuronide conjugates formed when the methyl-substituted hydrocarbon, DMBA, is metabolized by hamster embryo cell cultures can also be isolated by this procedure. They are extracted into ethyl acetate after, but not prior to, acidification of the medium (Table II).

Formation of BP-Phenol-Glucuronides in Hamster Embryo Cells

We had shown previously that in early-passage hamster embryo cells the formation of BP-phenol-glucuronides is a major pathway of BP metabolism (2). To determine if this remained a major pathway of hydrocarbon metabolism after prolonged subcultivation of these cells, BP metabolism was compared in cells at the 3rd, 10th, and 40th passages (Table III). In early-passage cells, the majority of the chloroform-extractable metabolites were dihydrodiols and β-glucuronidase treatment released mainly phenols and some quinones. At passage 10, the amount of BP metabolized was decreased, but the types and proportions of metabolites formed were similar to those at passage 3. At passage 40, less BP was metabolized than at passage 10 and the nature of the metabolites formed had changed; almost no dihydrodiols were formed and the only identifiable metabolites were the phenols released by β-glucuronidase treatment. The HPLC profiles of these samples are shown in Figure 2: peaks of 9-hydroxy BP, 3-hydroxy BP and BP-quinones were present in the β-glucuronidase-treated sample whereas no major metabolites were

TABLE II. Effect of pH on DMBA Metabolite Extraction

Extraction Conditions	Percent Radioactivity in Organic Solvent Phase
Medium + chloroform-methanol	66
β-Glucuronidase-treated medium + chloroform-methanol	83
Medium + ethyl acetate	65
Medium + 0.2 *M* acetate buffer (pH 4.5) + ethyl acetate	82
Medium + hydrochloric acid to pH 4.5 + ethyl acetate	81

TABLE III. BP Metabolism by Hamster Embryo Cell Cultures

		Percent of Original BP Recovered as				
Passage	β-Glucuronidase Treatment	Water-Soluble Metabolites	Diols	Phenols	Quinones	BP
3	–	71	11	2	1	15
	+	30	13	20	6	14
10	–	41	8	2	1	41
	+	18	10	16	2	43
40	–	28	2	1	2	57
	+	15	1	7	3	62

present in the control. Thus, formation of BP-phenol glucuronides remained a major pathway of BP metabolism in hamster embryo cells after 40 passages in culture and, at that passage level, only the analysis of the glucuronide conjugates contributed information about the pathways of BP metabolism.

Formation of BP-Phenol-Glucuronides in Cells from other Species

To determine if glucuronide conjugates can be formed when BP is metabolized by cells of other species, BP metabolism was studied in the cells described in Methods (Table IV). Both mouse cell lines, ME1109 and IT22, formed high proportions of water-soluble metabolites,and treatment with β-glucuronidase released mainly phenols and small amounts of quinones. BP-phenol-glucuronides were also major BP metabolites in the rat cell line WIRL-3C. In contrast, β-glucuronidase treatment of the medium from either cell of

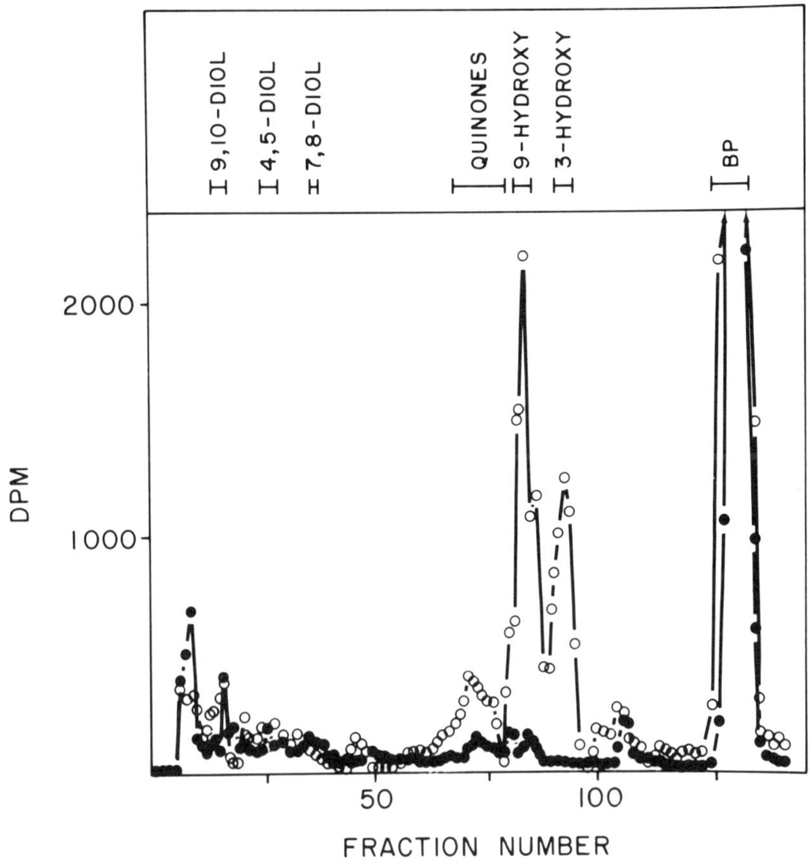

FIGURE 2. HPLC of BP metabolites in chloroform extracts of medium samples from a 40th-passage culture of hamster embryo cells exposed to 0.5 nmole [^{3}H] BP/ml medium for 24 hr. ○- - - -○, β-Glucuronidase-treated medium; ●—●, control medium incubated without enzyme. Samples were prepared and chromatographed as described in Methods. The elution positions of BP metabolite markers are at the top.

human origin, embryonic kidney or WI-38, did not change the amount of water-soluble metabolites or release any phenols. (Hydrocarbon metabolism in WI-38 cells is described in detail in ref. 4.). β-Glucuronidase treatment also did not affect the amounts of either the water-soluble or chloroform-extractable metabolites formed in bovine endothelial cells. The metabolites present in the medium from these cells contained a high proportion of free phenols, in contrast to the other cells studied in which phenols were detected only as conjugates. The bovine cells may be unable to conjugate BP-phenols to glucuronic acid.

TABLE IV. BP Metabolism By Cells From Other Species

Cells	β-Glucuronidase Treatment	Percent of Original BP Recovered as: Water-Soluble Metabolites	Diols	Phenols	Quinones	BP
Mouse						
ME1109	–	80	3	1	1	6
	+	61	4	11	3	6
IT22	–	86	1	0	0	8
	+	54	4	18	2	8
Rat						
WIRL-3C	–	56	9	2	1	29
	+	40	10	17	2	27
Human						
WI-38[a]	–	12	10	3	1	69
	+	14	10	3	1	68
Kidney	–	24	21	3	2	44
	+	22	20	3	2	46
Bovine						
Endothelial	–	20	3	10	2	56
	+	18	3	8	2	57

[a]After 6 days.

We conclude that formation of BP-phenol-glucuronides is a major pathway of BP metabolism in all cells tested from three rodent species. The formation of these conjugates was not detected in the human or bovine cells examined.

CONCLUSIONS

The results described above and in our previous studies demonstrate the following:

1. Formation of glucuronic acid conjugates of hydrocarbon phenols is a major pathway of metabolism for both methyl-substituted and unsubstituted hydrocarbons in rodent cell cultures.

2. Hydrocarbon-phenol glucuronides can be extracted into ethyl acetate from culture medium acidified to pH 4.5. If the unmetabolized hydrocarbon and its primary oxidation products are extracted first, ethyl acetate extraction after acidification will yield mainly hydrocarbon-glucuronides.

3. Hydrocarbon-phenol-glucuronides can be cleaved to free phenols by dilution of medium samples with 0.2 *M* acetate buffer (pH 4.5) and incubation with β-glucuronidase; the released phenols can be extracted into organic solvents and analyzed by TLC or HPLC.

4. The majority of the primary hydrocarbon oxidation products recovered by organic solvent extraction of medium from rodent cell cultures are dihydrodiols, whereas the majority of those released by β-glucuronidase treatment are phenols. Only by analysis of the conjugates formed can the pathways of hydrocarbon metabolism in these cells be determined.

ACKNOWLEDGMENTS

The authors wish to thank Drs. E. Levine and T. Shimada of the Wistar Institute and Dr. W. Nichols of the Institute for Medical Research, Camden, NJ, for providing cells; and Drs. A. Dipple and J. Tomaszewski of the Frederick Cancer Research Center, Frederick, MD, and the Chemical Repository of the National Cancer Institute for providing hydrocarbon metabolite markers.

This work was supported in part by grants CA 19948, CA 08936, CA 21778, and CA 10815 from the National Cancer Institute, Department of Health, Education and Welfare.

REFERENCES

1. Baird, W. M., Chemerys, R. A., Chern, C. J., and Diamond, L. (1978): Formation of glucuronic acid conjugates of 7,12-dimethylbenz[a] anthracene phenols in 7,12-dimethylbenz[a] anthracene-treated hamster embryo cell cultures. *Cancer Res.* 38:3432-3437.
2. Baird, W. M., Chern, C. J., and Diamond, L. (1977): Formation of benzo[a]pyrene-glucuronic acid conjugates in hamster embryo cell cultures. *Cancer Res.* 37:3190-3197.
3. Baird, W. M., and Diamond, L. (1976): Effect of 7,8-benzoflavone on the formation of benzo[a]pyrene-DNA-bound products in hamster embryo cells. *Chem.-Biol. Interact.* 13:67-75.
4. Baird, W. M., and Diamond, L. (1978): Metabolism and DNA binding of polycyclic aromatic hydrocarbons by human diploid fibroblasts, *Int. J. Cancer* 22:189-195.
5. Cohen, G. M., Haws, S. M., Moore, B. P., and Bridges, J. W. (1976): Benzo[a]pyren-3-yl hydrogen sulphate, a major ethyl acetate-extractable metabolite of benzo[a]pyrene in human, hamster and rat lung cultures. *Biochem. Pharmacol.* 25:2561-2570.
6. Cohen, G. M., and Moore, B. P. (1977): The metabolism of benzo[a] pyrene, 7,8-dihydro-7,8-dihydroxybenzo[a]pyrene and 9,10-dihydro-9, 10-dihydroxybenzo[a]pyrene by short-term organ cultures of hamster lungs. *Biochem. Pharmacol.* 26:1481-1487.
7. Cohen, G. M., and Moore, B. P. (1978): Metabolism of benzo[a]pyrene and its major metabolites by respiratory tissues. In: *Carcinogenesis, Vol.3.* P. W. Jones and R. I. Freudenthal, eds. pp. 325-339, Raven Press, New York.

8. Cohen, G. M., Moore, B. P., and Bridges, J. W. (1977): Organic solvent soluble sulphate ester conjugates of monohydroxybenzo[a]pyrenes. *Biochem. Pharmacol.* 26:551-553.
9. Croce, C. M., Knowles, B. B., and Koprowski, H. (1973): Preferential retention of the human chromosome C-7 in human-(thymidine kinase deficient) mouse hybrid cells. *Exp. Cell Res.* 82:457-461.
10. Diamond, L. (1971): Metabolism of polycyclic hydrocarbons in mammalian cell cultures. *Int. J. Cancer* 8:451-462.
11. Diamond, L., and Baird, W. M. (1977): Chemical carcinogenesis *in vitro*. In: *Growth, Nutrition and Metabolism of Cells in Culture,* vol. 3, G. H. Rothblat and V. J. Cristofalo, eds., pp. 421-470, Academic Press, New York.
12. Diamond, L., McFall, R., Tashiro, Y., and Sabatini, D. (1973): The WIRL-3 rat liver cell lines and their transformed derivatives. *Cancer Res.* 33:2627-2636.
13. Diamond, L., Sardet, C., and Rothblat, G. H. (1968): The metabolism of 7,12-dimethylbenz[a]anthracene in cell cultures. *Int. J. Cancer* 3:838-849.
14. Duncan, M., Brookes, P., and Dipple, A. (1969): Metabolism and binding to cellular macromolecules of a series of hydrocarbons by mouse embryo cells in culture. *Int. J. Cancer* 4:813-819.
15. Gentil, A., Lasne, C., and Chouroulinkov, I. (1977): Metabolism and cytotoxicity of 7,12-dimethylbenz[a]anthracene by hamster, rat and rabbit embryo cell cultures *Xenobiotica* 7:221-233.
16. Hayflick, L. (1965): The limited *in vitro* lifetime of human diploid cell strains. *Exp. Cell Res.* 37:614-636.
17. Heidelberger, C. (1973): Chemical oncogenesis in culture. *Adv Cancer Res.* 18:317-366.
18. Huberman, E., Sachs, L., Yang, S. K., and Gelboin, H. V. (1976): Identification of mutagenic metabolites of benzo[a]pyrene in mammalian cells. *Proc. Nat. Acad. Sci., U.S.* 73:607-611.
19. Huberman, E., Selkirk, J. K., and Heidelberger, C. (1971): Metabolism of polycyclic aromatic hydrocarbons in cell cultures. *Cancer Res.* 31: 2161-2167.
20. Jerina, D. M., and Daly, J. W. (1976): Oxidation at carbon. In: *Drug Metabolism–From Microbe to Man*, D. V. Parke and R. L. Smith, eds., pp. 13-32, Taylor and Francis, Ltd., London.
21. Levine, E. M., and Mueller, S. N. (1978): Cultured vascular endothelial cells as a model system for the study of cellular senescence. *Int. Rev. Cytol.*, in press.
22. Selkirk, J. K. (1978): Analysis of benzo[a]pyrene metabolism by high-pressure liquid chromatography. *Adv. Chromatog.* 16:1-36.
23. Sims. P. (1970): The metabolism of some aromatic hydrocarbons by mouse embryo cell cultures. *Biochem. Pharmacol.* 19:285-297.
24. Sims, P., and Grover, P. L. (1974): Epoxides in polycyclic aromatic hydrocarbon metabolism and carcinogenesis. *Adv. Cancer Res.* 20: 165-274.

Polynuclear Aromatic Hydrocarbons, edited by P.W. Jones and P. Leber. Ann Arbor Science Publishers, Inc. Ann Arbor, MI

HYDROXYLATION AND CONJUGATION AT THE BENZYLIC CARBON ATOM: A POSSIBLE MECHANISM OF CARCINOGENIC ACTIVATION FOR SOME METHYL-SUBSTITUTED AROMATIC HYDROCARBONS

E. Cavalieri, R. Roth and E. Rogan

Eppley Institute for Research in Cancer
University of Nebraska Medical Center
Omaha, Nebraska 68105

The complexity of polycyclic aromatic hydrocarbon (PAH) carcinogenesis can best be understood in terms of multiple mechanisms of activation arising from different reactive ultimate metabolites (6,10). The predominance of one mechanism over another may change, depending on the nature of the activating enzymes present in the target organs of various animals.

The data presently available indicate the importance of the "bay-region" diol epoxides as ultimate carcinogens of several PAH (19-21). However, it has been postulated that other reactive electrophilic intermediates, such as radical cations and benzylic carbonium ions of alkylated PAH generated via formation of benzylic esters, might be ultimate carcinogenic forms of PAH.

One-electron oxidation of PAH to radical cations has been postulated as an important pathway leading to carcinogenesis (9,18,32). This hypothesis is based on several lines of evidence which include the strong electron-donating properties of PAH and the presence *in vivo* of several potential oxidants in the form of metal ion-containing enzymes, such as cytochrome P-450 which was recently demonstrated to be a one-electron acceptor (24). Activation of the carcinogens *N*-hydroxy-2-acetylaminofluorene (1,2,17) and benzo[a]pyrene (B[a]P) (26) via the one-electron oxidation mechanism has been observed *in vitro* with horseradish peroxidase, as well as *in vivo* with mammary peroxidase (25) for the former compound.

Hydroxylation of the alkyl side chain in most tissues represents a major pathway in the metabolism of alkyl-substituted PAH (7,13,28,30). The conjugation of a hydroxy compound potentially involves glucuronidation, phosphorylation, acetylation or sulfation. These esters are generally stable and rapidly excreted, thus constituting a major detoxification pathway. A benzylic ester bearing a good leaving group, however, might be a potential ultimate carcinogen by generating a reactive carbonium ion capable of binding to cellular nucleophiles. The hydroxymethyl derivatives of PAH are generally carcinogenic with potencies varying from weaker to stronger, when compared to the parent compounds. For example the two monohydroxymethyl derivatives of 7,12-dimethylbenz[a]anthracene (DMBA) display a carcinogenicity in mice and rats lower or similar to DMBA depending on the route of administration, the animal tested and the type of tumor elicited (4, 5,15,31).

In the methylcholanthrene (MC) series, MC-2-OH is as carcinogenic as the parent compound by subcutaneous injection (27) and by repeated skin painting (8), while MC-1-OH is less potent than MC in both cases. BP-6-CH_2OH is a stronger inducer of subcutaneous sarcomas in rats (14) than BP-6-CH_3, but a weaker carcinogen when tested by repeated skin application in mice (7) (see also Results: Figure 3 and Table IV).

This paper reports carcinogenicity, mutagenicity and binding studies of benzylic esters of the 7-methylbenz[a]anthracene (BA-7-CH_3) and BP-6-CH_3 series with the purpose of evaluating their contribution to the carcinogenic activity of the parent compounds. If certain benzylic esters such as acetates, phosphates and/or sulfates are formed in target organs from the hydroxymethyl metabolites, the evidence thus far obtained suggest that these esters might be ultimate carcinogens of the parent compounds. In this regard, pyrrolic esters of pyrrolizidine alkaloids (12,22) and benzylic esters of safrole (3, 33,34) have been implicated as ultimate carcinogenic metabolites for these compounds.

RESULTS AND DISCUSSION

7-Substituted Benz[a]anthracene Derivatives

Carcinogenicity Experiments

Seven-week-old female Swiss mice from the Eppley Colony were treated in groups of 30 with BA-7-CH_3, BA-7-CH_2CH_3, BA-7-CH_2OH, BA-7-CHO and BA-7-CH_2OCOCH_3 at dose levels of 0.8 and 0.4 μmol in one drop (16.7 μl) of acetone twice weekly for 25 weeks. Control mice were treated with one drop of solvent. All animals were examined daily and the onset and progress of tumor growth were recorded. Mice were killed when moribund and survivors were sacrificed after 52 experimental weeks. Animals were autopsied and histological examination was performed on skin and other altered organs.

Table I shows the carcinogenicity of BA-7-CH_3 and some of its derivatives. BA-7-CH_2OCOCH_3 was found to be a stronger carcinogen than BA-7-CH_2OH (high dose, $p < 0.04$; low dose, $p < 0.063$), which in turn was more potent than the parent compound BA-7-CH_3 (high dose, $p < 0.2$; low dose, $p < 0.2$). The stronger activity of BA-7-CH_2OH relative to that of BA-7-CH_3 and the formation of BA-7-CH_2OH as a major metabolite of BA-7-CH_3 in mouse skin (unpublished results) suggest that BA-7-CH_2OH might be an important proximate carcinogen of BA-7-CH_3. BA-7-CHO elicited a slightly weaker carcinogenic activity than BA-7-CH_2OH, although not statistically significant at the two dose levels tested. This result, combined with the metabolic reduction of BA-7-CHO to BA-7-CH_2OH in mouse skin (unpublished results), suggests that the latter compound might be a proximate carcinogenic metabolite of BA-7-CHO. BA-7-CH_2CH_3 was found to be a very weak carcinogen when compared to BA-7-CH_3 (high dose, $p < 0.002$; low dose, $p < 0.002$).

ATP-Mediated Binding of 7-Hydroxymethylbenz[a]anthracene to DNA

Phosphate esters are among the potentially reactive benzylic esters of PAH. It has been reported that 6-hydroxymethylbenzo[a]pyrene (BP-6-CH_2-OH) and BA-7-CH_2OH-12-CH_3 bind efficiently to DNA *in vitro* in the presence of ATP alone (14,16,29), suggesting that binding might occur through the intermediacy of a phosphate ester. The formation of benzylic alcohols as major metabolites of methyl PAH and the ubiquitous presence of ATP in all tissues suggest that this mode of activation might be important *in vivo*.

We have studied the ATP-mediated binding of [^{14}C]BA-7-CH_2OH to DNA (Figure 1). A maximum level of binding was obtained with about 20 μM BA-7-CH_2OH. ADP mediated the binding half as well as ATP (Table II), while AMP effected no binding. BA-7-CH_2OH did not bind to DNA efficiently when the concentration of ATP was below 2 mM (Figure 2). These binding kinetics point to a complex reaction mechanism which needs further study for elucidation.

Binding of three hydroxymethyl derivatives of PAH to DNA is compared in Table III (see Table II for reaction conditions). Binding of BP-6-CH_2OH in the presence of ATP is 8 to 10 times higher than binding of BA-7-CH_2OH, while binding of BA-7-CH_2OH-12-CH_3 is three times higher than that of BA-7-CH_2OH. Table III also shows that comparable binding levels in the presence of ATP, ADP or AMP were obtained in our laboratory and that of Flesher.

Carcinogenicity of 6-Substituted Benzo[a]pyrene Derivatives

Nine-week-old female Swiss CD-1 mice in groups of 28 were treated with BP-6-CH_2OSO_3Na, BP-6-CH_2OH, BP-6-CH_2OCOCH_3, BP-6-CHO or BP-6-CH_3 at three dose levels of 0.8, 0.2, and 0.05 μmole in two drops (16.7 μl

TABLE I. Skin Neoplasms in Female Swiss Mice by 7-Substituted BA Derivatives

Group, Substance and dose	Effective No. of Mice[a]	Tumor Incidence (%)	Survival (mean ± S.D.) (wk)	No. and type of Skin Neoplasms: Tumor Appearance (mean ± S.D.) (wk)	No. of Papillomas/ No. of TBA[b]	No. of Keratotic Epitheliomas/ No. of TBA	No. of Carcinomas/ No. of TBA
0.8 μmole							
BA-7-CH_3	30	76.6	48 ± 6	34 ± 6	4/4	5/3	22/18
BA-7-CH_2CH_3	30	10.0	50 ± 6	37 ± 9	0	0	5/3
BA-7-CH_2OH	30	86.6	43 ± 9	35 ± 8	8/5	14/13	26/19
BA-7-CHO	29	72.4	46 ± 8	37 ± 7	1/1	8/7	19/17
BA-7-CH_2OCOCH_3	30	100	36 ± 6	26 ± 6	2/2	9/8	41/29
0.4 μmole							
BA-7-CH_3	30	66.7	48 ± 6	38 ± 7	4/3	4/3	16/13
BA-7-CH_2CH_3	29	6.9	44 ± 11	39 ± 17	1/1	1/1	0
BA-7-CH_2OH	29	82.8	44 ± 9	34 ± 7	6/6	15/13	16/12
BA-7-CHO	29	72.4	47 ± 9	36 ± 8	7/5	7/7	25/18
BA-7-CH_2OCOCH_3	30	93.3	43 ± 9	32 ± 8	6/5	19/15	31/23
Control	29	0	51 ± 5	–	0	0	0

[a]Mice histologically examined.
[b]Tumor-bearing animals.

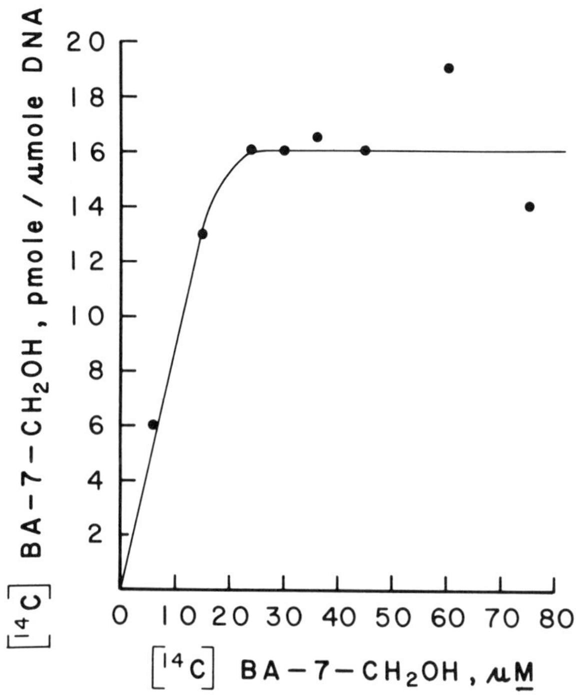

FIGURE 1. Effect of substrate concentration on ATP-mediated binding of [^{14}C]BA-7-CH_2OH. Incubations were conducted and DNA was purified as described in Table II.

each) of dioxane-dimethylsulfoxide (75:25) twice weekly for 40 weeks. Control mice were treated with two drops of solvent. All animals were examined daily and tumors recorded. Mice were killed when moribund and survivors were sacrificed after 70 experimental weeks.

The observed tumor incidence versus experimental weeks of the medium dose level of this experiment is shown in Figure 3, while the results of the entire experiment are presented in Table IV. BP-6-CH_2OSO_3Na was a significantly more potent carcinogen than BP-6-CH_2OH (high dose, $p < 0.001$; medium dose, $p < 0.0006$; low dose, $p < 0.028$). In this case, however, BP-6-CH_2OCOCH_3 did not significantly differ from BP-6-CH_2OH in terms of carcinogenicity at the three dose levels. This result is in contrast to the one presented above in which the activity of BA-7-CH_2OCOCH_3 is stronger than that of the corresponding alcohol. The weaker carcinogenicity of BP-6-CHO relative to that of BP-6-CH_2OH (high dose, $p < 0.028$; medium dose, $p < 0.016$) and the efficient metabolic reduction of BP-6-CHO to BP-6-CH_2OH in mouse

TABLE II. Binding of [^{14}C] BA-7-CH_2OH to DNA[a]

Cofactor	pmole/μmole DNA
ATP	12.7 ± 1.2[b]
ADP	6.3 ± 0.3
AMP	0.8 ± 0.02
None	0.9 ± 0.1

[a] Reaction mixtures containing 5 m*M* cofactor, 3 m*M* DNA, 3 m*M* $MgCl_2$ and 30 μ*M* [^{14}C] BA-7-CH_2-OH in 0.05 *M tris*, pH 7.0, were incubated at 37° for 1 hr. The DNA was purified by 2 extractions with water-saturated phenol, pH 8.0; 2 extractions with chloroform-octanol (10:1), and 2 precipitations in 75% ethanol.
[b] Standard error of measurement.

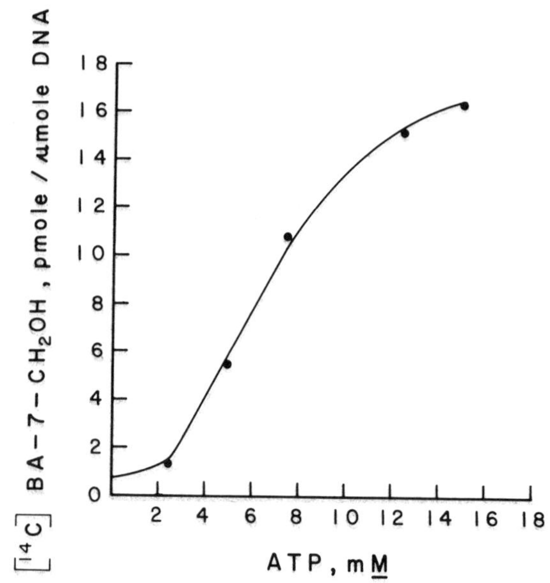

FIGURE 2. Effect of ATP concentration on ATP-mediated binding of [^{14}C] BA-7-CH_2OH. Incubations were conducted and DNA was purified as described in Table II.

skin (7) suggest that the latter compound might be a proximate carcinogenic metabolite of BP-6-CHO. BP-6-CH_3 was confirmed to be a stronger carcinogen than BP-6-CH_2OH (medium dose, $p < 0.035$, low dose, $p < 0.078$), as previously found (7). The lesser carcinogenicity of BP-6-CH_2OH relative to that of BP-6-CH_3 indicates that the former cannot function as a major carcin-

Table III. Binding of Hydroxymethyl PAH Derivatives to DNA

Cofactor	BA-7-CH_2OH	BP-6-CH_2OH[a]	BP-6-CH_2OH[b]	BA-7-CH_2OH-12-CH_3[c]
		pmole/μmoleDNA		
ATP	13	130	98	44
ADP	6	27	53	16
AMP	1	2	2	0

[a]See references 14, 29.
[b]See reference 26.
[c]See reference 16.

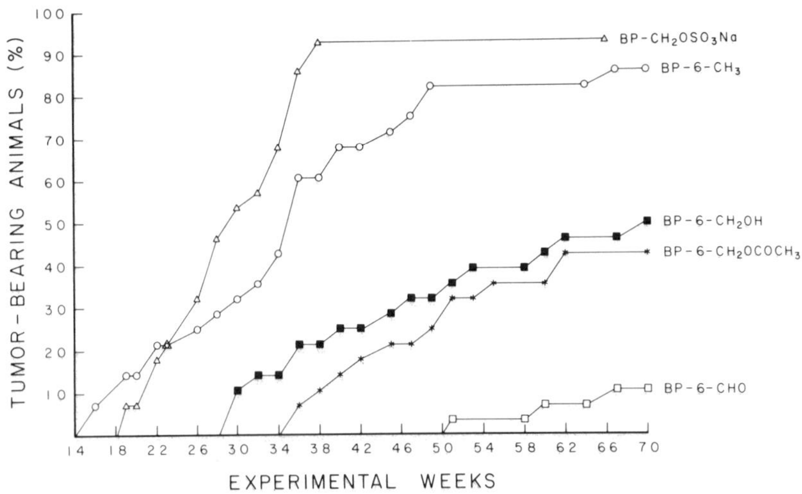

FIGURE 3. Rates of observed skin tumors in CD-1 Swiss mice repeatedly treated at 0.2-μmol dose level

ogenic metabolite of BP-6-CH_3, although BP-6-CH_2OH is a major metabolite of BP-6-CH_3 in mouse skin. It is not excluded, however, that this oxidative pathway might contribute to some extent to the BP-6-CH_3 activity.

Mutagenicity Studies of PAH Benzylic Esters

A series of PAH benzylic (and nonbenzylic) esters in DMSO were tested for mutagenicity (35), as described by Ames (23), in strain TA98, which was derived from TA1538, a strain sensitive to frameshift mutagens by the inclusion of a plasma factor causing error-prone recombinational repair. The compounds were tested without metabolic activation.

TABLE IV. Skin Tumors in Mice by 6-Substituted BP Derivatives

Substance and Dose	Tumor Incidence (%)	Survival (mean ± S.D.) (wk)	Age at First Tumor Appearance (mean ± S.D.) (wk)	Gross Tumor Observations: No. of Papillomas/No. of TBA[a]	Gross Tumor Observations: No. of Carcinomas/No. of TBA
0.8 μmole					
BP-6-CH_2OSO_3Na	100.0	42 ± 5	25 ± 7	10/5	28/27
BP-6-CH_2OH	92.3	58 ± 9	43 ± 12	17/12	15/13
BP-6-CH_2OCOCH_3	67.9	50 ± 12	36 ± 8	15/11	10/10
BP-6-CHO	51.9	59 ± 13	51 ± 9	6/5	9/9
0.2 μmole					
BP-6-CH_2OSO_3Na	96.3	47 ± 10	29 ± 6	10/7	28/25
BP-6-CH_2OH	50.0	63 ± 11	44 ± 13	9/7	8/7
BP-6-CH_2OCOCH_3	35.7	64 ± 7	44 ± 7	4/3	8/7
BP-6-CHO	10.7	68 ± 5	59 ± 8	3/3	0
BP-6-CH_3	78.6	56 ± 12	32 ± 10	13/8	20/19
0.05 μmole					
BP-6-CH_2OSO_3Na	82.1	64 ± 7	46 ± 10	10/9	18/17
BP-6-CH_2OH	25.0	68 ± 6	49 ± 5	3/2	5/5
BP-6-CH_2OCOCH_3	21.4	68 ± 4	56 ± 7	4/4	2/2
BP-6-CH_3	64.3	66 ± 5	45 ± 11	8/7	12/9
Control	0	70 ± 0	—	0	0

[a]Tumor-bearing animals.

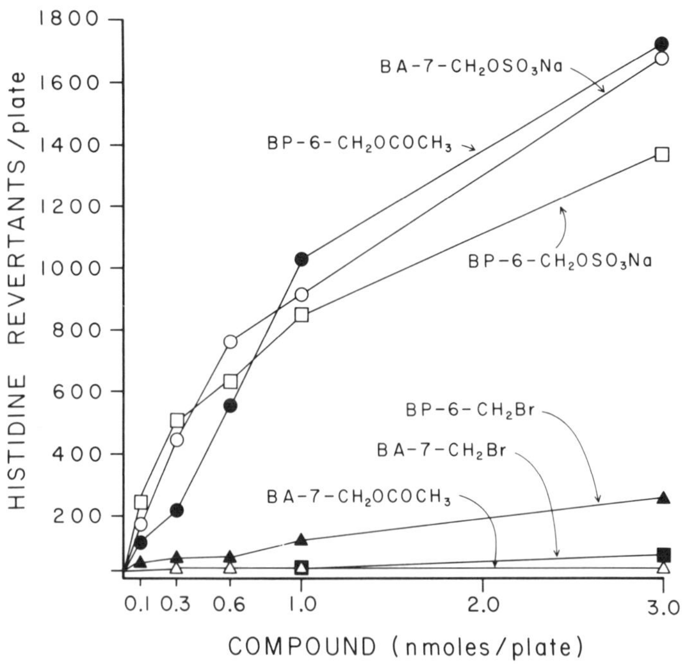

FIGURE 4. Relative mutagenicity of 6-benzo[a]pyrene and 7-benz[a]anthracene benzylic esters in *S. typhimurium* **TA 98.**

A portion of these results are summarized in Figure 4. The sulfate esters of BA-7-CH_2OH and BP-7-CH_2OH showed similarly high levels of mutagenicity at concentrations from 0.03 to 3.0 nmole/plate. No mutagenic activity for the nonbenzylic ester BA-7-$CH_2CH_2OSO_3Na$ was observed at concentrations up to 100 nmol/plate.

While BP-6-CH_2OCOCH_3 showed an activity comparable to that of the corresponding sulfate, BA-7-CH_2OCOCH_3 was only twice that of background at a concentration of 3 nmol/plate. The corresponding bromomethyl derivatives of BP and BA were much less mutagenic than the sulfates or the acetate ester of BP.

In the system used and in the absence of metabolic activation, BA-7-CH_3, BA-7-CH_2OH and BP-6-CH_3 were not mutagenic. However, at a concentration of 3.0 to 100 nmol/plate BP-6-CH_2OH was observed to give an approximately onefold increase in mutation rate over background which was determined to be real at the $p < 0.01$ level of significance.

The high mutagenic activity of BP-6-CH_2OSO_3Na correlated with its strong carcinogenicity on mouse skin (see above), whereas BP-6-CH_2OCOCH_3 showed high mutagenic activity and relatively low carcinogenicity. In the BA-7-CH_3 series, BA-7-CH_2OCOCH_3 was a relatively potent carcinogen and a very poor mutagen.

CONCLUSIONS

Hydroxylation of benzylic carbon atoms is a major metabolic pathway for alkyl-substituted PAH (7,13,28,30). The relatively high carcinogenicity of BA-7-CH_2OCOCH_3 and BP-6-CH_2OSO_3Na when compared to the respective alcohols offers an indication that these esters might contribute to the carcinogenicity of the alkyl-substituted PAH if conjugative enzymes are present in the target organs. If such esters are formed, their possible role in carcinogenesis is further substantiated by the high mutagenicity of the benzylic sulfate esters. For PAH the only data available concerning sulfate ester formation comes from the presence of BP-3-yl sulfate in human, hamster and rat lung cells cultured with BP (11), suggesting the presence of sulfotransferase activity in these biological systems.

The ATP-mediated binding of BA-7-CH_2OH to DNA occurs through the intermediacy of a phosphate ester, showing that this is one of the reactive benzylic esters.

Further experiments designed to demonstrate the presence of these esters in target tissues should help to assess their role in the carcinogenesis of alkyl-substituted PAH.

ACKNOWLEDGMENTS

We wish to thank Dr. J. Althoff for histological evaluation, Dr. P. Wislocki for mutagenicity assays and Dr. K. Patil for statistical analysis of data. This research was supported by Public Health Service contract NO1 CP33278 from the Division of Cancer Cause and Prevention of the National Cancer Institute, NIH.

REFERENCES

1. Bartsch, H., and Hecker, E. (1971): On the metabolic activation of carcinogen *N*-hydroxy-*N*-2-acetylamino-fluorene. III. Oxidation with horseradish peroxidase to yield 2-nitrosofluorene and *N*-acetoxy-*N*-2-acetyl-aminofluorene. *Biochim. Biophys. Acta* 237:567-578.

2. Bartsch, H., Miller, J. A., and Miller, E. C. (1972): *N*-Acetoxy-*N*-acetylaminofluorene and nitrosoarenes, one-electron non-enzymatic and enzymatic oxidation products of various carcinogenic aromatic acethydroxamic acids. *Biochim. Biophys. Acta* 273:40-51.
3. Borchert, P., Miller, J. A., Miller, E. C., and Shires, T. K. (1973): 1′-Hydroxysafrole, a proximate carcinogenic metabolite of safrole in the rat and mouse. *Cancer Res.* 33:590-600.
4. Boyland, E., and Sims, P. (1967): The carcinogenic action in mice of compounds related to benz[a]anthracene. *Int. J. Cancer* 2:500-504.
5. Boyland, E., Sims, P., and Huggins, C. (1965): Induction of adrenal damage and carcinogenesis with metabolites of 7,12-dimethylbenz[a]anthracene. *Nature* 207:816-817.
6. Cavalieri, E., Rogan, E., and Roth, R. (1979): Multiple mechanisms of activation in aromatic hydrocarbon carcinogenesis. In: *Free Radicals and Cancer,* R. A. Floyd, Ed. Marcel Dekker, Inc., New York (in press).
7. Cavalieri, E., Roth, R., Grandjean, C., Althoff, J., Patil, K., Liakus, S., and Marsh, S. (1978): Carcinogenicity and metabolic profiles of 6-substituted benzo[a]pyrene derivatives on mouse skin. *Chem.-Biol. Interact.* 22:53-67.
8. Cavalieri, E., Roth, R., Grandjean, C., Althoff, J., Patil, K., Marsch, S., and McLaughlin, D. (1978): Carcinogenicity and metabolic profiles of 3-methylcholanthrene oxygenated derivatives at the 1 and 2 positions. *Chem.-Biol. Interact.* 22:69-81.
9. Cavalieri, E., Roth, R., and Rogan, E. G. (1976): Metabolic activation of aromatic hydrocarbons by one-electron oxidation in relation to the mechanism of tumor initiation. In: *Carcinogenesis, Vol. 1,* R. I. Freudenthal and P. W. Jones, eds., pp. 181-190, Raven Press, New York.
10. Cavalieri, E., Roth, R., Rogan, E., Grandjean, C., and Althoff, J. (1978): Mechanisms of tumor initiation by polycyclic aromatic hydrocarbons. In: *Carcinogenesis, Vol. 3,* P. W. Jones and R. I. Freudenthal, eds., pp. 273-284, Raven Press, New York.
11. Cohen, G. M., Haws, S. M., Moore, B. P., and Bridges, J. W. (1976): Benzo[a]pyren-3-yl-hydrogen sulphate, a major ethyl acetate extractable metabolite of benzo[a]pyrene in human, hamster and rat lung cultures. *Biochem. Pharmacol.* 25:2561-2570.
12. Culvenor, C. C. J., Downing, D. T., Edgar, J. A., and Jago, M. V. (1969): Pyrrolizidine alkaloids as alkylating and antimitotic agents. *Ann. N.Y. Acad. Sci.* 163:837-847.
13. DiGiovanni, J., Slaga, T. J., Berry, D. L., and Juchau, M. R. (1977): Metabolism of 7,12-dimethylbenz[a]anthracene in mouse skin homogenates analyzed with high-pressure liquid chromatography. *Drug Metab. Dispos.* 5:295-301.
14. Flesher, J. W. Personal communication.
15. Flesher, J. W., and Sydnor, K. L. (1971): Carcinogenicity of derivatives of 7,12-dimethylbenz[a]anthracene. *Cancer Res.* 31:1951-1954.

16. Flesher, J. W., and Tay, L. K. (1978): Binding of 7-hydroxymethyl-12-methylbenz[a]anthracene to calf thymus DNA in the presence and absence of 3′-phosphoandenosine-5′-phosphosulfate generating system. *Proc. Fed. Am. Soc. Exp. Biol.* 31:749.
17. Floyd, R. A., and Soong, L. M. (1977): Obligatory free radical intermediate in the oxidative activation of the carcinogen *N*-hydroxy-2-acetylaminofluorene. *Biochim. Biophys. Acta.* 498:244-249.
18. Fried, J. (1974): One-electron oxidation of polycyclic aromatics as a model for the metabolic activation of carcinogenic hydrocarbons. In: *Chemical Carcinogenesis, Part A*, P. O. P. Ts'o and J. DiPaolo, eds., pp. 197-215, Marcel Dekker, New York.
19. Jerina, D. M., and Daly, J. W. (1976): Oxidation at carbon. In: *Drug Metabolism,* D. V. Parke and R. L. Smith, eds., pp. 15-33, Taylor & Francis, Ltd., London.
20. Jerina, D. M., Lehr, R., Schaefer-Ridder, M., Yagi, H., Karle, J. M., Thakker, D. R., Wood, A. W., Lu, A. Y. H., Ryan, D., West, S., Levin, W., and Conney, A. H. (1977): Bay region epoxides of dihydrodiols: A concept which explains the mutagenic and carcinogenic activity of benzo[a]pyrene and benz[a]anthracene. In: *Origins of Human Cancer, Book B,* H. H. Hiatt, J. D. Watson and J. A. Winsten, eds, pp. 639-658, Cold Spring Harbor Laboratory, NY.
21. Jerina, D. M., Lehr, R. E., Yagi, H., Hernandez, O., Dansette, P., Wislocki, P. G., Wood, A. W., Chang, R. L., Levin, W., and Conney, A. H. (1976): Mutagenicity of benzo[a]pyrene derivatives and the description of a quantum mechanical model which predicts the ease of carbonium ion formation from diol epoxides. In: *In Vitro Metabolic Activation and Mutagenesis Testing,* F. J. DeSerres, J. R. Fouts, J. R. Bend and R. M. Philpot, eds., pp. 159-177, North Holland Publishing Co., Amsterdam.
22. Mattocks, A. R., and White, I. N. H. (1971): The conversion of pyrrolizidine alkaloids to N-oxides and to dihydropyrrolizidine derivatives by rat-liver microsomes in vitro. *Chem.-Biol. Interact.* 3:383-396.
23. McCann, J., Springarm, N. E., Kabori, J., and Ames, B. N. (1975): Detection of carcinogens as mutagens: Bacterial tester strains with R factor plasmids. *Proc. Natl. Acad. Sci., U.S.* 72:979-983.
24. Peterson, J. A., White, R. E., Yasukochi, Y., Coomes, M. L., O'Keeffe, D. H., Ebel, R. E., Masters, B. S. S., Ballou, D. B., and Coon, M. J. (1977): Evidence that purified liver microsomal cytochrome P-450 is a one-electron acceptor. *J. Biol. Chem.* 252:4431-4434.
25. Reigh, D. L., Stuart, M., and Floyd, R. A. (1978): Activation of the carcinogen N-hydroxy-2-acetylaminofluorene by rat mammary peroxidase. *Experimentia* 34:107-108.
26. Rogan, E., Roth, R., and Cavalieri, E. (1979): Binding of aromatic hydrocarbons to DNA catalyzed by peroxidase and ATP. Chapter 44, this volume.

27. Sims, P. (1967): The carcinogenic activities in mice of compounds related to 3-methylcholanthrene. *Int. J. Cancer* 2:505-508.
28. Sims, P. (1970): Qualitative and quantitative studies on the metabolism of a series of aromatic hydrocarbons by rat-liver preparations. *Biochem. Pharmacol.* 19:795-818. (and references therein).
29. Tay, L. K., Sydnor, K. L., and Flesher, J. W. (1978): Binding of 6-acetoxymethylbenzo[a]pyrene and ATP-mediated binding of 6-hydroxymethylbenzo[a]pyrene to DNA. *Proc. Am. Assoc. Cancer Res.* 19:181.
30. Tierney, B., Hewer, A., Walsh, C., Grover, P. L., and Sims, P. (1977): The metabolic activation of 7-methylbenz[a]anthracene in mouse skin. *Chem.-Biol. Interact.* 18:179-193.
31. Wheatley, D. N., and Inglis, M. S. (1968): Mammary tumors induced in Sprague-Dawley female rats by 7,12-dimethylbenz[a]anthracene and its hydroxymethyl derivatives. *Brit. J. Cancer* 22:122-127.
32. Wilk, M., Bez, W., and Rochlitz, J. (1966): Neue Reaktionen der carcinogenen Kohlenwasserstoffe 3,4-Benzpyren, 9,10-Dimethyl-1,2-benzanthracene und 20-Methylcholanthren. *Tetrahedron* 22:2599-2608.
33. Wislocki, P. G., Borchert, P., Miller, J. A., and Miller, E. C. (1976): The metabolic activation of the carcinogen 1′-hydroxysafrole in vivo and in vitro and the electrophilic reactivities of possible ultimate carcinogens. *Cancer Res.* 36:1686-1695.
34. Wislocki, P. G., Miller, E. C., Miller, J. A., McCoy, E. C., and Rosenkranz, H. S. (1977): Carcinogenic and mutagenic activities of safrole, 1′-hydroxysafrole, and some known or possible metabolites. *Cancer Res.* 37:1883-1891.
35. Wislocki, P. G., Roth, R., and Cavalieri, E. (1979): High mutagenicity of esters of 6-hydroxymethylbenzo[a]pyrene and 7-hydroxymethylbenz[a]anthracene. (Submitted).

Polynuclear Aromatic Hydrocarbons,
edited by P. W. Jones and P. Leber.
Ann Arbor Science Publishers, Inc.
Ann Arbor, MI ©1979.

BAY-REGION ACTIVATION OF CARCINOGENIC POLYCYCLIC HYDROCARBONS

A. W. Wood, W. Levin, R. L. Chang,
H. Yagi,* D. R. Thakker,* R. E. Lehr,**
D. M. Jerina* and A. H. Conney

Department of Biochemistry and Drug Metabolism
Hoffmann-La Roche Inc.
Nutley, New Jersey 07110

*Laboratory of Bioorganic Chemistry
National Institute of Arthritis, Metabolism
and Digestive Diseases
National Institutes of Health
Bethesda, Maryland 20014

**Department of Chemistry
University of Oklahoma
Norman, Oklahoma 73019

The polycyclic aromatic hydrocarbons comprise a large class of environmental contaminants that exert their cytotoxic, mutagenic and carcinogenic effects only after metabolic activation to reactive metabolites (5,10,13,17, 26,37). Of the very large number of known and potential metabolites that can occur from the oxidative metabolism of polycyclic hydrocarbons, recent attention has focused on benzo-ring diol epoxides. These diol epoxides are formed via oxidation of a terminal benzo ring of the hydrocarbon to form an arene oxide, hydration of the arene oxide to form a *trans*-dihydrodiol and then oxidation of the adjacent olefinic double bond to form a diol epoxide. Cytochrome P450-dependent monooxygenases catalyze the two oxidative reactions while another microsomal enzyme, epoxide hydrase, catalyzes the formation of the dihydrodiol. Figure 1 summarizes the formation of the 7,8-diol-9,10-epoxide of the potent carcinogen benzo[a]pyrene (BP). In addition, this figure shows that a pair of diastereomeric diol epoxides are possible in which the benzylic hydroxyl group is either *cis* (diol-epoxide-1) or *trans* (diol epoxide-2) to the oxirane ring. Both diastereomeric "bay-region" diol epoxides of BP are among the most mutagenic compounds ever tested

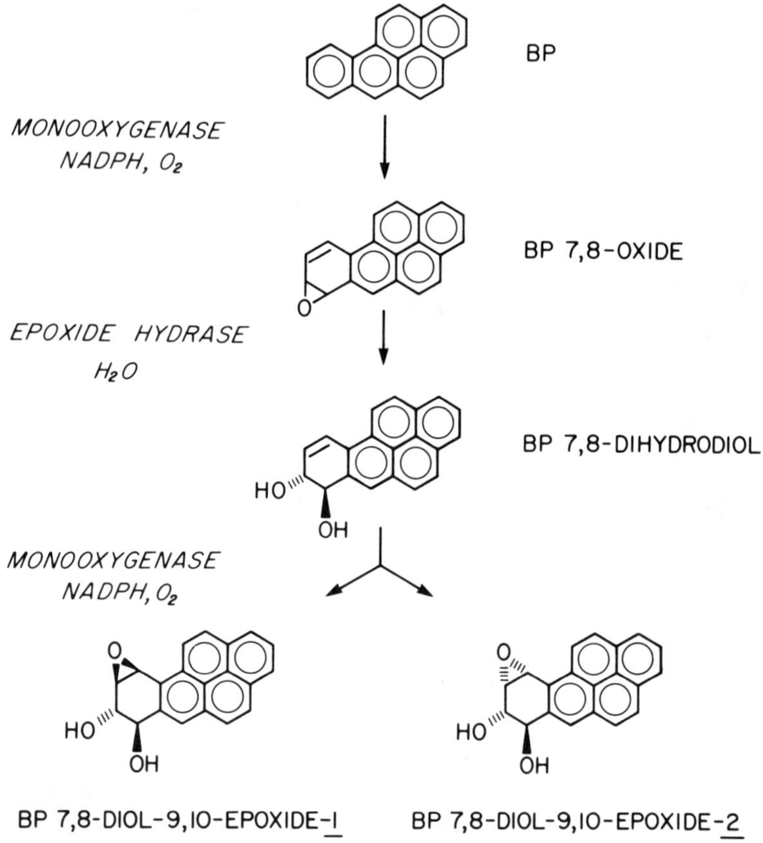

FIGURE 1. Metabolism of benzo[a]pyrene to the diastereomeric 7,8-diol-9,10-epoxides. Absolute stereochemistry is not implied.

(11,31,34,57) and BP 7,8-diol-9,10-epoxide-2 is also highly tumorigenic (2, 18,19,38). These findings and the high carcinogenicity of BP 7,8-oxide (28,49) and BP 7,8-dihydrodiol (3,18,27,29,40) indicate that a BP 7,8-diol-9,10-epoxide is the principal ultimate carcinogenic metabolite of BP. Several recent reviews have dealt extensively with the metabolic activation of benzo[a]pyrene (5,17,26).

The high biological activity of the BP 7,8-diol-9,10-epoxides suggested the possibility that diol epoxides were ultimate carcinogenic metabolites of other polycyclic hydrocarbons. Since most nonlinear hydrocarbons could form at least two isomeric diol epoxides, our laboratories sought to identify the key structural and chemical features of the BP 7,8-diol-9,10-epoxides. Comparison of the mutagenic activities of the two benzo-ring tetrahydroepoxides of benzo[a]pyrene at the 7,8- and 9,10-positions indicated that the H_4-9,10-epoxide had activity comparable to the 7,8-diol-9,10-epoxides while the H_4-7,8-epoxide was substantially less mutagenic (57). Furthermore, the H_4-9,10-epoxide was two orders of magnitude more mutagenic than the 9,10-arene oxide of BP which has a double bond in the 7,8-position (52). Thus the unique structural feature of the diol epoxide appears to be that the epoxide is on a saturated angular benzo ring and that it forms part of a bay region of the polycyclic hydrocarbon. The key chemical feature of such epoxides is that they are highly susceptible to attack by nucleophiles. The structures of the bay region diol epoxides of several polycyclic hydrocarbons are shown in Figure 2.

The high chemical reactivity of bay-region diol epoxides can be attributed to their unique electronic properties. Perturbational molecular orbital calculations, which predict π-electron energy changes, indicate that epoxides on saturated benzo rings which form part of the bay region of a hydrocarbon undergo ring opening to a carbonium ion much more easily than do nonbay-region epoxides (15,16,23). Furthermore, these calculations indicate that the more carcinogenic members of a series of polycyclic hydrocarbons tend to form carbonium ions from their bay-region diol epoxides more readily than the weaker carcinogens. Figure 3 shows the structure of a diol epoxide and tetrahydro epoxide, their resultant carbonium ions and the quantum mechanical model for estimating the π-electron energy change ($\Delta E_{deloc}/\beta$) on ring opening of the epoxides. These perturbational molecular orbital calculations (6) do not distinguish between the opening of diol epoxides or tetrahydro-epoxides to carbonium ions since they do not take into account the presence of the tetrahydro benzo ring.

In the present report we have attempted to establish whether a quantitative relationship exists between mutagenic activity and the calculated $\Delta E_{deloc}/\beta$ values by studying the mutagenicity of nine tetrahydroepoxides of five different polycyclic hydrocarbons in strain TA 100 of *S. typhimurium.* As shown in Figure 4, all nine tetrahydroepoxides induced mutations in a dose-dependent manner, and activities varied over a 200-fold range. As predicted by the molecular orbital calculations, the bay-region tetrahydro-epoxides of benzo[a]pyrene, benzo[a]anthracene, chrysene and phenanthrene were from 6 to 25 times more active than their nonbay-region counterparts.

The quantitative relationship between the mutagenic activities of the nine tetrahydroepoxides and the calculated values of $\Delta E_{deloc}/\beta$ is shown in Figure 5. A semilogarithmic plot was utilized since the logarithm of the rate of carbonium ion formation is expected to be linearly related to $\Delta E_{deloc}/\beta$ (6). Interestingly, the data in Figure 5 indicate that the logarithm of mutagenic activity also appears linearly related to $\Delta E_{deloc}/\beta$. A linear least-squares regression analysis of all nine data points produced a moderate correlation

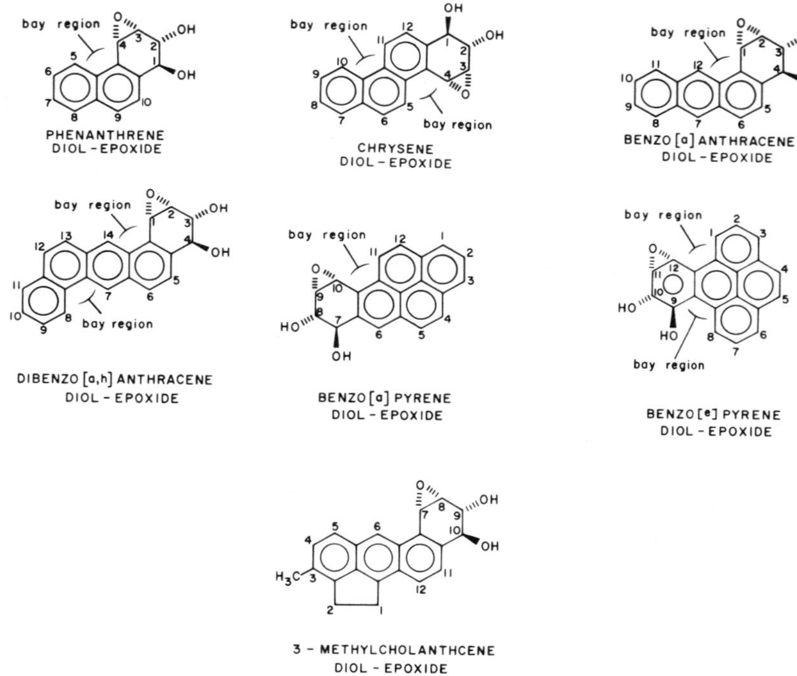

FIGURE 2. Relative stereochemistry of the bay-region diol epoxides of seven polycyclic hydrocarbons. Diol epoxides exist as diastereomeric pairs in which the epoxide oxygen is either *cis* (isomer 1 series) or *trans* (isomer 2 series) to the benzylic hydroxyl group (see Figure 1). The isomer 2 series is shown here since in the cases of benzo[a]pyrene and benzo[a]anthracene the diol epoxide 2 isomer is more tumorigenic than the diol epoxide 1 isomer.

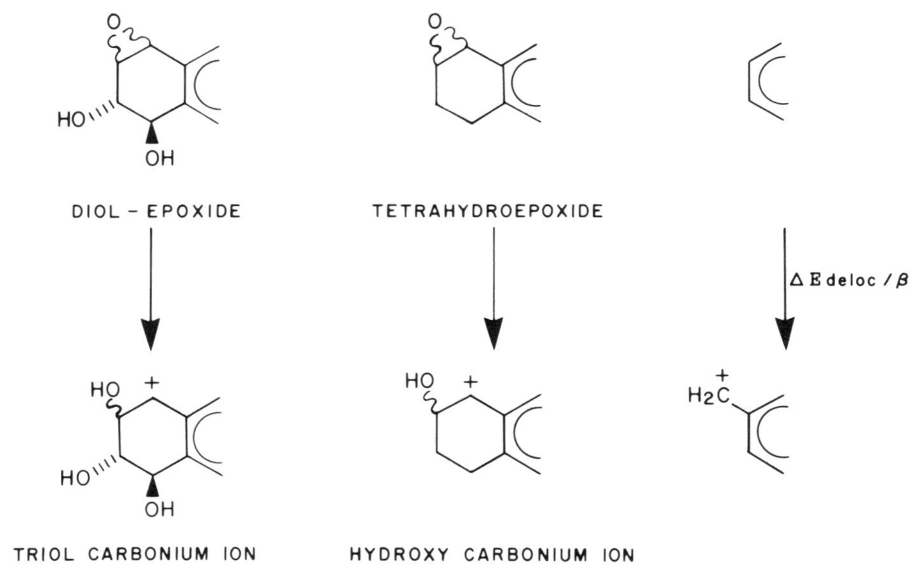

FIGURE 3. Structure of a diol epoxide and tetrahydroepoxide moiety, their carbonium ion products, and the quantum mechanical model for estimating the π-electron energy change (ΔE_{deloc}) on ring opening of such epoxides. Dewar perturbational molecular orbital calculations do not distinguish between the opening of a diol epoxide or a tetrahydroepoxide to a carbonium ion since these calculations do not take into account the presence of the tetrahydro benzo ring.

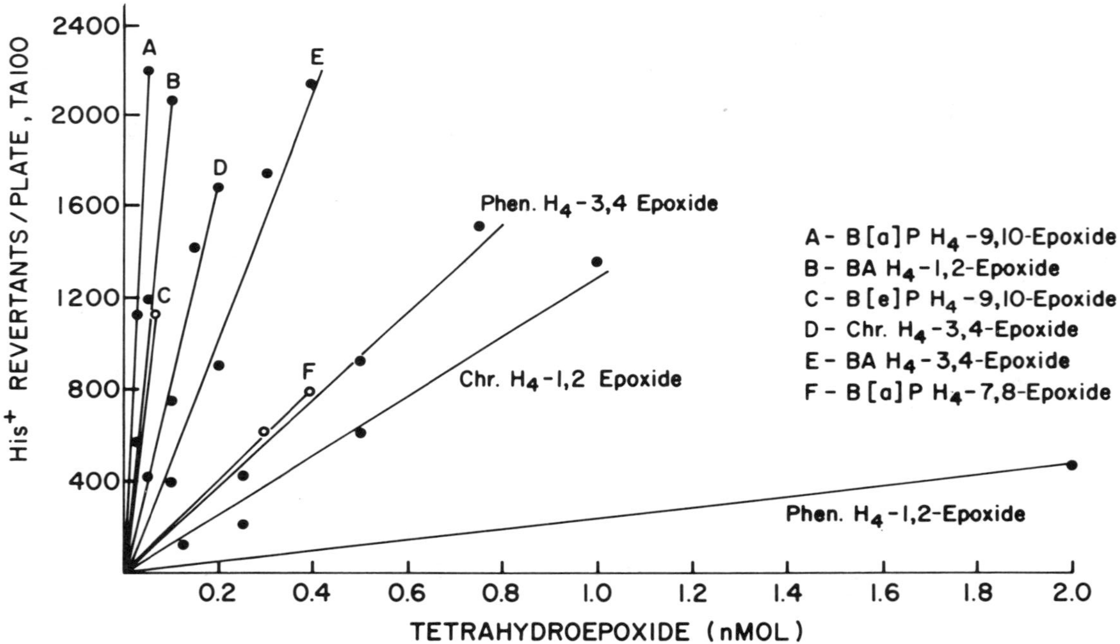

FIGURE 4. Intrinsic mutagenicity of nine polycyclic hydrocarbon tetrahydroepoxides in strain TA 100 of *Salmonella typhimurium.* Background mutation frequencies of 74 revertants per plate have been subtracted from the data, which are presented as the means of three replicate determinations. Mutagenicity of all the compounds was determined on the same day using an assay procedure described previously (50). Mutations induced by phenanthrene H_4-1,2-epoxide exhibited a linear dose response from 2 to 8 nmol of tetrahydroepoxide.

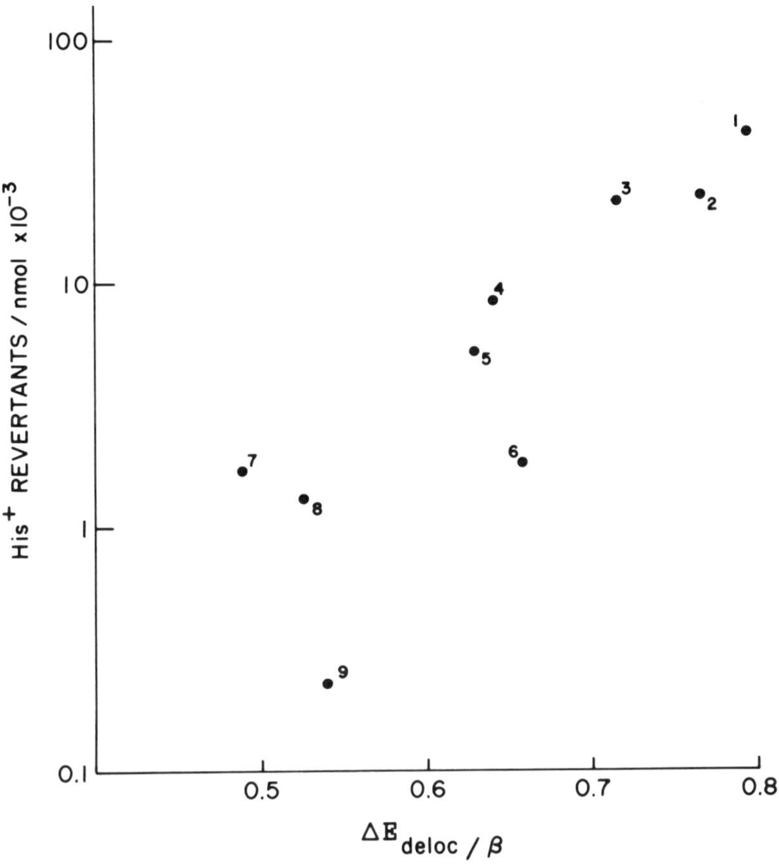

FIGURE 5. Relationship between mutagenic activity of the nine tetrahydroepoxides in strain TA 100 of *Salmonella typhimurium* and the value $\Delta E_{deloc}/\beta$ calculated by the perturbational molecular orbital method for the opening of these compounds to benzylic carbonium ions. Mutagenicity values are from Figure 4. 1, B[a]P H_4-9,10-epoxide; 2, BA H_4-1,2-epoxide; 3, B[e]P H_4-9,10-epoxide; 4, chrysene H_4-3,4-epoxide; 5, BA H_4-3,4-epoxide; 6, phenanthrene H_4-3,4-epoxide, 7, B[a]P H_4-7,8-epoxide; 8, chrysene H_4-1,2-epoxide; 9, phenanthrene H_4-1,2-epoxide.

(r = 0.74) which indicated that 55% of the variation in mutagenic activities was attributable to the different rates of carbonium ion formation. However inspection of the plot revealed that lower than predicted mutagenic activity of the two phenanthrene H_4-epoxides was primarily responsible for the deviation from a strictly linear relationship. When the data for the two phenanthrene H_4-epoxides was omitted, a correlation of 0.96 was obtained, suggesting that 92% of the 80-fold variation in mutagenic activities of the seven tetrahydroepoxides was predicted by the quantum mechanical calculations alone. Additional H_4-epoxides will need to be tested before the extent of the relationship between their mutagenic activities and $\Delta E_{deloc}/\beta$ is strictly defined. Recent studies comparing rates of solvolysis (47) and rates of reaction with the nucleophile *p*-nitrothiophenolate (14) have also verified the practicality of the quantum mechanical calculations.

We have emphasized that confirmation and practical application of the bay-region theory depends on both the demonstration that the appropriate dihydrodiol and diol epoxides of a number of polycyclic hydrocarbons possess unique mutagenic and tumorigenic activity and that these derivatives are actually formed in the species in which the parent hydrocarbons are carcinogenic (16). The remainder of this chapter is a summation of our mutagenicity and tumorigenicity studies with five of the polycyclic hydrocarbons shown in Figure 2. In separate chapters the metabolism of these hydrocarbons is reviewed (44) as is the biological activity and metabolism of benzo[e]pyrene (22).

PHENANTHRENE

Phenanthrene is the smallest angular polycyclic hydrocarbon and thus it is the prototype of the bay-region hydrocarbons. The data in Figure 4 indicate that phenanthrene H_4-3,4-epoxide, with its oxirane ring forming part of the bay region of the hydrocarbon, is almost ten times as mutagenic as the non-bay-region phenanthrene H_4-1,2-epoxide. The high mutagenicity of phenanthrene H_4-3,4-epoxide is consistent with the calculated energy change ($\Delta E_{deloc}/\beta$) expected for the formation of the corresponding carbonium ion. However, the mutagenicity data and the quantum mechanical calculations are apparently inconsistent with the observation that phenanthrene has little if any carcinogenic activity (35) and that phenanthrene 1,2-dihydrodiol, the immediate metabolic precursor of the bay-region diol epoxide of phenanthrene, is not metabolically activated to products mutagenic toward strain TA 100 of *S. typhimurium* (15). Preliminary experiments with the chemically synthesized diastereomeric phenanthrene 1,2-diol-3,4-epoxides indicate that they are substantially less mutagenic than the corresponding tetrahydroepoxide in both bacterial and mammalian cells. Since phenanthrene is a small polycyclic hydrocarbon, its diol epoxides are more polar than the diol epoxide of larger hydrocarbons. This increased polarity may result in a less efficient interaction with the bacterial DNA. Alternatively (or additionally), the negligible carcinogenicity of phenanthrene may result from a relatively small amount of the parent hydrocarbon being converted to phenanthrene 1,2-dihydrodiol and subsequently to the 1,2-diol-3,4-epoxide. Again, the polarity of the diol groups may make the dihydrodiol a poor substrate for the cytochrome P450 monooxygenase system. Tumorigenicity studies with phenanthrene and its derivatives are currently in progress.

CHRYSENE

Chrysene is a symmetric polycyclic hydrocarbon having two identical bay regions and three metabolically possible *trans*-dihydrodiols. The bay-region theory predicts that chrysene 1,2-dihydrodiol and chrysene 1,2-diol-3,4-epoxide are proximate and ultimate carcinogens, respectively, of chrysene. Perturbational molecular orbital calculations show a greater energy change when the bay region H_4-3,4-epoxide is converted to a carbonium ion ($\Delta E_{deloc}/\beta$ = 0.639) than when the nonbay-region H_4-1,2-epoxide is converted to its corresponding carbonium ion ($\Delta E_{deloc}/\beta$ = 0.526). The sixfold higher mutagenic activity of chrysene H_4-3,4-epoxide relative to chrysene H_4-1,2-epoxide (Figure 4) supports the calculations and provides experimental evidence for the high reactivity of the bay-region epoxide. Figure 6 summarizes the mutagenic (55) and skin tumor-initiating activity (25) of the parent hydrocarbon, the three chrysene dihydrodiols and chrysene H_4-1,2-diol–a derivative of chrysene 1,2-dihydrodiol which lacks a double bond in the 3,4-position and thus cannot be metabolized to a diol epoxide. For the mutagenicity experiments the compounds were incubated with strain TA 100 of *S. typhimurium* and a purified cytochrome P450-dependent monooxygenase system. The data clearly indicate the uniquely high mutagenic activity of the metabolic products of chrysene 1,2-dihydrodiol. Similarly this immediate metabolic precursor of the chrysene bay-region diol epoxides had the most tumorigenic activity on mouse skin. In these experiments 4.0 μmol of each compound was applied once to the backs of CD-1 female mice and the tumor promoter tetradecanoyl phorbol-13-acetate was applied twice weekly for 25 weeks. The low mutagenic and tumorigenic activity of chrysene H_4-1,2-diol indicates the importance of the olefinic double bond in the bay region of the molecule and suggests that chrysene 1,2-diol-3,4-epoxide is the biologically active metabolite formed from chrysene 1,2-dihydrodiol. We have recently synthesized the diastereomeric pair of chrysene 1,2-diol-3,4-epoxides and, as predicted from the quantum mechanical calculations and the biological activity of chrysene 1,2-dihydrodiol, both isomers have significant intrinsic mutagenic activity in both bacterial and mammalian cells. The bay-region diol epoxides are currently being evaluated for tumorigenic activity.

While chrysene is a relatively weak carcinogen, its 5-methyl derivative is a potent carcinogen found in cigarette smoke and elsewhere in our biosphere (7). Hecht *et al.* have studied the metabolism, mutagenicity and tumorigenicity of 5-methylchrysene and a number of its derivatives and concluded that

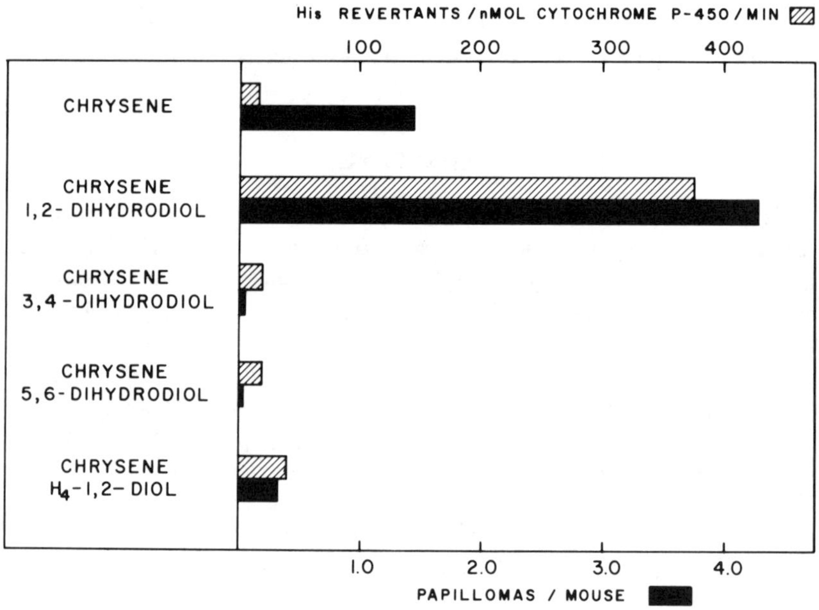

FIGURE 6. Mutagenicity (▨) and tumorigenicity (■) of chrysene, its three metabolically possible vicinal *trans*-dihydrodiols and chrysene H_4-1,2-diol. Mutagenic activity was determined by incubating 37.5 nmol of each compound with 2 x 10^8 bacteria of strain TA 100 of *Salmonella typhimurium* for 10 min with a highly purified and reconstituted cytochrome P450-dependent monooxygenase system, as described (55). Cytochrome P450 was purified from Aroclor 1254-pretreated Long Evans rats. Tumorigenic activity was determined by a single topical application of 4.0 μmol of each compound to CD-1 female mice followed by twice weekly applications of 16 nmol of tetradecanoyl-phorbol-13-acetate for 25 weeks, as described (24).

bay-region activation represents an important determinant of the biological activity of 5-methylchrysene (8,9).

BENZO[a]ANTHRACENE (BA)

BA, like chrysene, is a weakly carcinogenic tetracyclic hydrocarbon. A total of 5 *trans*-dihydrodiols of BA are metabolically possible and, except for the K-region BA 5,6-dihydrodiol, the dihydrodiols have adjacent olefinic double bonds that can be oxidized to form diol epoxides. Results of incubating BA and its five dihydrodiols with strain TA 100 of *S. typhimurium* in the presence of a highly purified microsomal monooxygenase system have been reported (54) and are summarized in Figure 7. BA 3,4-dihydrodiol, the immediate metabolic precursor of the bay-region diol epoxide of BA, was

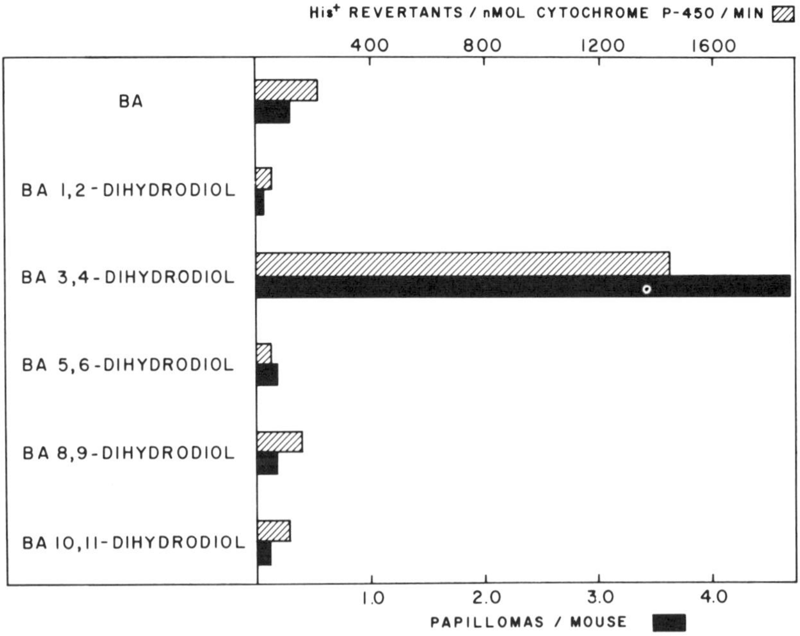

FIGURE 7. Mutagenicity (▨) and tumorigenicity (■) of BA and its five metabolically possible *trans*-dihydrodiols. Mutagenic activity was determined as described in Figure 6 except that the amount of hydrocarbon incubated was 12.5 nmol, incubation time was 15 min and cytochrome P450 was purified from 3-methylcholanthrene-pretreated Long Evans rats (54). Tumorigenic activity was determined as described in Figure 6 except that 2.0 μmol of each hydrocarbon was applied and the experiment was terminated after 20 weeks of promotion (53).

metabolically activated to products that were at least ten times more mutagenic than the metabolic products of BA or its four other dihydrodiols. Figure 7 also compares the skin tumor-initiating activities of a single 4.0 μmol dose of BA and its dihydrodiols; the data clearly indicate that BA 3,4-dihydrodiol is a proximate carcinogen of BA (53). The high tumorigenicity of BA 3,4-dihydrodiol has also been confirmed in the newborn mouse where this precursor of the bay-region diol epoxide was over 30 times more active in inducing lung adenomas than was BA and the other four BA dihydrodiols (48). More recently, resolution of racemic BA 3,4-dihydrodiol into the (-)- and (+)-enantiomers and assignment of their absolute stereochemistry revealed that the (-)-(3R,4R) isomer was at least fivefold more active as a tumor initiator on mouse skin than was the (+)-(3S,4S) isomer (24). Absolute stereochemical aspects of the mutagenicity and tumorigenicity of polycyclic hydrocarbon derivatives have been reviewed elsewhere (17).

Assessment of the inherent mutagenic activity of seven of the eight possible vicinal diol epoxide diastereomers of BA confirmed the unique biological activity of the bay-region 3,4-diol-1,2-epoxides (50, Table I). As was found for the bay region 7,8-diol-9,10-epoxides of BP, isomer 1 was more mutagenic than isomer 2 in *S. typhimurium* while the converse was true in the mammalian cells. When tested for their tumor-initiating activity on mouse skin, isomer 1 and 2 of BA 3,4-diol-1,2-epoxide were approximately 10- and 40-fold more tumorigenic, respectively, than the parent hydrocarbon BA (24). Thus the skin tumor-initiating activity of these two diastereomers paralleled the mutagenic activities observed with the mammalian cells. Slaga *et al.* (39) recently demonstrated the high skin tumor-initiating activity of the BA 3,4-diol-1,2-epoxide-2 relative to BA 1,2-diol-3,4-epoxide-2, the BA 8,9-diol-10,11-epoxides and BA 10,11-diol-8,9-epoxide-2. Thus there is now definitive proof that a bay-region diol epoxide is an ultimate carcinogen of BA. Given the efficiency at which BA 3,4-dihydrodiol is metabolized to mutagenic and tumorigenic products and the high inherent biological activity of the BA 3,4-diol-1,2-epoxides, it appears that BA is a weak carcinogen by virtue of its low rate of metabolism to the 3,4-dihydrodiol, as recently shown by Thakker *et al.* (43,44).

While BA is a weak carcinogen its two meso-anthracenic methyl-substituted derivatives, 7-methyl-BA and 7,12-DMBA are moderate and extremely potent carcinogens, respectively. Both of these alkyl-substituted BA derivatives have their 1,2,3,4-benzo ring intact and are capable of forming bay-region diol epoxides. Mutagenicity, DNA binding and tumorigenicity studies from the laboratories of Grover and Sims and their associates (4,32,45) have indicated that the 3,4-dihydrodiol of 7-methyl-BA is a proximate carcinogen of the parent hydrocarbon, and mutagenicity (30) and DNA-binding (12,33,46) studies suggest that 7,12-DMBA is also activated at the bay region.

DIBENZO[a,h]ANTHRACENE (DBA)

The first evidence that a chemically pure substance could induce cancer was reported in 1930 when Kennaway and Hieger demonstrated the carcinogenicity of DBA on mouse skin (20). DBA is a symmetrical pentacyclic hydrocarbon possessing two identical bay regions, and only the 1,2-, 3,4- and 5,6-dihydrodiols are metabolically possible. The metabolic activation of DBA, the three DBA dihydrodiols and DBA H_4-3,4-diol to mutagens have been examined using microsomes as well as a purified cytochrome P450 monooxygenase system as a metabolic activating system (56). Figure 8 summarizes the results with the purified monooxygenase system and shows that DBA 3,4-dihydrodiol, the immediate metabolic precursor of the bay-region 3,4-diol-1,2-epoxide of DBA was approximately three times more active than DBA

TABLE I. Mutagenicity of Benzo[a]anthracene Diol Epoxides in Bacterial and Mammalian Cells

	Strain TA100[a] *S. typhimurium*		Chinese Hamster[b] V79 Cells	
Compound	**Mutation Frequency**	**% Activity**	**Mutation Frequency**	**% Activity**
BA 1,2-Diol-3,4-epoxide				
Isomer 2	3	0.2	0.33	3
BA 3,4-Diol-1,2-epoxide				
Isomer 1	1,650	100	8.5	78
Isomer 2	850	52	10.9	100
BA 8,9-Diol-10,11-epoxide				
Isomer 1	16	1.0	0.06	0.6
Isomer 2	50	3.0	0.10	0.9
BA 10,11-Diol-8,9-epoxide				
Isomer 1	6	0.4	0.03	0.3
Isomer 2	55	3.0	0.21	2.0

[a]Mutation frequency expressed as histidine revertants/nmol/plate.
[b]Mutation frequency expressed as 8-azaguanine resistant colonies/nmol/10^5 surviving cells.

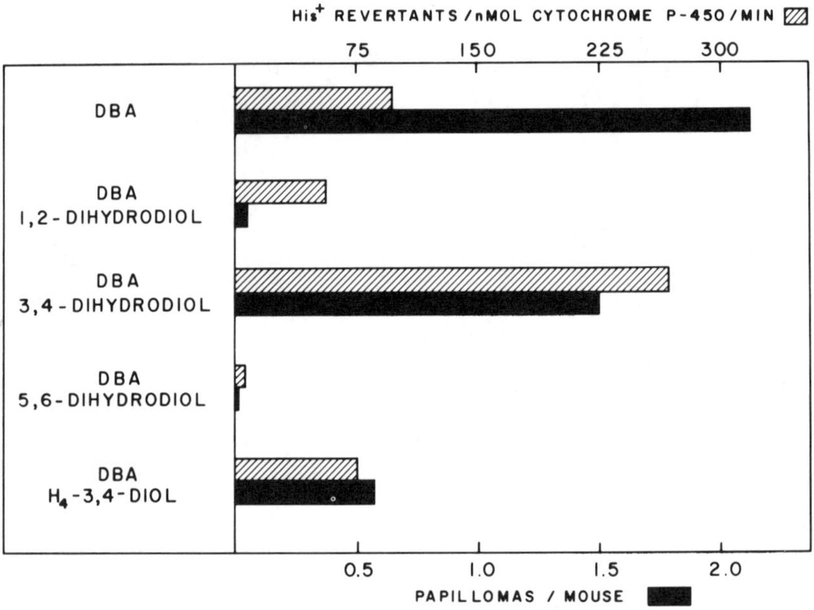

FIGURE 8. Mutagenicity (▨) and tumorigenicity (■) of DBA, its three metabolically possible *trans*-dihydrodiols and DBA H_4-1,2-diol. Mutagenic (56) and tumorigenic (1) activities were assessed by the protocols described in Figure 6 except that in the tumorigenicity experiments 0.16 μmol of hydrocarbon was applied to the mice.

and eight to fifty times more active than the other diols. The DBA derivatives had the same relative activites when they were incubated with microsomes from Aroclor-pretreated rats except that at relatively high substrate and enzyme concentrations DBA 5,6-dihydrodiol was activated to mutagenic products. Figure 8 also shows that DBA 3,4-dihydrodiol, in contrast to the 1,2- and 5,6-dihydrodiols, is a potent initiator of skin tumors (1). While DBA 3,4-dihydrodiol had 70% of the tumor-initiating activity of the parent hydrocarbon at a 0.16-μmol dose, a dose of 0.01 μmole of DBA 3,4-dihydrodiol induced almost twice as many tumors as did DBA. Metabolism studies (36,44) have indicated that rat liver converts substantial amount of DBA to the 3,4-dihydrodiol. If this is also the case in other tissues and species, it may partially explain why DBA and the 3,4-dihydrodiol have comparable tumorigenic activities. The high activity of DBA 3,4-dihydrodiol relative to DBA H_4-3,4-diol in both the mutagenicity and tumorigenicity models illustrates the importance of the isolated 1,2-double bond in the bay region of DBA and suggests that a substantial portion of the biological activity of DBA is mediated by its bay-region diol epoxide.

TABLE II. Relative Mutagenic Activities of 3-Methylcholanthrene Derivatives in Bacterial and Mammalian Cells

	Strain TA98 *S. typhimurium*		Strain TA100 *S. typhimurium*		Chinese Hamster V79 Cells
Compound	Microsomes	Purified Monooxygenase System	Microsomes	Purified Monooxygenase System	Microsomes
MC	16	5	7	<1	$\leqslant 5$
1-Keto MC	10	2	18	<1	30
1-Hydroxy MC	100	13	2	1	$\leqslant 5$
1-Hydroxy MC 9,10-dihydrodiol *a*	39	100	100	100	100
1-Hydroxy MC 9,10-dihydrodiol *b*	—	52	—	9	40
2-Keto MC	34	72	40	3	50
2-Hydroxy MC	24	9	3	<1	$\leqslant 5$
MC 11,12-Dihydrodiol	0	—	0	—	—
Cis-1,2-Dihydroxy MC	10	—	3	—	—
Trans-1,2-Dihydroxy MC	10	—	3	—	—

Note: Relative activities of the various MC derivatives are expressed as a percentage of the most active derivative in each test system. Since racemic 1-hydroxy-MC was used to form the 9,10-dihydrodiol, the two 1-hydroxy-MC 9,10-dihydrodiols, designated *a* and *b*, are presumed to be diastereomers in which the 1- and 10-hydroxyl groups are either *cis* or *trans.* At present nothing is known about the absolute stereochemistry of these isomers nor is it known whether the 1- and 10-hydroxyl groups are *cis* or *trans* in the major *a* isomer. Because relatively little of the *b* isomer was isolated from the incubations, it was not tested for metabolic activation with the bacteria in the presence of microsomes.

3-METHYLCHOLANTHRENE (MC)

As summarized above, the bay-region 3,4-diol-1,2-epoxide of benzo[a]-anthracene is an ultimate carcinogenic metabolite of benzo[a]anthracene. The potent carcinogen MC is an alkyl-substituted derivative of benzo[a]-anthracene (see Figure 2) in which the angular benzo ring and the bay region remain unsubstituted. We therefore sought to determine whether MC was metabolically activated in a manner analogous to benzo[a]anthracene, namely via formation of MC 9,10-diol-7,8-epoxide. Metabolism studies by Thakker *et al.* (41,42,44) demonstrated that rat liver enzymes form only small amounts of dihydrodiols but significant amounts of MC 9,10-dihydrodiol are formed from 1-HOMC, which is a major metabolite of MC. Large-scale incubations permitted the isolation of sufficient amounts of the 1-hydroxy-MC 9,10-dihydrodiol diastereometers, designated *a* and *b* (see legend to Table II), for detailed mutagenicity studies. Both microsomes and a purified mono-oxygenase system were utilized for activating systems and two strains of bacteria and a mammalian cell line were used to detect mutagenic activity. The results of these studies have been described (51) and Table II summarizes the activation of 1-hydroxy MC-9,10-dihydrodiol and several other known metabolites of MC.

The data indicate that a 1-hydroxy-MC-9,10-dihydrodiol undergoes metabolic activation to mutagenic products to a greater extent than does the parent hydrocarbon in all five of the test systems studied and that in four of the systems the immediate metabolic precursor of the bay-region diol epoxide of MC is activated to a greater extent than is any of the seven other MC metabolites tested. Additional evidence that MC is activated via a bay-region 9,10-diol-7,8-epoxide has recently been obtained from two independent studies on the nature of the DNA-bound adducts of MC obtained from mouse embryo cells treated in culture with MC or from mice treated topically with the parent hydrocarbon (21,46). Both studies indicated the covalently bound adducts were saturated in the 7,8,9,10-benzo ring of the hydrocarbon and had an intact anthracene nucleus. The studies in Table II also indicate that several other metabolites of MC can be metabolically activated to an appreciable extent. Unlike 1-hydroxy MC 9,10-dihydrodiol these other metabolites were not consistently activated and the expression of their mutagenic activity is highly dependent on both the source of the monooxygenase system and the target cell with which the mutation is detected.

REFERENCES

1. Buening, M., Levin, W., Wood, A. W., Chang, R. L., Ygi, H., Karle, J. M., Jerina, D. M. and Conney, A. H. (1979): Tumorigenicity of the

dihydrodiols of dibenzo[a,h] anthracene on mouse skin and in newborn mice. *Cancer Res.* 39:1310-1314.

2. Buening, M. K., Wislocki, P. G., Levin, W., Yagi, H., Thakker, D. R., Akagi, H., Koreeda, M., Jerina, D. M., and Conney, A. H. (1978): Tumorigenicity of the optical enantiomers of the diastereomeric benzo-[a] pyrene 7,8-diol-9,10-epoxides in newborn mice: Exceptional activity of (+)-7β,8α-dihydroxy-9α,10α-epoxy-7,8,9,10-tetrahydrobenzo[a]-pyrene. *Proc. Nat. Acad. Sci., U.S.*, 75:5358-5361.
3. Chouroulinkov, I., Gentil, A., Grover, P. L., and Sims, P. (1976): Tumour initiating activities on mouse skin of dihydrodiols derived from benz[a] pyrene. *Brit. J. Cancer* 34:523-532.
4. Chouroulinkov, I., Gentil, A., Tierney, B., Grover, P. L., and Sims, P. (1977): The metabolic activation of 7-methylbenz[a] anthracene in mouse skin: High tumor initiating activity of the 3,4-dihydrodiol. *Cancer Lett.* 3:247-253.
5. Conney, A. H., Levin, W., Wood, A. W., Yagi, H., Thakker, D. R., Lehr, R. E., and Jerina, D. M. (1978): Metabolism of polycyclic aromatic hydrocarbons to reactive intermediates with high biological activity. *Proceedings, Conference on Human Epidemiology and Laboratory Animal Correlations in Chemical Carcinogenesis,* (in press).
6. Dewar, M. J. S. (1969): *The Molecular Orbital Theory of Organic Chemistry,* pp. 214-217,304-305, McGraw-Hill, New York.
7. Hecht, S. S., Bondinell, W. E., and Hoffmann, D. (1974): Chrysene and methylchrysenes: Presence in tobacco smoke and carcinogenicity. *J. Nat. Cancer Inst.* 53:1121-1133.
8. Hecht, S. S., LaVoir, E., Mazzarese, R., Amin, S., Bedemko, V., and Hoffmann, D. (1978): 1,2-Dihydro-1,2-dihydroxy-5-methylchrysene, a major activated metabolite of the environmental carcinogen 5-methylchrysene. *Cancer Res.* 38:2191-2194.
9. Hecht, S., Mazzarese, R., Amin, S., LaVoir, E., and Hoffmann, D. (1979): The metabolic activation of the environmental carcinogen, 5-methylchrysene. Chapter 48, this volume.
10. Heidelberger, C. (1974): Chemical oncogenesis in culture. *Adv. Cancer Res.* 20:317-366.
11. Huberman, E., Sachs, L., Yang, S. K., and Gelboin, H. V. (1976): Identification of mutagenic metabolites of benzo[a] pyrene in mammalian cells. *Proc. Nat. Acad. Sci., U.S.* 73:607-611.
12. Ivanoivic, V., Geacintov, N. E., Jeffrey, A. M., Fu, P. P., Harvey, R. G., and Weinstein, I. B. (1978): Cell and microsome mediated binding of 7,12-dimethylbenz[a] anthracene to DNA studied by fluorescence spectroscopy. *Cancer Lett.* 4:131-140.
13. Jerina, D. M., and Daly, J. W. (1974): Arene oxides. A new aspect of drug metabolism. *Science* 185:573-582.
14. Jerina, D. M., and Lehr, R. E. (1977): The bay region theory: A quantum mechanical approch to aromatic hydrocarbon-induced carcinogenicity. In: *Microsomes and Drug Oxidations,* V. Ullrich, I. Roots, A. G.

Hildebrant, R. W. Estabrook and A. H. Conney, Eds., pp. 709-720, Pergamon Press, Oxford.

15. Jerina, D. M., Lehr, R., Schaefer-Ridder, M., Yagi, H., Karle, J. M., Thakker, D. R., Wood, A. W., Lu, A. Y. H., Ryan, D., West, S., Levin, W., and Conney, A. H. (1977): Bay region epoxides of dihydrodiols: A concept which may explain the mutagenic and carcinogenic activity of benzo[a]pyrene and benzo[a]anthracene. In: *Origins of Human Cancer,* H. H. Hiatt, J. D. Watson and J. A. Winsten, Eds., pp. 638-658, Cold Spring Harbor Laboratory, Cold Spring Harbor, NY.
16. Jerina, D. M., Lehr, R. E., Yagi, H., Hernandez, O., Dansette, P. M., Wislocki, P. G., Wood, A. W., Chang, R. L., Levin, W., and Conney, A. H. (1976): Mutagenicity of benzo[a]pyrene derivatives and the description of a quantum mechanical model which predicts the ease of carbonium ion formation from diol epoxides. In: *In Vitro Metabolic Activation in Mutagenesis Testing,* F. J. de Serres, J. R. Fouts, J. R. Bend and R. M. Philpot, Eds., pp. 159-177, Elsevier, Amsterdam.
17. Jerina, D. M., Yagi, H., Thakker, D. R., Karle, J. M., Mah, H. D., Boyd, D. R., Gadaginamath, G., Wood, A. W., Buening, M., Chang, R. L., Levin, W., and Conney, A. H. (1978): Stereoselective metabolic activation of polycyclic aromatic hydrocarbons. In: *Advances in Pharmacology and Therapeutics: Toxicology,* Vol. 9, Y. Cohen, Ed., pp. 53-62, Pergamon Press, Inc., New York.
18. Kapitulnik, J., Levin, W., Yagi, H., Jerina, D. M., and Conney, A. H. (1977): Benzo[a]pyrene 7,8-dihydrodiol is more carcinogenic than benzo[a]pyrene in newborn mice. *Nature* 266:378-379.
19. Kapitulnik, J., Wislocki, P. G., Levin, W., Yagi, H., Jerina, D. M., and Conney, A. H. (1978): Tumorigenicity studies with diol epoxides of benzo[a]pyrene which indicate that (±)*trans*-7β,8α-dihydroxy-9α,10α-epoxy-7,8,9,10-tetrahydrobenzo[a]pyrene is an ultimate carcinogen in newborn mice. *Cancer Res.* 38:354-358.
20. Kennaway, E. L., and Hieger, I. (1930): Carcinogenic substances and their fluorescence spectra. *Brit. Med. J.* 1:1044-1046.
21. King, H. W. S., Osborne, M. R., and Brookes, P. (1977): The metabolism and DNA binding of 3-methylcholanthrene. *Int. J. Cancer* 20: 564-571.
22. Lehr, R. E., Taylor, C. W., Kumar, S., Levin, W., Chang, R., Wood, A. W., Conney, A. H., Thakker, D. R., Yagi, H., Mah, H. D., and Jerina, D. M. (1979): Differences in metabolism provides a basis for the low mutagenicity and carcinogenicity of benzo[e]pyrene compared to benzo[a]pyrene. Chapter 3, this volume.
23. Lehr, R. E., Yagi, H., Thakker, D. R., Levin, W., Wood, A. W., Conney, A. H., and Jerina, D. M. (1978): The bay region theory of polycyclic aromatic hydrocarbon-induced carcinogenicity. In: *Carcinogenesis, Vol. 3,* P. W. Jones, and R. I. Freudenthal, Eds., pp. 231-239, Raven Press, New York.

24. Levin, W., Thakker, D. R., Wood, A. W., Chang, R. L., Lehr, R. E., Jerina, D. M. and Conney, A. H. (1978): Evidence that benzo[a]-anthracene 3,4-diol-1,2-epoxide is an ultimate carcinogen on mouse skin. *Cancer Res.* 38:1705-1710.
25. Levin, W., Wood, A. W., Chang, R. L., Yagi, H., Mah, H. D., Jerina, D. M., and Conney, A. H. (1978): Evidence for bay region activation of chrysene 1,2-dihydrodiol to an ultimate carcinogen. *Cancer Res.* 38:1831-1834.
26. Levin, W., Wood, A. W., Wislocki, P. G., Chang, R. L., Kapitulnik, J., Mah, H. D., Yagi, H., Jerina, D. M., and Conney, A. H. (1978): Mutagenicity and carcinogenicity of benzo[a]pyrene and benzo[a]pyrene derivatives. In: *Polycyclic Hydrocarbons and Cancer, Vol. 1: Environment, Chemistry, and Metabolism,* H. V. Gelboin and P.O.P. Ts'o, Eds., pp. 189-202, Academic Press, New York.
27. Levin, W., Wood, A. W., Wislocki, P. G., Kapitulnik, J., Yagi, H., Jerina, D. M., and Conney, A. H. (1977): Carcinogenicity of benzo-ring derivatives of benzo[a]pyrene on mouse skin. *Cancer Res.* 37:3356-3361.
28. Levin, W., Wood, A. W., Yagi, H., Dansette, P. M., Jerina, D. M., and Conney, A. H. (1976): Carcinogenicity of benzo[a]pyrene 4,5-, 7,8-, and 9,10-oxides on mouse skin. *Proc. Nat. Acad. Sci., U.S.* 73:243-247.
29. Levin, W., Wood, A. W., Yagi, H., Jerina, D. M., and Conney, A. H. (1976): Benzo[a]pyrene 7,8-dihydrodiol: A potent skin carcinogen when applied topically to mice. *Proc. Nat. Acad. Sci, U.S.* 73:3867-3871.
30. Malaveille, C., Bartsch, H., Tierney, B., Grover, P. L., and Sims, P. (1978): Microsome-mediated mutagenesis of the dihydrodiols of 7,12-dimethylbenz[a]anthracene: High mutagenic activity of the 3,4-dihydrodiol. *Biochem. Biophys. Res. Commun.* 83:1468-1473.
31. Malaveille, C., Kuroki, J., Sims, P., Grover, P. L., and Bartsch, H. (1977): Mutagenicity of isomeric diol-epoxides of benzo[a]pyrene and benz-[a]anthracene in *S. typhimurium* TA 98 and TA 100 and in V79 Chinese hamster cells. *Mutation Res.* 44:313-326.
32. Malaveille, C., Tierney, B., Grover, P. L., Sims, P., and Bartsch, H. (1977): High microsome-mediated mutagenicity of the 3,4-dihydrodiol of 7-methyl-benz[a]anthracene in *S. typhimurium* TA 98. *Biochem. Biophys. Res. Commun.* 75:427-433.
33. Moschel, R. C., Baird, W. M., and Dipple A. (1977): Metabolic activation of the carcinogen 7,12-dimethylbenz[a]anthracene for DNA binding. *Biochem. Biophys. Res. Commun.* 76:1092-1098.
34. Newbold, R. F., and Brookes, P. (1976): Exceptional mutagenicity of a benzo[a]pyrene diol epoxide in cultured mammalian cells. *Nature* 261:52-54.
35. Scribner, J. D. (1973): Tumor initiation by apparently noncarcinogenic polycyclic aromatic hydrocarbons. *J. Nat. Cancer Inst.* 50:1717-1719.
36. Sims, P. (1970): Qualitative and quantitative studies on the metabolism

of a series of aromatic hydrocarbons by rat liver preparations. *Biochem. Pharmacol.* 19:795-818.

37. Sims, P., and Grover, P. L. (1974): Epoxides in polycyclic aromatic hydrocarbon metabolism and carcinogenesis. *Advan. Cancer Res.* 20:165-274.
38. Slaga, T. J., Bracken, W. M., Gleason, G., Levin, W., Yagi, H., Jerina, D. M., and Conney, A. H. (1979): Marked differences in the skin tumor-initiating activities of the optical enantioners of the diastereomeric benzo[a]pyrene 7,8-diol-9,10-epoxides. *Cancer Res.* 39:67-71.
39. Slaga, T. J., Huberman, E., Selkirk, J. K., Harvey, R. G., Bracken, W. M. (1978): Carcinogenicity and mutagenicity of benz[a]anthracene diols and diol-epoxides. *Cancer Res.* 38:1699-1704.
40. Slaga, T. J., Viage, A., Berry, D. L., Bracken, W., Buty, S. G., and Scribner, J. D. (1976): Skin tumor initiating ability of benzo[a]pyrene 4,5-, 7,8-, and 7,8-diol-9,10-epoxides and 7,8-diol. *Cancer Lett.* 2: 115-122.
41. Thakker, D. R., Levin, W., Stoming, T. A., Conney, A. H., and Jerina, D. M. (1978): Metabolism of 3-methylcholanthrene by rat liver microsomes and a highly purified monooxygenase system with and without epoxide hydrase. In: *Carcinogenesis, Vol. 3,* P. W. Jones and R. I. Freudenthal, Eds., pp. 253-264, Raven Press, New York.
42. Thakker, D. R., Levin, W., Stoming, T. A., Conney, A. H., and Jerina, D. M. (1978): Metabolic formation of 1,9,10-trihydroxy-9,10-dihydro-3-methylcholanthrene. *J. Am. Chem. Soc.* 100:645-647.
43. Thakker, D. R., Levin, W., Yagi, H., Karle, J. M., Lehr, R. E., Ryan, D., Thomas, P. E., Conney, A. H., and Jerina, D. M. (1979): Metabolism of benzo[a]anthracene to its tumorigenic 3,4-dihydrodiol. *Mol. Pharmacol.* 15:138-153.
44. Thakker, D. R., Nordqvist, M., Yagi, H., Levin, W., Ryan, D., Thomas, P. E., Conney, A. H., and Jerina, D. M. (1979): Comparative metabolism of a series of polycyclic aromatic hydrocarbons by rat liver microsomes and purified cytochrome P450. Chapter 30, this volume.
45. Tierney, B., Hewer, A., Walsh, C., Grover, P. L., and Sims, P. (1977): The metabolic activation of 7-methylbenz[a]anthracene in mouse skin. *Chem.- Biol. Interact.* 18:179-193.
46. Vigny, P., Duqesne, M., Coulomb, H., Tierney, B., Grover, P. L., and Sims, P. (1977): Flourescence spectral studies on the metabolic activation of 3-methylcholanthrene and 7,12-dimethylbenz[a]anthracene in mouse skin. *FEBS Letters* 82:278-282.
47. Whalen, D. L., Ross, A. M., Yagi, H., Karle, J. M., and Jerina, D. M. (1978): Stereoelectronic factors in the solvolysis of bay region diol epoxides of polycyclic aromatic hydrocarbons. *J. Am. Chem. Soc.* 100:5218-5221.
48. Wislocki, P. G., Kapitulnik, J., Levin, W., Lehr, R., Schaefer-Ridder, M., Karle, J. M., Jerina, D. M., and Conney, A. H. (1978): Exceptional

carcinogenic activity of benzo[a]anthracene 3,4-dihydrodiol in the newborn mouse and the bay region theory. *Cancer Res.* 38:693-696.

49. Wislocki, P. G., Kapitulnik, J., Levin, W., Yagi, H., Jerina, D. M., and Conney, A. H. (1978): Tumorigenicity of benzo[a]pyrene 4,5-, 7,8-, 9,10- and 11,12-oxides in newborn mice. *Cancer Lett.* 5:191-197.

50. Wood, A. W., Chang, R. L., Levin, W., Lehr, R. E., Schaefer-Ridder, M., Karle, J. M., Jerina, D. M., and Conney, A. H. (1977): Mutagenicity and cytotoxicity of benzo[a]anthracene diol epoxides and tetrahydro epoxides: Exceptional activity of the bay region 1,2-epoxides. *Proc. Nat. Acad. Sci., U.S.* 74:2746-2750.

51. Wood, A. W., Chang, R. L., Levin, W., Thomas, P. E., Ryan, D., Stoming, T. A., Thakker, D. R., Jerina, D. M., and Conney, A. H. (1978): Metabolic activation of 3-methylcholanthrene and its metabolites to mutagens toward bacteria and mammalian cells. *Cancer Res.* 38:3398-3404.

52. Wood, A. W., Goode, R. L., Chang, R. L., Levin, W., Conney, A. H., Yagi, H., Dansette, P. M., and Jerina, D. M. (1975): Mutagenic and cytotoxic activity of benzo[a]pyrene 4,5-, 7,8-, and 9,10-oxides and the six corresponding phenols. *Proc. Nat. Acad. Sci., U.S.* 72:3176-3180.

53. Wood, A. W., Levin, W., Chang, R. L., Lehr, R. E., Schaefer-Ridder, M., Karle, J. M., Jerina, D. M., and Conney, A. H. (1977): Tumorigenicty of five dihydrodiols of benzo[a]anthracene on mouse skin: Exceptional activity of benzo[a]anthracene 3,4-dihydrodiol. *Proc. Nat. Acad. Sci., U.S.* 74:3176-3179.

54. Wood, A. W., Levin, W., Lu, A. Y. H., Ryan, D., West, S. B., Lehr, R. E., Schaefer-Ridder, M., Jerina, D. M., and Conney, A. H. (1976): Mutagenicity of metabolically activated benzo[a]anthracene 3,4-dihydrodiol: Evidence for bay region activation of carcinogenic polycyclic hydrocarbons. *Biochem. Biophys. Res. Commun.* 72:680-686.

55. Wood, A. W., Levin, W., Ryan, D., Thomas, P. E., Yagi, H., Mah, H. D., Thakker, D. R., Jerina, D. M., and Conney, A. H. (1977): High mutagenicity of metabolically activated chrysene 1,2-dihydrodiol: Evidence for bay region activation of chrysene. *Biochem. Biophys. Res. Commun;* 78:847-854.

56. Wood, A. W., Levin, W., Thomas, P. E., Ryan, D., Karle, J. M., Yagi, H., Jerina, D. M., and Conney, A. H. (1978): Metabolic activation of dibenzo[a,h]anthracene and its dihydrodiols to bacterial mutagens. *Cancer Res.* 38:1967-1973.

57. Wood, A. W., Wislocki, P. G., Chang, R. L., Levin, W., Lu, A. Y. H., Yagi, H., Hernandez, O., Jerina, D. M., and Conney, A. H. (1976): Mutagenicity and cytotoxicity of benzo[a]pyrene benzo-ring epoxides. *Cancer Res.* 36:3358-3366.

Polynuclear Aromatic Hydrocarbons, edited by P. W. Jones and P. Leber. Ann Arbor Science Publishers, Inc. Ann Arbor, MI ©1979.

STUDIES ON THE RELATIONSHIPS BETWEEN INDUCTION OF BIOTRANSFORMATION AND TUMOR-INITIATING ACTIVITY OF 7,12-DIMETHYLBENZ[a]ANTHRACENE IN MOUSE SKIN

J. DiGiovanni, D. L. Berry,* T. J. Slaga* and M. R. Juchau

Department of Pharmacology
School of Medicine
University of Washington
Seattle, Washington 98195

*Biology Division
Oak Ridge National Laboratory
Oak Ridge, Tennessee 37830

INTRODUCTION

7,12-Dimethylbenz[a]anthracene (DMBA) is a methyl-substituted polycyclic aromatic hydrocarbon (PAH) with a wide spectrum of biological activities. These include the production of mammary cancer (15) and adrenal necrosis (39) in rats as well as cytotoxicity to cells grown in culture (5,6). DMBA is the most potent mouse skin carcinogen of all tested PAH (13) and can induce tumors in mice (using the initiation-promotion regimen) with 500 pmoles administered as a single dose (33).

Biologic activation of the molecule is presumed to be essential for the observed toxic effects of DMBA as well as PAH in general. The microsomal enzyme system of mixed-function oxygenases has been implicated in the activation of PAH to toxic, mutagenic and/or carcinogenic metabolites. Investigators have provided evidence that the enzyme complex aryl hydrocarbon hydroxylase (AHH, E.C. 1.14.14.2) functions as a detoxification system (3,4), but Gelboin *et al.* (10,11) have proposed that the specific enzyme complex is also responsible for the activation of PAH. The enzyme complex has been found in most mammalian tissues including mouse skin in which it is highly inducible (11,34,40). The role of induction of mixed-function oxygenases in various target tissues as it relates to carcinogenesis by

PAH remains unresolved, however, since induction may affect the rate of detoxification of the hydrocarbon as well as the rate of activation.

Induction of microsomal enzymes in several tissues has been postulated as a mechanism for the anticarcinogenic effects of a wide variety of compounds. These include PAH (16,24,25,32), flavones (36-38), coumarins (9), phenothiazines (35) and phenobarbital (23,27). Induction of microsomal enzyme systems also has been implicated in the anticarcinogenic activity of 1,1,1-trichloro-2,2-*bis*-(*p*-chlorophenyl) ethane (DDT) (26) and various polychlorinated biphenyl mixtures (18,22). However, investigations such as these have not demonstrated conclusively that the inhibitory actions are a result of induction within the target tissue for a particular carcinogen or that the effect is on tumor initiation *per se*. In this investigation, several potent inducers of microsomal mixed-function oxygenases were chosen as tools for studying the relationships between induction of biotransformation of DMBA in mouse skin and tumor initiation in the same tissue. The inducers utilized included 2,3,7,8-tetrachlorodibenzo-*p*-dioxin (TCDD), Aroclor 1254 and several closely related derivatives. We have used the two-stage system of mouse skin tumorigenesis extensively in these studies for the following reasons: (1) this system allows one to study the effects of modifiers (*i.e.*, inducers, inhibitors, etc.) on initiation and promotion separately and (2) systemic metabolism of the carcinogen does not play a major role in the tumor response since only very small doses of the carcinogen are applied locally.

MATERIALS AND METHODS

Chemicals

DMBA, benzo[a]pyrene (B[a]P], 3-methylcholanthrene (MC), dibenz-[a,c]anthracene (DB[a,c]A) and N-methyl-N′-nitro-N-nitrosoguanidine (MNNG) were supplied by the Sigma Chemical Company, St. Louis, MO. [^{3}H]DMBA (10.4 Ci/mmole) was obtained from Amersham/Searle, Arlington, Heights, IL, and ± 7β,8α-dihydroxy-9α,10α-epoxy-7,8,9,10-tetrahydrobenzo-[a]pyrene (B[a]P-diol-epoxide) was a generous gift from the Carcinogenesis Program of the National Cancer Institute. 12-0-Tetradecanoylphorbol-13-acetate (TPA) was obtained from Dr. Peter Borchert, University of Minnesota, Minneapolis, MN. Aroclor 1254 was purchased from Analabs, Inc., North Haven, CT. TCDD was a gift from the Dow Chemical Co., Midland, MI (98.6% pure by glc, Lot # 851-144-2). 3,4,3′,4′-Tetrachloroazobenzene (TCAB) was generously supplied by Dr. A. Poland, McArdle Laboratory for Cancer Research, University of Wisconsin Medical Center, Madison, WI. 2,7-dichlorodibenzo-*p*-dioxin (DCDD), 3,4,3′,4′-tetrachlorobiphenyl (TCB) and 2,4,5,2′,4′,5′-hexachlorobiphenyl (HCB) were supplied by Analabs, Norwalk, CT.

Tumor Experiments

Female outbred Charles River CD-1 mice were purchased from Charles River Mouse Farms, North Wilmington, MA. Mice 7-9 weeks old were carefully shaved with surgical clippers 2 days prior to initial treatment, and only those mice in the resting phase of the hair cycle were used in the tumor experiments. Each experimental group contained 30 preshaved mice. All chemicals were applied to the shaved area in 0.2 ml acetone. One week after initiation, 10 μg TPA in 0.2 ml acetone was applied twice weekly. Promotion was continued for varying intervals and the incidences of papillomas and carcinomas were recorded weekly. Papillomas and carcinomas were removed at random for histologic verification.

Metabolism of DMBA by Mouse Epidermal Homogenates

The metabolite profile of DMBA was analyzed using epidermal homogenates (obtained from inducer-pretreated mice) as the enzyme source. Animals were treated with the following enzyme inducing agents: DMBA, MC and DB[a,c]A (200 nmoles/mouse, 18 hr prior to sacrifice); Aroclor 1254 (100 μg/mouse, 24 hr prior to sacrifice); and TCDD (1 μg/mouse, 72 hr prior to sacrifice). Preparation of epidermal homogenates, assay conditions and analysis of metabolites using high-pressure liquid chromatography (HPLC) were performed as previously described (7).

Effects of TCDD-Pretreatment on Covalent Binding of [^{3}H]DMBA to Epidermal Macromolecules

TCDD was applied topically to the shaved backs of CD-1 mice (1 μg/mouse) 3 days prior to application of 2.56 μg of ^{3}H-DMBA (10 μCi). Control animals were pretreated with 0.2 ml acetone. After topical application of ^{3}H-DMBA, mice were sacrificed 3 and 24 hr later. The epidermal material was removed by the heat-treatment method (34) and DNA, RNA and protein were extracted using the method described by Huberman and Sachs (14).

RESULTS AND DISCUSSION

Effects of Pretreatment with Enzyme Inducers on DMBA Metabolism *In Vitro*

The following enzyme inducers were used to determine both qualitative and quantitative effects on DMBA profiles: DMBA, MC, DB[a,c]A, Aroclor

1254 and TCDD. All inducers were applied topically as described under Materials and Methods. The results from these experiments are depicted in Figure 1. All metabolic profiles have been corrected for background and radioactivity occurring in heat-inactivated controls. In addition, the profiles have been normalized with respect to protein concentration and all other incubation conditions.

An examination of Figure 1 reveals a close similarity in the qualitative pattern of metabolite formation between all the metabolic profiles regardless of the inducer utilized. The order of inducing capacity at the doses utilized was as follows: TCDD > DB[a,c]A > MC > DMBA > Aroclor 1254. At the dose utilized, TCDD-pretreatment gave rise to profound increases in the rates of formation of hydroxylated products from DMBA. A comparison of profile C (using epidermal homogenates from MC-pretreated mice) and profile F (using epidermal homogenates from TCDD-pretreated mice) demonstrates greater proportionate increases in metabolites eluting with standard diols in experiments with tissues from TCDD-pretreated animals. We have found recently that the pattern of metabolite formation from DMBA in epidermal homogenates of untreated mice was qualitatively similar to those shown in Figure 1 with the possible exception of the TCDD profile (8). This finding is consistent with the notion that the relative content of P_1-450 (that form which is induced by MC), compared with other forms of P-450, may be large in the skin of uninduced mice when compared with liver (17). Since TCDD and Aroclor 1254 were found to be effective inducers of monooxygenases (TCDD >>> Aroclor 1254) in the skin and since both are present as contaminants in the environment, we decided to explore them further as possible tools for understanding the relationships between induction of biotransformation and tumor initiation in the skin.

Tumor-Initiating Activities of Aroclor 1254 and TCDD

TCDD (2 μg/mouse) and Aroclor 1254 (100 μg/mouse) in 0.2 ml acetone were applied alone as initiators to the backs of CD-1 mice. The dose of TCDD was based on the ability of this compound to induce AHH after topical administration to mice (2 μg/mouse/day) (31). Both "responsive" (C57BL/6N) and "nonresponsive" (DBA/2N) mice exhibited significant induction with this dose. The dose of Aroclor 1254 was based on the ability of this compound to induce drug-metabolizing enzymes in skin after a single topical application to rats (25 mg/kg) (2) and mice (as described above, 100 μg/mouse).

TCDD, when administered alone at a dose of 2 μg per mouse, showed very weak initiating activity after promotion for 32 weeks (0.1 papillomas/mouse, 14% of survivors with papillomas) as illustrated in Figure 2. At the dose utilized, TCDD produced lethality in approximately one-third of the animals

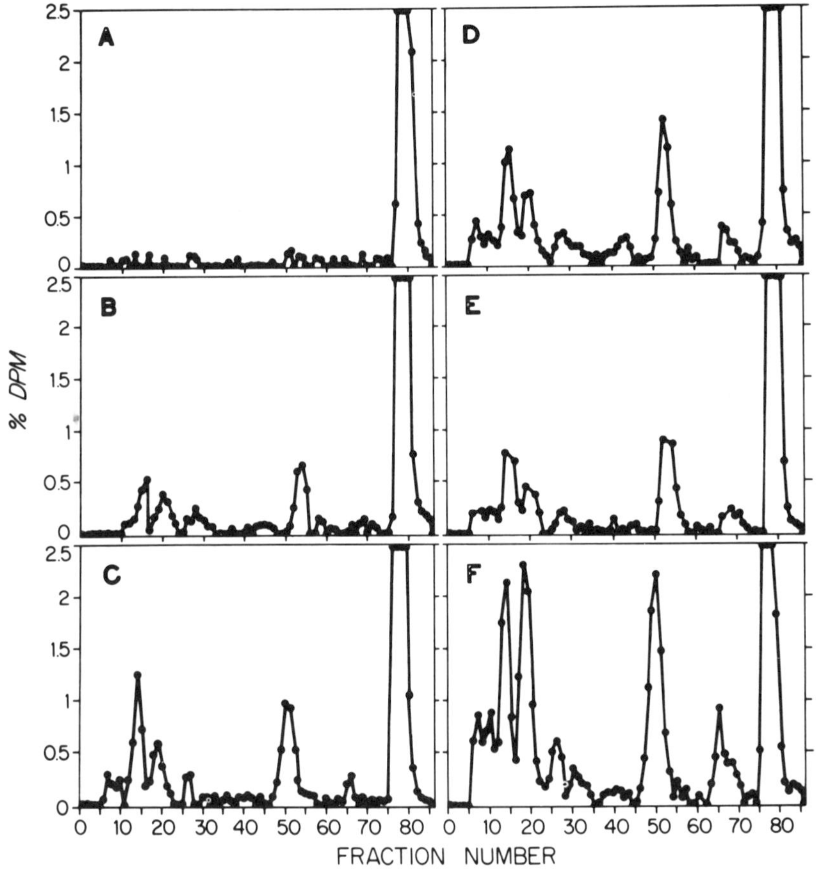

FIGURE 1. High-pressure liquid chromatographic profiles of [^{14}C]DMBA metabolites generated *in vitro* using epidermal homogenates from variously pretreated, female CD-1 mice. Panel A, heat-inactivated control; panel B, DMBA-pretreatment (200 nmol 18 hr prior to sacrifice); panel C, 3-methylcholanthrene-pretreatment (200 nmol 18 hr prior to sacrifice); panel D, dibenz[a,c]anthracene-pretreatment (200 nmol 18 hr prior to sacrifice); panel E, Aroclor 1254-pretreatment (100 μg 18 hr prior to sacrifice); and panel F, TCDD-pretreatment (1 μg 72 hr prior to sacrifice). All profiles have been normalized with respect to protein concentration. The peak fractions in which various synthetic standard compounds elute in this sytem are as follows: DMBA, 78; 5-hydroxy-DMBA, 72; 4-hydroxy-DMBA, 70; 2-hydroxy- and 3-hydroxy-DMBA, 61; 12-hydroxymethyl-7-methylbenz[a]anthracene, 57; 7-hydroxymethyl-12-methylbenz[a]anthracene, 52; 5,6-dihydro-5,6-dihydroxy-DMBA, 22; 7,12-dihydroxymethylbenz[a]anthracene, 19; 8,9-dihydro-8,9-dihydroxy-DMBA, 13. Further details of the methodology can be found in reference 7 as well as a complete listing of the sources for metabolic standards.

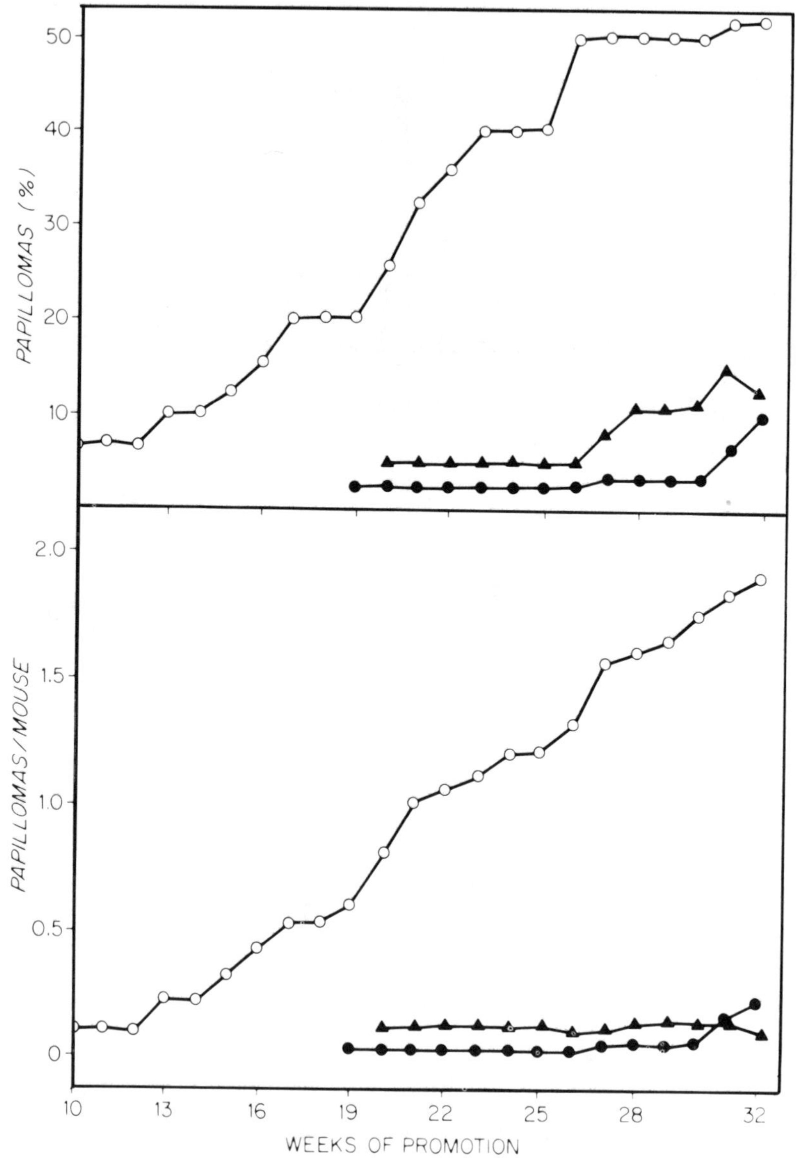

FIGURE 2. Tumor-initiating activity of TCDD and Aroclor 1254 in the two-stage, initiation-promotion system of mouse skin carcinogenesis. The animals in group 1 (○) were treated with 2.56 μg DMBA and promoted with twice weekly applications of 5 μg TPA for 32 weeks; group 2 (●), treated with 100 μg Aroclor 1254 and promoted with twice weekly applications of 5 μg TPA for 32 weeks; and group 3 (▲), treated with 2 μg TCDD and promoted as described above.

after 32 weeks. A higher dose of TCDD would have produced greater lethality and was therefore deemed impractical for the tumor experiments. Also shown in Figure 2 is the tumor-initiating activity of a 100-μg dose of Aroclor 1254. At this dose, Aroclor 1254 also displayed only very weak tumor-initiating capacity after promotion for 32 weeks (0.2 papillomas/mouse, 10% of survivors with papillomas). The group receiving tumor promoter (TPA) alone for 32 weeks displayed 0.06 papillomas per mouse (6% of survivors with papillomas). Since TCDD and Aroclor 1254 possessed little or no tumor-initiating or -promoting (1) properties in mouse skin using the two-stage initiation-promotion system of carcinogenesis, further experiments were designed to determine the effects of pretreatment with these two compounds on tumor initiation by DMBA.

Effects of TCDD on Tumor Initiation by DMBA and B[a]P

Table I illustrates the effects of topical pretreatment with TCDD on the initiation of tumors by DMBA. As presented in the table, the effect of TCDD was dependent on the time of pretreatment as well as dose. When applied 5 min prior to initiation with DMBA, a 2-μg topical dose of TCDD produced very little effect on the tumor response. However, with pretreatment times of 1, 3, 5 and 10 days, a 1-μg topical dose of TCDD resulted in 86, 91, 94 and

TABLE I. Inhibition by TCDD of DMBA Skin Tumor Initiation: Dependence on Time of Pretreatment and Dose[a]

Pretreatment (μg/mouse)	Weeks of Promotion	Pretreatment Time	Papillomas per Mouse[b]	Percent of Control[c]
Acetone	20	5 min	3.80	100
TCDD (2)	20	5 min	3.70	97
TCDD (1)	20	1 day	0.53	14
TCDD (1)	20	3 days	0.34	9
TCDD (1)	20	5 days	0.23	6
TCDD (1)	20	10 days	0.85	22
Acetone	24	3 days	3.83	100
TCDD (0.01)	24	3 days	0.65	17
TCDD (0.10)	24	3 days	0.31	8
TCDD (2)	24	3 days	0.14	4

[a]30 mice were used per experimental group. Pretreated mice were initiated with 2.56 μg of DMBA and promoted twice weekly, for either 20 or 24 weeks, with 10 μg of TPA.
[b]Average number of papillomas per mouse after 20 or 24 weeks of promotion.
[c]The average number of papillomas per mouse expressed as a percentage of the DMBA-initiated TPA-promoted groups for each period of promotion.

78% inhibition, respectively, in the number of papillomas formed per mouse. Maximum inhibition of tumor formation occurred when TCDD was applied 3 to 5 days prior to initiation with DMBA although significant inhibition was evident with the 10-day pretreatment. The time course of inhibition with TCDD corresponded with the time course of induction of monooxygenase activity in the skin (20), suggesting that the inhibitory effect might be related to enzyme induction.

To further analyze the inhibitory effect of TCDD, a dose-response study was designed. These results are also depicted in Table I. Three doses of TCDD were chosen; 0.01 μg, 0.10 μg and 2.0 μg/mouse and each dose was applied 3 days prior to initiation with DMBA. At doses of 0.01 μg, TCDD inhibited papilloma formation by approximately 80%, whereas a 0.1-μg dose of TCDD produced an approximately 92% inhibition. The 2-μg dose of TCDD was slightly more effective than the 0.1-μg dose. Except for the tumor experiments in which TCDD was applied at 2-μg doses, no toxicity was observed. Animals receiving 1 μg or less appeared normal in terms of weight gain and morphologic and histologic characteristics of the skin.

The effects of TCDD on tumor initiation with B[a]P also were investigated to determine the generality of this inhibitory response. These results are presented in Table II and indicate a similar inhibitory effect of TCDD on skin tumor initiation by B[a]P. However, the magnitude of the effect was slightly less than that observed with DMBA as the initiating agent. The same dependence on the time of pretreatment was noted. A 1-μg dose of TCDD, given 5 days prior to initiation with B[a]P, produced an approximately 65% inhibition in papilloma formation.

TABLE II. Effects of TCDD on Skin Tumor Initiation by B[a]P[a]

Pretreatment (μg/mouse)	Pretreatment Time	Papillomas per Mouse[b]	Percent of Control[c]
Acetone	5 min	2.30	100
TCDD (1)	1 day	1.90	83
TCDD (1)	3 days	1.00	43
TCDD (1)	5 days	0.80	35

[a]30 mice were used per experimental group. Pretreated mice were initiated with 100 nmoles of B[a]P and promoted with twice weekly applications of 10 μg of TPA. Promotion was continued for 20 weeks.

[b]Average number of papillomas per mouse after 20 weeks of promotion.

[c]Average number of papillomas per mouse expressed as a percentage of the B[a]P-initiated TPA-promoted group.

The unusually potent inhibitory action of TCDD on skin tumor initiation by DMBA and B[a]P stands in contrast to several reports by Kouri *et al.* (19,20). TCDD was found to enhance the formation of MC-induced sarcomas at the site of subcutaneous injection when the two compounds were administered concurrently to "nonresponsive" D-2 mice. This effect was attributed to the ability of TCDD to enhance the metabolism of MC in this system. However, they found that when given simultaneously with the initiator to C57BL/6N mice, TCDD did not exhibit an effect and in some cases actually inhibited tumorigenesis when the mice were pretreated. These results would tend to agree with those obtained in our experiments. Further work will be required to clarify the discrepancies between the results obtained with these two test systems (initiation promotion on mouse skin vs subcutaneous sarcoma formation) as well as metabolic differences between various mouse strains and tissues.

Inhibitory Effects of Chlorinated-Biphenyl- and -Dioxin-Related Inducers on Tumor Initiation by DMBA

A series of chlorinated-biphenyl- and dibenzo-*p*-dioxin-related inducers of AHH were tested for their ability to modify the initiating activity of DMBA or B[a]P. As shown in Table III, Aroclor 1254 (100 μg/mouse) exhibited little or no effect on tumorigenesis when given 5 min prior to initiation with DMBA. However, when administered in doses of 100 μg/mouse at 18 hr and 3 days prior to initiation with DMBA, Aroclor 1254 inhibited the formation of tumors by 45% and 12%, respectively. A 625-μg dose of Aroclor 1254, applied 3 days prior to initiation with DMBA, produced a 29% inhibition in papilloma formation. The effects of pretreatment with Aroclor 1254 on B[a]P tumor initiation were not tested. The inhibitory effects of Aroclor 1254 and TCDD correlated with the magnitude of monooxygenase induction as illustrated in Figure 1. A 1-μg topical dose of TCDD produced two-to-six-fold greater increases in hydroxylated metabolites from DMBA compared with a 100-μg dose of Aroclor 1254. Examination of the tumor experiments (Tables I and II) indicates that TCDD was approximately ninefold more effective at inhibiting tumor initiation with DMBA than Aroclor 1254 at the same topical doses utilized in the induction study.

Polychlorinated biphenyl (PCB) mixtures such as Aroclor 1254 display a spectrum of enzyme induction exhibiting properties similar to both phenobarbital and MC. However, it is now known that PCBs are mixtures of isomers and each possesses inducing properties that fit one of two well-established categories (*i.e.*, phenobarbital- or MC-type) (12). TCB and HCB are pure PCB isomers possessing MC- and phenobarbital-type inducing properties, respectively (12, 28). These two compounds were chosen for use in tumor-initiation experiments to further analyze the effects of Aroclor 1254 on

TABLE III. Effects of Various Polychlorinated Biphenyls and Derivatives of TCDD on Skin Tumor Initiation by DMBA and B[a] P[a]

Pretreatment (μg/mouse)	Weeks of Promotion	Initiator	Pretreatment Time	Papillomas per Mouse	Percent of Control
Acetone	20	DMBA	5 min	3.80	100
Aroclor 1254 (100)	20	DMBA	5 min	3.60	95
Aroclor 1254 (100)	20	DMBA	18 hr	2.1	55
Aroclor 1254 (100)	20	DMBA	3 days	3.3	88
Aroclor 1254 (625)	20	DMBA	18 hr	2.70	71
DCDD (100)	20	DMBA	3 days	3.44	91
TCAB (10)	20	DMBA	3 days	4.73	125
Acetone	24	DMBA	3 days	3.83	100
3,4,3′,4′-TCB (625)	24	DMBA	3 days	0.40	10
2,4,5,2′,4′,5′-HCB (625)	24	DMBA	3 days	3.36	88
Acetone	24	B[a] P	3 days	3.27	100
3,4,3′,4′-TCB (625)	24	B[a] P	3 days	4.06	124
2,4,5,2′,4′,5′-HCB (625)	24	B[a] P	3 days	2.56	78

[a]Pretreated mice were initiated with either 10 nmoles (experiments 1-7) or 100 nmoles (experiments 8-10) of DMBA or 200 nmoles of B[a] P (experiments 11-13). 30 mice were used for each experimental group.

tumor initiation by DMBA. The results of these experiments also are shown in Table III. TCB (625 μg/mouse) given three days prior to initiation with DMBA produced an approximately 90% inhibition in the formation of papillomas. When given 3 days prior to initiation with B[a] P, however, TCB had no effect or slightly enhanced the tumor response. HCB pretreatment (625 μg/mouse, 3 days prior to initiation) had a slight inhibitory effect on formation of papillomas following initiation with DMBA or B[a] P. It is interesting to note that TCB and TCDD are approximate isostereomers; however, TCB is over 400-fold less potent than TCDD for inducing AHH (28). In addition, TCB is a nearly planar molecule, similar to TCDD. As shown in Table III, TCB (at an approximately 600-fold higher dose than TCDD) was quite effective at inhibiting tumor initiation by DMBA but not B[a] P. There is no explanation at this time for the differences observed between DMBA and B[a] P.

Finally, several other TCDD-related derivatives were tested for their effects on tumor initiation by DMBA. DCDD and TCAB were applied at 100-μg and 10-μg doses, respectively, 3 days prior to initiation with DMBA. DCDD had little effect and TCAB slightly enhanced tumor initiation with DMBA. DCDD lacks the ability to induce AHH in the chick embryo (29) and this could explain the inability to inhibit tumor initiation by DMBA. The ED_{50} for induction of AHH in chick embryo by TCAB is approximately 6.45 that of TCDD (30). The fact that TCAB did not inhibit tumor initiation by DMBA, even at a dose tenfold greater than TCDD, may be explained by the ability of the skin to reduce the azo linkage in the molecule, thus rendering the compound inactive as an inducer.

Effects of Pretreatment with TCDD on Tumor Initiation by MNNG and B[a]P-Diol-Epoxide

MNNG, which does not require metabolic activation for its mutagenic or tumor-initiating effects (13), and B[a]P-diol-epoxide, which is thought to be the ultimate carcinogenic metabolite of B[a]P (21), were utilized as tumor initiators to further test the hypothesis that pretreatment with TCDD inhibited tumor initiation by virtue of its ability to induce monooxygenase enzymes of the skin. These results are presented in Table IV. Initiation with DMBA (100 nmole/mouse) was included in the experiments as a positive control as was the effect of pretreatment with TCDD (1 μg/mouse, 3 days prior to initiation). Pretreatment with TCDD (1 μg/mouse, 3 days prior to initiation) enhanced MNNG initiation significantly and inhibited tumor initiation by B[a]P-diol-epoxide significantly. The fact that TCDD significantly inhibited tumor initiation by B[a]P-diol-epoxide ($P < 0.001$) was surprising. This finding suggested that TCDD could alter tumor initiation by B[a]P-diol-epoxide through a mechanism other than enzyme induction. Further studies will be required to understand this interesting result.

Reduction in ^{3}H-DMBA Bound to Epidermal Macromolecules Following TCDD-Pretreatment

To understand further the inhibitory effects of pretreatment with Aroclor 1254 and TCDD on skin tumor initiation by DMBA, the effects of TCDD on covalent binding of ^{3}H-DMBA to mouse epidermal DNA, RNA and protein were investigated. These experiments are summarized in Table V. TCDD was applied topically to CD-1 mice (1 μg/mouse) 3 days prior to application of ^{3}H-DMBA. After topical application of ^{3}H-DMBA, mice were sacrificed 3 and 24 hr later. Under these conditions (which mimicked the tumor experiments described above), TCDD-pretreatment was found to significantly reduce the covalent binding of DMBA to epidermal DNA (60-70%) and RNA

TABLE IV. Effects of TCDD Pretreatment on Tumor Initiation by MNNG and B[a]P-Diol-Epoxide[a]

Pretreatment[b] (μg/mouse)	Initiator (μmoles/mouse)	Papillomas per Mouse	Percent of Control
Acetone	DMBA (0.1)	6.46	100
TCDD (1)	DMBA (0.1)	0.53	8
Acetone	MNNG (2.0)	1.30	100
TCDD (1)	MNNG (2.0)	2.80	215
Acetone	B[a]P-diol-epoxide (0.4)	1.10	100
TCDD (1)	B[a]P-diol-epoxide (0.4)	0.08	7

[a]30 mice were used per experimental group. One week after initiation, mice received 10 μg of TPA twice weekly for 18 weeks. These experiments are still in progress.
[b]TCDD was applied 3 days prior to initiation.

(45-55%). Interestingly, TCDD pretreatment had no effect on the amount of hydrocarbon covalently bound to epidermal proteins.

CONCLUSIONS

1. A series of enzyme inducers was utilized to determine the effects of pretreatment on the qualitative and quantitative aspects of DMBA metabolism in a target tissue. As judged by HPLC profiles, the pattern of metabolite

TABLE V. Effect of Pretreatment with TCDD on Covalent Binding of ^{3}H-DMBA to Epidermal Macromolecules[a]

Treatment	Hydrocarbon Bound to Macromolecule (pmoles/mg x 10^2)		
	DNA	RNA	Protein
DMBA			
3 hr[b]	4.79	2.88	69.5
24 hr	12.63	7.16	38.8
DMBA + TCDD[c]			
3 hr	1.78	1.56	63.8
24 hr	3.52	3.16	39.4

[a]Mice were treated topically with 10 nmoles of ^{3}H-DMBA (10 μCi) and sacrificed 3 and 24 hr later.
[b]Time of sacrifice after application of ^{3}H-DMBA; 40 mice were used for each experimental group.
[c]Mice in this group received 1 μg of TCDD, applied topically, 72 hours prior to application of ^{3}H-DMBA.

formation in mouse epidermal homogenates was quite similar regardless of the inducer utilized with the possible exception of TCDD. The order of inducing capacity was as follows: TCDD > DB[a,c] A > MC > DMBA > Aroclor 1254.

2. TCDD and Aroclor 1254, at 2-μg and 100-μg topical doses per mouse, were only very weak tumor initiators in the skin.

3. Pretreatment with TCDD and Aroclor 1254 effectively inhibited tumor initiation by DMBA. TCDD was extremely potent in this regard. In addition, pretreatment with TCDD effectively inhibited tumor initiation by B[a] P.

4. The anticarcinogenic effects of TCDD and Aroclor 1254 appeared to correlate with the magnitude, as well as the time course, of induction of monooxygenase enzymes of the skin. Furthermore, the quantity of ^{3}H-DMBA bound to DNA and RNA (but not to protein) in the presence and absence of TCDD pretreatment correlated well with the tumor response under similar conditions.

5. These experiments suggested that pretreatment with TCDD and Aroclor 1254 both resulted in increased rates of inactivation of DMBA and B[a] P relative to the rates of activation in mouse skin.

ACKNOWLEDGMENTS

Research was supported by grants HD-04839 from the National Institute of Child Health and Human Development, USPHS, National Institutes of Health, USPHS, Training Grant GM 00109, by Grant CA-20076 from the National Cancer Institute, USPHS, and by the U.S. Department of Energy under contract with Union Carbide Corporation. By acceptance of this article, the publisher or recipient acknowledges the U.S. Government's right to retain a nonexclusive, royalty-free license in and to any copyright covering this article.

REFERENCES

1. Berry, D. L., DiGiovanni, J., Juchau, M. R., Bracken, W. M., Gleason, G. L., and Slaga, T. J. (1978): Lack of tumor-promoting ability of certain environmental chemicals in a two-stage mouse skin tumorigenesis assay. *Res. Commun. Chem. Path. Pharmacol.* 20:101-107.
2. Bickers, D. R., Kappas, A., and Alvares, A. P. (1974): Differences in inducibility of cutaneous and hepatic drug metabolizing enzymes and cytochrome P-450 by polychlorinated biphenyl and 1,1,1-trichloro-2,2-bis (*p*-chlorophenyl) ethane (DDT). *J. Pharmacol. Exp. Ther.* 188: 300-309.

3. Burki, K., Kiebelt, A. G., and Bresnick, E. (1973): Induction of aryl hydrocarbon hydroxylase in mouse tissues from a high and low cancer strain and their F_1 hybrids. *J. Nat. Cancer Inst.* 50:369-380.
4. Conney, A. H. (1967): Pharmacological implication of microsomal enzyme induction. *Pharmacol. Rev.* 19:317-366.
5. Diamond, L. (1971): Metabolism of polycyclic hydrocarbons in mammalian cell cultures. *Int. J. Cancer* 8:451-462.
6. Diamond, L., Sardet, C., and Rothblat, G. H. (1968): The metabolism of 7,12-dimethylbenz[a]anthracene in cell cultures. *Int. J. Cancer* 3:838-849.
7. DiGiovanni, J., Slaga, T. J., Berry, D. L., and Juchau, M. R. (1977): Metabolism of 7,12-dimethylbenz[a]anthracene in mouse skin homogenates analyzed with high pressure liquid chromatography. *Drug Metab. Dispos.* 5:295-301.
8. DiGiovanni, J., Slaga, T. J., Berry, D. L., and Juchau, M. R. (1979): Inhibitory effects of environmental chemicals on polycyclic aromatic hydrocarbon carcinogenesis. In: *Modifiers of Chemical Carcinogenesis,* T. J. Slaga, Ed., Raven Press, New York (in press).
9. Feuer, G., and Kellen, J. A. (1974): Inhibition and enhancement of mammary tumorigenesis by 7,12-dimethylbenz(a)anthracene in female Sprague-Dawley rat. *Int. J. Clin. Pharmacol.* 9:62-69.
10. Gelboin, H. V., Wiebel, F. J., and Diamond, L. (1970): Dimethylbenz[a]anthracene tumorigenesis and aryl hydrocarbon hydroxylase in mouse skin. Inhibition by 7,8-benzoflavone. *Science* 170:169-171.
11. Gelboin, H. V., Wiebel, F. J., and Kinoshita, N. (1972): Microsomal aryl hydrocarbon hydroxylases: On their role in polycyclic hydrocarbon carcinogenicity and toxicity and the mechanism of enzyme induction. *Biochem. Soc. Symp.* 34:103-133.
12. Goldstein, J. A., Hickman, P., Bergman, H., McKinney, J. D., and Walker, M. P. (1977): Separation of pure polychlorinated biphenyl isomers into two types of inducers on the basis of induction of cytochrome P-450 or P-448. *Chem.-Biol. Interac.* 17:69-87.
13. Heidelberger, C. (1975): Chemical carcinogenesis. *Ann. Rev. Biochem.* 44:79-121.
14. Huberman, E., and Sachs, L. (1977): DNA binding and its relationship to carcinogenesis by different polycyclic hydrocarbons. *Int. J. Cancer* 19:122-127.
15. Huggins, C., Grand, L. C., and Brillantes, F. P. (1961): Mammary cancer induced by a single feeding of polynuclear hydrocarbons, and its suppression. *Nature* 189:204-207.
16. Huggins, C. B., Grand, L., and Fukunishi, R. (1964): Aromatic influences on the yields of mammary cancers following the administration of 7,12-dimethylbenz[a]anthracene. *Proc. Nat. Acad. Sci.* 51:737-742.

17. Kahl, G. F., Kahl, R., Kumaki, K., and Nebert, D. W. (1976): Association of the Ah locus with the specific changes in metyrapone and ethylisocyanide binding to mouse liver microsomes. *J. Biol. Chem.* 251:5397-5407.
18. Kimura, N. T., Kanematsu, T., and Baba, T. (1976): Polychlorinated biphenyl(s) as a promoter in experimental hepatocarcinogenesis in rats. *Z. Krebsforsch.* 87:257-266.
19. Kouri, R. E. (1976): Relationship between levels of aryl hydrocarbon hydroxylase activity and susceptibility to 3-methylcholanthrene and benzo(a)pyrene-induced cancers in inbred strains of mice. In: *Carcinogenesis: A Comprehensive Survey,* R. Freudenthal and P. Jones, eds., pp. 139-151, Raven Press, New York.
20. Kouri, R. E., Rude, T. H., Joglekar, R., Dansette, P. M., Jerina, D. M., Atlas, S. A., Owens, I. S., and Nebert, D. W. (1978): 2,3,7,8-Tetrachlorodibenzo-*p*-dioxin as a cocarcinogen causing 3-methylcholanthrene-initiated tumors in mice genetically "nonresponsive" at Ah locus. *Cancer Res.* 38:2777-2783.
21. Lehr, R. E., Yagi, H., Thakker, D. R., Levin, W., Wood, A. W., Conney, A. H., and Jerina, D. M. (1978): The bay region theory of polycyclic aromatic hydrocarbon-induced carcinogenicity. In: *Carcinogenesis, Vol. 3,* P. W. Jones and R. E. Freudenthal, Eds., Raven Press, New York.
22. Makimura, S., Aoe, H., Sugihara, S., Hirao, K., Arai, M., and Ito, N. (1974): Inhibitory effect of polychlorinated biphenyl on liver tumorigenesis in rats treated with 3′-methyl-4-dimethylaminoazobenzene, N-2-fluorenylacetamide and diethylnitrosamine. *J. Nat. Cancer Inst.* 53:1253-1257.
23. Mclean, A. E. M., and Marshall, A. (1971): Reduced carcinogenic effects of aflatoxin in rats given phenobarbitone. *Brit. J. Exp. Path.* 52:322-329.
24. Meechan, R. J., McCafferty, D. E., and Jones, R. S. (1953): 3-Methylcholanthrene as an inhibitor of hepatic cancer induced by 3′-methyl-4-dimethylaminobenzene in the diet of the rat: A determination of the time relationships. *Cancer Res.* 13:802-806.
25. Miller, E. C., Miller, J. A., Brown, R. R., and McDonald, J. C. (1958): On the protective action of certain polycyclic aromatic hydrocarbons against carcinogenesis by aminoazo dyes and 2-acetylaminofluorene. *Cancer Res.* 18:469-477.
26. Okey, A. B. (1972): Dimethylbenz[a]anthracene-induced mammary tumors in rats: Inhibition by DDT. *Life Sci.* 11:833-843.
27. Peraino, C., Fry, R. J. M., and Staffeldt, E. (1971): Reduction and enhancement by phenobarbital of hepatocarcinogenesis induced in the rat by 2-acetylaminofluorene. *Cancer Res.* 31:1506-1512.

28. Poland, A., and Glover, E. (1977): Chlorinated biphenyl induction of aryl hydrocarbon hydroxylase activity: a study of the structure-activity relationships. *Mol. Pharmacol.* 13:924-938.
29. Poland, A., Glover, E., and Kende, A. S. (1976): Stereospecific, high affinity binding of 2,3,7,8-tetrachlorodibenzo-*p*-dioxin by hepatic cytosol. *J. Biol. Chem.* 251:4936-4946.
30. Poland, A., Glover, E., Kende, A. S., Decamp, M., and Giandomenico, C. M. (1976): 3,4,3′,4′-Tetrachloroazoxybenzene and azobenzene: Potent inducers of aryl hydrocarbon hydroxylase. *Science* 194:627-630.
31. Poland, A. P., Glover, E., Robinson, J. R., and Nebert, D. W. (1974): Genetic expression of aryl hydrocarbon hydroxylase activity: Induction of monooxygenase activities and cytochrome P_1-450 formation by 2,3,7,8-tetrachloro-dibenzo-*p*-dioxin in mice genetically "nonresponsive" to other aromatic hydrocarbons. *J. Biol. Chem.* 249:5599-5606.
32. Richardson, H. L., Stier, A. R., and Borsos-Nachtnebel, E. (1952): Liver tumor inhibition and adrenal histologic responses in rats to which 3′-methyl-4-dimethylaminoazobenzene and 20-methylcholanthrene were simultaneously adminsitered. *Cancer Res.* 12:356-361.
33. Slaga, T. J., Bowden, G. T., Scribner, J. D., and Boutwell, R. K. (1974): Dose-response studies on the ability of 7,12-dimethylbenz[a] anthracene and benz[a] anthracene to initiate skin tumors. *J. Nat. Cancer Inst.* 53:1337-1340.
34. Thompson, S., and Slaga, T. J. (1976): Mouse epidermal aryl hydrocarbon hydroxylase. *J. Invest. Dermatol.* 66:108-111.
35. Wattenberg, L. W., and Leong, J. L. (1967): Inhibition of 9,10-dimethylbenz[a] anthracene (DMBA) induced mammary tumorigenesis. *Fed. Proc.* 26;692.
36. Wattenberg, L. W., and Leong, J. L. (1968): Inhibition of the carcinogenic action of 7,12-dimethylbenz(a)anthracene by beta-napthoflavone. *Proc. Soc. Exp. Med. Biol.* 128:940-943.
37. Wattenberg, L. W., and Leong, J. L. (1970): Inhibition of the carcinogenic action of benzo[a] pyrene by flavones. *Cancer Res.* 30:1922-1925.
38. Wattenberg, L. W., Page, M. A., and Leong, J. L. (1968): Induction of increased benzpyrene hydroxylase activity by flavones and related compounds. *Cancer Res.* 28:934-937.
39. Wheatley, D. N., Hamilton, A. G., Currie, A. R., Boyland, E., and Sims, P. (1966): Adrenal necrosis induced by 7-hydroxymethyl-12-methylbenz[a] anthracene and its prevention. *Nature* 211:1311-1312.
40. Wiebel, F. J., Leutz, B. S., and Gelboin, H. V. (1975): Aryl hydrocarbon (benzo[a] pyrene) hydroxylase: A mixed function oxygenase in mouse. *J. Invest. Dermatol.* 64:184-189.

Polynuclear Aromatic Hydrocarbons, edited by P. W. Jones and P. Leber. Ann Arbor Science Publishers, Inc. Ann Arbor, MI ©1979.

AB INITIO QUANTUM MECHANICAL CHARACTERIZATION OF THE 7β,8α,9α-TRIHYDROXY-7,8,9-TRIHYDROBENZO[a]PYRENE CARBONIUM ION

Lester L. Shipman

Chemistry Division
Argonne National Laboratory
Argonne, Illinois 60439

INTRODUCTION

Benzo[a]pyrene (Figure 1) is activated by a three-step metabolic process (7) to a powerful mutagen and carcinogen, its 7,8-diol-9,10 epoxide (6). This metabolite or the corresponding 7,8,9-trihydro-triol carbonium ion (Figure 2) has been shown to attack and bind covalently to the 2-amino group of guanosine (4,5,8,9,14). The stereochemical structure of the guanosine adduct has been determined (4,5,8,9,14) and that stereochemistry has been employed in the present calculations. In particular, the 7β,8α,9α-trihydroxy-7,8,9-trihydrobenzo[a]pyrene carbonium ion (triol-carbonium ion; Figure 2) has been characterized using the *ab initio* molecule fragment floating spherical Gaussian orbital (MF-FSGO) method. The total π-electronic charge distribution and the lowest unoccupied molecular orbital (LUMO) are reported in this paper. The LUMO has particular significance for the mechanism of attack on guanosine. In a previous and related study, (11) the MF-FSGO method was used to characterize the parent molecule, benzo[a]pyrene.

MOLECULAR GEOMETRY

Before a Hartree-Fock-type wave function can be computed for a molecule, the nuclear geometry must be selected. In the present study, standard bond lengths and bond angles were chosen from the Tables of Interatomic Distances and Configurations in Molecules and Ions (13). The pyrene moiety was taken to have an idealized geometry with all C-C and C-H bond lengths

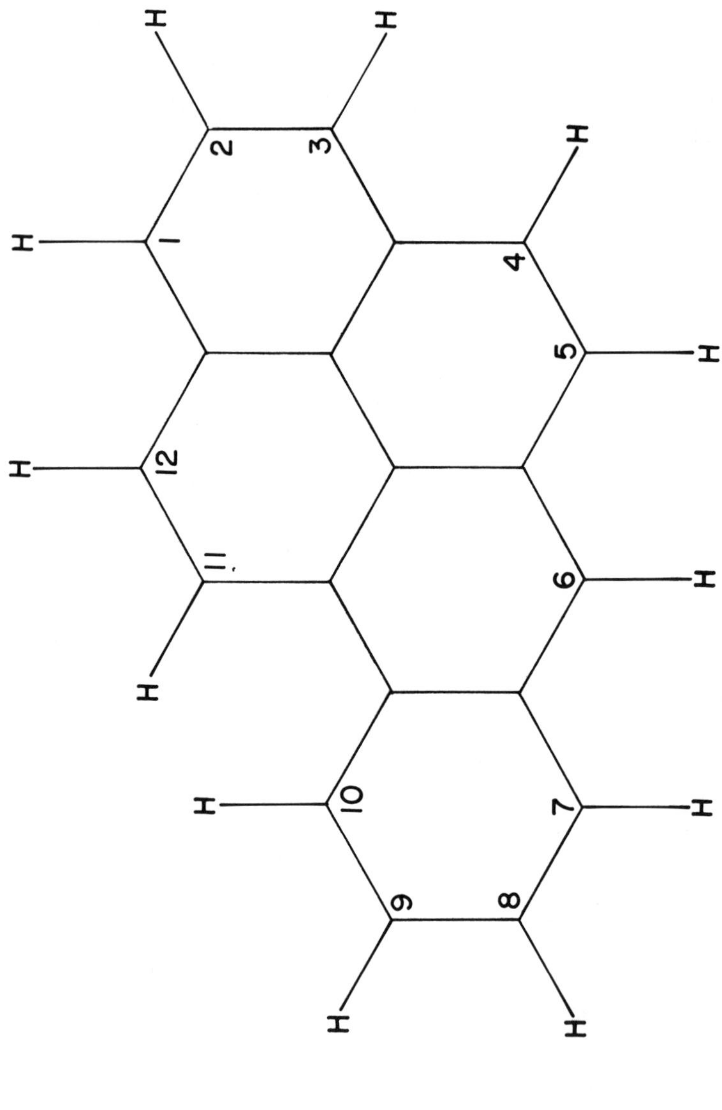

FIGURE 1. Chemical structure and atomic numbering system for benzo[a]pyrene.

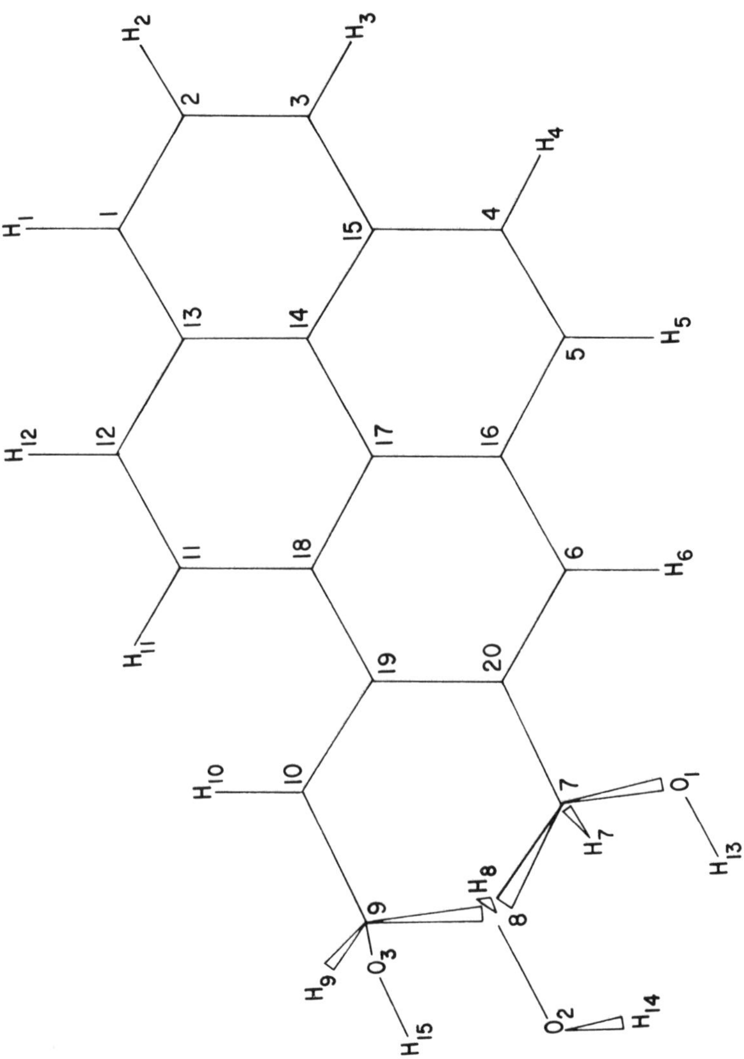

FIGURE 2. Chemical structure and atomic numbering system for 7β,8α,9α,-trihydroxy-7,8,9-trihydrobenzo[a] pyrene carbonium ion.

fixed at their standard values of 1.394 and 1.084 (Å), respectively, and all C-C-C and C-C-H bond angles were fixed at the idealized value of 120°. As for the reduced ring, all atoms except the C_7, C_8, C_9 and C_{10} carbon atoms in the ring were placed according to the standard parameters given in Table I. Just one of the 27 possible conformational minima with respect to rotations about the three C-O bonds was selected for this calculation. The positions of the C_7, C_8, C_9 and C_{10} carbon atoms of the reduced ring were determined by allowing their coordinates to vary in a least-squares procedure until the sum of the squares of the deviations between the computed and the standard geometry parameters (shown in Table I) was minimized. In the sum of squares, distance deviations (in units of $Å^2$) were given double the weighting of the angle deviations (in units of $degrees^2$). The final computed parameters that minimize the sum of squares are given in Table I. In the

TABLE I. Geometry Parameters[a] for the Reduced Ring

Parameter	Standard Value	Optimized Value
C7-C20	1.505	1.496
C7-C8	1.537	1.531
C8-C9	1.537	1.539
C9-C10	1.510	1.519
C10-C19	1.394	1.400
∢C7-C20-C19	120	118.3
∢C8-C7-C20	120	117.3
∢C7-C8-C9	109.5	109.2
∢C8-C9-C10	109.5	111.8
∢C9-C10-C19	120	123.1
∢C10-C19-C20	120	121.1
C7-01	1.426	
C8-02	1.426	
C9-03	1.426	
01-H13	0.97	
02-H14	0.97	
03-H15	0.97	
C7-H7	1.073	
C8-H8	1.073	
C9-H9	1.073	
C10-H10	1.084	
∢H7-C7-01	109.5	
∢C7-01-H13	109.5	
∢H8-C8-02	109.5	
∢C8-02-H14	109.5	
∢H9-C9-03	109.5	
∢C9-03-H15	109.5	

[a]Bond lengths and bond angles are given in (Å) and degrees, respectively.

minimization, atoms C_7, C_9 and C_{10} were constrained to lie in the pyrene plane so that π-conjugation would be maintained, but atom C_8 was allowed to vary in all three orthogonal directions, giving a "pucker" to the reduced ring at the C_8 position. The final cartesian coordinates of all 38 atoms determined by the procedures above are given in Table II. Note that C_8 lies approximately 0.6 Å out of the pyrene plane.

TABLE II. Atomic Positions for the Triol-Carbonium Ion[a]

Label	X[b]	Y[b]	Z[b]
C1	0.0	0.0	0.0
C2	1.207	-0.697	0.0
C3	1.207	-2.091	0.0
C4	0.0	-4.182	0.0
C5	-1.207	-4.879	0.0
C6	-3.622	-4.879	0.0
C7	-6.146	-4.892	0.0
C8	-7.346	-4.146	0.588
C9	-7.396	-2.725	0.0
C10	-6.027	-2.065	0.0
C11	-3.622	-0.697	0.0
C12	-2.414	0.0	0.0
C13	-1.207	-0.697	0.0
C14	-1.207	-2.091	0.0
C15	0.0	-2.788	0.0
C16	-2.414	-4.182	0.0
C17	-2.414	-2.788	0.0
C18	-3.622	-2.091	0.0
C19	-4.829	-2.788	0.0
C20	-4.829	-4.182	0.0
H1	0.0	1.084	0.0
H2	2.146	-0.155	0.0
H3	2.146	-2.633	0.0
H4	0.939	-4.724	0.0
H5	-1.207	-5.963	0.0
H6	-3.622	-5.963	0.0
H7	-6.383	-5.132	-1.019
H8	-7.247	-4.090	1.655
H9	-8.069	-2.129	0.587
H10	-5.977	-0.982	0.0
H11	-4.560	-0.155	0.0
H12	-2.414	1.084	0.0
H13	-6.806	-6.609	0.716
H14	-8.515	-5.737	0.629
H15	-8.755	-3.201	-1.350
O1	-5.983	-6.097	0.745
O2	-8.548	-4.841	0.260
O3	-7.874	-2.796	-1.342

[a]See text for an explanation of how the coordinates were derived.
[b]Units are Å.

MOLECULAR FRAGMENT FLOATING SPHERICAL GAUSSIAN ORBITAL METHOD

The *ab initio* MF-FSGO method has been reviewed in detail elsewhere (1,2). This method is *ab initio* in the sense that all electrons (σ and π) and all nuclei are explicitly included in the calculations and all integrals are explicitly considered. The wave function is of the Hartree-Fock type, *i.e.*, a single Slater determinantal wave function. For the particular case of the triol-carbonium ion, the wave function is constructed from 79 doubly occupied molecular orbitals (MO), each of which is a linear combination of 115 basis functions. Each basis function is either a single floating spherical Gaussian orbital (FSGO) or a subtractive linear combination of two FSGO placed 0.11 Å apart to form a *p*-type Gaussian basis function. Altogether, there are 132 FSGO. The linear coefficients of each basis function in each MO are determined by minimizing the total energy with respect to the linear coefficients. This minimization is accomplished by the iterative self-consistent-field (SCF) procedure of Hall and Roothaan (3,10). A total of 22,247,785 two-electron integrals were evaluated, but only those 3,798,194 integrals that had an absolute value greater than 10^{-10} Hartree ($\sim 6.3 \times 10^{-8}$ kcal/mol) were retained and utilized in the SCF procedure.

π-ELECTRONIC CHARGE DISTRIBUTION

Of particular interest is the distribution of electrons in the π-system of the triol-carbonium ion and the corresponding distribution of the net positive charge. The total π-populations and π-bond orders are shown in Figure 3. All populations and bond orders have been computed by performing a population analysis on the symmetrically orthogonalized (12) version of the original basis functions. The number alongside each atom is the number of electrons in the symmetrically orthogonalized π *p*-orbital on that particular atom and the numbers in parentheses between atoms are the bond orders between the symmetrically orthogonalized π *p*-orbitals on the atoms. The bond orders have the following chemical meaning; bond orders of -1, 0 and +1 correspond to a full antibond, a nonbond and a full bond between the atoms. The fractional positive bond orders shown in Figure 3 correspond to the fraction of a full π-bond. For example, the C_{10}-C_{19} bond order is 0.777 of an ethylene-like π-bond. The average π-bond order and standard deviation are 0.581 and 0.111, respectively. The only bond orders more than one standard deviation below the average are the C_{18}-C_{19} and C_{19}-C_{20} bond orders. The only bond orders greater than one standard deviation above the average are the C_{10}-C_{19}, C_{11}-C_{12}, C_4-C_5 and C_6-C_{20} bond orders. Obviously, the pyrene π-system acts effectively to delocalize the cationic charge away from C_{10}.

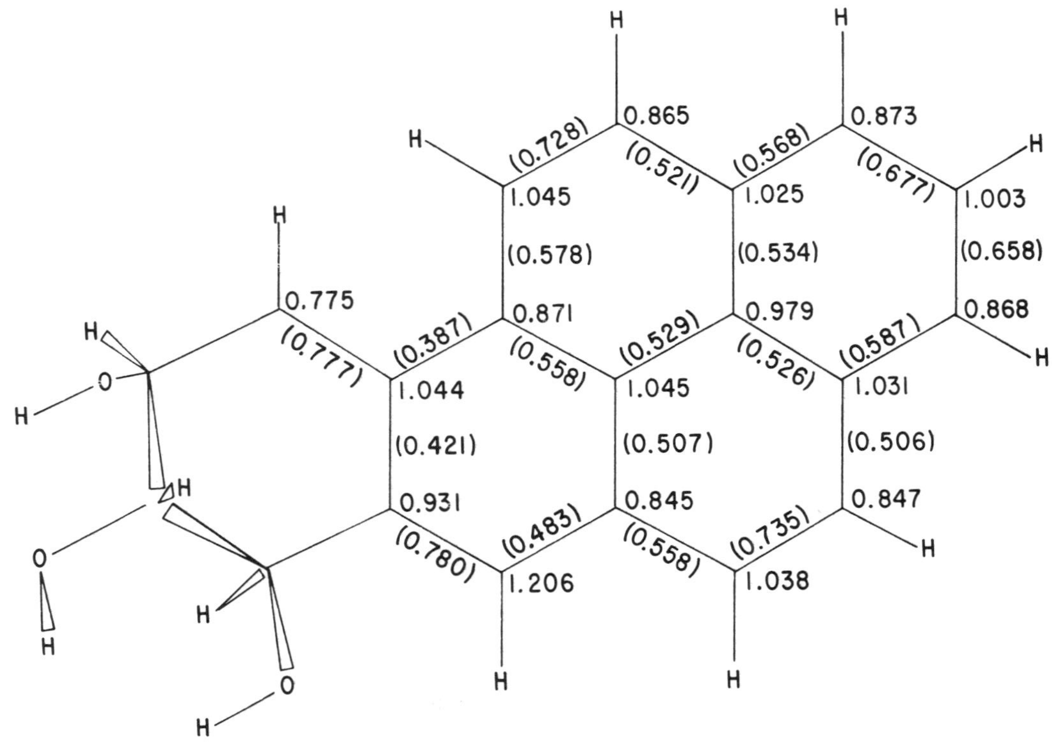

FIGURE 3. π-populations and bond orders for the triol-carbonium ion. The numbers alongside the atoms correspond to the number of electrons in the π p-orbital on the atom and the numbers in parentheses between atoms correspond to the bond orders between the atoms.

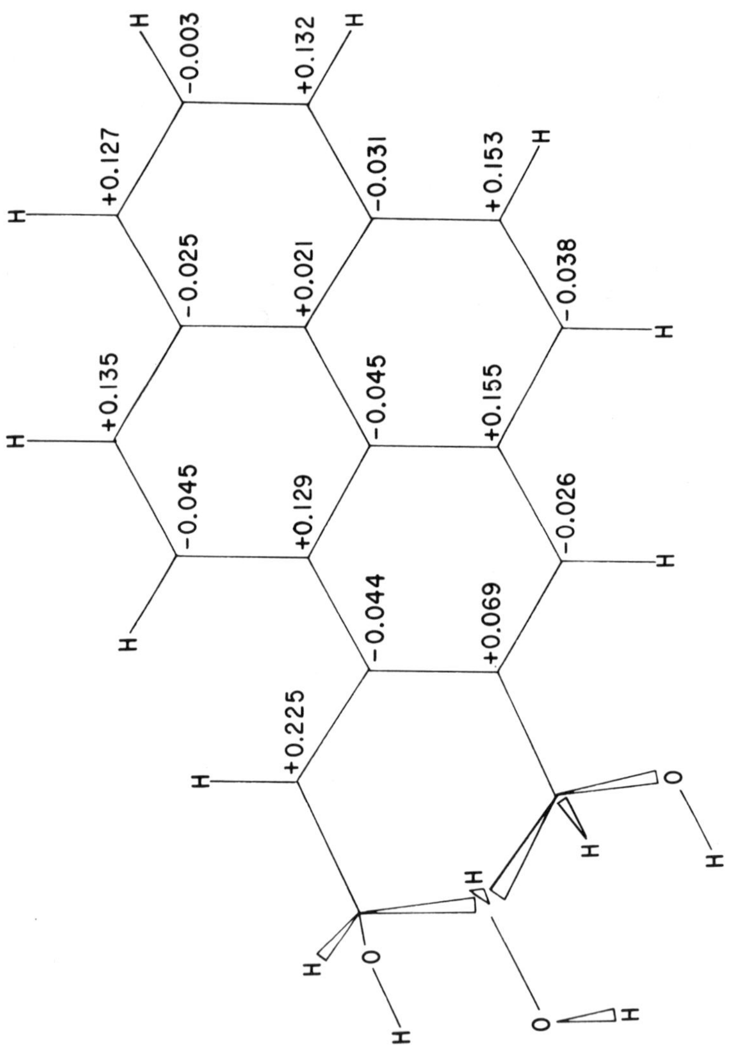

FIGURE 4. Net π-charges for the triol-carbonium ion. The numbers alongside the atoms correspond to the net π charge on the atom.

The distribution of positive charge is perhaps more clearly seen in Figure 4 where a pattern of alternating positive and negative charges is apparent. The average charge and the standard deviation are 0.052 and 0.092, respectively. The only net charges more than one standard deviation below the average are C_{11}, C_{17} and C_{19}. The only net charges greater than one standard deviation above the average are C_4, C_{10} and C_{16}. It is interesting that for a molecule with a net charge of +1, some of the atoms have net π-charges that are negative. The total net π-charge obtained by summation over the net π-charges of the 17 atoms in the π-system is +0.89; thus the reduced ring and the σ-system of pryene, taken together, have a net charge of +0.11. This indicates that a simpler π-only quantum mechanical method such as the Hückel theory cannot properly account for the charge distribution in the triol-carbonium ion.

LOWEST UNOCCUPIED MOLECULAR ORBITAL

The lowest unoccupied molecular orbital (LUMO) has a special significance for the chemical reactivity of the triol-carbonium ion. The LUMO is the unoccupied MO which most readily accepts additional electron density into the molecule and therefore may figure prominently in the electrophilic attack on another molecule such as guanosine. The LUMO is shown in Figure 5; the numbers alongside the atoms are the coefficients of the symmetrically orthogonalized π *p*-orbitals on the atoms. Clearly, the LUMO is strongly delocalized with the largest contribution coming from C_{10} followed by C_{16}, C_{18}, C_4, C_3, C_1, C_{20} and C_{12} in decreasing order. The lone pair of electrons in the 2-amino group of guanosine react with the C_{10} atom on the triol-carbonium ion to form a covalent bond. The lone pair on the 2-amino group is a part of the guanine π-system. This suggests that the initial reaction between guanosine and the triol-carbonium ion may proceed via an intermediate π-π charge transfer complex in which electron density is transferred from the higher occupied π-MO of guanine to the lower unoccupied π-MOs of the triol-carbonium ion.

ACKNOWLEDGMENTS

This work was performed under the auspices of the U.S. Department of Energy and was supported by contract W-31-109-Eng-38. Accordingly, the U.S. Government retains a nonexclusive, royalty-free license to publish or reproduce the published form of this contribution, or allow others to do so, for U.S. Government purposes.

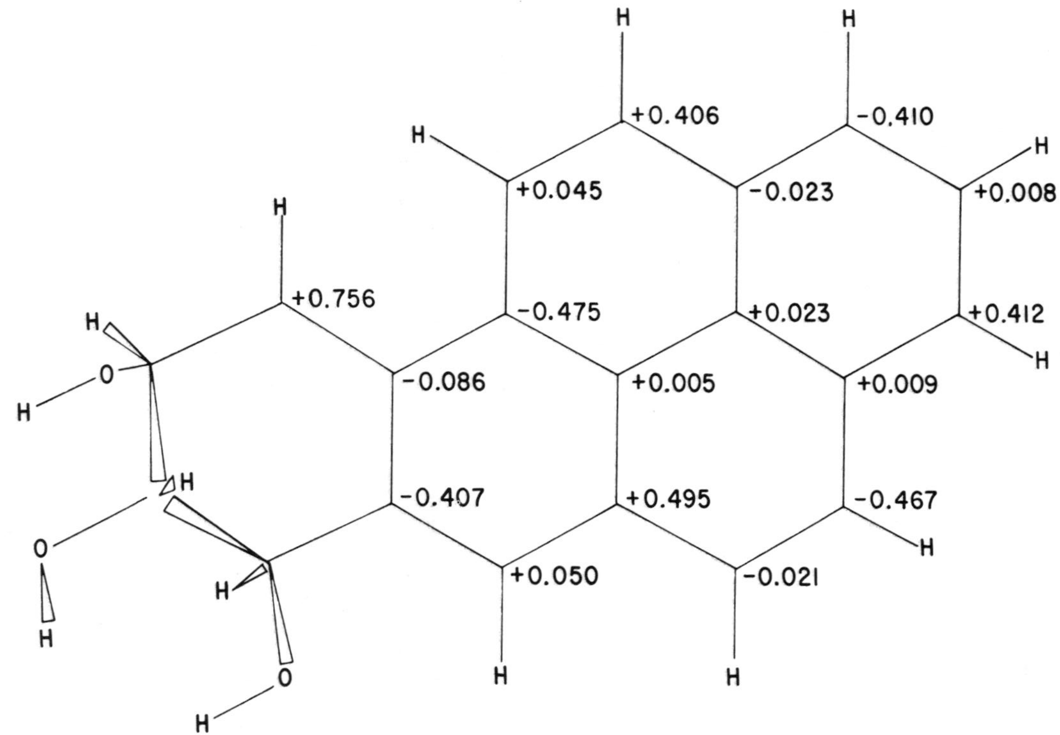

FIGURE 5. The lowest unoccupied molecular orbital (LUMO) of the triol-carbonium ion. The number alongside each atom corresponds to the coefficient of the symmetrically orthogonalized π p-orbital on the atom. The LUMO is normalized to 2 electrons, *i.e.*, the sum of the squares of the coefficients of the symmetrically orthogonalized basis function equals 2.

REFERENCES

1. Christoffersen, R. E. (1972): *Ab initio* calculations on large molecules. *Adv. Quantum Chem.* 6:333-393.
2. Christoffersen, R. E., Spangler, D., Hall, G. G., and Maggiora, G. M. (1973): *Ab initio* calculations on large molecules using molecular fragments. Evaluation and extension of initial procedures. *J. Am. Chem. Soc.* 95:8526-8536.
3. Hall, G. G. (1951): The molecular orbital theory of chemical valency. VIII. A method of calculating ionization potentials. *Proc. Roy. Soc.* A205:541-552.
4. Jeffrey, A. M., Jennette, K. W., Blobstein, S. H., Weinstein, I. B., Beland, F. A., Harvey, R. G., Kasai, H., Miura, I., and Nakanishi, K. (1976): Benzo[a]pyrene-nucleic acid derivative found *in vivo:* Structure of a benzo[a]pyrene-tetrahydrodiol epoxide-guanosine adduct. *J. Am. Chem. Soc.* 98:5714-5715.
5. Jeffrey, A. M., Weinstein, I. B., Jennette, K. W., Grzeskowiak, K., Nakanishi, K., Harvey, R. G., Autrup, H., and Harris, C. (1977): Structures of benzo[a]pyrene-nucleic acid adducts formed in human and bovine bronchial explants. *Nature* 5626:348-350.
6. Jerina, D. M., Yagi, H., Levin, W., and Conney, A. H. (1977): Carcinogenicity of Benzo[a]pyrene. In: *Drug Design and Adverse Reactions,* Alfred Benzon Symposium X, pp. 261-275, Munksgaard.
7. Levin, W., Wood, A. W., Lu, A. Y. H., Ryan, D., West, D., Conney, A. H., Thakker, D. R., Yagi, H., and Jerina, D. M. (1977): Role of purified cytochrome P-450 and epoxide hydrase in the activation and detoxification of benzo[a]pyrene. In: *Drug Metabolism Concepts,* ACS Symposium No. 44, pp. 99-126, American Chemical Society.
8. Nakanishi, K., Kasai, H., Cho, H., Harvey, R. G., Jeffrey, A. M., Jennette, K. W., and Weinstein, I. B. (1977): Absolute configuration of a ribonucleic acid adduct formed *in vivo* by metabolism of benzo[a]pyrene. *J. Am. Chem. Soc.* 99:258-260.
9. Osborne, M. R., Beland, F. A., Harvey, R. G., and Brookes, P. (1976): The reaction of (±)-7α,8β-dihydroxy-9β,10β-epoxy-7,8,9,10-tetrahydrobenzo[a]pyrene with DNA. *Int. J. Cancer* 18:362-368.
10. Roothaan, C. C. J. (1951): New developments in molecular orbital theory. *Rev. Mod. Phys.* 23:69-89.
11. Shipman, L. L. (1978): *Ab initio* quantum mechanical characterization of the ground electronic state of benzo[a]pyrene. Implications for the mechanism of polynuclear aromatic hydrocarbon oxidation to epoxides by cytochrome P-450. In: *Carcinogenesis, Vol. 3,* P. W. Jones and R. K. Freudenthal, Eds., pp. 139-144, Raven Press, New York.
12. Shipman, L. L., and Christoffersen, R. E. (1972): *Ab initio* calculations on large molecules using molecular fragments. Initial P-matrices for SCF calculations. *Chem. Phys. Lett.* 15:469-474.

13. *Tables of Interatomic Distances and Configurations in Molecules and Ions* (1965): The Chemical Society, Burlington House, London.
14. Weinstein, I. B., Jeffrey, A. M., Jennette, K. W., Blobstein, S. H., Harvey, R. G., Harris, C., Autrup, H., Kasai, H., and Nakanishi, K. (1976): Benzo[a]pyrene diol epoxides as intermediates in nucleic acid binding *in vitro* and *in vivo*. *Science* 193:592-595.

Polynuclear Aromatic Hydrocarbons,
edited by P. W. Jones and P. Leber.
Ann Arbor Science Publishers, Inc.
Ann Arbor, MI ©1979.

METABOLISM OF BENZO[a]PYRENE TO OXIDATIVE AND CONJUGATIVE METABOLITES BY ISOLATED MAMMALIAN HEPATOCYTES

C. A. Jones, B. P. Moore, G. M. Cohen and J. W. Bridges
Department of Biochemistry
University of Surrey
Guildford, Surrey, England

Benzo[a]pyrene (BP) is an important member of the class of chemical compounds termed polycyclic aromatic hydrocarbons (PAH) which produce serious cytotoxic, mutagenic and carcinogenic effects in many different species and tissues (15,24). An essential prerequisite for the induction of these toxic changes is the biotransformation of PAH into electrophilic metabolites which interact with cellular material to initiate the biological response. In particular, there are good correlations between the carcinogenic potency of the metabolites and the extent of their binding to DNA (3,5).

The metabolism of BP is highly complex. It is converted by the microsomal mixed-function oxidases and epoxide hydratase(s) to several different phenols, quinones, epoxides and dihydrodiols (1,6,18,30,31). Some of these metabolites may be converted to their respective glucuronide, sulfate or glutathione conjugates or, alternatively, they may be recycled through the oxidase system leading to the production of further activated species (16,35). Thus the generation and possible accumulation of active intermediates is dependent on the balance in the reactions leading to the formation of primary metabolites and their further metabolism by either oxidative or conjugation mechanisms. Other factors which may also influence the fate of active intermediates within the cell include cell membrane permeability, the degree of binding to intracellular organelles and the subsequent rate of release of the free and conjugated metabolites from the cell into the extracellular environment. Some of these factors can only be adequately investigated at the intact cell level. Freshly isolated adult mammalian hepatocytes which have retained their full capacity for catalyzing both the phase 1 oxidative reactions and the phase 2 conjugative reactions are a valuable tool in studying the overall metabolic fate of a xenobiotic such as BP. In addition, the conservation of the biochemical and structural organization of the cell, in a manner analogous to that within the intact organ, enables the consequences of the xenobiotic metabolism on the normal physiological processes of the cell to be investigated concomitantly.

In order to understand the organ and species selectivity in PAH carcinogenesis, it is necessary to investigate the metabolic fate of these compounds in both susceptible and nonsusceptible tissues from a variety of species. The metabolic profiles formed by hepatic microsomal preparations from a number of species have been compared (12,16,22), as well as those from human epithelial cells (11), lymphocytes (27) and embryonic cell cultures (28).

In the present study, isolated viable hepatocytes have been prepared from the rat, ferret and hamster and used in the study of the phase 1 and phase 2 metabolism of BP. We were particularly interested in the distribution of metabolites between the cells and the extracellular medium since those metabolites with a higher affinity for the intracellular environment are potentially more likely to interfere with normal cellular functions.

MATERIALS AND METHODS

Preparation of Isolated Hepatocytes

Male Wistar albino rats (60-100 g), bred at the University of Surrey, male Syrian hamsters (60-80 g) obtained from the Coombehurst Breeding Establishment, Basingstoke, Hampshire, and male polecat or albino ferrets (0.7-1.0 kg) obtained from Procellus Ltd., Sussex, were used in the study.

Hepatocytes were isolated from the livers by collagenase/hyaluoronidase digestion of the finely sliced tissue (13). The viability of the cell preparations was assessed by the degree of exclusion of trypan blue (26) and was routinely greater than 85%. The hepatocytes were diluted to 2×10^6 viable cells/ml in Leibovitz L-15 medium containing 2 m*M* glutamine and 10% fetal calf serum.

Metabolism of BP and Metabolites

^{3}H-BP was purified according to the method of DePierre *et al.* (10) and diluted to the appropriate concentration with unlabeled BP (Koch-Light Ltd., England). Unlabeled or ^{3}H-labeled metabolites were prepared by the method of Sims (29) and their identity confirmed by their UV spectra, fluorescence excitation and emission spectra and their chromatographic properties in several solvent systems.

Samples (1 ml) of hepatocyte suspension were pipetted into 10-ml conical flasks and incubated with [^{3}H]-BP (80 μM) or [^{3}H]-3-OH-BP (7.5 μM)

or [^{3}H]-7,8-diol (4 μM) or [^{3}H]-9,10-diol (7.0 μM) at 37° in a shaking water bath (90 oscillations/min). After timed intervals, the cells and medium were separated by centrifugation at 0° as described previously (20) and extracted with ethyl acetate (2 x 1 vol). The ethyl acetate extracts were dried over anhydrous sodium sulfate, concentrated to dryness on a rotary evaporator and then redissolved in a small volume of ethyl acetate containing appropriate reference compounds. The extracts were examined by TLC in a selection of solvent systems. The metabolites were detected by inspection of the wet plates in UV light (254 nm) and quantified by radioscanning on a Varian-Berthold scanner and by liquid scintillation counting (7).

The water-soluble metabolites remaining in the aqueous medium after organic extraction were subjected to hydrolysis with either β-glucuronidase (ketodase) or arylsulfatase (with D-saccharic acid-1,4-lactone to inhibit β-glucuronidase) in 0.1 *M* acetate buffer (pH 5.0) or 0.1 *M* acetate buffer (pH 5.0) alone (20). Deconjugated products were extracted with ethyl acetate and the products examined by TLC as described above.

RESULTS AND DISCUSSION

Metabolism of BP by Rat Hepatocytes

BP was rapidly metabolized by the hepatocytes. Initially the majority of products formed were ethyl acetate-soluble (65% of metabolites in medium after 25 min incubation), but later the products underwent further metabolism to water-soluble conjugates (78% of metabolites in medium after 90 min).

Ethyl Acetate-Soluble Metabolites

Figures 1 and 2 show the pattern of ethyl acetate-soluble metabolites obtained in the cells and extracellular medium during the course of the incubation. The metabolite profile within the hepatocytes differed in several important respects from that seen in the extracellular medium. The major intracellular metabolites were phenols but the levels measured in the extracellular medium were very low. In the TLC separation systems used, it was not possible to resolve the separate phenols produced by the hepatocytes. Similar intracellular and extracellular differences in metabolite profiles have also been observed with hamster embryo cells (23) and in a recent study using isolated hepatocytes from 3-methylcholanthrene-pretreated rats (4). Quinones were also dominant in the intracellular metabolite profile. The quinones may have arisen metabolically either directly from BP or the phenols or by autooxidation during the metabolite isolation procedure. The quinones may

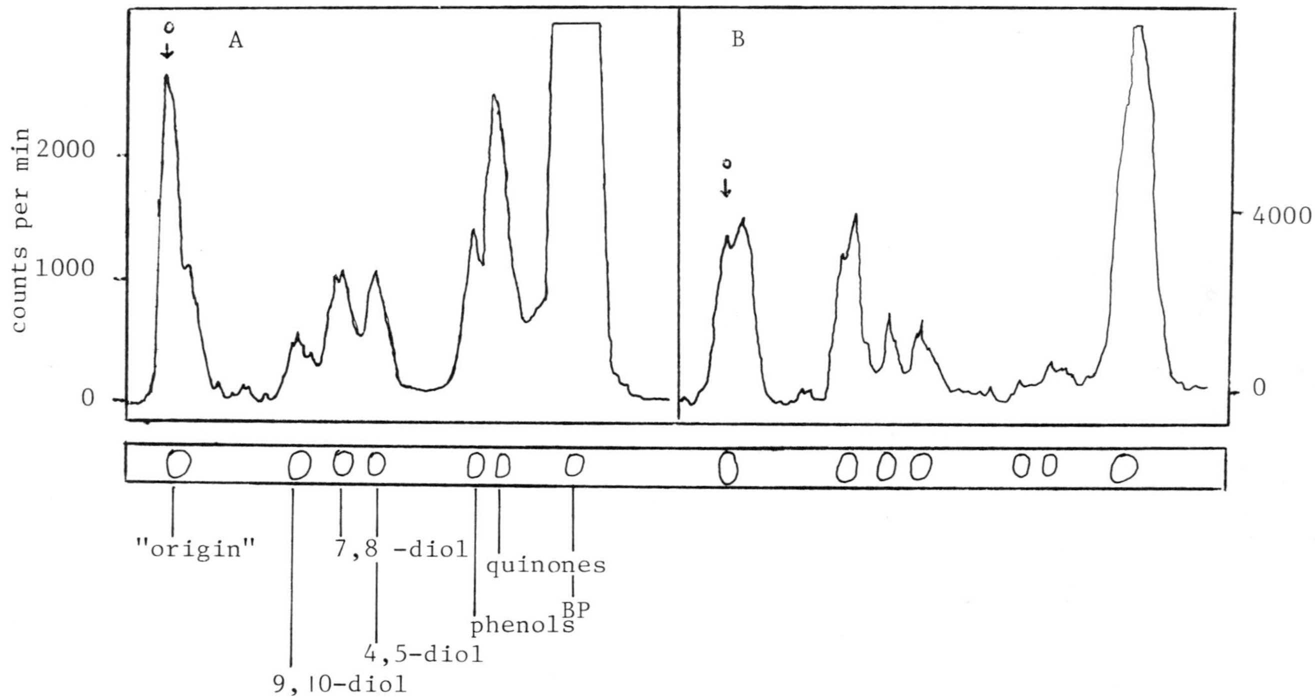

FIGURE 1. Tracings of radioscans of thin-layer radiochromatograms of ethyl acetate-soluble metabolites: (A) the hepatocytes and (B) the extracellular medium. Rat hepatocytes (2×10^6 cell/ml) were incubated with ^{3}H-BP (80 μm) for 90 min. The cells and extracellular medium were separated, extracted with ethyl acetate and the ethyl acetate-soluble radioactive products were separated as described previously (20).

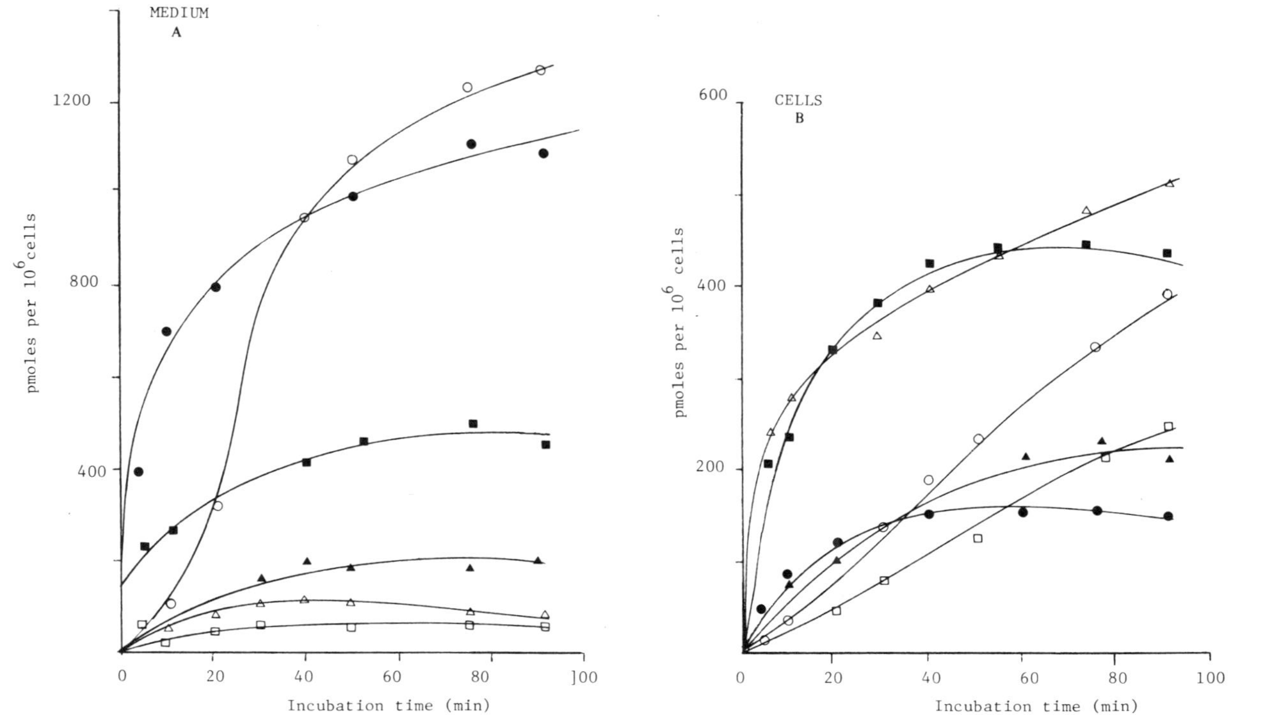

FIGURE 2. Distribution of ethyl acetate-soluble metabolites of BP between (A) the extracellular medium and (B) the hepatocytes. Following incubation of ^{3}H-BP with hepatocytes, the ethyl acetate-soluble metabolites were separated and quantified as described previously (2). 3-OH-BP △—△; 4,5-diol ■—■; 7,8-diol ▲—▲; 9,10-diol ●—●; metabolite(s) at or near origin ○—○.

have arisen metabolically either directly from BP or the phenols or by autooxidation during the metabolite isolation procedure. The quinones may undergo reduction and conjugation with sulfate (25) and this would facilitate their removal from the cells.

The extent to which different diols egressed from the cells into the medium differed markedly (Figures 1 and 2). 9,10-Diol egressed rapidly into the extracellular medium whereas 4,5- and 7,8-diol were distributed more evenly between cells and medium. Relatively higher levels of 7,8-diol predominated intracellularly. When the diol distribution is expressed per unit volume of cells or medium (Table I), the relative affinities of the diols for either environment can be better assessed. It is unlikely that these distributional differences can be explained simply in terms of their relative partition coefficients which, when calculated by the Hansch method (14), are essentially the same. Therefore the differences in diol retention may be related to their affinities for intracellular binding proteins or lipid membranes. Preliminary results show that each diol possesses approximately the same affinity for ligandin but 9,10-diol exhibited a considerably reduced affinity (less by a factor of ten) for egg lecithin or egg lecithin:cholesterol (1:1) liposomes and for protein A (33). As the principal potential intracellular binding material within the hepatocytes is the phospholipid component of the membranes (9), one would predict that both 4,5- and 7,8-diol should be better retained within the cells; this was found to be true in our studies (Figure 1).

A considerable amount of radioactivity remained at the origin in the radiochromatograms of the ethyl acetate extracts. This material also demonstrated a higher affinity for the cells. A major product in this material was identified as the sulfate conjugate of 3-OH-BP (6) but the other products present were not resolved. These probably included other phenolic sulfate esters, polyhydroxylated phenols and tetrahydrotetrols.

Water-Soluble Metabolites

The water-soluble products in the extracellular medium were subjected to enzymic hydrolysis with β-glucuronidase and the primary metabolites released were extracted into ethyl acetate. The principal hydrolysis products were 3-OH-BP and 4,5-diol with trace amounts of 7,8-diol (Figure 3). A significant amount of radioactivity was associated with a product migrating close to 9,10-diol in the chromatograms (region I on Figure 3). This product corresponded to that described as the 9,10-catechol or 9,10-dihydroxy-BP in a microsomal study by Booth and Sims (1). However, in a more recent study by Thakker *et al.* (32), a further metabolite of 9,10-diol was identified as a triol, 1 (3), 9,10-trihydroxy-9,10-dihyrobenzo[a]pyrene. As the fluorescence and UV spectra of the metabolite formed in the hepatocytes

TABLE I. Concentration of Metabolites in Cells and Extracellular Medium

Time (min)	3-Hydroxbenzo[a]-pyrene nmole ml^{-1}		Material at origin nmole ml^{-1}		9,10-Dihydrodiol nmole ml^{-1}		4,5-Dihydrodiol nmole ml^{-1}		7,8-Dihydrodiol nmole ml^{-1}	
	M	C	M	C	M	C	M	C	M	C
10	0.12	66.2	0.22	6.5	1.41	21.2	0.53	55.6	0.12	17.7
30	0.18	82.0	1.00	30.4	2.48	30.3	1.17	90.7	0.33	28.4
90	0.16	118.3	2.50	96.3	2.09	35.3	0.91	88.3	0.41	48.9

Note: Benzo[a]pyrene was incubated with isolated rat hepatocytes. The metabolites in the cells (C) and extracellular medium. (M) were separated and estimated as described in the text. The amounts of metabolites are expressed per ml of extracellular medium or intracellular volume. In the later calculation 2 x 10^6 hepatocytes occupied a packed volume of 8 μl based on the assumption that each hepatocyte was a sphere of mean diameter 22 μl) in an incubation containing 1 ml extracellular medium.

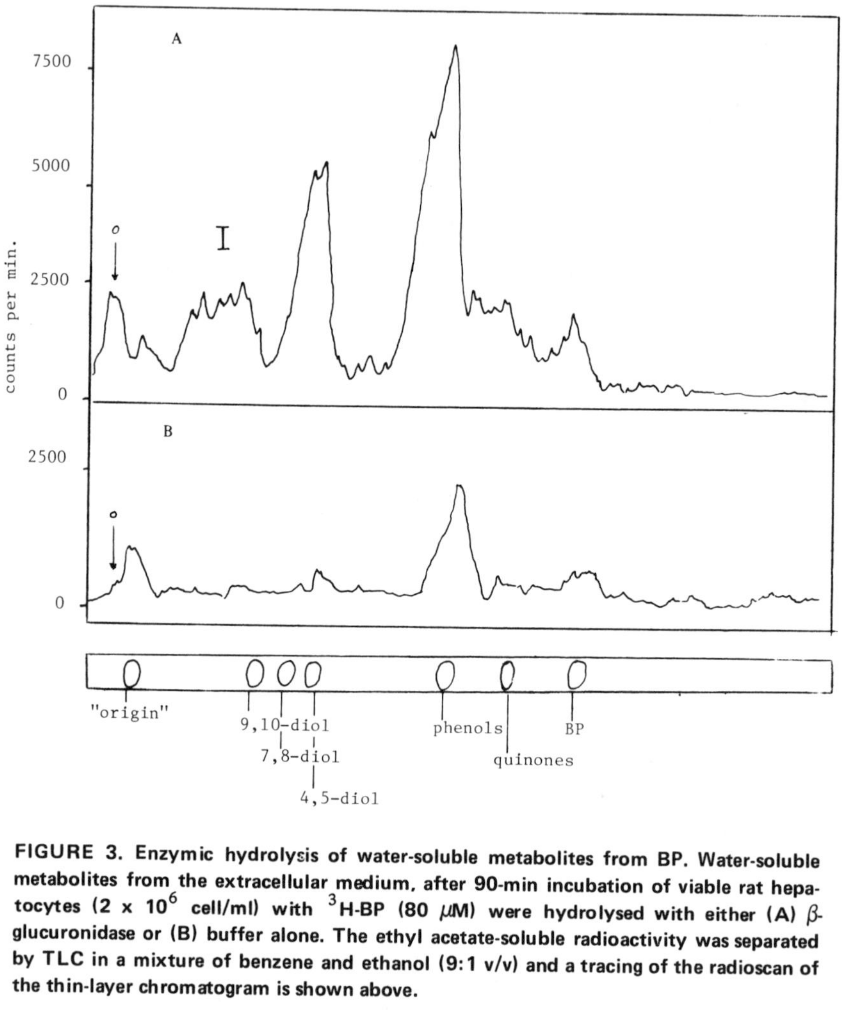

FIGURE 3. Enzymic hydrolysis of water-soluble metabolites from BP. Water-soluble metabolites from the extracellular medium, after 90-min incubation of viable rat hepatocytes (2×10^6 cell/ml) with ^{3}H-BP (80 μM) were hydrolysed with either (A) β-glucuronidase or (B) buffer alone. The ethyl acetate-soluble radioactivity was separated by TLC in a mixture of benzene and ethanol (9:1 v/v) and a tracing of the radioscan of the thin-layer chromatogram is shown above.

corresponded to that of the catechol and the triol, further characterization is necessary befores its identity can be confirmed.

Two further metabolites were detected following protein precipitation of the aqueous samples and TLC in butanol-propanol-2 MNH_3 (2:1:1). These may be glutathione conjugates but were not identified. They contributed approximately 6% of the added radioactivity after 60 min incubation. Low levels of radioactivity were covalently bound to the intracellular constituents at the end of the incubation. The proportion of bound material increased with time and therefore was attributed to BP metabolites but was not further investigated. In order to obtain a better understanding of the overall metabolic fate of BP in isolated rat hepatocytes, we have also studied the metabolic fate of some of the major metabolites of BP.

Metabolism of 7,8-diol, 9,10-diol and 3-OH-BP by Isolated Rat Hepatocytes

Significant differences in the distribution of these metabolites between hepatocytes and extracellular medium were observed (Figure 4).

7,8-Diol

More of the radioactivity associated with 7,8-diol or its metabolites was found intracellularly than with either 9,10-diol or 3-OH-BP at all times (Figure 4). All three metabolites were converted to both ethyl acetate-soluble and water-soluble products (Table II). Most of the ethyl acetate-soluble radioactivity in both the cells and medium was associated with the unmetabolized 7,8-diol (Figure 5). The major ethyl acetate-soluble metabolite, accounting for 3% of the radioactivity in the extract at 30 min was identified as the 7,8,9,10-tetrol on the basis of its UV and fluorescence spectra. It was found in both cells and medium.

Following enzymic hydrolysis of the water-soluble products in the medium, 7,8-diol and tetrol were released. Both were conjugated primarily with glucuronic acid. More recently, we have used HPLC to analyze the metabolic products formed on incubation of 7,8-diol with rat hepatocytes. In agreement with our earlier studies, the results showed the formation of both tetrol and the glucuronide conjugate of 7,8-diol. However, an uncharacterized product was formed which eluted just prior to BP and was also further metabolized to a glucuronide conjugate. We are currently investigating the nature of this relatively nonpolar product.

It is interesting to note the formation of tetrol thus indicating that hepatocytes are capable of generating the diol-epoxide, 7,8-dihydro-7,8-dihydroxy-BP-9,10-oxide. The possibility exists that 7,8-diol retained within hepatocytes may act as a store of slowly released metabolite, capable of being metabolized to the ultimate carcinogen in other tissues. However, it appears more likely that the liver can efficiently metabolize 7,8-diol either via this unidentified product or through direct conjugation, and that the diol-epoxide generated can also be efficiently further metabolized to tetrol and conjugated so as to facilitate its removal from the cells. In this manner, the liver is protected from the possible toxic consequences of diol-epoxide accumulation within the cells.

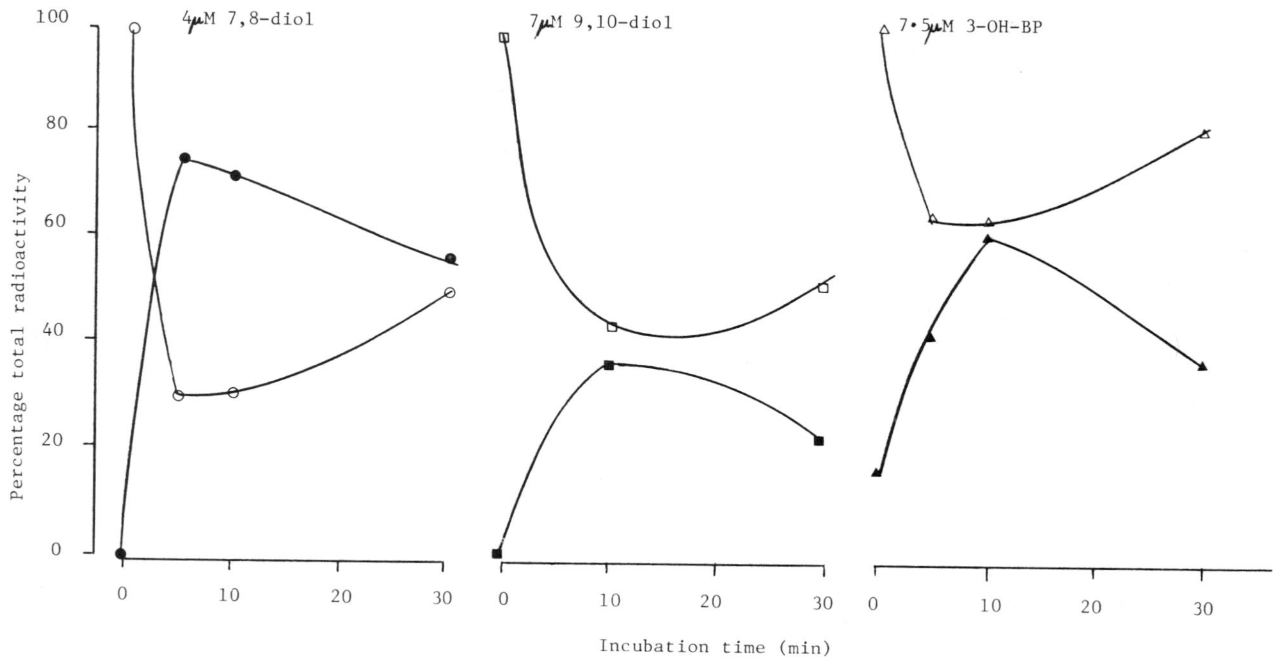

FIGURE 4. Time course of the distribution of radioactivity associated with hepatocytes (closed symbols) and extracellular medium (open symbols) on incubation of rat hepatocytes (2×10^6 cell/ml) with each metabolite. At various times, the incubation was stopped, the cells and extracellular medium separated and the total radioactivity in both determined.

TABLE II. Overall Metabolism of 9,10-diol, 7,8-diol and 3-OH-BP to Organic-Soluble and Aqueous Metabolites

	9,10-diol	7,8-diol	3-OH-BP
Total Metabolites	90%	3%	70%
Organic soluble metabolites	8%	3%	12%
Aqueous metabolites	82%	57%	58%

Note: Values refer to the percentage of metabolites formed in the total incubate from each substrate, during a 60-min incubation.

9,10-Diol

9,10-Diol was extensively metabolized to both ethyl acetate-soluble and water-soluble products (Table II). The major ethyl acetate-soluble metabolite cochromatographed with a metabolite with similar spectral characteristics to either 9,10-dihydroxy-BP (1) or 1, (3), 9, 10-trihydroxy-9,10-dihydro-BP (32) but only small amounts of tetrol were formed (Figure 6). After 60 min incubation, greater than 90% of 9,10-diol had been metabolized via the catechol/triol (Figure 6). Most of this metabolite (60%) was present in the medium as a sulfate conjugate and 19% as a glucuronide conjugate but no detectable glucuronide conjugate of 9,10-diol was observed. The major metabolic fate of 9,10-dihydrodiol in isolated rat hepatocytes is therefore initial metabolism to either catechol and/or triol and then conjugation of the product(s), with only small amounts of tetrol being formed. In contrast to this, 7,8-diol was metabolized more slowly but primarily via a biologically reactive diol epoxide to a tetrol, and then further conjugated. Thus these results with isolated hepatocytes confirm and extend earlier observations of Booth and Sims (1) showing similar differences in microsomal metabolism of 7,8- and 9,10-dihydrodiols. Both these studies suggest that the higher biological activity of 7,8-dihydrodiol than 9,10-dihydrodiol may be due to its preferential metabolism via a diol epoxide. It was of interest that isolated rat hepatocytes were capable of relatively rapid further metabolism of 9,10-dihydrodiol in contrast to our studies with short-term organ culture of rodent lung and trachea (8) when 9,10-dihydrodiol is only metabolized very slowly.

3-OH-BP

3-OH-BP was metabolized primarily by rat hepatocytes to both glucuronic acid and sulfate conjugates with lesser amounts of at least one quinone, BP-3,6-quinone, and some uncharacterized polyhydroxylated products (triols, diols, etc.) formed, after 60 min incubation, approximately 70-80%, 20%, 3% and 1% of the final products were accounted for by the glucuronide conjugate, sulfate ester, quinone and polyhydroxylated products, respectively.

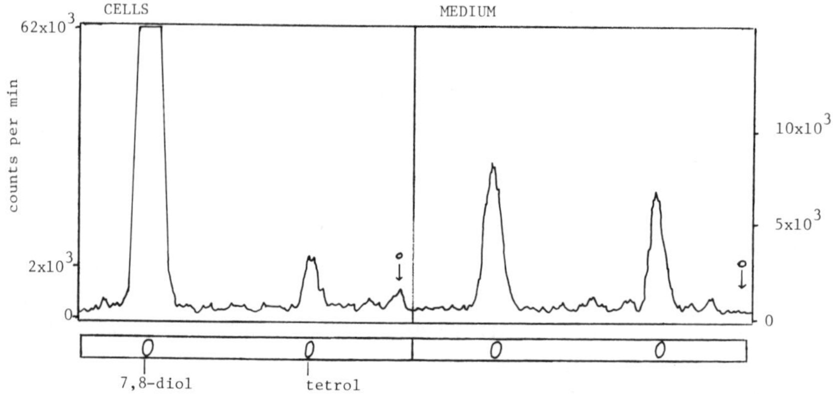

FIGURE 5. Tracings of radioscans of thin-layer chromatograms of ethyl acetate-soluble metabolites in rat hepatocyte extracellular medium from an incubation with 7,8-diol. Rat hepatocytes (2 x 10^6 cell/ml) were incubated with 7,8-diol (4 μM) for 30 min. The cells and medium were separated and extracted with ethyl acetate and the ethyl acetate-soluble radioactive products were separated by TLC in chloroform and methanol (4:1 v/v).

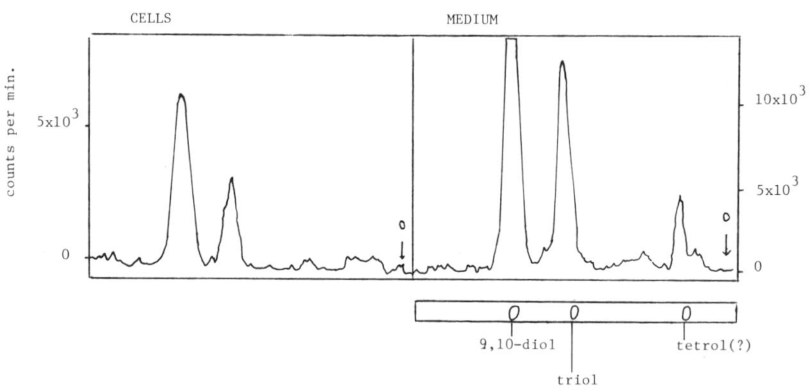

FIGURE 6. Tracings of radioscans of thin-layer radiochromatograms of ethyl acetate-soluble metabolites in rat hepatocytes and extracellular medium from an incubation with 9,10-diol. Rat hepatocytes (2 x 10^6 cell/ml) were incubated with ^{3}H-9,10-diol (7.0 μM) for 30 min. The cells and medium were separated, extrracted with ethyl acetate-soluble radioactive products were separated by TLC in chloroform and methanol (4:1 v/v).

While most of the radioactivity associated with 3-OH-BP or its metabolites was in the extracellular medium throughout the incubation (Figure 4), unmetabolized 3-OH-BP accumulated rapidly in the cells and at the termination of the incubation, the remaining unmetabolized phenol was located almost entirely intracellularly (Figure 7). The glucuronic acid conjugate egressed rapidly from the cells on formation (Figure 6). However, a lag phase in the onset of conjugation was observed which might be attributed to a degree of latency in the glucuronyl transferase enzymes or to a temporary shortage of the necessary cofactors, namely UDPGA at the enzymic site, but this delay was not in the release of the conjugate, once formed, from the cells. The ethyl acetate-soluble sulfate, on the other hand, persisted as the major conjugate within the cells (Figure 7). The polyhydroxylated metabolites were not further characterized but they constituted less than 1% of the total metabolites near the origin when the extent of sulfate conjugate present was more accurately determined following enzymic hydrolysis with sulfatase.

Metabolism of BP by Hamster and Ferret Hepatocytes

Significant qualitative differences were observed in the organic solvent-soluble metabolites generated by hamster and ferret hepatocytes. The major ethyl acetate-soluble metabolites in the hamster were quinones (Figure 8), most probably the 3,6-quinone, whereas only small amounts were produced by the rat or ferret. As the metabolite isolation procedures were similar in all species used, it appears that in the hamster at least, some enzymic action may be involved.

4,5-Diol was the major diol formed by hamster hepatocytes but not by other species (Table III). This agrees with the results of microsomal studies (2,28,34). Approximately sevenfold more 4,5-diol was produced relative to 7,8-diol and to 9,10-diol formation, which were formed in approximately equal amounts. The ratio of diol formation in human liver microsomes (27) resembles that observed in rat and ferret hepatocytes whereas the ratio in rhesus monkey liver microsomes (17,22) more closely resembles that of hamster hepatocytes.

A different distribution of metabolites between cells and extracellular medium was observed in ferret hepatocytes (Figure 9) compared to both hamster (Figure 8) and rat hepatocytes (Figure 3). Thus, at the end of 60 min of incubation with ferret hepatocytes, the unconjugated phenols were distributed evenly between the cells and medium, approximately 70% of 7,8-diol formed and greater than 80% of the other ethyl acetate-soluble metabolites had egressed into the extracellular medium (Figure 9). The lack of persistence of the phenols and 7,8-diol within the ferret hepatocytes relative to that in

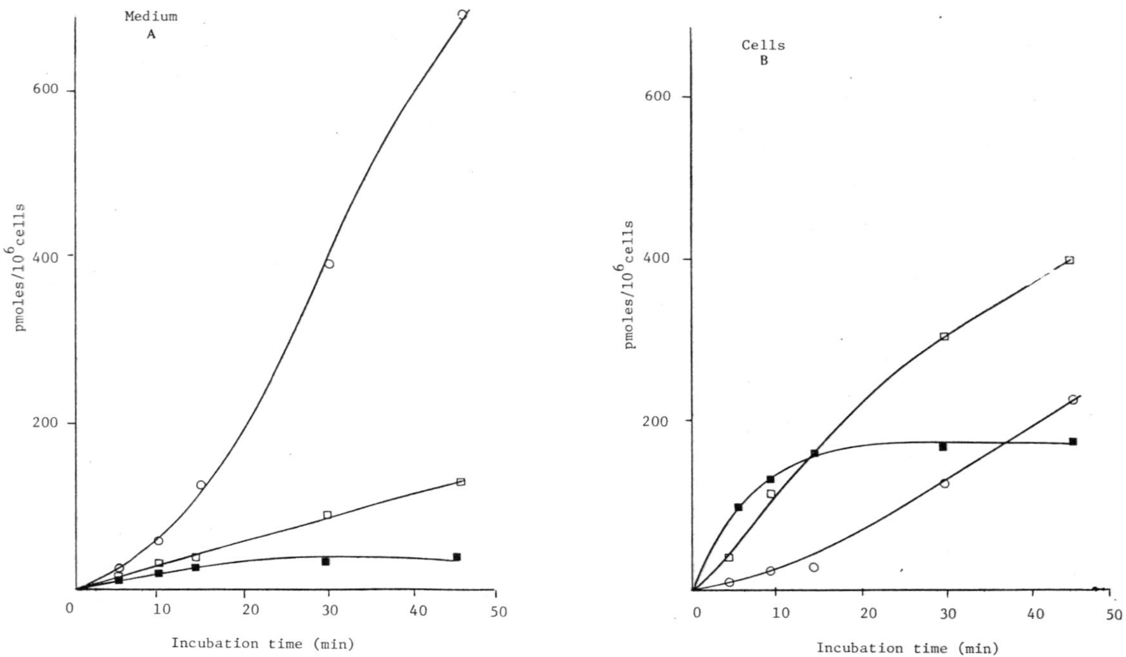

FIGURE 7. Distribution of 3-OH-BP (■—■), its glucuronide conjugate (○—○) and its sulfate ester (□—□) between (A) extracellular medium and (B) hepatocytes. Hepatocytes (2×10^6 cell/ml) were incubated with BP (80 μM), the cells and medium separated and the amounts of glucuronide and sulfate conjugates were determined by the fluorescence method of Nebert and Gelboin. The amounts of glucuronide and sulfate conjugates were determined from the 3-OH-BP present following enzymic hydrolysis with either β-glucuronidase (with saccharo-1,4,-β-lactone) or aryl sulfatase, respectively.

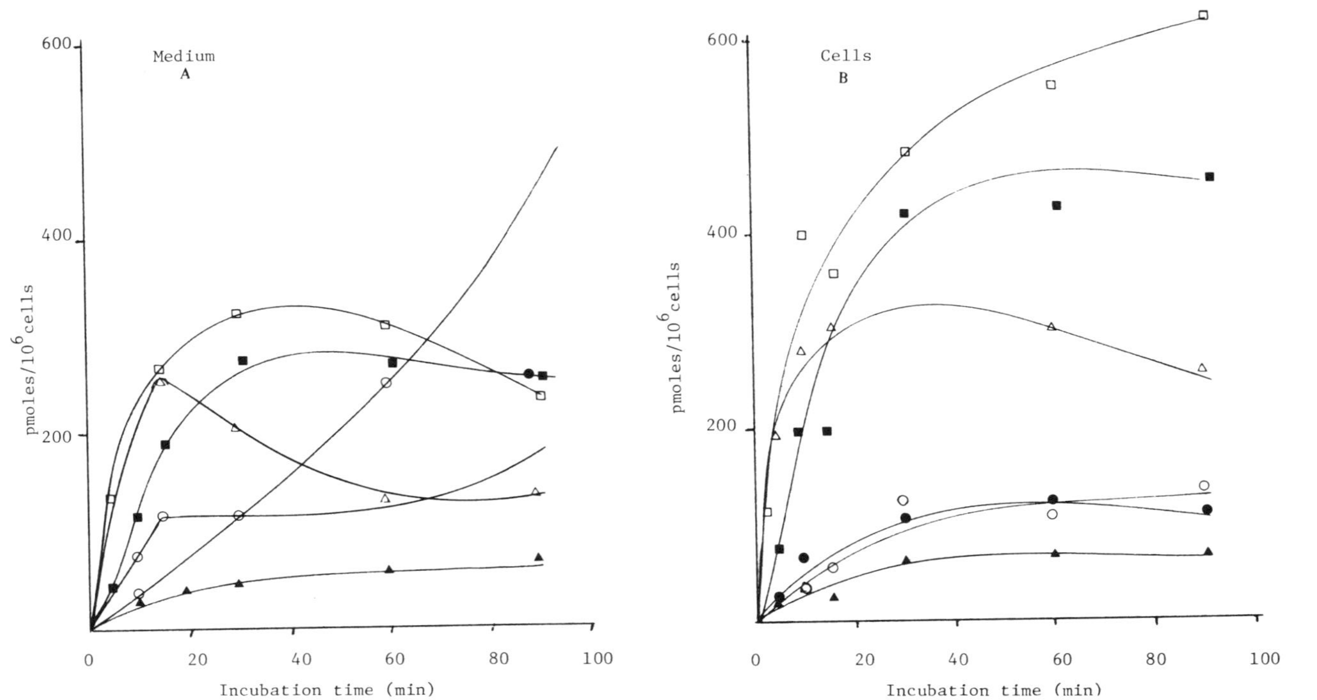

FIGURE 8. Distribution of ethyl acetate-soluble metabolites of BP between (A) extracellular medium and (B) hamster hepatocytes. Following incubation of ^{3}H-BP with hamster hepatocytes (2 x 10^6/ml) the ethyl acetate-soluble metabolites were separated and quantified as described previously (22). 3-OH-BP △—△; 4,5-diol ■—■; 7,8-diol ▲—▲; 9,10-diol ●—●; metabolites at or near origin ○—○.

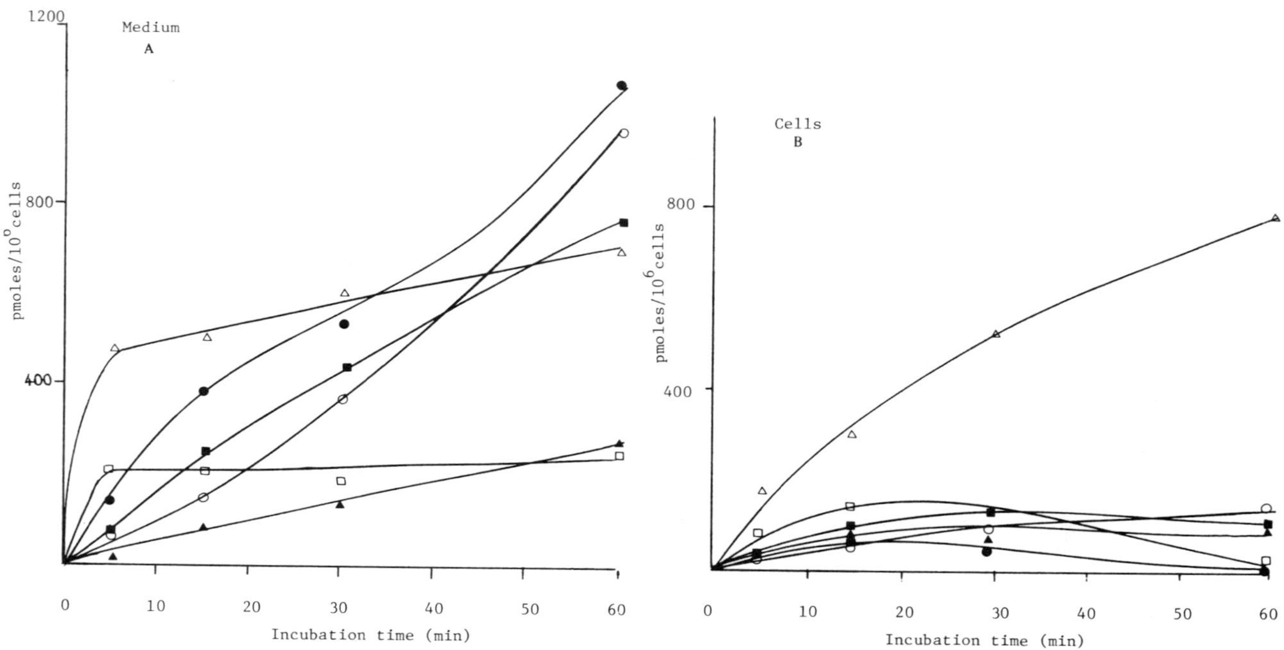

FIGURE 9. Distribution of ethyl acetate-soluble metabolites of BP between (A) extracellular medium and (B) ferret hepatocytes. Following incubation of ^{3}H-BP with ferret hepatocytes (2 x 10^6 cell/ml) the ethyl acetate-soluble metabolites were separated and quantified as described previously (22). 3-OH-BP △—△; 4,5-diol ■—■; 7,8-diol ▲—▲; 9,10-diol ●—●; metabolites at or near origin ○—○.

TABLE III. Relative Percentage of Each Diol in the Total Incubate at 60 Minutes

	4,5-diol	7,8-diol	9,10-diol
Rat	30%	13%	53%
Hamster	65%	12%	23%
Ferret	37%	17%	46%

Note: For each diol, the amount of free and further metabolized products were summed together and expressed as a percentage of the total amount of all diols formed.

the hamster and rat hepatocytes was not related to a difference in cell viability as this was checked using several biochemical parameters (19). Ferret hepatocytes may differ from rat hepatocytes in some respect, possibly the availability of binding proteins or certain membrane charcteristics may enable the primary metabolites to egress more readily from the cells without further metabolism.

Hamster and ferret hepatocytes metabolized BP to water-soluble metabolites. The ferret was relatively less efficient at performing the conjugative reactions (Table IV). Qualitatively the conjugates produced by the three species were similar, *i.e.*, the glucuronide conjugates of 3-OH-BP and 4,5-diol and the sulfate ester of 3-OH-BP being the major conjugates identified. Low levels of 7,8-diol glucuronide were also measured.

In conclusion, it is evident that isolated hepatocytes metabolize BP via the cytochrome P-450-epoxide hydrase linked system to a variety of potential toxic intermediates but which on the whole are efficiently and extensively detoxified to excretable water-soluble products by the conjugation processes. Speculatively, it appears possible that these intermediates escaping unconjugated from the liver *in vivo* could be taken up by other lipid-rich tissues. The intact hepatocytes have been demonstrated to retain their organ specificity in their capacity to activate carcinogens in a manner analogous to the organ *in vivo*. In a recent paper (21), liver cells were unable to activate BP, a nonhepatocarcinogen to mutagenic metabolites in a cell-mediated assay, whereas DMN and aflatoxin were shown to be highly mutagenic in this system. Thus the use of isolated hepatocytes to study the metabolism and activation of carcinogens would seem to have several advantages over those systems involving cell lines or bacterial systems.

SUMMARY

The metabolism of benzo[a]pyrene has been investigated in adult hepatocytes isolated from the rat, hamster and ferret. The metabolites produced

TABLE IV. Amounts of Phenols and Dihydrodiols Produced from BP in a 60-Minute Incubation by Rat, Hamster and Ferret Hepatocytes

	Amounts of Metabolite		
Metabolite	**Rat**	**Hamster**	**Ferret**
9,10-diol	1.68 (20%)	0.31 (8%)	1.14 (19%)
7,8-diol	0.50 (6%)	0.15 (4%)	0.42 (7%)
4,5-diol	1.35 (16%)	1.09 (50%)	0.93 (15%)
Phenols	4.89 (58%)	2.14 (58%)	3.61 (59%)
Total/Free Metabolites	3.50 (42%)	2.25 (60%)	4.11 (68%)
Total of Conjugates	4.92 (58%)	1.44 (40%)	1.98 (32%)
Overall Total	8.42 (100%)	3.69 (100%)	6.09 (100%)

Note: Each amount is expressed in nmoles/10^6 cells, with its percentage of the total metabolites given in parenthesis. For each value, the proportion of metabolite in the cells and medium were added together.

initially were mainly ethyl acetate-soluble, comprising mainly phenols, diols and quinones. As the incubation proceeded, the primary metabolites underwent further metabolism to more polar products. Water-soluble glucuronic acid conjugates of the phenols and 4,5-diol and ethyl acetate-soluble sulfate ester conjugates of the phenols were formed predominantly. Qualitatively similar metabolites were formed by rat and ferret hepatocytes, but quantitatively, metabolism in the ferret was less extensive. In the hamster, a significantly greater proportion of 4,5-diol and quinone was produced. The further metabolism of some primary products by rat hepatocytes was investigated in separate incubations. 7,8-Diol formed mainly glucuronic acid conjugates as well as some tetrol which was subsequently conjugated with glucuronic acid. 9,10-Diol did not form conjugates directly but was metabolized to a catechol/triol product which was then conjugated with both glucuronic acid and sulfate.

Significant differences in the distribution of metabolites between the cells and extracellular medium were observed (Figure 10). Phenols tended to accumulate intracellularly and only low amounts were released into the medium. Sulfate ester conjugates of monohydroxybenzo[a]pyrenes accumulated intracellularly to a slightly lesser degree. 4,5-Diol and 7,8-diol were distributed more evenly between cells and medium, whereas 9,10-diol was found mainly in the medium. Low amounts of radioactivity bound covalently to cellular macromolecules. These studies have further demonstrated the potential value of intact hepatocytes for the comparative study of BP metabolism in different species.

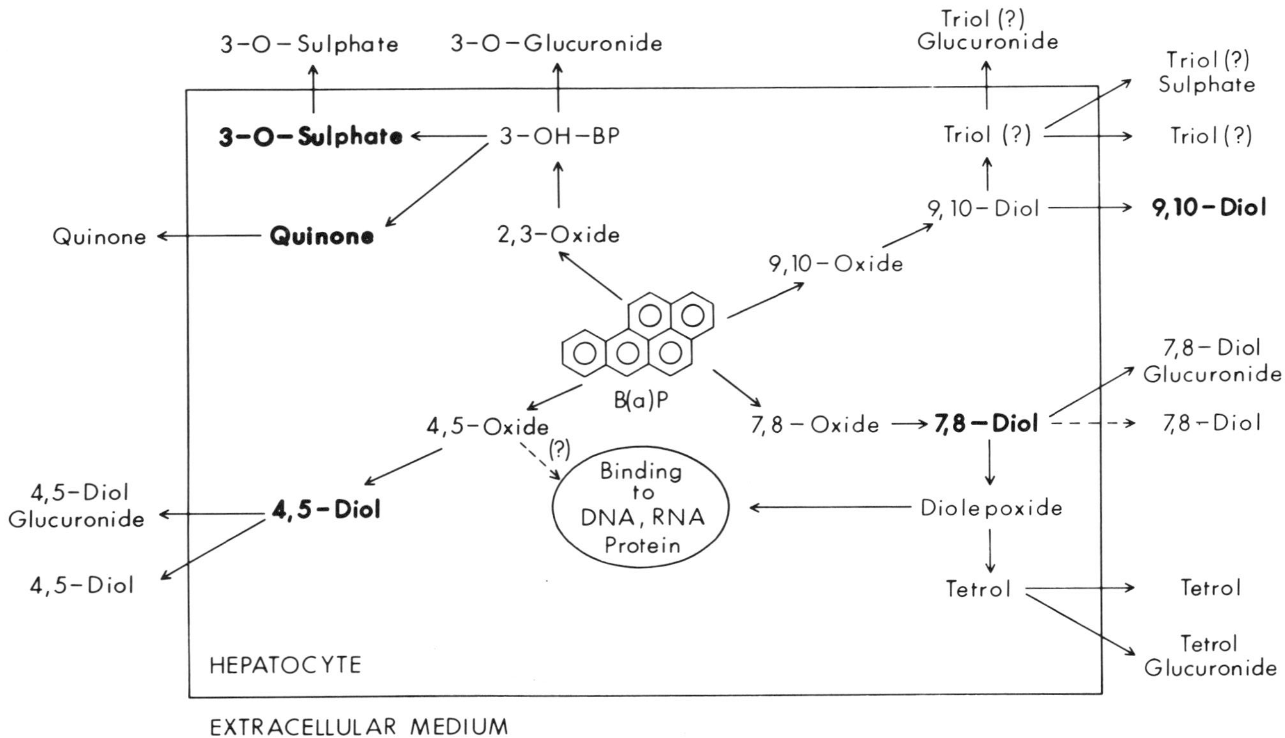

FIGURE 10. Distribution of the major metabolites of B[a]P between the hepatocytes and extracellular medium.

REFERENCES

1. Booth, J., and Sims, P. (1976): Different pathways involved in the metabolism of the 7,8- and 9,10-dihydrodiols of benzo[a]pyrene. *Biochem. Pharmacol.* 25:979-980.
2. Borgen, A., Darvey, H., Castagnoli, N., Crocker, T. T., Rasmussen, R. E., and Wang, I. Y. (1973): Metabolic conversion of benzo[a]pyrene by Syrian hamster liver microsomes and binding of metabolites to deoxyribonucleic acid. *J. Med. Chem.* 16:502-506.
3. Brookes, P., and Lawley, P. D. (1964): Evidence for the binding of polynuclear aromatic hydrocarbons to the nucleic acids of mouse skin. *Nature* 202:781-784.
4. Burke, D. M., Vadi, H., Jernstrom, B., and Orrenium, S. (1977): Metabolism of benzo[a]pyrene with isolated hepatocytes and the formation and degradation of DNA-binding derivatives. *J. Biol. Chem.* 252:6424-6431.
5. Buty, S. G., Thompson, S., and Slaga, T. J. (1976): The role of epidermal aryl hydrocarbon hydroxylase in the covalent binding of polycyclic hydrocarbon to DNA and its relationship to tumor initiation. *Biochem. Biophys. Res. Commun.* 70:1102-1108.
6. Cohen, G. M., Haws, S. M., Moore, B. P., and Bridges, J. W. (1976): Benzo[a]pyren-3-yl hydrogen sulphate, a major ethyl acetate extractable metabolite of benzo[a]pyrene in human, hamster and rat lung cultures. *Biochem. Pharmacol.* 25:2561-2570.
7. Cohen, G. M., and Moore, B. P. (1976): Metabolism of benzo[a]pyrene by different portions of the respiratory tract. *Biochem. Pharmacol.* 25:1623-1629.
8. Cohen, G. M., and Moore, B. P. (1977): Metabolism of benzo[a]pyrene, 7,8-dihydro-7,8-dihyroxybenzo[a]pyrene and 9,10-dihydro-9,10-dihydroxybenzo[a]pyrene by short-term organ cultures of hamster lung. *Biochem. Pharmacol.* 26:1481-1487.
9. Davies, W. W., Krahl, M. E., and Clowes, G. H. A. (1942): Solubility of carcinogenic and related hydrocarbons in water. *J. Am. Chem. Soc.* 64:108-110.
10. De Pierre, J. W., Moron, M. S., Johannesen, K. A. M., and Ernsker, L. (1975): A reliable, sensitive and convenient radioactive assay of benzpyrene monooxygenase. *Anal. Biochem.* 63:470-484.
11. Fox, C. H., Selkirk, J. K., Price, F. M., Croy, R. G., Sanford, K. K., and Cottler-Fox, M. (1975): Metabolism of benzo[a]pyrene by human epithelial cells *in vitro. Cancer Res.* 35:3551-3557.
12. Freudenthal, R. I., Hundley, S. G., and Cattaneo, S. M. (1978): A comparison of the metabolites of benzopyrene by lung mixed function oxidase from rat, rhesus, and humans. In: *Carcinogenesis, Vol. 3,* P. W. Jones and R. I. Freudenthal, Eds., Raven Press, New York.
13. Fry, J. R., Jones, C. A., Wiebkin, P., Bellemann, H. P., and Bridges, J. W. (1976): The enzymic isolation of adult rat hepatocytes in a functional and viable state. *Anal. Biochem.* 71:341-350.

14. Hansch, C. (1972): Quantitative relationships between liophilic character and drug metabolism. *Drug Metab. Rev.* 1:1-14.
15. Heidelberger, C. (1975): Chemical carcinogenesis. *Ann. Rev. Biochem.* 44:79-121.
16. Holder, G. M., Yagi, H., Jerina, D. M., Levin, W., Lu, A. Y. H., and Conney, A. H. (1975): Metabolism of benzo[a]pyrene. Effect of substrate concentration and 3-methylcholanthrene pretreatment on hepatic metabolism by microsomes from rats and mice. *Arch. Biochem. Biophys.* 170:557-566.
17. Hundley, S. G., and Freudenthal, R. I. (1977): High-pressure liquid chromatography analysis of benzo[a]pyrene metabolism by liver and lung microsomal enzymes from rhesus liver and lung. *Cancer Res.* 37:244-249.
18. Jerina, D. M., and Daly, J. W. (1974): Arene oxides: A new perspective of drug metabolism. *Science* 185:573-582.
19. Jones, C. A., Fry, J. R., and Bridges, J. W. (1978). In press.
20. Jones, C. A., Moore, B. P., Cohen, G. M., Fry, J. R., and Bridges, J. W. (1978): Studies on the metabolism and excretion of benzo[a]pyrene in isolated adult rat hepatocytes. *Biochem. Pharmacol.* 27:693-702.
21. Langenbach, R., Freed, H. J., Raveh, D., and Huberman, E. (1978): Differences in metabolic activation of the carcinogens aflatoxin B_1 and benzo[a]pyrene to mutagens by fibroblasts and liver cells in culture. *Nature.*
22. Leber, P., Kerchner, G., and Freudenthal, R. I. (1978): A comparison of benzo[a]pyrene metabolism by primates, rats, and miniature swine. In: *Polynuclear Aromatic Hydrocarbons: Chemistry, Metabolism and Carcinogenesis,* R. Freudenthal and P. W. Jones, Eds., Raven Press, New York.
23. MacLeod, M., Cohen, G. M., and Selkirk, J. K. Manuscript submitted.
24. National Academy of Sciences (1972): *Biological Effects of Atmospheric Pollutants, Particular Polycyclic Organic Matter.* Washington, D.C.
25. Nemoto, N., Takayama, S., and Gelboin, H. V. (1977): Enzymic conversion of benzo[a]pyrene phenols, dihydrodiols, and quinones to sulphate conjugates. *Biochem. Pharmacol.* 26:1825-1829.
26. Paul, J. (1972): *Cell and Tissue Culture,* 4th ed. p. 91, E & S Livingstone, Edinburgh.
27. Selkirk, J. K., Croy, R. G., Whitlock, J. P., Jr., and Gelboin, H. V. (1975): *In vitro* metabolism of benzo[a]pyrene by human liver microsomes and lymphocytes. *Cancer Res.* 35:3651-3655.
28. Selkirk, J. K., Croy, R. G., Wiebel, F. S., and Gelboin, H. V. (1976): Differences in benzo[a]pyrene metabolism between rodent liver microsomes and embryonic cells. *Cancer Res.* 36:4476-4479.
29. Sims, P. (1970): Qualitative and quantitative studies on the metabolism of a series of aromatic hydrocarbons by rat liver preparations. *Biochem. Pharmacol.* 19:795-818.
30. Sims, P., and Grover, P. L. (1974): Epoxides in polycyclic aromatic

hydrocarbon metabolism and carcinogenesis. *Adv. Cancer Res.* 20: 165-274.

31. Sims, P., Grover, P. L., Swaisland, A., Pal, K., and Hewer, A. (1974): Metabolic activation proceedes via by a diol-epoxide. *Nature* 252: 326-328.
32. Thakker, D. R., Yagi, H., Lehr, R. E., Levin, W., Buening, M., Lu, A. Y. H., Chang, R. L., Wood, A. W., Conney, A. H., and Jerina, D. M. (1978): Metabolism of *trans*-9,10-dihydroxybenzo[a]pyrene occurs primarily via arylhydroxylation rather than formation of a diol epoxide. *Mol. Pharmacol.* 14:502-513.
33. Tipping, E., Ketterer, B., Bridges, J. W., Cohen, G. M., and Moore, B. P. (1978). In preparation.
34. Wang, I. Y., Rasmussen, R. E., and Crocker, T. T. (1974): The metabolism of benzo[a]pyrene by microsomes from tissues of pregnant and fetal hamsters. *Life Sciences* 15:1291-1300.
35. Wiebel, F. J. (1975): Metabolism of monohydroxybenzo[a]pyrenes by rat liver microsomes and mammalian cells in culture. *Arch. Biochem. Biophys.* 168:609-621.

Polynuclear Aromatic Hydrocarbons,
edited by P. W. Jones and P. Leber.
Ann Arbor Science Publishers, Inc.
Ann Arbor, MI ©1979.

COMPARATIVE MICROBIAL AND MAMMALIAN CELL *IN VITRO* BIOASSAY ASSESSMENT OF FOSSIL FUEL-GENERATED RESPIRABLE PARTICULATES

James P. Crowley, Anthony J. Dennis, Thomas J. Facklam and Werner L. Margard

Biomedical Sciences Section
Battelle Columbus Laboratories
505 King Avenue
Columbus, Ohio 43201

INTRODUCTION

Human exposure to potentially harmful environmental materials has prompted the development and application of short-term *in vitro* bioassays which are designed to measure the cytotoxic, mutagenic and carcinogenic potential of these materials. Most of these assays have been developed over a period of years for specific applications other than as screening assays for potentially harmful environmental materials. Many have been developed for the study of the molecular mechanisms of various normal and abnormal biological processes. The modification of these assays and the application to complex environmental samples not only will lead to greater insight into the types of harmful agents present in the environment, but also to the molecular basis for their action.

Studies of mutation and DNA repair rates in microorganisms such as *Salmonella typhimurium* and *Escherichia coli* indicate that approximately 90% of chemical carcinogens cause mutation or DNA damage in these systems (6, 8). However, at this time, no single *in vitro* test system provides 100% correlation between mutagenicity and carcinogenicity (6). Also, the more complex nucleic acid organization, metabolic and repair mechanisms of mammalian cells and integrated profiles of both microbial and mammalian cell test systems insure a sufficient data base for primary evaluation of sample materials and reduce the risk of false results due to metabolic variation.

The application of a Primary Screening Profile to determine possible harmful effects of fossil fuel combustion products (FFCP) is of primary concern due to the possible high rate of human exposure in industrialized countries. One of the most important forms of FFCP, in terms of potential biological activity, is respirable-range particulate flyash. These particles have long residency times in the atmosphere, and consist of ferrous silica particles with surface condensations of heavy metal salts, polycyclic aromatic hydrocarbons and numerous other chemical species as a result of the combustion process and environmental interactions (10). Organic extracts of airborne particulates have been demonstrated to be cytotoxic and mutagenic in mammalian cell and microbial test systems, respectively (2,9). Recently, it has been demonstrated that an aqueous extract of respirable flyash particulates emitted from the stack of a coal-fired power plant was mutagenic in the *Salmonella* test system (1). In the present study, we report the biological effects of several different flyash particulates and aqueous extracts of those particulates in a regimen of microbial and mammalian cell systems.

MATERIALS AND METHODS

Sample Collection

Samples were collected from a stack post downstream from the electrostatic precipitator (ESP). Fly ash, designated as sample 1, was generated from pulverized low-sulfur Kentucky coal burned in a conventional cyclone-type burner. Fly ash designated as sample 2 was generated from an oil-fired power plant with no ESP, and consisted of pure Venezualian oil with a high concentration of vanadium (approximately 400 ppm). Sample 3 particulates were generated from a plant using low-sulfur Montana coal with limestone scrubbers and ESP.

In all cases, samples were collected using a high-volume cyclone-type sampler. Stack gas and dust were drawn into a bag collector using a centrifugal blower, with the stack dust being captured in a Teflon-coated fiberglass cloth bag. The sampling vessel was heated to a temperature above the dew point of the stack moisture (200°F) so that condensation would neither contaminate nor alter the sample. The sample collection probe was comprised of two sections of 0.5 in. i.d. stainless-steel pipe filtered with a 1 in. i.d. hood-type nozzle.

Sample Extraction

The fly ash samples were sized to exclude particles greater than 10 microns in diameter and transported in dark glass containers. On reception they were extracted with double-distilled demineralized water and incomplete Minimum Essential Medium with Earle's Salts (MEM) or Basal Medium

Eagles with Earle's Salts (BME) medium specific for the mammalian cell system utilized. Ten grams of each sample was extracted with 40 ml of water at ambient temperature for 2 hr with constant stirring. The resulting mixture was centrifuged at 1,000 g for 10 min and the supernatant filter sterilized with a 0.20-μ Nalgene (TM) filter unit.

In Vitro Bioassay Profile

The general test profile adapted for this study is shown in Figure 1. Selected tests from this profile (*e.g.*, a minimum of one from each group) were used to measure specific types of biological activities of the three parent samples and subsequent extracts.

Rabbit Alveolar Macrophage Assay

This assay system utilizes primary alveolar macrophages acquired through lung lavage and is used to define acute cellular toxicity of airborne particulates (11). Briefly, macrophages were harvested from healthy 1.5-2.0 kg New Zealand white rabbits sacrificed by intravenous injection of 0.5 ml of T61 (TM) euthanizing solution. A tracheotomy was performed and 30 ml of sterile prewarmed 0.14 *N* saline infused for 15 min. Five additional sequential lavage cycles were performed with the lavage fluid being filtered through sterile cheesecloth into a 250-ml polypropylene centrifuge bottle.

The pooled suspension was centrifuged at 265 x g for 15 min, resuspended in 10 ml of sterile 0.85% saline, and a differential count conducted using phase microscopy. The cell preparations contained a minimum of 95% alveolar macrophages with no appreciable red cell contamination. The cell suspension were placed in 60-mm tissue culture plates, each containing 4 ml of complete M199 medium. After 2 hr, the medium was decanted and 5 ml of fresh medium added to each plate.

The test material was diluted to the desired concentration with M199 medium and 1-ml aliquots added to each plate. Initially, three concentrations of test material were used with three replicates per dose. The plates were incubated in a humidified 95% air, 5% CO_2 atmosphere at 37°C for 20 hr. The plates were rotated for the first hour to insure uniform exposure of the test material. At the end of the incubation period, the following parameters were measured as a function of exposure to the varying concentration levels:

1. total cell count
2. cell viability trypan blue (0.4%)
3. intracellular ATP concentration–Dupont 760 Luminescence Biometer

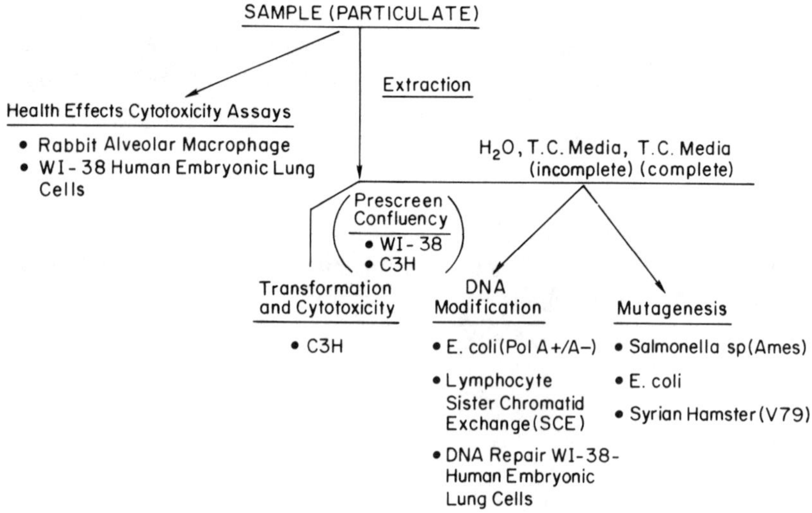

FIGURE 1. ***In vitro*** **bioassay profile of respirable particlates.**

Prescreen Confluency Assay (PSC)

This is a rapid cell toxicity assay system using any type of contact-dependent mammalian cell line, *e.g.* C3H10T1/2 mouse fibroblast, WI38 human lung fibroblast, BALB/c 3T3 fibroblast, etc, developed in our laboratory. This assay allows for rapid determination of toxicity of large numbers of materials using broad concentration ranges. The results of this assay were used to calculate the toxic dose 50 (TD_{50}) or the mean lethal dose (MLD_0). These values were initially determined in order to use the nontoxic concentration ranges for subsequent mutagenesis and carcinogenesis assays.

The assay utilizes Falcon (No. 3008) multiwell tissue culture plates. Each plate has 24 wells, each measuring 26 mm i.d. and 10 mm in depth. The wells are situated in 6 x 4 row configurations. Subconfluent monolayers of cells were established in the plate by seeding 5 x 10^3 cells per well and allowing them to attach and replicate for 24 hr. Subsequently, the growth medium was removed and replaced with fresh growth medium containing a known concentration of the material. Six of the wells (or one row) were assigned as control wells receiving no test material in the replacement medium. The cells were exposed for 24 hr after which the material was removed by aspiration, the cells washed in sterile saline, refed with fresh medium, and the plates reincubated. When the monolayers in the control wells reached 90-100% confluency (approximately 6 days), the cells were fixed with methanol and stained with

Giemsa, May-Gruenwald Giemsa or crystal violet. The plates then were scored on the basis of percentage of surface area covered by the cell monolayer relative to the control wells. The toxic dose 50 (TD_{50}) was expressed as a function of the concentration of the test material which would yield 50% confluency.

DNA Repair in WI38 Human Embryonic Lung Fibroblasts

This assay measures the amount of DNA damage in cells exposed to a subtoxic sample concentration gradient. The incorporation of a radioactive DNA precursor (^{3}H-thymidine) was used as a measure of repairable DNA damage (5), A liver microsomes activation system also was used as described.

Exposure Method

Sixty-mm plates at 6 X replicates per dose were seeded with 2×10^5 cells/plate and grown to confluency. The cells were then sequentially exposed to 0.6 m*M* hydroxyurea for 2 hr, then a mixture of the test sample and ^{3}H Tdr for 24 hr. Appropriate control cells were exposed to the test sample and pulsed with ^{3}HTdr.

DNA Isolation

All plates were washed three times in phosphate-buffered saline (PBS, pH 7.2) and subsequently lysed with, 1 *N* NaOH (2 ml/plate). The resulting solutions were neutralized with 1 *N* HCl (0.2 mg/plate). Subsequently, 2.2 ml of cold 10% trichloracetic acid (TCA) was added to each plate and then the plate was incubated at 4°C for 1 hr. The solutions were filtered through a GF/A filter and washed three times with TCA. The filters were dried for 8-12 hr and counted in a Nuclear Chicago Mark III Liquid Scintillation Counter.

ESCHERICHIA COLI POLYMERASE ASSAY

The *E. coli* pol A+/pol A- assay was performed according to the method of Rosencranz *et al.* (8). The assay was designed to score nucleic acid damage in mutant strains of the bacterium *Escherichia coli* which are deficient in nucleic acid repair enzymes (DNA polymerase).

Mammalian microsomal mixtures for use in metabolizing parent forms of carcinogens/mutagens to "proximate" carcinogens/mutagens were prepared as described in the (Ames) mutagenesis assay.

Liquid media (medium$_{HA}$) supplemented with 5 μg of thymine per ml was used for the cultivation of the *E. coli* stock cultures. The composition of this medium was as follows: K_2HPO_4,07%; KH_2PO_4, 0.3%; Na citrate, 0.5%

$MgSO_4\ 7H_2O$, 0.01%; $(NH_4)_2SO_4$, 0.1%; casein hydrolysate, 0.2%; and glucose, 0.2%. The glucose and thymine were sterilized separately and added aseptically.

The basal agar medium for the assay was prepared by the addition of 1.5% agar to the above basal minimal salt medium and poured into standard petri dishes in 25-ml volumes. The top agar overlays were prepared by the addition of 0.75% agar to the basal medium.

This test was performed in duplicate for both the parent *E. coli* W3110 thy- (pol A+) and its derivative p3478 (pol A_1-) for each concentration of sample. The determination was made in nonactivated and S-9 mixture activated preparations. The assay was conducted as follows: a 0.1-ml aliquot of exponentially growing bacterial culture of each *E. coli* strain was added to 2 ml of molten overlay medium-soft agar. The overlay agar then was poured onto the surface of a basal HA agar plate and slightly swirled to ensure agar coverage of the plate. After the soft agar solidified, a 0.8 cm i.d. well was cut into the agar of each plate. The plug was removed and the bottom of the well was sealed with 0.1 ml of molten agar.

For metabolic-activation assays, 50-μ aliquot of S-9 mixture was added to the well. A 50μl aliquot at the desired concentration of the test material was immediately added to the well and the plate incubated. For nonactivated assays, a 50μl aliquot of the solvent was added to the well immediately preceeding the addition of the same volume of the test material. All plates were subsequently incubated at 37°C for 24 hr. Following incubation, all plates were observed for the presence of a zone of inhibition surrounding the well. The diameter of this zone was measured, averaged and recorded.

AMES *SALMONELLA TYPHIMURIUM* MUTAGENESIS ASSAY

The complete test system utilizing five tester strains was carried out according to procedures described by Ames (6). Specifically, the plate-incorporation test with, and without, metabolic activation was performed on aqueous extracts of the three parent samples.

The S-9 microsomal mix contained per ml: S-9 (0.15 ml), $MgC1_2$ (8 μM), KCL (33 μM), glucose-6-phosphate (5 μM). Stock solutions of NADP (0.1 M) and glucose-6-phosphate were prepared with sterile water, aliquoted in appropriate amounts and maintained in a Revo freezer. The stock salt solutions were prepared, autoclaved and refrigerated. The S-9 mix was prepared fresh daily and maintained on ice before and during use.

MAMMALIAN CELL CLONAL TRANSFORMATION ASSAY (C3H10T1/2)

This assay was performed as described by Heidelberger *et al.* (7). It utilizes the C3H10T1/2 embryonic fibroblast (supplied by Dr. C. Heidelberger) as an indicator of malignant transformation of cells following postconfluence inhibition of cell growth.

The C3H10T1/2 cell line was maintained in BME containing 10% heat-inactivated fetal calf serum (GIBCO) and no antibiotics. Cells were grown in 25-cm^2 tissue culture flasks and subcultured weekly. The cells were subcultured using 0.1% Trypsin and reseeded into 25-cm^2 flasks at 5 x 10^4 viable cells/flask. The cells were routinely monitored for morphological consistency in terms of symmetry of mitotic figures, uniformity of size, degree of granulation, clarity of cytoplasmic boundary and saturation density.

The transformation assay was performed using 12 replicate plates per dose of test material and transformation was scored according to the method described by Heidelberger.

MEMBRANE PROTEIN STUDIES

C3H10T1/2 fibroblasts grown in 75-cm^2 falcon flasks were exposed to water extracts (300 mg/ml) of fly ash samples 1 and 3 for 24 and 48 hr. Parallel groups of cells were exposed to DMBA (2.5 μg/ml) for the same time period, while a fourth group was left untreated. The cells were harvested at the appropriate time intervals and the membrane proteins labeled with ^{125}I, as follows.

The confluent monolayers of C3H10T1/2 cells were washed three times with PBS followed by addition of 0.4 ml of a 200-mg/ml Lactoperoxidase solution of 50 μl of a 100-μg/ml Glucoseoxidase solution. Four milliliters of phosphate-buffered saline with 5 m*M* glucose was then added along with 5 μCi of ^{125}I-Carrierfree. The reaction was allowed to proceed for 10 min at ambient temperature after which the cells were scraped from the flask, placed in a centrifuge tube and washed 3 times with PBS by centifugation. The final pellet was resuspended in 0.3 ml of a 0.2-m*M* solution of phenylmethyl sulfonyl fluoride in the electrophoresis buffer which contains 0.1% SDS and 1% 2-mercaptoethanol.

Electrophoresis was performed with bromphenol blue as a tracking dye in 3.9% stacking and 7.5% resolving gel (4). Subsequent to drying, the gels were sliced (2 mm/slice) and counted in a Nuclear Chicago gamma counter.

RESULTS AND DISCUSSION

Acute Cellular Toxicity

In this screening mode it was necessary to measure multiple parameters in order to obtain an adequate data base for comparative analysis. The measurement of cell numbers or survival presented in Figure 2 show that the oil-based

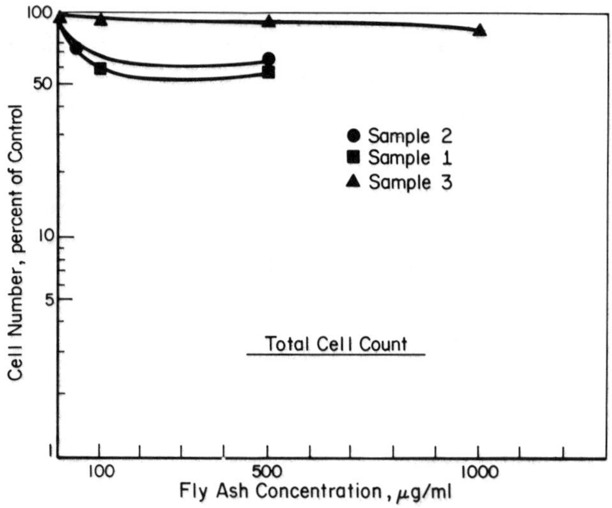

FIGURE 2. Rabbit alveolar macrophage response to coal-generated fly ash exposure.

and one coal-based fly ash induced similar toxicity profiles while the third coal fly ash sample did not.

Viability measurements presented in Figure 3 show a differential effect of the fly ash samples on alveolar macrophages. The oil-based fly ash elicited the most toxic repsonse with an LC_{50} of approximately 100 μg/ml, with fly ash sample 3 again exhibiting the least toxic profile. Measurement of intracellular ATP (seen in Figure 4) also indicates a differential response among the three fly ashes similar to the pattern in Figure 3.

Comparison of the cell survival, viability and ATP results indicate that fly ash samples 1 and 2 had similar effects relative to cell survival, but differential effects in membrane integrity (viability) and intracellular ATP concentration in the surviving cell population. The minimal cytotoxic response elicited by fly ash sample 3 indicates it is the least toxic in this system, while the concomitant decrease in intracellular ATP levels may be due to the energy-dependent phagocytosis of the fly ash particulates.

The dose-response curves of the three samples in the PSC are presented in Figure 5. The results indicate that samples 1 and 2 did contain water-soluble cytotoxic components, while sample 3 again elicited a minimal cytotoxic response. The dose-response information obtained in this test was used to establish the "low" toxic dose used in subsequent assays assessing potential genetic toxicity.

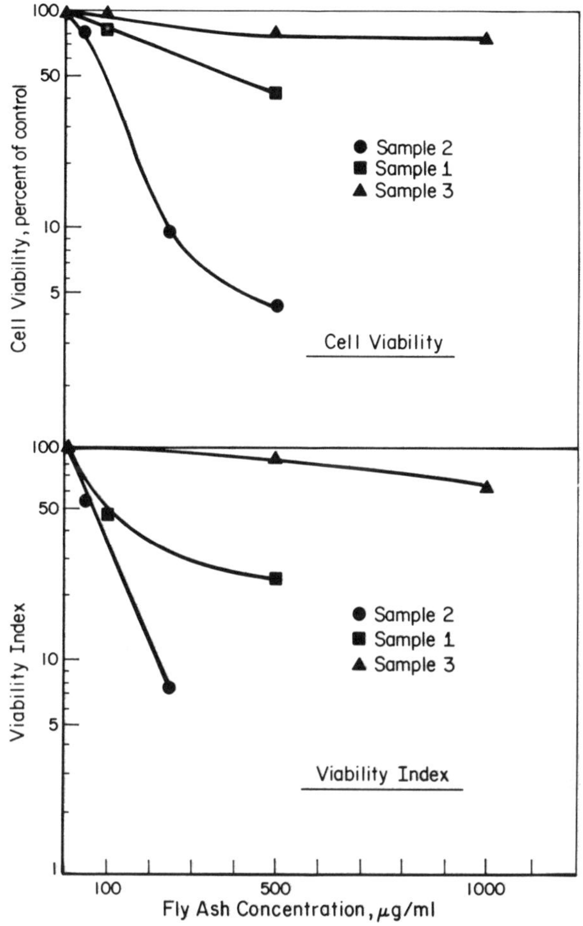

FIGURE 3. Viability of rabbit alveolar macrophages exposed to coal-generated fly ash.

PRIMARY ASSESSMENT OF NUCLEIC ACID EFFECTS

The ability of several aqueous-based extracts to damage DNA was monitored in the *E. coli* polymerase system as described by Rosenkrantz (8). Although there were inconsistent differential effects between activated and nonactivated samples, as shown in Figure 6, extracts of fly ash samples 1 and 2 did exhibit preferential toxicity to the polymerase negative strain, indicating damage specific to DNA. Extracts of fly ash sample 3 (not shown) did not produce any response in this system.

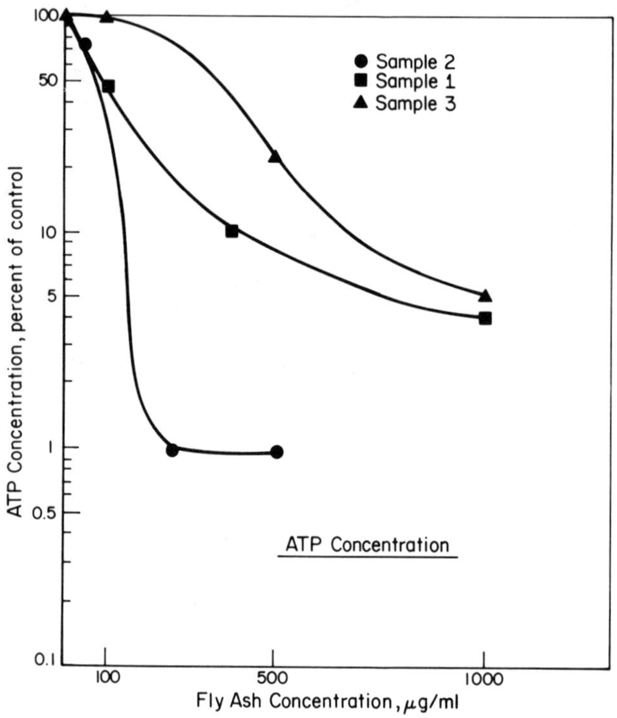

FIGURE 4. Effect of fly ash exposure on intracellular ATP concentration in rabbit alveolar macrophage.

Results of the uptake of ^{3}H-Tdr into the DNA of WI-38 human embryonic lung fibroblasts exposed to the water extracts are shown in Figure 7. Extracts of sample 2 did produce significant incorporation over control levels, while extracts of samples 1 and 3 did not increase incorporation of ^{3}H-Tdr over control levels. The combined results indicate that both samples 1 and 2 did contain water-soluble moieties having potential DNA-damaging capability.

Microbial Mutagenesis

None of the aqueous extracts elicited a positive response in the Ames plate incorporation test. The tester organisms did respond to positive control chemical listed in Table I.

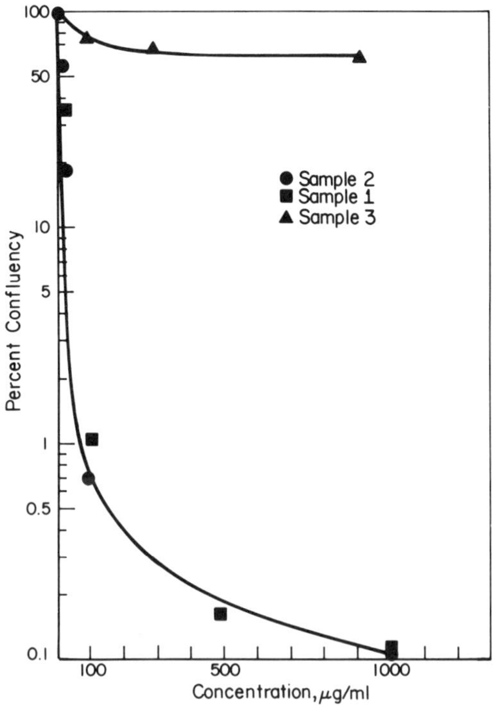

FIGURE 5. Cytotoxicity (PSC) in WI-38 human embryonic lung cells exposed to fly ash–water extract.

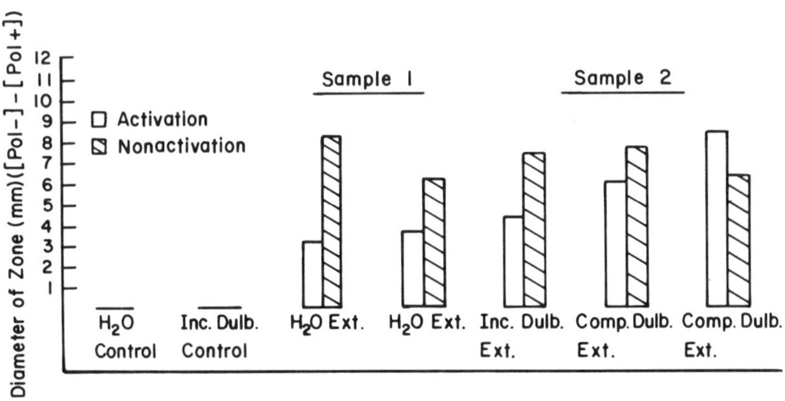

FIGURE 6. DNA modifying capability of fly ash extracts.

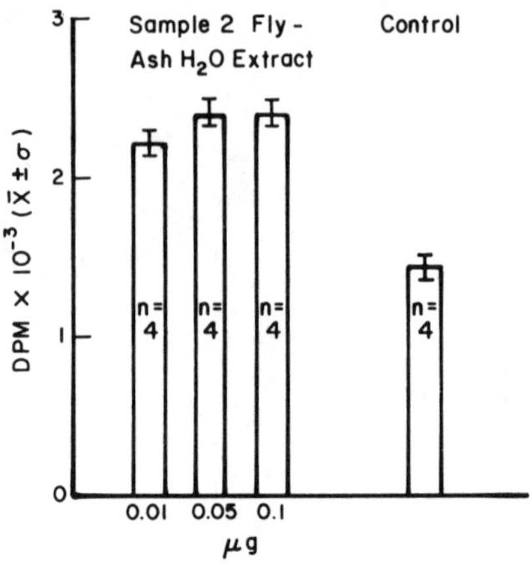

FIGURE 7. Uptake of ^{3}HTdR in WI-38 human embryonic lung fibroblasts exposed to fly ash-water extract.

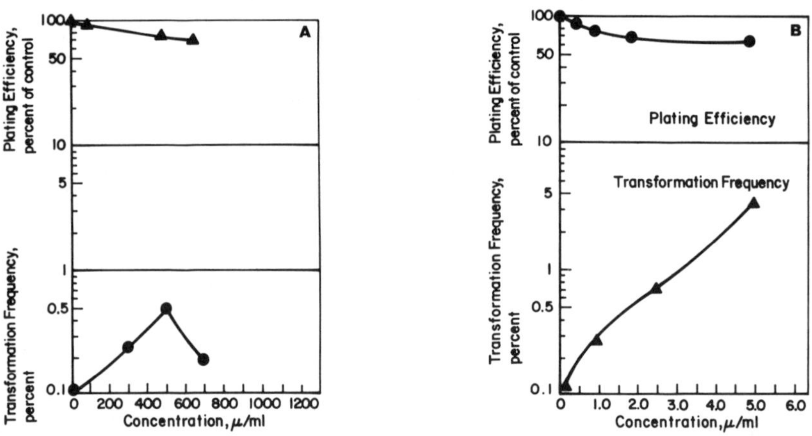

FIGURE 8. (A) C3H transformation frequency and cytotoxicity as a function of exposure to fly ash-water extract (sample 1); (B) Cytotoxicity and transformation frequency of DMBA as a function of concentration.

Mammalian Cell Clonal Transformation Assay

None of the sample extracts exhibited statistically significant positive responses in the C3H10T1/2 transformation assay. One sample, as shown in Figure 8, did exhibit an initial linear increase in transformation frequency which changed discontinuously to a negative slope at a concentration >500 μg/ml.

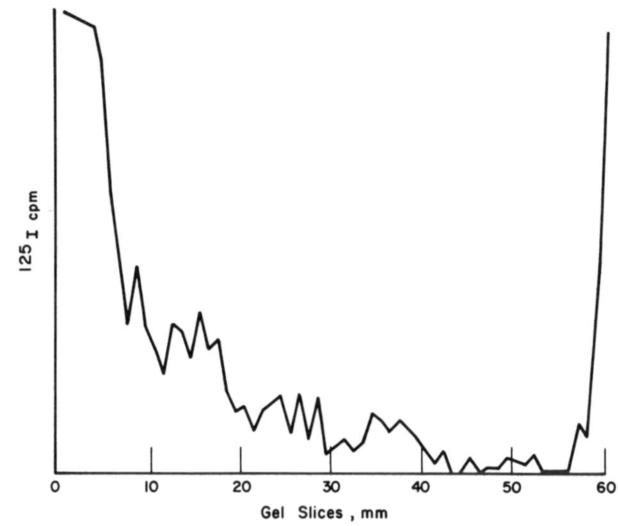

FIGURE 9. Membrane protein profile of C3H10T1/2 fibroblasts.

The slope change may be a reflection of increasing cytotoxicity; however, further exposure studies were necessary to define the results.

Parallel positive-control exposures to dimethylbenzanthracene (DMBA) did produce a linear increase in transformation frequency confirming the ability of the cells to undergo *in vitro* transformation, while negative control (unexposed) cells had a low background or spontaneous transformation frequency of approximately 0.1%.

Based on the overall results, sample 1 was judged as being highly suspect and was used in secondary *in vitro* studies designed to further study exposure on a molecular level.

Membrane Protein Studies

The membrane protein profile of the untreated control cells showed little variation from experiment to experiment using the same passage cells. A representative profile is shown in Figure 9. The membrane protein profiles of the DMBA exposed cells are shown in Figure 10. As shown, a significant change in the labeling pattern was evident after a 24-hr exposure and continuing through the 48-hr period. The cells exposed to the fly ash sample 1 water extract also exhibited a change in the labeling pattern as shown in Figure 11, while the cells exposed to a water extract from fly ash sample 3 did not exhibit a change in labeling pattern compared to the untreated control cells (Figure 12).

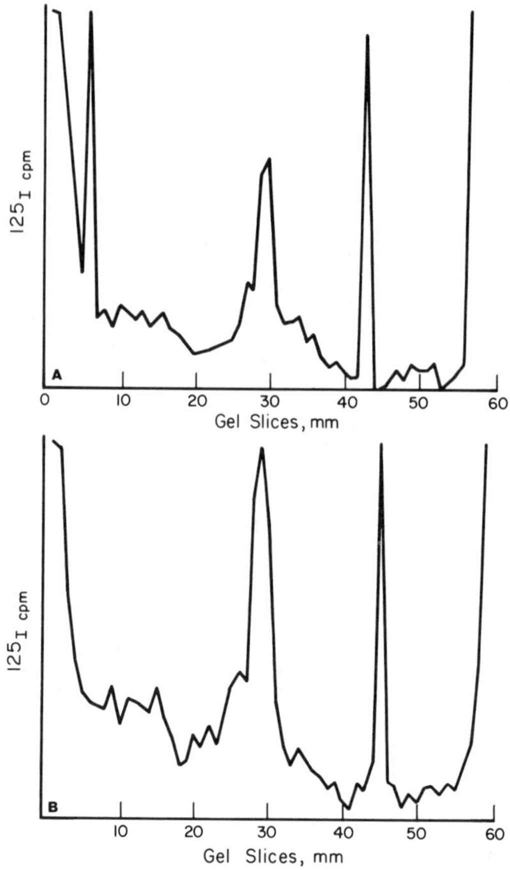

FIGURE 10. (A) Membrane protein profile of C3H10T1/2 fibroblasts exposed to DMBA for 24 hr; (B) Membrane protein profile of C3H10T1/2 fibroblasts exposed to DMBA for 48 hr.

These "pretransformation" membrane changes correlate closely with the differential cytotoxic and genetic toxicity results noted for the three fly ash samples. This data is not sufficient to determine whether the quantative membrane changes are due to increased protein synthesis or exposure of previously unavailable labeling sites. There are indications that these changes are heritable and that further cell kinetic studies are prerequisite to our understanding of these effects.

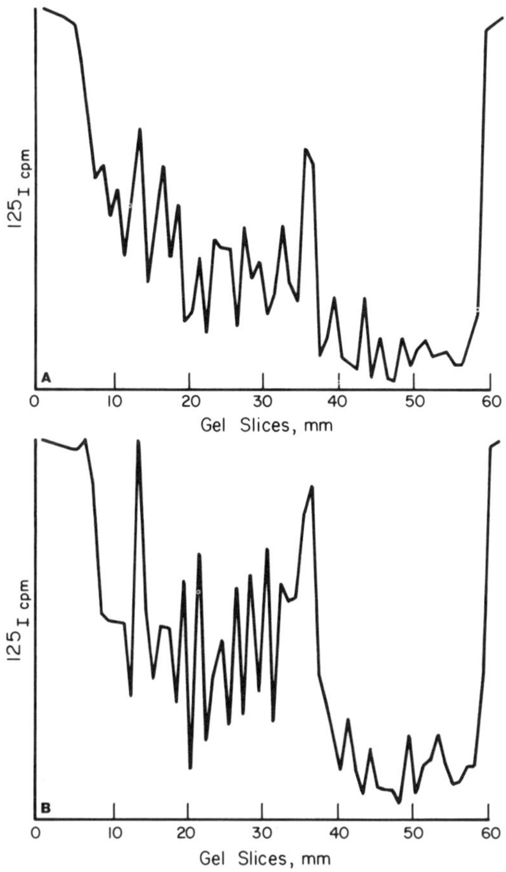

FIGURE 11. (A) Membrane protein profile of C3H10T1/2 fibroblasts exposed to sample 1 fly ash-water extract for 24 hr; (B) Membrane protein profile of C3H10T1/2 fibroblasts exposed to sample 1 fly ash-extract for 48 hr.

TABLE I. Positive Control Chemicals

Indicator Strain	Nonactivation Assays	Activation Assays
TA-1535	sodium azide	2-aminoanthracene
TA-1537	quinacrine hydrochloride	2-aminoanthracene
TA-1538	2-nitrofluorene	2-aminofluorene
TA-98	2-nitrofluorene	2-aminofluorene
TA-100	2-nitrofluorene	2-aminofluorene

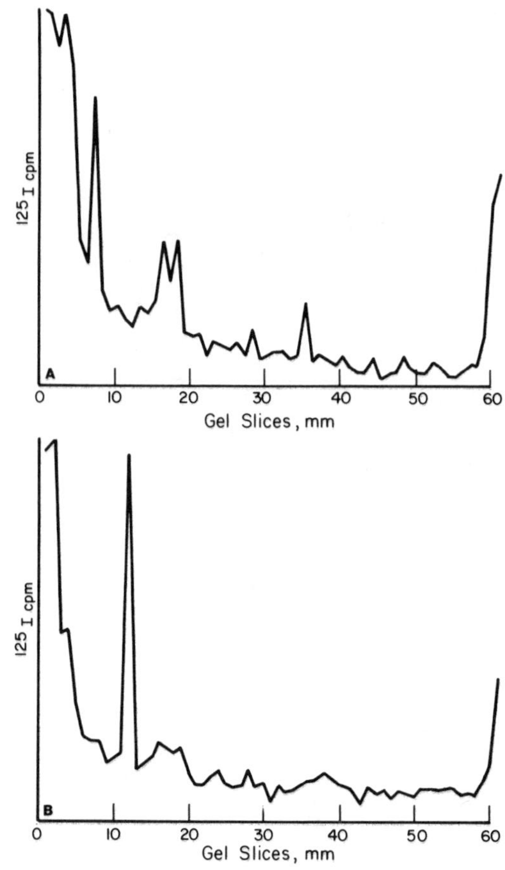

FIGURE 12. (A) Membrane protein profile of C3H10T1/2 fibroblasts; (B) Membrane protein profile of C3H10T1/2 fibroblasts exposed to sample 3 fly ash-water extract for 24 hr.

CONCLUSIONS

While negative microbial (Ames) mutation results were obtained, a battery of *in vitro* tests did identify two parent samples which contained water-soluble moieties having both cytotoxic and genetic toxicity potential. The variable results obtained in detecting DNA damage may be a reflection of mutagen specificity or inherent constraints present in the test systems.

Observed activity of a particular compound may be due to related factors such as diffusion, membrane transport and protein binding which may vary in different test systems.

Specifically, a noncomparable result between the *E. coli* PolA+A- assay and the WI-38 DNA repair assay may not be an indication of lack of activity. The effective dose or chemical configuration of componenets of complex mixtures at the nucleic acid level for these two widely varying cell types could be significantly different. Additionally, the mechanisms of these two processes vary greatly in complexity even though mucleic acid effects form the basis for the endpoint of both assays.

Similar considerations apply to the analysis of cytotoxicity data. It is theoretically possible that both synergistic and antagonistic chemical reactions occur in these complex environmental mixtures in the physiological environment. This again might alter the effects observed in one or more types of assays which would be sensitive to the compound modifications. Thus, it is mandatory that a spectrum of assays be applied for primary assessment of any complex environmental material. The results obtained from a primary screening profile may be applied in choosing further *in vivo* or *in vitro* test systems having specific endpoints in order to define multicellular exposure effects on subcellular mechanisms.

ACKNOWLEDGMENTS

This research was supported in part by the Electric Power Research Institute (EPRI). We would like to thank Ms. Ann Aukerman, Ms. Sue Gruebel, Mr. Tim Hutson and Ms. Joan Pitman for their expert technical assistance.

REFERENCES

1. Chrisp, G. E., Fisher, G. L., and Lammert, J. E. (1978): Mutagenicity of filtrates from respirable coal fly ash. *Science* 199:73-75.
2. Dreesen, D. R., Gladney, E. S., Owens, J. W., Perkins, B. L., Wienke, C. L., and Wangen, L. F. (1977): Comparison of levels of trace elements extracted from fly ash and levels found in effluent waters from a coal-fire power plant. Environmental Science and Technology (11)10:1017-1019.
3. Gordon, R. J., Bryan, R. J., Rhim, J. S., Demoise, E., Wolford, R. G., Freeman, A. E., and Huebner, R. J. (1973): Transformation of rat and mouse embryo cells by a new class of carcinogenic compounds isolated from particles in city air. *Int. J. Cancer* 12:223-232.
4. Laemmli, U. K. (1970): Cleavage of structural proteins during the assembly of the head of bacteriophage T4. *Nature* 227:(August 15).
5. Lieberman, M. W., Baney, R. N., Lee, R. E., Sell, S., and Farber, E. (1971): Studies of DNA repair in human lymphocytes treated with proximate carcinogens and alkylating agents. *Cancer Res.* 31:1297-1307.

6. McCann, J., and Ames, B. N. (1976): Detection of carcinogens as mutagens in the *Salmonell*/microsome test: Assay of 300 chemicals: Discussion. *Proc. Nat. Acad. Sci., U. S.* 73:950-954.
7. Reznikoff, C. A., Bertraus, J. S., Brankow, D. W., and Heidelberger, C. (1973): Quantitative and/qualitative studies of chemical transformation of cloned C3H mouse embryo cells sensitive to postconfluence inhibition of cell division. *Cancer Res.* 33:3239-3249.
8. Rosenkranz, H. S., Gutter, B., and Speck, W. T. (1976): Mutagenicity and DNA modifying activity: A comparison of two microbial assays. *Mutat. Res.* 41:61-70.
9. Sheridan, M. A., Axler, D. A., Dennis, A. J., and Mitchell, R. I. (1974): Transformation of nonvirus producing BALB/3T3 cells by an environmental extract. In: *Experimental Lung Cancer Carcinogenesis and Bioassays,* E. Karbe and J. F. Park, Eds., Springer-Verlag, Berlin.
10. Tomatis, L., Agthe, C., Bartsch, H., Huff, J., Ruggero, M., Saracci, R., Walker, E., and Wilbourn, J. (1978): Evaluation of the carcinogenicity of chemicals: A review of the monograph program of the International Agency for Research on Cancer (1971 to 1977). *Cancer Res.* 38:877-885.
11. Waters, M. D., Vaughan, T. O., Abernethy, D. J., Garland, H. R., Cox, C. C., and Coffin, D. L. (1975): Toxicity of platinum (IV) salts for cells of pulmonary origin. *Environ. Health Pers.* 12:45-56.

Polynuclear Aromatic Hydrocarbons,
edited by P. W. Jones and P. Leber.
Ann Arbor Science Publishers, Inc.
Ann Arbor, MI ©1979.

IMMUNOCHEMICAL CHARACTERIZATION OF RAT LIVER CYTOCHROMES P450 AND EPOXIDE HYDRASE

Paul E. Thomas, Dene Ryan and Wayne Levin

Department of Biochemistry and Drug Metabolism
Hoffmann-La Roche Inc.
Nutley, New Jersey 07110

MULTIPLE FORMS OF CYTOCHROME P450

Properties of Polyspecific Antibodies

Three components function in the liver microsomal monooxygenase system: NADPH-cytochrome P450 reductase, phospholipid and cytochrome P450 (14,15). The substrate specificity of the system is determined by the cytochrome P450 component (15,17). The presence of multiple forms of cytochrome P450 with different but often overlapping substrate specificities explains the apparent broad metabolic activity of the monooxygenase system (15,17). Many factors influence the tissue levels of individual forms of cytochrome P450 such as sex, age, nutrition, hormonal levels and pretreatment with inducers (3,5,20). Understanding the regulation of these forms is important because the tissue levels of the enzymes determine the *in vivo* capacity to metabolically activate and inactivate many carcinogens, mutagens, drugs and xenobiotics assimilated by an organism. Clearly, tissue levels of a particular form of cytochrome P450 may be a critical determinant of whether a compound is carcinogenic, mutagenic or toxic in that tissue. The lack of specific assays for a single form of cytochrome P450 has hindered the elucidation of the regulation of multiple forms of cytochrome P450. Antibodies are highly specific and therefore uniquely suited as probes to identify and discriminate between functionally similar proteins such as the multiple forms of cytochrome P450. We shall also show in this chapter that the specificities of these antibodies can be used to quantitate, localize and purify both epoxide hydrase and cytochrome P450.

We have prepared antibodies against highly purified hepatic cytochrome P450 and P448 isolated from phenobarbital (PB)-treated rats (33) and 3-

methylcholanthrene (MC)-treated rats (32), respectively. Although the cytochromes P450 were highly purified, they contained more than one form of cytochrome P450 and the antibodies obtained were not monospecific. These antibodies are nevertheless highly specific for cytochrome P450 and differentially inhibit the metabolism of xenobiotics in microsomes (34) and the purified reconstituted system (32,33) depending on the substrate metabolized and the source of cytochrome P450. We have been able to demonstrate the presence of at least six forms of cytochrome P450 in the livers of rats with our two antibody preparations (33). These antibodies do not react with other microsomal enzymes, but do recognize more than one form of rat liver microsomal cytochrome P450 and are therefore not monospecific. If immunochemical approaches are to be highly useful for quantitation, localization, etc., the antibodies must be monospecific.

Monospecific Antibodies: Purification and Location of Cytochrome P450

Monospecific antibody to the major form of hepatic cytochrome P450 from PB-treated rats has been prepared from polyspecific antibody by immunoabsorption with solid-phase partially purified cytochrome P450 from MC-treated rats (31). This monospecific antibody is directed against the major form of cytochrome P450 present in PB-treated rats (PB-$P450_b$), and does not recognize other forms of cytochrome in liver microsomes from either PB- or MC-treated rats was made monospecific for the major form in these rats (MC-$P450_c$) by immunoabsorption with solid-phase partially purified cytochrome P450 from PB-treated rats. It is important to note that the immunoabsorbants used are low in the homologous antigen but posses the necessary cross-reacting forms of cytochrome P450. Using these monospecific and polyspecific antibodies, we have obtained immunodiffusion evidence that PB-$P450_b$, MC-$P450_c$ and two additional forms of P450 are not immunochemically related (31).

The monospecific anti-MC-$P450_c$ has also been useful in establishing the immunochemical identity of the nuclear and microsomal sources of MC-$P450_c$ (30). We have also obtained immunohistochemical evidence for the *intranuclear* localization of a significant amount of the total nuclear MC-$P450_c$ (1). This intracellular location of the monooxygenase system may be important for carcinogen or mutagen activation at sites adjacent to the primary target (DNA) for these reactive intermediates.

Since these monospecific antibodies (anti-MC-$P450_c$ and anti-PB-$P450_b$) recognize only a single form of cytochrome P450, they should be ideal tools for examining the properties and amounts of a single form of the enzyme in the presence of many forms. For example, when covalently bound to Sepharose, these monospecific antibodies can selectively remove the corresponding form of P450 from a crude microsomal preparation in a single step.

The cytochrome P450 can subsequently be recovered from the immunoaffinity column by elution with a chatropic salt such as 4 *M* KSCN. Some of our results using this procedure with the two monospecific antibodies to recover single forms of cytochrome P450 from detergent-solubilized microsomes are shown in Figure 1. The prominent band in well 3 of the sodium

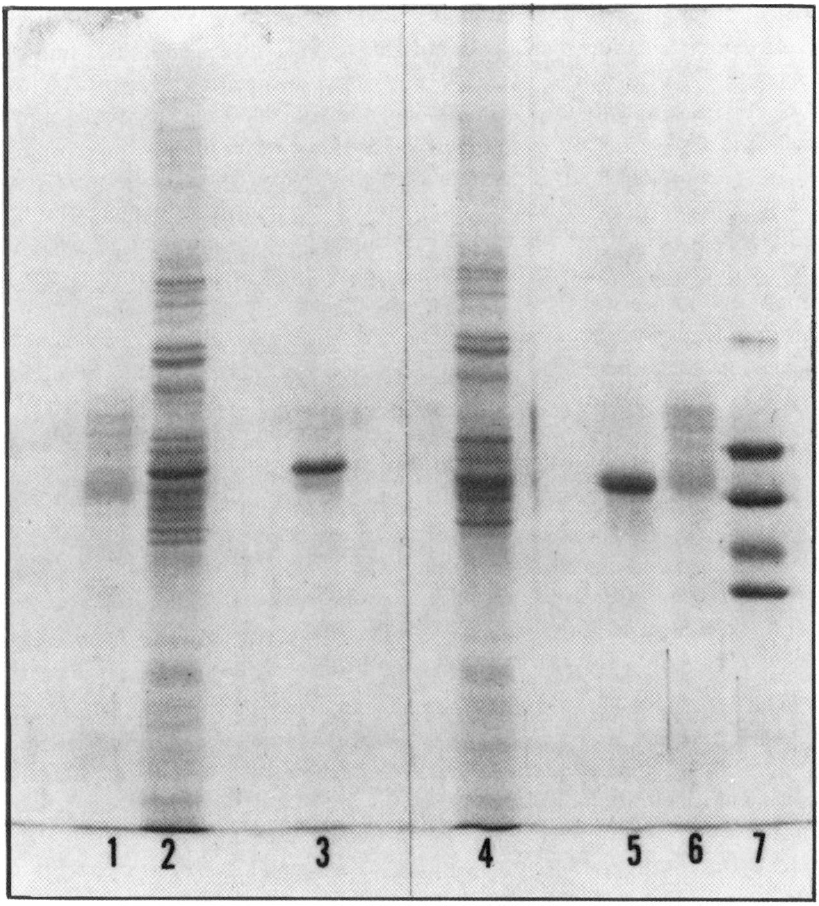

FIGURE 1. SDS PAGE results of one step immunoaffinity purification of PB-P450$_b$ and MC-P450$_c$ from rat liver microsomes. See text for details. Samples: (1) 4 *M* KSCN elution of a control IgG column, 0.6 μg; (2) liver microsomes from PB-treated rats, 6 μg; (3) immunoaffinity purified PB-P450$_b$ from microsomes in well 2, 1 μg; (4) liver microsomes from MC—treated rats, 6 μg; (5) immunoaffinity purified MC-P450$_c$ from microsomes in well 4, 1 μg; (6) 4 *M* KSCN elution of a control IgG column, 0.6 μg; (7) protein standards with molecular weights of 58,000, 53,000, 45,000 and 41,000 (top to bottom).

dodecyl sulfate (SDS) polyacrylamide gel electrophoresis (PAGE) pattern is MC-P450$_c$ purified from liver microsomes of MC-treated rats (well 2). The minor contaminants present with MC-P450$_c$ in well 3 are the result of antibody leaching from the immunoaffinity column. Well 1 contains the protein removed with 4 *M* KSCN from a control immunoaffinity column which had not previously been exposed to any antigen. The minor bands in well 3 correspond in mobility with those in well 1. In the same manner, PB-P450 (well 5) can be purified in a single immunoaffinity column step from liver microsomes of PB-treated rats (well 4). Both immunoaffinity purified proteins coelectrophorese with the corresponding form of P450 purified by conventional techniques (results not shown). While these forms of cytochrome P450 purified by immunoaffinity chromatography are catalytically inactive, the technique is nevertheless useful. Forms of P450 purified by immunoaffinity chromatography can be subjected to a variety of physicochemical analyses such as SDS-PAGE (shown in Figure 1) or amino acid sequencing. Heterogeneity of cytochrome P450 from various tissues, organelles or animals with differing xenobiotic pretreatment could be assessed by the physicochemical properties of the immunoaffinity purified protein. The specificity of these antibodies could also be used to study the turnover of a single form of cytochrome P450. Finally, we have used immunoaffinity chromatography to remove unwanted forms of cytochrome P450, epoxide hydrase, etc. from partially purified preparations when there is no other readily available technique.

Quantitation of Individual Forms of Cytochrome P450 in Rat Liver Microsomes

No routine techniques have been successful in reliably quantitating the amounts of different forms of cytochrome P450 in microsomal preparations, although estimates have been made using substrate binding spectra (35). The electrophoretic profiles of microsomal preparations on SDS-polyacrylamide gels are difficult to interpret, since microsomal proteins other than cytochromes P450 are known to have similar minimum molecular weights (2,18, 21,29). Different cytochromes P450 have overlapping substrate specificities which exclude an accurate quantitation of these enzymes based on catalytic activity.

Figure 2 shows the quantitation of two different forms of cytochrome P450 in different microsomal preparations by a radial immunodiffusion assay with the monospecific antibodies described above. The assay has been adapted (31) from a previously described method (19) based on the diffusion of the antigen into a gel saturated with antibody. Diffusion of the antigen ceases when the antibody has been consumed. The area of the resultant

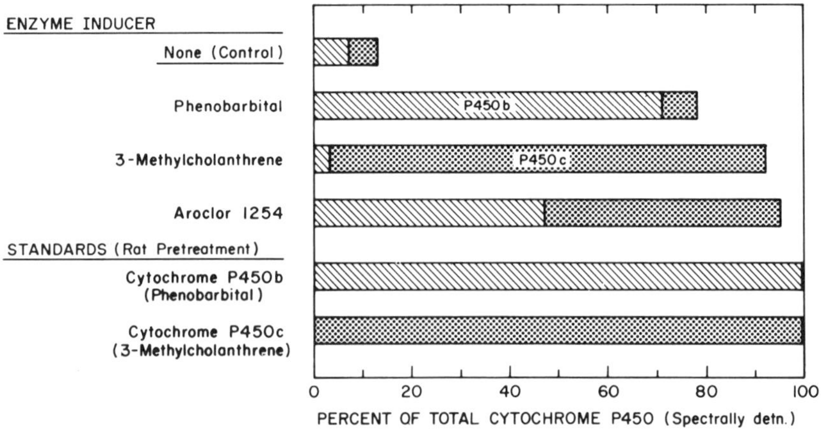

FIGURE 2. Radial immunodiffusion assay of two forms of cytochrome P450 in liver microsomes from rats pretreated with different enzyme inducers.

precipitin ring is directly proportional to the amount of the antigen. The amount of each antigen in different microsomal preparations can be determined since both electrophoretically homogeneous standards (26) and microsomes are proportional to enzyme concentration over approximately a 16-fold range (31).

The amount of each cytochrome P450 is expressed in Figure 2 as a percentage of the total microsomal cytochrome P450 determined spectrally (25). Approximately 70% of the total cytochrome P450 in microsomes from PB-treated rats is immunochemically identical to PB-$P450_b$. Similarly, about 90% of the microsomal cytochrome P450 from MC-treated rats is MC-$P450_c$. Clearly, both inducers preferentially induce a single form of cytochrome P450. Aroclor 1254, a polychlorinated biphenyl mixture, induces approximately equal amounts of the two forms of cytochrome P450, which agrees with earlier results showing that Aroclor 1254 induces forms of cytochrome P450 indistinguishable from those induced by a mixture of MC and PB (27).

Liver microsomes from untreated rats (control) contain less than 10% of either form of cytochrome P450 (Figure 2). Thus, more than 80% of the cytochrome P450 in microsomes from control rats is unrelated to the major forms of P450 induced by PB and MC. It is unknown whether this uncharacterized 80% is composed of one or several forms of cytochrome P450, but the results of SDS-PAGE suggests the latter possibility. These results are consistent with previous observations showing that partially purified cytochrome P450 preparations from control, MC- and PB-treated rats have different substrate specificities (17).

Immunochemical Analyses of Three Forms of Rat Liver Microsomal Cytochrome P450

A third form of cytochrome P450, designated $P450_a$, has recently been isolated from liver microsomes of rats treated with PB (PB-P450), MC (MC-$P450_a$) and Aroclor 1254 (Aro-$P450_a$) (26). Cytochrome $P450_b$ has been isolated from both PB-treated rats (PB-$P450_b$) and Aroclor 1254-treated rats (Aro-$P450_b$). Cytochrome $P450_c$ has been obtained from MC-treated rats (MC-$P450_c$) and Aroclor 1254-treated rats (Aro-$P450_c$). Each form of cytochrome P450 shows a single protein-staining band on SDS gels with minimum molecular weights of 48,000 ($P450_a$), 52,000 ($P450_b$) and 56,000 ($P450_c$). The three forms of cytochrome P450 also differ in their substrate specificities, spectral properties and protease digestion patterns (26).

Figure 3 shows double diffusion analyses of the purified cytochromes P450 using antibodies against three forms of the enzyme. The peripheral wells in each plate are filled with an electrophoretically homogenous form of cytochrome P450, as described in the legend. The antibody in the center well of plate A is directed against Aro-$P450_a$. In the center wells of plates B and C are the monospecific antibodies to PB-$P450_b$ and MC-$P450_c$, respectively. Plate A shows that cytochromes $P450_a$ isolated from Aroclor 1254-(well 1), PB- (well 2) or MC- (well 3) treated rats are immunochemically identical. Neither cytochrome $P450_b$ nor $P450_c$ react with this antibody. As shown in plate B, cytochrome $P450_b$ from PB-treated rats is antigenically identical to that enzyme isolated from Aroclor 1254-treated rats. The monspecific antibody to cytochrome $P450_b$ does not recognize either cytochrome $P450_a$ or $P450_c$. Likewise, MC-$P450_c$ is immunologically indistinguishable from Aro-$P450_c$ (plate C), and further, cytochromes $P450_a$ and $P450_b$ do not cross react with the monospecific antibody to MC–$P450_c$. Therefore, cytochromes $P450_a$, $P450_b$ and $P450_c$ are distinct proteins as judged by their immunochemical properties.

Multiple forms of some enzymes arise from post-translational processing. If post-translational processing were responsible for the generation of these three forms of cytochrome P450, they would necessarily have large domains of homology in their amino acid sequence. Since antibodies are directed against several distinct and spatially distant surface determinants, the absence of shared immunochemical determinates between these three forms of cytochrome P450 implies the absence of significant regions of surface homology. Consequently, it is unlikely that these three forms of cytochrome P450 arise by post-translational processing. Alternatively, one would have to postulate such extensive processing that all antigenic sites were removed and a new set exposed, but this would not be consistent with the small differences of 4,000 daltons in minimum molecular weight between the different forms.

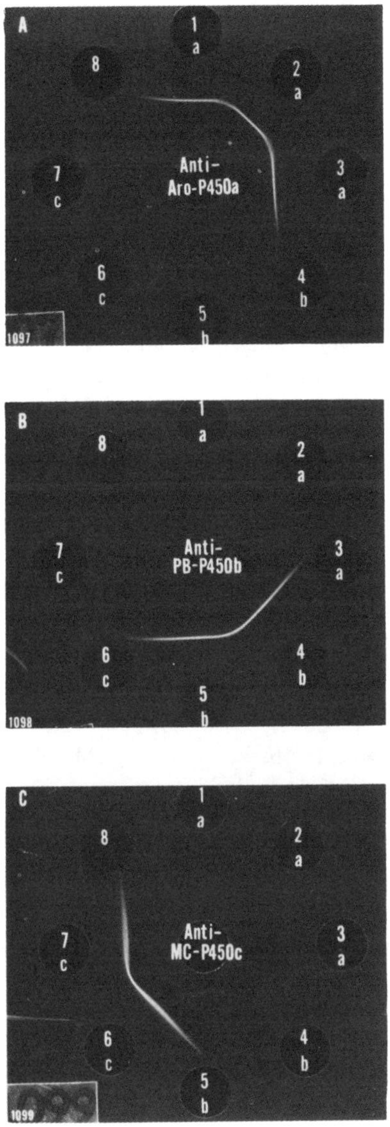

FIGURE 3. Immunodiffusion analysis of three forms of cytochrome P450 using antibodies specific for each form. The center well of plate A contains antibody against Aro-P450$_a$. The center wells in plates B and C contain monospecific antibody against PB-P450$_b$ and MC–P450$_c$, respectively. The peripheral wells of all three plates contain the following forms of purified cytochrome P450 in wells numbered: (1) 3.3 μM Aro-P450; (2) 2.9 μM PB-P450$_a$, (3) 3.3 μM MC-P450$_a$, (4) 3.8 μM Aro-P450$_b$, (5) 4.2 μM PB-P450$_b$, (6) 4.2 μM Aro-P450$_c$, (7) 4.2 μM MC-P450$_c$, (8) phosphate-buffered saline.

Therefore, each of the three forms of cytochrome P450 are probably products of distinct genes.

EPOXIDE HYDRASE

Immunochemical Quantitation of Epoxide Hydrase

Similar to the hepatic monooxygenase system, microsomal epoxide hydrase has a broad substrate specificity (9,13,22) and is capable of activating and/or inactivating various carcinogens, mutagens and other toxic substances (8,28,36). Our laboratory has purified and characterized rat liver epoxide hydrase (18) and recently obtained antibodies to the enzyme from four animal species (11). Interestingly, the antibodies do not inhibit the catalytic activity of the enzyme. The results of immunodiffusion studies indicated that liver microsomal epoxide hydrase from untreated rats is immunochemically identical to the enzyme in hepatic microsomes from PB-, MC- or pregnenolone 16α-carbonitrile-treated rats (11).

The catalytic activity of epoxide hydrase is dependent on assay conditions (13) including substrate, lipid and protein concentration. For example, the ratio of the specific activity of the purified enzyme to microsomes varies from 13 for dibenzo[a,h]anthracene 5,6-oxide to 45 for styrene 7,8-oxide or naphthalene 1,2-oxide (13). Therefore, catalytic activity cannot be used to accurately quantitate epoxide hydrase content in microsomal preparations.

As shown in Figure 4, the highly specific antibodies can be used to determine the amount of epoxide hydrase in liver microsomal preparations and hyperplastic nodule microsomes from 2-acetylamino fluorene (AAF) pretreated rats by a radial immunodiffusion assay. The values obtained for pure epoxide hydrase and microsomes are proportional to enzyme concentration over a large range. The results of the quantitation are compared with the catalytic activity of epoxide hydrase in Table I. These results indicate that the amount of epoxide hydrase in liver microsomes from phenobarbital-treated rats is increased threefold over control values, as judged by the hydration of octene oxide or the radial immunodiffusion assay. This data demonstrates that phenobarbital treatment of rats results in an increase in the amount of epoxide hydrase enzyme rather than a modulation of catalytic activity.

This laboratory, in collaboration with Drs. M. J. Griffin and D. E. Kizer, has recently identified the PN antigen (23,24) present in hyperplastic nodules from hepatic microsomes of rats fed AAF as epoxide hydrase (10). Both the antibody to PN antigen and the antibody to epoxide hydrase were used to establish this protein as epoxide hydrase in conjunction with electrophoretic profiles and catalytic activity (10). As shown in Table I, the amount of epoxide hydrase increases five- to seven-fold after feeding 0.05% AAF for 1 week to rats.

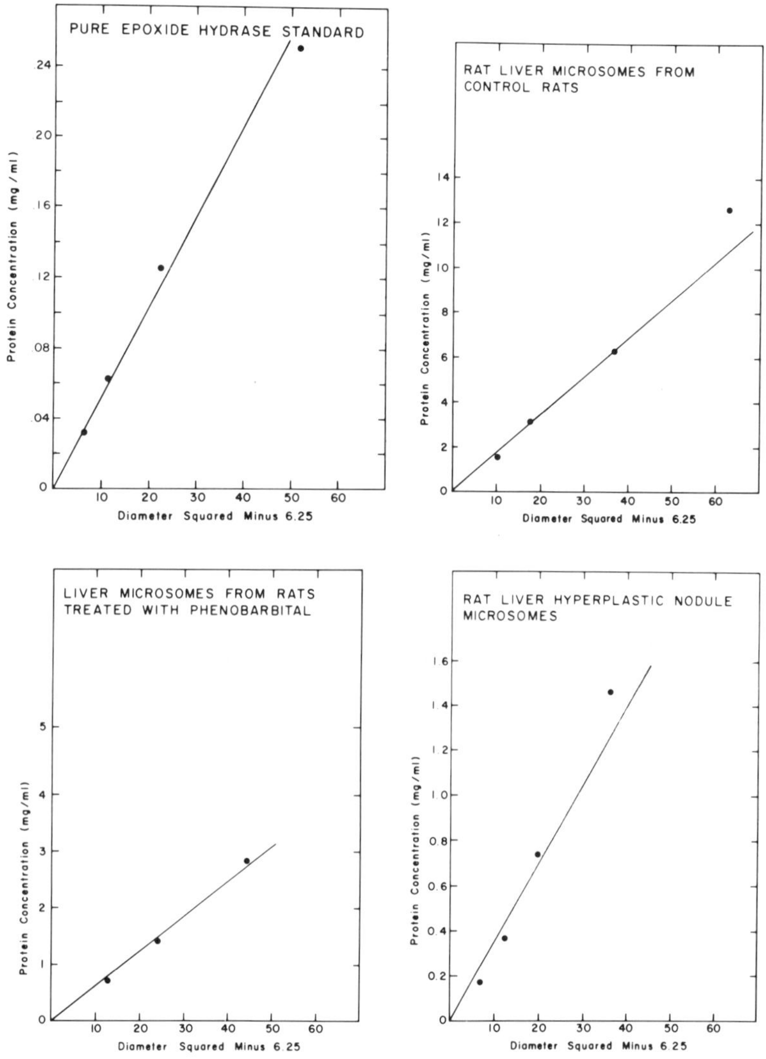

FIGURE 4. Radial immunodiffusion results using gels impregnated with antibody to epoxide hydrase. The abscissa gives the square of the immunodiffusion ring diameter in mm at equilibrium minus the square of the well diameter. The protein concentration on the ordinate is total protein and was estimated by the method of Lowry *et al.* (12) using bovine serum albumin as standard.

TABLE I. A Comparison of the Radial Immunodiffusion Quantiation of Epoxide Hydrase with the Catalytic Activity in Liver Microsomes and Hyperplastic Nodules from Drug-Treated Rats

Drug Pretreatment	Epoxide Hydrase (mg/mg protein)	Epoxide Hydrase Activity Octene Oxide (nmole/min/mg protein)
Long Evans Rat Liver Microsomes		
Control (Untreated)	0.015	16.3
Phenobarbital	0.045	45.2
Sprague Dawley Rat Microsomes		
Control (Untreated)	0.018	13.4
2-Acetylaminofluorene (1 week)	0.084	85.6
Hyperplastic Nodules	0.079	93.2

Note: The concentration of epoxide hydrase was determined from the results in Figure 5. Catalytic activity was assayed by a radiometric method at 37° (9).

When rats are given AAF in a discontinuous feeding schedule for 13 to 16 weeks, they develop hyperplastic nodules which have been characterized as precursor lesions of hepatomas (cf. ref. 4). As shown in Table I, hyperplastic nodules contain six- to ten-fold more epoxide hydrase than the liver microsomes of untreated rats. Once hyperplastic nodules have developed in the livers of AAF-fed rats, additional AAF in the diet is unnecessary for the maintenance of elevated epoxide hydrase levels in these hyperplastic nodules and hepatomas (10). Thus, for the most part, hepatic epoxide hydrase activity in phenobarbital- or AAF-treated rats and in hyperplastic nodules is the result of increased tissue concentrations of the enzyme. These results make it unnecessary to postulate the presence of an altered epoxide hydrase in hyperplastic nodules although this possibility cannot be excluded.

Immunochemical Properties of Epoxide Hydrase in Other Tissues and Species

Figure 5 shows the cross reaction of the antibody to rat liver epoxide hydrase with microsomal preparations from kidney and lung which were solubilized with detergents. It is evident from this immunodiffusion analysis (Figure 5) that the epoxide hydrase present in these extrahepatic tissues is identical to the liver microsomal enzyme. The epoxide hydrase in isolated nuclei and nuclear envelope has also been shown by this laboratory to be immunochemically indistinguishable from the microsomal enzyme (30).

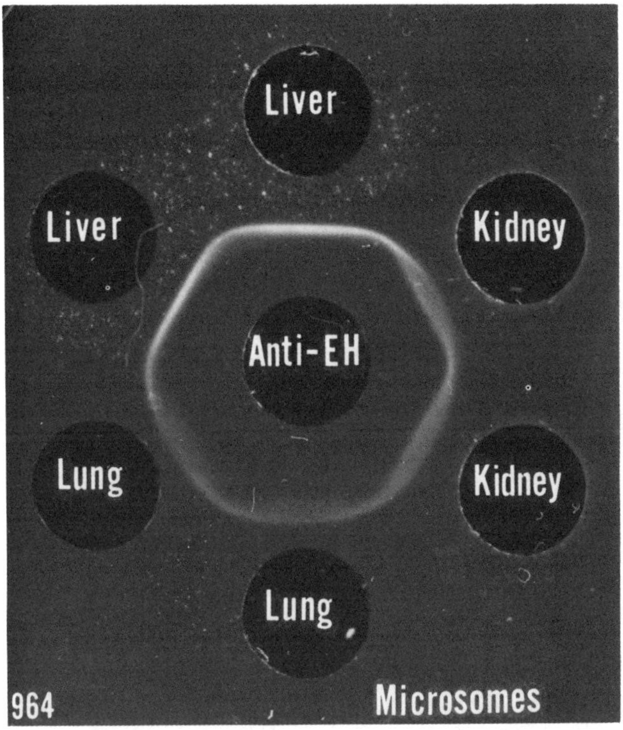

FIGURE 5. Double immunodiffusion plate comparing microsomal epoxide hydrase derived from different tissues of the rat. The antibody (7.5 mg/ml IgG) to epoxide hydrase in the center well was made in sheep against electrophoretically homogeneous rat liver microsomal epoxide hydrase. The peripheral wells contained detergent solubilized (0.2 mg Emulgen 911 and 1.0 mg cholate per mg protein) microsomal fractions derived from the indicated tissues of untreated rats. Microsomal protein concentrations were: liver, 23 mg/ml; kidney, 26 mg/ml; lung, 30 mg/ml.

Immunodiffusion analyses are highly sensitive assays which can detect alterations in antigenic sites. For example, even the same antigen is immunochemically nonidentical if it is partially heat denatured (7). Therefore, since the epoxide hydrases from various tissues are immunochemically identical, much of the knowledge about liver metabolism should also be applicable to other rat tissues.

Antibody to rat liver epoxide hydrase was also used to investigate the immunochemical properties of this enzyme in other species. The epoxide hydrase from mouse liver microsomes (strains C57BL/6J and DBA/2J) has been reported to be antigenically identical to the rat liver enzyme using

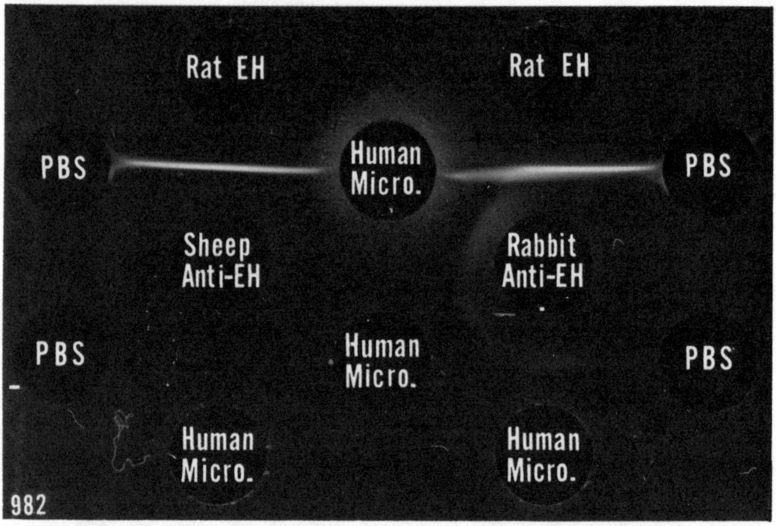

FIGURE 6. A comparison of the reactivity of anti-rat epoxide hydrase made in sheep with that made in rabbits against human microsomal epoxide hydrase. The protein concentrations of sheep and rabbit IgG fractions were 7.5 and 10 mg/ml, respectively. The pure rat epoxide hydrase in the uppermost wells was present at 0.25 mg/ml. The human microsomes (final protein concentration 35 mg/ml) were detergent-solubilized (0.2 mg Emulgen 911 and 0.5 mg cholate per mg protein). The wells marked "PBS" contained phosphate-buffered saline.

antibody preparations from sheep (11). The antibody prepared in sheep, however, cross reacts poorly with microsomal preparations from hamster and guinea pig and not at all with rabbit or human liver microsomes (11,16).

Figure 6 shows a comparison of the cross reactivity of human liver microsomes with the antibody to rat liver epoxide hydrase obtained from sheep and rabbit. Although the antibody made in sheep does not recognize human liver microsomal epoxide hydrase (11,16), the preparation from rabbits does cross react weakly with the human enzyme. The double diffusion analysis (Figure 6) indicates that human liver epoxide hydrase is immunochemically partially identical to the purified rat liver enzyme. Dr. B. Burchell has also found that antibody to rat liver epoxide made in rabbits cross reacts with the human liver enzyme (personal communication). The human liver epoxide hydrase is not recognized by antibodies to the rat liver enzyme made in donkey or goat as well as sheep. The differences in reactivity of the human liver epoxide hydrase with antibody preparations from different species is not unusual, since these antibodies may recognize different antigenic sites on the enzyme used as the antigen. Differences in reactivities of these various antibodies could be utilized as tools to investigate the antigenic structure of epoxide hydrase.

IMMUNOCHEMICAL COMPARISON AND ANTIGENICITY OF FIVE DRUG–METABOLIZING ENZYMES

The immunodiffusion plate in Figure 7 compares the immunochemical properties of five drug-metabolizing enzymes which our laboratory has purified from rat liver microsomes. The crossing of these immunoprecipitin bands in Figure 7 shows that these five antigens are not immunochemically related. These conclusions confirm and extend the results contained in Figure 3 where it was shown in a different fashion that $P450_a$, $P450_b$ and $P450_c$ are immunochemically unique. This absence of immunochemical identity among these five proteins is surprising especially for the three forms of cytochrome P450 which are functionally similar.

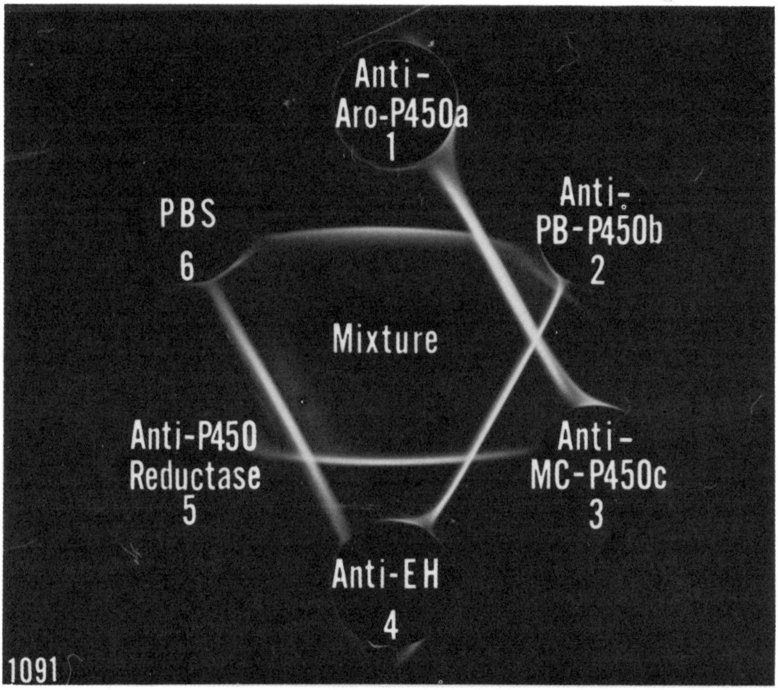

FIGURE 7. A double diffusion plate comparing the immunochemical properties of five enzymes involved in carcinogen and drug metabolism. The antibodies in the peripheral wells were IgG fractions with the indicated protein concentrations and species of origin: (1) 50 mg/ml, sheep; (2) 25 mg/ml, rabbit; (3) 50 mg/ml, rabbit; (4) 7.5 mg/ml, sheep; (5) 50 mg/ml, sheep; (6) phosphate-buffered saline. The central well contained a mixture of the appropriate purified antigens.

All the antigens in Figure 7, except NADPH-cytochrome P450 reductase, have been examined for antigenicity in several rabbits and in sheep. The antigenicity of NADPH-cytochrome P450 reductase has been examined in one sheep only. From our experience, these five rat liver microsomal proteins can be arranged in descending order of antigenicity: epoxide hydase >>PB-$P450_b$ > MC-$P450_c$ ≅ Aro-$P450_a$ ≅ NADPH-cytochrome P450 reductase. The high antigenicity of epoxide hydrase is illustrated in other ways. For example, when producing antibody to certain protein mixtures containing epoxide hydrase, we have found a disproportionally large amount of antibody to this enzyme in the immune sera. It is of interest that epoxide hydrase has been identified as the PN antigen of hepatocarcinogen-induced preneoplastic nodules of liver (10). The high antigenicity of epoxide hydrase coupled with its induction by liver carcinogens could play a role in the immune surveillance system which may be responsible for the spontaneous regression of certain preneoplastic lesions.

ACKNOWLEDGMENTS

We thank Drs. M. J. Griffin and D. E. Kizer for the liver and hyperplastic nodule microsomes from Sprague Dawley rats. We also thank Ann Marie Williams for her assistance in the preparation of this manuscript.

REFERENCES

1. Bresnick, E., Boraker, D., Levin, W., and Thomas, P. E. (1978): Intranuclear localization of hepatic cytochrome P448 by an immunochemical method. *Mol. Pharmacol.* (in press).
2. Burchell, B. (1977): Studies on the purification of rat liver uridine diphosphate glucuronyltransferase. *Biochem. J.* 161:543-549.
3. Conney, A. H. (1967): Pharmacological implications of microsomal enzyme induction. *Pharmacol. Rev.* 19:317-366.
4. Farber, E. (1973): Hyperplastic liver nodules. In: *Methods in Cancer Research,* Vol. 8, H. Busch, Ed., pp. 345-375, Academic Press, New York.
5. Gillette, J. R., Davis, D. C., and Sasame, H. A. (1972): Cytochrome P450 and its role in drug metabolism. *Ann. Rev. Pharmacol.* 12:57-84.
6. Griffin, M. J., and Kizer, D. E. (1978): Purification and quantitation of preneoplastic antigen from hyperplastic nodules and normal liver. *Cancer Res.* 38:1136-1141.
7. Jefferis, R., Mathews, J. B., and Bayley, P. M. (1978): Studies of human IgD myeloma proteins. Conformational changes induced in the Fcδ fragment on heating or expousre to acid pH. *Immunochemistry* 15:19-25.

8. Jerina, D. M., and Daly, J. W. (1974): Arene oxides: A new aspect of drug metabolism. *Science* 185:573-582.
9. Jerina, D. M., Dansette, P. M., Lu, A. Y. H., and Levin, W. (1977): Hepatic microsomal epoxide hydrase: A sensitive radiometric assay for hydration of arene oxides of carcinogenic aromatic hydrocarbons. *Mol. Pharmacol.* 13:342-351.
10. Levin, W., Lu, A. Y. H., Thomas, P. E., Ryan, D., Kizer, D. E., and Griffin, M. J. (1978): Identification of epoxide hydrase as the preneoplastic antigen in rat liver hyperplastic nodules. *Proc. Nat. Acad. Sci., U.S.* 75:3240-3243.
11. Levin, W., Thomas, P. E., Korzeniowski, D., Seifried, H., Jerina, D. M., and Lu, A. Y. H. (1978): Liver microsomal epoxide hydrase: Activation, inhibition and immunological properties of the purified enzyme. *Mol. Pharmacol.* 14:1107-1120.
12. Lowry, O. H., Rosebrough, N. J., Farr, A. L., and Randall, R. J. (1951): Protein measurement with the Folin-phenol reagent. *J. Biol. Chem.* 193:265-275.
13. Lu, A. Y. H., Jerina, D. M., and Levin, W. (1977): Liver microsomal epoxide hydrase: Hydration of alkene and arene oxides by membrane-bound and purified enzymes. *J. Biol. Chem.* 252:3715-3723.
14. Lu, A. Y. H., Junk, K. W., and Coon, M. J. (1969): Resolution of the cytochrome P450-containing ω-hydroxylation system of liver microsomes into three components. *J. Biol. Chem.* 244:3714-3721.
15. Lu, A. Y. H., and Levin, W. (1974): The resolution and reconstitution of the liver microsomal hydroxylation system. *Biochim. Biophys. Acta.* 344:205-240.
16. Lu, A. Y. H., Levin, W., Thomas, P. E., Jerina, D. M., and Conney, A. H. (1978): Enzymological properties of purified liver microsomal cytochrome P450 system and epoxide hydrase. In: *Carcinogenesis, Vol. 3,* P. W. Jones and R. I. Freudenthal, Eds., pp. 243-252, Raven Press, New York.
17. Lu, A. Y. H., Levin, W., West, S. B., Jacobson, M., Ryan, D., Kuntzman, R., and Conney, A. H. (1973): Reconstituted liver microsomal enzyme system that hydroxylates drugs, other foreign compounds, and endogenous substrates: VI. Different substrate specificities of the cytochrome P450 fractions from control- and phenobarbital-treated rats. *J. Biol. Chem.* 248:456-460.
18. Lu, A. Y. H., Ryan, D., Jerina, D. M., Daly, J. W., and Levin, W. (1975): Liver microsomal epoxide hydrase: Solubilization, purification, and characterization. *J. Biol. Chem.* 250:8283-8288.
19. Mancini, G., Carbonara, A. O., and Heremans, J. F. (1965): Immunochemical quantitation of antigens by single radial immunodiffusion. *Immunochemistry* 2:235-254.
20. Mannering, G. J. (1971): Properties of cytochrome P450 as affected by environmental factors: Qualitative changes due to administration of

polycyclic hydrocarbons. *Metab. Clin. Exp.* 20:228-245.

21. Nakayasu, H., Mihara, K., and Sato, R. (1978): Purification and properties of a membrane-bound aldehyde dehydrogenase from rat liver microsomes. *Biochem. Biophys. Res. Commun.* 83:697-703.
22. Oesch, F. (1973): Mammalian epoxide hydrases: Inducible enzymes catalyzing the inactivation of carcinogenic and cytotoxic metabolites derived from aromatic and olefinic compounds. *Xenobiotica* 3:305-340.
23. Okita, K., and Farber, E. (1975): An antigen common to preneoplastic hepatocyte populations and to liver cancer induced by N-2-fluorenylacetamide, ethionine, or other hepatocarcinogens. *GANN Monograph on Cancer Research,* No. 17, pp. 283-299, University of Tokyo Press, Japan.
24. Okita, K., Kligman, K. H., and Farber, E. (1975): A new common marker for premalignant and malignant hepatocytes induced in the rat by chemical carcinogens. *J. Nat. Cancer Inst.* 54:199-202.
25. Omura, T., and Sato, R. (1964): The carbon monoxide-binding pigment of liver microsomes. II. Solubilization, purification, and properties. *J. Biol. Chem.* 239:2379-2385.
26. Ryan, D., Thomas, P. E., Korzeniowski, D., and Levin, W. (1979): Separation and characterization of highly purified forms of liver microsomal cytochrome P450 from rats treated with polychlorinated biphenyls, phenobarbital and 3-methylcholanthrene. *J. Biol. Chem.* 254:1365-1374.
27. Ryan, D. E., Thomas, P. E., and Levin, W. (1977): Properties of purified liver microsomal cytochrome P450 from rats treated with the polychlorinated biphenyl mixture Aroclor 1254. *Mol. Pharmacol.* 13:521-532.
28. Sims, P., and Grover, P. L. (1974): Epoxides in polycyclic aromatic hydrocarbon metabolism and carcinogenesis. *Adv. Cancer Res.* 20:165-274.
29. Strittmatter, P., Spatz, L., Corcoran D., Rogers, M. J., Setlow, B., and Redline, R. (1974): Purification and properties of rat liver microsomal stearyl coenzyme A desaturase. *Proc. Nat. Acad. Sci., U.S.* 71:4565-4569.
30. Thomas, P. E., Korzeniowski, D., Bresnick, E., Bornstein, W. A., Kasper, C. B., Fahl, W. E., Jefcoate, C. R., and Levin, W. (1979): Hepatic cytochrome P448 and epoxide hydrase: Enzymes of nuclear origin are immunochemically identical with those of microsomal origin. *Arch. Biochem. Biophys.* 192:22-26.
31. Thomas, P. E., Korzeniowski, D., Ryan, D., and Levin, W. (1979): Preparation of monospecific antibodies against two forms of rat liver cytochrome P450 and quantitation of these antigens in microsomes. *Arch. Biochem. Biophys.* 192:524-532.

32. Thomas, P. E., Lu, A. Y. H., Ryan, D., West, S. B., Kawalek, J., and Levin, W. (1976): Multiple forms of rat liver cytochrome P450: Immunochemical evidence with antibody against cytochrome P448. *J. Biol. Chem.* 251:1385-1391.
33. Thomas, P. E., Lu, A. Y. H., Ryan, D., West, S. B., Kawalek, J., and Levin, W. (1976): Immunochemical evidence for six forms of rat liver cytochrome P450 obtained using antibodies against purified rat liver cytochrome P450 and P448. *Mol. Pharmacol.* 12:746-758.
34. Thomas, P. E., Lu, A. Y. H., West, S. B., Ryan, D., Miwa, G. T., and Levin, W. (1977): Accessibility of cytochrome P450 in microsomal membranes: Inhibition of metabolism by antibodies to cytochrome P450. *Mol. Pharmacol.* 13:819-831.
35. Werringloer, J., and Estabrook. R. W. (1975): Heterogeneity of liver microsomal cytochrome P450: The spectral characterization of reactants with reduced cytochrome P450. *Arch. Biochem. Biophys.* 167: 270-286.
36. Wood, A. W., Levin, W., Lu, A. Y. H., Yagi, H., Hernandez, O., Jerina, D. M., and Conney, A. H. (1976): Metabolism of benzo[a]pyrene and benzo[a]pyrene derivatives to mutagenic products by highly purified hepatic microsomal enzymes. *J. Biol. Chem.* 251:4882-4890.

Polynuclear Aromatic Hydrocarbons, edited by P. W. Jones and P. Leber. Ann Arbor Science Publishers, Inc. Ann Arbor, MI © 1979.

BIOACTIVATION OF POLYCYLIC AROMATIC HYDROCARBONS IN THE AORTA: EVIDENCE FOR A ROLE IN THE GENESIS OF ATHEROSCLEROTIC LESIONS

Mont R. Juchau, James A. Bond,
***Richard M. Kocan and *Earl P. Benditt**

Department of Pharmacology
*Department of Pathology
University of Washington
School of Medicine
Seattle, Washington 98195

The sequelae of atherosclerosis are widely recognized to be the major cause of death in the United States as well as in most of the western world. Cardiovascular diseases (of which atherosclerosis is highly prominent) afflict more than 28 million people and are responsible for approximately 1 million deaths per year in the U.S. alone (24). In spite of the widespread recognition of the importance of the disease, very little is known concerning the molecular, biochemical or cellular mechanism(s) whereby atheromatous lesions are initiated. Currently it is widely accepted that the principal response in the artery wall associated with the genesis of lesions is the focal proliferation and accumulation of smooth muscle cells (myointimal cells) within the affected intima, the innermost layer of the vessel (16,29,41,46). The unknown factor in the etiology is the mechanism(s) whereby a wide variety of stimuli can initiate and maintain the proliferative behavior of myointimal cells. Two principal therories have guided recent research directed toward delineating such etiologic mechanisms. The first is commonly referred to as the "response-to-injury" or "lipid-insudation" hypothesis and contends that any repeated or chronic injury to the endothelium (mechanical, chemical or biochemical) will result in inflammation and associated platelet aggregation with secondary proliferation of myointimal cells (38). The second and most recent hypothesis is referred to as the "monoclonal theory" and is based upon analyses of human plaques obtained at autopsies (4,8). It was discovered that the cells obtained from single plaques in the arterial walls of black females [heterozygotes for glucose 6-phosphate dehydrogenase (G6PD)] expressed only one or the other of two forms of G6PD while the surrounding non-plaque tissue expressed both forms of the enzyme. The results of these experiments were later confirmed in independent investigations (33,47) and suggested the possibility that all or nearly all the cells within the plaque were derived from a single precursor cell. Elimination of cells with one form of the enzyme within plaques via selection seemed unlikely since that process should have resulted in the appearance of only one or the other form of G6PD (but not both) in plaques examined at random. A second alternative appeared more likely, namely that a genetic (mutation) or epigenetic event originally resulted in a selective advantage for a single precursor smooth muscle cell and its progeny. This process would be analogous to that proposed for the initiation and development of tumors (28). Thus, the most likely conclusion [although others are feasible (47)] resulting from the observations was that atherosclerotic lesions and tumors could be initiated by similar stimuli. Indeed, benign smooth-muscle tumors (leiomyomas) of the uterus (26) and other tumors (15) are assumed to exhibit monoclonality as the result of similar observations. By contrast, many common cell proliferations seen in embryogenesis, maintenance and repair are multicellular in character.

The implications of the monoclonal theory for the role of environmental chemicals in atherosclerosis are immediately obvious. Environmental factors are presumed to be responsible for 60-90% of all human malignancies and of these factors, chemicals appear to play a major role (27). Probably the most common environmental chemical contaminants are the PAH (14), many of which are known to be promutagens and procarcinogens and, in light of the evidence for the monoclonal hypothesis discussed above [and in greater detail in recent reviews (5-7)], are candidates as potential etiological factors in atheromatous diseases. Thus it becomes incumbent upon investigators to determine whether, to what extent and by what mechanisms PAH may play a role in the etiology of atherosclerosis.

EPIDEMIOLOGIC DATA

One of the first questions encountered is whether epidemiologic data exist that could support the hypothesis that PAH are causative factors in atheromatous diseases. It is well known that cigarette smoking is a highly important risk factor (24,40,44,52, and references therein) and it has been estimated (49) that about 70% of arteriosclerosis obliterans is causally associated with cigarette smoking. Although carbon monoxide (20), cadmium (48) and other factors in cigarette smoke (44) have been implicated as etiologic factors, only recently has attention been given to PAH. In view of the monoclonal theory and the importance of PAH in cigarette smoke as tumorigenic agents (1) it would appear that considerable attention should now be given to these ubiquitous chemicals in terms of their role in atherogenesis. Particularly pertinent to this were the observations of Yamasaki and Ames (55), who showed that cigarette smokers had elevated levels of mutagens in the urine. We have succeeded in confirming their observations and also have found that the mutagens appear almost immediately after the commencement of smoking but also disappear very rapidly after cessation (unpublished results). PAH may act not only as tumor initiators (presumably via somatic cell mutagenesis) but also as promoters, "accelerators" and otherwise as cocarcinogens. Since cigarette smoking is a major risk factor for a large number of defined neoplastic diseases, the commonality of cigarette smoking as an etiologic factor for both atherosclerosis and cancer could be regarded not only as a piece of indirect corroborative evidence for the role of PAH in atherosclerosis but, perhaps, for the monoclonal hypothesis as well.

Do cancer and atherosclerosis share other epidemiologic associations? Aside from cigarette smoking, high blood pressure and high concentrations of serum cholesterol are regarded as major risk factors for atherosclerosis (24). Although the latter two are not commonly regarded as major risk factors for neoplastic diseases, it is well to emphasize that these factors are uniquely confined to the cardiovascular system. As such, they could possibly play roles in the "promotion" or "acceleration" of atheromatous lesions by mechanisms akin to those now only beginning to be understood for the promotion and development of tumors of the lung, skin, liver, etc. (27). In this connection, it is well to emphasize that virtually any type of injury to tissues has long been known to facilitate the development of tumors, either prior to or following exposure to tumorigenic agents (45).

An interesting observation that could interrelate the monoclonal hypothesis with hypertension as a risk factor and PAH as causative agents was made by Pero *et al.* (34) who found, in measurements of N-acetoxyacetylamino-fluorene-induced unscheduled DNA synthesis (excision-repair) in human lymphocytes, that the repair synthesis showed a linear increase with the blood pressure of the individuals under study. They speculated on the basis of the results obtained that the individuals with hypertension have a greater potential for accumulating DNA damage because of an increased chemical reactivity of lymphocytes to carcinogen exposure.

Urban living and industrialization also represent common factors for atherosclerosis and cancer (32). PAH are known to be more heavily concentrated in urban than rural environments due to their presence in automobile exhausts, factory smoke and other products of combustion common to industrialization and urban environments; this fact is likewise consistent with the idea that PAH may be acting as chemical atherogens. It is believed that the PAH are probably the most widespread of environmental contaminants (14) and that urban environments contain the highest concentrations of such contaminants.

Diets high in animal (saturated) fat and cholesterol are regarded by most scientists as risk factors for atheromatous diseases (32) and high beef/high fat diets also have been associated with cancers of the colon, pancreas, kidney, breast, ovary, endometrium and prostate gland (54). This information suggests a common etiology for these diseases. The lipophilic nature of PAH and the fact that alterations in the fat content of diets can have profound effects upon the metabolism of PAH (25) as well as on the conversion of procarcinogens to mutagens (13) are therefore also pertinent to the discussion. Rose *et al.* (37) found that individuals with coronary heart disease and blood cholesterol levels below an expected level exhibited a positive correlation with mortality from colon cancer while those with elevated serum cholesterol levels exhibited a negative correlation. An implication is that the cholesterol levels may determine the probability of colon cancer vs atherosclerosis (11). This is important since it is well known that atherosclerosis is not extensive in subjects dying from most malignant diseases, except cancer of the lung and prostate (43). Also, significant positive correlations between incidences of cancers of the colon, breast, ovary and endometrium and cigarette smoking do not appear to occur (53).

In summary, it can be stated that while the available epidemiologic data do not provide definitive proofs of a role of PAH in the genesis of atherosclerosis, most of the data are compatible with the suggestion of such a role in human populations. Common associations of atherogenic and tumorigenic diseases with cigarette smoking, urban living, industrialization, high fat diets and slow development/long latency point toward a common etiology. In addition, none of the epidemiologic data collected to date would directly contradict or discredit this idea.

EXPERIMENTAL STUDIES *IN VIVO*

The best evidence for the role of PAH in the genesis of atheromatous lesions is derived from experiments with animals. Albert *et al.* (2) were able to demonstrate that both 7,12-dimethylbenz[a]anthracene (DMBA) and benzo[a]pyrene (B[a]P) could increase the incidence and size of atherosclerotic lesions when administered intramuscularly to chickens at relatively high doses (25 mg/kg and 50 mg/kg, respectively) over a period of 9 to 18 weeks. Based on observations of autoradiographic labeling with tritiated thymidine, the lesions produced appeared to be proliferative in character. Neither DMBA nor B[a]P produced elevations in serum cholesterol; slight but statistically insignificant decreases were observed. In their experiments, PAH produced proliferative lesions in the abdominal aorta whereas cholesterol administration commonly produce fatty lesions in the thoracic aorta. Importantly, the lesions produced by the PAH could not be distinguished morphologically from lesions that occurred spontaneously at an older age.

These results appeared to confirm some heretofore little-noted observations at the National Cancer Institute that were published much earlier. White and White (51) reported that the treatment of cystine-deficient mice with methylcholanthrene (MC) (0.2% in ether applied topically) resulted in the appearance of sclerotic lesions of the aorta and other large arteries. No lesions were observed when diets contained sufficient quantitites of cystine. Interestingly, the mice on the low-cystine diets developed considerably less leukemias than those on cystine-supplemented diets such that an inverse correlation between sclerosis and leukemia was observed in the experiments. The authors interpreted this observation to indicate that the vascular damage produced by MC was unrelated to its carcinogenic properties. When the dosage of MC was decreased from 45 to 20 applications (50) the incidence of sclerosis was not significantly lower (94% vs 80%) but the latent period was increased from 80 days to 114 days.

Because of the observed relationships between atherosclerotic plaque cells and leiomyoma cells (30) it is also interesting to note that administration of DMBA to pregnant rats resulted in the appearance of uterine leiomyosarcomas in 2 of the 31 (6.5%) rats treated with the PAH (42). Adenocarcinomas, which were the most frequently occurring tumors, appeared in 6 of 31 (19.4%) rats treated. Further implications of these observations are discussed below.

EXPERIMENTAL STUDIES *IN VITRO*

In accordance with the above discussion, the question as to *whether* PAH can be implicated in the etiology of atherosclerotic lesions appears to have been answered in the affirmative. The next questions pertain to concentration-effect or dose-response relationships, applicability to human populations and mechanisms of action. To our knowledge, no further studies on dose-response relationships have been reported and an appreciation of the applicability to human populations must await much more extensive epidemiologic investigations and studies in human tissues. In our laboratories we have begun to address the question of mechanisms. Considerations of the chemistry of PAH as well as a wealth of experimental data obtained to date both indicate that biotransformation of these substances is requisite for their elicitation of biological effects. It is well established that phenolic metabolites of PAH exhibit considerable cytotoxicity and that certain epoxides and diol-epoxides possess marked mutagenic and carcinogenic potential in contrast to the absence of these properties in the unmetabolized parent hydrocarbons. Such metabolites are generated via monooxygenation in cytochrome P-450-dependent enzyme systems which occur in a wide variety of tissues and cells.

In mammals, the highest concentrations of P-450-dependent monooxygenase activities occur in hepatic and adrenocortical endoplasmic reticulum and adrenocortical mitochondria. The liver is the most important quantitative site of PAH monooxygenation and conceivably could generate reactive metabolites that would produce cytotoxic or mutagenic effects in arterial walls. (The exact role of bioactivation of PAH in terms of the mechanisms of their tumor-promoting or other cocarcinogenic properties has not been delineated.) However, the possibility that the *particular* metabolites that could elicit effects in arterial walls would be rapidly inactivated by enzymatic or nonenzymatic mechanisms seems likely in view of the known chemical instability of metabolites now presumed to act as ultimate mutagens and carcinogens as well as a number of efficient enzymatic tissue components that mediate the inactivation of reactive intermediates via further oxidation, glucuronidation, sulfation, glutathione conjugation and perhaps other reactions. The current widespread interest in the bioactivation of PAH and other procarcinogens in the cellular nucleus is indicative of the concern for the importance of bioactivation within the target sites (35,39). Therefore, the question regarding the relative contributions of bioactivation/inactivation in arterial vs nonarterial tissues to the observed atherogenic effects of PAH seems highly pertinent at present.

In an approach to these problems, some very basic questions needed first to be answered: Do PAH distribute into the cellular elements of arteries? Do arterial cells contain enzymes or enzyme systems (particularly monooxygenases) that will metabolize PAH? What is the nature of the biotransforming enzymes? Are the arterial smooth-muscle ("target") cells capable of effecting metabolic conversion of PAH? What kinds of metabolites are formed? Will some form covalent bonds with macromolecules? Are some of the intermediary metabolites mutagenic?

Evidence for a positive answer to the first question rests with the following considerations: The lipophilic nature of PAH leads to the expectation that they would distribute readily to all body tissues; secondly, administration of PAH by any of a number of different routes evokes enzyme induction in nearly every mammalian tissue examined thus far, including the arteries (10, 22); and thirdly, studies on the plasma distribution of PAH (5,7) show that they are closely associated with plasma lipoproteins including the low density lipoproteins (LDL). These LDL are taken up by smooth-muscle cells. Further studies in our laboratories (unpublished) have shown that MC, DMBA and B[a]P each associate preferentially with human plasma lipoproteins although MC appears to be more closely associated with LDL than either DMBA or B[a]P. The distribution profiles of MC and cholesterol were remarkably similar. The quantitative extent to which given doses of PAH (administered by specific routes) accumulate in smooth-muscle cells of arterial walls, however, remains a subject for more intensive future investigations.

The answer to the second question likewise has been answered in the affirmative (9,10,22) for arteries of humans, monkeys, rabbits, and chickens. Aortic homogenates from each of these species exhibited low but readily measurable monooxygenase activities with B[a]P and DMBA as substrates except that biotransformation of B[a]P in chicken aortas/(Table1) was not detectable except after pretreatment with an inducing agent (MC). Activities were comparable to those observed in mouse skin, a known target organ for the tumorigenic effects of PAH. Cytochrome P-450 also was detected in preparations of primate aortas. Epoxide hydratase (E.C. 4.2.1.63) activity was measurable in homogenates of the arterial walls of chickens and rabbits with B[a]P-4,5-oxide as substrate (unpublished) but, to our knowledge, other biotransformation reactions involving PAH in aortic tissues have not been investigated. The monooxygenase system that biotransforms B[a]P exhibited characteristics similar to those observed in several other extrahepatic tissues. Pretreatment with MC, polychlorinated biphenyls (Aroclor 1254) or 5,6-benzoflavone resulted in marked increases in activity whereas pretreatment with phenobarbital or pregnenolone-16α-carbonitrile produced no significant effects. Unexpectedly, fortified preparations of rabbit aortic homogenates produced much higher quantities of alkali-extractable metabolites on addition of hematin to reaction vessels. As much as 28-fold increases were observed (10). It was later found, however, that several extrahepatic tissues of rabbits and, to a lesser extent, other species exhibit the same phenomenon (31) indicating that the response to hematin is not peculiar to aortic tissues. The mechanism of this interesting effect of hematin upon extrahepatic monooxygenases is currently under study in our laboratory.

We have recently obtained evidence that the hypothetical target cells (arterial smooth-muscle cells) exhibit the capacity to convert PAH to various metabolites (unpublished). Cells obtained from human fetal aortas were cultured and incubated with DMBA. The metabolites were extracted with ethyl acetate and separated with high-pressure liquid chromatography. The major metabolite(s) formed cochromatographed with 12-hydroxymethyl-7-methylbenz[a]anthracene (12-OHM-7-MBA) which was interesting in view of previous experiments with fresh human fetal tissues (23). Homogenates of fetal lung, liver, kidney and adrenal gland each generated DMBA metabolites of which the major peak coeluted with 12-OHM-7-MBA. Other metabolites detected in incubations of DMBA with cultured smooth-muscle cells were 7-OHM-12-MBA, 7,12-diOHMBA and DMBA-trans-8,9-diol. Treatment of the cell cultures with benzanthracene (13 μM, final concentration) resulted in an increased production of the ethyl-acetate-soluble metabolites. Details of these experiments are to be published at a later date. Clearly, the biotransformation of PAH by aortic and arterial smooth-muscle cells should be studied in detail.

TABLE I. Effects[a] of Pretreatment of Rabbits and Chickens on the Aortic Conversion in vitro of B[a]P and DMBA to Oxidized Metabolites

Pretreatment	Rabbits		Chickens	
	B[a]P	DMBA	B[a]P	DMBA
Aroclor 1254	464	-[b]	>100	123
3-Methylcholanthrene	286	677	>100	183
5,6-Benzoflavone	153	-	>100	193
Pregnenolone-16α-carbonitrile	114	-	ND[c]	29
Phenobarbital	107	-	ND	117
Control values	0.42 ± 0.26	2.3 ± 1.4	ND	16.9 ± 4.1

[a]Control values are mean specific activities (pmoles/mg protein/min) ± S. E. of triplicate determinations on 5 pools of aortic homogenates (5 aortas/pool). Other values are percentages of corresponding control values determined from a single pool. For DMBA, total ethyl-acetate extractable metabolites (separated on HPLC) are calculated; for B[a]P, total base-extractable phenolic metabolites are calculated. Details of the methodology have been submitted for publication.

[b]Blanks signify that the experiments were not performed.

[c]ND signifies not detectable.

Production of mutagenic and/or cytotoxic metabolites of PAH within arterial smooth-muscle cells could represent a highly significant contribution to the initiation of atheromatous lesions *in vivo* and studies along these lines may provide a much greater understanding of the initiation process.

Thus far, evidence has accumulated to indicate that the primary B[a]P metabolites formed in aortic homogenates of rabbits are phenolic compounds that cochromatograph (HPCL) with 3-hydroxy-B[a]P and 9-hydroxy-B[a]P (10). Phenolic metabolites are known to exhibit considerable cytotoxicity. Other metabolites were not detected but with further characterization of aortic enzyme systems and the development of more sensitive assay procedures it is to be expected that other metabolites will be found. The major metabolite formed on incubation of chicken aortic homogenates with DMBA was 7-hydroxymethyl-12-methylbenz[a]anthracene. A second major metabolite peak did not chromatograph with any of the available standards. 12-Hydroxymethyl-7-methylbenzanthracene and 7,12-dihydroxymethylbenz[a]anthracene also were tentatively identified on the basis of retention times. Interestingly, the aortas of chickens exhibited much higher activity than rabbit aortas with DMBA as substrate but the converse was true with B[a]P as substrate (Table I).

That some of the aortic metabolites of B[a]P and DMBA will form covalent bonds with macromolecules also has been demonstrated in preliminary

TABLE II. Effects of Additions of 7,8-Benzoflavone (7,8-BF) *in vitro* and Pretreatment with MC on the Conversion of DMBA to Metabolites Covalently Bound to DNA in the Presence of Chicken Aortic Homogenates and Cofactors for Monooxygenation

Pretreatment	Inhibitor	DNA Binding[a]
1. None	None	0.38
2. None	7,8-BF, 10^{-6}M[c]	0.18
3. None	7,8-BF, 10^{-5}M	0.12
4. None	7,8-BF, 10^{-4}M	0.15
5. Methylcholanthrene[b]	None	1.98

[a]Values are expressed as fmoles of DMBA metabolite(s) bound/μg DNA/mg protein/min. Experiments 1-4 were performed on the same homogenate pool (5 aortas/pool); experiment 5 on a separate pool of 5 aortas.

[b]Chickens were given a single dose of MC in corn oil (40 mg/kg 3.p.) 48 hr prior to sacrifice.

[c]As final concentrations in reaction vessels.

studies (unpublished). Aortic homogenates from MC-pretreated chickens catalyzed the formation of metabolites that formed covalent bonds with calf thymus DNA when incubated together and assayed in the system described by Gelboin (17) and Grover and Sims (18) as slightly modified (12). When compared with chickens treated with vehicle only, pretreatment of the chickens with MC markedly increased the amounts of PAH covalently bound to DNA (Table II) and 7,8-benzoflavone added *in vitro* inhibited the binding. Activities observed were comparable to those commonly detected with mouse epidermis, a common target organ for PAH-initiated tumors. In still further studies, we have been able to demonstrate (9) that aortic enzymes would catalyze the conversion of DMBA to metabolites that produced back mutations in *Salmonella typhimurium* tester strains TA-98, TA-100 and TA-1538 when tested in the systems originally designed by Ames and his colleagues (3). Low activities again were observed in these systems but revertant colony counts on test plates were significantly higher statistically than on control plates incubated without PAH, without cofactors for the monooxygenase reactions or without either. Therefore, it appears that mutagenic as well as cytotoxic metabolites of PAH can be generated in aortic tissues. Details of these studies will be published in a separate paper.

OTHER CORROBORATIVE DATA

One of the interesting aspects of questions pertaining to the possible role and mechanism of action of PAH in the etiology of atheromatous lesions is the relationship between uterine leiomyomas and atheromatous lesions. Uterine leiomyomas are benign smooth-muscle tumors that occasionally metastasize although these metastasizing tumors lack histologic criteria of malignancy (14). Metastasizing leiomyomas reportedly behave much like low-grade sarcomas. The apparent close similarities between normal arterial and uterine smooth-muscle cells (19) as well as between smooth-muscle cells of atheromas and leiomyomas (30) would indicate that rigorous attention should be given to parallels between the etiology and pathogenesis of uterine leiomyomas vs atheromas. As discussed above, both kinds of lesions exhibit monotypism with respect to G6PD and the known tendency for both to become fibrotic and calcified (30) further heightens the analogy. Like atheromatous lesions, leiomyomas occur frequently. They are the most common tumors in women and are responsible for at least one-third of all gynecologic admissions to hospitals (36). As mentioned above, DMBA is capable of producing an increased incidence of both smooth-muscle tumors of the uterus as well as sclerotic lesions. In spite of these correlations, no statistical evidence that links cigarette smoking to uterine smooth-muscle tumors has yet been found (53). It seems possible that more careful analyses could reveal a statistically significant association. The observations do show, however, that the proliferation of smooth-muscle cells of arterial walls observed in atherosclerosis exhibits parallelisms with smooth-muscle-cell tumors occurring in other parts of the body, thus providing corroborative evidence for the theory that tumors and atheromas can have common etiologies. The exact relationship of PAH to the etiology of human atherosclerosis, however, must await more detailed further investigations.

SUMMARY AND CONCLUSIONS

Evidence that the bioactivation of PAH in the aorta may play a role in the etiology of atheromatous lesions may be summarized as follows:

1. PAH are capable of producing sclerotic lesions in arterial walls of experimental animals.
2. Epidemiologic data are consistent with the idea that PAH may also be causative etiologic factors in human populations.
3. PAH require metabolism via monooxygenation for production of mutagenic, carcinogenic and cytotoxic effects.
4. Evidence exists that bioactivation of PAH within the target cells may play a critical role in determining their biologic effects. Target cells for atherogenesis (arterial smooth-muscle cells) are capable of mediating the monooxygenation of PAH.
5. Evidence exists that atheromas arise as a result of mutations (monoclonal origin). Arterial cells will convert PAH to active mutagens in bacterial systems.
6. A large number of analogies exist between atheromatous lesions and various tumorous lesions. The role of the bioactivation of PAH in tumorigenesis is well-documented.

7. Damage to arterial cells is believed to play a major role in the development of sclerotic lesions (response to injury). Arterial cells possess the capability of converting PAH to metabolites with cytotoxic properties.

ACKNOWLEDGMENTS

Original research by the authors was supported by NIH Grants HL-03174, HD-04839, GM-00109, GM-13543 and National Foundation (March of Dimes) Grant CRBS-250.

REFERENCES

1. Akin, F. J., Snook, M. E., Severson, R. E., Chamberlain, W. J., and Walters, D. B. 91976): Identification of polynuclear aromatic hydrocarbons in cigarette smoke and their importance as tumorigens. *J. Nat. Cancer Inst.* 57:191-195.
2. Albert, R. E., Vanderlaan, M., Burns, F. J., and Nishizumi, M. (1977): Effect of carcinogens on chicken atherosclerosis. *Cancer Res.* 37: 2232-2235.
3. Ames, B. N., McCann, J., and Yamasaki, E. (1975): Methods of detecting carcinogens and mutagens with the *Salmonella*/mammalian-microsome mutagenicity test. *Mutation Res.* 31:347-364.
4. Benditt, E. P. (1974): Evidence for a monoclonal origin of human atherosclerotic plaques and some implications . *Circulation* 50:650-652.
5. Benditt, E. P. (1976): Implications of the monoclonal character of human atherosclerotic plaques. *Beitr. Path.* 158:405-416.
6. Benditt, E. P. (1977): The origin of atherosclerosis. *Scientific Am.* 236:74-86.
7. Benditt, E. P. (1978): The monoclonal theory of atherogenesis. In: *Athersclerosis Reviews*, R. Paoletti and A. M. Gotto, Jr., Eds., pp. 77-85, Raven Press, New York.
8. Benditt, E. P. and Benditt, J. M. (1973): Evidence for a monoclonal origin of human atherosclerotic plaques. *Proc. Nat. Acad. Sci., U.S.* 70:1753-1756.
9. Bond, J. A., Jones, A. H. and Juchau, M. R. (1978): Biotransformation and bioactivation of 7,12-dimethylbenz[a]anthracene (DMBA) in aortas of rabbits and chickens. *Pharmacologist* 20:266.
10. Bond, J. A., Omiecinski, C. J., and Juchau, M. R. (1979): Kinetics, activation and induction of aortic monooxygenases: Biotransformation of benzo[a]pyrene. *Biochem. Pharmacol.* 28:305-312.
11. Broitman, S. A., Vitale, J. J., Vavrousek-Jakuba, E., and Gottlieb, L. S. (9177): Polyunsaturated fat, cholesterol and large bowel tumorigenesis. *Cancer* 40:2455-2463.
12. Buty, S. G., Thompson, S., and Slaga, T. J. (1976): The role of epidermal aryl hydrocarbon hydroxylase in the covalent binding of polycyclic

hydrocarbons to DNA and its relationship to tumor initiation. *Biophys. Res. Commun.* 70:1102-1108.
13. Castro, C. E., Felkner, I. C., and Yang, S. P. (1978): Dietary lipid-dependent activation of the carcinogen N-2-fluorenylacetamide in rats as monitored by Salmonella typhimurium. *Cancer Res.* 38:2836-2841.
14. Clark, D. H., and Weed, J. C. (1977): Metastasizing leiomyoma: A case report. *Am. J. Obstet. Gynecol.* 127:672-673.
15. Fialkow, R. J. (1974): The origin and development of human tumors studied with cell markers. *New England J. Med.* 291:26-35.
16. Florentin, R. A., and Nam, S. C. (1968): Dietary induced atherosclerosis in miniature swine. *Exp. Mol. Path.* 8:263-301.
17. Gelboin, H. V. (1969): A microsome-dependent binding of benzo[a] pyrene to DNA. *Cancer Res.* 29:1272-1276.
18. Grover, P. L., and Sims, P. (1968): Enzyme-catalyzed reactions of polycyclic hydrocarbons with deoxyribonucleic acid and protein *in vitro*. *Biochem. J.* 110:159-160.
19. Haust, M. D., Heras, J. L., and Harding, P. G. (1977): Fat-containing uterine smooth muscle cells in "toxemia": Possible relevance to atherosclerosis. *Science* 195:1353-1354.
20. Heliovaara, M., Karvonen, M. J., Vilhunen, R., and Punsar, J. (1978): Smoking, carbon monoxide and atherosclerotic diseases. *Brit. Med. J.* 1:268-270.
21. Jones, P. W., and Freudenthal, R. I. (1978): In: *Carcinogenesis vol. 3, A* P. W. Jones and R. I. Freudenthal, Eds., p. vii, Raven Press, New York.
22. Juchau, M. R., Bond, J. A., and Benditt, E. P. (1976): Aryl hydrocarbon hydroxylase and cytochrome P-450 in the aorta: Possible role in atherosclerosis. *Proc. Nat. Acad. Sci., U.S.* 73:3723-3728.
23. Juchau, M. R., Namkung, M. J., Jones, A. H., and DiGiovanni, J. (1978): Biotransformation and bioactivation of 7,12-dimethylbenz[a] anthracene in human fetal and placental tissues: Analyses of HPLC profiles and studies with Salmonella typhimurium. *Drug Metab. Dispos.* 6: 273-282.
24. Kolata, G. B., and Marx, J. L. (1976): Epidemiology of heart disease: Searches for causes. *Science* 194:509-512.
25. Lambert, L,, and Wills, E. D. (1977): The effect of dietary lipids on 3, 4-benzo[a]pyrene metabolism in the hepatic endoplasmic reticulum. *Biochem. Pharmacol.* 26:1423-1427.
26. Linder, D., and Gartler, S. M. (1965): Glucose-6-phosphate dehydrogenase mosaicism: Utilization as a cell marker in the study of leiomyomas. *Science* 150:67-69.
27. Miller, E. C. (1978): Some current perspectives on chemical carcinogenesis in humans and experimental animals. *Cancer Res.* 38:1479-1496.
28. Mondal, S., and Heidelberger, C. (1970): *In vitro* malignant transformation by methylcholanthrene of the progeny of single cells derived from C34 mouse prostate. *Proc. Nat. Acad. Sci., U.S.* 65:219-225.

29. Moss, N. S., and Benditt, E. P. (1970): The ultrastructure of spontaneous and experimentally induced arterial lesions. II. The spontaneous plaque in the chicken. *Lab. Invest.* 23:231-145.
30. Moss, N. S., and Benditt, E. P. (1975): Human atherosclerotic plaque cells and leiomyoma cells, *Am. J. Path.* 78:175-190.
31. Omiecinski, C. J., Bond, J. A., and Juchau, M. R. (1978): Stimulation by hematin of monooxygenase activity in extra-hepatic tissues from rats, rabbits and chickens. *Biochem. Biophys. Res. Commun.* 83:1004-1011.
32. Peacock, P. B. (1973): Atherosclerotic heart disease and the environment. *Trans. N.Y. Acad. Sci.* 35:631-635.
33. Pearson, T. A., Wang, A., Solez, K., and Hepinstall, R. H. (1975): Clonal characteristics of fibrous plaques and fatty streaks from human aortas. *Am. J. Path.* 81:379-387.
34. Pero. R. W., Bryngelsson, C., Mitelman, F., Thulin, T., and Norden, A., (1976): High blood pressure related to carcinogen-induced unscheduled DNA synthesis, DNA carcinogen binding and chromosomal aberrations in human lymphocytes. *Proc. Nat. Acad. Sci., U.S.* 73:2496-2500.
35. Pezzuto, J. M., Lea, M. A., and Yang, C. S. (1976): Binding of metabolically activated benzo[a]pyrene to nuclear macromolecules. *Cancer Res.* 36:3647-3653.
36. Robbins, S. L. (1974): *Pathologic Basis of Disease,* W. B. Saunders Company, Philadelphia.
37. Rose, G., Blackburn, H., Keys, A., Taylor, H. L., Kannel, W. B., Paul, O., Reid, D. D., and Stamler, J. (1974): Colon cancer and blood cholesterol. *Lancet* 1:181-183.
38. Ross, R., and Harker, L. (1976): Hyperlipidemia and atherosclerosis: Chronic hyperlipidemia initiates and maintains lesions by endothelial cell desquamation and lipid accumulation. *Science* 193:1094-1098.
39. Sakai, S., Reinhold, C. E., Wirth, P. J., and Thorgeirsson, S. S. (1978): Mechanism of *in vitro* mutagenic activation and covalent binding of N-hydroxy-2-acetylaminofluorene in isolated liver cell nuclei from rat and mouse. *Cancer Res.* 38:2058-2067.
40. Schettler, G., Nüssel, E., and Buchholz, L. (1978): Epidemiologic research in western Europe. In: *Atherosclerosis Reviews* R. Paoletti and A. M. Gotto, Jr. Eds., pp. 201-211, Raven Press, New York.
41. Scott, R. F., Jarmolych, J., Fritz, K. E., Imai, H., Kim, D. N., and Morrison, E. S. (1970): In: *Atherosclerosis: Proceedings of the Second International Symposium* R. J. Jones, Ed., p. 50, Springer-Verlag, New York.
42. Sekiya, S., Takamizawa, H., Wang, F., Takane, T., and Kuwata, T. (1972): *In vivo* and *in vitro* studies on uterine adenocarcinoma of the rat induced by 7,12-dimethylbenz[a]anthracene. *Am. J. Obstet. Gynecol* 113:691-695.

43. Sternby, N. H., (1976): Atherosclerosis and malignant tumors. *Bull. World Health Organ.* 53:555-561.
44. Strong, J. P., and Richards, M. L. (1976): Cigarette smoking and atherosclerosis in autopsied men. *Atherosclerosis* 23:451-476.
45. Suss, R., Kinzel, V., and Scribner, J. (1973): *Cancer: Experiments and Concepts*, Springer-Verlag, New York.
46. Thomas, W. A., Florentin, R. A., Nam, S. C., Kim, D. N., Jones, R. M., and Lee, K. T. (1968): Pre-proliferative phase of atherosclerosis in swine fed cholesterol. *Arch. Pathol* 86:621-643.
47. Thomas, W. A., Janakidevi, K., Reiner, J. M., and Florentin, R. A. (1978): Some aspects of population dynamics of arterial smooth muscle cells in atherogenesis. In: *Athersclerosis Reviews* R. Paoletti and A. M. Gotto, Jr., Eds., pp. 87-95, Raven Press, New York.
48. Voors, A. W., Shuman, M. S., and Gallagher, P. N. (1975): Atherosclerosis and hypertension in relation to some trace elements in tissues. *World Rev. Nutr. Diet.* 20:299-323.
49. Weiss, N. S. (1972): Cigarette smoking and arteriosclerosis obliterans: An epidemiologic approach. *Am. J. Epidemiol.* 95:17-25.
50. White, J., Mider, G. B., and Heston, W. E. (1942-3): Note on the comparison of dosage of methylcholanthrene on the production of leukemia and sclerotic lesions in strain Dilute Brown mice on a restricted cystine diet. *J. Nat. Cancer Inst.* 3:453-454.
51. White, F. R., and White, J. (1941): The effect of dietary cystine on the reaction of Dilute Brown mice to methylcholanthrene. *J. Nat. Cancer Inst.* 2:95-97.
52. Wilhemsson, C., Elmfeldt, D., Vedin, J. A., Tibblin, G., and Wilhelmsson, L. (1975): Smoking and myocardial infarction. *Lancet* 1975:415-429.
53. Williams, R. R., and Horm, J. W. (1977): Association of cancer sites with tobacco and alcohol consumption and socioeconomic status of patients: Interview study from the third national cancer survey. *J. Nat. Cancer Inst.* 58:525-547.
54. Wynder, E. L. (1976): Nutrition and cancer. *Fed. Proc.* 35:1309-1316.
55. Yamasaki, E., and Ames, B. N. (1977): Concentration of mutagens from urine by adsorption with the non-polar resin XAD-2: Cigarette smokers have mutagenic urine. *Proc. Nat. Acad. Sci., U.S.* 74:3555-3559.

Polynuclear Aromatic Hydrocarbons,
edited by P. W. Jones and P. Leber.
Ann Arbor Science Publishers, Inc.
Ann Arbor, MI © 1979.

NUCLEAR ARYL HYDROCARBON HYDROXYLASE AND ITS ROLE IN THE ACTIVATION OF BENZO[a]PYRENE

Chung S. Yang and John M. Pezzuto[1]

Department of Biochemistry
CMDNJ-New Jersey Medical School
Newark, New Jersey 07103

It is generally agreed that most chemical carcinogens require metabolic activations before exerting their carcinogenic actions in animals (6,16). Of great current interest is the subcellular site of the activation process. It is usually assumed that carcinogens are activated by the monooxygenase or aryl hydrocarbon hydroxylase (AHH)[2] system of the endoplasmic reticulum and then enter into the cell nucleus to react with target molecules. However, the AHH system also exists in the nuclear envelope (9-11,25). It may be proposed that due to its close proximity to the genetic apparatus, the nuclear AHH is the major site of carcinogen activation and the metabolism by the endoplasmic reticulum would result in detoxification.

THE ROLES OF NUCLEI AND MICROSOMES IN THE ACTIVATION OF BENZO[a]PYRENE

The above two hypotheses were tested in a model system which consists of isolated nuclei and [^{3}H]benzo[a]pyrene (BP) (18). After they were incubated in the presence of NADPH the nuclei were washed. The covalent binding of [^{3}H]BP to nuclear DNA, RNA, histones and nonhistone proteins were analyzed by procedures involving stepwise enzyme digestions and solvent extractions (18). It was observed that an incubation mixture containing 30 μM BP and nuclei corresponding to 3.5 mg of protein per ml, and a

---

[1] Present address: Department of Chemistry, Massachusetts Institute of Technology, Cambridge, Massachusetts 02139.

[2] Abbreviations used are: AHH, aryl hydrocarbon hydroxylase; BP, benzo[a]pyrene; MC, 3-methylcholanthrene; PB, phenobarbital; MC (PB) nuclei or microsomes, nuclei or microsomes isolated from MC-(PB) treated rats.

30-min incubation at 37° to be optimal conditions for assaying the covalent binding of BP to nuclear macromolecules. Under these conditions, low-level binding of BP to nuclear macromolecules was observed with control nuclei (Table I). Treatment of the animals with 3-methylcholanthrene (MC) increased the AHH of the nuclear envelope by six times and increased the level of binding two to six times. Addition of rat liver microsomes increased the binding of BP metabolites to macromolecules of control nuclei by 10- to 15-fold. Microsomes also enhanced the binding to MC nuclear macromolecules by two- to fivefold. In the presence of microsomes, the systems with control nuclei and MC nuclei had comparable levels of carcinogen bound to DNA, RNA, histone and the nonhistone fractions (Table I, Experiments 3 and 6).

TABLE I. Binding of [^{3}H]BP Metabolites to Macromolecules of Liver Nuclei[a]

Experiment	NADPH	DNA	RNA	Histone	Nonhistone
1. Control nuclei (AHH activity, 33 units)	-	9.3	71.6	11.6	36.6
2. Control nuclei	+	18.1	106.4	17.8	37.7
3. Control nuclei + liver microsomes	+	236.2	1228.0	218.8	531.1
4. MC nuclei (AHH activity, 210 units	-	9.5	62.3	10.7	29.3
5. MC nuclei	+	117.7	357.7	87.8	130.0
6. MC nuclei + liver microsomes	+	260.4	1345.0	269.5	613.5
7. MC nuclei	-	10.2	67.3	15.7	47.8
8. MC nuclei + 8 nmoles P-448[b]	-	11.7	70.7	15.1	38.5

[a]Each incubation contained liver nuclei (7 mg protein) or liver nuclei plus MC liver microsomes (4 mg protein) in 2 ml. Binding was expressed as pmoles of BP bound per mg of macromolecules. One unit of AHH activity corresponded to the formation of 1 pmole of 3-hydroxybenzo[a]pyrene per min per mg protein. The microsomes had an AHH activity of 5,400 units.
[b]Partially purified cytochrome P-448 from MC rat liver microsomes (29).

The binding values observed in the absence of NADPH in Experiments 1, 4, 7 and 8 of Table I are believed to be background values, rather than due to any one-electron transfer from the carcinogen to ferric cytochrome P-450 as has been suggested by others (23). Several lines of observations are in support of this conclusion.

1. It was consistently observed during the course of this study that MC nuclei which have a higher cytochrome P-450 content did not bind a higher level of [^{3}H]BP than control nuclei in the absence of NADPH.
2. The addition of microsomes to nuclei in the absence of NADPH did not result in a higher level of binding.
3. The addition of partially purified cytochrome P-448 also did not result in significant increases in the binding (Table I, Experiment 8).
4. These background values were lower if the incubation was carried out at 0-4°, a temperature less favorable than 37° for the hydrophobic binding of [^{3}H]BP to macromolecules.
5. Reduction of ferric cytochrome P-450 was not observed, as judged by the absorption spectrum in the presence of CO, on the addition of BP to microsomes or solubilized cytochrome P-448 under anaerobic conditions.

The binding of metabolically activated [^{3}H]BP to the components of rat lung nuclei was also studied (18). The AHH activity of the lung nuclear envelope was very low but inducible by the treatment with MC. Very low or no binding with control nuclei and low levels of binding with MC nuclei were observed. However, the addition of liver microsomes resulted in binding values comparable to those obtained with liver nuclei. The addition of lung microsomes also increased the level of binding, although they were much less effective than liver microsomes. In all the studies, the levels of BP binding to RNA, nonhistone protein, DNA and histone were parallel, and followed the order RNA > nonhistone proteins > DNA ⩾ histones. One factor that may have contributed to the different levels of binding is the accessibility of the macromolecules to the activated form of BP; the RNA and nonhistone proteins are probably more exposed than the DNA and histones in the chromatin.

The above experiments demonstrated that a carcinogen activated by the microsomes can enter into the nucleus and react with its components. On the other hand, the nucleus also possesses enzyme systems which can activate carcinogens proximate to the target tissue. On the basis of these analyses, it was suggested that both the endoplasmic reticulum and nuclear envelope were potentially important sites of carcinogen activation (18). Consistent with this concept, increased binding of BP to nuclear DNA in the presence of microsomes has also been reported by Alexandrov *et al.* (1) and Jernstrom *et al.* (8). On the other hand, Rogan and Cavalieri (22) and Vaught and Bresnick (26) have observed that upon addition of microsomes to isolated nuclei, the level of bound BP was reduced. It may be inferred from the last two reports that the metabolic role of microsomes is one of detoxification.

In order to resolve the existing conflict and further substantiate the roles of microsomes and the nuclear envelope in carcinogen activation, we have examined the effect of added microsomes on BP binding to nuclear macromolecules under a variety of conditions.

The effect of BP concentration on the binding to the DNA of MC nuclei is shown in Figure 1. A 30-min incubation time was used in this experiment. In agreement with the results of Vaught and Bresnick (26), the addition of microsomes reduced the levels of BP binding when 1 or 2 μM BP were used. With 30 μM BP, however, the microsomes increased the level of binding, consistent with our previous observation (18). Furthermore, when equal amounts of denatured microsomes were added to the incubation systems, the

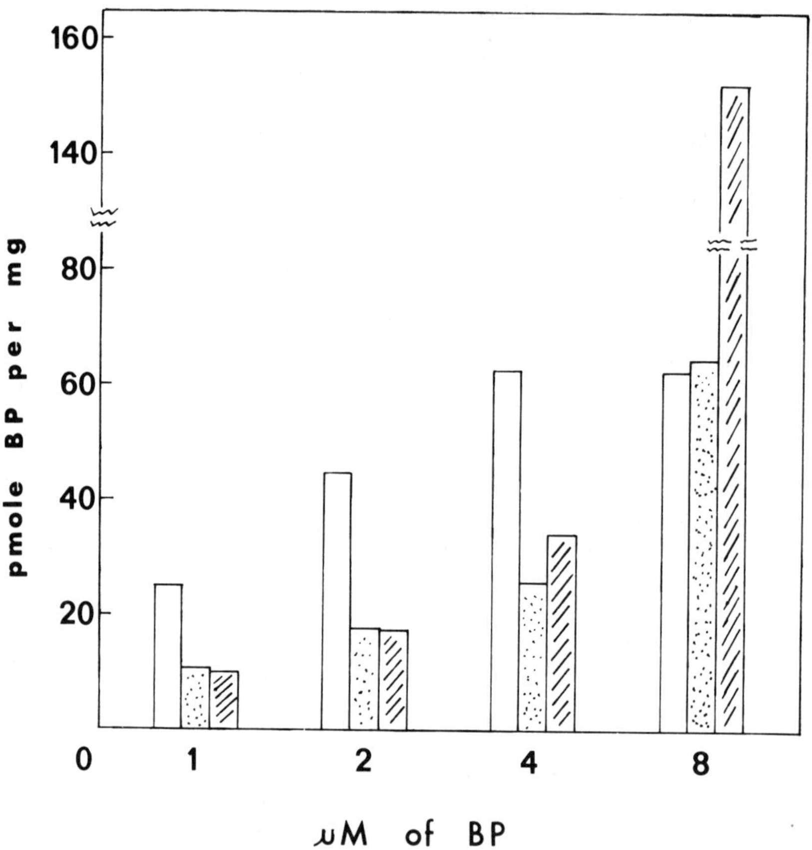

FIGURE 1. Effect of benzo[a]pyrene concentration on the binding to nuclear DNA. Each incubation contained MC nuclei (5 mg protein) or MC nuclei plus native or denatured MC microsomes (4 mg protein) in 2 ml. [^{3}H]BP was added in 0.02 ml of acetone, and the incubation was at 37° for 30 min. Binding is expressed as pmoles of [^{3}H]BP bound per mg of DNA. □, MC nuclei; ▣, MC nuclei plus denatured microsomes; ▨, MC nuclei plus microsomes. The data were replotted from reference 19.

levels of binding were reduced in the experiments with 1, 2 and 4 μM of BP. The reduced levels of binding were also reflected in control values (no NADPH added). These results suggested that due to the physical binding of BP to microsomes, the quantity of BP available for metabolism by the nuclei was reduced. This resulted in a lower covalent binding of BP to nuclear macromolecules with the addition of microsomes when lower BP concentrations were used. A comparison between the levels of binding observed with denatured and active microsomes indicates that the metabolism of BP by active microsomes contributed to the total BP binding to nuclear components. Such an effect is not apparent with low BP concentrations but becomes clear as the concentration of BP increased.

The effect of incubation time on the binding of [^{3}H] BP metabolites to nuclear macromolecules is shown in Figure 2. MC nuclei, active or denatured microsomes, 2 μM [^{3}H] BP and NADPH were incubated for the indicated time at 37°. With short incubation times (2 to 8 min) the level of binding was greater when native MC microsomes were included. The difference diminished with an incubation time of 12 min and disappeared with 30 min of incubation. Microsomes also increased the binding of BP to nuclear components with short incubation times when the experiment was repeated with 1 μM BP.

The binding of BP metabolites to the macromolecules of control nuclei was also studied. The levels of binding in the presence of active microsomes were compared with those obtained in the presence of denatured microsomes. At all the BP concentrations tested, control microsomes were able to increase the binding. Addition of MC microsomes resulted in much higher levels of binding, comparable to those obtained with MC nuclei. It is therefore apparent that metabolism by microsomes can increase the level of BP metabolites bound to nuclear components, even when low concentrations of BP are used in the incubation system.

The distribution of [^{3}H] BP physically bound to the components in the incubation system was studied in a separate experiment. With 2 μM BP in the mixture, a large portion of BP (68%) was bound to microsomes, 22% to nuclei and 10% was in the supernatant. In the presence of NADPH, the fraction of BP physically associated with microsomes decreased to 35% and that with nuclei decreased to 14%. About half of the BP molecules were found in the supernatant. Similar results were also observed with 30 μM [^{3}H] BP (19).

The above results suggest that the microsomes have two opposite effects on the metabolism of BP: Effect A, the metabolic activation of BP by microsomes; and Effect B, the physical binding of BP to microsomes causing a decrease in the effective concentration of BP to nuclear enzymes. The K_m of the MC-induced aryl hydrocarbon hydroxylase system has been reported to have values ranging from 0.3 to 2.4 μM BP (2,14,21). With 30 μM BP added to incubations containing microsomes, the decreased BP concentration should still be much higher than the K_m; therefore, Effect A outweighs Effect B and an enhanced level of binding is seen. With lower BP concentrations, when the metabolism is limited by substrate, *e.g.*, 2 *μM*, Effect B outweighs Effect A and decreased binding is seen. Under these conditions, the metabolism by microsomes should still contribute to the binding. This is demonstrated by the experiments with shorter incubations (Figure 2). Such

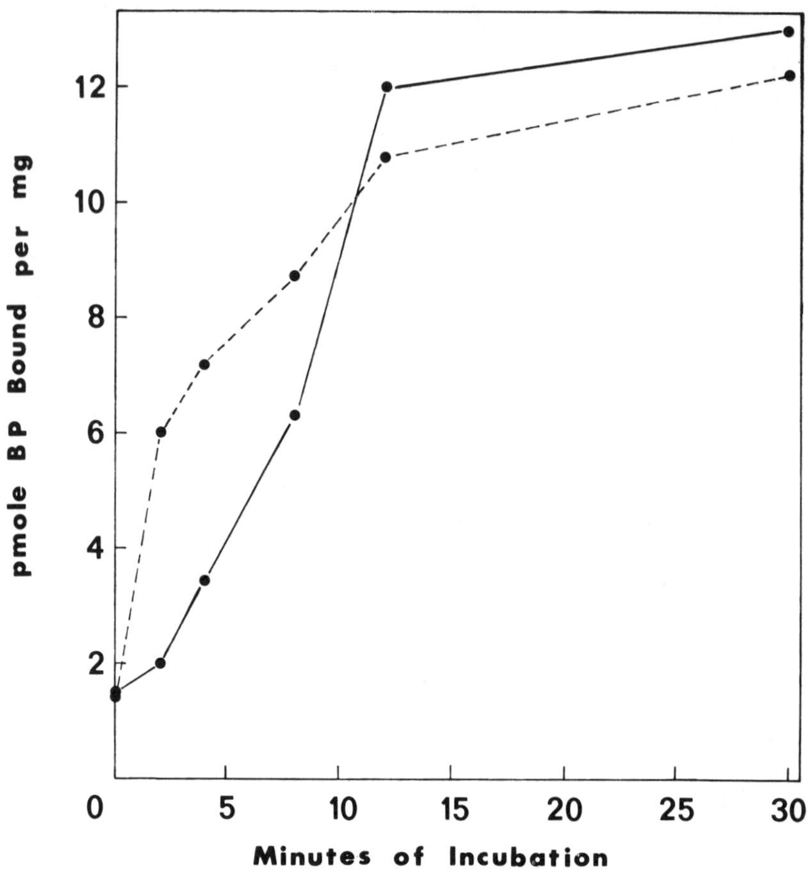

FIGURE 2. Effect of incubation time on the binding of $[^3H]$BP metabolites to nuclear DNA. MC nuclei (5 mg protein) were incubated with native or denatured MC microsomes (4 mg protein) at 37° for the indicated time. The concentration of $[^3H]$BP was 2 *μM*. The zero time samples were kept on ice. ●– –●, with native microsomes; ●——●, with denatured microsomes. The data were replotted from reference 19.

an effect is less apparent with longer incubations, due possibly to the exhaustion of the substrate BP (Figures 1 and 2). With control nuclei, which have very low AHH activity, Effect A outweighs Effect B even at low BP concentrations when either control or MC microsomes are used.

PROPERTIES OF THE NUCLEAR AND MICROSOMAL MONOOXYGENASE ENZYMES

In assessing the role of the nuclear monooxygenase system in carcinogen activation, an important question is whether the properties of this system are different from those of the microsomal monooxygenase. We have used several methods of nuclear isolation; Kasper's method (10) gave us the cleanest preparation. Two criteria were used to examine the purity of the nuclear preparation: (1) the absence of endoplasmic membrane contamination as determined by electron microscopy (Figure 3), and (2) the consistency of the cytochrome P-450 content when a sample of isolated nuclei was subjected to a second discontinuous sucrose gradient centrifugation. Our nuclear preparations satisfied these criteria; nevertheless, there is not proof that the nuclear samples were completely free from microsomal contamination. The CO difference spectra of reduced nuclei and microsomes are shown in Figure 4. Both the nuclei and microsomes from PB-treated rats had absorption peaks at 450 nm; whereas, those from MC-treated animals had peaks at 448 nm. The nuclei and microsomes of control animals also had peaks at 450 nm (spectra not shown). In some preparations, the position of absorption peaks of the microsomes and nuclei of the same animals could differ by as much as 1 nm, but on the average a consistent difference between the nuclei and microsomes was not observed in fifteen preparations. Similar results were observed both in Long Evans and Sprague Dawley rats weighing 70-100 g. These results are in agreement with those reported by Bresnick *et al.* (4), but are different from those of Rogan *et al.* (24) who reported that the position of the absorption peaks of the nuclei differed from those of the microsomes by 2 nm. The reason for this conflict is not apparent.

The contents of the monooxygenase enzymes in nuclei and microsomes are shown in Table II. Both MC and PB treatments increased the nuclear cytochrome P-450 content. This is similar to the induction of microsomal cytochrome P-450, although a quantitative difference existed. On the other hand, the nuclear NADPH-cytochrome P-450 reductase was not responsive to the treatment by either MC or PB; the latter was an inducer of microsomal reductase. The AHH activity was induced sevenfold by MC in the nuclei vs fourfold in the microsomes when it was expressed on a protein basis. When expressed on the basis of pmole of cytochrome P-450, the AHH activities in the nuclei were similar to those in the microsomes. The result agrees with the concept that treatment with PB induces a form of cytochrome P-450 which has lower AHH activity than those in the control or MC microsomes. These results, however, are different from those obtained by Kasper and co-workers with rat liver nuclear envelope (5,9,11). In a series of thorough studies, these authors observed that PB treatment did not increase the cytochrome P-450 content of the nuclear envelope. With control and PB-treated rats, the AHH activity (based on per nmole cytochrome P-450) of the nuclear envelope was found to be much lower than the corresponding activity in microsomes (5).

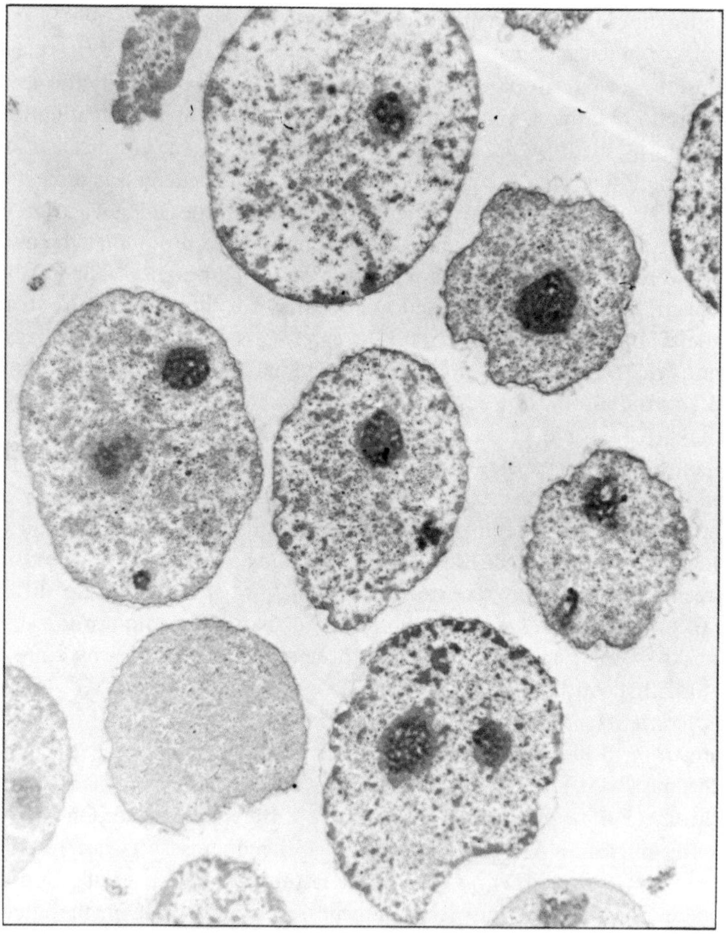

FIGURE 3. Electron micrograph of MC liver nuclei. The nuclei were fixed in glutaraldehyde and postfixed in osmium tetroxide (x 3,300).

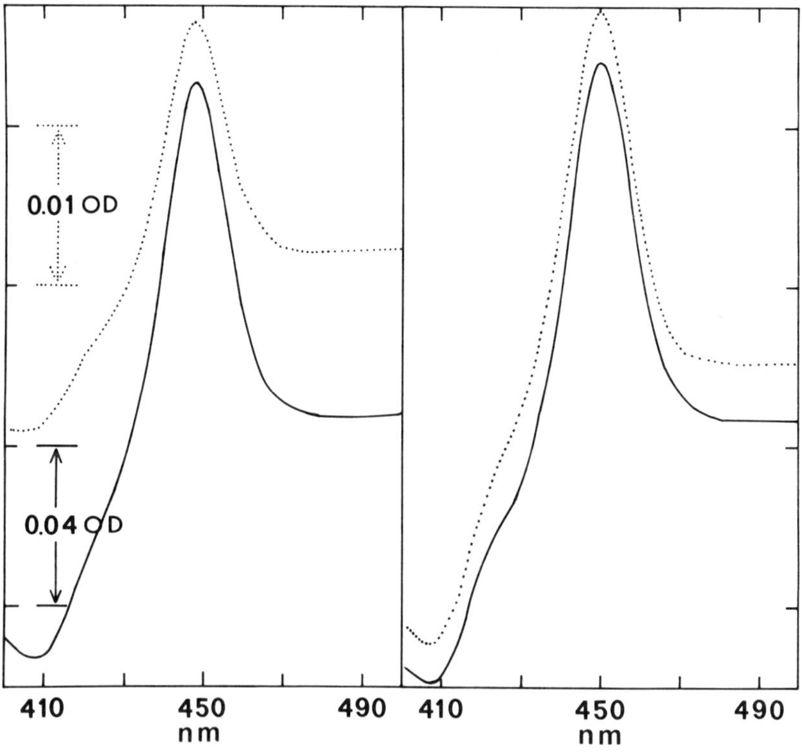

FIGURE 4. Carbon monoxide-difference spectra of nuclei and microsomes in the presence of dithionite. Samples were suspended in 0.25 *M* sucrose at concentrations (mg protein) indicated. Left panel: . . . . ., MC nuclei, 1.3 mg/ml; ———, MC microsomes, 0.6 mg/ml. Right panel: . . . . ., PB nuclei, 2.2 mg/ml; ———, PB microsomes 0.5 mg/ml.

The difference between the results obtained with whole nuclei and those with nuclear envelope would be resolved if the monooxygenase enzymes and the AHH activity is present not only in the nuclear envelope but also in the interior of the nucleus. This possibility has been put forth by Bresnick *et al.* (4). Nevertheless, more work is needed to illustrate the applicability of the histofluorescence technique (4). To demonstrate the existence of nuclear membranous structure in addition to the nuclear envelope is also essential, since membrane phospholipids are required for the monooxygenase activities (28). The data in Table III, however, do not support the hypothesis that the AHH activity is present inside the nucleus. Treatment of the nuclei with 0.5% Triton X-100, which is known to remove the nuclear envelope, abolished the nuclear AHH activity. About 90% of the original nuclear protein was

TABLE II. Induction of Monooxygenase Enzymes in Nuclei and Microsomes[a]

Sample	Cytochrome P-450	Reductase	Aryl Hydrocarbon Hydroxylase	
	pmol/mg	nmole/min/mg	pmole/min/mg	pmole/min/pmol P-450
Control Nuclei	30 ± 16	10.6 ± 2.1	22 ± 12	0.73
MC Nuclei	82 ± 25	10.5 ± 0.15	158 ± 53	1.92
PB Nuclei	80 ± 26	11.1 ± 2.5	27 ± 16	0.34
Control Microsomes	890 ± 210	178	635 ± 75	0.71
MC Microsomes	1,500 ± 250	182	2,533 ± 310	1.69
PB Microsomes	2,000 ± 340	322	853 ± 135	0.42

[a]Male Long Evans rats (70-100 g body weight) were injected, i.p., daily with 25 mg of 3-methylcholanthrene or 75 mg of phenobarbital per kg of body weight, for 3 or 4 days. Control animals received no treatment. The monooxygenase enzymes were assayed according to previous methods (29,30). Means and standard deviation values were obtained from 5 to 7 preparations of nuclei and 3 to 7 preparations of microsomes. The microsomal reductase was from one preparation of microsomes.

TABLE III. Effect of Triton on Nuclear Aryl Hydrocarbon Hydroxylase Activity[a]

Sample	AHH Activity pmole/min/mg
1. MC nuclei	218
2. MC nuclei after Triton wash	0.4
3. Dialyzed supernatant of the wash	0
4. MC microsomes	2,585
5. MC microsomes + Triton treated nuclei	3,459
6. MC microsomes + dialyzed supernatant of the wash	43

[a]A suspension of MC nuclei (11 mg protein/ml) was washed with 0.5% Triton X-100. The supernatant of the wash was dialyzed and concentrated and was used in Experiments 3 and 6. The nuclei were washed again with a pH 7.5 buffer containing 0.25 *M* sucrose, 50 m*M* tris·HCl, 25 m*M* KCl and 5 m*M* $MgCl_2$ to remove Triton. The washed nuclei were used in Experiments 2 and 5 at a concentration of 0.5 mg protein per ml. MC microsomes were used in Experiments 4-6, at 0.01 mg protein per ml; the AHH activity in these experiments was expressed on the basis of per mg microsomal protein.

recovered after the two-step washing and the nuclei retained their spherical shape. The lack of AHH activity of the washed nuclei cannot be attributed to the inhibitory action of the Triton. Upon mixing these nuclei with a small amount of microsomes, the microsomal AHH activity was not inhibited. The nuclear AHH activity was also not recovered in the supernatant of the wash, possibly due to the inhibition by the residue of Triton in the sample. In a separate experiment, Triton X-100 was found to stimulate the nuclear AHH activity slightly at low concentrations and inhibit the activity at high concentrations. In considering the differences between the results obtained with whole nuclei and with nuclear envelope, it is still possible that the differences are due to either the microsomal contamination of the nuclear preparations or due to the procedures of preparing the nuclear envelope. We have observed that sonicating a nuclear suspension (5 mg protein per ml) at ice temperature with a Sonifier cell disruptor (microtip) at 50 watts for two 15-sec intervals caused a 52% denaturation of cytochrome P-450 in PB nuclei.

Analogous to microsomes, rat liver nuclei can activate BP to r-7, t-8-dihydroxy-t-9,10-oxy-7,8,9,10-tetrahydrobenzo[a]pyrene (diol epoxide I) (20), a proposed proximate or ultimate carcinogen (7,12,13,27). With nuclei from control, MC- and PB-treated rats, the pattern of BP metabolites was similar to that of the corresponding microsomes, although quantitatively reduced (20). Similar results concerning BP metabolism were also observed by several other groups (1,3,5,8) with either nuclei or nuclear envelope. Thus, the nuclear and microsomal monooxygenase systems are grossly similar in the production of metabolites of BP, although the nuclear system is quantitatively less and qualitatively different from the microsomal system in certain respects (5).

SITES OF ACTIVATION AND TRANSPORT OF CARCINOGENS IN THE CELL

The problem of carcinogen activation and detoxification appears to depend on the interplay of a variety of factors such as the rates of metabolism of a carcinogen and its metabolites by activating and detoxification enzymes, the rates of transport of these compounds between organelles, and the chemical reactivity of these metabolites. Several possible mechanisms and routes of carcinogen activation are shown in Figure 5 (17).

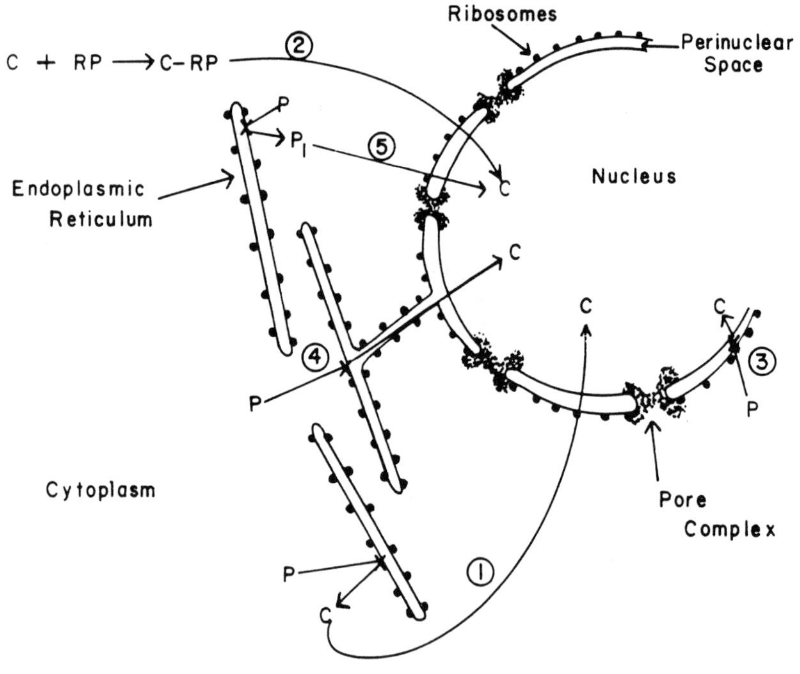

FIGURE 5. A graphic representation of some possible routes of carcinogen activation leading to binding with nuclear macromolecules. The abbreviations used are as follows: P, procarcinogen; P_1, proximate carcinogen; C, ultimate carcinogen; RP, receptor protein; X, metabolism by the mixed-function oxidase system. The transport routes (1 to 5) and activation steps are described in the text.

1. The carcinogen is activated by the AHH of the endoplasmic reticulum, travels through the cytoplasm and enters the nucleus.
2. The carcinogen activated by the endoplasmic reticulum binds to a receptor protein, and the complex is transported into the nucleus (15).
3. The carcinogen is activated by the monooxygenase of the nuclear envelope and enters the nucleus.
4. The carcinogen is activated by the endoplasmic reticulum at a point where it continues with the nuclear envelope. The activated carcinogen then moves through the lumen into the nucleus.
5. The procarcinogen is metabolized to a proximate carcinogen by the endoplasmic reticulum and to the ultimate carcinogen by the nuclear envelope (8).

At the present time, there is no good experimental result for assessing the relative importance of each mechanism. Since the pattern of BP metabolites produced by the nuclei is similar to that produced by microsomes, it appears that the site of carcinogen activation is determined not by the specificity of the enzyme but by the rates of metabolism and transport of the carcinogen and its metabolites in the cell.

ACKNOWLEDGMENTS

This work was supported by Grant CA-16788 from the National Cancer Institute.

REFERENCES

1. Alexandrov, K., Brookes, P., King, H. W. S., Osborne, M. R., and Thompson, M. H. (1974): Comparison of the metabolism of benzo[a]pyrene and binding to DNA caused by rat liver nuclei and microsomes. *Chem.-Biol. Interact.* 12:269-277.
2. Alvares, A. P., Schilling, G., Garbut, A., and Kuntzman, R. (1970): Studies on the hydroxylation of 3,4-benzpyrene by hepatic microsomes. Effect of albumin on the rate of hydroxylation of 3,4-benzpyrene. *Biochem. Pharmacol.* 19:1449-1455.
3. Bresnick, E., Stoming, T. A., Vaught, J. B., Thakkar, D. R., and Jerina, D. M. (1977): Nuclear metabolism of benzo[a]pyrene and (±)-*trans*-7,8-dihydroxy-7,8-dihydrobenzo[a]pyrene. *Arch. Biochem. Biophys.* 183:31-37.
4. Bresnick, E., Vaught, J. B., Chuang, A. H. L., Stoming, T. A., Bockman, D., and Mukhtar, H. (1977): Nuclear aryl hydrocarbon hydroxylase and interaction of polycyclic hydrocarbons with nuclear components. *Arch. Biochem. Biophys.* 181:257-269.
5. Fahl, W. E., Jefcoate, C. R., and Kasper, C. B. (1978): Characteristics of benzo[a]pyrene metabolism and cytochrome P-450 heterogeneity

in rat liver nuclear envelope and comparison to microsomal membrane. *J. Biol. Chem.* 253:3106-3113.

6. Heidelberger, C. (1975): Chemical carcinogenesis. *Ann. Rev. Biochem.* 44:79-121.
7. Huberman, E., Sachs, L., Yang, S. K., and Gelboin, H. V. (1976): Identification of mutagenic metabolites of benzo[a]pyrene in mammalian cells. *Proc. Nat. Acad. Sci., U.S.* 73:607-611.
8. Jernstrom, B., Vadi, H., and Orrenius, S. (1976): Formation in isolated rat liver microsomes and nuclei of benzo[a]pyrene metabolites that bind to DNA. *Cancer Res.* 36:4107-4113.
9. Kasper, C. (1971): Biochemical distinctions between the nuclear and microsomal membranes from rat hepatocytes. *J. Biol. Chem.* 246: 577-581.
10. Kasper, C. B. (1974): Isolation and properties of the nuclear envelope. *Methods Enzymol.* 31:279-292.
11. Khandwala, A. S., and Kasper, C. B. (1973): Preferential induction of aryl hydrocarbon hydroxylase activity in rat liver nuclear envelope by 3-methylcholanthrene. *Biochem. Biophys. Res. Commun.* 54: 1241-1246.
12. Lehr, R. E., Yagi, H., Thakker, D. R., Levin, W., Wood, A. W., Conney, A. H., and Jerina, D. M. (1978): The bay region theory of polycyclic aromatic hydrocarbon-induced carcinogenicity. In: *Carcinogenesis, Vol. 3,* P. W. Jones and R. I. Freudenthal, Eds., pp. 231-241, Raven Press, New York.
13. Levin, W., Wood, A. W., Chang, R. L., Slaga, T. J., Yagi, H., Jerina, D. M., and Conney, A. H. (1977): Marked differences in the tumor-initiating activity of optically pure (+)- and (-)-*trans*-7,8-dihydroxy-7,8-dihydrobenzo[a]pyrene on mouse skin. *Cancer Res.* 37:2721-2725.
14. Lu, A. Y. H., and West, S. (1972): Reconstituted liver microsomal enzyme system that hydroxylates drugs, other foreign compounds, and endogenous substrates. III. Properties of the reconstituted 3,4-benzpyrene hydroxylase system. *Mol. Pharmacol.* 8:490-500.
15. Mainigi, K. D., and Sorof, S. (1977): Evidence for a receptor protein of activated carcinogen. *Proc. Nat. Acad. Sci., U.S.* 74:2293-2296.
16. Miller, E. C. (1978): Some current perspectives on chemical carcinogenesis in humans and experimental animals. *Cancer Res.* 38:1479-1496.
17. Pezzuto, J. M. (1977): Studies on the metabolic activation of benzo[a]-pyrene and subsequent binding to nuclear macromolecules. PhD Thesis, pp. 81-82, College of Medicine and Dentistry of New Jersey, Newark.
18. Pezzuto, J. M., Lea, M. A., and Yang, C. S. (1976): Binding of metabolically activated benzo[a]pyrene to nuclear macromolecules. *Cancer Res.* 36:3647-3653.
19. Pezzuto, J. M., Lea, M. A., and Yang, C. S. (1977): The role of microsomes and nuclear envelope in the metabolic activation of benzo[a]-pyrene leading to binding with nuclear macromolecules. *Cancer Res.* 37:3427-3433.

20. Pezzuto, J. M., Yang, C. S., Yang, S. K., McCourt, D. W., and Gelboin, H. V. (1978): Metabolism of benzo[a]pyrene and (-)-*trans*-7,8-dihydroxy-7,8-dihydrobenzo[a]pyrene by rat liver nuclei and microsomes. *Cancer Res.* 38:1241-1245.
21. Robie, K. M., Cha, Y. -N., Talcott, R. E., and Schenkman, J. B. (1976): Kinetic studies of benzpyrene and hydroxybenzpyrene metabolism. *Chem. -Biol. Interact.* 12:285-297.
22. Rogan, E. G., and Cavalieri, E. (1974): 3-Methylcholanthrene-inducible binding of aromatic hydrocarbons to DNA in purified rat liver nuclei. *Biochem. Biophys. Res. Commun.* 58:1119-1126.
23. Rogan, E. G., Mailander, P., and Cavalieri, E. (1976): Metabolic activation of aromatic hydrocarbons in purified rat liver nuclei: Induction of enzyme activities and binding to DNA with and without monooxygenase-catalyzed formation of active oxygen. *Proc. Nat. Acad. Sci., U.S.* 73:457-461.
24. Rogan, E., Roth, R., and Cavalieri, E. (1978): Enzymology of polycyclic hydrocarbon binding to nucleic acids. In; *Carcinogenesis Vol. 3,* P. W. Jones and R. Freudenthal, Eds., pp. 265-271, Raven Press, New York.
25. Sikstrom, R., Lanoix, J., and Bergeron, J. J. M. (1976): An enzymatic analysis of a nuclear envelope fraction. *Biochim. Biophys. Acta* 448: 88-102.
26. Vaught, J., and Bresnick, E. (1976): Binding of polycyclic hydrocarbons to nuclear components in vitro. *Biochem. Biophys. Res. Commun.* 69:587-591.
27. Weinstein, I. B., Jeffrey, A., Jennette, K., Blobstein, H., Harvey, R., Harris, C., Autrup, H., Kasai, H., and Nakanishi, K. (1976): Benzo[a]pyrene diol epoxides as intermediates in nucleic acid binding in vitro and in vivo. *Science* 193:592-595.
28. Yang, C. S. (1977): The organization and interaction of monooxygenase enzymes in the microsomal membrane. *Life Sci.* 21:1047-1058.
29. Yang, C. S., and Strickhart, F. S. (1975): Interactions between solubilized cytochrome P-450 and hepatic microsomes. *J. Biol. Chem.* 250: 7968-7972.
30. Yang, C. S., Strickhart, F. S., and Kicha, L. P. (1978): Interaction between NADPH-cytochrome P-450 reductase and hepatic microsomes. *Biochim. Biophys. Acta* 509:326-337.

Polynuclear Aromatic Hydrocarbons, edited by P. W. Jones and P. Leber. Ann Arbor Science Publishers, Inc. Ann Arbor, MI ©1979.

SUBSTRATE SPECIFICITY OF ARYL HYDROCARBON (BENZO[a]PYRENE) HYDROXYLASE

F. B. Thomas,[1] D. M. Thomas, D. S. Shewach and N. B. Furlong

Department of Biochemistry
The University of Texas System Cancer Center
M. D. Anderson Hospital and Tumor Institute
Houston, Texas 77030

Presented here is a conceptual model for the metabolism of benzo[a]-pyrene (BP)[2] and other polycyclic aromatic hydrocarbons by the mixed-function monooxygenase system. Our research efforts in this area began in 1974, when there occurred two major developments in the understanding of BP carcinogenicity which particularly affected the course of our research. First, the K-region epoxide was being supplanted by an increasing body of evidence implicating the 7,8-dihydrodiol-9,10-epoxide as the suspected active form of BP (1,11,14,16,27). Secondly, high-pressure liquid chromatography (HPLC) was being applied to the identification and quantification of BP metabolites (7,8,20,21,28,29).

The structure of BP is shown in Figure 1. The molecule is essentially planar and consists of five coupled benzene rings. One can surmise much of the basic chemistry of this compound from looking at its structure.

---

[1]Present address: Health, Safety & Environment–Toxicology, Shell Oil Company, P.O. Box 2463, One Shell Plaza, Houston, Texas 77001.

[2]Abbreviations used: BP, benzo[a]pyrene; 1-BPOH, 1-hydroxybenzo[a]pyrene; other phenols abbreviated similarly; BP-7,8-dihydrodiol, *trans*-7,8-dihydro-7,8-dihydroxy-benzo[a]pyrene; other dihydrodiols abbreviated similarly; BP-7,8-dihydrodiol-9,10-oxide, *trans*-7,8-dihydro-7,8-dihydroxybenzo[a]pyrene-9,10-oxide; other dihydrodiol epoxides abbreviated similarly; P, bond order as calculated by Hückel method; ΔP_{arom}, difference between calculated bond order value and theoretical value (0.667) of pure benzenoid C=C bond.

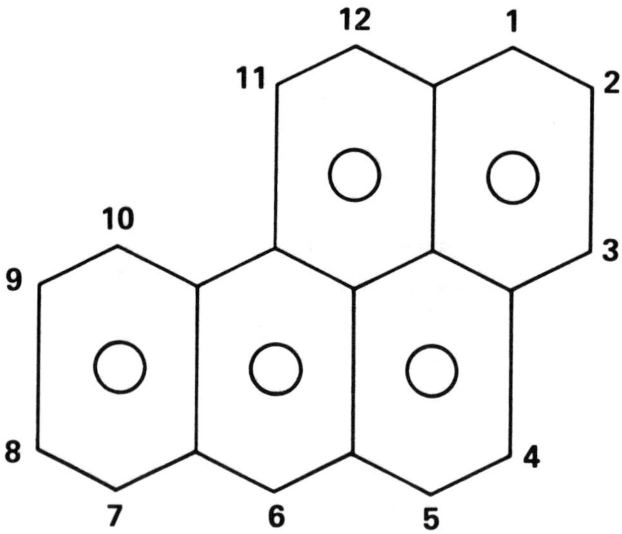

FIGURE 1. Structure and numbering of benzo[a]pyrene.

1. Like benzene, one might expect this compound to be fairly insoluble in water. Several of our early studies show that this is, in fact, the case. BP, when added to an aqueous solution, forms hydrophobic complexes with a variety of macromolecules, especially protein. We found, for example, that in order to assay BP metabolism accurately it was necessary to include bovine serum albumin to prevent loss of BP to the glass wall of our reaction vessel (24).
2. With 20 carbon atoms, one would anticipate BP to have a fairly high boiling point; reportedly this compound boils at 310-312°C. The implication of this is that BP is not volatile at normal temperatures, and when emitted into the air, it readily forms microparticulates or adsorbs to the surface of a particle.
3. BP is highly lipid-soluble, such that it can be eluted from a particle or macromolecule and dissolve into the membranes of a target cell.
4. The electrons that form the double bonds in BP cannot be localized to any one carbon atom, but migrate over the entire system. The delocalization of the π electrons imparts to the BP molecule some degree of stability and makes it chemically unreactive.

Nonetheless, it is clear that in many living tissues BP is readily metabolized to a variety of derivatives. The major pathway of BP metabolism involves a two-electron oxidation catalyzed by aryl hydrocarbon hydroxylase (AHH, E.C.1.14.14.2). This activity is located in the cellular endoplasmic reticulum and nuclear membrane, and is functionally dependent on both molecular oxygen and NADPH (3). AHH converts BP to a series of BP epoxides. From previous studies of aromatic hydrocarbon metabolism, two general rules-of-thumb seem to be applicable in this reaction: (1) epoxidation does not involve a carbon atom at a ring juncture, and (2) epoxidation does not involve a carbon atom where a substituent group (such as an hydroxyl or a methyl group) is attached. This means then that of all the C=C double bonds in the BP molecule, only seven are accessible for epoxidation—those at position 1-2, 2-3, 4-5, 7-8, 8-9, 9-10 and 11-12.

With advent of the HPLC method, a substantial improvement in the separation and quantification of BP metabolites was possible. The technique resolves the metabolites of BP into three dihydrodiol, two quinone and two phenol fractions. The identities of the BP derivatives composing each of these fractions have, to a large extent, been determined (7,8,20,21,28,29). In this regard it was interesting to note that these seven positions are not equally epoxidated. It is clear, for example, that the major BP metabolite is 3-BPOH, which we assume is formed from the opening of the 2,3-oxirane ring. By comparison the K-region epoxide at the 4,5-position is formed to a much smaller extent, while no evidence for epoxidation at the 8,9 position has been obtained. These observations led us to consider the theoretical basis for this apparent site-specificity of AHH. One possibility is that unknown steric factors are involved in the interaction between BP and AHH. An alternative hypothesis is that there may be small, but important, differences in the electronic structure of the C=C double bonds of the BP molecule.

QUANTUM CHEMICAL STUDIES

Quantum chemical methods have been applied to the polycyclic aromatic hydrocarbons for many years. Perhaps the best known result of such efforts is the classical K- and L-region hypothesis of the Pullmans (17). More recently, Jerina *et al.* (11) have used quantum chemical methods to predict the ease of formation of carbonium ions from diol epoxides. We had available in our laboratory a computer program written by Dr. Bill Novak (15). Using this program we were able to calculate a variety of molecular orbital indices by means of the Hückel linear-combination-of-atomic-orbitals method, the simplest of the quantum chemical approaches. It soon became apparent that the C=C double bonds in the BP molecule do indeed differ in their electronic structure.

One of the molecular orbital parameters examined was bond order (P). Essentially the P-value gives us a measure of how single-bonded or double-bonded a particular chemical bond is, such that a perfect single bond is defined to have a P-value of 0, while a perfect double bond has a value of 1. Table I displays the calculated P-values for the accessible C=C double bonds

TABLE I. Correlation of Bond Order (P) with the Percentage of Metabolic Epoxidation of Benzo[a]pyrene as Reported by Yang *et al.* (29) and Holder *et al.* (7).

				Control (%)		3-MC (%)
Bond	P	ΔP_{arom}	Metabolite	Yang *et al.*	Holder *et al.*	Holder *et al.*
C_8=C_9	0.609	-0.058	—	—	—	—
C_2=C_3	0.658	-0.009	3-BPOH	36	26.6	24.8
C_1=C_2	0.680	0.013	—	—	—	—
C_9=C_{10}	0.710	0.043	9-BPOH	3-13 } 18-38	6.1 } 19.9	2.3 } 21.8
			9,10-BP dihydrodiol	15-25	13.8	19.5
C_7=C_8	0.721	0.054	7, 8-BP dihydrodiol	12-14	9.7	11.8
C_{11}=C_{12}	0.754	0.087	11,12-BP dihydrodiol	—	—	—
C_4=C_5	0.784	0.117	4, 5-BP dihydrodiol	8	9.2	8.7

P = Bond Order, ΔP_{arom} = Difference Between Calculated Bond Order Value and Theoretical Value (0.667) of Pure Benzenoid C=C Bond of the BP molecule. Also included are estimates of the extent of metabolism at each of these bonds by the AHH activity of untreated and 3-methylcholanthrene-induced microsome preparations as determined by HPLC methods and reported in the literature (7,29). There is a general correlation between bond-order values and the percentage of metabolites formed at each bond. In subsequent studies made on BP and other polycyclic aromatic hydrocarbons, we have found that this general correlation can be improved by looking at how closely the P-value of a particular bond approximates that of the theoretical aromatic double-bond P-value of 0.667 (ΔP_{arom}). Our interpretation of these data is that aryl hydrocarbon hydroxylase preferentially epoxidates benzenoid double bonds with a P-value of 0.667. As the calculated bond-order value of a particular bond deviates from this optimal P-value, the probability of its epoxidation by AHH decreases. Thus the $C_2{=}C_3$ double bond of BP with a bond-order value of 0.658, deviates from the preferred P-value by only -0.009, and is readily metabolized to BP-2,3-epoxide. This epoxide subsequently rearranges via the NIH shift mechanism to form 3-BPOH. The bond at $C_4{=}C_5$ on the other hand assumes a more alkene character with a bond-order value of 0.784, and is epoxidated to a less efficient extent.

In explaining these observations to people who are not familiar with molecular orbital terminology, we have found it convenient to note that the bond-order value correlates with bond length. It then becomes easy to conceptualize that the active site of AHH prefers bonds of a benzenoid bond length of approximately 1.39 Å. As bonds assume a more alkene character, that is become shorter in length, they fit less efficiently in this active site and are epoxidated to a lesser extent. It likewise becomes possible to explain that the bond at $C_8{=}C_9$, which is larger than the benzenoid length of 1.39 Å, is too large to fit into the active site, thus explaining the lack of epoxidation at this position (7).

In order to evaluate the electronic structure of the various BP epoxides, we assumed the cyclicized oxygen to be a type of ether linkage and utilized the Coulomb and resonance integral estimates suggested by Pullman and Pullman (17). While such a treatment must certainly contain some error for these strained-ring structures, its use is consistent with the general approach of the Hückel method and has proven satisfactory in predicting the relative chemical properties of a series of BP epoxides.

Epoxidation of BP results in a loss of aromaticity and is reflected in our calculations as a decrease in resonance energy. As shown in Figure 2, the various BP epoxides are predicted by our calculations to differ in their relative chemical stabilities. The least stable of these is BP-8,9-oxide which, as mentioned above, is not believed to be formed metabolically since previous workers have failed to identify the stable oxepin tautomer as a BP metabolite (7).

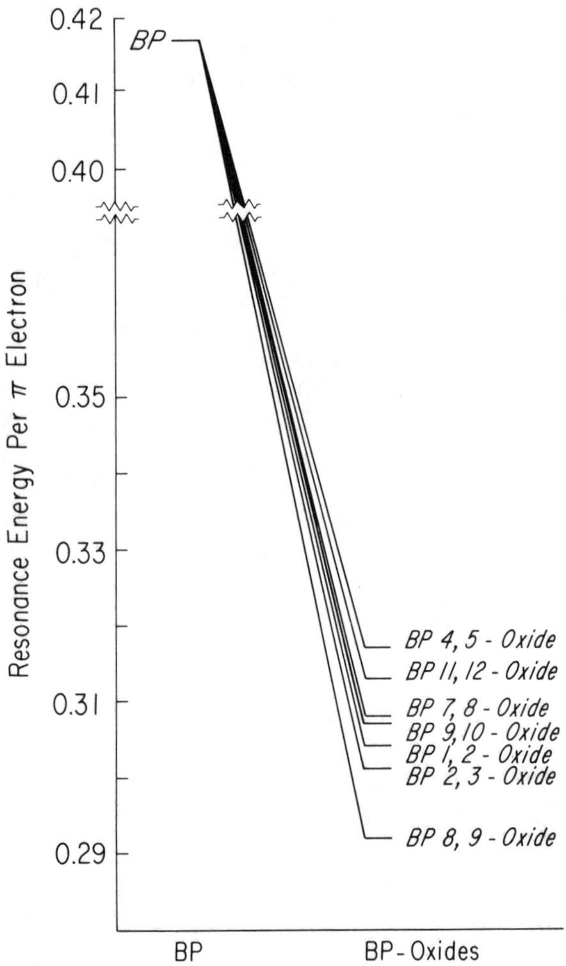

FIGURE 2. Relative stabilities of the epoxide derivatives of benzo[a]pyrene as predicted by their calculated resonance energy values.

BP-2,3-oxide is also predicted to be unstable by these calculations. Structural constaints established by adjacent rings presumably prevent rearrangement to an oxepin. Instead, the 2,3-epoxide spontaneously isomerizes to a phenol, 3-BPOH, via a nonenzymatic mechanism known as the NIH shift (13). Phenol formation is also expected from the BP epoxides of intermediate stabilities–BP-1,2-oxide, BP-9,10-oxide and BP-7,8-oxide which would be expected to rearrange primarily to 1-BPOH, 9-BPOH and 7-BPOH, respectively. This prediction has, in fact, been confirmed by the studies of Holder *et al.* (7), who found that BP-9,10-oxide and BP-7,8-oxide could be extracted from an *in vitro* reaction mixture, but were unstable during HPLC. In comparison, BP-4,5-oxide, predicted by the Hückel calculations to be the most stable of the BP epoxides, was found to be both extractable and stable during HPLC.

The stability of BP epoxides is thought to be an important consideration in the metabolism, toxicity and carcinogenicity of this hydrocarbon. Metabolically derived BP epoxides, with the exception of BP-2,3-oxide, appear to have physiological half-lives of sufficient length to allow varying degrees of interaction with cellular components. Epoxide hydratase, a microsomal activity, converts BP epoxides to the corresponding *trans*-dihydrodiols. The microsomal content of this enzyme is inducible by polycyclic hydrocarbons and, in 3-methylcholanthrene-treated systems, can cause a significant shift in the HPLC distribution of BP metabolites away from the phenolic fractions (7). The interaction of BP epoxides with a cytosol enzyme, gluatatione-S-epoxide transferase, yields water-soluble glutathione conjugates (3). BP epoxides have also been shown to covalently react with cellular macromolecules such as DNA, RNA and protein, and it is thought that these reactions may be critical to the mechanisms of cytotoxicity and carcinogenicity (27).

It must be reemphasized that in applying the Hückel method we have made several broad assumptions; we feel that the generality of our observations must be confirmed in the future by more sophisticated quantum chemical methods applied to a larger series of polycyclic aromatic hydrocarbons. Methods applicable to large molecular systems like BP have only recently become available, and in this regard we have initiated collaborative studies with the Department of Chemistry at the University of Texas at Austin. To date, these studies, using self-consistent field methods, have essentially reproduced the values obtained by the simple Hückel method. Similarly, X-ray crystallographic estimates of bond lengths in crystalline BP are in general agreement with our Huckel predictions (4,10). Therefore, we offer our observations as a useful conceptual model. Specifically it has led us to postulate and to test certain hypotheses, and to make what we believe are important observations concerning the metabolism of BP.

CHARACTERISTICS OF AHH SUBSTRATES

The first hypothesis relates to the characteristics of the substrates of AHH. With the realization that AHH seems to prefer a benzenoid-type bond, we postulated that the best substrate for AHH should be benzene with six perfect benzenoid double bonds. To test this hypothesis we developed a quantitative radioassay of BP metabolism which is not sensitive to interference from the fluorescence of other polynuclear aromatic hydrocarbons

(24). Essentially the method involves the timed spotting of the metabolic reaction mixture on a silica gel TLC plate, and utilizes the mobile phase of the chromatographic system to extract BP and its metabolites away from a protein target molecule which remains precipitated at the origin. While getting around the fluorescence problem, this assay also has a potential advantage in that all BP derivatives are spotted (*i.e.*, quantitation is not influenced by the differential extraction efficiencies of BP metabolites). Therefore, quantitative estimates of BP metabolism can be made from changes in the radioactivity associated with any of several chromatographically distinct bands or by that no longer extractable from the macromolecular target.

As shown in Figure 3, benzene did not inhibit BP activation even at concentrations up to 100 times that of BP in the reaction mix. Similarly naphthalene, phenanthrene and anthracene failed to inhibit BP metabolism. Significant inhibition was seen only with compounds the size of benz[a]-anthracene and larger. When these larger polycyclic hydrocarbons were present in equimolar concentrations relative to BP, approximately 50% inhibition of AHH activity was observed and we have been unable to identify any one compound as being the better competitive inhibitor of BP metabolism.

The significance of these observations is that there must exist at least two arene-specific epoxidases in the microsomal mixed-function monooxygenase system. The microsomal activity generally referred to as AHH only metabolizes large polycyclic aromatic hydrocarbons of four or more rings. Therefore, a second arene-specific epoxidase must exist to account for the oxidative metabolism of smaller polycyclic aromatic hydrocarbons.

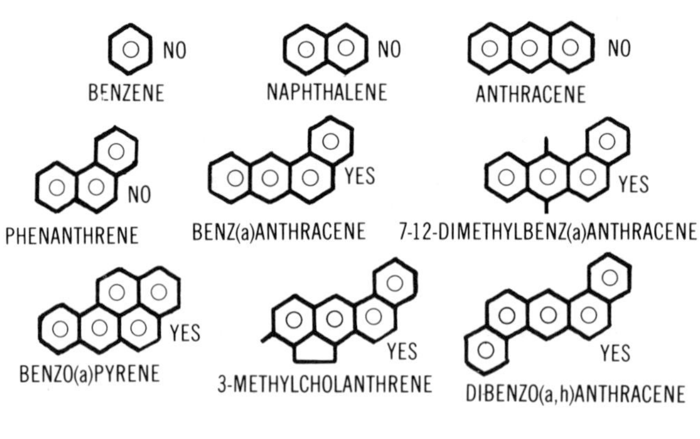

FIGURE 3. Inhibition of benzo[a]pyrene metabolism by polycyclic aromatic hydrocarbons.

Planarity of AHH substrates does not appear to be an important factor. 7,12-Dimethylbenz[a]anthracene, for example, is known to be significantly nonplanar, yet effectively inhibits epoxidation. Likewise, we have observed that 17β-estradiol is a competitive inhibitor of BP metabolism. This steroid may be considered essentially a cycloalkane-substituted phenol. It is, of course, nonplanar and is interesting because of the available C_1=C_2 bond (calculated P-value = 0.668). 2-Hydroxylation is a recognized metabolic pathway for this compound and may involve an epoxide intermediate. Surprisingly, the synthetic estrogen diethylstilbestrol is not effective as an inhibtor of BP metabolism. Thus, while its structure is sufficiently similar to the natural steroid to allow its interaction with cellular estrogen receptors, it is not sufficiently similar to replace estradiol as a substrate for AHH.

STUDIES ON BP-7,8-DIHYDRODIOL

Because of the suggestions that "bay-region" diol epoxides may be of critical importance in the carcinogenic process, we have used the Hückel program to approximate the electronic structure of BP-7,8-dihydrodiol. As shown in Table 2, the C=C double bonds are not equivalent in terms of their calculated bond orders. On the basis of our previous observations, we would

TABLE 2. Calculated Bond Order (P) Values for Benzo[a]pyrene-7,8-dihydrodiol.

Bond	P	ΔP_{arom}
C_8=C_9	0.273	-0.394
C_1=C_2	0.669	0.002
C_2=C_3	0.669	0.002
C_{11}=C_{12}	0.767	0.100
C_4=C_5	0.769	0.102
C_9=C_{10}	0.853	0.186

P = Bond Order, ΔP_{arom} = Difference Between Calculated Bond Order Value and Theoretical Value (0.667) of Pure Benzenoid C=C Bond

predict that the most likely sites of epoxidation should be the equivalent bonds at C_1=C_2 and C_2=C_3. The important C_9=C_{10} bond assumes a substantial alkene character (P = 0.853) and is predicted to be the least-likely site for epoxidation by AHH. Figure 4 shows the calculated resonance energies for the various epoxide derivatives of BP-7,8-dihydrodiol. It can be seen that although the bay-region epoxide, BP-7,8-dihydrodiol-9,10-oxide, is the least-likely product of metabolism by AHH, it is predicted to be the most stable of the possible diol-epoxides. Such chemical stability may be an important consideration, since our current models of carcinogenesis require the

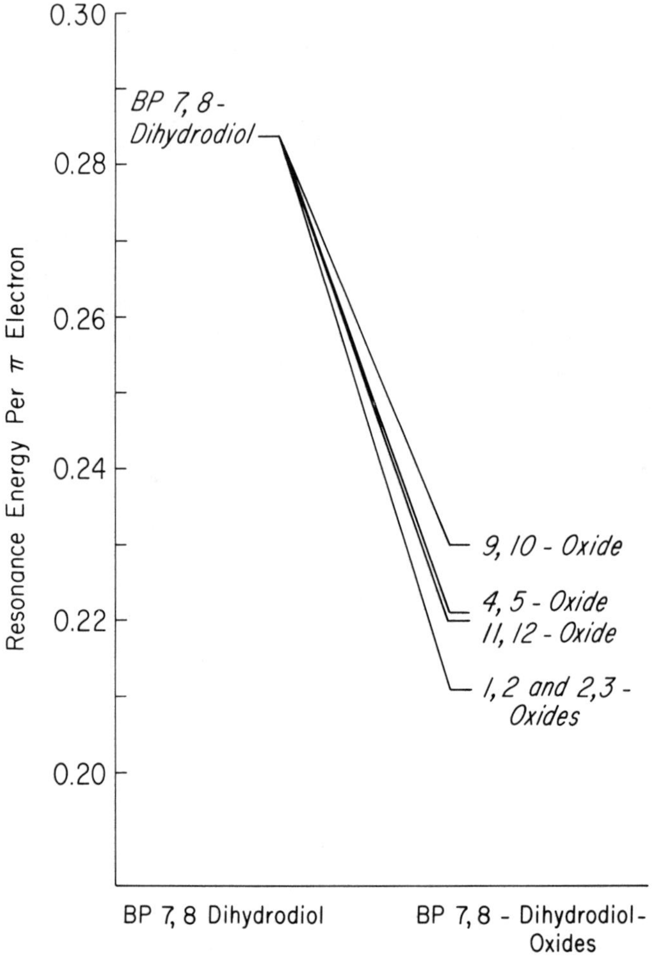

FIGURE 4. Relative stabilities of the epoxide derivatives of benzo[a]pyrene-7,8-dihydrodiol as predicted by their calculated resonance energy values.

"ultimate" carcinogenic derivative to migrate from the membrane-bound sites of activation, then to react instantaneously with critical macromolecular targets.

That neoplastic transformation should be initiated by a minor derivative of BP-7,8-dihydrodiol, itself a minor metabolite, seems counterintuitive and has prompted us to consider several alternative explanations. The first possibility, of course, is that our observations of a correlation between bond order and metabolic epoxidation are spurious, and that the site preference of

AHH is controlled by other factors. A second possibility is that BP-7,8-dihydrodiol-9,10-oxide reaches and reacts with critical macromolecular targets with high efficiency. A third speculation assumes that the epoxidation of BP-7,8-dihydrodiol at the C_9=C_{10} bond is mediated by an enzyme which is distinct from AHH and which preferentially attacks "alkene" double bonds.

The heterogeneity of the mixed-function monooxygenase system is well known (5,19,25,26), and the postulation of at least two general classes of microsomal epoxidases differing in the types of C=C bonds they attack is attractive from a functional point of view. An alkene-specific epoxidase, for example, would be expected to be involved in the epoxidation of substances like vinyl chloride, aflatoxin B_1 and cholesterol. Two arene-specific epoxidases, as discussed above, allow for the oxidative metabolism of a wide variety of aromatic substrates. It appears that the cellular microsomes contain enzymes from both classes, although exposure to certain natural or xenobiotic compounds may selectively induce one or more specific members of each. Thus, the hepatic microsomal monooxygenase system from untreated rats and that from rats pretreated with 3-methylcholanthrene are similar in that the predominating enzymes are from the class of arene-specific epoxidases. In contrast, the HPLC distribution of the BP-dihydrodiols formed by a phenobarbital-induced rat liver microsome preparation shows a relationship with P-values which is suggestive of the induction of an alkene-specific epoxidase (7). This latter observation provided us with an opportunity to examine the possibility than an alkene-specific epoxidase is involved in the formation of BP-7,8-dihydrodiol-9,10-oxide.

As shown in Table III, we have examined the relative efficiencies of a 3-methylcholanthrene- and a phenobarbital-induced rat liver microsome preparation to activate BP and BP-7,8-dihydrodiol to derivatives capable of reverting *Salmonella typhimurium* strain TA100. It can be seen that whereas

TABLE III. Differential Activation of BP-7,8-dihydrodiol by Microsomal Enzyme Preparations

Compound	Preparation	#Revertants/10^8 cells of TA100
DMSO (Control)	W/O Activation	479
	3-MC Microsomes	408
	PB Microsomes	557
BP	W/O Activation	450
	3-MC Microsomes	3,074
	PB Microsomes	805
BP-7,8-dihydrodiol	W/O Activation	550
	3-MC Microsomes	2,308
	PB Microsomes	1,929

the 3-methylcholanthrene-induced preparation is several times more efficient at activating BP than is the phenobarbital-induced preparation, both are equally effective at activating BP-7,8-dihydrodiol. While there are still other possible explanations for these observations, we believe that these mutagenicity data are consistent with an hypothesis that an alkene-specific epoxidase is responsible for the conversion of BP-7,8-dihydrodiol to its highly reactive bay-region epoxide.

CONCLUSIONS

First, we have used a relatively simple molecular orbital model to characterize the electronic properties of BP and have found a qualitative correlation between the degree of aromaticity and the extent of metabolic epoxidation by AHH. We have also been able to predict the relative stabilities of the resulting BP epoxides and, therefore, to predict the probability of non-enzymatic rearrangement or further metabolism. Although not presented today, we have found that this method can be applied successfully to other polycyclic aromatic hydrocarbons for which limited metabolism data are available.

Secondly, we have found that the size of a polycyclic aromatic hydrocarbon, but not its planar character, is an important determinant of suitability as a substrate of AHH. Only compounds the size of benz[a]anthracene and larger are epoxidated by AHH, and we postulate the existence of a second arene-specific epoxidase to account for the epoxidation of smaller aromatics. Such observations may be particularly relevant to understanding the complex interactions seen with mixtures of polycyclic aromatic hydrocarbons (6,9,22,23,26,30).

Thirdly, we applied the Hückel model to BP-7,8-dihydrodiol and on the basis of our calculations would predict that BP-7,8-dihydrodiol-9,10-oxide should be the least-likely metabolic derivative if AHH is involved in the second epoxidation step. We postulate the involvement of a distinct epoxidase with a preference for bonds having substantial alkene character, and have obtained some preliminary support for this hypothesis in observations using *Salmonella typhimurium.*

Finally, we propose as a conceptual model, that at least three epoxidase activities (*i.e.,* 2 arene-specific and 1 alkene-specific) are present in the mixed-function monooxygenase system. The metabolite profile that one observes represents the concerted action of these activities on appropriate C=C bonds in the substrate molecules. Pretreating an animal with certain substances may specifically induce one or more of these epoxidases and can result in a significant change in the types of metabolites formed, as well as the nature and intensity of the biological effects observed following exposure to polycyclic aromatic hydrocarbons. As methods of isolating and characterizing the various components of the cytochrome P-450 system improve, we hope the general applicability of our observations will be affirmed.

ACKNOWLEDGMENTS

The authors wish to thank Mrs. Florence Schmidt, Mrs. Carmen Gustafson and Mrs. Alisa Klingler for their assistance in the preparation of this manuscript. This work was supported in part by Grant G-120 from the Robert A. Welch Foundation. Computer time was generously provided in part by the Department of Biomathematics through Grant CA 11430. F. B. Thomas was a recipient of a Rosalie B. Hite Postdoctoral Fellowship.

REFERENCES

1. Borgen, A., Darvey, H., Castagnoli, N., Crocker, T. T., Rasmussen, R. E., and Wany, I. Y. (1973): Metabolic conversion of benzo[a]pyrene by Syrian hamster liver microsomes and binding of metabolites to deoxyribonucleic acid. *J. Med. Chem.* 16:502-506.
2. Boyland, E. (1969): The biochemistry of aromatic hydrocarbon amines and urethane. In: *Jerusalem Symposia on Quantum Chemistry and Biochemistry,* Vol. I, E. D. Bergmann and B. Pullman, Eds., pp. 25-44, Israel Academy of Sciences and Humanities, Jerusalem.
3. Conney, A. H. (1967): Pharmacological implications of microsomal enzyme induction. *Pharmacol. Rev.* 19:317.
4. Glusker, J. P. Zacharias, D. E., Carrell, H. L., Fu, H. P., and Harvey, R. G. (1976): Molecular structure of benzo[a]pyrene 4,5-oxide. *Cancer Res.* 36:3951-3957.
5. Haugen, D. A., van der Hoeven, T. A., and Coon, M. J. (1975): Purified liver microsomal cytochrome P-450. Separation and characterization of multiple forms. *J. Biol. Chem.* 250:3567-3570.
6. Hill, W. T., Stranger, D. W., Pizzo, A., Riegel, B., Shubik, P., and Wartman, W. B. (1951): Inhibition of 9,10-dimethyl-1,2-benzanthracene skin carcinogenesis in mice by polycyclic hydrocarbons. *Cancer Res.* 11:892-897.
7. Holder, G., Yagi, H., Dansette, P., Jerina, D. M., Levin, W., Lu, A. Y. H., and Conney, A. H. (1974): Effect of inducers and epoxide hydrase on the metabolism of benzo[a]pyrene by liver microsomes and a reconstituted system: Analysis by high pressure liquid chromatography. *Proc. Nat. Acad. Sci., U.S.* 71:4356-4360.
8. Holder, G. M., Yagi, H., Jerina, D. M., Levin, W., Lu, A. Y. H., and Conney, A. H. (1975): Metabolism of benzo[a]pyrene. Effect of substrate concentration and 3-methylcholanthrene pretreatment on hepatic metabolism by microsomes from rats and mice. *Arch. Biochem. Biophys.* 170:557-566.

9. Huggins, C., Grand, L., and Fukunishi, R. (1964): Aromatic influences on the yields of mammary cancers following administration of 7,12-dimethylbenz[a]anthracene. *Proc. Nat. Acad. Sci., U.S.* 51:737-742.
10. Iball, J., Scrimgeour, S. N., and Young, D. W. (1976): 3,4-Benzo[a]-pyrene (a new refinement). *Acta Cryst.* B32:328-330.
11. Jerina, D. M., Lehr, R. E., Yagi, H., Hernandez, O., and Dansette, P. M. (1976): Mutagenicity of benzo[a]pyrene derivatives and the description of a quantum mechanical model which predicts the ease of carbonium ion formation from diol epoxides. In: *In Vitro Metabolic Activation in Mutagenesis Testing,* F. J. de Serres, J. R. Fouts, J. R. Bend and R. M. Philpot, Eds., pp. 159-177, Elsevier/North-Holland Biomedical Press, Amsterdam.
12. Kapitulnik, J., Wislocki, P. G., Levin, W., Yagi, H., Thakker, D. R., Akagi, H., Koreeda, M., Jerina, D. M., and Conney, A. H. (1978): Marked differences on the carcinogenic activity of optically pure (+)- and (-)-trans-7,8-dihydroxy-7,8-dihydrobenzo[a]pyrene in newborn mice. *Cancer Res.* 38:2661-2665.
13. Kasperek, G. J., and Bruice, T. C. (1972): The mechanism of the aromatization of arene oxides. *J. Am. Chem. Soc.* 94:198-202.
14. King, H. W. S., Thompson, M. H., and Brookes, P. (1975): The benzo-[a]pyrene deoxyribonucleoside products isolated from DNA after metabolism of benzo[a]pyrene by a rat liver microsomes in the presence of DNA. *Cancer Res.* 34:1263-1269.
15. Novak, B., and Furlong, N. B. (1969): Computer programs in the biomedical sciences: I. A computer program for the rapid calculation of molecular orbital indices. *Texas Reports Biol. Med.* 27:1041-1055.
16. Osborne, M. R., Thompson, M. H., Tarry, E. M., Beland, F. A., Harvey, R. G. and Brookes, P. (1976): The reaction of 7,8-dihydro-7,8-dihyroxybenzo[a]pyrene-9,10-oxide with DNA in relation to the benzo[a]pyrene-DNA products isolated from cells. *Chem. -Biol. Interact.* 13:343-348.
17. Pullman, B., and Pullman, A. (1960): Some electronic aspects of biochemistry. *Rev. Mod. Phys.* 32:428-433.
18. Rogan, E. G., Mailander, P., and Cavalieri, E. (1976): Metabolic activation of aromatic hydrocarbons in purified rat liver nuclei: induction of enzyme activities and binding to DNA with and without monooxygenase-catalyzed formation of active oxygen. *Proc. Nat. Acad. Sci., U.S.* 73:457-461.
19. Ryan, D., Lu, A. Y. H., West, S., and Levin, W. (1975): Multiple forms of cytochrome P-450 in phenobarbitol- and 3-methylcholanthrene-treated rats. *J. Biol. Chem.* 250:2157-2163.
20. Selkirk, J. K., Croy, R. G., and Gelboin, H. V. (1975): Isolation by high pressure liquid chromatography and characterization of benzo[a]pyrene-4,5-epoxide as a metabolite of benzo[a]pyrene. *Arch. Biochem. Biophys.* 168:322-326.

21. Selkirk, J. K., Croy, R. G., and Gelboin, H. V. (1976): High-pressure liquid chromatographic separation of 10 benzo[a]pyrene phenols and the identification of 1-phenol and 7-phenol as new metabolites. *Cancer Res.* 36:922-926.
22. Slaga, T. J., and Boutwell, R. K. (1977): Inhibition of the tumor-initiating ability of the potent carcinogen 7,12-dimethylbenz[a]-anthracene by the weak tumor initiator 1,2,3,4-dibenzanthracene. *Cancer Res.* 37:128-133.
23. Steiner, P. E., and Falk, H. L. (1951): Summation and inhibition effects of weak and strong carcinogenic hydrocarbons: 1,2-dibenzanthracene, chrysene, 1,2,5,6-dibenzanthracene and 20-methylcholanthrene. *Cancer Res.* 11:56-63.
24. Thomas, F. B., and Furlong, N. B. (1976): A simple radioassay of benzo[a]pyrene and activation: Observation on the covalent interactions of benzo[a]pyrene with protein. *Anal. Biochem.* 72:546-551.
25. Thomas, P. E., Lu, A. Y. H., Ryan, D., West, S. B., Kawalek, J., and Levin, W. (1976): Multiple forms of rat liver cytochrome P-450. Immunochemical evidence with antibody against cytochrome P-448. *J. Biol. Chem.* 251:1385-1391.
26. Wiebel, F. J., Lentz, J. C., Diamond, L., and Gelboin, H. V. (1971): Aryl hydrocarbon (benzo[a]pyrene) hydroxylase in microsomes from rat tissues: differential inhibition and simulation by benzoflavones and organic solvents. *Arch. Biochem. Biophys.* 144:76-86.
27. Wood, A. W., Wislocki, P. G., Chang, R. L., Levin, W., Lu, A. Y. H., Yagi, H., Hernandez, O., Jerina, D. M., and Conney, A. H. (1976): Mutagenicity and cytotoxicity of benzo[a]pyrene benzo-ring epoxides. *Cancer Res.* 36:3358-3366.
28. Yang, S. K., Roller, P. P., and Gelboin, H. V. (1977): Enzymatic mechanism of benzo[a]pyrene conversion to phenols and diols and an improved high-pressure liquid chromatographic separation of benzo-[a]pyrene derivatives. *Biochemistry* 16:3680-3687.
29. Yang, S. K., Selkirk, J. K., Plotkin, E. V., and Gelboin, H. V. (1975): Kinetic analysis of the metabolism of benzo[a]pyrene to phenols, dihydrodiols, and quinones by high-pressure liquid chromatography compared to analysis by aryl hydrocarbon hydroxylase assay, and the effect of enzyme induction. *Cancer Res.* 35:3642-3650.
30. Zytkovicz, T. H., Moses, H. L., and Spelsberg, T. C. (1978): Competitive binding and metabolism of several carcinogenic polycyclic aromatic hydrocarbons by AKR mouse embryo cells in culture. *Cancer Res.* 38:2307-2313.

Polynuclear Aromatic Hydrocarbons, edited by P. W. Jones and P. Leber. Ann Arbor Science Publishers, Inc. Ann Arbor, MI ©1979.

BINDING OF AROMATIC HYDROCARBONS TO DNA CATALYZED BY PEROXIDASE AND BY ATP

E. Rogan, R. Roth and E. Cavalieri

Eppley Institute for Research in Cancer
University of Nebraska Medical Center
Omaha, Nebraska 68105

Biological activation of polycyclic aromatic hydrocarbons (PAH) has been thought to occur primarily by oxygenation reactions. The most thoroughly elucidated pathway is that of the metabolism of benzo[a]pyrene (BP) to the BP 7,8-diol-9,10 epoxides by cytochrome P-450 and epoxide hydrase.

Activation of PAH by an oxidation involving loss of one electron rather than an oxygenation to form epoxides has also been postulated (4) and studied in isolated rat liver nuclei and on mouse skin (15,16). Peroxidases, including those from horseradish (HRP) and rat mammary gland, have been shown to activate N-hydroxy-2-acetylaminofluorene by one-electron oxidation (1, 2,9-11,13). Therefore, we have begun to use HRP as a possible model system for activation of PAH by one-electron oxidation (14).

A nonenzymatic ATP-mediated activation of hydroxymethyl derivatives of PAH, which are the major metabolites of methyl-substituted PAH, has also been discovered (8,17). This reaction appears to bind PAH to DNA through the methyl group.

In this chapter are reported studies which describe some aspects of PAH activation by HRP and by ATP.

RESULTS AND DISCUSSION

Activation by HRP

One tool to study the mechanism of PAH activation has been use of PAH doubly labeled with ^{14}C and tritium at specific positions. In these experiments the loss of tritium from a position is an indication (not conclusive, however) of binding of DNA to that position. We have studied the HRP-catalyzed activation and binding of such PAH to DNA. Previously reported

TABLE I. Horseradish Peroxidase/H_2O_2 Catalyzed Binding of BP to DNA[a]

Position of tritium in $[^3H,^{14}C]$ BP	$[^{14}C]$ BP bound, pmole/μmole DNA		$^3H/^{14}C$, % original ratio
6	47.9 ± 7.2[b]	(6)	5.7 ± 1.3
1, 3, 6	79.5 ± 9.1	(7)	81.4 ± 1.7
7	53.0 ± 11.9	(12)	88.4 ± 1.7
8	58.3 ± 3.0	(10)	99.2 ± 2.1
Control Incubations	0.2 ± 0.1	(6)	

[a]Reaction mixtures containing 0.2 mg HRP, Type II, 4 m*M* DNA, 30 μM $[^3H,^{14}C]$ BP, 5×10^{-4} *M* H_2O_2 and Sörenson buffer, pH 7.0, to 1 ml were incubated at 37° for 30 min in the dark. The reaction was stopped and DNA purified as previously described. Control incubations did not contain H_2O_2.

[b]Standard error of measurement. The number in parentheses is the number of determinations.

experiments (15) using $[^3H, ^{14}C]$ BP with tritium at the 6; 1,3,6; and 7 positions showed that virtually all the tritium is lost from the 6 position, indicating binding of DNA to that position. However, a small (12%) loss of tritium from the 7 position was observed, raising some questions about interpretation of these results. We have now completed studies of HRP-catalyzed binding of $[8\text{-}^3H,^{14}C]$ BP to DNA (Table I) and found that no loss of tritium occurs from the 8 position. Thus these results indicate more strongly that activation of BP by HRP results in binding of BP at its 6 position to DNA. The small loss of tritium from the 7 position may result from a close proximity to the substituent at the 6 position.

While the double-label experiments show binding of DNA to the 6 position of BP, they do not identify a particular mechanism of activation. While the 6 position of BP is the most reactive, if activation occurs by one-electron oxidation (16), it is still possible that another mechanism could occur involving activation at the 6 position. The most likely reaction would be formation of BP-6-OH, which has been shown to bind covalently to DNA through the BP-6-oxy radical (12,18).

In order to learn whether or not metabolites of BP are formed during incubation of $[^{14}C]$ BP with the HRP/H_2O_2 activation system, reaction mixtures were examined by high-pressure liquid chromatography (HPLC). Reaction mixtures containing 0.2 mg/ml HRP, 4 m*M* DNA, 30 μM $[^{14}C]$ BP and 5×10^{-4} *M* H_2O_2 in Sörenson buffer, pH 7.0, were incubated at 37° for 30 min. Under these conditions approximately 2% of the BP is bound to the DNA. The following 0.1-ml reaction mixtures were incubated and analyzed by HPLC:

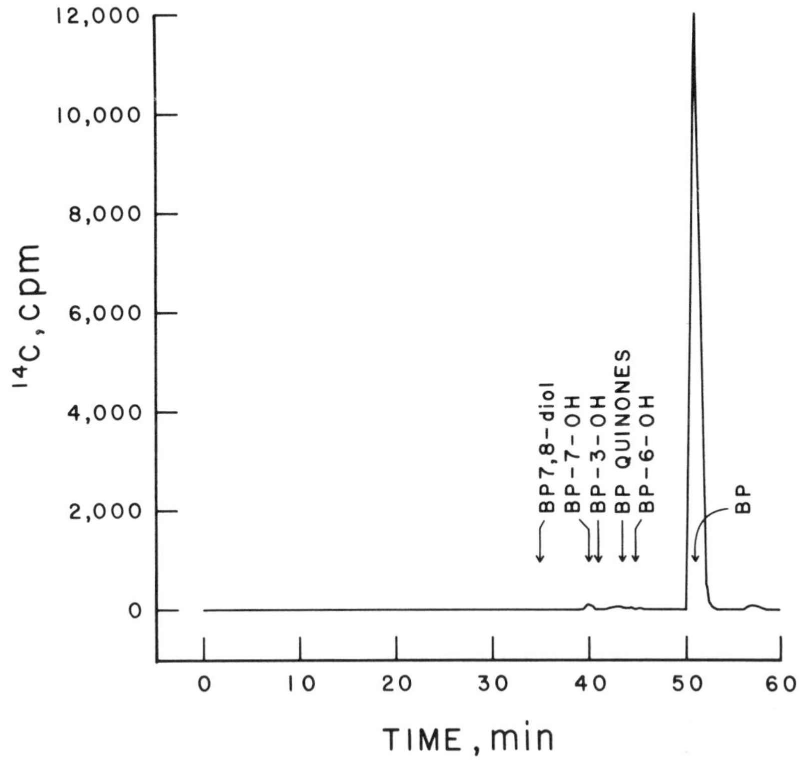

FIGURE 1. HPLC analysis of possible metabolites of [^{14}C] BP formed in HRP-activated binding to DNA. Samples were prepared and analyzed as described in the text.

1. [^{14}C]BP and H_2O_2, evaporated under nitrogen and dissolved in methanol
2. [^{14}C]BP, HRP and H_2O_2, evaporated under nitrogen and dissolved in methanol
3. [^{14}C]BP, DNA, HRP and H_2O_2, evaporated under nitrogen and dissolved in methanol
4. [^{14}C]BP, DNA, HRP and H_2O_2, extracted with 0.5 ml benzene (washed twice with 0.1 ml water, dried over sodium sulfate and evaporated under nitrogen) and dissolved in methanol
5. [^{14}C]BP, DNA, HRP and H_2O_2, extracted with 0.5 ml 4:1 ethyl acetate-acetone (washed twice with 0.1 ml water, dried over sodium sulfate and evaporated under nitrogen) and dissolved in methanol

Figure 1 shows the results of HPLC analysis of a sample prepared by method 5 above. HPLC was conducted on a 50-cm x 0.9-cm micropack ODS column at 45°. A 30-min linear gradient of 45-99% methanol in water was used. Standards were detected simultaneously by UV (257 nm) and fluorescence, and metabolites by radioactivity. Recovery of ^{14}C through the extraction procedure and HPLC was 100%. BP and BP-6-OH were run as standards with each sample. HPLC of samples prepared by methods 1-4 above gave exactly the same results as in Figure 1. The only significant radioactivity observed corresponded to BP. Thus neither BP-6-OH nor BP quinone formed from BP-6-OH was observed. The sensitivity of detection of possible metabolites was greater than the level of BP binding to DNA. For instance in Figure 1, 46,400 cpm of [^{14}C] BP was recovered. Since 2% of the BP is bound to DNA under these reaction conditions, a minimum of 930 cpm (2%) of metabolite could be expected. Thus these results demonstrate that the HRP-catalyzed binding of BP is not occurring by formation of BP-6-OH and provide further support for the hypothesis that HRP activates BP by one-electron oxidation.

Activation of MC by HRP was also studied using MC tritiated at the 1; 1, 5,6; 2; and 6 positions. Loss of tritium was observed from all these positions and it was found that significant amounts of polar metabolites were formed in the presence of H_2O_2. Thus the double-label technique cannot be used to study the mechanism of activation of MC by HRP.

ATP-Mediated Activation of PAH

Binding of hydroxymethyl derivatives of PAH to DNA in the presence of ATP alone has only recently been described (6,8,17). Activation of hydroxymethyl PAH by formation of a reactive ester was suggested by the fact that sarcomas were induced in rats by BP-6-CH_2OH and BP-6-CH_2OAc with a potency similar to BP, while BP-6-CH_3 was weaker and BP-6-CH_2OCH_3 did not induce tumors (6). 7-Hydroxy-methyl-12-methyl-benz[a]anthracene (BA-7-CH_2OH-12-CH_3) and BA-7-CH_2OAc-12-CH_3 induced sarcomas in rats, while BA-7-CH_2OCH_3-12-CH_3 was very weak (3,6,7). We have confirmed the ATP-mediated binding of BP-6-CH_2OH to DNA previously reported by Tay *et al.* (17) and extended these studies to BA-7-CH_2OH (5).

As previously reported (17), binding of [^{14}C] BP-6-CH_2OH is mediated by both ATP and ADP, but ATP is far more effective (Table II). The inability of AMP to activate BP-6-CH_2OH for binding to DNA indicates the need for a high-energy phosphate bond to drive the reaction.

Under our standard reaction conditions, the reaction proceeds for more than 90 min before slowing down (Figure 2). The amount of BP-6-CH_2OH bound to DNA is proportional to the concentration of BP-6-CH_2OH (Figure 3) and $MgCl_2$ at limiting levels of these compounds. However, the response to ATP is quite different (Figure 4), showing a lag in binding at low levels and then a greater than proportional response to added ATP until a saturation level is reached. The effect of ATP is more complex than a reaction second-order with respect to ATP. Further studies are required to elucidate the role of ATP and the exact mechanism of this reaction.

TABLE II. Binding of [^{14}C] BP-6-CH_2OH to DNA[a]

Cofactor	pmole/μmole DNA
ATP	130 ± 8[b]
ADP	26 ± 2
AMP	1.7 ± 0.1
None	0.9 ± 0.02

[a]Reaction mixtures containing 5 m*M* cofactor, 3 m*M* DNA, 3 m*M* $MgCl_2$ and 26 μ*M* (10 μg/ml) [^{14}C] BP-6-CH_2OH in 0.05 *M* Tris, pH 7.0, were incubated at 37° for 1 hr. The DNA was purified by 2 extractions with water-saturated phenol, pH 8.0; 2 extractions with chloroform-octanol (10:1), and 2 precipitations in 75% ethanol.
[b]Standard error of measurement.

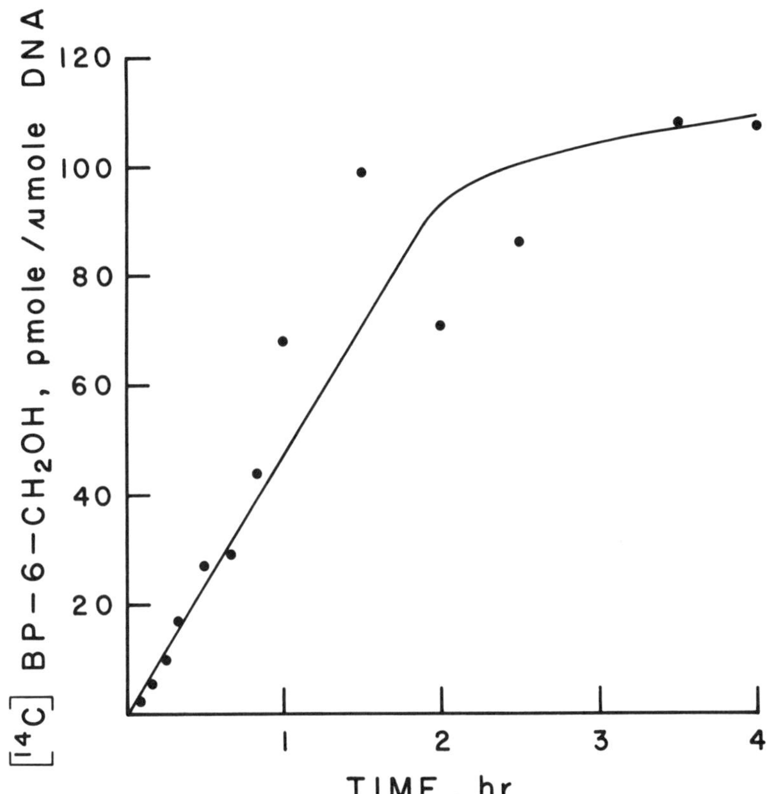

FIGURE 2. Time course of ATP-mediated binding of [^{14}C] BP-6-CH_2OH to DNA. Incubations were conducted and DNA was purified as described in Table II.

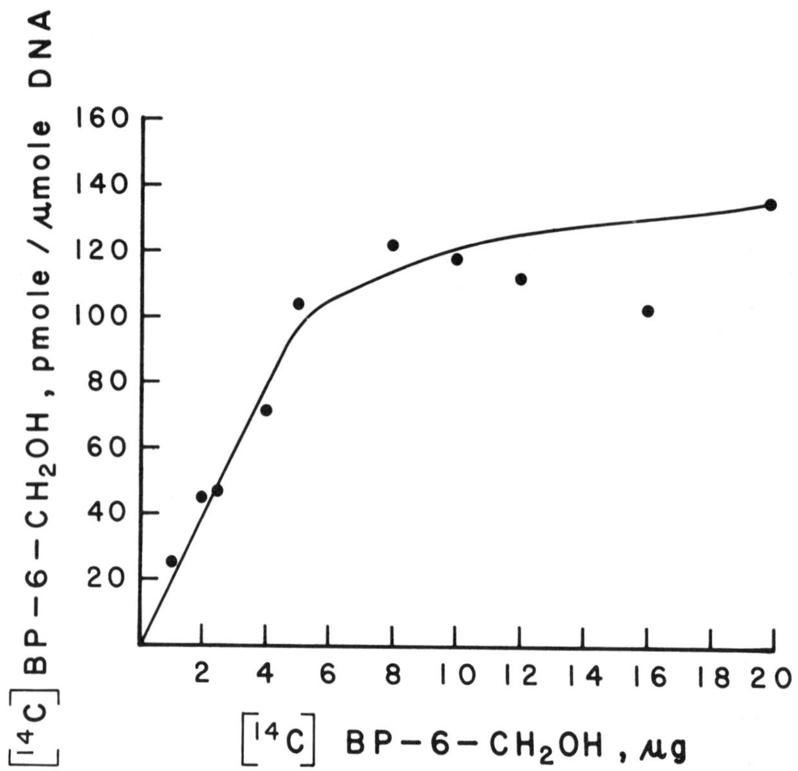

FIGURE 3. Effect of substrate concentration on ATP-mediated binding of [^{14}C]BP-6-CH_2OH to DNA. Incubations were conducted and DNA was purified as described in Table II, except that reaction mixtures contained 0.1 m*M* EDTA.

ATP-mediated binding of BP-6-CH_2OH to DNA occurs at the 6-methyl position. This is shown by the fluorescence spectra of BP-6-CH_2OH bound to DNA (Figure 5). DNA-bound BP-6-CH_2OH (300 pmole/µmole DNA) was prepared and purified under standard conditions (see Table II) except that 10 m*M* ATP was used. BP-6-CH_2Br (0.1 mg) was reacted with 1 mg DNA by shaking 15 min in the dark and the DNA purified as above (see Table II). Fluorescence spectra of these two DNA (approximately 1 m*M*) and free BP-6-CH_3 (about 100 n*M*) in 33% ethanol were then recorded on an Aminco-Bowman spectrophotofluorometer. The emission spectrum (excitation at 300 nm) of DNA-bound BP-6-CH_2OH was almost identical to those of DNA-bound BP-6-CH_2Br and free BP-6-CH_3. The DNA-bound BP-6-CH_2OH showed maxima at 412 and 432 nm, while DNA-bound BP-6-CH_2Br had maxima at 412 and 434 nm, with a shoulder at 462 nm, and free BP-6-CH_3

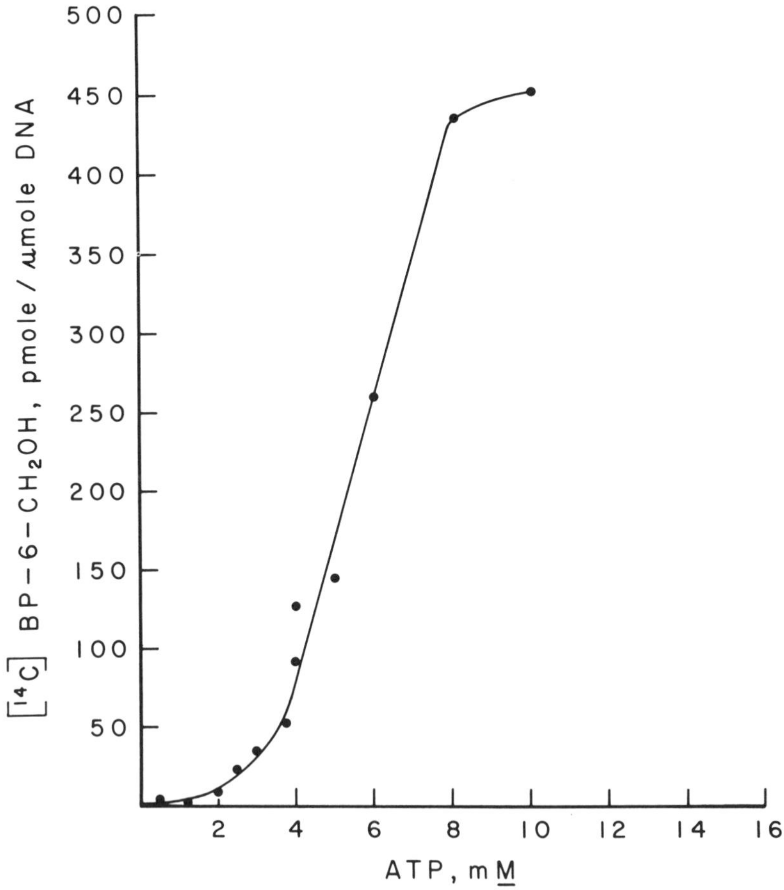

FIGURE 4. Effect of ATP concentration on ATP-mediated binding of [^{14}C]BP-6-CH_2OH to DNA. Incubations were conducted and DNA was purified as described in Table II.

exhibited maxima at 412 and 434 nm and a shoulder at about 462 nm. The excitation spectrum (emission at 415 nm) of DNA-bound BP-6-CH_2OH showed peaks at 302 and about 385 nm, while DNA-bound BP-6-CH_2Br had peaks at 302, 380 and 398 nm and free BP-6-CH_3 had peaks at 300 and 392 nm and a shoulder about 265 nm. These spectra are very similar, all showing a major peak at 300-302 nm. Comparison of the spectra of DNA-bound BP-6-CH_2OH, DNA-bound BP-6-CH_2Br and free BP-6-CH_3, which have identical major peaks, demonstrates that the ATP-mediated binding occurs at the 6-methyl position.

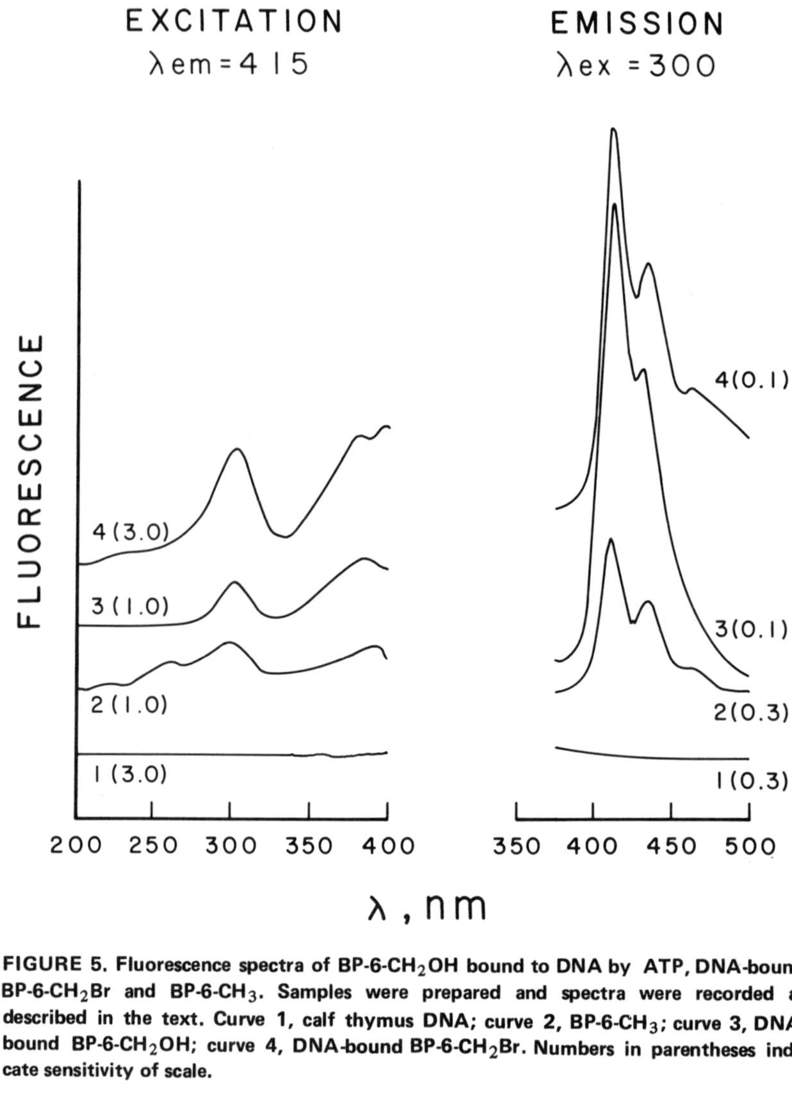

FIGURE 5. Fluorescence spectra of BP-6-CH_2OH bound to DNA by ATP, DNA-bound BP-6-CH_2Br and BP-6-CH_3. Samples were prepared and spectra were recorded as described in the text. Curve 1, calf thymus DNA; curve 2, BP-6-CH_3; curve 3, DNA-bound BP-6-CH_2OH; curve 4, DNA-bound BP-6-CH_2Br. Numbers in parentheses indicate sensitivity of scale.

CONCLUSIONS

Evidence has been presented for metabolic activation of PAH by two mechanisms which occur in addition to the highly explored metabolism to epoxides. Peroxidase is capable of catalyzing the binding of PAH to DNA; for BP the evidence presently available indicates that binding occurs at the 6 position and activation may occur by one-electron oxidation. This mechanism is particularly interesting because peroxidase in rat mammary cells has been shown to catalyze one-electron oxidation of N-hydroxy-2-acetylaminofluorene (13). Since HRP-activated binding of BP appears to occur by one-electron oxidation, nucleoside-bound PAH prepared in this system may be useful for comparison with PAH nucleosides obtained *in vivo*. Such comparisons may provide indications that one-electron oxidation is an activating mechanism in some target organs.

Binding of hydroxymethyl derivatives of PAH to DNA is catalyzed by ATP, presumably by formation of a reactive phosphate-ester intermediate. Binding has been shown to occur at the 6-methyl group of BP-6-CH_2OH. Methyl derivatives of PAH are predominantly metabolized to hydroxymethyl derivatives, but experiments must be conducted to determine their possible role in carcinogenesis.

ACKNOWLEDGMENTS

This research was supported by Public Health Service Contract NO1 CP33278 from the Division of Cancer Cause and Prevention of the National Cancer Institute, NIH.

REFERENCES

1. Bartsch, H., and Hecker, E. (1971): On the metabolic activation of the carcinogen N-hydroxy-N-2-acetylaminofluorene. III. Oxidation with horseradish peroxidase to yield 2-nitrosofluorene and N-acetoxy-N-2-acetylaminofluorene. *Biochim. Biophys. Acta*. 237:567-578.
2. Bartsch, H., Miller, J. A., and Miller, E. C. (1972): N-Acetoxy-N-acetylaminoarenes and nitrosoarenes. One-electron non-enzymatic and enzymatic oxidation products of various carcinogenic aromatic acethydroxamic acids. *Biochim. Biophys. Acta.* 273:40-51.
3. Boyland, E., Sims, P., and Huggins, C. (1965): Induction of adrenal damage and carcinogenesis with metabolites of 7,12-dimethylbenz[a]-anthracene. *Nature* 207:816-817.
4. Cavalieri, E., Roth, R., and Rogan, E. G. (1976): Metabolic activation of aromatic hydrocarbons by one-electron oxidation in relation to the mechanism of tumor initiation. In: *Carcinogenesis, Vol. 1*, R. Freudenthal and P. W. Jones, Eds., pp. 181-190, Raven Press, New York.
5. Cavalieri, E., Roth, R., and Rogan, E. (1979): Hydroxylation and conjugation at the benzylic carbon atom: A possible mechanism of carcinogenic activation for some methyl-substituted aromatic hydrocarbons. Chapter 34, this volume.
6. Flesher, J. W. Personal communication.

7. Flesher, J. W., and Sydnor, K. L. (1971): Carcinogenicity of derivatives of 7,12-dimethylbenz[a]anthracene. *Cancer Res.* 31:1951-1954.
8. Flesher, J. W., and Tay, L. K. (1978): Binding of 7-hydroxymethyl-12-methylbenz[a]anthracene to calf thymus DNA in the presence and absence of 3′-phosphoadenosine-5′-phosphosulfate generating system. *Proc. Fed. Am. Soc. Exp. Biol.* 37:749.
9. Floyd, R. A., and Soong, L. M. (1977): Obligatory free radical intermediate in the oxidative activation of the carcinogen N-hydroxy-2-acetylaminofluorene. *Biochim. Biophys. Acta.* 498:244-249.
10. Floyd, R. A., Soong, L. M., and Culver, P. L. (1976): Horseradish peroxidase/hydrogen peroxide-catalyzed oxidation of the carcinogen N-hydroxy-N-acetyl-2-aminofluorene as affected by cyanide and ascorbate. *Cancer Res.* 36:1510-1517.
11. Floyd, R. A., Soong, L. M., Walker, R. N., and Stuart, M. (1976): Lipid hydroperoxide activation of N-hydroxy-N-acetylaminofluorene via a free radical route. *Cancer Res.* 36:2761-2767.
12. Nagata, C., Tagashira, Y., and Kodama, M. (1974): Metabolic activation of benzo[a]pyrene: Significance of the free radical. In: *Chemical Carcinogenesis, Part A*, P. O. P. Ts'o and J. A. DiPaolo, Eds., pp. 87-111, Marcel Dekker, New York.
13. Reigh, D. L., Stuart, M., and Floyd, R. A. (1978): Activation of the carcinogen N-hydroxy-2-acetylaminofluorene by rat mammary peroxidase. *Experientia* 34:107-108.
14. Rogan, E., Katomski, P., Roth, R., and Cavalieri, E. (1978): Binding of benzo[a]pyrene and 3-methylcholanthrene to DNA by horseradish peroxidase. *Proc. Fed. Am. Soc. Exp. Biol.* 37:750.
15. Rogan, E., Roth, R., and Cavalieri, E. (1978): Enzymology of polycyclic hydrocarbon binding to nucleic acids. In: *Carcinogenesis, Vol. 3;* P. W. Jones and R. I. Freudenthal, Eds., pp. 265-271, Raven Press, New York.
16. Rogan, E., Roth, R., Katomski, P., Benderson, J., and Cavalieri, E. (1978): Binding of benzo[a]pyrene at the 1,3,6 positions to nucleic acids in vivo on mouse skin and in vitro with rat liver microsomes and nuclei. *Chem.-Biol. Interact.* 22:35-51.
17. Tay, L. K., Sydnor, K. L., and Flesher, J. W. (1978): Binding of 6-acetoxymethylbenzo[a]pyrene and ATP-mediated binding of 6-hydroxymethylbenzo[a]pyrene to DNA. *Proc. Am. Assoc. Cancer Res.* 19:181.
18. Ts'o, P. O. P., Caspary, W. J., Cohen, B. I., Leavitt, J. C., Lesko, S. A., Lorentzen, R. J., and Schechtman, L. M. (1974): Basic mechanisms in polycyclic hydrocarbon carcinogenesis. In: *Chemical Carcinogenesis Part A,* P. O. P. Ts'o and J. A. DiPaolo, Eds., pp. 113-147, Marcel Dekker, New York.

Polynuclear Aromatic Hydrocarbons,
edited by P. W. Jones and P. Leber.
Ann Arbor Science Publishers, Inc.
Ann Arbor, MI ©1979.

SYNTHESIS AND METABOLISM OF CYCLOPENTA [c,d] PYRENE

A. Gold, J. Schultz and E. Eisenstadt

Department of Physiology and
Department of Microbiology
Harvard School of Public Health
665 Huntington Avenue
Boston, Massachusetts 02115

Polycyclic aromatic hydrocarbons (PAH) are widespread environmental contaminants with potential for carcinogenicity in humans (11). Increasing human exposure to PAH through fossil-fuel utilization has stimulated interest in the biological activity of these compounds. Biological activity requires that the parent hydrocarbons be metabolically activated to reactive species (2). Studies on the structural basis of PAH activity have heretofore emphasized polycyclics containing the "bay-region" feature (9) as exemplified by the area enclosed by the 4,5 positions of phenanthrene (Figure 1). Diol epoxides adjacent to the bay region appear to be the most active metabolites of bay-region PAH.

We have recently reported (4) that a widely occurring non-bay-region PAH, cyclopenta[c,d]pyrene (CPP; Figure 2), is highly mutagenic for in the presence of liver-metabolizing enzymes. Our studies on metabolism and mutagenicity require CPP in greater quantities than can be conveniently extracted from soots (6) or synthesized by a recently published pyrolytic method (8). We, therefore, embarked on a new synthetic route. This synthesis has been reported elsewhere as a brief communication (7).

Formation of the c,d-fused ring by intramolecular Friedel-crafts acylation of suitably functionalized pyrene was selected as an approach to the synthesis. Cyclization of 1-pyrenylacetic acid, a readily available candidate (12) for internal Friedel-Crafts condensation, could not be achieved either by polyphosphoric acid or by treating with $AlCl_3$ the acid chloride (Figure 3). The carbonium ions and corresponding changes in delocalization energy (ΔE_{π};3) resulting from electrophilic attack at peripheral ring positions of pyrene are shown in Figure 4. The low reactivity at C(4) may be a reflection

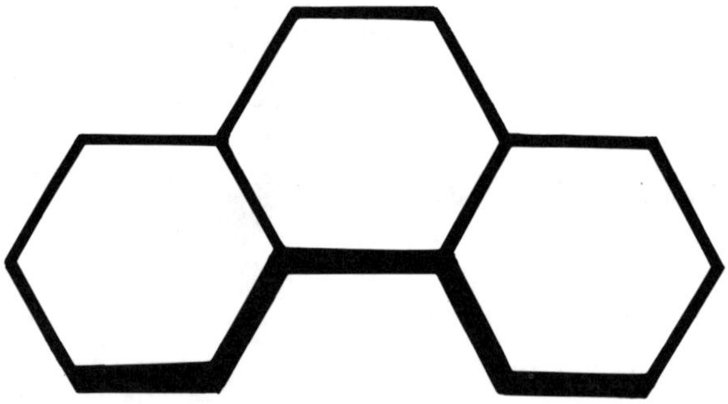

FIGURE 1. Bay region indicated by heavy outline.

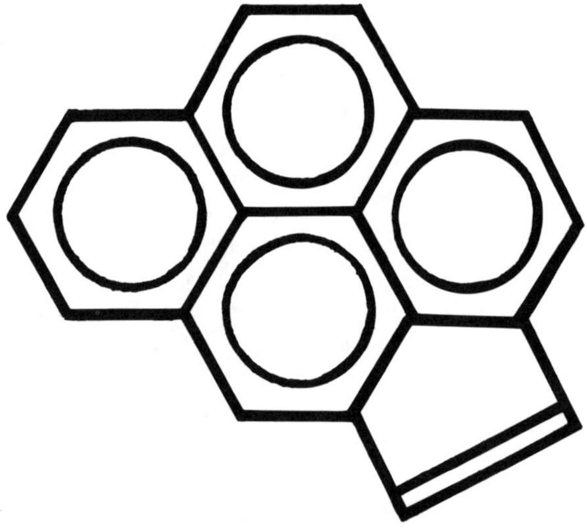

FIGURE 2. Cyclopenta[c,d] pyrene.

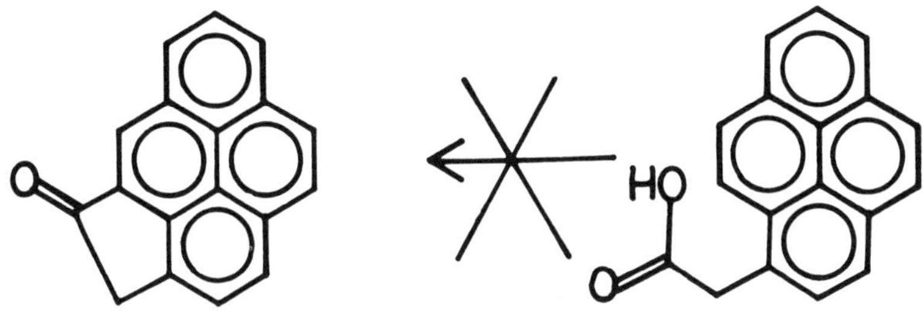

FIGURE 3. Unsuccessful cyclization of 1-substituted pyrene.

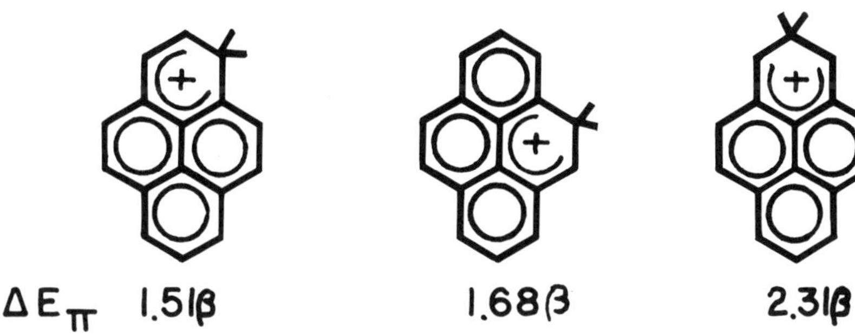

FIGURE 4. Relative reactivities of pyrene ring positions towards electrophilic attack as estimated by PMO ΔE_{π} energies in units of β.

of the ordering of the ΔE_{π} values. The exceptionally high reactivity of pyrene to electrophilic attack at C(1) has been well-documented (12); however, the relative reactivities at C(2) and C(4), although considerably less than at C(1), cannot be established from available experimental evidence.

The expected reactivity at C(1) suggested that internal Friedel-Crafts cyclization of 4-pyrenylacetic acid would be successful. 4-Pyrenylacetic acid was synthesized by a published procedure (5) modified at several points to improve yields (Figure 5).

Functionalization of pyrene at C(4) was accomplished by acylation of 1,2, 3,6,7,8-hexahydropyrene with acetyl chloride and $AlCl_3$ in nitromethane (3 hr, room temperature). Chromatography of the crude reaction mixture in benzene on silica yielded pure 4-acyl-1,2,36,7,8-hexahydropyrene (84%). The published procedure (12) utilizing acetic anhydride and $AlCl_3$ in benzene yielded considerable amounts of benzophenone in addition to the desired product.

Aromatization of the 4-acyl hexahydropyrene proceeded smoothly by the reported method (5) to give 4-acetylpyrene (85%) which was almost quantitatively converted by Willgerodt oxidation (5) to 4-pyrenylthioacetamide. The thioacetamide was purified by recrystallization from HOAc-chlorobenzene and acid hydrolyzed to give 4-pyrenylacetic acid. The yield (83%) is much improved compared to that reported (5) for hydrolysis of crude thioacetamide with ethanolic KOH (34%): 4-Pyrenylthioacetamide (2 g) was dissolved in boiling HOAc (30ml) and, after addition of conc, HCL (15ml), refluxed for 2 hr. 4-Pyrenylacetic acid, precipitated by cooling and dilution with additional conc. HCl (15 ml) was collected, partitioned between 5% KOH and CH_2Cl_2, reprecipitated by acidification of the aqueous layer and recrystallized from chlorobenzene after decolorization with activated carbon.

4-Pyrenylacetic acid was cyclized to the desired ketone, 3,4-dihydrocyclopenta[c,d]pyren-3-one, either by treating it with polyphosphoric acid or by treating its acid chloride with $AlCl_3$. The yield with $AlCl_3$ was identical to that with polyphosphoric acid; however, the $AlCl_3$-cyclized product was more easily purified and this procedure was therefore adopted for routine use: 4-Pyrenylacetic acid (0.4 g), covered with $SOCl_2$ and one drop of pyridine, was stirred for 1 hr under N_2. After evaporating the excess $SOCl_2$ *in vacuo*, the residue was covered with nitromethane and $AlCl_3$ (0.24 g) was added. The reaction was stirred for 1 hr at room temperature and then partitioned between H_2O and CH_2Cl_2. The organic layer was extracted with 5% KOH, dried over Na_2SO_4, evaporated and the residue chromatographed on silica. A highly fluorescent yellow band, eluted with benzene, was collected and recrystallized from EtOH to yield the ketone as pale yellow neddles (0.065 g, 16%), mp 214°, V_{co} (KBr) 1701 cm^{-2}; UV (Figure 6); mass spectrum, $M.^+$, 242.0735 (expected for $C_{18}H_{10}O$, 242.0732); NMR ($CDCl_3$, 60 MHz), singlet (2 H), 383 δ, aromatic multiplet (8 H) 7.8-8.5 δ. An explanation for the poor yield is not apparent. Dilution of the reaction mixture, reverse addition or use of $SnCl_4$ as catalyst did not improve the yield.

The ketone was qualitatively (by UV) reduced by stirring overnight in Et_2O with a 25% molar excess of LAH and hydrolyzing with a small amount of 6 *N* HCl. Pure alcohol could not be isolated by evaporation of the Et_2O because of decomposition induced by HCl dissolved in the Et_2O, but a mass spectrum of the slightly discolored crude alcohol was obtained: $M.^+$, 244; base peak, m/e 226 ($M.^+$-H_2O). The peak at m/e 226 is likely to be a fragment resulting from dehydration of the alcohol rather than the molecular ion of cyclopenta[c,d]pyrene because of the absence of the doubly charged molecular ion at m/e 113 which appears in the mass spectrum of cyclopenta[c,d] pyrene on the same instrument under identical condictions.

The alcohol was dehydrated without isolation from the hydrolysis reaction. After addition of benzene to the reaction mixture, the organic layer was

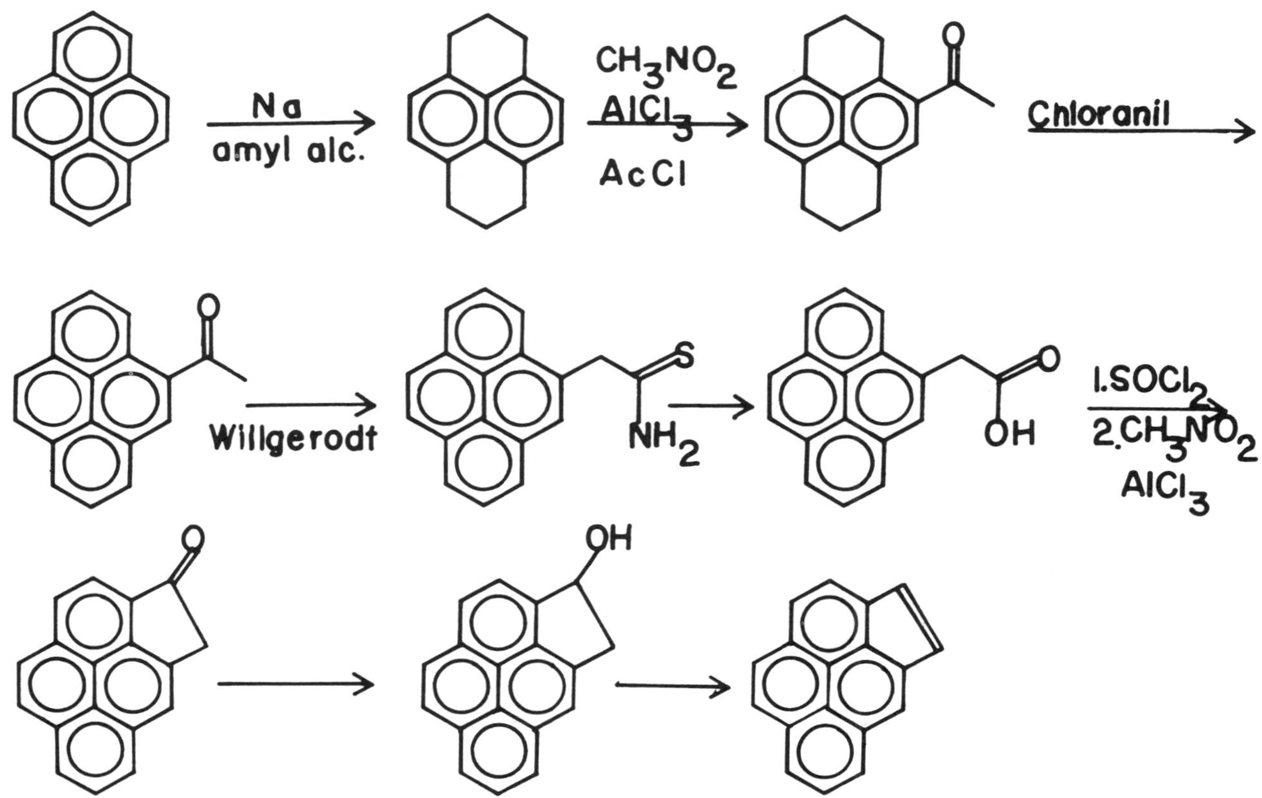

FIGURE 5. Synthetic route to cyclopenta[c,d] pyrene.

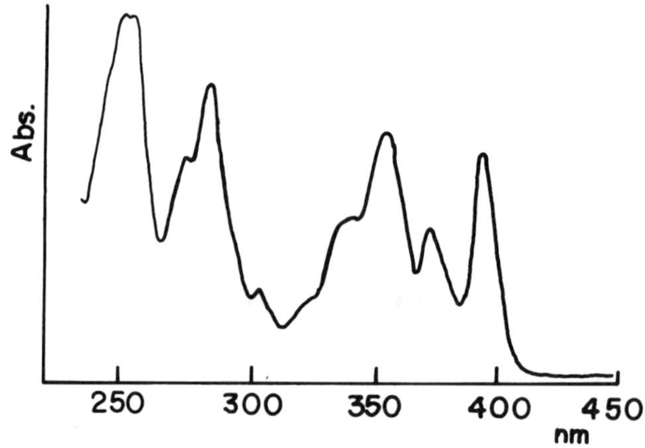

FIGURE 6. UV spectrum (CH_2Cl_2): 3,4 dihydrocyclopenta[c,d] pyren-3-one.

separated and dried over Na_2SO_4 and a tenfold excess (by weight of ketone) of 200 mesh basic alumina activity grade I (ICN Pharmaceuticals) was added. The Et_2O was removed by distillation and the reaction refluxed for 2 hr. Filtration and evaporation of the benzene yielded brilliant orange crystalline cyclopenta[c,d] pyrene (63%) which was pure by the following criteria: mp. 176° (6,8,13); UV (Figure 7); mass spectrum and NMR (Figure 8).

The high degree of localization of the 3,4 double bond on the ethylene bridge suggested by the Hückel bond order ($P_{34} = 0.787$) is borne out by the high field position of the H_3-H_4 AB quartet in the NMR (Figure 8) and the ease of catalytic hydrogenation (6,13) of cyclopenta[c,d] pyrene to the corresponding 3,4-dihydro derivative. The 3,4 double bond should, therefore, be an important site in the metabolism of the compound. Indications from preliminary work on separation and identification of metabolites appears to confirm this prediction. Ethyl acetate-acetone extracts (10) of cyclopenta[c, d] pyrene incubated for 30 min at 37° with liver microsomes prepared from rats treated with Aroclor 1254 (1) were worked up by preparative HPLC and four fractions, a-d (Figure 9), were collected. Fraction b, a major fraction by weight, was a colorless crystalline solid with a pyrene-like UV (Figure 10) [indicating saturation of C(3), C(4)] and a molecular weight of 260 by mass spectrometry. Both UV and mass spectrum (M^{+} 260, fragments at m/e 242, 213, 106.5) are consistent with a mixture of *cis* and *trans* 3,4-dihydrodiols.

The UV and mass spectrum of d revealed that this fraction contained unmetabolized cyclopenta[c,d] pyrene.

The relative retention time on reverse phase HPLC for fraction c is consistent with its containing phenolic and quinonoid metabolites. The UV

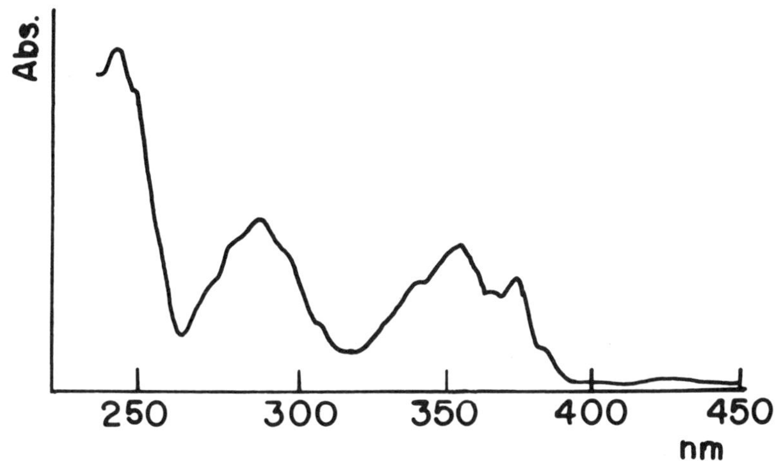

FIGURE 7. UV spectrum (CH_2Cl_2): cyclopenta[c,d] pyrene.

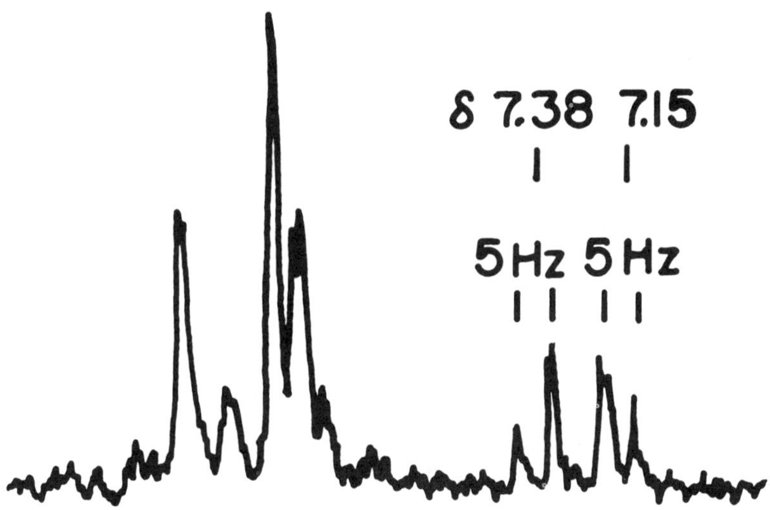

FIGURE 8. 60 MHz NMR ($CDCl_3$): cyclopenta[c,d] pyrene.

spectrum of this fraction (Figure 11) is consistent with the presence of phenol(s) and the mass spectrum of this fraction contains peaks expected for phenolic derivatives ($M^{+\cdot}$ 242, fragments at m/e 213 and 106.5). No peaks corresponding to quinonoid metabolites were evident in the mass spectrum of fraction c.

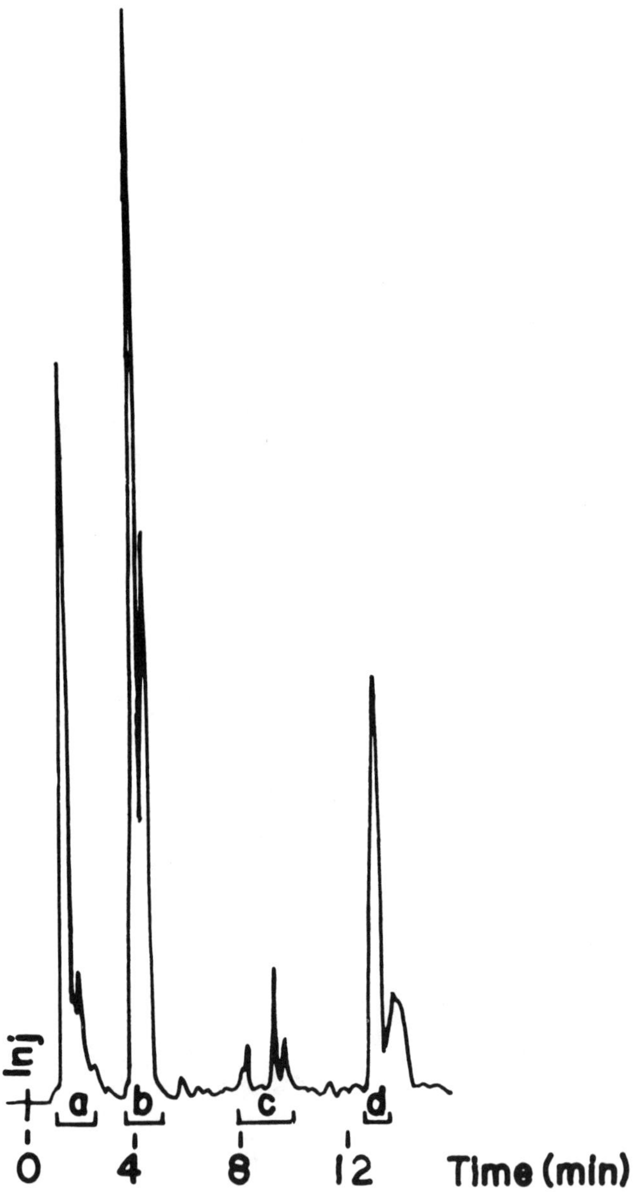

FIGURE 9. HPLC of EtOAc-soluble metabolites of cyclopenta[c,d] pyrene after incubation for 30 min at 37°C with liver microsome preparation: Reverse-phase chromatography on C_{18} ODS column, 2.6 mm x 25 cm; eluant, 30% CH_3CN- 70% H_2O, with 5%/min gradient to 100% CH_3CN; flowrate 1 ml/min and fixed wavelength UV detector at 254 nm.

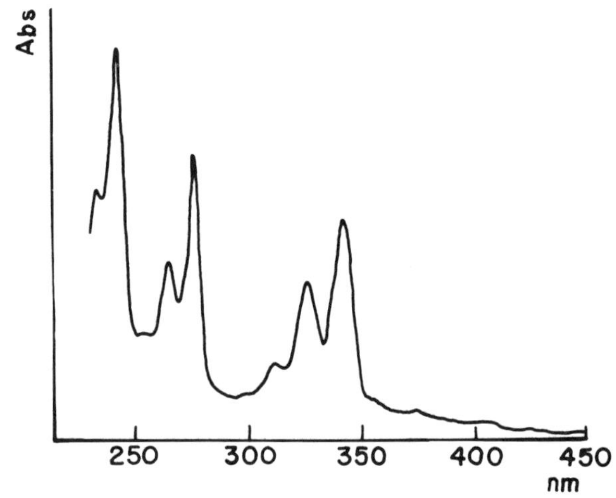

FIGURE 10. UV spectrum (CH_2Cl_2): fraction *b* of cyclopenta[c,d] pyrene metabolites.

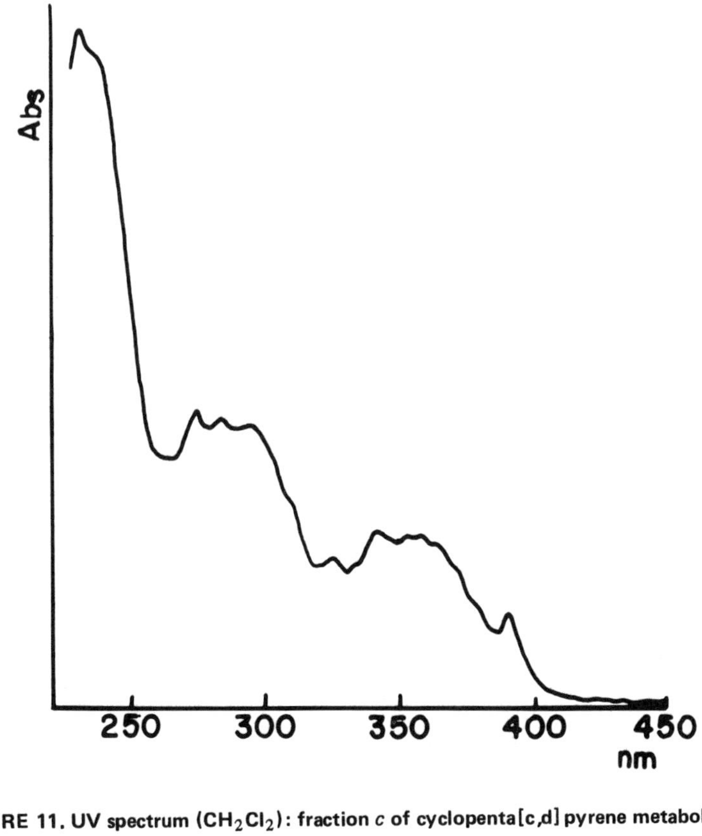

FIGURE 11. UV spectrum (CH_2Cl_2): fraction *c* of cyclopenta[c,d] pyrene metabolites.

The material in fraction a had a pyrene like UV spectrum and the mass spectrum shows major peaks at m/e 260, 242 and 215.

The preliminary identification of 3,4-dihydrodiols as major components of the metabolic reaction supports the prediction (4) that a mutagenic metabolite of cyclopenta [c,d] pyrene is the 3,4-oxide. Cyclopenta [c,d] pyrene 3,4-oxide can be made in one enzymatic step, whereas transformation of benzo [a] pyrene to to the 7,8-dihydrodiol-9,10-oxide requires three enzymatically catalyzed steps. This could explain the relatively small quantity of S-9 preparation required to optimize activity–approximately one-tenth that required for benzo [a] pyrene (4).

ACKNOWLEDGMENTS

Supported by American Cancer Society, Massachusetts Division, Grant #1485-C-1 and by Biomedical Research Support Grant #5-SO7-RR-05446-16 from the NIH.

REFERENCES

1. Ames, B. N., McCann, J., and Yamasaki, E. (1975): *Muta. Res.* 31:347.
2. Borgen, A., Darvey, H., Castognoli, N., Crocker, T. T., Rasmussen, R. E., and Wang, I. Y. (1973): *J. Med. Chem.* 16:502.
3. Dewar, M. J. S., and Dougherty, R. C. (1975): *The PMO Theory of Organic Chemistry,* pp. 197-391, Plenum Press, New York).
4. Eisenstadt, E., and Gold, A. (1978): Proc. Nat. Acad. Sci., U.S. 75: 1667
5. Gerasimenko, Yu. E., and Shevchuk, I. N. 91968): *J. Org. Chem., USSR* 4:2120
6. Gold, A. (1975): *Anal. Chem.* 47:1469
7. Gold, A., Schultz, J., and Eisenstadt, E. (1978): *Tetrahedron Lett.* (in press).
8. Jacob, J., and Grimmer, G. (1977): *Z. Bakteriol.,* Abt. 1:Orig., Reihe B 165:305.
9. Jerina, D. M., Lehr, R. E., Yagi, H., Hernandez, O., Dansette, P. M., Wislocki, P. G., Wood, A. W., Chang. R. L., Levin, W., and Conney, A. H. (1976): In: *In Votro Metabolic Activation in Mutagenesis Testing* F. J. de Serres, J. R. Fouts, J. E. Bend and R. M. Philpot Eds., pp 159-178, Elsevier, Amsterdam.
10. Kinoshita, N., Sherars, B., and Gelboin, H. V. (1973): *Cancer Res.* 33: 1937.
11. Sims, P., and Grover, P. L. (1974): *Adv. Cancer Res.* 20:165.
12. Vollmann, H., Becker, H., Corell, M., and Streek, H. (1937): *Ann.* 531:1.
13. Wallcave. L., Nagel, D. L., Smith, J. W., and Waniska, R. D.(1975): *Environ. Sci. Technol.* 9:143

Polynuclear Aromatic Hydrocarbons, edited by P. W. Jones and P. Leber. Ann Arbor Science Publishers, Inc. Ann Arbor, MI ©1979.

A COMPARISON OF THE MUTAGENICITY, TUMOR-INITIATING ACTIVITY AND COMPLETE CARCINOGENICITY OF POLYNUCLEAR AROMATIC HYDROCARBONS

Edmond LaVoie, Victoria Bedenko, Norio Hirota, Stephen S. Hecht and Dietrich Hoffmann

Naylor Dana Institute for Disease Prevention
American Health Foundation
Valhalla, New York 10595

Mutagenic activity towards *Salmonella typhimurium* and tumor-initiating activity and complete carcinogenicity on mouse skin have all been employed to determine the mutagenic and carcinogenic potential of polynuclear aromatic hydrocarbons (PAH). Several reviews have been reported on the advantages and significance of bioassays of PAH for tumor initiation and complete carcinogenicity on mouse skin (6,24,25,29). While these *in vivo* bioassays are generally accepted as effective indicators of the carcinogenic potential of PAH, tumorigenicity varies in different strains of mice. Comparisons between mutagenicity and carcinogenicity of PAH have been reviewed in several studies (3,7,19,20). The carcinogenicity data in these studies are often limited to complete carcinogenicity or are derived from a number of laboratories and different strains and species of animals. The studies on tumor initiation and complete carcinogenicity reported here were carried out within this institute using female Swiss albino (Ha/ICR) mice. These data, therefore, provide a uniform basis for the comparison of mutagenicity with carcinogenic activity.

The diverse structures and complexity of various PAH have been shown in earlier studies to elicit major differences in mutagenic potential in the *Salmonella*/microsomal assay. These results are not unexpected in view of the variations in the number of sites available for metabolic activation among PAH. In contrast, the lack of solubility of higher PAH in this *in vitro* assay could also reduce the effectiveness of detecting their mutagenic potential. Test compounds were, therefore, divided into several groups based on the number of fused rings incorporated in their structure. Thus, the potential and limitations of the *S. typhimurium* mutagenicity assay to suggest the carcinogenic potential of structurally related PAH were evaluated both qualitatively and quantitatively.

MUTAGENICITY ASSAYS

Mutagenicity studies were performed using *S. typhimurium* tester strains TA 100 (TA 1535/pKM101) and TA 98 (TA 1538/pKM101) provided by Dr. Bruce Ames of the University of California, Berkeley. The S-9 fraction routinely employed for microsomal activation was obtained from the livers of male Fisher-344 rats weighing 300-350 g which had been treated five days prior to sacrifice with 500 mg/kg Aroclor 1254 (Analabs, Inc.). The S-9 fraction was prepared by centrifugation of the liver homogenate (25% in 0.15 *M* KCl) at 9,000 g for 15 min as described by Ames *et al.* (1). The S-9 fraction was filter-sterilized at 4°C using a 600-ml Millipore pressure-filtration apparatus equipped with a 0.45-μm Swinnex filter. Each microsomal preparation was checked for sterility on nutrient agar prior to storage at -80°C. The S-9 mix contained, per ml, potassium phosphate buffer (100 μmoles) pH 7.4, 8 μmoles of $MgCl_2$, 1.65 μmoles of KCl, 5 μmoles of glucose-6-phosphate, and 4 μmoles of NADP and 0.5 ml of S-9 fraction.

The procedure of Ames *et al.* (2) was employed in performing these assays. In summary, various concentrations of PAH in 50 μl of dimethyl sulfoxide (DMSO) were added to 0.1 ml of an overnight nutrient-broth culture of the bacterial tester strain. After addition of 200 μl of S-9 mix and 2 ml of molten top agar at 45°C, the contents were mixed and poured on minimal glucose agar plates. Percent survivors for all compounds assayed was determined by employing dilutions of bacterial broth under identical conditions with the exception that excess histidine was added to the top agar. No significant toxicity towards the bacterial tester strains was observed for the compounds assayed (Tables I-VII).

Minor impurities in the test sample can have a significant effect on its mutagenic response (8). All the compounds used in this study were analyzed for purity using a Waters Associates Model ALC/GPC-202 high-speed liquid chromatograph equipped with a Model 440 UV detector and a Microbondapak/C_{18} column. Only the test samples which were at least 99% pure under these analytical conditions were employed in this study.

Compounds were considered mutagenic when the histidine revertant colonies (His^+ rev) obtained at any dose level assayed were at least 1.8 times that of the solvent control. Solvent controls (50 μl of DMSO) were employed at the beginning and end of every group of compounds assayed.

TABLE I. Bioassays on Anthracene, Phenanthrene, Fluorene and Pyrene

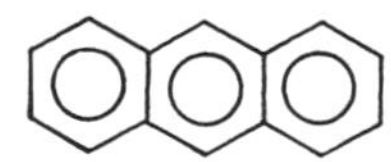

(1)

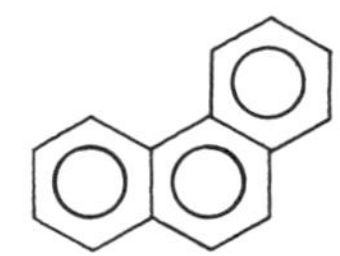

(2)

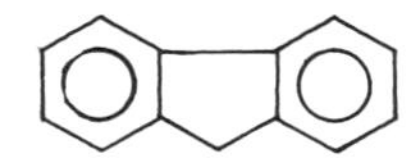

(3)

(4)

	Mutagenicity[a]		Complete Carcinogenicity[b] (27)		
	Dose (μg)	His^+ rev	% Sol	% TBA	Blanding Index (5)
Anthracene (1)	100	40	1.0	0	(0)
Phenanthrene (2)	100	50	–	–	–
Fluorene (3)	100	20	–	–	–
Pyrene (4)	100	30	0.1	0	(0)

[a] All compounds were tested at doses ranging from 10-100 μg per plate. Data listed represent the average number of revertant colonies per plate at 100 μg which have been corrected from DMSO control revertants. Average DMSO control value in this study was 105 revertants per plate. Revertant colonies are corrected to two significant figures.

[b] The data represent the results of application of the percent solutions in acetone listed three times a week with a No. 5 camel's hair brush to groups of 5 mice. % TBA indicates the percentage of tumor-bearing animals. Blanding indices in parentheses are extrapolated values.

TUMOR INITIATION

Compounds tested as tumor initiators were applied to the skin of groups of Swiss albino female mice (Ha/ICR). When the animals had entered the second telogen phase of the hair cycle, ten doses of the hydrocarbon to be tested were applied on alternate days to the shaved backs of the mice. Ten days after the last initiator dose, promotion was begun by application of 2.5 μg of tetradecanoyl phorbol acetate or 2.5% croton oil in acetone. The method of application was either by biopette or by skin painting with a No. 5 camel's hair brush.

COMPLETE CARCINOGENICITY

Assays for complete carcinogenicity were performed employing groups of mice of the same strain and sex as those used in the tumor-initiation studies (Swiss albino female mice, Ha/ICR). Applications were begun during the second telogen phase of the hair cycle and were continued three times weekly for the duration of the experiment. Test solutions were applied in doses of 100 μl either by No. 5 camel's hair brush or a biopette.

RESULTS AND DISCUSSION

The mutagenicity and complete carcinogenicity of anthracene, phenanthrene and fluorene are outlined in Table I. The lack of mutagenic activity at doses ranging from 10-100 μg per plate for these compounds correlates with the generally accepted view of their inactivity as complete carcinogens. Because of the limited number of compounds assayed in this series of PAH, the potential of the *Salmonella*/microsomal assay systems to qualitatively, or possibly quantitatively, reflect the tumorigenicity of these less complex PAH cannot be adequately determined. In screening methylated three-ring PAH, 9-methylfluorene and 1,9-dimethylfluorene were found to be relatively potent mutagens towards TA 100 in the presence of rat liver homogenate (Figure 1). The relative tumor-initiating activity and complete carcinogenicity of these methylated PAH are currently being assayed in order to provide a better understanding of the efficiency of the mutagenicity assay in respect to potential carcinogens or noncarcinogens within PAH having three or fewer fused-ring systems.

The relationship of mutagenicity toward *S. typhimurium* TA 100 with the carcinogenicity of tetracyclic PAH was thoroughly investigated. Pyrene was found to be inactive as a mutagen which is in agreement with its lack of carcinogenic activity (Table I). The mutagenic activity of chrysene and its methylated analogs, however, failed to provide a complete quantitative or

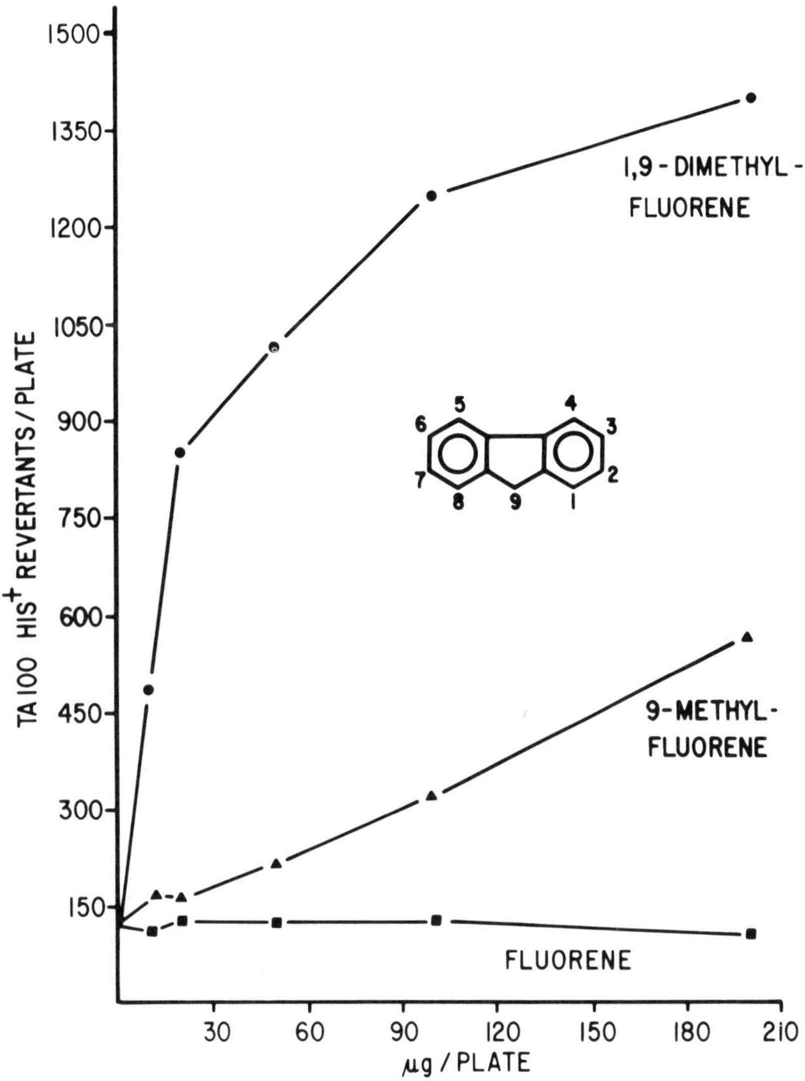

FIGURE 1. The mutagenic activity of 9-methylfluorene, and 1,9-dimethylfluorene towards *S. typhimurium* TA 100. The 1-, 2-, 3- and 4-methylfluorenes and 2,3- and 9,9-dimethylfluorenes were as inactive as their parent compound when assayed in the presence of rat liver homogenate and *S. typhimurium* TA 100.

TABLE II. Bioassays on Chrysene and Methylchrysenes

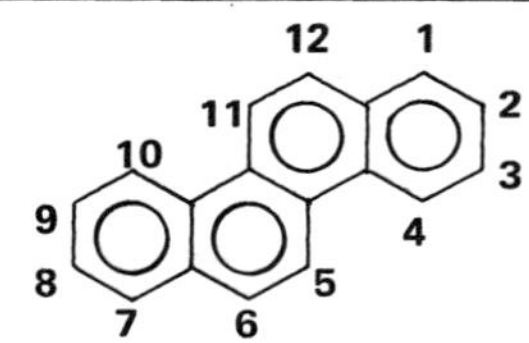

	Mutagenicity[a]	Tumor Initiation[b]		Complete Carcinogenicity[b,c] (9,13)		
	His$^+$ rev	% TBA	T/A	% TBA	T/A	Blanding Index
Chrysene	270	50	1.0	20[e]	0.5	(16)
1-CH_3	150	30	0.3	0	0	(0)
2-CH_3	90	40	0.7	55[d]	1.1	22
3-CH_3	150	70	1.3	25[e]	0.3	(20)
4-CH_3	120	35	0.5	15	0.3	(15)
5-CH_3	360	85	4.8	100	4.5	90
6-CH_3	40	30	0.6	15	0.3	(15)
Acetone control (100 μl)		0	0	0	0	(0)

[a]All compounds were listed at doses ranging from 5-100 μg per plate. Data listed represent the average number of revertants per plate at 20 μg (maximal dose-response observed for all compounds) which have been corrected from DMSO control revertants. Average DMSO control value in this study was 125 revertants per plate. Revertant colonies are corrected to two significant figures.

[b]All compounds were assayed with an initiator dose of 10 x 100 μg with promotion by thrice weekly applications of 2.5-μg TPA in acetone for 20 weeks using 20 mice per group. % TBA represents the percentage of tumor-bearing animals. T/A is the average number of tumors per animal.

[c]Compounds were assayed with groups of 20 mice at a dose of 100 μg three times weekly. Blanding indices in parentheses are extrapolated values.

[d]$p < 0.01$, compared to acetone control.

[e]$p < 0.05$, compared to acetone control.

qualitative correlation (Table II). While 6-methylchrysene was not mutagenic under these assay conditions, it is a marginal tumor initiator and a complete carcinogen. The observed mutagenic activity for 1-methylchrysene, which possesses tumor-initiating potential similar to 6-methylchrysene, does preclude a quantitative correlation among isomeric PAH which are weakly tumorigenic. Mutagenicity assays of these compounds employing *S. typhimurium* TA 98 did not provide an improved correlation with carcinogenic activity on mouse skin. The most noteworthy feature of this comparative study of the methylchrysenes is that 5-methylchrysene is most mutagenic and most carcinogenic. The possibility, therefore, that mutagenic activity within an isomeric series of PAH might distinguish between potent and weak or noncarcinogens was further investigated.

The known fluorinated analogs of 5-methylchrysene provide not only an isomeric but also isosteric series of PAH with significant variations in carcinogenicity. These fluorinated 5-methylchrysene analogs are compared for mutagenic activity toward *S. typhimurium* TA 100, tumor-initiating activity, and complete carcinogenicity (Table III). If one views the mutagenic activity of these compounds at a single dose of 10 μg, a correlation between the observed mutagenic activity and carcinogenicity would appear to exist. This direct correlation, however, rapidly weakened when a range of doses was employed in the mutagenicity assay (Table III).

Expressing the mutagenic activity of fluorinated 5-methylchrysenes as the maximum number of revertants per nmol where a linear response is observed may appear to provide some quantitative correlation between mutagenicity toward *S. typhimurium* TA 100 and carcinogenic activity. The implications of this study would tend to indicate that mutagenic potency expressed in this fashion could be used as a criterion for selecting representative members of a series of isomeric PAH for *in vivo* bioassay. The weakness of this assumption can be seen in the comparison of the mutagenic activity of fluoranthene and several of its methyl and dimethyl analogs with tumor-initiating activity (Table IV). In this, study 1-methylfluoranthene and 7-methylfluoranthene would be considered prime candidates for determining the potential of alkylfluoranthenes as tumor initiators. However, both of these isomeric methylfluoranthenes were noncarcinogenic (Table IV). Employing mutagenic activity toward *S. typhimurium* to limit *in vivo* bioassays of PAH would have misrepresented the carcinogenic hazard within this series of compounds.

The use of the *S. typhimurium* TA 98 failed to provide an improved indication of carcinogenic potential for both the methylated fluoranthenes and fluorinated 5-methylchrysenes. The utility of a general screen of PAH for mutagenic activity toward *S. typhimurium* in view of these data becomes limited to its use as an indicator of a possible carcinogen. Since the weak carcinogen, 6-methylchrysene, was not found to be significantly mutagenic

TABLE III. Comparison of the Bioactivity of 5-Methylchrysene and its Fluorinated Derivatives

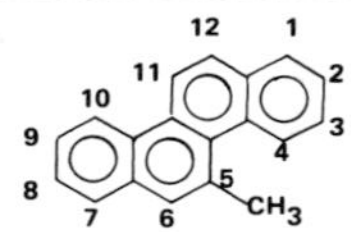

	Mutagenicity[a]		Maximum	Tumor Initiation[b]		Complete Carcinogenicity[b,c]		
	Dose (μg)	His⁺ rev	rev/nmol	% TBA	T/A	% TBA	T/A	Blanding Index
5-methyl-chrysene	10	330	49	68[d,f]	2.0[f]	80[d]	1.0	31
	20	360						
1-F	10	190	22	10	0.1	5	0	(0)
	20	440						
3-F	10	220	12	0	0	0	0	(0)
	50	610						
6-F	10	350	35	80[d]	4.1	100[d]	3.2	32
	100	510						
7-F	10	730	73	60[d]	0.9	70[d]	1.0	30
	20	800						
9-F	10	410	41	45[d]	0.8	75[d]	1.0	26
	100	460						
11-F	10	250	41	45[d]	0.6	80[d]	1.0	33
	20	300						
12-F	10	120	18[f]	18[f]	0.2[f]	25[e]	0.2	(17)
	100	400						
Acetone control (100 μl)			0	0	0	0	0	(0)

[a]All compounds were tested at doses ranging from 5-100 μg per plate. Data listed represent the average revertants per plate (corrected from DMSO controls) at 10μg and at the highest mutagenic response level observed. The average DMSO control value in this study was 130 revertants per plate. Revertant colonies are corrected to two significant figures.

[b]Compounds were assayed using groups of 20 mice at an initiation dose of 10 x 3.0 μg with promotion by thrice weekly applications of 2.5-μg TPA for 20 weeks. % TBA represents the percentage of tumor-bearing animals. T/A is the average number of tumors per animal.

[c]Compounds were assayed using groups of 20 mice at a dose of 5 μg applied three times weekly. Blanding indices in parentheses are extrapolated values.

[d]$p < 0.01$, compared to acetone control.

[e]$p < 0.05$, compared to acetone control.

[f]Represents the average of three separate assays.

TABLE IV. The Mutagenicity and Tumor-Initiating Activity of Fluoranthene and its Methylated Derivatives

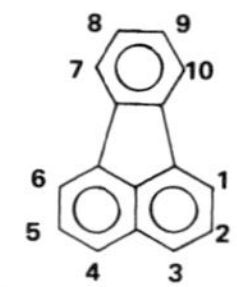

	Mutagenicity[a]		Maximum	Tumor Initiation[b] (15)	
	Dose (μg)	His^+ rev	rev/nmol	% TBA	T/A
Fluoranthene[c]	200 100	140 180	11	3.3	0
1-CH_3	200 10	120 320	22	0	0
2-CH_3[c]	200 100	220 220	7.3	30[d]	0.5
3-CH_3	200 100	280 310	5.7	30[d]	0.5
7-CH_3	200 10	210 350	35	6.7	0.1
8-CH_3	200 100	70 80	15	3.3	0.1
2,3-Dimethyl	200 100	120 120	3.5	47[d]	0.8
7,8-Dimethyl	200 20	120 140	12	43[d]	0.8
7,10-Dimethyl	200 100	70 80	1.5	–	–
8,9-Dimethyl	200 50	60 60	4.4	13	0.1
Acetone Control (100 μl)	–	–	–	3.3	0.1

[a]All compounds were tested at doses ranging from 10-200 μg per plate. Data listed represent the average number of revertants per plate (corrected from DMSO controls) at 200 μg and at the dose in which the highest mutagenic response was observed. The average DMSO control value in this study was 124 revertants per plate. Revertant colonies are corrected to two significant figures.

[b]Compounds were assayed at an initiation dose of 10 x 100 μg with promotion by thrice weekly application of a 2.5% croton oil solution in acetone (average dose 3.8 mg) with a No. 5 camel's hair brush for 20 weeks using groups of 30 mice. % TBA represents the percentage of tumor-bearing animals. T/A is the average number of tumors per animal.

[c]Fluoranthene and 2-methylfluoranthene were assayed as complete carcinogens using groups of 20 and 10 mice, respectively. These compounds were applied thrice weekly at a dose of 100 μg for one year. While 2-methylfluoranthene gave a 60% incidence of tumors and 1.2 tumors per animal, fluoranthene was inactive.

[d]$p < 0.01$, compared to acetone control.

under the assay conditions employed, the reliability of even this limited view of the significance of mutagenic PAH is reduced.

In contrast to mutagenicity, the tumor-initiating activity of the methylchrysenes did provide a reasonable agreement with their complete carcinogenicity. A similar indication of the activity of the fluorinated analogs of 5-methylchrysene as complete carcinogens was obtained when assayed at a dose of 10 x 3.0 μg for tumor-initiating activity.

The mutagenicity and complete carcinogenicity of several benzofluoranthenes are outlined in Table V. While there exist significant differences in the complete carcinogenic activity of these PAH, mutagenic activity toward *S. typhimurium* TA 100 was not found to differ significantly. The tumor-initiating activity of this series of PAH on mouse skin is currently being studied.

The mutagenic activities of a few additional pentacyclic PAH are listed in Table VI. Benzo[a]pyrene (B[a]P) and its methylated analogs represent the only PAH in this study where an improved correlation between carcinogenicity and mutagenicity was observed when *S. typhimurium* TA 98 was substituted for TA 100 employing this general protocol. While the number of compounds in this comparison is limited, their mutagenic and tumorigenic potential are in agreement. The tumor-initiating activity of B[e]P and perylene is currently being assayed on mouse skin for comparison with the mutagenicity data shown in Table VI.

The tumor-initiating activity, complete carcinogenicity and mutagenic potential toward *S. typhimurium* TA 100 of several hexacyclic PAH are listed in Table VII. The mutagenic activity of the compounds in this series varied only slightly with the bacterial tester strain employed. Comparison of the relative mutagenic activity of these compounds employing a wide range of doses would indicate only the likelihood of dibenzo[e,l]pyrene being noncarcinogenic.

The tumor-initiation potential of these compounds did correlate well with their complete carcinogenicity. The tumor-initiating activity of these compounds did detect, as would be expected, the three complete carcinogens within this series of PAH. Of the six PAH which were not complete carcinogens, only naphtho[2,3-e]pyrene had significant tumor-initiating activity, $p < 0.05\%$.

The inability of the *Salmonella*/microsomal mutagenicity assay to provide data which are of comparable significance to tumor-initiating activity may be associated with the selection of the metabolic activation system employed in this study. The use of a microsomal suspension from mouse skin or possibly mouse liver could produce a significant alteration in both the number and ratio of *in vitro* metabolites formed. Even under these more specific metabolic activation conditions, however, the metabolic profile obtained would

TABLE V. The Mutagenicity and Complete Carcinogenicity of Benzofluoranthenes

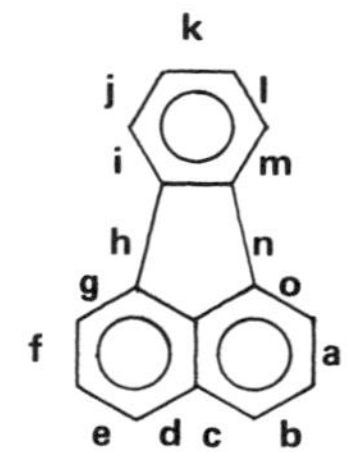

	Mutagenicity[a]		Complete Carcinogenicity[b] (28)		
	Dose (μg)	His^+ rev	% Sol	% TBA	Blanding Index
Benzo[b] fluoranthene	200	460	0.5%	100	110
	100	440	0.1%	85	36
Benzo[j] fluoranthene	200	130	0.5%	95	73
	50	240	0.1%	100	55
Benzo[k] fluoranthene	200	270	0.5%	10	—
	100	320	0.1%	0	(0)
Benzo[g,h,i] fluoranthene	200	280	0.5%	0	(0)
	20	570	0.1%	0	(0)

[a] All compounds were tested at doses ranging 5-200 μg per plate. Data listed represent the average number of revertants (corrected from DMSO controls) at 200 μg and at the dose in which the highest mutagenic response was observed. The average DMSO control value in this study was 124 revertants per plate. Revertant colonies are corrected to two significant figures.

[b] Compounds were tested by application of the acetone solutions listed three times a week to groups of 20 mice with a No. 5 camel's hair brush. Blanding indices in parentheses are extrapolated values.

[c] $p < 0.01$, compared to benzo[g,h,i] fluoranthene.

TABLE VI. Bioassays on Benzopyrenes and Perylene

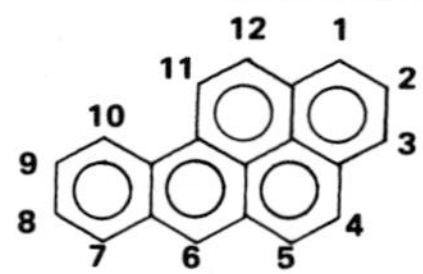

B[a]P

B[e]P

Perylene

	Mutagenicity[a]			Tumor Initiation[b]		Complete Carcinogenicity[b,c] (10,26-28)		
	Dose (μg)	TA100	TA98	% TBA	T/A	% TBA	T/A	Blanding Index
B[a]P	10	380	450	40[d]	0.5	95[d]	4.3	78
	20	320	480					
10-Methyl B[a]P	10	200	330	20[e]	0.2	–	–	–
	20	150	410					
7,10-Dimethyl B[a]P	10	150	130	0	0	–	–	–
	20	260	180					
B[e]P	10	130	20	–	–	15		(16)
	20	120	20					
Perylene	10	170		–	–	–	–	–
	20	210	70					
Acetone control (100 μl)	–	–	–	0	0	0	0	(0)

[a]All compounds were listed at doses ranging from 2-200 μg per plate. Data listed represent the number of reverants per plate (corrected from DMSO controls) at 10 and 20 μg. The average DMSO control values in this study were 103 and 52 revertants for TA100 and TA98, respectively. Revertant colonies are corrected to two significant figures.

[b]Tumor initiation was performed by application of 10 x 10 μg of each compound. Promotion was performed by thrice weekly applications of 2.5 μg TPA for 20 weeks. % TBA represents the percentage of tumor-bearing animals. T/A is the number of tumors per animal.

[c]Compounds were tested by applying thrice weekly a 0.1% solution in acetone with a No. 5 camel's hair brush. Blanding indices in parentheses are extrapolated values.

[d]$p < 0.01$, compared to acetone control.

[e]$p < 0.05$, compared to acetone control.

TABLE VII. Comparison of the Bioactivity of Various Six-Membered PAH

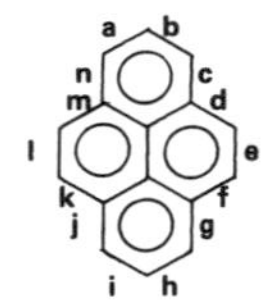

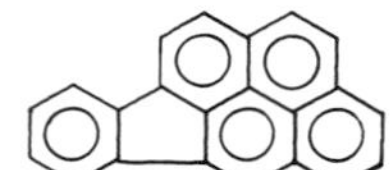

Indeno[1,2,3-c,d] pyrene

Benzo[g,h,i] perylene

	Mutagenicity[a]		Tumor Initiation[b] (16)		Complete Carcinogenicity[c] (16)		
	Dose (μg)	His$^+$ rev	% TBA	T/A	% TBA	T/A	Blanding Index
Dibenzo[a,e] pyrene	100	170	33[d]	0.8	45[d]	0.6	(26)
	20	120					
Dibenzo[a,h] pyrene	100	120	70[d]	3.5	75[d]	2.9	48
	20	120					
Dibenzo[a,i] pyrene	50	130	40[d]	1.8	80[d]	1.0	33
	20	60					
Dibenzo[c,d & j,k] pyrene	100	160	6.7	0.1	0	0	(0)
	20	170					
Dibenzo[e,l] pyrene	100	30	0	0	0	0	(0)
	20	40					
Naphtho [2,3-e] pyrene	100	280	30[d]	0.5	0	0	(0)
	20	270					
Indeno[1,2,3-c,d] pyrene	50	220	17	0.3	0	0	(0)
	20	160					
Benzo[g,h,i] perylene	100	100	6.7	0.1	0	0	(0)
	20	110					
Dioxane (Solvent control)	0		6.7	0.1	0	0	(0)

[a]All compounds were tested at doses ranging from 1 to 100 μg per plate. Data listed represent the average revertants per plate (corrected from DMSO controls) at 20 μg and at the dose in which the highest mutagenic response was observed. The average DMSO control value for this study was 148 revertants per plate. Revertant colonies are corrected to two significant figures.

[b]Compounds were assayed at an initiation dose of 10 x 25 μg in dioxane with promotion by thrice weekly applications of 2.5% croton oil in acetone. % TBA represents the percentage of tumor-bearing animals. T/A is the average number of tumors per animal.

[c]Compounds were assayed by application to groups of 20 mice of 0.1% solutions in dioxane thrice weekly with a No. 5 camel's hair brush. Blanding indices in parentheses are extrapolated values.

[d]$p < 0.01\%$, compared to dioxane control.

most likely not be identical to that of the intact cell (4,21). Several K-region epoxides of PAH, which appear to contribute marginally to the carcinogenicity of the parent compound, have been shown to possess significant mutagenic activity towards *S. typhimurium* (22). In view of these data, it is more likely that mutagenic activity in this assay system in the presence of various microsomal preparations would only be an indication of the ability of PAH to form alkylating agents. Since metabolic activation of several PAH has been shown to involve a multistage process for the formation of the ultimate carcinogenic agent (12,17,18) the *S. typhimurium* mutagenic assay system may simply be unable to distinguish this possible metabolite from other reactive metabolites formed *in vitro.*

Mutagenicity towards *S. typhimurium* has, however, been shown to be an effective tool for the identification of the metabolically activated species of carcinogenic PAH. In recent studies in which the mutagenic activity of the procarcinogens and their suspect ultimate carcinogenic forms were compared to their carcinogenic potential, a good correlation was obtained (17,18). The ability of mutagenicity towards *S. typhimurium* to qualitatively and quantitatively correlate with carcinogenic potential within metabolites of carcinogenic PAH may reflect its effectiveness as an indicator of carcinogenic potential when single-stage metabolic activation processes are involved.

CONCLUSIONS

Mutagenicity toward *S. typhimurium* has proved to be an effective assay system for several practical and theoretical studies in environmental carcinogenesis. The potential of mutagenic activity towards *S. typhimurium* to be an indicator of tumorigenic activity has not been adequately studied for PAH with three or less fused-ring systems. The lack of significant mutagenic activity of some of these less complex PAH is associated with their noncarcinogenic response on mouse skin. However the available data are insufficient to assess the correlation of active mutagens of this series of PAH with their carcinogenic potential. Studies on the tumor-initiating potential of 9-methyl- and 1,9-dimethylfluorene and on the mutagenicity of methylphenanthrenes are in progress to determine the qualitative and quantitative limitations of this bacterial assay system within these series of PAH.

The data obtained from a general mutagenicity screen of 40 PAH which contain four or more fused-ring systems have indicated only five compounds (including one false negative) as inactive mutagens. Overall, of 27 tumorigenic PAH assayed for mutagenicity (including very weakly or marginally active tumorigenic compounds), 26 were positive (96%). Of 11 nontumorigenic PAH, only 4 (36%) were not mutagenic. Since mutagenicity is more likely to reflect tumor-initiating activity than complete carcinogenicity, inactive complete carcinogens which have not been assayed for tumor-initiating activity are not included in this analysis. Quantitation of the response of mutagenic PAH which were structurally related failed to provide a reliable indication of their relative complete carcinogenicity or tumor-initiating activity.

The *Salmonella* mutagenicity assay was not effective in detecting PAH which are noncarcinogenic. This is in contrast to the tumor-initiation studies in which no inactive tumor initiator exhibited activity as a complete carcinogen. In addition, all complete carcinogenic PAH were tumor initiators when assayed in both bioassay systems. Since the target tissue and the strain of mice employed were identical, this correlation is not surprising. Although the number of PAH assayed for tumor initiation and complete carcinogenicity is limited in this study, 21% (4/19) of the PAH assayed were inactive as complete carcinogens.

In view of these results, general screening of PAH employing the standard *Salmonella*/mammalian microsomal mutagenicity assay system should be performed accepting the following criteria for interpretation of the data.

1. All PAH which are assayed for mutagenicity should be screened at several dose levels prior to establishing their relative activity.
2. Mutagenicity assays of various mixtures of PAH without adequate analytical determination of their relative composition should not be viewed as relevant to their carcinogenic hazard.
3. Quantitative comparisons of the mutagenic potential of PAH are not reliable indicators of relative carcinogenic potential. Priority, if required, for *in vivo* testing should be limited to a distinction between mutagenic and nonmutagenic PAH.

ACKNOWLEDGMENTS

These studies were supported by National Cancer Institute Grant CA 17613 and National Cancer Institute Contract No. 1 CP-55666.

We express our deep appreciation to Dr. Hans D. Sauerland of Ruetgerswerke A. G. Duisburg, Federal Republic of Germany, for samples of several polycyclic aromatic hydrocarbons.

REFERENCES

1. Ames, B. N., Duston, W. E., Yamasaki, E., and Lee, F. D. (1973): Carcinogens are mutagens: A simple test system combining liver homogenates for activation and bacteria for detection. *Proc. Nat. Acad. Sci., U.S.* 70:2281-2285.
2. Ames, B. N., McCann, J., and Yamasaki, E. (1975): Methods for detecting carcinogens and mutagens with the *Salmonella*/mammalian microsome mutagenicity test. *Mutat. Res.* 31:347-364.

3. Andrews, A. W., Thibault, L. H., and Lijinsky, W. (1978): The relationship between carcinogenicity and mutagenicity of some polynuclear hydrocarbons. *Mutat. Res.* 51:311-318.
4. Bigger, C. A. H., Tomazewski, J. E., and Dipple, A. (1978): Differences between products of binding of 7,12-dimethylbenz[a]anthracene to DNA in mouse skin and in a rat liver microsomal system. *Biochem. Biophys. Res. Commun.* 80:229-235.
5. Blanding, F. H., King, W. H. Jr., Priestly, W. Jr., and Rehner, J. Jr. (1951): Property of high-boiling petroleum products. *Arch. Ind. Hyg.* 4:39-49.
6. Bock, F. G. (1964): Early effects of hydrocarbons on mammalian skin. In: *Progress in Experimental Tumor Research, Vol. 4.* F. Homburger, Ed., pp. 126-168, Karger/Hafner Publishing Co., Inc., New York.
7. Coombs, M. M., Dixon, C., and Kissonerghis, A. (1976): Evaluation of the mutagenicity of compounds of known carcinogenicity, belonging to the benz[a]anthracene, chrysene, and cyclopenta[a]phenanthrene series, using Ames' test. *Cancer Res.* 36:4525-4529.
8. Donahue, E. V., McCann, J., and Ames, B. N. (1978): Detection of mutagenic impurities in carcinogens and noncarcinogens by high-pressure liquid chromatography and the *Salmonella*/microsome test. *Cancer Res.* 38:431-438.
9. Hecht, S. S., Bondinell, W. E., and Hoffmann, D. (1974): Chrysene and methylchrysenes: Presence in tobacco smoke and carcinogenicity. *J. Nat. Cancer Inst.* 53:1121-1133.
10. Hecht, S. S., Hirota, N., and Hoffmann, D. (1978): Comparative tumor initiating activity of 10-methylbenzo[a]pyrene, 7,10-dimethylbenzo-[a]pyrene and benzo[a]pyrene. *Cancer Lett.* 5:179-183.
11. Hecht, S. S., Hirota, N., Loy, M., and Hoffmann, D. (1978): Tumor-initiating activity of fluorinated 5-methylchrysenes. *Cancer Res.* 38: 1694-1968.
12. Hecht, S. S., LaVoie, E., Mazzarese, R., Amin, S., Bedenko, V., and Hoffmann, D. (1978): 1,2-Dihydro-1,2-dihydroxy-5-methylchrysene, a major activated metabolite of the environmental carcinogen 5-methylchrysene. *Cancer Res.* 38:2191-2194.
13. Hecht, S. S., Loy, M., and Hoffmann, D. (1976): On the structure and carcinogenicity of the methylchrysenes. In: *Carcinogenesis, Vol. 1,* R. I. Freudenthal and P. W. Jones, Eds., pp. 325-340, Raven Press, New York.
14. Hecht, S. S., Loy, M., Mazzarese, R., and Hoffmann, D. (1978): Synthesis and mutagenicity of modified chrysenes related to the carcinogen, 5-methylchrysene. *J. Med. Chem.* 21:38-44.
15. Hoffmann, D., Rathkamp, G., Nesnow, S., and Wynder, E. L. (1972): Fluoranthenes: Quantitative determination in cigarette smoke, formation by pyrolysis and tumor initiating activity. *J. Nat. Cancer Inst.* 49:1165-1175.

16. Hoffmann, D., and Wynder, E. L. (1966): Beitrag zur carcinogenen Wirkung von Dibenzopyrenen. *Z. Krebsforsch.* 68:137-149.
17. Lehr, R. E., Yagi, H., Thakker, D. R., Levin, W., Wood, A. W., Conney, A. H., and Jerina, D. M. (1978): The bay region theory of polycyclic aromatic hydrocarbon-induced carcinogenicity. In: *Carcinogenesis, Vol. 3*, P. W. Jones and R. I. Freudenthal, Eds., pp. 231-242, Raven Press, New York.
18. Levin, W., Thakker, D. R., Wood, A. W., Chang, R. L., Lehr, R. E., Jerina, D. M., and Conney, A. H. (1978): Evidence that benzo[a]-anthracene-3,4-diol-1,2-epoxide is an ultimate carcinogen on mouse skin. *Cancer Res.* 38:1705-1710.
19. McCann, J., Choi, E., Yamasaki, E., and Ames, B. N. (1975): Detection of carcinogens as mutagens in the *Salmonella*/microsome test: assay of 300 chemicals. *Proc. Nat. Acad. Sci., U.S.* 72:5135-5139.
20. Purchase, I. F. H., Longstaff, E., Ashby, J., Styles, J. A., Anderson, D., Lefevre, P. A., and Westwood, F. R. (1978): An evaluation of 6 short-term tests for detecting organic chemical carcinogens. *Brit. J. Cancer* 37:873-959.
21. Schmeltz, I., Tosk, J., and Williams, G. M. (1978): Comparison of the metabolic profiles of benzo[a]pyrene obtained from primary cell cultures and subcellular fractions derived from normal and methylcholanthrene induced rat liver. *Cancer Lett.* 5:81-89.
22. Sims, P., and Grover, P. L. (1974): Epoxides in polycyclic aromatic hydrocarbon metabolism and carcinogenesis. *Adv. Cancer Res.* 10:165-267.
23. Van Duuren, B. L. (1969): Tumor-promoting agents in two-stage carcinogenesis. In: *Progress in Experimental Tumor Research*, Vol. 11, F. Homburger, Ed., pp. 31-68, Karger/Hafner Publishing Co., Inc., New York.
24. Van Duuren, B. L., Sivak, A., Goldschmidt, M., Katz, C., and Melchionne, S. (1970): Initiating activity of aromatic hydrocarbons in two-stage carcinogenesis. *J. Nat. Cancer Inst.* 44:1167-1173.
25. Weisburger, J. H. and Weisburger, E. K. (1967): Tests for chemical carcinogens. In: *Methods in Cancer Research,* Vol. 1, H. Busch, Ed., pp. 307-398, Academic Press, New York.
26. Wynder, E. L. Fritz, L., and Furth, N. (1957): Effect of concentration of benzopyrene in skin carcinogenesis. *J. Nat. Cancer Inst.* 19:361-370.
27. Wynder, E. L., and Hoffmann, D. (1959): A study of tobacco carcinogenesis: The role of higher polycyclic hydrocarbons. *Cancer* 12:1079-1086.
28. Wynder, E. L., and Hoffmann, D. (1959): The carcinogenicity of benzofluoranthenes. *Cancer* 12:1194-1198.
29. Wynder, E. L., and Hoffmann, D. (1970): The epidermis and the respiratory tract as bioassay systems in tobacco carcinogenesis. *Brit. J. Cancer* 24:574-587.

Polynuclear Aromatic Hydrocarbons, edited by P. W. Jones and P. Leber. Ann Arbor Science Publishers, Inc. Ann Arbor, MI © 1979.

CHEMICAL CYTOTOXICITY: A CANCER PROMOTER

G. D. Griffin, T. D. Jones and P. J. Walsh

Health and Safety Research Division
Oak Ridge National Laboratory
Oak Ridge, Tennessee 37830

The fundamental biochemical mechanisms by which various chemicals induce cancer in animals and humans are still shrouded in mystery. Attempts to assess the carcinogenicity (in a quantitative manner) of chemicals and chemical mixtures which have not been tested in long-term animal studies must therefore depend on essentially empirical relationships which can be discovered between cancer and other biological effects produced by the same chemical. The use of short-term biological tests which use chemicals to induce mutagenesis in various cultured cells is an example of attempts to correlate an early biologic effect (mutation) with carcinogenesis. The molecular basis for this correlation (that chemicals interact with DNA to produce cancer and/or mutation) seems reasonable and, indeed, a 85-90% correlation (on a qualitative basis) between the carcinogenic activity and the mutagenicity of a given chemical has been reported (14). Quantitatively, however, it has been much more difficult to relate the extent of mutagenicity to the potency of the chemical as a carcinogen. As Huberman notes: "no correlation has yet been found between mutagenic and carcinogenic activities of a number of potent carcinogens" (9). We have recently studied radiation-induced cytotoxicity vs cancer incidence data (11). These relationships usually indicate a linear correlation between cytotoxicity and cancer incidence from zero cell-killing to almost total necrosis in a given organ. DiPaolo *et al.* (7) previously reported "that cytotoxicity is related to concentration and potency of the carcinogen." It is well-known, however, that cytotoxicity and cellular transformation are not directly related processes (5,7,9).

The degree of carcinogenicity of various chemicals, in our risk estimates, is hypothesized to be directly related to the degree of increased cell-proliferation and this proliferation can be estimated from cytotoxicity data (11). Cellular proliferation can be envisioned as a homeostatic response of the surviving cells to the initial chemical insult which produced the observed cytotoxicity. We have attempted to determine if a linear relationship could be established between chemically induced cytotoxicity and cancer incidence in animals exposed to the same chemical. The available data are not sufficient to produce a strong confirmation or refutation of our approach to chemical carcinogenesis, but based on the composite literature on oncology (1,2,3,16) our model may be useful. We have examined the cytotoxicity data of a number of polycyclic aromatic hydrocarbons (PAH) and have found a linear relationship between cellular cytotoxicity and cancer. Cytotoxicity data might be useful, therefore, in estimating the carcinogenic risk of various chemicals.

If one considers the multitude of interactions which a PAH or its metabolite can undergo with various macromolecular species such as protein, RNA and DNA, it may be possible to sometimes correlate carcinogenesis with DNA-PAH interaction, yet the degree of carcinogenicity can often be found to correlate with binding to RNA or protein instead of DNA (4,13). Cytotoxicity, on the other hand, can arise from interaction of chemicals with a multitude of intracellular molecular species. The cytotoxic effect of carcinogens may be thought of as a promotional stimulus which can "drive" a previously small and slowly proliferating population of altered cells into an actively proliferating tumor cell mass (Figure 1). In this model, the initial event which transforms cells is coupled temporally to a proliferation stimulus, and one explanation for this stimulus could be a regulatory response by the organism's cells to the cellular loss caused by cytotoxicity. Although loss of transformed cells by chemical cytotoxicity might be expected, some work suggests that these cells are more resistant to chemical toxicity (5), so that the proportion of altered cells killed may be zero or much less than in the case of normal cells. This model envisions the proliferative stimulus (whether mediated by hormones, cell growth factors or loss of contact inhibition) to be a decisive factor which enables the tumor cells to escape from cellular control.

Chemical insults in an organism occur in different tissues which vary greatly in their sensitivity and the length of time they are in contact with the chemical. An understanding of the metabolism of some of these chemicals is critical in determining the cancer risk to a specific cellular population. Benzo [a] pyrene (BP) can serve as a model compound in this regard. In order for BP to show any significant cytotoxicity, it must be metabolized (10). A number of lines of evidence suggest that BP must be metabolized to a so-called "proximal carcinogen" in order to be carcinogenic (10,12). This metabolic activation seems to occur through epoxide formation by the enzyme aryl hydrocarbon hydroxylase (AHH). The cytotoxicity of BP toward various cell lines has been related to AHH activity, in that cell lines with higher AHH activity are killed to a greater degree by BP (8). Figure 2 presents cytotoxicity data for BP and a number of its metabolites on V79 hamster cell cultures. The

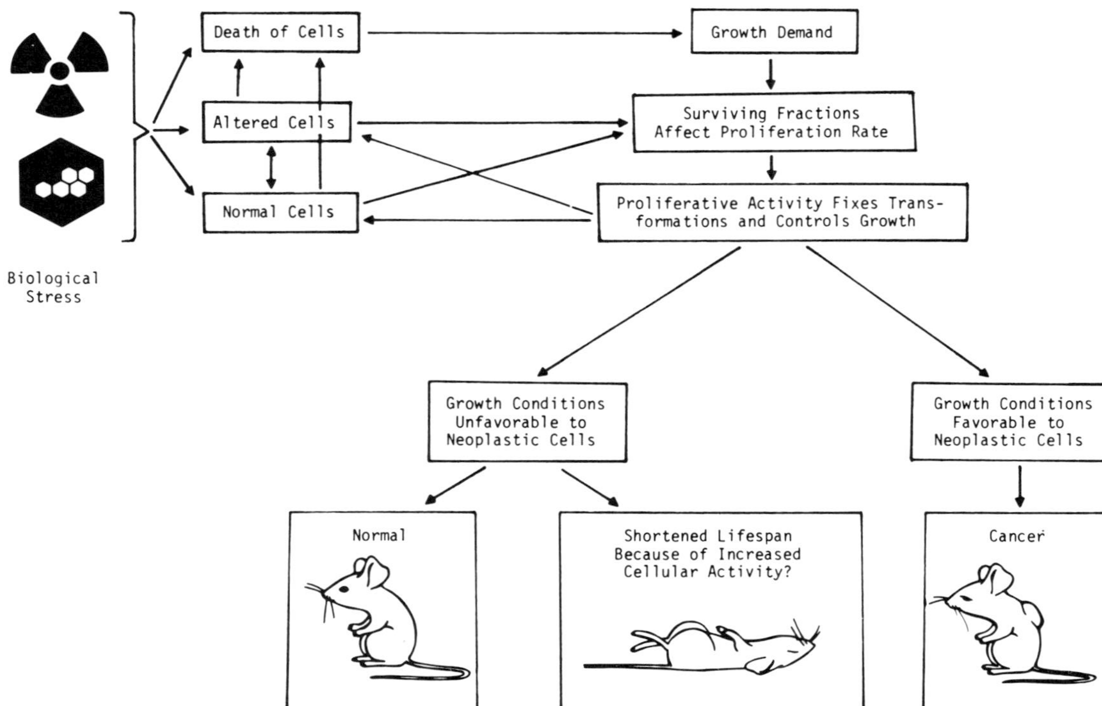

FIGURE 1. Diagrammatic representation of cellular responses to various insults, and possible role of these responses to cancer development.

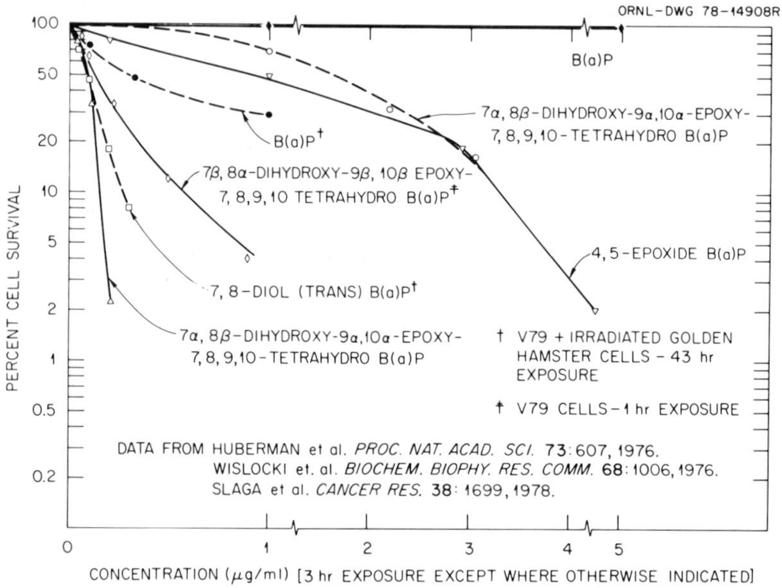

FIGURE 2. The cytotoxicity of benzo[a]pyrene and various benzo[a]pyrene metabolites toward V79 hamster lung cells or V79 cells and irradiated golden hamster cells combined.

cytotoxicity data in Figure 2 may be divided into three general classes. The dihydrodiol epoxides of BP and the 7,8-diol metabolite show a marked cytotoxic effect and the cell survival decreases linearly with dose. If BP is tested on V79 cells alone, it shows no cytotoxicity because it is not metabolized. In a culture system containing V79 cells plus metabolizing cells, however, BP exhibits a cytotoxicity curve that has a concave-upward component. Other metabolites, such as the 4,5-epoxide of BP, display a cytotoxicity which results in a curve that is concave downward.

These results might be interpreted as follows:

1. The concave-upward curve indicates that after metabolism which results in increasing cytotoxicity at low BP doses, a concentration of BP is reached which saturates the metabolic capacities of the cells, so that the cellular death rate declines.
2. Concave-downward cytotoxicity curves may reflect cellular ability to metabolize these compounds to nontoxic products, if the dose of BP is not too high. Higher doses overwhelm the metabolic rate of conversion to nontoxic compounds, and the accumulation of lethal metabolites causes increased death rate.

3. Straight-line cytotoxicity relationships could be generated by BP derivatives which do not require metabolism for toxicity or which are almost totally converted to a cytotoxic form.

The same types of cytotoxicity relationships can be seen in Figure 3, which shows the cytotoxic effect of BP on various cell culture systems. BP exhibits a wide range of toxicity toward different types of cultured cells. The cytotoxicity of BP alone shows the concave-upward shape seen in Figure 2, while toxicity of the 4-5-epoxide is quite different, depending on whether the target cell is human or hamster. In the case of human fibroblast cells, the epoxide at low concentrations is highly cytotoxic, while increases in hamster V79 lung cell death rates only occur at higher epoxide concentrations.

A variety of data from radiation studies have led us to the conclusion that cell survival can be inversely linearly related to cancer incidence in many cases of radiation-induced cancer (11), and we postulate that a fundamental mechanism which produces this empirical relationship may be the same one as mentioned before—namely, the cytotoxicity caused by radiation calls forth a proliferative response, which allows the transformed cells to proceed toward tumor development. We hypothesize that such relationships may hold for carcinogenic chemicals as well, and preliminary data on incidence of BP-induced tumors in mice (15) seems to support this idea (Figure 4).

Using this cytotoxicity-cancer relationship, it is possible to construct a model to estimate carcinogenic risk of exposure to various levels of PAH based on cytotoxicity data for these compounds (Figure 5). Several salient points of this model should be emphasized. Cell survival is related to various doses of radiation, and these doses of radiation are then correlated with excess malignancies produced in atomic bomb survivors. (It should be noted that human excess malignancies are expressed as the ratio of observed over expected cases, so that on a percentage basis, the number 1.1 would correspond to 10% excess malignancies.) *In vitro* data on the cytotoxicity of BP toward various mammalian cells can then be expressed and correlated with each other simply on the basis of cytotoxicity. Cell survival *in vivo* is prererred, when available, but usually only shifts the slope of cancer vs cytotoxicity line through the zero-dose or spontaneous cancer rate. Data on incidence of BP-induced tumors in mice can also be fit into this general scheme, by relating dose of BP that causes a certain incidence of mouse tumors to the excess human malignancy line.

It can be seen from Figure 5 that, although the data are derived from many different studies, there is a rather good agreement between the doses of BP which cause a 10% incidence of mouse tumors (0.10 μg/g body weight) and those doses of BP *in vitro* which result in 80% cell survival. The excess human malignancies in A-bomb survivors at doses

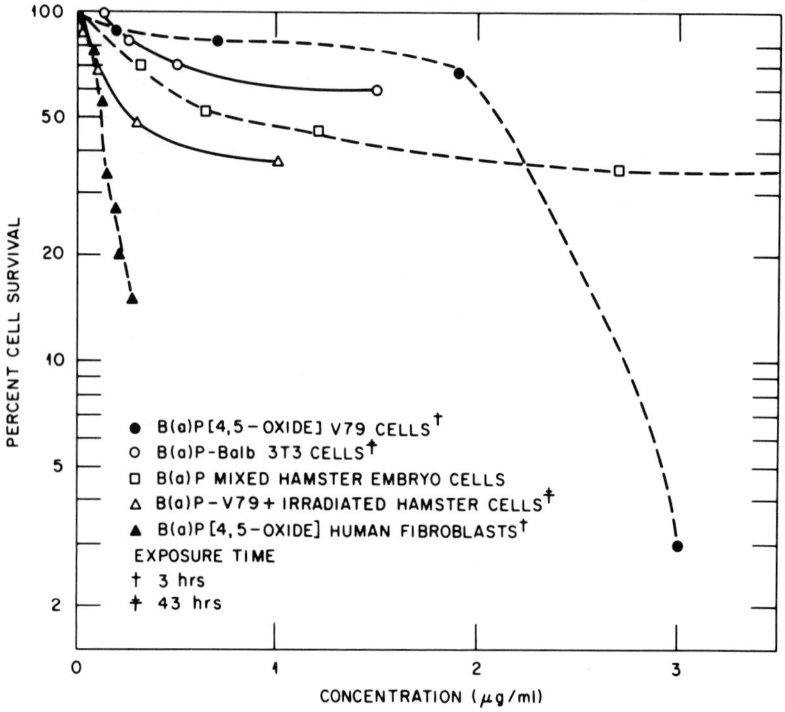

FIGURE 3. The cytotoxic activity of benzo[a]pyrene toward various mammalian cultured cells.

of radiation which give 80% cell survival is approximately 10%. Likewise, concentrations of BP which give 25% tumor incidence in mice (~0.20 μg/g) agree reasonably with those doses of BP which give 50% cell survival in *in vitro* systems (and 50% cell survival in radiation studies can be correlated with 25% excess cancer in A-bomb survivors). From a study of Figure 5, it is apparent that all of the *in vitro* data as well as the mouse tumor-incidence data do not exhibit a linear dose-response relationship in that neither cytotoxicity nor tumor incidence is linearly correlated with BP dose. This does not invalidate the model and only serves to reflect the exceeding complexity of both the *in vivo* and *in vitro* systems used in formulation of this scheme. The important points

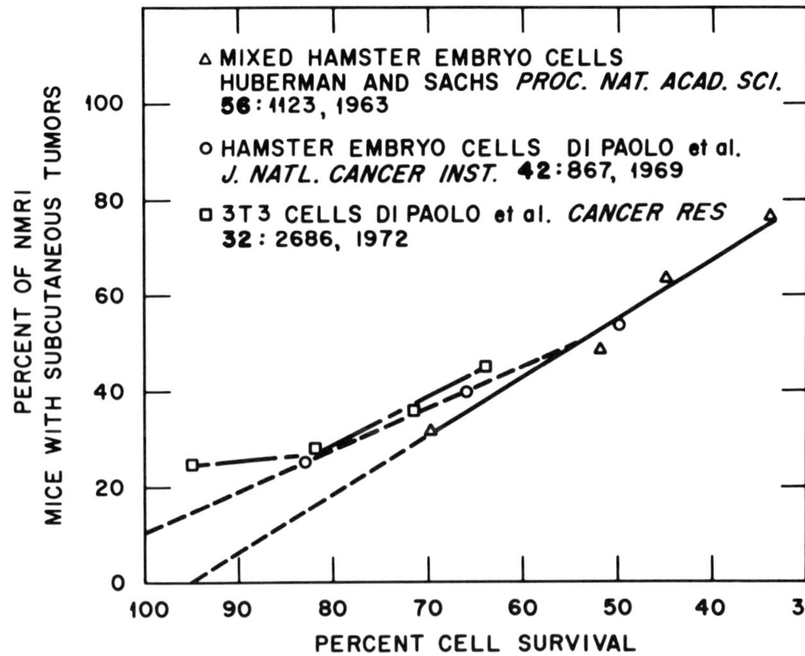

FIGURE 4. Relationship of subcutaneous tumor incidence induced by benzo[a]pyrene in mice to the cytotoxic activity of the same chemical toward various mammalian cells *in vitro.*

of the model are the linear correlation between cancer incidence and cytotoxicity, and the use of *in vitro* cytotoxicity data to estimate cancer risk.

All these risk estimates are based on the assumption that the insult is delivered in a time interval that is shorter than the turnover rate of the cells at risk. Insults of longer duration may produce fewer cancers because of biological repair processes. The present state-of-the-art is insufficient to estimate this decrease and protection guidelines can be derived assuming acute-exposure conditions. The degree of overestimation is expected to be similar for radiation and chemical pollutants. The validity of this model probably could be better tested if dose rates to the cells at risk in animals could be estimated. It is difficult to ascertain the actual dose to target tissues with any degree of precision. Figure 5 illustrates another important consideration, namely the sensitivity of various cell lines to the cytotoxic effect of BP. The 3T3 fibroblast line is much more resistant than the other

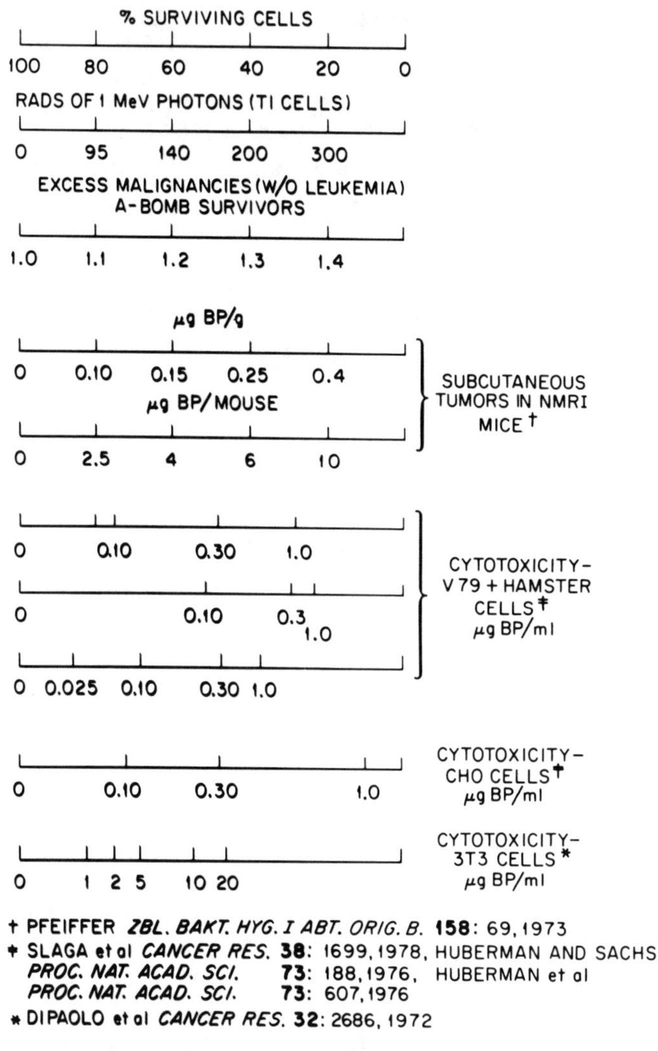

FIGURE 5. Model system used to estimate cancer risk from *in vitro* cytotoxic data. Excess human malignancies are related to doses of radiation received in the atom bomb exposures and these levels of radiation then can be associated with a certain degree of cell killing. Benzo[a]pyrene cytotoxicity in *in vitro* cell culture systems is then plotted in relation to the radiation-generated cytotoxicity data. The incidence of benzo[a]pyrene-induced tumors in mice is also plotted as a function of chemical dose, relating the 10% tumor incidence caused by a certain dose of benzo[a]pyrene to the 10% excess human malignancy point, etc.

lines used in the *in vitro* studies illustrated. For best risk estimation, it is probably important to match sensitivities of different cell types for the chemicals of interest, at least so far as it is possible to do this.

Even with the many parameters which ought to be, but are not now included in the present model (mainly due to lack of basic knowledge), we feel this system may be very useful in terms of coming to grips with the problem of risk assessment for chemical carcinogenesis. Hopefully the basic data to extend, modify or correct the model will be forthcoming.

ACKNOWLEDGMENTS

This research was sponsored by the Division of Biomedical and Environmental Research, U.S. Department of Energy under Contract W-7405-eng-26 with the Union Carbide Corporation.

By acceptance of this article, the publisher or recipient acknowledges the U.S. Government's right to retain a nonexclusive, royalty-free license in and to any copyright covering the article.

REFERENCES

1. Albert, R. E., Burns, F. J., and Heinboch, R. D. (1967): The association between chronic radiation damage of the hair follicles and tumor formation in the rat. *Rad. Res.* 30:590-599.
2. Arcos, J. C. (1974): French research on chemical carcinogenesis in the last quarter century with special reference to the contributions of A. Lacassagne and N. O. Buu-hoi. In: *Chemical Carcinogenesis, Part A,* P. O. P. Ts'o and J. A. DiPaolo, Eds., pp. 5-46, Marcel Dekker, Inc., New York.
3. Bryan, W. R. (1960): A reconsideration of the nature of the neoplastic reaction in the light of recent advances in cancer research. *J. Nat. Cancer Inst.* 24:221-251.
4. Buu-Hoi, N. P., Croisy-Delcey, M., Jacquignon, P., Chayet, C., Dandel, P., Gachelin, F., Jacquier, A., Lataillade, G., and Moreau, J. (1974): Action of halogenomethyl hydrocarbon derivatives on *in vitro* DNA and RNA synthesis, and on tRNA properties. Relationship with carcinogenic potency of methyl parent hydrocarbons. In: *Chemical Carcinogenesis, Part A,* P. O. P. T'so and J. A. DiPaolo, Eds., pp. 159-175, Marcel Dekker, Inc., New York.
5. Chen, T. T., and Heidelberger, C. (1969): Quantitative studies on the malignant transformation of mouse prostate cells by carcinogenic hydrocarbons *in vitro. Int. J. Cancer* 4:166-178.
6. DiPaolo, J. A. (1974): Quantitative Aspects of *in vitro* chemical carginogenesis. In: *Chemical Carcinogenesis, Part B,* P. O. P. T'so and J. A. DiPaolo, Eds., pp. 443-455, Marcel Dekker, Inc., New York.

7. DiPaolo, J. A., Donovan, P. J., and Nelson, R. L. (1971): *In vitro* transformation of hamster cells by polycyclic hydrocarbons: Factors influencing the number of cells transformed. *Nature* 230:240-242.
8. Gelboin, H. V., Wiebel, F. J., and Kinoshita, N. (1974): Aryl hydrocarbon hydroxylase: Regulation and role in polycyclic hydrocarbon action. In: *Chemical Carcinogenesis, Part A,* P. O. P. T'so and J. A. DiPaolo, Eds., pp. 309-351, Marcel Dekker, Inc., New York.
9. Huberman, E. (1975): Mammalian cell transformation and cell-mediated mutagenesis by carcinogenic hydrcarbons. *Mutat. Res.* 29:285-291.
10. Huberman, E., Sachs, L., Yang, S. K., and Gelboin, H. V. (1976): Identification of mutagenic metabolites of benzo[a]pyrene in mammalian cells. *Proc. Nat. Acad. Sci. U.S.* 73:607-611.
11. Jones, T. D., Jacobs, D. G., Auxier, J. A., and Kerr, G. D. (1979): Risk of environmental cancer based on promotion from cytotoxicity. In: *Proceedings of Conference on Neutron from Electron Medical Accelerators,* National Bureau of Standards, Gaithersburg, Md.
12. Kapitulnik, J., Levin, W., Conney, A. H., Jagi, H., and Jerina, D. M. (1977): Benzo[a]pyrene-7,8-dihydrodiol is more carcinogenic than benzo[a]pyrene in newborn mice. *Nature* 266:378-380.
13. Levin, W., Conney, A. H., Yagi, H., and Jerina, D. M. (1978): Binding of benzo[a]pyrene 7,8-diol-9,10-epoxides to DNA, RNA and protein of mouse skin occurs with high stereoselectivity, *Science* 199:778-781.
14. McCann, J., Choi, E., Yamasaki, E., and Ames, B. N. (1975): The detection of carcinogens as mutagens in the *Salmonella* microsome test: Assay of 300 chemicals, *Proc. Nat. Acad. Sci. U.S.* 72:5135-5139.
15. Pfeiffer, E. H. (1973): Investigation on the carcinogenic burden by air pollution in man. VII' Studies on the oncogenetic interactions of polycyclic aromatic hydrocarbons. *Z. Bakteriol. [Orig. B],* 158:69-83.
16. United Nations Scientific Committee on the Effects of Atomic Radition (1977): Experimental radiation carcinogenesis H. The role of cell proliferation. In: *Sources and Effects of Ionizing Radiation,* pp. 577-578, United Nations, New York.

Polynuclear Aromatic Hydrocarbons, edited by P. W. Jones and P. Leber. Ann Arbor Science Publishers, Inc. Ann Arbor, MI ©1979.

ON THE METABOLIC ACTIVATION OF 5-METHYLCHRYSENE

Stephen S. Hecht, Robert Mazzarese, Shantilal Amin, Edmond LaVoie and Dietrich Hoffmann

Naylor Dana Institute for Disease Prevention
American Health Foundation
Valhalla, New York 10595

Numerous studies in recent years have indicated the importance of specific angular-ring dihydrodiol epoxides in the metabolic activation of various polynuclear aromatic hydrocarbons including benzo[a]pyrene, benz[a]-anthracene and chrysene (5,12,16,19,20,27,28,30). In the benz[a]anthracene and chrysene systems, the activation of methylated derivatives has also been of interest since certain isomers are strongly carcinogenic, in contrast to the parent compounds. Research on benz[a]anthracene, 7-methylbenz[a]-anthracene and 7,12-dimethylbenz[a]anthracene has indicated that, in each case, a 3,4-dihydrodiol-1,2-epoxide may be a major ultimate carcinogen (3,13,21,23,29). Thus, in the benz[a]anthracene system, the bay-region hypothesis of polynuclear aromatic hydrocarbon carcinogenicity appears to apply in the same way to the methylated and parent compounds. In the chrysene series, studies on synthetic chrysene dihydrodiols showed that 1,2-dihydro-1,2-dihydroxychrysene, which can form a bay-region 1,2-dihydrodiol-3,4-epoxide, was the most mutagenic and most tumorigenic isomer (20,31). Substitution of a methyl group at position 5 of chrysene (see Figure 1) converts the weakly carcinogenic parent compound to a strong carcinogen, 5-MeC (7,10). In addition, this substitution renders the hydrocarbon unsymmetric and two nonequivalent angular rings with bay regions become available for possible activation, providing an interesting contrast to the benz-[a]anthracene system. To study the activation of 5-MeC, which is the most carcinogenic of the methylchrysene isomers (7,10), a series of fluorinated 5-MeC derivatives was synthesized for bioassay on mouse skin. The results of these bioassays and of parallel studies on the metabolism of 5-MeC and the fluorinated compounds have led to the conclusion that activation takes place predominantly in the angular (1-4) ring adjacent to the 5-methyl group.

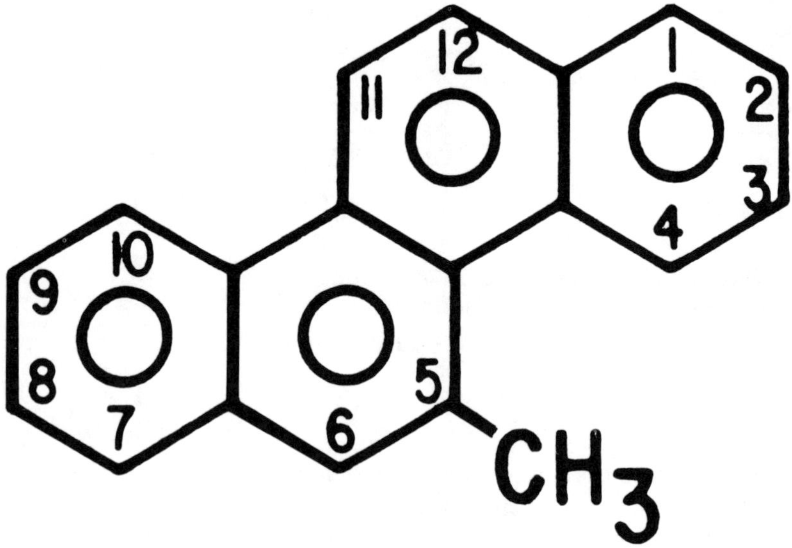

FIGURE 1. Structure of 5-methylchrysene.

BIOASSAYS OF FLUORINATED 5-METHYLCHRYSENES

A series of monofluorinated 5-MeC derivatives was synthesized. Fluorine was substituted at the 1,3,6,7,9,11, or 12-position of 5-MeC to give the seven compounds, 1-fluoro-5-methylchrysene (1-F-5-MeC), 3-F-5-MeC, 6-F-5-MeC, 7-F-5-MeC, 9-F-5-MeC, 11-F-5-MeC and 12-F-5-MeC. Each compound was synthesized by an unambiguous route, as described previously (11). This fluorine-probe approach has also been used in studies on other carcinogens, including 7-methylbenz[a]anthracene, 7,12-dimethylbenz[a]anthracene, 2-acetylaminofluorene, and dibenzo[a,i]pyrene (1,22,24) and the results have generally been consistent with current concepts on the metabolic activation of these compounds.

The fluorinated compounds were tested on mouse skin as tumor initiators with promotion by tetradecanoyl phorbol acetate. (TPA), and as complete carcinogens. For bioassays as initiators, groups of 20 Ha(ICR) female outbred Swiss mice were treated with initiating doses (30 or 100 μg) of the fluorinated compounds or 5-MeC itself with promotion by three times weekly application of 2.5 μg of TPA for 20 weeks. For complete carcinogenicity, the compounds were applied in acetone solution (0.005%) three times weekly for 72 weeks.

The results of the assays as tumor initiators are presented in Table I (8). Both 3-F-5-MeC and 1-F-5-MeC were significantly less active than 5-MeC as

TABLE I. Tumor-Initiating Activity of Fluorinated Derivatives of 5-MeC[a]

Compound	Initiating Dose (μg)	% Tumor-Bearing Animals (compound/5-MeC)	Tumors per Animal (compound/5-MeC)
1-F-5-MeC	30	10/60[b]	0.1/1.5
	100	40/80[b]	0.6/1.9
3-F-5-MeC	30	0/60[b]	0/1.5
	100	5/80[b]	0/1.9
6-F-5-MeC	30	80/95	4.1/3.0
	100	85/80	2.8/5.2
7-F-5-MeC	30	60/50	0.9/1.4
	100	90/80	4.8/4.7
9-F-5-MeC	30	45/50	0.8/1.4
	100	85/80	3.2/4.7
11-F-5-MeC	30	45/50	0.6/1.4
	100	65/80	2.5/4.7
12-F-5-MeC	30	18/67[c]	0.2/1.8[c]
	100	77/82[c]	2.1/4.9[c]
5-MeC[c]	30		
	100		

[a]20 Ha/ICR ♀ mice per group.
[b]Statistically significant, $p < 0.01$.
[c]Average of 3 experiments.

tumor initiators, as judged by the percentage of tumor-bearing animals and by the number of tumors per animal. Reduced activity compared to 5-MeC was also observed for 12-F-5-MeC at the 30-μg dose level, but not when a dose of 100 μg was applied. The other fluorinated 5-MeC derivatives showed no decreased activity as tumor initiators compared to 5-MeC.

The results of the assays as complete carcinogens are shown in Figures 2 and 3. Neither 1-F-5-MeC nor 3-F-5-MeC showed carcinogenic activity under the conditions of this assay, whereas 5-MeC gave 80% tumor-bearing animals. Significantly diminished activity was also observed for 12-F-5-MeC. The only compound showing greater activity than 5-MeC was 6-F-5-MeC, whereas 7-F-5-MeC, 9-F-5-MeC and 11-F-5-MeC were as active as the unfluorinated hydrocarbon. These results are in good agreement with those presented in Table I. The comparative tumor-initiating, complete carcinogenic and mutagenic activities of these compounds and a number of other polynuclear aromatic hydrocarbons are discussed in detail in another chapter of this volume.

Thus, substitution of fluorine at position 1, 3 or 12 of 5-MeC inhibited tumorigenicity. Substitution of fluorine at the 6-position increased activity, in contrast to positions 7,9 or 11 where no significant change was observed. These results suggest that the 12, 1, and 3-positions are involved in metabolic

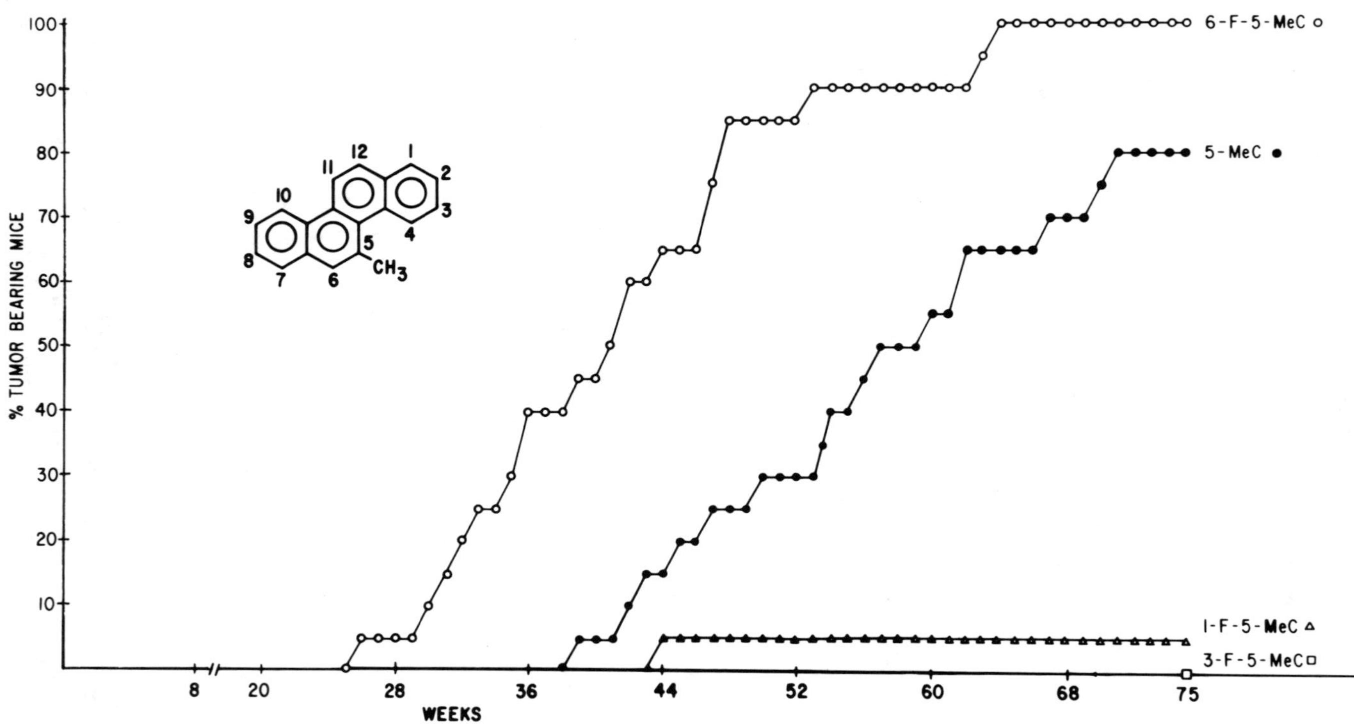

FIGURE 2. Carcinogenicity on mouse skin of 1-F-5-MeC, 3-F-5-MeC, 6-F-5-MeC, and 5-MeC. Each compound (5 μg) was applied to mouse skin three times weekly.

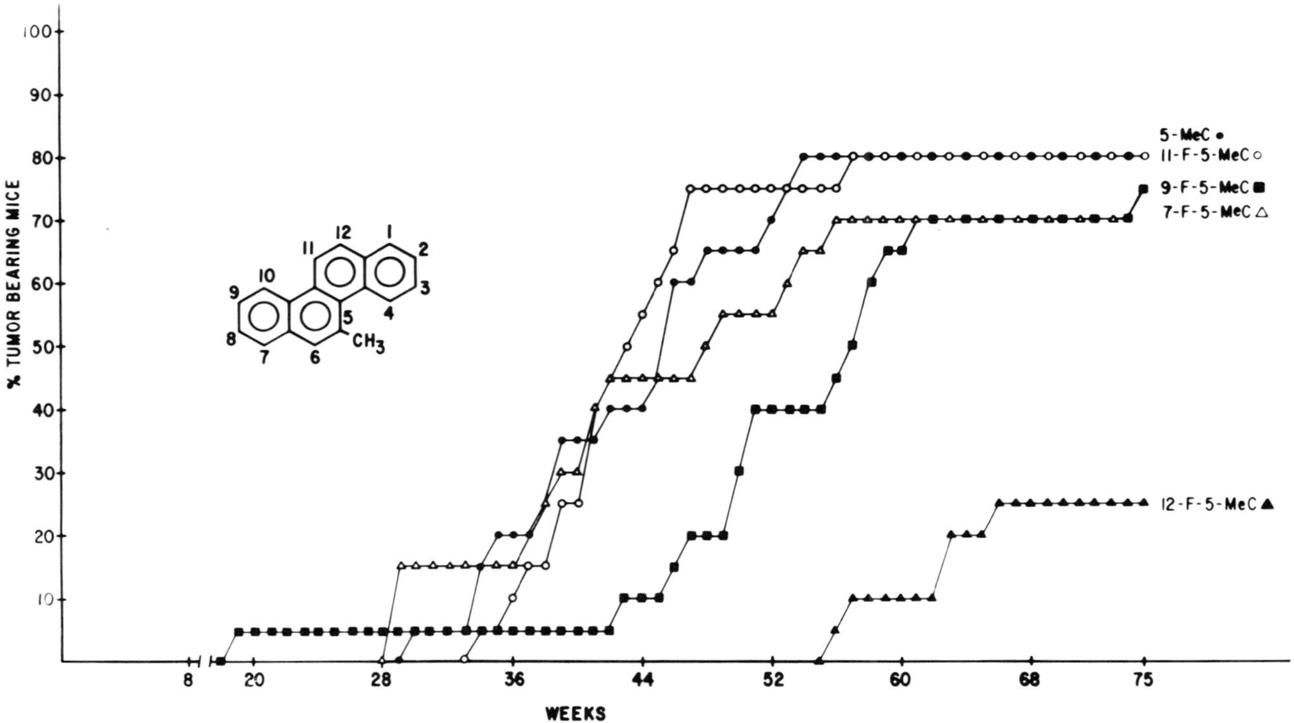

FIGURE 3. Carcinogenicity on mouse skin of 7-F-5-MeC, 9-F-5-MeC, 11-F-5-MeC, 12-F-5-MeC and 5-MeC. Each compound (5 μg) was applied to mouse skin three times weekly.

activation of 5-MeC in mouse skin whereas the 6,7,9 and 11-positions are not involved. *In vitro* metabolic studies on these fluorinated compounds were useful in determining the effects of fluorine substitution; these results will be described in the section on the relationship of metabolism to mutagenicity and tumorigenicity of 5-MeC derivatives.

METABOLISM OF 5-MeC *IN VITRO*

The metabolism of 5-MeC was studied using the 9,000 x g supernatant from the livers of Aroclor-treated rats (9). Metabolites were examined by reverse-phase high pressure liquid chromatography (HPLC); a typical chromatogram is shown in Figure 4. Peaks A-I were collected and their mass spectra

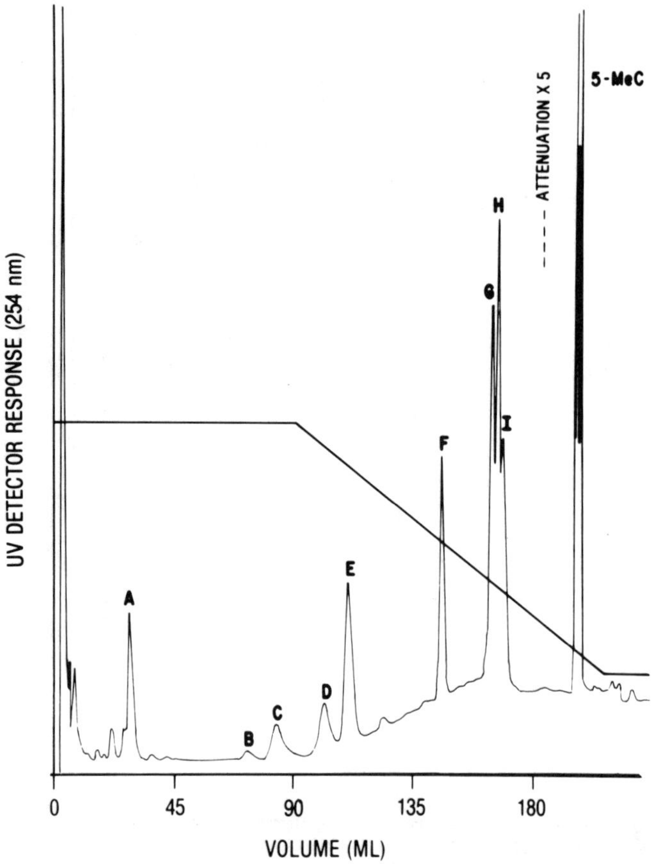

FIGURE 4. HPLC chromatogram of metabolites formed from 5-MeC in incubations with Aroclor-induced rat liver 9,000 x g supernatant. See Table II for identity of peaks A-I.

and ultraviolet spectra were determined. From these data, it was concluded that peaks A, D and E (which also disappeared when incubations were done in the presence of trichloropropylene oxide) were dihydrodiols. Peak F was identified as 5-hydroxymethylchrysene by comparison to a reference sample, and peaks G-I were 5-Me-chrysenols.

For characterization of the dihydrodiols and 5-Me-chrysenols, a series of 5-Me-chrysenols was synthesized, as summarized in Figures 5 and 6. For 1-, 2- and 3-hydroxy-5-MeC (**11-13**), the phosphonium salt **1** from 1-bromomethyl-3-methylnaphthalene was condensed with the appropriate methoxybenzaldehyde **2-4**, using the Wittig reaction. The resulting alkenes **5-7** were photocyclized to the methoxy derivatives **8-10** which were converted to 5-Me-chrysenols **11-13** by reaction with BBr_3. Since photolysis of alkene **6** could have given a mixture of 2-methoxy-5-MeC (**9**) and 4-methoxy-5-MeC, an independent synthesis of 2-hydroxy-5-MeC (**12**) was developed, as outlined in Figure 5b. The product of this synthesis was identical to that obtained in the photochemical synthesis of **12**, indicating that cyclization of **6** gave exclusively 2-methoxy-5-MeC (**9**), as expected on steric grounds. The syntheses of 6-, 7-, 8- and 12-hydroxy-5-MeC (**27-29** and **33**) followed similar routes, as shown in Figure 6a,b. These syntheses provided reference standards which could be distinguished by their HPLC retention volumes and/or their ultraviolet spectra.

The dihydrodiols A, D and E (Figure 4) were each dehydrated and the resulting 5-Me-chrysenols identified by comparison to the synthetic standards. Dihydrodiol A gave predominantly 9-hydroxy-5-MeC, D gave 7-hydroxy-5-MeC (93%) and 8-hydroxy-5-MeC (7%), and E gave 1-hydroxy-5-MeC (95%) and 2-hydroxy-5-MeC (5%). Since the ultraviolet spectrum of A agreed well with a published spectrum of 3,4-dihydro-3,4-dihydroxychrysene (17), it was assigned the structure 9,10-dihydro-9,10-dihydroxy-5-MeC. A shoulder preceding A was tentatively identified as the 3,4-dihydrodiol. From the dehydration and ultraviolet spectral data, peaks D and E were identified as 7,8-dihydro-7,8-dihydroxy-5-MeC and 1,2-dihydro-1,2-dihydroxy-5-MeC, respectively. The formation of predominantly 7-hydroxy-5-MeC in the dehydration of the 7,8-dihydrodiol and of 1-hydroxy-5-MeC in the dehydration of the 1,2-dihydrodiol indicate that the carbonium ions at positions 2 and 8 (allylic) were more stable than those at positions 1 and 7 (benzylic). Similar results were obtained in the dehydration of 1,2-dihydroxynaphthalene (14).

The 5-Me-chrysenols G, H and I were identified as 9-hydroxy-5-MeC, 7-hydroxy-5-MeC and 1-hydroxy-5-MeC, respectively. A shoulder preceding G was tentatively identified as 3-hydroxy-5-MeC. Of course, these assignments are based on the available standards and do not exclude the presence of other 5-Me-chrysenols.

The yields of the metabolites, as determined in incubations with 5-MeC-^{14}C are summarized in Table II. When protein concentration was decreased

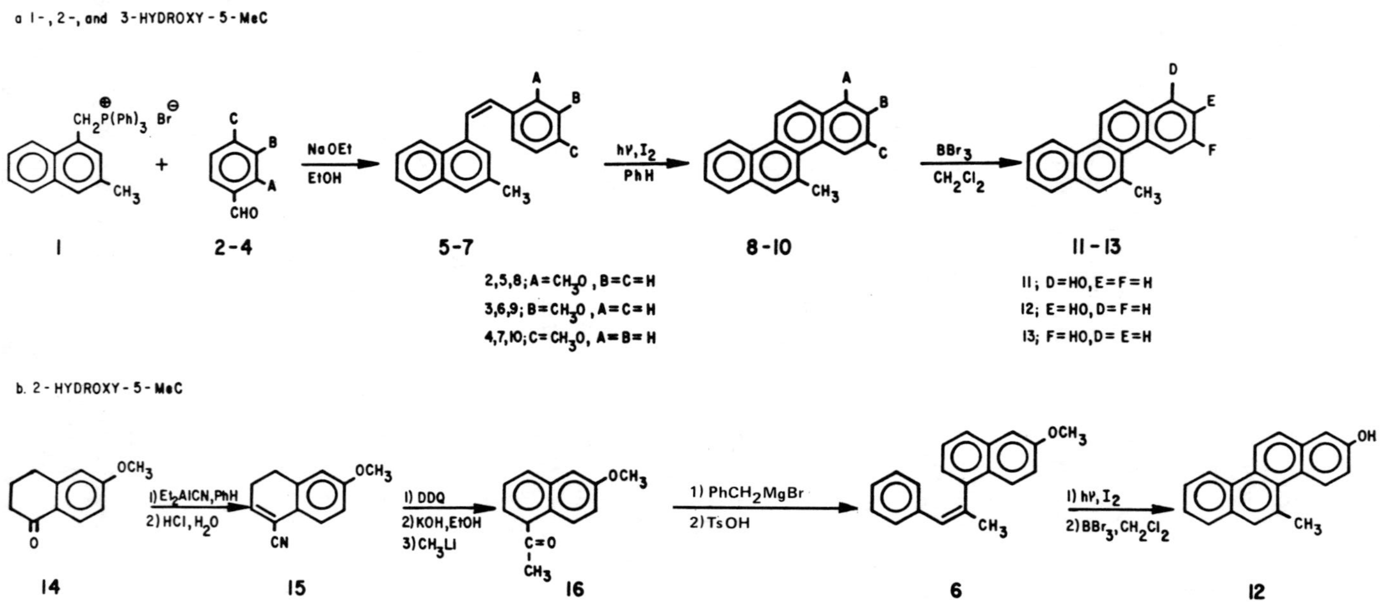

FIGURE 5. (a) Synthesis of 1-hydroxy-5-MeC, 2-hydroxy-5-MeC and 3-hydroxy-5-MeC. (b) Alternate synthesis of 2-hydroxy-5-MeC.

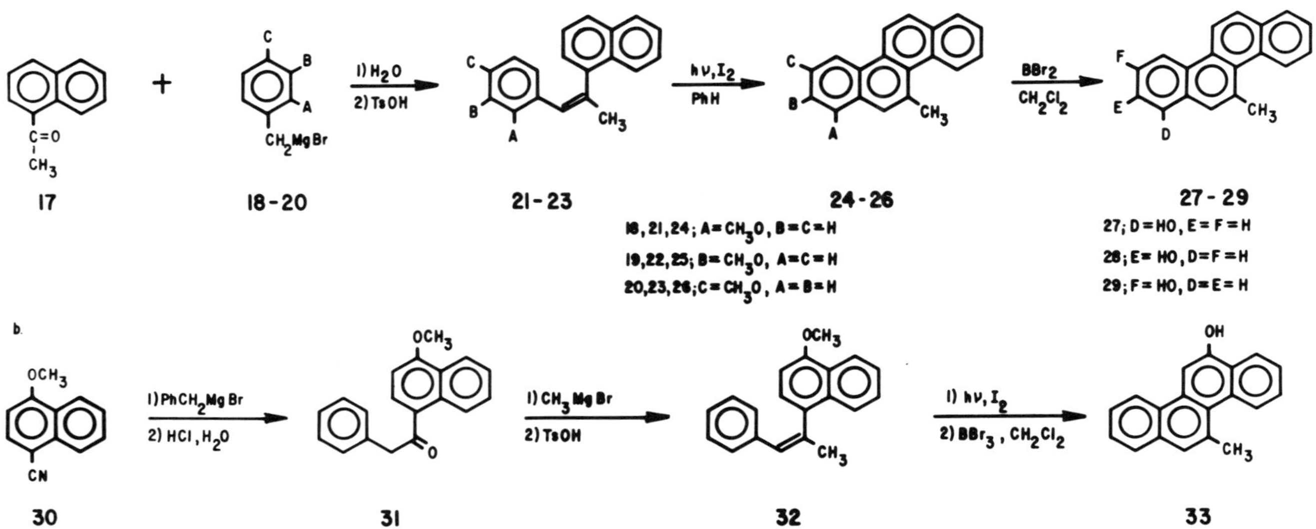

FIGURE 6. (a) Synthesis of 7-hydroxy-5-MeC, 8-hydroxy-5-MeC and 9-hydroxy-5-MeC; (b) Synthesis of 12-hydroxy-5-MeC.

from one-half to one-eighth of the usual amount, there was a corresponding decrease in metabolite formation, but the relative amounts of each metabolite remained essentially constant. It is evident from the data in Table II that the major sites of initial oxidation of 5-MeC were the 1,2-, 7,8- and 9,10-bonds, and the methyl group. The initial products of ring oxidation were presumably epoxides. Dihydrodiol formation was the preferred pathway from the 1,2-epoxide while 5-Me-chrysenol formation predominated from the 7,8-epoxide (see Table II). These differences may be due to differing rates of rearrangement of the 1,2-epoxide vs the 7,8-epoxide to 5-Me-chrysenols. Since the electron-releasing ability of the 5-methyl group would tend to stabilize a developing carbonium ion at the 8-position, but not at the 2-position, the 7,8-epoxide may undergo an NIH shift more readily to give 7-hydroxy-5-MeC than would the 1,2-epoxide to give 2-hydroxy-5-MeC (see Figure 7). Other factors such as rates of hydration to dihydrodiols may also be involved.

TABLE II. Mutagenicity of Metabolites of 5-MeC

			His$^+$ revertant/plate[c]	
Peak[a]	Structure	nmol[b]	Assay 1[d]	Assay 2[e]
A	5-MeC-9,10-diol[f]	5.7	110 ± 11[g]	
B	Unknown	0.37		110 ± 11
C	Unknown	1.4	562 ± 12	101 ± 2
D	5-MeC-7,8-diol	2.7		199 ± 38
E	5-MeC-1,2-diol	7.2		373 ± 63
F	5-Hydroxymethylchrysene	7.3	136 ± 14	
G	9-OH-5-MeC	4.1		
H	7-OH-5-MeC	6.1	148 ± 15	
I	1-OH-5-MeC	2.8		

[a]See Figure 4.
[b]From 103 nmol 5-MeC, 5.4 mg protein, 20 min.
[c]Values observed without subtraction of negative control.
[d]From 3.10 μmol 5-MeC. Negative control, DMSO, 100 ± 5 revertants/plate.
[e]From 6.20 μmol 5-MeC. Negative control, DMSO, 86 ± 15 revertants/plate.
[f]9,10-Dihydro-9,10-dihydroxy-5-methylchrysene.
[g]Mean ± S.D.

RELATIONSHIP OF METABOLISM TO MUTAGENICITY AND TUMORIGENICITY OF 5-MeC DERIVATIVES

Metabolites formed from 5-MeC *in vitro* were assayed for mutagenicity towards *S. typhimurium* TA100 with activation to determine which were proximate mutagens (9). HPLC peaks were collected and two separate assays were done as summarized in Table II. In the first assay, peaks A, B-E, F and

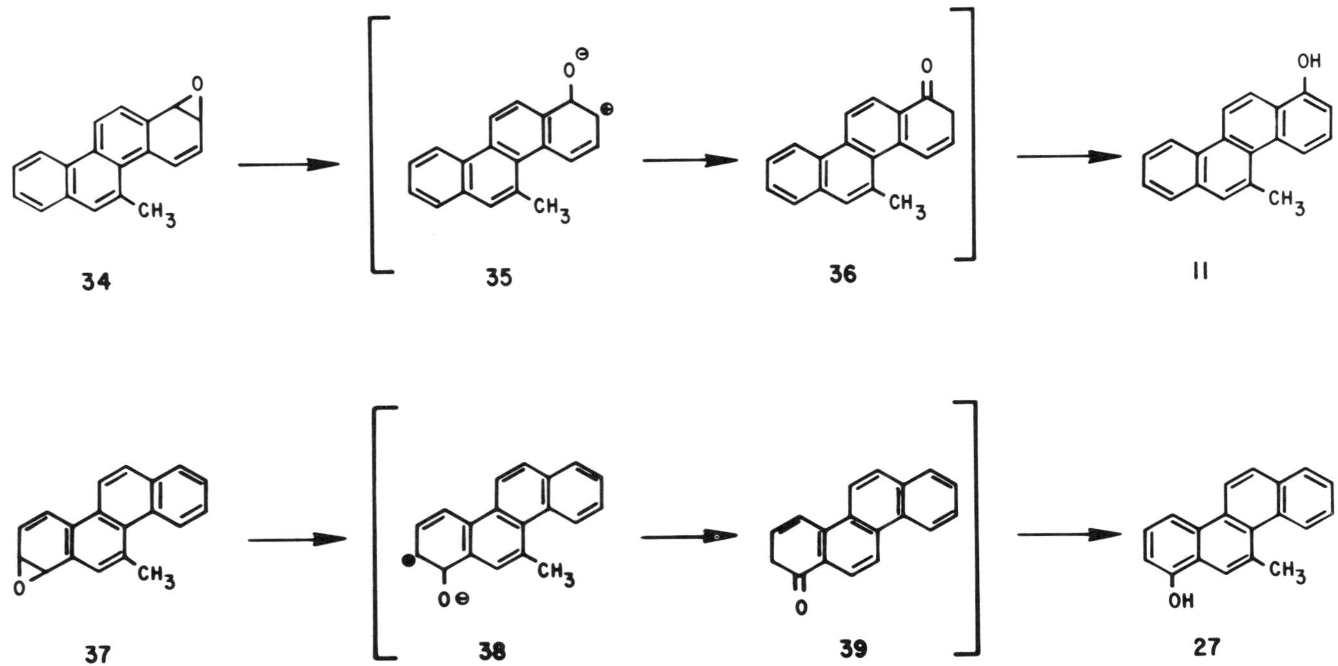

FIGURE 7. Proposed NIH shifts of the 1,2-epoxide (34) and the 7,8-epoxide (37) of 5-MeC. The electron-releasing effect of the methyl group should stabilize intermediate 38, but not 35.

G-I (Figure 4) were tested. Significant mutagenicity was observed for peaks B-E. In the second assay, peaks B, C, D and E were each tested for mutagenicity. Peak E was the most mutagenic with some activity also observed for D. Thus, the 1,2-dihydrodiol was the major metabolite capable of further oxidation to a mutagen. The lower mutagenicity observed for the 7,8-dihydrodiol was probably a reflection of the quantity tested rather than of structural features, since identical aliquots of all metabolic fractions were used for the assays. These results are in agreement with studies on the mutagenicity and tumorigenicity of the 1,2-, 3,4- and 5,6-dihydrodiols of chrysene which indicated greatest activity for the 1,2-dihydrodiol and support the bay-region hypothesis (20,31). Formation of a 1,2-dihydrodiol-3,4-epoxide is likely to be a major activation pathway for both chrysene and 5-MeC. The bay-region hypothesis also predicts the 7,8-dihydrodiol-9,10-epoxide of 5-MeC to be an active form but metabolic formation of this dihydrodiol is apparently not favored (15).

The results of the bioassays for carcinogenicity of the fluorinated 5-MeC derivatives also support the concept that 5-MeC is activated predominantly via a 1,2-dihydrodiol-3,4-epoxide. The lack of activity of 1-F-5-MeC and 3-F-5-MeC on mouse skin was presumably a reflection of low rates of formation of the 1,2-dihydrodiol, whereas, in 7-F-5-MeC and 9-F-5-MeC, diminished metabolism would be expected in the 7-10 ring. These assumptions were supported, at least partially, by a study of the *in vitro* metabolism of the fluorinated compounds using rat liver 9,000 x g supernatants. The HPLC traces obtained from 1-F-5-MeC, 3-F-5-MeC, and 7-F-5-MeC are shown in Figure 8a-c. UV spectra for the major peaks or groups of peaks numbered 1-4 in each case corresponded to the 7,8- or 1,2-dihydrodiols (peaks 1 or 2), hydroxymethyl derivatives (3) and chrysenols (4). Based on the relative retention volumes for peaks 1 and 2 in the three chromatograms, and on the order of elution of dihydrodiols from 5-MeC, peaks 1 and 2 in each chromatogram were tentatively assigned as the 7,8-dihydrodiols and 1,2-dihydrodiols, respectively. Thus, substitution of fluorine at the 1 and 3 positions apparently blocked formation of the 1,2-dihydrodiol, whereas substitution at the 7-position inhibited formation of the 7,8-dihydrodiol, as well as chrysenols. These tentative assignments were not possible for 9-F-5-MeC and 11-F-5-MeC.

Substitution of fluorine at the "peri" positions (6 and 12) of 5-MeC also had pronounced effects on tumorigenicity; 6-F-5-MeC was more tumorigenic than 5-MeC while 12-F-5-MeC was less active. HPLC traces from incubation of these two compounds with rat liver 9,000 x g supernatants are shown in Figure 9a, b. In the case of 12-F-5-MeC, a peak (numbered 1) corresponding in relative retention volume and UV spectrum to the 7,8-dihydrodiol was observed, but formation of the 1,2-dihydrodiol was inhibited. Thus, the major activation pathway in the 1-4 ring was inhibited by fluorine at position 12.

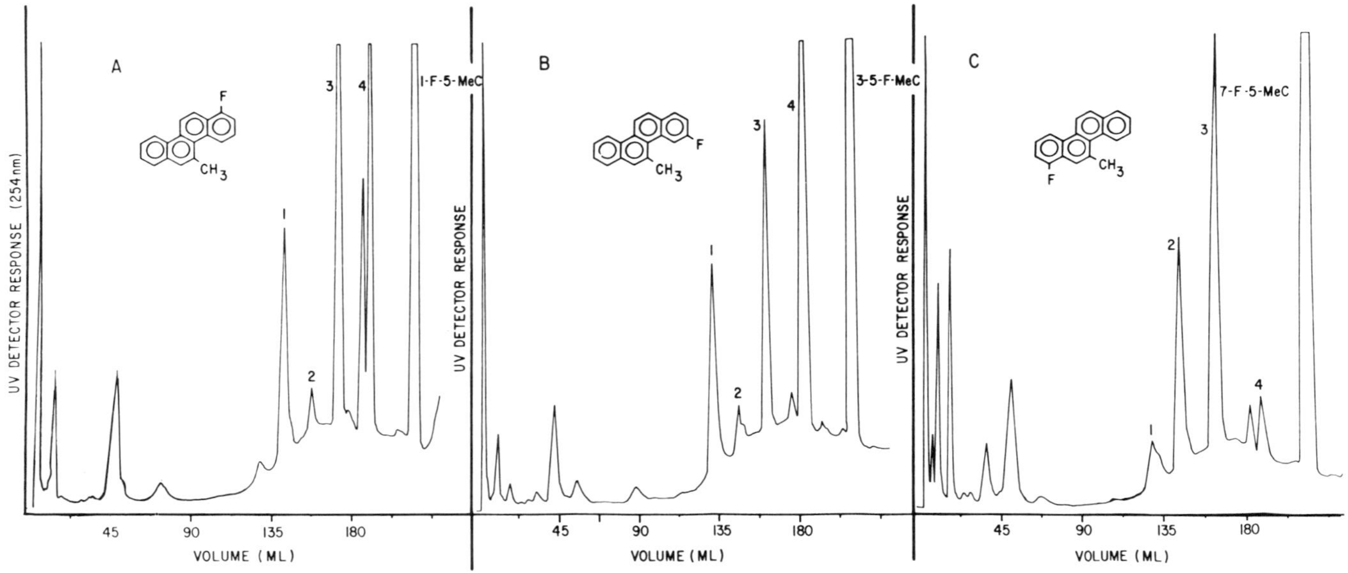

FIGURE 8. HPLC chromatograms of metabolites formed from (A) 1-F-5-MeC; (B) 3-F-5-MeC; and (C) 7-F-5-MeC in incubations with Aroclor-induced rat liver 9,000 x g supernatant. The peaks numbered 1-4 correspond to the 7,8-dihydrodiol (1), 1,2-dihydrodiol (2), hydroxymethylchrysene (3) and chrysenols (4).

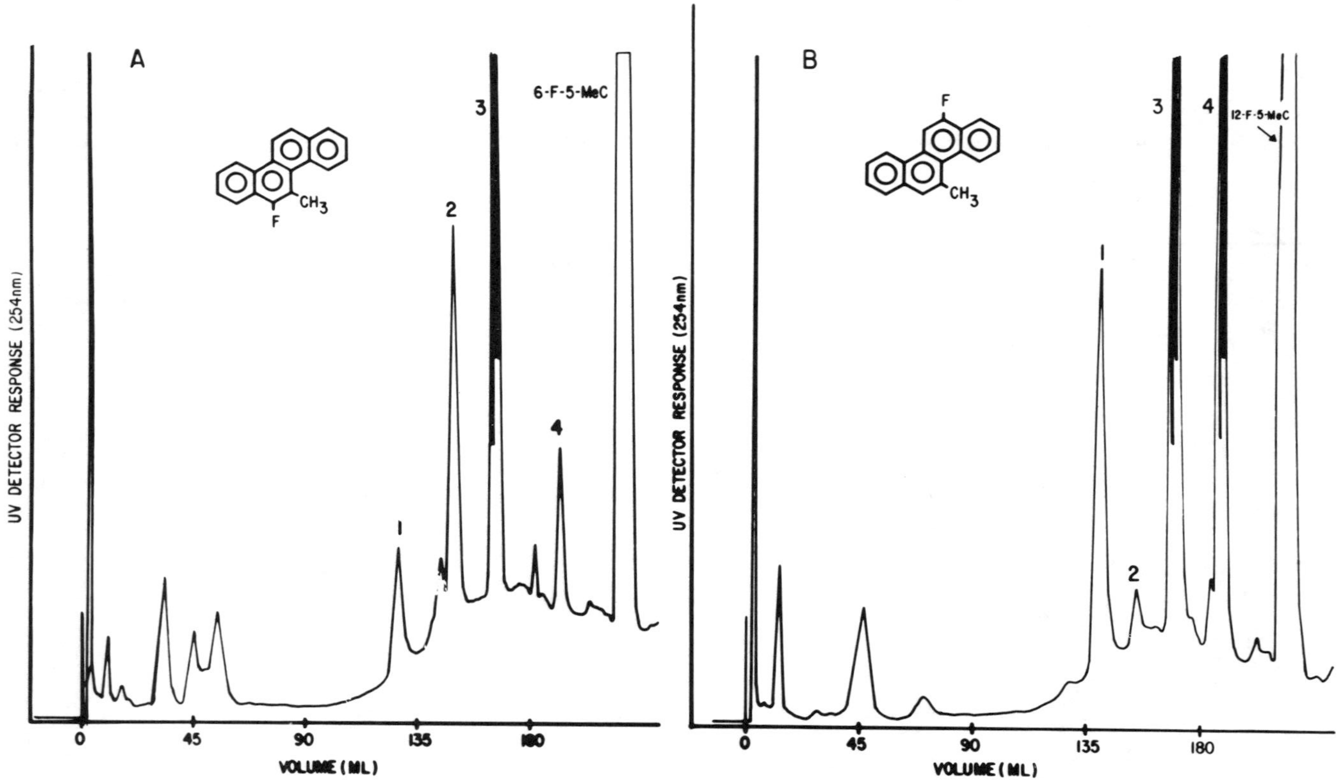

FIGURE 9. HPLC chromatograms of metabolites formed from (A) 6-F-5-MeC and (B) 12-F-5-MeC in incubations with Aroclor-induced rat liver 9,000 x g supernatant. The peaks numbered 1-4 correspond to the 7,8-dihydrodiol (1), 1,2-dihydrodiol (2), hydroxymethylchrysene (3) and chrysenols (4).

For 6-F-5-MeC, a peak (numbered 2) having the proper relative retention volume and UV spectrum for the 1,2-dihydrodiol was present, but formation of chrysenols was minimal. Since chrysenols derive mainly from the 7-10 ring, fluorine at the 6-position apparently inhibited oxidation in the 7-10 ring. This may partially explain the greater activity of 6-F-5-MeC, compared to 5-MeC, since chrysenol formation is a detoxification pathway. Thus, substitution of fluorine at a peri position inhibits metabolism in the adjacent ring with the tumorigenicity of the resulting compound depending on whether an activation or detoxification process was blocked. While the mechanism of this peri inhibition of metabolism is not known, the simplest explanation is that the electronegativity of the peri fluorine atom affects positions in the adjacent ring. Another possibility is that a peri fluorine atom inhibits enzyme-substrate complex formation.

STRUCTURAL CHARACTERISTICS FAVORING CARCINOGENICITY OF METHYLATED PAH

The structural requirements favoring carcinogenicity in the methylchrysene series are a bay-region methyl group and a free peri position adjacent to an unsubstituted angular ring. Among the methylchrysene isomers, only 5-MeC has these features, and it is the most carcinogenic isomer. The reasons for certain of these requirements are evident from the data discussed above. An unsubstituted angular ring is necessary for bay-region dihydrodiol-epoxide formation and substitution at the adjacent peri position can inhibit formation of the requisite proximate dihydrodiol. The presence of the methyl group in the bay region adjacent to the angular ring leads to a deformation from planarity and to an electronic distribution which might result in increased rates of dihydrodiol formation in that ring, although further evidence is needed on this point. The presence of the methyl group in this bay region is clearly important, since all the methylchrysene isomers have a free angular ring and adjacent peri position but only 5-MeC is a strong carcinogen.

As shown in Figure 10, these structural requirements may operate in other systems. In both the dimethylbenz[a]anthracene and cyclopenta[a]-phenanthrene systems, the presence of a bay-region methyl group and a free peri position adjacent to an unsubstituted angular ring results in increased carcinogenicity. The isomers shown in Figure 10, 7,12-dimethylbenz[a]-anthracene and 15,16-dihydro-11-methylcyclopenta[a]phenanthren-17-one are among the most active in these series (2,4,24,25). Similar considerations apply to many of the trimethylbenz[a]anthracenes, although 4,7,12-trimethylbenz[a]anthracene, which induces sarcomas in rats, appears to be an exception (6,24,25). Other strong carcinogens fulfilling these structural requirements are 11-methylbenzo[a]pyrene and 7-methyldibenz[a,h]anthracene (18,26).

To further test the above generalization, 5,11-dimethylchrysene (44) was synthesized, as outlined in Figure 11. Bioassays for tumor-initiating activity are currently in progress. Preliminary *in vitro* metabolic studies showed that formation of the hydroxymethyl derivative was favored, indicating that the additional methyl group may facilitate detoxification.

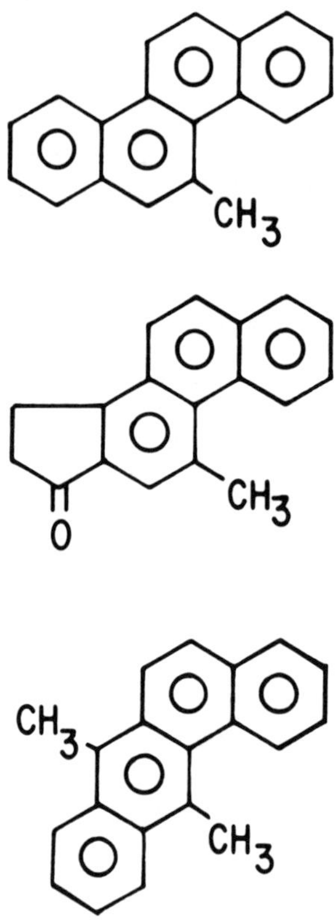

FIGURE 10. Structures of 5-MeC, 7,12-dimethylbenz[a] anthracene and 15,16-dihydro-11-methylcyclopenta[a] phenanthrene-17-one. Each compound has a bay-region methyl group and free peri position adjacent to an unsubstituted angular ring and each is more carcinogenic than other isomers in the same series.

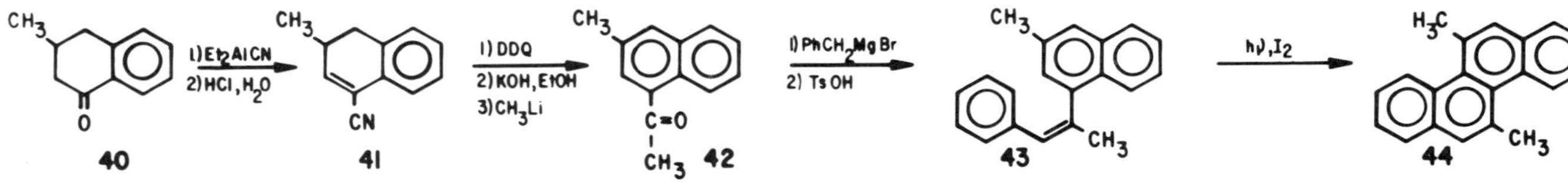

FIGURE 11. Synthesis of 5,11-dimethylchrysene.

ACKNOWLEDGMENTS

This study, number 15 in a series entitled "A Study of Chemical Carcinogenesis," was supported by Grant CA-012376 from the National Cancer Institute. Stephen S. Hecht is a recipient of the National Cancer Institute Research Career Development Award 5K04 CA-00124.

REFERENCES

1. Boger, E., O'Malley, R. F., and Sardella, D. J. (1976): Active sites in dibenzopyrenes: Synthesis and studies of 3-fluoro- and 2,10-difluorobenzo[rst]pentaphene. *J. Fluorine Chem.* 8:513-525.
2. Butenandt, A., and Dannenberg, H. (1953): Studies on the carcinogenicity of methyl derivatives of 1,2-cyclopentenophenanthrenes. *Arch. Geschwultsforsch.* 6:1-7.
3. Chouroulinkov, I., Gentil, A., Tierney, B., Grover, P. L., and Sims, P. (1977): The metabolic activation of 7-methylbenz[a]anthracene in mouse skin: High tumor-initiating activity of the 3,4-dihydrodiol. *Cancer Lett.*. 3:247-253.
4. Coombs, M. M., Bhatt, T. S., and Croft, C. J. (1973): Correlation between carcinogenicity and chemical structure in cyclopenta[a]phenanthrenes. *Cancer Res.* 33:832-837.
5. Freudenthal, R. I., and Jones, P. W., Eds. (1978): *Carcinogenesis, Vol. 3,* Raven Press, New York, p. 487.
6. Harvey, R. G., and Dunne, F. B. (1978): Evidence for multiple regions of metabolic activation of carcinogenic hydrocarbons. *Nature* 273:566-568.
7. Hecht, S. S., Bondinell, W. E., and Hoffmann, D. (1974): Chrysene and methylchrysenes: Presence in tobacco smoke and carcinogenicity. *J. Nat. Cancer Inst.* 52:1121-1133.
8. Hecht, S. S., Hirota, N., Loy, M., and Hoffmann, D. (1978): Tumor initiating activity of fluorinated 5-methylchrysenes. *Cancer Res.* 38: 1694-1698.
9. Hecht, S. S., LaVoie, E., Mazzarese, R., Amin, A., Bedenko, V., and Hoffmann, D. (1978): 1,2-Dihydro-1,2-dihydroxy-5-methylchrysene, a major activated metabolite of the environmental carcinogen, 5-methylchrysene. *Cancer Res.* 38:2191-2194.
10. Hecht, S. S., Loy, M., and Hoffmann, D. (1976): On the structure and carcinogenicity of the methylchrysenes. In: *Carcinogenesis, Vol. 1,* R. I. Freudenthal and P. W. Jones, Eds. pp. 325-340, Raven Press, New York.
11. Hecht, S. S., Loy, M., Mazzarese, R., and Hoffmann, D. (1978): Synthesis and mutagenicity of modified chrysenes related to the carcinogen, 5-methylchrysene. *J. Med. Chem.* 21:38-44.
12. Huberman, E., Sachs, L., Yang, S. K., and Gelboin, H. V. (1976): Identification of mutagenic metabolites of benzo[a]pyrene in mammalian cells. *Proc. Nat. Acad. Sci., U.S.* 73:607-611.

13. Ivanovic, V., Geacintov, A. M., Jeffrey, A. M., Fu, P. P., Harvey, R. G., and Weinstein, I. B. (1978): Cell and microsome mediated binding of 7,12-dimethylbenz[a]anthracene to DNA studied by fluorescence spectroscopy. *Cancer Lett.* 4:131-140.

14. Jeffrey, A. M., Yeh, H. J. C., Jerina, D. M., Patel, T. R., Davey, J. F., and Gibson, D. J. (1975): Initial reactions in the oxidation of naphthalene in *Pseudomonus putida. Biochemistry* 14:575-584.

15. Jerina, D. M., and Daly, J. W. (1976): Oxidation at carbon. In: *Drug Metabolism–from Microbe to Man.* D. V. Parke and R. L. Smith, Eds., pp. 13-32, Taylor and Francis Ltd., London.

16. Kapitulnik, J., Levin, W., Conney, A. H., Yagi, H., and Jerina, D. M. (1977): Benzo[a]pyrene 7,8-dihydrodiol is more carcinogenic than benzo[a]pyrene in newborn mice. *Nature* 266:378-380.

17. Karle, J. M., Mah, H. D., Jerina, D. M., and Yagi, H. (1977): Synthesis of dihydrodiols from chrysene and dibenzo[a,h]anthracene. *Tetrahedron Lett.* 4021-4024.

18. Lacassagne, A., Zajdela, F., Buu-Hoi, N. P., Chalvet, O., and Daub, G. H. (1968): High carcinogenic activity of mono-, di-, and trimethylbenzo-[a]pyrene. *Int. J. Cancer* 3:238-243.

19. Levin, W., Thakker, D. R., Wood, A. W., Chang, R. L., Lehr, R. E., Jerina, D. M., and Conney, A. H. (1978): Evidence that benz[a]-anthracene-3,4-diol-1,2-epoxide is an ultimate carcinogen on mouse skin. *Cancer Res.* 38:1705-1710.

20. Levin, W., Wood, A., Chang, R. L., Yagi, H., Mah, H. D., Jerina, D. M., and Conney, A. H. (1978): Evidence for bay region activation of chrysene 1,2-dihydrodiol to an ultimate carcinogen. *Cancer Res.* 38:1831-1834.

21. Malaveille, C., Bartsch, H., Tierney, B., Grover, P. L., and Sims, P. (1978): Microsome-mediated mutagenicities of the dihydrodiols of 7,12-dimethylbenz[a]anthracene: High mutagenic activity of the 3,4-dihydrodiol. *Biochem. Biophys. Res. Commun.* 83:1468-1473.

22. Miller, E. C., Fletcher, T. L., Margreth, A., and Miller, J. A. (1962): The carcinogenicities of derivatives of fluorene and biphenyl: Fluoro derivatives as probes for active sites in 2-acetylaminofluorene. *Cancer Res.* 22:1002-1014.

23. Moschel, R. C., Baird, W. H., and Dipple, A. (1977): Metabolic activation of the carcinogen 7,12-dimethylbenz[a]anthracene for DNA binding. *Biochem. Biophys. Res. Commun.* 76:1092-1098.

24. Newman, M. S. (1976): Carcinogenic activity of benz[a]anthracenes. In: *Carcinogenesis, Vol. 1,* R. I. Freudenthal and P. W. Jones, Eds., pp. 203-207, Raven Press, New York.

25. Pataki, J., and Huggins, C. (1969): Molecular site of substituents of benz[a]anthracene related to carcinogenicity. *Cancer Res.* 29:506-509.

26. Shear, M. J., and Leiter, J. (1940): Studies in Carcinogenesis XIV. 3-Substituted and 10-substituted derivatives of 1,2-benzanthracene. *J. Nat. Cancer Inst.* 1:303-306.

27. Sims, P., Grover, P. L., Swaisland, A., Pal, K., and Hewer, A. (1974): Metabolic activation of benzo[a]pyrene proceeds by a diol-epoxide. *Nature* 252:326-328.
28. Slaga, T. J., Huberman, E., Selkirk, J. K., Harvey, R. G., and Bracken, W. M. (1978): Carcinogenicity and mutagenicity of benz[a]anthracene diols and diol-epoxides. *Cancer Res.* 38:1699-1704.
29. Tierney, B., Hewer, A., Walsh, C., Grover, P. L., and Sims, P. (1977): The metabolic activation of 7-methylbenz[a]anthracene in mouse skin. *Chem. -Biol. Interact.* 18:179-193.
30. Weinstein, I. B., Jeffrey, A. M., Jennette, K. W., Blobstein, S. H., Harvey, R. G., Harris, A., Autrup, H., Kasai, H., and Nakanishi, K. (1976): Benzo[a]pyrene diol epoxides as intermediates in nucleic acid binding *in vitro* and *in vivo. Science* 193:592-595.
31. Wood, A. W., Levin, W., Ryan, D., Thomas, P. E., Yagi, H., Mah, H. D., Thakker, D. R., Jerina, D. M., and Conney, A. H. (1977): High mutagenicity of metabolically activated chrysene 1,2-dihydrodiol: Evidence for bay region activation of chrysene. *Biochem. Biophys. Res. Commun.* 78:847-854.

Polynuclear Aromatic Hydrocarbons, edited by P. W. Jones and P. Leber. Ann Arbor Science Publishers, Inc. Ann Arbor, MI © 1979.

TUMOR–INITIATING ACTIVITIES OF VARIOUS DERIVATIVES OF BENZ[a]ANTHRACENE AND 7,12–DIMETHYLBENZ[a]ANTHRACENE IN MOUSE SKIN

T. J. Slaga, G. L. Gleason, J. DiGiovanni,* D. L. Berry, M. R. Juchau,* P. P. Fu,† K. B. Sukumaron† and R. G. Harvey†

Biology Division
Oak Ridge National Laboratory
Oak Ridge, Tennessee 37830

*Department of Pharmacology
School of Medicine
University of Washington
Seattle, Washington 98195

†Ben May Laboratory for Cancer Research
University of Chicago
Chicago, Illinois 60637

INTRODUCTION

Current information indicates that polycyclic aromatic hydrocarbons (PAH) exert their toxic, mutagenic and carcinogenic activities after they have been metabolically activated by target cells to reactive epoxides (1,11,13,23, 31). The results obtained from *in vivo* and *in vitro* binding (3,6,17,22,24,27, 40), mutagenicity (12,26,41,42), metabolism (31,38,39,49) and carcinogenicity (4,18-20,34,36,37) studies have led to the conclusion that BP-7,8-diol is a proximate carcinogenic metabolite of BP, and the BP diol epoxide is an ultimate carcinogenic metabolite of BP. Recent results concerning the strong carcinogenicity of BP-7β,8α-diol-9α10α-epoxide in newborn mice (16) and in mouse skin (33) strongly indicate that it is the ultimate carcinogenic metabolite of BP.

Since diol epoxides may be responsible for the carcinogenicity of PAH other than BP, we have undertaken the testing of diols and diol epoxides as well as other derivatives of PAH for skin tumor initiation in a two-stage system of tumorigenesis. In addition, since activation of methylated PAH may involve the side-chain methyl group, the skin tumor-initiating activity of various side-chain derivatives of methylated PAH are also being determined. Specifically, in this report we are comparing the skin tumor initiation of various

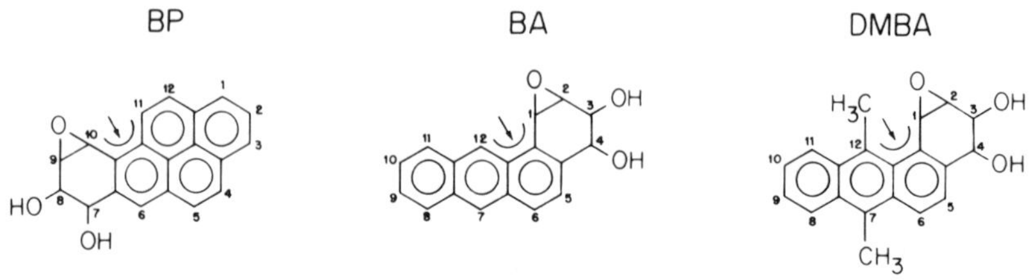

FIGURE 1. Bay-region diol-epoxides.

derivatives of a nonmethylated PAH, BA as well as a methylated PAH, DMBA. The data suggests that "bay-region" diol epoxides may be important in BA and DMBA carcinogenicity in mice; this supports the theory proposed by Jerina and co-workers which predicts that diol epoxides in the bay region occurs in a PAH when an angularly fused benzo ring is present (14). The bay regions of BP, BA and DMBA are shown in Figure 1.

MATERIALS AND METHODS

Chemicals

BA was purchased from Aldrich Chemical Co., Milwaukee, WI., and DMBA from Sigma Chemical Co., St. Louis, M., with both being more than 99% pure. TPA was obtained from Dr. Peter Borchert, University of Minnesota, Minneapolis, MN. The DMBA andBA diols and diol epoxides were synthesized and purified as previously described (8,10). The DMBA and BA diol epoxides were applied topically in tetrahydrofuran (distilled over $LiAlH_4$ and stored over sodium wire). The following compounds were synthesized by the method described in the cited reference; 7-OHM-12MBA (2), 7-OHM-12-MBA-5,6-oxide (30), 7-CHO-12-MBA (29), 12-CHO-7-MBA (29), 7,12-diCHOBA (29), 7-BrME-12-MBA (2) and 12-BrME-7-MBA (29). The 1-, 2-, 5- and 11-fluoro DMBA, the 1-, 2- and 5-CH_3DMBA and the 1-, 2-, 3-, 4-, 5-, 9- and 10-OHDMBA were generous gifts from Professor M. Newman, Ohio State University, Columbus. The above compounds plus DMBA and BA were applied topically in spectroquality acetone. In some cases other reactive compounds such as 7-OHM-12-MBA-5,6-oxide and the bromo derivatives of DMBA were applied in tetrahydrofuran. All hydrocarbons were consistently prepared under yellow light immediately before use. The mice were treated topically with the above compounds under subdued light with the time lapse between preparation of the above solutions and animal treated being less than one-half hour. TPA was prepared in stock solutions and kept in a freezer until use. Mice received twice weekly applications of 10 μg of TPA 1 week after treatment.

TABLE I. Skin Tumor-Initiating Activities of BP, BA and BA Diols after TPA Tumor Promotion

Initiator	No. of Mice[a]	Papillomas/Mouse at:[b] 15 wk	26 wk	Mice with tumors (%) at:[c] 15 wk	26 wk
BP	29	2.0	5.3	60	92
BA	30	0.7	1.2	37	57
BA 1,2-diol	30	0.2	0.4	13	33
BA 3,4-diol	28	3.2	4.7	80	85
BA 5,6-diol	29	0.1	0.2	10	14
BA 8,9-diol	30	0.2	0.8	18	46
BA 10,11-diol	29	0.1	0.2	10	14
BA control[d]	0.1	0	0	0	0
TPA control[e]	29	0	0.1	0	6

Note: All the BA compounds were applied at a dose of 2 μmoles, whereas BP was given at 0.2 μmole. Each of these treatments was followed 1 week later by twice weekly applications of 10 μg of TPA.

[a]Surviving at the 26th week after promotion.

[b]Total number of papillomas divided by the total number of surviving mice.

[c]Percentage of surviving mice with papillomas.

[d]Treated once with 2 μmoles of BA, not followed by promotion.

[e]Treated with TPA only, twice weekly for 26 weeks.

Tumor Experiments

Female CDl mice were purchased from Charles River Farms, North Wilmington, MA. Mice, 7- to 9-weeks-old, were shaved with surgical clippers 2 days before treatment and only those in the resting phase of the hair cycle were used. In the tumor experiments, groups of 30 animals received a single topical application of the test compound, followed 1 week later by twice weekly applications of TPA. The incidence of both papillomas and carcinomas was recorded weekly; papillomas and carcinomas were removed at random for histological verification.

RESULTS AND DISCUSSION

Carcinogenicity of BA Derivatives

Shown in Table I are the skin tumor-initiating activities of BA and the 1,2- 3,4-, 5,6-, 8,9- and 10,11-diols of BA. It is quite evident that BA 3,4-diol was the most active tumor initiator when compared with BA and its other diols. After 26 weeks of promotion with TPA, BA 3,4-diol had induced tumors in

TABLE II. Skin Tumor-Initiating Activities of BA and BA Diol Epoxides after TPA Tumor Promotion

Initiator	No. of Mice[a]	Papillomas/Mouse at:[b]		Mice with Tumors (%) at:[c]	
		15 wk	26 wk	15 wk	26 wk
BA	30	0.5	1.0	38	52
BA 1,2-diol-3,4-epoxide I	28	0	0.1	0	10
BA 3,4-diol-1,2-epoxide I	27	4.2	5.0	100	100
BA 8,9-diol-10,11-epoxide I	29	0.3	0.6	25	40
BA 8,9-diol-10,11-epoxide II	28	0	0.2	0	20
BA 10,11-diol-8,9-epoxide I	29	0.1	0.2	7	17

Note: All the compounds were applied at a dose of 2 μmoles and were followed 1 week later by twice weekly applications of 10 μg of TPA.

[a]Surviving at the 26th week after promotion.

[b]Total number of papillomas divided by the total number of surviving mice.

[c]Percentage of surviving mice with papillomas.

85% of the animals, with an average of 4.7 papillomas per mouse, whereas BA induced tumors in 57% of the animals, with approximately one papilloma per mouse. The 1,2- and 8,9-diols of BA had weak tumor-initiating activity, whereas the other diols were extremely weak. Although BA 3,4-diol at a dose level of 2 μmoles is a good tumor initiator when compared with BA, its tumor-initiating activity is less impressive when compared with BP at a dose of 0.2 μmole (Table I).

The incidence of skin tumors in female CDl mice after a single topical application of BA, BA 3,4-diol-1,2-epoxide I, BA 1,2-diol-3,4-epoxide I, BA 10,11-diol-8,9-epoxide I, BA 8,9-diol-10,11-epoxide I or BA 8,9-diol-10,11-epoxide II, followed by twice weekly applications of TPA, is shown in Table II. BA 3,4-diol-1,2-epoxide I was found to be approximately five times more active than BA. This should be noted as a first example of a reactive intermediate of a PAH being more active as a tumor initiator on mouse skin than the parent hydrocarbon. BA 8,9-diol-10,11-epoxide I had weak tumor-initiating activity in mouse skin, whereas the other diol epoxides were extremely weak in this regard.

Wood *et al.* recently reported that BA 3,4-diol-1,2-epoxides, in which the epoxide forms part of the bay region, were more mutagenic to bacterial cells than were the other diol epoxides (44,46). Wood *et al.* also reported that BA-3,4-diol was more tumorigenic toward mice than were BA or the four other possible *trans* diols of BA (45). In our studies we found that both the 3,4-diol and the 3,4-diol-1,2-epoxide of BA are much stronger skin tumor initiators than BA or the other possible diols and diol epoxides of BA (35). In addition we reported that BA 3,4-diol was a potent mutagen in mammalian

TABLE III. Tumor-Initiating Ability of Various 7- and 12-Substituted Derivatives of DMBA[a]

Hydrocarbon	Dose[b]	No. of Mice[c]	Time of 1st Tumor[d]	Papillomas/ Mouse[e]	Mice with Tumors (%)[f]
DMBA	10	29	8	4.50	90
7-OHM-12-MBA-5,6-oxide	200	31	24	0.10	10
7-OHM-12-MBA-5,6-oxide	400	30	15	0.17	17
7-CHO-12-MBA	740	28	9	6.10	84
12-CHO-7-MBA	740	28	11	0.75	54
7,12-diCHOBA	704	30	15	0.75	52
7-BrM-12-MBA	600	30	7	6.80	96
12-BrM-7-MBA	600	29	7	7.20	87
7-OHM-12-MBA	400	29	12	1.80	48
12-OHM-7-MBA	200	29	9	1.73	73

[a]Thirty mice were used per experiment and all mice were promoted one week after initiation with twice weekly applications of 10 μg TPA for 30 weeks.
[b]Doses of all hydrocarbons are expressed in nmoles.
[c]Numbers of mice surviving at the 30th week.
[d]Time to appearance of first tumor in weeks.
[e]Total number of papillomas divided by total number of surviving mice.
[f]Percent of surviving mice with papillomas.

cells (35).

It should be emphasized that in most *in vivo* animal tumorigenic models, as well as in *in vitro* cell transformation systems, BA is generally considered to be noncarcinogenic (32). Previously we reported that, although not a complete skin carcinogen, BA is a weak tumor initiator in a two-stage tumorigenic system (32).

Carcinogenicity of DMBA Derivatives

The skin tumor-initiating activities of various substituted derivatives of DMBA are shown in Table III. Both 7-OHM-12MBA and 12-OHM-7MBA were found to be strong tumor initiators but less active than DMBA. 7-BrM-12-MBA, 12-BrM-7-MBA and 7CHO-12-MBA were also strong tumor-initiators but once again less so than DMBA. 12-CHO-7-MBA and 7,12-diCHO-BA were less active than 7-CHO-12-MBA, whereas 7-OHM-12-MBA-5,6-oxide was found to be essentially inactive (Table III). It is of interest to point out that 7-CHO-12-MBA was found to be nearly as potent as DMBA in producing cancer in the rat whereas, as stated above, 7-CHO-12-MBA was found to have strong skin tumor-initiating activity although of a lesser degree than DMBA. In addition, Pataki and Huggins (28) stated that when two or three methyl

TABLE IV. Skin Tumor-Initiating Activities of Various Fluoro-and Methyl-substituted DMBA after TPA Promtion[a]

DMBA and Derivative	No. of Mice[b]	Papillomas per Mouse[c]	Mice with Tumors (%)[d]
Control (only DMBA initiation)[e]	30	0	0
Control (only TPA promotion)[f]	29	0.1	6
DMBA	28	9.1	100
1-CH_3DMBA	30	0.03	3
2-CH_3DMBA	30	0.10	7
5-CH_3DMBA	29	0.40	34
1-fl DMBA	30	0.03	3
2-fl DMBA	29	0.10	10
5-fl DMBA	28	0.20	15
11-fl DMBA	29	8.2	100
6,8-dimethyl BA	29	0.7	40
3,9-dimethyl BA	29	0.07	3
5,7-dimethyl BA	28	0.03	3

[a]DMBA and DMBA derivatives were applied at a dose of 200 nmoles and were followed one week later by twice weekly applications of 10 μg of TPA.
[b]Surviving at the 30th week after promotion.
[c]Total number of papillomas divided by total number of surviving mice.
[d]Percentage of surviving mice with tumors.
[e]These mice were only initiated with 200 nmoles of DMBA.
[f]These mice were only promoted twice weekly with 10 μg of TPA for 30 weeks.

groups were added to BA, in positions 6, 7, 8 and 12 (any combination), the resulting products were equally carcinogenic in the rat. However, we found that 6,8-dimethyl BA as well as 3,9-dimethyl BA and 5,7-dimethyl BA had very weak skin tumor-initiating activity when compared to DMBA (Table IV).

Direct evidence for determining the importance of the bay-region diol-epoxide of DMBA is difficult because of the hindrance of the side-chain methyl groups during synthesis. However, the influence on carcinogenicity of the substitution of methyl or fluoro groups in the bay region would provide indirect evidence. Table IV shows that either a methyl or fluoro substitution to the 1, 2 and 5 positions almost completely destroyed the skin tumor-initiating activity whereas a fluoro group in position 9 or 11 did not. Harvey and Dunne (9) recently reported that DMBA fluorinated at positions 1 and 2 also drastically reduced the sarcomagenic activity of DMBA when injected into rats although fluorination at position 11 did not. These results suggest that, indeed, the bay region of DMBA is involved in the metabolic activation of DMBA into a carcinogen. In addition, our data implicate the 5 position in such an activation. Moschel *et al.* (25) have recently presented fluorescence

TABLE V. Skin Tumor-Initiating Activities of Various Derivatives of DMBA After TPA Promotion[a]

DMBA and Derivatives	No. of Mice[b]	Papillomas per Mouse[c]	Mice with Tumors (%)[d]
Control (only DMBA initiation)[e]	30	0	0
Control (only TPA promotion)[f]	29	0.1	6
DMBA	28	8.86	100
1-OH	29	0.10	10
2-OH	30	0.07	7
3-OH	30	0.14	14
4-OH	28	0.17	14
5-OH	29	0.10	10
9-OH	30	0.40	24
10-OH	30	0.47	30
3,4-diol[g]	30	15.30	100
5,6-diol	29	0.10	10
8,9-diol	29	0.21	18
8,9-diol-10,11-epoxide	28	0.30	20

[a]The DMBA and DMBA derivatives were applied topically at a dose of 200 nmoles and were followed one week later by twice weekly applications of 10 μg of TPA.
[b]Surviving at the 20th week after promotion.
[c]Total number of papillomas divided by total number of surviving mice.
[d]Percentage of surviving mice with tumors.
[e]These mice were only initiated with 200 nmoles of DMBA.
[f]These mice were only promoted twice weekly with 10 μg of TPA for 30 weeks.
[g]Tumor response after 15 weeks of promotion for 100 nmoles of DMBA-3,4-diol.

data which implicates the bay-region diol epoxide of DMBA in the interaction of DMBA with cellular DNA. In addition, binding of DMBA to cells in culture analyzed by LH20 chromatography also suggests that a bay-region diol epoxide is involved in DMBA binding (7).

Table V shows the relative skin tumor-initiating activities of the various derivatives of DMBA. The 1-, 2-, 3-, 4- and 5-OH DMBA had weak activity. The 5,6- and 8,9-diols of DMBA were also found to be inactive and the (±) DMBA 8β,9α-diol-10α,11α-epoxide, a non-bay-region diol epoxide, had only weak skin tumor-initiating activity. The difficult synthesis of the bay-region diol of DMBA has been recently achieved (unpublished results). In the early stages of an ongoing tumor experiment, DMBA 3,4-diol has approximately 10 times the skin tumor-initiating activity as DMBA (Table V). Lack of tumorigenicity of DMBA 8,9-diol and DMBA 8β,9α-diol-10α,11α-epoxide, the results concerning the fluoro and methyl substitution of DMBA, the extreme potency of DMBA 3,4-diol in initiating skin tumors plus studies on binding of

DMBA to cellular DNA after metabolism (7,25) strongly implicate the bay-region diol epoxide of DMBA in its carcinogenic activity. The results in this report plus those recently published (4,5,12,13,15,16,18-21,26,32-37,43-46, 48) provide evidence that bay-region diol epoxides are involved in BP, BA, chrysene, 7-methylbenz[a]anthracene, dibenz[a,h]anthracene and DMBA carcinogenicity and mutagenicity. Again, this data supports the theory proposed by Jerina and co-workers, which predicts that diol-epoxides in the bay region are the principle determinants of PAH carcinogenicity (14,15).

ACKNOWLEDGMENTS

This work was supported in part by NIH Grants CA-20076 and CA-11968 and by the Division of Biomedical and Environmental Research, U.S. Department of Energy, under Contract W-7405-eng-26 with the Union Carbide Corporation.

REFERENCES

1. Ames, B. N., Dursten, W. E., Yamasaki, E., and Lee, F. D. (1973): Carcinogens are mutagens: A simple test system combining liver homogenates for activation and bacteria for detection. *Proc. Nat. Acad. Sci.* 70:2281-2885.
2. Boyland, E., and Sims, P. (1965): Metabolism of polycyclic compounds. The metabolism of 7,12-dimethylbenz[a]anthracene by rat liver homogenates. *Biochem. J.* 95:780-787.
3. Borgen, A., Darvey, H., Castagnoli, N., Crocker, T. T., Rasmussen, R. E., and Wang, I. Y. (1973): Metabolic conversion of benzo[a]pyrene by syrian hamster liver microsomes and binding of metabolites to DNA. *J. Med. Chem.* 16:502-506.
4. Chouroulinkov, I., Gentil, A., Grover, P. L., and Sims, P. (1976): Tumor-initiating activities on mouse skin of dihydrodiols derived from benzo[a] pyrene. *Brit. J. Cancer* 34:523-532.
5. Chouroulinkov, I., Gentil, A., Tierney, B., Grover, P., and Sims, P. (1977): The metabolic activation of 7-methylbenz[a]anthracene in mouse skin: High tumor-initiating activity of the 3,4-dihydrodiol. *Cancer Lett.* 3:247-253.
6. Dandel, P., Duquesne, M., Vigny, P., Grover, P. L., and Sims, P. (1975): Fluorescence spectral evidence that benzo[a]pyrene-DNA products in mouse skin arise from diol-epoxides. *FEBS Lett.* 57:250-253.
7. Dipple, A., and Nebzydoski, J. A. (1978): Evidence for the involvement of a diol-epoxide in the binding of 7,12-dimethylbenz[a]anthracene to DNA in cells in culture. *Chem.-Biol. Interact.* 20:17-26.

8. Fu, P. P., and Harvey, R. G. (1977): Synthesis of the diols and diol epoxides of carcinogenic hydrocarbons. *Tetrahedron Lett.* 2059-2062.
9. Harvey, R. G., and Dunne, F. A. (1978): Multiple regions of metabolic activation of carcinogenic hydrocarbons. *Nature* 273:566-567.
10. Harvey, R. G., and Sukumaran, K. B. (1977): Synthesis of the A-ring diols and diol epoxides of BA. *Tetrahedron Lett.* 2387-2390.
11. Heidelberger, C. (1975): Chemical carcinogenesis. *Ann. Rev. Biochem.* 44:79-121.
12. Huberman, E., Sachs, L., Yang, S. D., and Gelboin, H. V. (1976): Identification of mutagenic metabolites of benzo[a]pyrene in mammalian cells. *Proc. Nat. Acad. Sci.* 73:607-611.
13. Jerina, D. M., and Daly, J. W. (1974): Arene oxides. A new aspect of drug metabolism. *Science* 185:573-582.
14. Jerina, D. M., and Daly, J. W. (1977): Oxidation at Carbon. In: *Drug Metabolism*, D. V. Parke and R. L. Smith, Eds., pp. 15-33, Taylor & Francis, Ltd., London.
15. Jerina, D. M., Lehr, R. E., Yagi, H., Hernandez, O., Dansette, P. M., Wislocki, P. G., Wood, A. W., Chang, R. L., Levin, W., and Conney, A. H. (1977): Mutagenicity of benzo[a]pyrene derivatives and the description of a quantum mechanical model which predicts the ease of carbonium ion formation from diol epoxides. In: *In Vitro Metabolic Activation and Mutagenesis Testing*, F. J. de Serres, J. R. Fouts, J. R. Bend and R. M. Philpot, Eds., pp. 159-177, Elsevier/North Holland Biomedical Press, Amsterdam.
16. Kapitulnik, J., Levin, W., Conney, A. H., Yagi, H., and Jerina, D. M. (1977): Benzo[a]pyrene 7,8-dihydrodiol is more carcinogenic than benzo[a]pyrene in newborn mice. *Nature* 266:378-380.
17. King, H. W. S., Osborne, M. R., Beland, F. A., Harvey, R. G., and Brooks, P. (1976): (±)-7α, 8β-dihydroxy-9β, 10β-epoxy-7,8,9,10-tetrahydrobenzo[a]pyrene is an intermediate in the metabolism and binding to DNA of benzo[a]pyrene. *Proc. Nat. Acad. Sci., U.S.* 73:2679-2681.
18. Levin, W., Wood, A. W., Chang, R. L., Slaga, T. J., Yagi, H., Jerina, D. M., and Conney, A. H. (1977): Marked differences in the tumor-initiating activity of optically pure (+)- and (–)-trans-7,8-dihydroxy-7,8-dihydrobenzo[a]pyrene on mouse skin. *Cancer Res.* 37:2721-2725.
19. Levin, W., Wood, A. W., Yagi, H., Dansette, P. M., Jerina, D. M., and Conney, A. H. (1976): Carcinogenicity of benzo[a]pyrene 4,5-, 7,8- and 9,10-oxides on mouse skin. *Proc. Nat. Acad. Sci., U.S.* 73:243-247.
20. Levin, W., Wood, A. W., Yagi, H., Jerina, D. M., and Conney, A. H. (1976): (±)-trans-7,8-dihydroxy-7,8-dihydrobenzo[a]pyrene: A potent skin carcinogen when applied topically to mice. *Proc. Nat. Acad. Sci., U.S.* 73:3867-3871.

21. Malaveille, C., Bartsch, H., Grover, P. L., and Sims, P. (1975): Mutagenicity of non-K-region diol and diol epoxides of benzo[a] anthracene and benzo[a] pyrene in *S. typhimurium* TA100. *Biochem. Biophys. Res. Commun.* 66:693-700.
22. Meehan, T., Straub, K., and Calvin, M. (1976): Elucidation of hydrocarbon structure in an enzyme catalyzed benzo[a] pyrene-poly (G) covalent complex. *Proc. Nat. Acad. Sci., U.S.* 73:1437-1441.
23. Miller, E. C., and Miller, J. A. (1974): Biochemical mechanisms of chemical carcinogensis. In: *Molecular Biology of Cancer,* H. Busch, Ed., pp. 377-402, Academic Press, Inc., New York.
24. Moore, P. D., Koreeda, M., Wislocki, P. G., Levin, W., Conney, A. H., Yagi, H., and Jerina, D. M. (1977): *In vitro* reactions of the disastereomeric 9,10-epoxides of (+)- and (-)-trans-7,8-dihydroxy-7,8-dihydrobenzo[a] pyrene with polyguanylic acid and evidence for formation of an enantiomer of each diastereomeric 9,10-epoxide from benzo[a]-pyrene in mouse skin. In: *Concepts in Drug Metabolism No. 44,* D. M. Jerina, Ed., pp. 127-154, Washington, D.C.
25. Moschel, R. C., Baird, W. M., and Dipple, A. (1977): Metabolic activation of the carcinogen 7,12-dimethylbenz[a] anthracene for DNA binding. *Biochem. Biophys. Res. Commun.* 76:1092-1098.
26. Newbold, R. F., and Brooks, P. (1976): Exceptional mutagenicity of a benzo[a] pyrene diol epoxide in cultured mammalian cells. *Nature* 261:52-54.
27. Osborne, M. R., Thompson, M. H., Tarmy, E. M., Beland, F. A., Harvey, R. G. and Brooks, P. (1976): The reaction of 7,8-dihydro-7,8-dihydroxybenzo[a] pyrene-9,10-oxide with DNA in relation to the benzo[a] pyrene DNA products isolated from cells. *Chem. -Biol. Interact.* 13:343-348.
28. Pataki, J., and Huggins, C. B. (1969): Molecular site of substituents of benzo[a] anthracene related to carcinogenicity. *Cancer Res.* 29:506-509.
29. Pataki, J., Wlos, R., and Cho, Y-J. (1971): Adrenocorticolytic derivatives of benz[a] anthracene. *J. Med. Chem.* 11:1083-1086.
30. Sims, P. (1973): Epoxy derivatives of aromatic polycyclic hydrocarbons. The preparation and metabolism of epoxides related to 7,12-dimethylbenz[a] anthracene. *Biochem. J.* 131:405-413.
31. Sims, P., and Grover, P. L. (1974): Epoxides in polycyclic aromatic hydrocarbon metabolism and carcinogenesis. *Adv. Cancer Res.* 20: 165-274.
32. Slaga, T. J., Bowden, G. T., Scribner, J. D., and Boutwell, R. K. (1974): Dose-response studies on the ability of 7,12-dimethylbenz[a] anthracene to initiate skin tumors. *J. Nat. Cancer Inst.* 53:1337-1340.
33. Slaga, T. J., Bracken, W. J., Gleason, G., Levin, W., Yagi, H., Jerina, D. M., and Conney, A. H. (1979): Marked differences in the skin tumor-initiating activities of the optical enantiomers of the diastereomeric benzo[a] pyrene 7,8-diol-9,10-epoxides. *Cancer Res.* 39:67-71.

34. Slaga, T. J., Bracken, W. M., Viaje, A., Levin, W., Yagi, H., Jerina, D. M., and Conney, A. H. (1977): Comparison of the tumor-initiating activities of benzo[a]pyrene arene oxides and diol-epoxides. *Cancer Res.* 37:4130-4133.
35. Slaga, T. J., Huberman, E., Selkirk, J. K., Harvey, R. G., and Bracken, W. M. (1978): Carcinogenicity and mutagenicity of benz[a]anthracene diols and diol-epoxides. *Cancer Res.* 38:1699-1704.
36. Slaga, T. J., Viaje, A., Berry, D. L., Bracken, W., Buty, S. G., and Scribner, J. D. (1976): Skin tumor initiating ability of benzo[a]pyrene 4,5-, 7,8-, and 7,8-diol-9,10-epoxides and 7,8-diol. *Cancer Lett.* 2: 115-122.
37. Slaga, T. J., Viaje, A., Bracken, W. M., Berry, D. L., Fischer, S. M., Miller, D. R., and LeClerc, S. M. (1977): Skin tumor intiating ability of benzo[a]pyrene-7,8-diol-9,10-epoxide (anti) when applied topically in tetrahydrofuran. *Cancer Lett.* 3:23-30.
38. Thakker, D. R., Yagi, H., Akagi, H., Koreeda, M., Lu, A. Y. H., Levin, W., Wood, A. W., Conney, A. H., and Jerina, D. M. (1977): Metabolism of benzo[a]pyrene. VI. Stereoselective metabolism of benzo[a]pyrene and benzo[a]pyrene 7,8-dihydrodiol to diol epoxides. *Chem. -Biol. Interact.* 16:281-300.
39. Thakker, D. R., Yagi, H., Lu, A. Y. H., Levin, W., Conney, A. H., and Jerina, D. M. (1976): Metabolism of benzo[a]pyrene: Conversion of (+)-trans-7,8-dihydroxy-7,8-dihydrobenzo[a]pyrene to highly mutagenic 7,8-diol-9,10-epoxides. *Proc. Nat. Acad. Sci., U.S.* 73:3381-3385.
40. Weinstein, I. B., Jeffrey, A. M., Jennette, K. W., Blobstein, S. H., Harvey, R. G., Harris, C., Autrup, H., Kasai, H., and Nakanishi, K. (1976): Benzo[a]pyrene diol epoxides as intermediates in nucleic acid binding *in vitro* and *in vivo*. *Science* 193:592-595.
41. Wislocki, P. G., Wood, A. W., Chang, R. L., Levin, W., Yagi, H., Hernandez, O., Dansette, P. M., Jerina, D. M., and Conney, A. H. (1976): Mutagenicity and cytotoxicity of benzo[a]pyrene oxides, phenols, quinones and dihydrodiols in bacterial and mammalian cells. *Cancer Res.* 36:3350-3357.
42. Wislocki, P. G., Wood, A. W., Chang, R. L., Levin, W., Yagi, H., Hernandez, O., Jerina, D. M., and Conney, A. H. (1976): High mutagenicity and toxicity of a diol epoxide derived from benzo[a]pyrene. *Biochem. Biophys. Res. Commun.* 68:1006-1012.
43. Wislocki, P. G., Wood, A. W., Chang, R. L., Levin, W., Yagi, H., Hernandez, O., Jerina, D. M., and Conney, A. H. (1976): High mutagenicity and toxicity of a diol epoxide derived from benzo[a]pyrene. *Biochem. Biophys. Res. Commun.* 68:1006-1012.
44. Wood, A. W., Chang, R. L., Levin, W., Lehr, R. E., Schaefer-Ridder, M., Karle, J. M., Jerina, D. M., and Conney, A. H. (1977): Mutagenicity and cytotoxicity of benz[a]anthracene diol epoxides and tetrahydro-epoxides: Exceptional activity of the bay region 1,2-epoxides. *Proc. Nat. Acad. Sci., U.S.* 74:2746-2750.

45. Wood, A. W., Levin, W., Chang, R. L., Lehr, R. E., Schaefer-Ridder, M., Karle, J. M., Jerina, D. M., and Conney, A. H. (1977): Tumorigenicity of five dihydrodiols of benz[a]anthracene on mouse skin: Exceptional activity of benz[a]anthracene-3,4-diol. *Proc. Nat. Acad. Sci., U.S.* 74:3176-3179.
46. Wood, A. W., Levin, W., Lu, A. Y. H., Ryan, D., West, S. B., Lehr, R. E., Schaefer-Ridder, M., Jerina, D. M., and Conney, A. H. (1976): Mutagenicity of metabolically activated benz[a]anthracene-3,4-dihydrodiol. Evidence for bay region activation of carcinogenic polycyclic hydrocarbons. *Biochem. Biophys. Res. Commun.* 72:680-686.
47. Wood, A. W., Levin, W., Ryan, D., Thomas, P. E., Yagi, H., Moh, H. D., Thakker, D. R., Jerina, D. M., and Conney, A. H. (1977): High mutagenicity of metabolically activated chrysene-1,2-dihydrodiol. Evidence for bay region activation of chrysene. *Biochem. Biophys. Res. Commun.* 78:847-854.
48. Wood, A. W., Levin, W., Thomas, P. E., Ryan, D., Karle, J. M., Yagi, H., Jerina, D. M., and Conney, A. H. (1978): Metabolic activation of dibenzo[a,h]anthracene and its dihydrodiols to bacterial mutagens. *Cancer Res.* 38:1967-1973.
49. Yang, S. K., McCourt, D. W., Roller, P. P., and Gelboin, H. V. (1976): Enzymatic conversion of benzo[a]pyrene leading predominantly to the diol-epoxide r-7, t-8-dihydroxy-t-9,10-oxy-7,8,9,10-tetrahydrobenzo[a]pyrene through a single enantiomer of r-7, t-8-dihydroxy-7,8-dihydrobenzo[a]pyrene. *Proc. Nat. Acad. Sci., U.S.* 73:2594-2598.

Polynuclear Aromatic Hydrocarbons,
edited by P. W. Jones and P. Leber.
Ann Arbor Science Publishers, Inc.
Ann Arbor, MI ©1979.

HUMAN PLACENTAL ARYL HYDROCARBON HYDROXYLASE: GENETICS AND ENVIRONMENTAL INFLUENCES

Olavi Pelkonen, Niilo T. Kärki, Pekka Korhonen,
Maila Koivisto, Risto Tuimala and Antti Kauppila

Departments of Pharmacology, Microbiology,
Pediatrics and Obstetrics & Gynecology
University of Oulu
Oulu, Finland

INTRODUCTION

The placenta is the unique organ of pregnancy, whose role as an effective barrier protecting the fetus from harmful effects, including foreign chemical substances, has to be revised in light of research done in the last two decades. It has been found that most organic and inorganic compounds pass the placenta easily and reach the fetus. If, however, such compounds were metabolized, even in part, by the placenta, this would offer partial protection. On the other hand, because some compounds must be metabolized before becoming harmful, placental metabolism may actually serve, in some cases, to enhance toxicity.

Welch *et al.* (33,34) found that placentas from smoking mothers metabolized benzo[a]pyrene (BP), a well-known carcinogen, whereas placentas from nonsmokers were unable to do so; these findings were soon confirmed (9,24,28). The variability in enzyme activity was extremely high, even in placentas from mothers reportedly smoking a similar number of cigarettes (24,34). Although without doubt partly a consequence of differential exposure to cigarette smoke, this extreme variability raises the question of the genetic background. Genetic regulation with respect to the inducibility of aryl hydrocarbon (benzo[a]pyrene) hydroxylase (AHH) has been extensively studied in inbred strains of mice (23,31) and there are a number of studies showing that AHH in humans is, to a large extent, subject to genetic control, although the mode of inheritance is probably multigenic, not monogenic as in some inbred mouse strains (2,20,25,26). These earlier studies prompted us to try to elucidate the relative roles of genetic background and environment in the regulation of AHH activity in the placenta.

METABOLISM OF XENOBIOTICS IN THE PLACENTA

Components of the electron transport chain typical of the xenobiotic-oxidizing system, cytochrome P-450 and its reductase, have been found in human placental microsomes (3,16,22). As a first implication, this ought to indicate that the placenta is able to oxidize xenobiotics. The properties of the placental cytochrome P-450 system have been extensively studied by Juchau's group (10,11,12,14,30) and need not be reviewed here. Nevertheless, a few remarks may be called for in the context of this report. As is shown below, the substrate specificity of the placental monooxygenase resembles that of cytochrome P-448 or P_1-450 associated monooxygenase in the liver and yet the spectral properties of the placental cytochrome do not exhibit any "spectral shift to the blue." Also, there is no increase in the specific content of cytochrome P-450 in conjunction with the increase in AHH activity. These discrepancies may be explained by assuming that the monooxygenase associated with cytochrome P-448 or P_1-450 is not actually a single entity but consists of several closely related cytochromes, some of which exhibit spectral shifts while others do not (1). At the same time the turnover number for BP in the form of cytochrome P-450 preferentially metabolizing BP may be so high that the increase in this form cannot be detected spectrally or electrophoretically. However, what interests us is the substrate specificity and activity of the placental drug-metabolizing systems. Table I is a compilation of studies on the metabolism of xenobiotics by the human placenta. The metabolism of some substrates (benzo[a]pyrene, N-methylaniline, 3-methyl-4-monomethylaminoazobenzene and 7-ethoxycoumarin) by the placenta from smoking mothers seems to be clearly established, and some authors have found considerable oxidative metabolism of some other compounds (*e.g.*, pentobarbital, amphetamine, hexobarbital), but most studies with the same and other substrates have given negative results. An especially interesting finding has been that, according to most results, xenobiotics metabolized preferentially by control or phenobarbital-treated rat liver are not metabolized at all by the human placenta. Data on substrate specificity are further confirmed by analysis of the spectral interactions in placental microsomes (Table II). Compounds known to produce Type I spectral change with rat liver microsomes do not give any spectrum with placental microsomes. Since Type I spectral interaction is thought to reflect the enzyme-substrate complex, whereas the Type II spectrum is produced by ferric hemochromogen formation, typical compounds for producing Type II interactions give typical spectra with placental microsomes.

TABLE I. Metabolism *in vitro* of Xenobiotics by the Human Placenta

Compound	Metabolic Pathway	Activity	References
Benzo[a]pyrene	Hydroxylation (3,9)	Inducible	9,24,28,33,34
3-Methyl-4-mono-methylaminoazo-benzene	N-Demethylation	Inducible	34
7-Ethoxycoumarin	D-Deethylation	Inducible	8
Zoxazolamine	Tritium release	Inducible	19
2-Acetylaminofluorene	Hydroxylation (3,5,7)	Inducible	13
	Hydroxylation (N,1)	Present	13
N-Methylaniline	Disappearance	Present	28
Dimethylaniline	N-Demethylation	Present	18
Coumarin	Hydroxylation (7)	Not detectable	8,16
Aminopyrine	N-Demethylation	Not detectable	11,16
		Present	18
Ethylmorphine	N-Demethylation	Not detectable	4,21
Meperidine	N-Demethylation	Not detectable	32
Codeine	O-Demethylation	Not detectable	11
Mescaline	O-Demethylation	Not detectable	11
p-Nitroanisole	O-Demethylation	Not detectable	3,4
Diazepam	N-Demethylation	Not detectable	7
	Hydroxylation	Not detectable	7
Aniline	Hydroxylation (4)	Not detectable[a]	3,11,16
Acetanilide	Hydroxylation (4)	Not detectable	3
Hexobarbital	Disappearance	Not detectable	11
		Present	21
Pentobarbital	Hydroxylation	Not detectable	34
	Disappearance	Present	32
Biphenyl	Hydroxylation (4)	Not detectable	4
Amphetamine	Disappearance	Not detectable	3,11
		Present	

[a]Present in soluble fraction, but activity probably catalyzed contaminating hemoglobin (15).

In their reviews of placental drug metabolism, Juchau *et al.* (12,14,16) express the view that the placenta can oxidize only a very limited number of foreign compounds through a limited number of pathways. They consider it probable that the low rates of the drug oxidation reactions observed in the placenta may be explicable in terms of high substrate specificities for the placental enzymic components in contrast to the low substrate specificities of the corresponding liver systems (16). Another possible explanation is based on the multiplicity of cytochromes P-450. It is probable that the liver possesses multiple cytochromes P-450, whereas the placenta possesses only a few.

TABLE II. Spectral Interactions of Xenobiotics with Human Placental Microsomes

Compound[a]	Peak (nm)	Trough (nm)	Type	Reference
Type I Compounds				
Aminopyrine		No spectrum		3
Hexobarbital		No spectrum		3
Pentobarbital		No spectrum		3
Naphthalene		No spectrum		3
Desmethylimipramine		No spectrum		30
Type II Compounds				
Aniline	430	408	II	3
	435	416	II	30
n-Octylamine	424	407,439	II, atypical	17
Metyrapone	426	395	II	3
Nicotinamide	426	408	II	3
	435	416	II	30
p-Aminophenol	434	411	II	3

[a]Xenobiotics were grouped into Type I and Type II compounds according to spectral changes they cause with rat liver microsomes.

Recent work suggests that an analogy appears to exist between the substrate specificity of cytochrome P-448 or P_1-450, a polynuclear aromatic hydrocarbon-inducible hemoprotein, and the ability of the human placenta to oxidize xenobiotics. Compounds which have been unequivocally shown to be metabolized by the human placenta (from smoking mothers) benzo[a]-pyrene, 3-methyl-4-monomethylaminobenzene, N-methylaniline and 7-ethoxycoumarin, are also preferential substrates for cytochrome P-448 or P_1-450, as evidenced by their induced metabolism after exposure of the subject to polynuclear aromatic hydrocarbons. The work of Pelkonen (27) also shows that placental AHH resembles the enzyme from 3-methylcholanthrene-treated rat liver, at least with respect to its preferential inhibition by α-naphthoflavone and other compounds. On the other hand, *p*-nitroanisole, for example, which is shown to be metabolized preferentially by cytochrome P-448 in rat liver, is not a substrate for the placental enzyme (3,16).

INDUCTION OF PLACENTAL AHH ACTIVITY BY MATERNAL SMOKING

Maternal cigarette smoking induces placental AHH activity from the last weeks of the first trimester onwards (9,28), attaining its highest level at term. A correlation has been demonstrated between placental AHH activity and the number of cigarettes smoked daily (24), even though a much more conspicuous feature of placental AHH activity is its extreme variability (9,24,

33,34). Welch *et al.* (33) demonstrated, for example, that the range of AHH activity in placentas from mothers who smoked the same number of cigarettes daily could vary as much as 70-fold. Table III shows recent results obtained by the present authors on the correlation between placental AHH activity and maternal smoking. A certain increase in mean AHH activity appears with increasing numbers of cigarettes smoked. There are also a couple of confusing features, however. Firstly, in the whole population of smoking mothers there were only two who reportedly smoked over 20 cigarettes per day. This small number of heavy smokers is not in accordance with epidemiological studies or with earlier surveys in Finland and the U.S. (24,34) and is, perhaps, indicative of a change in the general population's awareness of the harmful effects of cigarette smoke, leading to a "guilt complex." Hence, the reports on the number of cigarettes smoked per day are more distorted (underestimated) than those given by other smokers. We feel, therefore, that the reported levels of smoking are highly dubious, and that the difference between the groups smoking 1-5 and 6-10 cigarettes per day is particularly unreliable. Secondly, there were several cases of nonsmokers in whom there was an increased AAH activity in the placenta. This may be also associated with the same "guilt complex" but there may also be other explanations, such as exposure to polynuclear aromatic hydrocarbons in the home or at work. These possibilities remain to be explored in the future, however.

The variability in AHH activity was in any event extreme, and this naturally leads to questions concerning the relative contributions of environmental factors and genetic background: (1) the variable exposure of the placenta (and fetus) to cigarette smoke constituents, and (2) genetic difference between placentas in their susceptibility to such inducing action. The role of both these factors may then be partially distorted by the mother's inherent or environmentally modified ability to metabolize or eliminate the constituents of cigarette smoke before they reach the placenta or fetus. Evaluation of the genetic contribution to the variation in placental AHH activity is very difficult, but it may be possible to make some progress by using placental cultures or lymphocyte cultures from the fetus (our preliminary work along these lines is described below). Maternal lymphocyte cultures may be used for the evaluation of any possible maternal influence. The evaluation of the real exposure of the placenta to cigarette smoke may also be difficult, since as far as we know there are no reliable indicators for measuring exposure over extended time periods. Measurements of the blood levels of cigarette smoke constituents such as CO or nicotine are perhaps too transient and too dependent on day-to-day variations in smoking to reflect the true exposure to inducing substances. As a first approximation, we hypothesized that AHH activity in the placenta may be used to measure the actual exposure of the placenta and fetus to the effects of maternal smoking.

Table III. Relationship Between the Reported Number of Cigarettes Smoked Daily and Placental AHH Activity

Cigarettes[a] Daily	n	AHH Activity (arbitrary fluorescence units)							
		<7	7 - 20	21 - 50	51 - 100	101 - 200	>200	Mean $\pm$ S.D.	Range
0	177	164	7	5	1	0	0	7	$<$ 7 - 73
1 - 5	39	7	8	16	3	4	1	49 $\pm$ 61	$<$ 7 - 243
6 - 10	39	7	7	10	6	6	3	80 $\pm$ 88	$<$ 7 - 343
11 - 15	7	0	0	2	2	1	2	127 $\pm$ 143	23 - 298
16 - 20	4	0	1	1	1	0	1	146 $\pm$ 222	13 - 405
>20	2	0	1	0	0	1	0	94	13 - 174
All Smokers	91	14	17	29	12	12	7	70 $\pm$ 82	$<$ 7 - 405

[a]According to clinical history. It must be stressed that this figure is highly unreliable.

This seemed a reasonable hypothesis since, even allowing for a substantial genetic contribution, placental AHH activity must also reflect the actual exposure in a large population. The investigation of this point led to an interesting preliminary observation, which may be significant in the context of differential exposure (Table IV), namely that the newborn infants from pregnancies in which the placental AHH was positive weighed about 400 grams less than those of nonsmoking mothers, whereas those from AHH-negative pregnancies were of the same size as those of nonsmoking mothers. It must be noted that there were no significant differences in the reported number of cigarettes smoked between the AHH-positive and AHH-negative groups. Although it is possible that the differences in placental AHH inducibility and fetal response to maternal smoking are due to genetic factors, a more likely explanation, and one which may be adopted for the time being, is that despite the same number of cigarettes smoked, the actual exposure was different in the AHH-positive and AHH-negative groups. Another, totally hypothetical yet reasonable explanation is that maternal metabolism of cigarette smoke constituents in the negative cases is so effective that insufficient amounts reach the placenta to induce AHH activity or the fetus to cause any reduction in birth weight.

TABLE IV. Maternal Cigarette Smoking, Placental AHH Activity and Birth Weight

Group	n	Cigarettes Per Day	Birth Weight (G) Mean ± S.D.	Birth Weight (G) Reduction
Nonsmokers	139	0	3662 ± 410	
Smokers	85	7.7 ± 4.0	3423 ± 488[a]	- 239
AHH Positive	55	7.8 ± 4.1	3266 ± 413[a]	- 396
AHH Negative	17	7.3 ± 0.9	3714 ± 546	+ 52
"Early Quitters"[b]	13	6.7 ± 3.4	3710 ± 428	+ 48

[a]Significantly different ($p < 0.05$) from nonsmokers.
[b]Mothers quit smoking after the first trimester of pregnancy.

INDUCTION OF AAH ACTIVITY IN PLACENTAL EXPLANT CULTURES

Perhaps the best and most direct way to study the genetic background of AHH inducibility in the placenta is to obtain placenta cultures, expose them to polynuclear aromatic hydrocarbon inducers under *in vitro* conditions and measure AHH activity in these and in control cultures. Placental cells are relatively difficult to culture however; meanwhile,we have attempted to study placental explant cultures. Methods have recently been developed by Trump,

Harris and others (5) for maintaining pieces of human tissue in culture for extended periods of time and it has been shown that these explants can metabolize various carcinogens (5). The covalent binding of BP to DNA in cultured human bronchus, for example, varies about 75-fold (6). We therefore attempted to culture pieces of human term placentas under similar conditions. These can be maintained in culture for several weeks, and the culture medium becomes positive for the pregnancy test, for example, indicating excretion of HCG. The cultured placenta is able to metabolize BP, and AHH activity is induced, when it is exposed to benzo[a]anthracene, even though the magnitude of induction is much smaller than in lymphocytes and fetal cells, only from 1.2- to 2-fold, and has no correlation with AHH activity in the corresponding placental homogenates. The reason for this small extent of induction is not clear. One may hypothesize that there is either a specific type of regulation of AHH inducibility in *in vitro* placental cultures, or else the culture methods used do not allow the maximal induction to occur. At the same time the considerable experimental variation lessens the viability of using AHH inducibility in placental explant cultures in the evaluation of the genetic background.

INDUCIBILITY OF AHH ACTIVITY IN MITOGEN-TREATED LYMPHOCYTES FROM MATERNAL AND CORD BLOOD

Since it has been suggested that the inducibility of AHH activity in lymphocytes may reflect the genetic constitution of an individual with respect to sensitivity to environmental PAH, we thought it appropriate to study the use of AHH inducibility in cord blood lymphocytes as a measure of the genetic susceptibility of the neonate to the effects of maternal smoking. We also wished to check if this parameter could be associated with the condition of newborn infants of smoking and nonsmoking mothers. The results of our preliminary studies are shown in Table V. It was found that both basal and induced levels were much higher in the cord blood lymphocytes than in the maternal lymphocytes, so that in the mother/neonate pair P.R. in Table V, for example, the AHH activity in the cord blood lymphocytes was 34-fold (basal) and 56-fold (induced) higher than in the maternal lymphocytes. Another significant finding was the great variability in absolute levels with 6.1-fold variations in absolute basal levels in maternal lymphocytes and 14.7-fold in the induced levels and corresponding 10.3- and 8.4-fold fluctuations in basal and induced levels in the cord lymphocytes. Despite the large difference in absolute activities between the maternal and cord blood lymphocytes, the ratios of induction were surprisingly similar varying from 1.5 to 6.8 in the maternal lymphocytes and from 1.9 to 6.9 in the cord blood lymphocytes. The reason for the observation that AHH activities are higher

TABLE V. Aryl Hydrocarbon Hydroxylase Activity in Lymphocytes from Smoking and Nonsmoking Mothers and from Cord Blood

		Aryl Hydrocarbon Hydroxylase Activity (arbitrary fluorescence units/10^6 cells/30 min)					
		Maternal Lymphocytes			Cord Blood Lymphocytes		
Subject	Age (yr)	Basal	BA-Induced	Ratio	Basal	BA-Induced	Ratio
Smokers							
R.K.	18	34.5	177.0	5.1	129.0	354.0	2.7
K.L.	27	40.0	242.0	6.1	88.0	170.5	1.9
P.H.	34	9.0	35.0	3.9	47.0	239.0	5.1
H.L.	33	55.0	324.0	5.9	111.0	634.0	5.7
Mean		34.6	194.5	5.2	93.8	394.4	3.9
S.E.M.		9.6	61.1	0.4	17.9	102.1	0.9
Nonsmokers							
K.I.	31	23.0	73.5	3.2	62.5	434.0	6.9
K.R.	41	27.0	96.0	3.6	156.0	794.0	5.1
N.S.	22	15.0	22.0	1.5	69.0	294.0	4.3
P.R.	–	13.0	25.5	2.0	445.0	1,435.0	3.2
T.E.	–	19.0	130.0	6.8	43.0	240.0	5.6
M.P.	25	25.0	108.0	4.3	76.5	449.0	5.4
H.I.	32	48.0	144.0	3.0	133.5	714.0	5.3
Mean		24.3	89.6	3.5	140.8	622.9	5.1
S.E.M.		4.4	18.1	0.7	53.0	155.7	0.4

in the cord blood than in the maternal lymphocytes is not known at this moment, but the similarity of the induction ratios suggests that the induction capacities are regulated in a quantitatively similar manner in the two cases and can thus be used, at least tentatively, for measuring the inherent responsiveness of an individual to PAH. If the inheritance of AHH activity and inducibility in lymphocytes were multigenic, one could predict that there should be no significant relationship between the induction ratios in maternal and cord blood lymphocytes in such a small population. As is shown in Figure 1, there is indeed no correlation here. With respect to the correlation between lymphocytes and AHH activity in placental homogenates, the population is far too small to enable any conclusions to be drawn. Theoretically, one should take into consideration both the actual exposure to cigarette smoke constituents and the genetic background. Thus, being unable to measure the actual exposure, one cannot hope to reach a perfect correlation between AHH inducibility in lymphocytes and AHH activity in placental homogenates *in vitro* (even assuming that both parameters are regulated in a similar fashion). Nevertheless, one should be able to find a relatively strong correlation, as mean placental AHH activity should be higher and the scatter of values wider in the "high lymphocyte inducibility" group than in that characterized by "low lymphocyte inducibility." The situation may be even more complex, however, for if we assume that high inducibility in maternal lymphocytes is associated with efficient overall maternal metabolism, then a smaller amount of cigarette smoke constituents would reach the placenta and fetus and less induction should be found. Resolution of all these possibilities is a matter which must await further research.

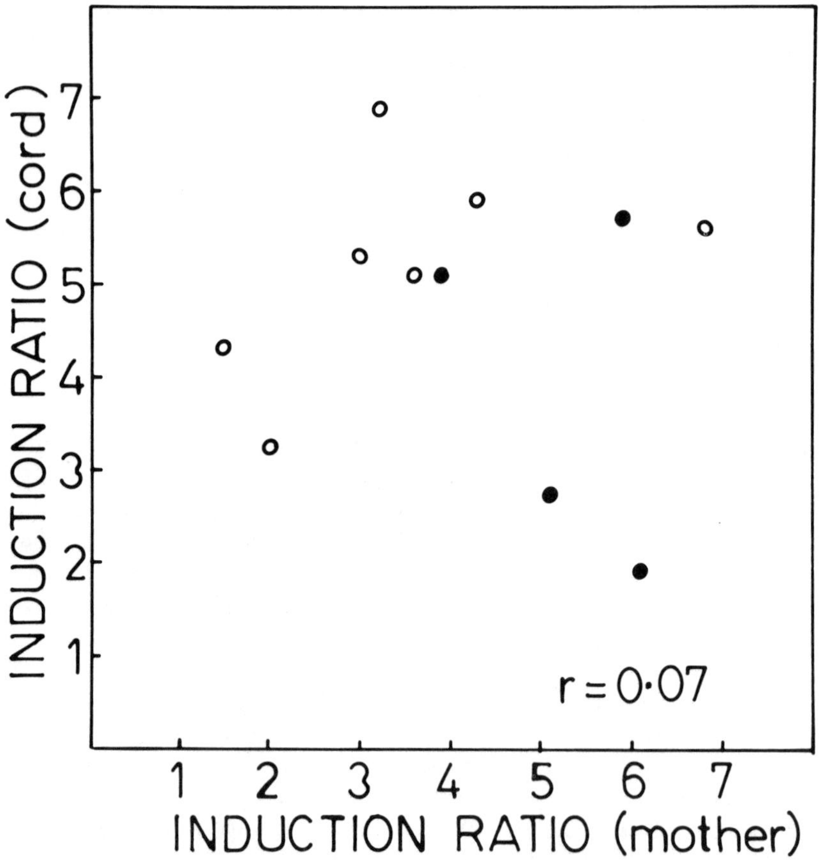

FIGURE 1. Correlation between maternal and cord blood lymphocytes in aryl hydrocarbon hydroxylase inducibility. Closed circles represent smokers, open circles represent nonsmokers.

SUMMARY

The extreme variability of placental AHH activity as a response to maternal smoking raises the question of possible genetic and environmental contributions. The finding that elevated AHH activity in the placenta is associated with a reduction in birth weight points to the possibility that placental AHH activity may be an indicator for the true exposure of the fetus (and placenta) to the effects of maternal smoking. On the other hand, the study of AHH inducibility in placental explant cultures has not succeeded in showing any genetic susceptibility for induction in the placenta, probably because of methodological difficulties. The variation in AHH induction in cord blood lymphocytes is similar to that in maternal lymphocytes, suggesting that it may be possible to use cord blood lymphocytes in a similar manner to lymphocytes from adults in measuring genetically determined interindividual variability in the metabolism of polynuclear aromatic hydrocarbons.

ACKNOWLEDGMENTS

We thank Drs. Daniel W. Nebert and Snorri S. Thorgeirsson for valuable discussions concerning portions of this work, and we appreciate the expert technical and secretarial assistance of Ms. Ritva Saarikoski and Ms. Leena Pyykkö.

REFERENCES

1. Atlas, S. A., Boobis, A. R., Felton, J. S. Thorgiersson, S. S., and Nebert, D. W. (1977): Ontogenetic expression of polycyclic aromatic compound-inducible monooxygenase activities and forms of cytochrome P-450 in the rabbit. Evidence of temporal control and organ specificity of two genetic regulatory systems. *J. Biol. Chem.* 252:4712-4721.
2. Atlas, S. A., Vesell, E. S., and Nebert, D. W. (1976): Genetic control of interindividual variations in the inducibility of aryl hydrocarbon hydroxylase in cultured human lymphocytes. *Cancer Res.* 36:4619-4630.
3. Bergheim, P., Rathgen, G. H., and Netter, K. J. (1973): Interaction of drugs and steroids with human placental microsomes. *Biochem. Pharmacol.* 22:1633-1645.
4. Chakraborty, J., Hopkins, R., and Parke, D. V. (1971): Biological oxygenation of drugs and steroids in the placenta. *Biochem. J.* 125:15P.
5. Harris, C. (1976): Chemical carcinogenesis and experimental models using human tissues. *Beitr. Path.* 158:389-404.
6. Harris, C. C., Autrup, H., Connor, R., Barrett, L. A., McDowell, E. M., and Trump, B. F. (1976): Interindividual variation in binding of benzo[a]-pyrene to DNA in cultured human bronchi. *Science* 194:1067-1069.
7. Idänpään-Heikkilä, J. E., Jouppila, P. I., Puolakka, J. O., and Vorne, M.S. (1971): Placental transfer and fetal metabolism of diazepam in early human pregnancy. *Am. J. Obstet. Gynecol.* 109:1011-1016.

8. Jacobsson, M., Levin, W., Poppers, P. J., Wood, A. W., and Conney, A. H. (1974): Comparison of the O-dealkylation of 7-ethoxycoumarin and the hydroxylation of benzo[a]pyrene in human placenta. *Clin. Pharmacol. Ther.* 16:701-710.
9. Juchau, M. R. (1971): Human placental hydroxylation of 3,4-benzypyrene during early gestation and at term. *Toxicol. Appl. Pharmacol.* 18:665-675.
10. Juchau, M. R. (1972): Mechanisms of drug biotransformation reactions in the placenta. *Fed. Proc.* 31:48-51.
11. Juchau, M. R., and Dyer, D. C. (1972): Pharmacology of the placenta, *Ped. Clin. N. Am.* 19:65-79.
12. Juchau, M. R., Lee, Q. H. Louviaux, G. L., Symms, K. G., Krasner, J., and Yaffe, S. J. (1973): Oxidation and reduction of foreign compounds in the human placenta and fetus. In: *Fetal Pharmacology*, L. Boreus, Ed., pp. 321-333, Raven Press, New York.
13. Juchau, M. R., Namkung, M. J., Berry, D. L., and Zachariah, P. K. (1975): Oxidative biotransformation of 2-acetylaminofluorene in fetal and placental tissues of humans and monkeys: Correlations with aryl hydrocarbon hydroxyase activities. *Drug. Metab. Disp.* 3:494-503.
14. Juchau, M. R., Pedersen, M. G., Fantel, A. G., and Shepard, T. H. (1973): Drug metabolism by placenta. *Clin. Pharmacol. Ther.* 14:673-679.
15. Juchau, M. R., and Symms, K. G. (1972): Aniline hydroxylation in the human placental mechanistic aspects. *Biochem. Pharmacol.* 21:2053-2064.
16. Juchau, M. R., and Zachariah, P. K. (1975): Comparative studies on the oxidation and reduction of drug substrates in human placental versus rat hepatic microsomes. *Biochem. Pharmacol.* 24:227-233.
17. Juchau, M. R., Zachariah, P. K., Colson, J., Symms, K. G., Krasner, J., and Yaffe, S. J. (1974): Studies on human placental carbon monoxide-binding cytochromes. *Drug. Metab. Disp.* 2:79-86.
18. Kamataki, T., Kitada, M., and Kitagawa, H. (1973): Difference in the substrate specificities of drug metabolizing enzymes in human term placenta and fetal liver. *Chem. Pharm. Bull.* 21:2329-2333.
19. Kapitulnik, J., Levin, W., Poppers, P. J., Tomaszwski, J. E., Jerina, D. M., and Conney, A. H. (1976): Comparison of the hydroxylation of zoxazolamine and benzo[a]pyrene in human placenta: Effect of cigarette smoking. *Clin. Pharmacol. Ther.* 20:557-564.
20. Kellermann, G., Luyten-Kellermann, M., and Shaw, C. R. (1973): Metabbolism of polycyclic aromatic hydrocarbons in cultured human leukocytes under genetic control. *Humangenetik* 20:257-263.
21. Kyecombe, D., Franklin, C., and Turner, P. (1973): Drug-metabolizing enzymes in the human placenta, their induction and repression. *Lancet* 1:405-406.
22. Meigs, R. A., and Ryan, K. J. (1968): Cytochrome P-450 and steroid biosynthesis in the human placenta. *Biochim. Biophys. Acta* 165:476-482.
23. Nebert, D. W., Robinson, J. R., Niwa, A. Kimaki, K., and Poland, A. P. (1975): Genetic expression of aryl hydrocarbon hydroxylase activity in the mouse. *J. Cell. Physiol.* 85:393-414.
24. Nebert, D. W., Winker, J., and Gelboin, H. V. (1969): Aryl hydrocarbon hydroxylase activity in human placenta from cigarette smoking and nonsmoking women. *Cancer Res.* 29:1763-1769.

25. Okuda, T., Vesell, E. S., Plotkin, E., Tarone, R., Bast, R. C., and Gelboin, H. V. (1977): Interindividual and intraindividual variations in aryl hydrocarbon hydroxylase in monocytes from monozygotic and dizygotic twins. *Cancer Res.* 37:3904-3911.
26. Paigen, B., Gurtoon, H. L., Minowada, J., Houten, I., Vincenti, R., Paigen, K., Parker, N. B., Ward, E., and Hayner, N. T. (1977): Questionable relation of aryl hydrocarbon hydroxylase to lung cancer risk. *New England J. Med.* 297:346-350.
27. Pelkonen, O. (1977): Differential inhibition of aryl hydrocarbon hydroxylase in human fetal liver, adrenal gland and placenta. *Acta Pharmacol. Toxicol.* 41:306-316.
28. Pelkonen, O., Jouppila, P., and Karki, N. T. (1972): Effect of maternal cigarette smoking in 3,4-benzpyrene and N-methyl-aniline metabolism in human fetal liver and placenta. *Toxicol. Appl. Pharmacol.* 23:399-407.
29. Pelkonen, O., and Karki, N. T. (1973): 3,4-Benzpyrene and aniline are hydroxylated by human fetal liver but not by placenta at 6-7 weeks of fetal age. *Biochem. Pharmacol.* 22:1538-1540.
30. Symms, K. G., and Juchau, M. R. (1973): Stabilization, solubilization and some properties of cytochrome P-450 present in $CaCl_2$-precipitated human placental microsomes. *Life Sci.* 13:1221-1230.
31. Thorgeirsson, S. S. and Nebert, D. W. (1977): The *Ah* locus and the metabolism of chemical carcinogens and other foreign compounds. *Adv. Cancer Res.* 25:149-193.
32. Van Petten, G. R., Hirsch, G. H., and Cherrington, A. D. (1968): Drug-metabolizing activity of the human placenta. *Can. J. Biochem.* 46:1057-1061.
33. Welch, R. M., Harrison, Y. E., Conney, A. H., Poppers, P. J., and Finster, M. (1968): Cigarette smoking stimulatory effect on metabolism of 3,4-benzpyrene by enzymes in human placenta. *Science* 160:541-542.
34. Welch, R. M., Harrison, Y. E., Gommi, B. W., Poppers, P. J., Finster, M., and Conney, A. H. (1969): Stimulatory effect of cigarette smoking on the hydroxylation of 3,4-benzpyrene and the N-demethylation of 3-methyl-4-monomethylaminoazobenzene by enzymes in human placenta. *Clin. Pharm. Ther.* 10:100-109.

Polynuclear Aromatic Hydrocarbons, edited by P. W. Jones and P. Leber. Ann Arbor Science Publishers, Inc. Ann Arbor, MI ©1979.

ARYL HYDROCARBON HYDROXYLASE IN HUMAN CANCER POPULATIONS

Marilyn S. Arnott, Toshio Yamauchi† and Dennis A. Johnston*

Department of Biology
*Department of Biomathematics
The University of Texas System Cancer Center
M. D. Anderson Hospital and Tumor Institute
6723 Bertner
Houston, Texas 77030

†The University of Texas Health Science Center
Medical School at Houston
Texas Medical Center
Houston, Texas 77030

INTRODUCTION

The work of Elizabeth and James Miller in the 1940s on aromatic amines was responsible for a unifying hypothesis concerning structure-activity relationships for the wide variety of chemical structures known to be carcinogenic (20,21). The hypothesis stated first that most carcinogenic chemicals must be metabolized to the active forms and, second, that the reactive species is a strong electrophile. The concept is summarized in Figure 1. The ingested or inhaled chemical undergoes one or a series of reactions leading to a proximate carcinogenic form. In the case of the prototype carcinogenic polycyclic aromatic hydrocarbon (PAH), benzo[a]pyrene (BP), one such metabolite is the 7,8-dihydrodiol.

At various points along the metabolic pathways, or at the point of the proximate carcinogen itself, the compound may be detoxified. Ultimately, conjugation to glutathione, sulfuric acid or glucuronic acid produces non-reactive hydrophilic metabolites which are excreted. On the other hand, the diol can undergo another oxidation to form the 7,8-diol-9,10-epoxide, which can combine with electron-rich portions of cellular macromolecules and lead to carcinogenesis.

In 1969 Grover and Sims (8) and, independently, Gelboin (6) demonstrated that microsomes are required for the binding of PAH to macromolecules. The enzyme system responsible for this binding is known variously as

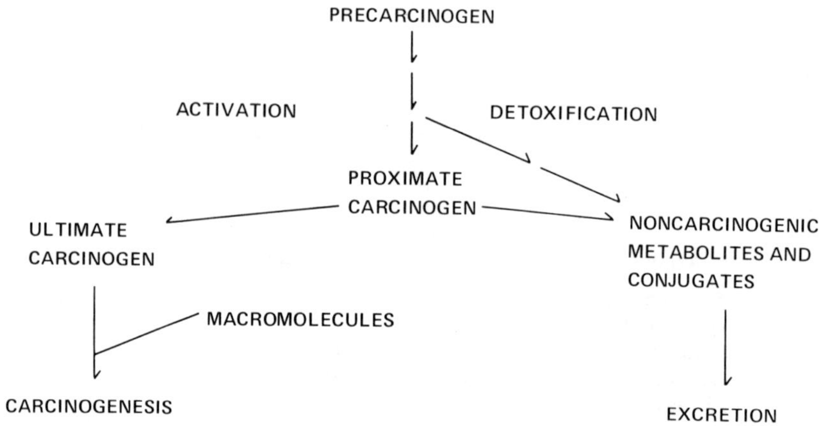

FIGURE 1. Generalized scheme for metabolism of chemical carcinogens.

the mixed-function oxygenases, drug-metabolizing enzymes, benzpyrene hydroxylase and aryl hydrocarbon hydroxylase (AHH).

These multicomponent enzymes are membrane-bound electron transfer systems with broad substrate specificity. One of their hallmarks is that in the presence of substrates, the components and activities are inducible.

The components include a cytochrome named for its wavelength of maximum absorption observed in its carbon monoxide difference spectrum, an enzyme which reduces the cytochrome, and a lipid. Several cytochromes exist (30) which differ in their spectral characteristics, in their substrate specificities and in their response to specific inducers.

Two prototype inducers of the mixed-function oxygenases are phenobarbital, which causes the synthesis of cytochrome P-450 and 3-methylcholanthrene (3-MC), which causes the synthesis of a different cytochrome, denoted P-448 or P_1-450. It is this cytochrome which is a component of AHH.

A parameter commonly used to define the AHH characteristics of a tissue is the inducibility: the ratio of AHH activity in a tissue which has been exposed to a hydrocarbon, such as 3-MC, to the activity in a comparable tissue which has not been exposed.

Genetics of AHH Inducibility and Tumor Susceptibility

Studies on Inbred Mice

In 1972 and 1973, Thomas *et al.* (28,29) and Gielen *et al.* (7) reported the hepatic AHH inducibility of several inbred strains of mice. About half the strains were inducible—or responsive to the hydrocarbon—and the other half were noninducible. Among the commonly used laboratory animals, the C57BL/6 (B6), C57L, C3H/He (C3) and Balb/C animals are inducible, while DBA/2 (D2), DBA/1, AKR and SWR animals are noninducible.

The genetic control of AHH inducibility was then examined. Crossing two noninducible strains gave all noninducible offspring, as expected. Likewise, crosses between various inducible strains gave inducible offspring. The outcome of crosses between inducible and noninducible animals, however, depended on which strains were used. When the inducible B6 was crossed with the noninducible D2, the offspring resembled the B6 parent, suggesting the inducibility is a dominant trait. However, when the inducible C3 was crossed with D2, the offspring had inducibility intermediate between the two parental strains, suggesting inducibility is an additive trait. Further studies on the F_2 generations and backcrosses were consistent with these modes of inheritance.

But the picture is even more complex. Nebert later reported that in the crosses between one substrain of the noninducible AKR and the B6, noninducibility appeared to be dominant. Such studies have led to the suggestion that AHH inducibility is under the control of at least three alleles at each of two nonlinked loci, plus a structural gene for cytochrome P-450 (23,26).

Although the overall control of AHH inducibility is quite complex, one can pick specific strains in which the genetics are fairly straightforward. This has proven to be a powerful tool for investigating the importance of carcinogen metabolism in susceptibility to cancer, as shown by the studies of Richard Kouri (17,19). He measured AHH inducibility in several hundred mice, both inbred strains and hybrids of those strains. The mice were classified as high, intermediate and low inducible animals. Subcutaneous injection of carcinogens such as 3-MC and BP produced fibrosarcomas. The percent of mice which developed tumors divided by the latency (the number of days after injection before a tumor developed) gives a parameter called the carcinogenic index. The correlation between carcinogenic index and AHH inducibility was excellent.

Kouri has improved on these experiments by examining the occurrence of lung lesions after intratracheal instillation of 3-MC. In this type of experiment, one can examine tumor formation and AHH inducibility in the same tissue. In addition, the tumors are carcinomas rather than sarcomas, which is more like the situation in humans. The inducible strain (B6) and the F_1 generation all had a high percentage of tumors compared to the noninducible strain (D2) and to the inducible strain which had received no 3-MC. The backcross to the inducible parent gave animals with high AHH inducibility and a high percentage of tumors; the backcross to the noninducible parent gave both inducible and noninducible animals. Those animals which were inducible for AHH had three times the incidence of lung lesions observed in the animals which were noninducible.

Under these experimental conditions, then, AHH-inducible animals exhibit increased susceptibility to 3-MC- and BP-induced fibrosarcomas and lung lesions.

AHH Inducibility in Humans

Such experimental animal models suggest that AHH inducibility could be an important determinant of susceptibility to hydrocarbon-induced tumors in humans. The possibility of testing this hypothesis presented itself in 1972 when a method was developed (3,33) for measuring AHH activity in cultured human lymphocytes. The original method involved dextran sedimentation of whole blood to obtain the leukocytes. A more purified population of lymphocytes is now generally obtained from centrifugation of the blood over Hypaque-Ficoll. The lymphocytes are stimulated with mitogens to transform the cells to lymphoblasts, rich in endoplasmic reticulum. Half the cultures from an individual are induced with 3-MC 24 hr before harvest, while the other half receive solvent alone. The cells are harvested and assayed for AHH activity by measuring the fluorescence of the alkali-soluble metabolites of BP (22).

Using this system, Kellermann *et al.* (13) surveyed the inducibility in a population of 353 healthy donors. He reported that the inducibilities fell into three groups: low, intermediate and high. The trimodal distribution suggested that two alleles at a single locus were responsible for the genetic control of AHH inducibility. Family studies were consistent with this concept (14,15).

Kellermann *et al.* (16) then measured the AHH inducibility on a group of 50 lung cancer patients along with 85 normal controls and 46 patients with other types of cancer. Of the normal and cancer controls, about 45% were in the low inducible group, 45% intermediate and 10% high. The lung cancer group, however, was skewed to the higher inducibilities: 30% high, 66% intermediate and only 4% low. The interpretation of these results was that the higher inducibility in cultured lymphocytes from lung cancer patients reflected a higher level of AHH in target cells of the respiratory tract. Persons with higher AHH levels were thus hypothesized to be at greater risk of developing lung cancer when exposed to hydrocarbon-rich cigarette smoke.

The studies of Kellermann *et al.* can be summarized as follows:

1. Variation in AHH inducibility exists in humans.
2. The variation in inducibility is under genetic control.
3. The trimodal distribution of inducibilities suggests a single gene control.
4. AHH inducibility is higher in lung cancer patients than in the normal population.

Because of the importance of these findings, we attempted to confirm and extend these studies. A computerized system was developed for information retrieval pertaining to the normal donors and cancer patients. A number of variables were followed, including age, sex, smoking history and family history of cancer. For the cancer patients, the type and stage of cancer were included, as well as any type of therapy they might have received. Computerization also facilitated the use of nonparametric statistics and analyses of variance and covariance. Over 700 normal donors and 600 cancer patients, including approximately 250 lung cancer patients, have been tested.

Because changes in the culture and assay methods caused some significant changes in the enzyme activities and inducibilities, the data are considered in three separate populations. Among the normal donors in all three populations, no differences in AHH inducibility were observed between male and female subjects, or between smokers and nonsmokers. The age of the donor, however, had a significant effect. Older donors exhibited lower enzyme activities and inducibilities. This is probably a reflection of reduced mitogen response in lymphocytes obtained from older individuals (2,10,24,27). Because of this age effect, comparisons between various cancer groups and normals include only individuals over the age of forty.

The distribution of AHH inducibilities in the normal controls from Population II is shown in Figure 2. The unimodal distribution is representative of every normal population we have examined. No evidence for a trimodal distribution, as reported by Kellermann, has been observed.

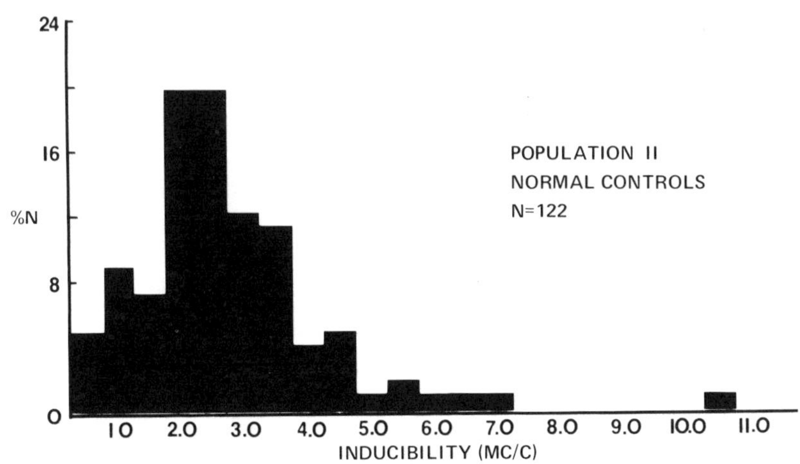

FIGURE 2. Distribution of AHH inducibilities observed in cultured lymphocytes obtained from a population of 122 healthy subjects over 40 years of age. %N is percent of total subjects. Inducibility is the AHH activity in methylcholanthrene- (MC) induced cells divided by the AHH activity in noninduced, or control (C), cells.

A unimodal distribution could either reflect the involvement of multiple gene loci in the control of AHH inducibility, or could simply result from the variability in enzyme measurements. Repeated tests on normal donors indicate that 15% is a conservative estimate of the variation observed in measurements on lymphocytes from a single donor, cultured on different days. Using the frequencies and mean inducibilities of the three groups reported by Kellermann, and a 15% variation about each mean, computer simulations of trimodal distributions are indistinguishable from a single log-normal distribution for populations of fewer than 200 individuals. Thus, with the present limitations in the lymphocyte system, it is unlikely that a single locus effect could be observed without testing hundreds of individuals.

Recent support for genetic control of the enzyme system in humans comes from the report of Atlas *et al.* (1), who observed a high heritability index for AHH inducibility in twin studies. However, recalling the complexity of genetic control of AHH inducibility observed in inbred strains of mice, it seems likely that the enzyme is under polygenic control in the human population.

Since chi-square and Kolmogorov-Smirnov tests of normality showed that the data were not normally distributed, nonparametric statistical methods were used to determine the significance of observed differences. In comparing the cancer populations to the normals, we relied on the Mann-Whitney test for comparing variables between two groups, and the Kruskal-Wallis one-way analysis of variance for comparing variables between two groups. Differences were considered significant when $p < 0.05$ was obtained.

In the first population, AHH was measured in cultured leukocytes obtained from dextran sedimentation of whole blood. The results are summarized in Table I for 80 healthy donors, 25 lung cancer patients and 78 patients with other types of cancer. The noninduced activity was the same for all

TABLE I. Population I

		AHH units/10^6 cells		Inducibility
Diagnosis	N	C	MC	MC/C
Normal Controls	80	0.19 ± 0.01[a]	0.42 ± 0.03	2.40 ± 0.10
Cancer Controls	78	0.21 ± 0.02	0.47 ± 0.03	2.42 ± 0.10
Lung Cancer	25	0.20 ± 0.02	0.56 ± 0.07	2.70 ± 0.22

Note: AHH units are defined as fluorescence equivalent to 1 pmole 3-hydroxybenzo[a]-pyrene per minute.

[a]Mean ± standard error.

three groups, but the induced activity for the lung cancer patients was higher than for the normal controls (p = .07). This resulted in a slightly higher inducibility for the lung cancer patients (p = 0.13). The cancer control group was indistinguishable from the normal controls.

For all subsequent studies, lymphocytes were isolated by centrifugation on Ficoll-Hypaque. The first study done in this manner included 122 normal donors, 100 cancer controls, 54 lung cancer patients and 13 oropharyngeal cancer patients. As shown in Table II, the noninduced activity was significantly lower for all the cancer patient groups than for the normal controls. No significant difference was observed in the induced activity between any of the cancer patient groups and the normal population. However, the inducibility for the lung cancer group was significantly higher than for the normal controls. Furthermore, the higher inducibility observed for the oropharyngeal cancer group, compared to the normals, was close to significance (p = 0.056).

After reevaluation of the pH optimum of the lymphocyte enzyme, the pH of the assay was changed from 7.4 to 8.5. This modification resulted in higher activities for all groups in Population III.

TABLE II. Population II

		AHH units/10^6 cells		Inducibility
Diagnosis	N	C	MC	MC/C
Normal Controls	122	0.36 ±0.03[a]	0.87 ±0.05	2.90 ±0.13
Cancer Controls	100	0.31[b] ±0.03	0.88 ±0.08	3.38[b] ±0.17
Lung Cancer	54	0.24[b] ±0.03	0.81 ±0.08	4.06[b] ±0.30
Oropharyngeal Cancer	13	0.20[b] ±0.03	0.69 ±0.16	3.90 ±0.34

Note: AHH units are defined as fluorescence equivalent to 1 pmole 3-hydroxybenzo[a]-pyrene per minute.
[a]Mean ± standard error.
[b]Significantly different from the normal population ($p < 0.05$).

Using this method, additional samples from oropharyngeal cancer patients were obtained. As shown in Table III, the inducibility for this group was significantly higher than for the normal controls. AHH inducibilities of lung cancer patients who had undergone chemo- or radiotherapy were also examined. The most striking difference between these patients and the normal

TABLE III. Population III

Diagnosis	N	AHH units/10^6 cells C	AHH units/10^6 cells MC	Inducibility MC/C
Normal Controls	71	0.49 ± 0.05[a]	1.55 ± 0.20	3.27 ± 0.17
Cancer Controls	41	0.49 ± 0.06	1.58 ± 0.22	3.53 ± 0.29
Lung Cancer				
+ Therapy	25	0.38 ± 0.06	0.72 ± 0.14	2.12 ± 0.29
Oropharyngeal Cancer	12	0.36 ± 0.06	1.46 ± 0.24	4.42[b] ± 0.36

Note: AHH units are defined as fluorescence equivalent to 1 pmole 3-hydroxybenzo[a]-pyrene per minute.
[a]Mean ± standard error.
[b]Significantly different from the normal population ($p < 0.05$).

controls was a marked reduction in the growth of the lymphocytes. Using morphological criteria, the normal controls exhibited 45% blastogenesis (100 x number of lymphoblasts harvested ÷ number of lymphocytes cultured), while the lung cancer patients undergoing therapy exhibited only 23% blastogenesis. This was the only group which showed a significantly decreased capacity for lymphocyte growth. Although the activities and inducibilities were not significantly different from the normal population, they tended to be lower. The myelosuppressive action of virtually all chemo- and radiotherapeutic agents precludes the use of these patients for any study involving cultured lymphocytes.

The major difficulty in this work (1,9,18) is our inability to control or even properly measure variations in the degree of mitogen response. Although we count cells, measure protein content and attempt to normalize the enzyme activities accordingly, we have no way of knowing how many of those cells have been maximally stimulated with mitogen and are synthesizing AHH. The reproducibility of the method is good on any given day, but comparison between days is less satisfying. However, if a group of donors is measured on one day, and the same group a week or month later, the individuals generally maintain their same relative order. In other words, a high inducer generally remains a high inducer.

With this in mind, we tried a different experimental design in which we measured the AHH inducibility from lung cancer patients and their spouses, as approximate age- and environment-matched controls. The results of this study are shown in Table IV. Once again, the patients exhibited lower noninduced activity, the induced levels were the same, and the inducibility was significantly higher for the patients than for the spouses.

Of some concern is the fact that the higher inducibility in lung and oropharyngeal cancer patients generally was the result of a lower noninduced

TABLE IV. Population IV (Lung Cancer Patient-Spouse)

Subject	N	AHH units/10^6 cells C	AHH units/10^6 cells MC	Inducibility MC/C
Patient	24	$0.06^a \pm 0.01^b$	0.40 ± 0.07	$6.39^c \pm 0.81$
Spouse	24	0.08 ± 0.01	0.42 ± 0.07	4.94 ± 0.68

[a]Significantly different from spouse ($p = 0.011$), paired t test.
[b]Mean ± standard error.
[c]Significantly different from spouse ($p = 0.007$), paired t test.

activity, rather than a higher induced activity. Two possible explanations come to mind.

Several studies have shown that the hepatic enzyme present in male rats induced with hydrocarbons is a different enzyme from that seen in non-induced animals (12,25,30,34). Both enzymes metabolize BP, but by different pathways. The noninduced enzyme produces mainly the 4,5-epoxide, while the hydrocarbon-induced enzyme forms relatively more epoxides at the 7,8 and 9,10 positions. Thus, the noninduced enzyme may be regarded as a detoxifying enzyme, whereas the induced enzyme metabolizes BP to an ultimate carcinogen, the 7,8-diol-9,10-epoxide. The AHH inducibility, then, can be viewed as a ratio of activating to detoxifying enzymes, and represents the very crucial balance between activating and detoxifying pathways.

If this holds true for the lymphocytes, then whether high AHH inducibility results from a high level of activating enzyme or a low level of detoxifying enzyme is unimportant. In either case, the balance of metabolism has been shifted toward activation of the carcinogen. It should be noted, however, that studies on rat lung and kidney (34) indicate that the detoxifying enzyme may not be present in extrahepatic tissues. Hence, this interpretation may not be valid for lymphocytes.

Perhaps a more appropriate explanation comes from the many studies on mitogen response of lymphocytes from cancer patients. Using the conventional method of ^{3}H-thymidine incorporation, several workers have reported impairment of mitogen response in lymphocytes from lung cancer patients in particular (4,11) and from cancer patients in general (5,31). Other investigators, however, observed no difference in ^{3}H-thymidine incorporation between lymphocytes from recently diagnosed, untreated lung cancer patients and age-matched controls (2,32). The latter study, by Whitcomb and Parker, is particularly relevant to the AHH question. Under culture conditions in which ^{3}H-thymidine incorporation was the same for lung cancer patients and controls, these investigators demonstrated decreased protein synthesis in the cultured lymphocytes from the lung cancer patients.

If this is borne out in our studies, then the noninduced enzyme level may be serving as a measure of the ability of the lymphocytes to synthesize protein. The inducibility in the lymphocytes from cancer patients could then be viewed as a measure of induced activity, corrected for a decreased capacity of the cancer patients' lymphocytes to synthesize protein. The higher inducibility in lymphocytes then, may reflect a higher induced enzyme level in the target tissue. Studies are in progress to test this possibility.

CONCLUSION

The studies reported here lead to the following summary of AHH inducibility in humans:

1. Individuals differ in their AHH inducibility.
2. The inducibility differences observed between people are under genetic control.
3. The unimodal distribution of AHH inducibility in the normal population may stem in part from day-to-day variability in the lymphocyte system, but also suggests polygenic control of the enzyme system.
4. Lung and oropharyngeal cancer patients exhibit higher AHH inducibility than the normal population.

To make a statement about the role of AHH in susceptibility to certain cancers on the basis of these data alone would be unwise. Obviously, a cause and effect relationship cannot be inferred from these experiments. The most direct way of establishing that point would be to measure AHH inducibility on a population at risk for developing lung cancer before the appearance of any signs of the disease. But, until appropriate measures of mitogen response are developed, so that results can be compared with better precision between laboratories and between days, this type of study is not feasible. For the time being, then, the human lymphocyte AHH results must be viewed in the context of what has been shown very clearly in experimental animal systems: that AHH inducibility is one important determinant of susceptibility to hydrocarbon-induced tumors. The data presented here are consistent with this hypothesis.

ACKNOWLEDGMENTS

The authors sincerely appreciate the generous support of Dr. Charles R. Shaw; the expert technical work of Ms. Betty Bruce, Ms. Kathy Slaughter, Mr. Richard Branum, Mr. Robert Daimler and Mr. Timothy Jenkins; the data management provided by Ms. Peggy Wright, Ms. Sheryl Tatar and Ms. Yvonne Freedman; and the manuscript preparation by Ms. Diane Hicks. This work was supported by Grant GM-15597 from the National Institute of General Medical Sciences, National Cancer Institute Conract NO1 CP 55604, and grants from The Council for Tobacco Research-U.S.A., Inc.

REFERENCES

1. Atlas, S. A., Vesell, E. S., and Nebert, D. W. (1976): Genetic control of interindividual variation in the inducibility of aryl hydrocarbon hydroxylase in cultured human lymphocytes. *Cancer Res.* 36:4619-4630.
2. Barnes, E. W., Farmer, A., Penhale, W. J., Irvine, W. J., Roscoe, P., and Horne, N. W. (1975): Phytohemagglutinin-induced lymphocyte transformation in newly presenting patients with primary carcinoma of the lung. *Cancer* 36:187-193.
3. Busbee, D. L., Shaw, C. R., and Cantrell, E. T. (1972): Aryl hydrocarbon hydroxylase induction in human leukocytes. *Science* 178: 315-316.
4. Ducos, J., Migueres, J., Colombies, P., Kessous, A., and Poujoulet, N. (1970): Lymphocyte response to P.H.A. in patients with lung cancer. *Lancet* i:1111-1112.
5. Garrioch, D. B., Good, R. A., and Gatti, R. A. (1970): Lymphocyte response to P.H.A. in patients with non-lymphoid tumors. *Lancet* i:618.
6. Gelboin, H. V. (1969): Microsome-dependent binding of benzo[a]-pyrene to DNA. *Cancer Res.* 29:1271-1276.
7. Gielen, J. E., Goujon, F. M., and Nebert, D. W. (1972): Genetic regulation of aryl hydrocarbon hydroxylase induction. II. Simple Mendelian expression in mouse tissues *in vivo. J. Biol. Chem.* 247:1125-1137.
8. Grover, P. L., and Sims, P. (1969): Enzyme-catalyzed reactions of polycyclic hydrocarbons with deoxyribonucleic acid and proteins *in vitro. Biochem. J.* 110:159-160.
9. Gurtoo, H. L., Minowada, J., Paigen, B., Parker, N. B., and Hayner, N. T. (1977): Factors influencing the measurement and reproducibility of aryl hydrocarbon hydroxylase activity in cultured human lymphocytes. *J. Nat. Cancer Inst.* 59:787-798.
10. Hagen, C., and Froland, A. (1972): Lymphocyte transformation and myocardial infarction. *Brit. Med. J.* 1:445.
11. Han, T., and Takita, H. (1972): Immunologic impairment in bronochogenic carcinoma: A study of lymphocyte response to phytohemagglutinin. *Cancer* 30:616-620.
12. Holder, G., Yagi, H., Dansette, P., Jerina, D. M., Levin, W., Lu, A. Y. H., and Conney, A. H. (1972): Effects of inducers and epoxide hydrase on the metabolism of benzo(a)pyrene by liver microsomes and a reconstituted system: Analysis by high pressure liquid chromatography. *Proc. Nat. Acad. Sci., U.S.* 71:4356-4360.
13. Kellermann, G., Cantrell, E., and Shaw, C. R. (1973): Variation in extent of aryl hydrocarbon hydroxylase induction in cultured human lymphocytes. *Cancer Res.* 33:1654-1656.

14. Kellermann, G., Luyten-Kellermann, M., and Shaw, C. R. (1973): Genetic variation of aryl hydrocarbon hydroxylase in human lymphocytes. *Am. J. Human Genet.* 25:327-331.
15. Kellermann, G., Luyten-Kellermann, M., and Shaw, C. R. (1973): Metabolism of polycyclic aromatic hydrocarbons in cultured human leukocytes under genetic control. *Humangenetik* 20:257-263.
16. Kellerman, G., Shaw, C. R., and Luyten-Kellermann, M. (1973): Aryl hydrocarbon hydroxylase inducibility and bronchogenic carcinoma. *New England J. Med.* 289:934-937.
17. Kouri, R. E. (1976): Relationship between levels of aryl hydrocarbon hydroxylase activity and susceptibility to 3-methylcholanthrene and benzo[a]pyrene-induced cancers in inbred strains of mice. In: *Carcinogenesis, Vol. 1,* R. I. Freudenthal and P. W. Jones, Eds., Raven Press, New York.
18. Kouri, R. E., Imblum, R. L., and Prough, R. A. (1978): Measurement of aryl hydrocarbon hydroxylase and NADH-dependent cytochrome C reductase activities in mitogen-activated human lymphocytes. In: *Proceedings of the Third International Symposium on the Detection and Prevention of Cancer,* H. Nieburgs, Ed., Marcel-Dekker, New York.
19. Kouri, R. E., and Nebert, D. W. (1978): Genetic regulation of susceptibility to polycyclic hydrocarbon-induced tumors in the mouse. In: *Cold Spring Harbor Conference on Cell Proliferation,* Vol. 4, The Origin of Human Cancer, pp. 811-835.
20. Miller, E. C., and Miller, J. A. (1947): The presence and significance of bound aminoazo dyes in the livers of rats fed *p*-dimethylaminoazobenzene. *Cancer Res.* 32:1073-1081.
21. Miller, J. A. (1970): Carcinogenesis by chemicals: An overview. G. H. A. Clowes Memorial Lecture. *Cancer Res.* 30:559-576.
22. Nebert, D. W., and Gelboin, H. V. (1968): Substrate-inducible microsomal aryl hydrocarbon hydroxylase in mammalian cell culture. I. Assay and properties of induced enzyme. *J. Biol. Chem.* 243:6242-6249.
23. Niwa, A., Kumaki, K. and Nebert, D. W. (1975): Genetic expression of aryl hydrocarbon hydroxylase activity in the mouse. *Arch. Biochem. Biophys.* 166:559-564.
24. Pisciotta, A. V., Westring, D. W., Deprey, C. and Walsh, B. (1967): Mitogenic Effect of Phytohaemagglutinin at different ages. *Nature* 215:193-194.
25. Rasmussen, R. E., and Wang, I. Y. (1974): Dependence of specific metabolism of benzo[a]pyrene on the inducer of hydroxylase activity. *Cancer Res.* 34:2290-2295.
26. Robinson, J. R., Considine, N., and Nebert, D. W. (1974): Genetic expression of aryl hydrocarbon hydroxylase induction. Evidence for the involvement of other genetic loci. *J. Biol. Chem.* 249:5851-5859.
27. Sample, W. F., Gertner, H. R., and Chretien, P. B. (1971): Inhibition of phytohemagglutinin-induced *in vitro* lymphocyte transformation by serum from patients with carcinoma. *J. Nat. Cancer Inst.* 46:1291-1297.

28. Thomas, P. E., and Hutton, J. J. (1973): Genetics of aryl hydrocarbon hydroxylase induction in mice: Additive inheritance in crosses between C3H/HeJ and DBA/2J. *Biochem. Genet.* 8:249-267.
29. Thomas, P. E., Kouri, R. E., and Hutton, J. J. (1972): The genetics of aryl hydrocarbon hydroxylase induction in mice: A single gene difference between C57BL/6J and DBA/2J. *Biochem. Genet.* 6:157-168.
30. Thomas, P. E., Lu, A. Y. H., Ryan, D., West, S. B., Kawalek, J., and Levin, W. (1975): Multiple forms of rat liver cytochrome P-450. Immunochemical evidence with antibody against cytochrome P-448. *J. Biol. Chem.* 251:1385-1391.
31. Trubowitz, S., Masek, B., and Del Rosario, A. (1966): Lymphocyte response to phytohemagglutinin in Hodgkin's disease, lymphatic leukemia and lymphosarcoma. *Cancer* 19:2019-2023.
32. Whitcomb, M. E., and Parker, R. L. (1977): Abnormal lymphocyte protein synthesis in bronchogenic carcinoma. *Cancer* 40:3014-3018.
33. Whitlock, J. P., Cooper, H. L., and Gelboin, H. V. (1972): Aryl hydrocarbon (benzopyrene) hydroxylase is stimulated in human lymphocytes by mitogens and benz[a]anthracene. *Science* 177:618-619.
34. Wiebel, F. J., Leutz, J. C., Diamond, L., and Gelboin, H. V. (1971): Aryl hydrocarbon (benzo[a]pyrene) hydroxylase in microsomes from rat tissues: differential inhibition and stimulation by benzoflavones and organic solvents. *Arch. Bioch. Biophys.* 144:78-86.
35. Wislocki, P. G., Wood, A. W., Chang, R. L., Levin, W., Yagi, H., Hernandez, O., Jerina, D. M., and Conney, A. H. (1976): High mutagenicity and toxicity of a diol epoxide derived from benzo[a]pyrene. *Biochem. Biophys. Res. Comm.* 68:1006-1012.

Polynuclear Aromatic Hydrocarbons, edited by P. W. Jones and P. Leber. Ann Arbor Science Publishers, Inc. Ann Arbor, MI ©1979.

CELLULAR REPAIR OF DNA DAMAGE INDUCED BY 7,12-DIMETHYLBENZ[a] ANTHRACENE AND ITS FLUORO ANALOGS *IN VITRO*

Steven M. D'Ambrosio, F. Bernard Daniel and Ronald W. Hart

Departments of Pharmacology and Radiology
College of Medicine
The Ohio State University
Columbus, Ohio 43210

Polycyclic aromatic hydrocarbons (PAH) like 7,12-dimethylbenz[a]-anthracene (DMBA) in the presence of a microsomal activation system damage cellular DNA (9,10). Unless removed or correctly repaired the resulting base alterations can lead to cell death, mutation and transformation. For example, DMBA-induced damage can result in morphological transformation *in vitro* and *in vivo* (7,21) and mutagenesis *in vitro* (13,14,17). Replacement of the hydrogen on the two or five carbon of DMBA with a fluorine or a methyl group dramatically reduces both the binding (6), and the carcinogenic (11,22,26) and mutagenic potential (12) of these hydrocarbons. However, the 4-methyl and 11-fluoro analogs of DMBA are strong carcinogens (11).

DMBA and all its fluoro analogs to some extent damage cellular DNA (6). Although 2-fluoro- and 5-fluoro-DMBA produce 96% and 64% less damage respectfully than does their parent compound DMBA (6), these fluoro analogs are at least 100 to 1,000 times less carcinogenic than DMBA (11, 22,26). These hydrocarbons, like many other chemical carcinogens, produce many different types of adducts with DNA, RNA and protein (10). As has been observed with damage produced by methylating agents (16,19,20) and N-acetoxy-acetylaminofluorene (AAAF) (13,18,19), some of the resulting adducts are rapidly repaired while others persist. Although it may be thought that the lesions entering replication are probably mutagenic, it has not been substantiated that all lesions entering replication are mutagenic or carcinogenic (8,10). It is not known whether persistent lesions, those not recognized by DNA repair excision enzymes, or the potentially excisable ones which are not excised before entering DNA replication are the ones responsible for the biological consequences of the PAH.

In this report we analyze:

1. the DNA repair capacity of mammalian cells to repair DNA damage induced by structurally similar but biologically different PAH;
2. The postreplication repair capacity of Syrian hamster embryo (SHE) cells for DNA damage induced by DMBA and 11-fluoro-DMBA and their less carcinogenic analogs 2-fluoro-DMBA and 5-fluoro-DMBA; and
3. the excision repair [bromodeoxyuridine (BrdUrd) photolysis] capacity of V79 cells for DNA damage induced using a SHE cell activation layer.

We find that:

1. the postreplication repair capacity parallels both the binding and the carcinogenic and mutagenic potential of the PAH;
2. V79 cells repair little of the damage induced by these compounds; and
3. there is little difference in prereplication repair of damage induced by DMBA or its fluoro analogs.

MATERIAL AND METHODS

Postreplication Repair

Cultures of SHE were obtained by disassociation of 12-day-old embryos. Cultures were maintained in modified minimal essential medium (MEM) (GIBCO) supplemented with 10% fetal calf seum (FCS) (Flow), 100 units penicillin (GIBCO), 100 μg/ml streptomycin (GIBCO), and 2.5 units/ml Fugazone (GIBCO). Cultures were maintained at 37°C and frozen at passage 2 in liquid nitrogen. Thawed cells were grown to confluency at which time they were trypsinized and seeded to a density of 300,000 in 60-cm^2 Corning tissue dishes. Six hours later, after cells had attached to the plates, they were starved prior to radiolabeling by incubation in MEM containing 3% calf serum. Twelve to eighteen hours later, parental DNA was labeled by incubating starved cells in MEM containing 10% FCS and 0.1 μCi/ml [^{14}C] thymidine (New England Nuclear, spec. act. 50 mCi/mmole). Cells were relabeled 24 hr later, treated with PAH and postreplication repair assayed as schematically shown in Figure 1 and described below and elsewhere (4, 5,18).

Cells prelabeled with [^{14}C] thymidine were treated in MEM containing 10% FCS with thin-layer chromatographically pure DMBA or one of its fluorinated analogs. The PAH were dissolved in dimethylsulfoxide (DMSO) and added to the culture medium such that the final concentration of DMSO was less than 0.1%. Cells were incubated with the compounds for 6 hr, since it has been shown that at least 80% of total maximum DNA binding occurs with all compounds within this time period (6).

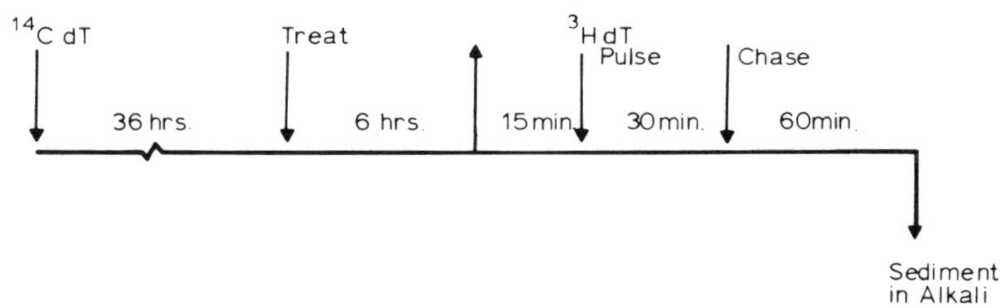

FIGURE 1. Experimental protocol for assaying postreplication repair in SHE cells. Cells were incubated with [^{14}C] thymidine (dT) for 36 hr to label parental DNA and the medium changed. DMBA or fluoro-DMBA dissolved in DMSO was added to the plates and incubated at 37°C for 6 hr. The medium was changed and after 15 min cells were pulse-labeled with [^{3}H] thymidine for 30 min and chased in radioactive medium for 1 hr.

Following treatment of cells with either DMBA or fluoro-DMBA, the cells were washed with MEM and 15 min later were pulse-labeled with 20 μCi/ml of [^{3}H] thymidine (Amersham Searle Corp., spec. act. 50 Ci/mmole) for 30 min to label newly synthesizing (daughter) DNA. After pulse-labeling the medium was replaced with a nonradioactive one containing 4 μg/ml thymidine and the pulse-labeled DNA was chased for 1 hr. The cells were washed with a NaCl-ethylenediaminetetraacetic acid (EDTA) solution (25) and exposed to 2,000 R of X-ray to disassociate aggregates formed during centrifugation. Cells were scraped from the dishes with a rubber policeman and resuspended in 0.5 ml and 50 μl containing 30,000 cells and 3,000 cpm of each isotope were lysed in a 0.2 ml solution of 1 *N* NaOH, 10 m*M* EDTA layered on top of a 5-20% alkaline sucrose gradient. The DNA was sedimented in a Beckman 50.1 rotor by centrifugation at 30,000 rpm for 120 min in either a Sorval OTD-65 or Beckman L50 ultracentrifuge. Fractions were collected onto paper strips and the radioactivity of each fraction determined as previously described.

Prereplication Repair

Chinese hamster V79 cells were grown in Dulbecco's modified Earle's salts medium supplemented with 10% FCS, 100 units penicillin, 100 μg/ml streptomycin and 2.5 units/ml Fugazone. Cultures were maintained and labeled with 0.1 μCi/ml [^{14}C] thymidine (spec. act. 50 mCi/mmole) or [^{3}H] thymidine (spec. act. 7 Ci/mmole) as described above. After labeling cells were trypsinized and 200,000 cells, containing about 0.5 cpm/cell, were seeded on top of 300,000 SHE cells attached to 60-cm^2 Corning tissue dishes. Thus the PAH was activated by the SHE cells and the repair of the resultant DNA damage studied in V79 cells.

The protocol for studying excision repair in V79 cells is schematically shown in Figure 2. Details of the BrdUrd photolysis assay have been described elsewhere (23,24). Cells were incubated with a final concentration of 2 m*M* hydroxyurea to inhibit semiconservative replication 30 min before treatment. Cells labeled with [^{3}H] thymidine were incubated with 1 μg/ml DMBA or fluoro-DMBA dissolved in DMSO, and BrdUrd at a final concentration of 10^{-4} *M*. Cells labeled with [^{14}C] thymidine were incubated with DMSO and thymidine as a control. After 20 hr the cells were washed twice with the NaCl-EDTA solution and exposed to 2,000 R X-ray. The cells were scraped from the plate and the ^{3}H and ^{14}C-labeled cells mixed together. The mixture in 700 μl NaCl-EDTA was exposed to light from two mercury lamps passed through a 12.4 nm Bandpass filter (1) allowing the selective passage of 313-nm radiation. Exposure of cells to 313-nm light photolyses the BrdUrd incorporated into DNA during repair replication (23,24). After irradiation the cells were lysed, the DNA was centrifuged and the molecular weights determined as described above and elsewhere.

RESULTS

Postreplication Repair

The size of newly synthesized (daughter) DNA in SHE cells treated with DMBA or 5-fluoro-DMBA was determined by sedimentation in alkaline sucrose. The gradient profiles of DNA from cells treated for 6 hr with either DMBA or 5-fluoro-DMBA are shown in Figure 3. These profiles show that the pulse-chased DNA from cells treated with DMBA is smaller in size than the DNA from cells treated with 5-fluoro-DMBA at the same concentration (5 μg/ml) or not treated. Correspondingly, the gradient profile of daughter DNA from cells treated with 5-fluoro-DMBA more closely resembles that of

DNA from untreated cells. The size of parental DNA was not appreciably affected by treatment of cells with either DMBA or 5-fluoro-DMBA.

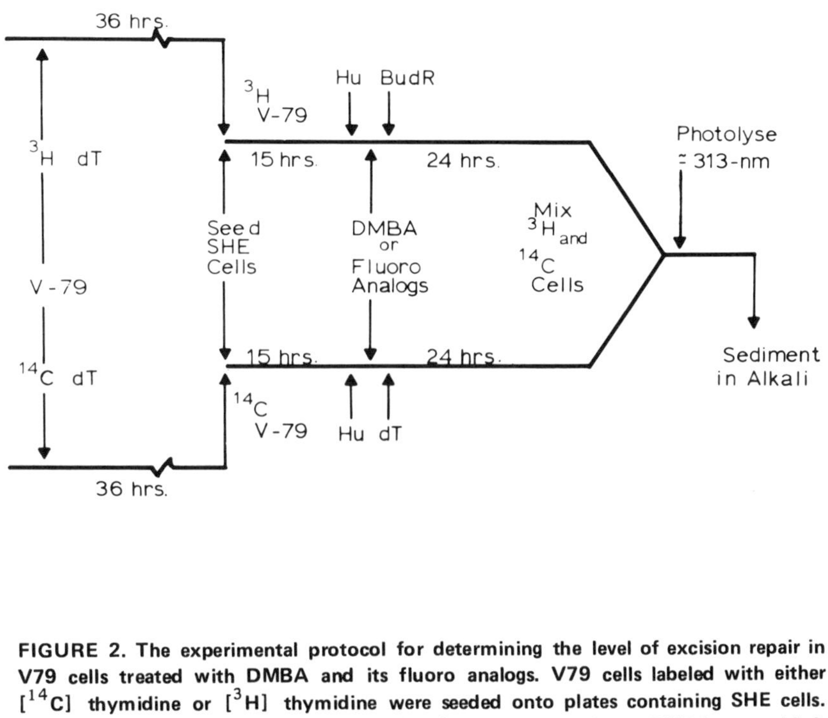

FIGURE 2. The experimental protocol for determining the level of excision repair in V79 cells treated with DMBA and its fluoro analogs. V79 cells labeled with either [^{14}C] thymidine or [^{3}H] thymidine were seeded onto plates containing SHE cells. After the cells had attached, hydroxyurea (HU) and DMBA or fluoro-DMBA were added. Bromodeoxyuridine (BudR) was added to the plates containing ^{3}H-labeled cells while thymidine was added to the plates containing ^{14}C-labeled cells. The ^{3}H- and ^{14}C-labeled cells were mixed, exposed to 313-nm light and sedimented in alkaline sucrose.

The amount of postreplication repair is estimated by measuring the percentage of pulse-labeled DNA sedimenting as high-molecular-weight DNA, *i.e.*, the percentage of ^{3}H-labeled DNA sedimenting under ^{14}C-labeled DNA peak (the fractions from the bottom of the gradient that include 40% of the

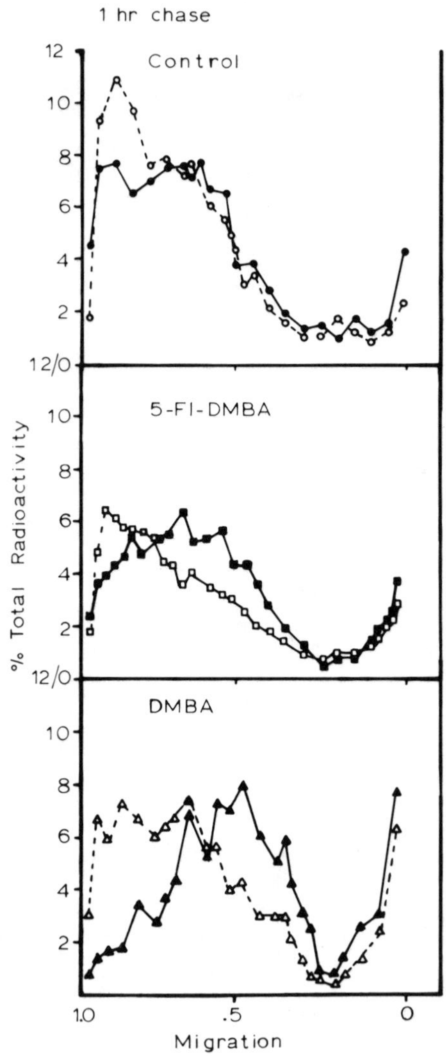

FIGURE 3. Alkaline sucrose gradient profiles (sedimentation to the left) of DNA from SHE cells treated with 5-fluoro-DMBA or DMBA and control cells. Cells were treated with 5 μg/ml compound, pulse-labeled and chased (closed symbols) as described in Figure 1. Parental DNA (open symbols) was labeled with [^{14}C] thymidine.

^{14}C-label) (4,5). When we use this method of analysis, we find that 15% and 13% of the pulse-chased DNA cosedimented as high-molecular-weight DNA following DMBA and 11-fluoro-DMBA treatments, respectively, while 28% and 26% cosedimented after 2-fluoro-DMBA and 5-fluoro-DMBA treatments, respectively, and 28% after no treatment (Table I).

TABLE I. Effect of DMBA and Fluoro-DMBA on Pulse-Chased DNA in SHE Cells

PAH[a]	Concentration (μg/ml)	% Pulse-Labeled DNA Sedimenting as High-Molecular-Weight[b]
DMBA	5	15.0
	1	14.0
2-fluoro-DMBA	1	27.5
5-fluoro-DMBA	5	26.0
	1	25.0
11-fluoro-DMBA	1	12.5
None	–	28.0

[a]Cells were treated with PAH for 6 hr, pulse-labeled for 0.5 hr and chased for 1 hr.
[b]High-molecular-weight DNA was taken as the cumulative fractions from the bottom of the gradient that included 40% of the parental ^{14}C-radioactivity.

The differences observed in pulse-chased DNA sedimenting as high-molecular-weight DNA following DMBA and selected fluoro-DMBA treatments could be due to different levels of these various PAH bound to DNA. Results obtained in this laboratory and published in this volume (6) indicate that there are differences in the amount of PAH bound to SHE DNA following DMBA and fluoro-DMBA treatments. For example, cells treated with 5-fluoro-DMBA bind 64% less PAH than cells treated with DMBA; cells treated with 11-fluoro-DMBA bind 2.4 times more PAH than cells treated with DMBA. The differences observed in the pulse-chased DNA could be due to: (1) the differences observed in the amount of PAH bound following treatment with DMBA and its fluoro analogs; (2) the removal of damage induced by the PAH; and/or (3) the types of adducts entering DNA replication. In order to determine whether the differences observed in the pulse-chased DNA were due to the levels of PAH bound, we treated cells with 5-fluoro-DMBA and DMBA at 1 and 5 μg/ml. The results showed that increasing the concentration of 5-fluoro-DMBA did not decrease the amount of pulse-chased DNA sedimenting as high-molecular-weight DNA, as would be expected if the differences were due only to levels of PAH bound to DNA. Further, there is very little difference between the pulse-chased DNA from cells treated with 11-fluoro-DMBA and DMBA, although the level of PAH bound after 11-fluoro-DMBA treatment has been reported to be 2.4 times greater than DMBA treatment (6).

Excision Repair

The ability of cells to remove damage in DNA resulting from PAH treatment by excision repair was indirectly investigated using the bromodeoxyuridine photolysis technique (24). In this technique the regions of DNA excised after PAH treatment are marked with BrdUrd incorporated during the treatment and repair period (23). At the end of this period cells are exposed to 313-nm light and the regions of DNA containing BrdUrd are photolysed. Sedimentaion in alkaline sucrose reveals the average molecular weight of the DNA. Thus the number of breaks and, therefore, the number of repair sites can be calculated. Table II shows the results of BrdUrd photolyses of V79 cells treated with 1 μg/ml DMBA, 5-fluoro-DMBA and 11-fluoro-DMBA. As is shown in Table II there is apparently little difference in the values of $\Delta(1/M_w)$ [$\Delta(1/M_w) = 1/M_{w\,\mathrm{BrdUrd}} - 1/M_{w\,\mathrm{thymidine}}$] between cells treated with DMBA and its fluoro analogs. Often the DNA of cells treated with chemicals like PAH (6) and AAAF (5) containing aromatic rings is photosensitive to 313-nm light. The $1/M_{w\,\mathrm{thymidine}}$ of DNA from treated cells is smaller than that of control cells. We found a slight reduction in the $1/M_{w\,\mathrm{thymidine}}$ of cells treated with DMBA, 5-fluoro-DMBA and 11-fluoro-DMBA. This reduction is reflected in the values of $\Delta 1/M_w$ shown in Table II. However, the reduction in $1/M_{w\,\mathrm{thymidine}}$ following treatment with 2-fluoro-DMBA was too large to accurately calculate $\Delta 1/M_w$.

TABLE II. Prereplication Repair Capacity in V-79 Cells[a]

Compound[b]	$\Delta(1/M_w)$[c]
DMBA	0.31×10^8
5-fluoro-DMBA	0.41×10^8
11-fluoro-DMBA	0.36×10^8
Control	0.16×10^8

[a]Repair was measured using BrdUrd photolysis (See Methods).
[b]Cells were treated with 1 μg/ml PAH and incubated with BrdUrd for 24 hr.
[c]$\Delta(1/M_W)$ $(1/M_{W\,\mathrm{BrdUrd}} - 1/m_{W\,\mathrm{thymidine}})$

DISCUSSION

Most chemicals that damage cellular DNA are mutagenic and/or carcinogenic (9,10). Chemicals like PAH usually produce a multitude of different types of DNA damage and it is not known whether one or more types of such damage are responsible for the biological consequences of chemical carcinogens. The mutagenic and carcinogenic endpoints resulting from DNA damage probably depend on the ability of the agent to cause base mispairing (8) and/or the ability of the cell to insert the correct base during DNA replication upon a damaged template (9). There are wide differences in the responses of animal strains, tissues and cells to chemical carcinogens as well as in the biological effectiveness of structurally similar compounds. The former probably depends upon the repair potential (9), while the latter upon the type of metabolite and damage made (10).

It has been shown (11,12,15,22,26) that replacement of a hydrogen atom on DMBA with a fluorine or a methyl group greatly modifies the mutagenic and carcinogenic potential of this PAH. The level of binding of the structurally similar analogs of DMBA to DNA and the amount of pulse-chased DNA sedimenting as high-molecular-weight DNA parallels their biological potential (Table III). Even though the mutagenic and carcinogenic potential of 2-fluoro-DMBA is approximately 0.001 that of DMBA, the reduction in the level binding is only 0.04. The comparatively smaller differences in the total amount of PAH bound to DNA following 2-fluoro-DMBA, 5-fluoro-DMBA and DMBA may account for the comparatively larger differences in the biological effects observed between these PAH. However, the differences observed in the pulse-chased DNA cannot be totally accounted for by the levels of binding since: (1) a fivefold increase in treatment concentration of 5-fluoro-DMBA does not significantly change the amount of pulse-chased DNA sedimenting as high-molecular-weight DNA; and (2) there is little difference between the pulse-chased DNA from cells treated with 11-fluoro-DMBA and DMBA although 11-fluoro-DMBA binds 2.4 times more PAH than DMBA (6). Also, it appears that only a small percentage of the damage made by DMBA and its fluoro analogs is removed by excision repair. The lack of significant differences between the excision repair of DMBA and its fluoro analogs suggests that most of the damage made by these PAH enters DNA replication. Therefore, consistent with similar studies on other chemical carcinogens (2,14,17,20), it appears that the differences observed with DMBA and its fluoro analogs are not simply due to the total number of adducts produced but to the levels of specific types of adducts entering DNA replication. The nature of these adducts is now under investigation.

TABLE III. Comparison of DNA Repair to Biological Potential of PAH

Compound	Carcinogenic Potential	Mutagenic Potential[12]	Repair Prerep[a]	Repair Postrep[b]	Binding[6]
DMBA	+[11,22,26]	+	0.31	0.45	1.00
2-fluoro-DMBA	-[11,26]	-	-	0.98	0.04
5-fluoro-DMBA	weak[22,26]	weak	0.41	0.89	0.36
11-fluoro-DMBA	+[11,22,26]	+	0.36	0.47	2.40

[a]See Table II.
[b]See Table I.

ACKNOWLEDGMENTS

We thank Professor M. S. Newman for the DMBA monofluoro analogs. We wish to thank R. E. Gibson, C. C. Kung and R. Clark-Crane for their excellent technical assistance. These studies were supported by Grants No. R-805337 from the Environmental Protection Agency, No. R01-CA-21371 from the National Cancer Institute and No. 221182 from the Ohio State University.

REFERENCES

1. Ahmed, F. E., Robb, R. C., and Hart, R. W. (1977): New arrangement for high-intensity 313-nm light. *Rev. Sci. Instrum.* 48:1442-1444.
2. Amacher, D. E., and Lieberman, M. W. (1977): Removal of acetylaminofluorene from the DNA of control and repair-deficient human fibroblasts. *Biochem. Biophys. Res. Comm.* 74:285-290.
3. Brookes, P., and Lawley, P. D. (1964): Evidence for the binding of polynuclear hydrocarbons to the nucleic acids of mouse skin. Relation between carcinogenic power of hydrocarbons and their binding to deoxyribonucleic acid. *Nature* 202:781-784.
4. D'Ambrosio, S. M., and Setlow, R. B. (1976): Enhancement of postreplication repair in Chinese hamster cells. *Proc. Nat. Acad. Sci., U.S.* 73:2396-2400.
5. D'Ambrosio, S. M., and Setlow, R. B. (1978): Defective and enhanced postreplication repair in classical and variant xeroderma pigmentosum cells treated with N-acetoxy-acetylaminofluorene. *Cancer Res.* 38:1147-1153.
6. Daniel, F. B., Wong, L., Oravec, C., Cazer, F. D., Wang, A., D'Ambrosio, S. M., Hart, R. W., and Witiak, D. T. (1978): Biochemical studies on the metabolism and DNA-binding of DMBA and some of its monofluoro derivatives of varying carcinogenicity. Chapter 56, this volume.
7. DiPaolo, J. A., Nelson, R. L., and Donovan, P. J. (1971): Morphological oncogenesis and karyological characteristics of Syrian hamster embryo cells transformed *in vitro* by carcinogenic polycyclic hydrocarbons. *Cancer Res.* 31:1118-1127.

8. Drake, J. W., and Blatz, R. H. (1976): The biochemistry of mutagenesis. *Ann. Rev. Biochem.* 45:11-37.
9. Hart, R. W., D'Ambrosio, S. M., Ng, K., and Modak, S. P. (1979): Longevity, stability and DNA repair. *Mech. Aging Dev.* 9:203-223.
10. Hart, R. W., Hall, K. Y., and Daniel, F. B. (1978): DNA repair and mutagenesis in mammalian cells. *Photochem. Photobiol.* 28:131-155.
11. Harvey, R. G., and Dunne, F. B. (1978): Multiple regions of metabolic activation of carcinogenic hydrocarbons. *Nature* 273:566-568.
12. Huberman, E. (1978): Unpublished data, Oak Ridge National Laboratory.
13. Huberman, E., and Sachs, L. (1974): Mutability of different genetic loci in mammalian cells by metabolically activated carcinogenic polycyclic hydrocarbons. *Proc. Nat. Acad. Sci., U.S.* 73:188-192.
14. Huberman, E., and Sachs, L. (1976): Cell-mediated mutagenesis of mammalian cells with chemical carcinogen. *Int. J. Cancer.* 13:326-333.
15. Huggins, C. B. (1978): Communicated through M. S. Newman (The Ohio State University), University of Chicago.
16. Kleihues, P., and Margison, G. P. (1974): Carcinogenicity of N-methyl-N-nitrosourea: Possible role of excision repair of O-Methylguanine from DNA. *J. Nat. Cancer Inst.* 53:1839-1841.
17. Krahn, D. F., and Heidelberger, C. (1977): Liver homogenate-mediated mutagenesis in Chinese hamster V79 cells by polycyclic aromatic hydrocarbons and aflatoxins. *Mut. Res.* 46:27-44.
18. Lehmann, A. R. (1974): Postreplication repair in mammalian cells. *Life Sciences.* 15:2006-2016.
19. Magee, R. N., NiColl, J. W., Pegg, A. E., and Swann, P. F. (1975): Alkylating intermediates in nitrosamine metabolism. *Biochem. Soc. Trans.* 3:62-65.
20. Marginson, G. P., and Kleihues, P. (1975): Preferential accumulation of O-Methylguanine in rat brain deoxyribonucleic acid during repetitive administration of N-Methyl-N-Nitrosourea. *Biochem. J.* 148:521-525.
21. Marquardt, H., Sodergren, J. E., Sims, P., and Grover, P. L. (1974): Malignant transformation *in vitro* of mouse fibroblasts by 7,12-dimethylbenz[a]anthracene and by their K-region derivatives. *Int. J. Cancer.* 13:304-310.
22. Miller, J. A., and Miller, E. C. (1978): Personal Communication, McArdle Laboratory, University of Wisconsin, Madison.
23. Regan, J. D., and Setlow, R. B. (1974): Two forms of repair in the DNA of human cells damaged by chemical carcinogens and mutagens. *Cancer Res.* 34:3318-3325.
24. Regan, J. D., Setlow, R. B., and Ley, R. D. (1971): Normal and defective repair of damaged DNA in human cells: A sensitive assay utilizing the photolysis of bromodeoxyuridine. *Proc. Nat. Acad. Sci., U.S.* 64:708-712.
25. Setlow, R. B., Regan, J. D., German, J., and Carrier, W. L. (1969): Evidence that xeroderma pigmentosum cells do not perform the first step in the repair of ultraviolet damage to their DNA. *Proc. Nat. Acad. Sci., U.S.* 64:1035-1041.
26. Slaga, T. J., Gleason, G., DiGiovanni, J., Berry, D. L., Juchau, M. R., Yu, P. P., Sukumaron, K. B., and Harvey, R. G. (1978): Tumor initiating activities of various derivatives of benz[a]anthracene and 7,12-dimethylbenz[a]anthracene in mouse skin. Chapter 49, this volume.

Polynuclear Aromatic Hydrocarbons, edited by P. W. Jones and P. Leber. Ann Arbor Science Publishers, Inc. Ann Arbor, MI ©1979.

INTERACTION OF (±)BENZO[a]PYRENE-7β,8α-DIOL 9α,10α-EPOXIDE WITH FRACTIONATED EUKARYOTIC DNA

Radhakrishnan Iyer,[1] Larry L. Triplett,* Thomas J. Slaga* and John Papaconstantinou*

The University of Tennessee–
Oak Ridge Graduate School of Biomedical Sciences

*Biology Division
Oak Ridge National Laboratory
Oak Ridge, Tennessee 37830

INTRODUCTION

Polycyclic aromatic hydrocarbons (PAH) exert their carcinogenic and mutagenic effects only after they are metabolized to chemically reactive species. These reactive metabolites can undergo covalent interactions with cellular RNA, DNA and protein (5,13), as well as with synthetic copolymers of nucleic acids and proteins. It has been hypothesized that these covalent interactions, particularly with DNA, may be an essential initial step in the process of chemical carcinogenesis (14,26).

Studies of the binding of metabolites of benzo[a]pyrene to DNA have provided evidence that benzo[a]pyrene-7,8-diol-9,10-epoxide (BPDE) is responsible for most of the bound adducts (4,10,29). Recently, the (±)benzo-[a]pyrene-7β,8α-diol-9α,10α-epoxide was synthesized (12,34) and shown to be the metabolite responsible for the binding of benzo[a]pyrene to nucleic acids *in vivo* (16,29) and that this interaction is mainly with the guanine residues (90%) and less with adenine and cytosine residues (2-5%) (25). In addition, this metabolite has been shown to be a strong tumor initiator in mouse skin (30), a potent carcinogen when injected into newborn mice (19, 20) and a potent mutagen (15,27).

[1]Postdoctoral investigator supported by Subcontract 3322 from the Biology Division of Oak Ridge National Laboratory to the University of Tennessee.

The significance of the interaction of benzo[a]pyrene-diol-epoxide with guanine in the mechanism of chemical carcinogenesis is not clear. In view of the apparent importance of this carcinogen-DNA interaction, we were interested in determining several parameters of this reaction that might contribute to an understanding of the mechanism of chemical carcinogensis. Since the carcinogen is known to react with the guanine residues in DNA, one of our major aims was to determine if there were any "hot spots" in eukaryotic DNA where the diol epoxide showed a preferential binding.

It is well known that the eukaryotic genome is quite complex and consists of localized regions of (G+C)-rich and (A+T)-rich DNA sequences. Within the genome of a specific species, for example, it is possible to find both (A+T)-rich and (G+C)-rich satellite DNA (8,9,11,23,32,33) as well as structural genes (2,3,6,7,21,31). With the development of molecular cloning techniques, a wealth of information on the structure and sequence of specific structural genes is now emerging. These specific DNA sequences will enable detailed studies of the role of (G+C) content as well as sequence specificity in the interaction of carcinogens with DNA. Our ability to carry out the present studies on the interaction of carcinogens with various fractions of mouse liver and calf thymus DNA was facilitated by the recent development of techniques in our laboratory for the enrichment of (G+C)-rich and (A+T)-rich DNA sequences (Papaconstantinou, DuBrul and Triplett, in preparation). We present here a preliminary report on the interaction of benzo[a]pyrene-diol-epoxide with fractionated mouse liver and calf thymus DNA.

MATERIALS AND METHODS

Chemicals

[^{14}C] 7β,8α-dihydroxy,9α,10α-epoxy-7,8,9,10- tetrahydrobenzo[a]pyrene (anti-BPDE) was kindly supplied to us by the Carcinogenesis Program of the National Cancer Institute. All other chemicals were reagent-grade.

Preparation of DNA

DNA was isolated from mouse liver nuclei (Sencar and BALB/c) and calf thymus nuclei according to the method of Marmur (24). The DNA was dialyzed into 0.1 *M* Na_2SO_4 and frozen. Spectral values of the DNA preparations were: $A_{260}/A_{280} \geqslant 2.0$ and $A_{260}/A_{230} \geqslant 1.9$. DNA preparations had a molecular weight of 10-14 x 10^6 daltons as determined by agarose gel electrophoresis.

Preparative Cs_2SO_4 Density Gradient Centrifugation

DNA, in 0.1 *M* Na_2SO_4, was fractionated by isopycnic density-gradient centrifugation in Ag^+-Cs_2SO_4 at pH 7 and various metal ion concentrations (17,28,33). Cacodylate buffer (0.1 *M*) was used to maintain a pH of 7. Silver perchlorate (1×10^{-3} *M*) and DNA were reacted at the desired r_f values, where r_f is the molar ratio of Ag^+ to DNA phosphorus, and were calculated on the premise that 1 mg of DNA has an A_{260} value of 20 and that DNA phosphorus is 10% of the DNA concentration. Saturated Cs_2SO_4 (EM Laboratories Inc., Elmsford, NY) was diluted with the components of the gradient plus 0.1 *M* Na_2SO_4 to give a final density of 1.485-1.50 g/cm^3, as determined by the refractive index. The samples were overlaid with mineral oil and centrifuged at 35,000 rpm in a Spinco 50 Ti angle rotor for 66 hr at 25°C.

After centrifugation, the mineral oil was removed; the gradient solution was pumped through a 0.3-cm ISCO flow cell with Fluorinert FC-48 (ISCO) or saturated Cs_2SO_4 and collected in 0.15-ml fractions. The fractions were divided into six pools in the case of mouse DNA and nine pools in the case of calf thymus DNA (see Figure 1A and B). The pooled fractions were dialyzed extensively against 5 *M* NaCl to remove the silver and then against water.

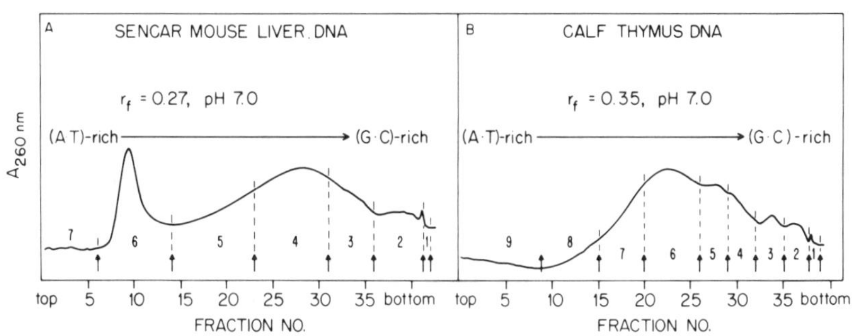

FIGURE 1. A profile of (A) Sencar mouse DNA and (B) calf thymus DNA formed by Ag^+-Cs_2SO_4 isopycnic density-gradient centrifugation. The fractions were pooled and dialyzed to remove Ag^+ and Cs_2SO_4. (See Materials and Methods.)

Reaction of Anti-BPDE with Mouse and Calf DNA Fractions

A solution approximately 20 μg of the DNA fractions in 1.5 ml water was treated with [^{14}C]anti-BPDE (29.4 mCi/mmol, 0.36 μl) at a molar concentration of BPDE:dNTP (1:100). Aliquots (0.2 ml) of the reaction mixture were withdrawn at 1, 5, 10, 15, 20, 30 and 40 min and precipitated with 2.5 ml EtOH and 0.5 ml KCl (4%). The precipitated DNA was stored in a freezer at -20°C for 4 hr, after which it was pelleted and extensively washed with ethanol until no significant radioactivity could be detected in the supernatant. The pellet was taken up in 0.8 ml of 1 x SSC (1 *M* NaCl, 1 *M* Na-citrate), and the radioactivity was counted by liquid scintillation spectroscopy.

Sephadex LH-20 Chromatography

The DNA-BPDE adduct was digested enzymatically to the nucleo*tide* level, according to the procedure of Baird and Brooks (1) and chromatographed on an LH-20 column, 2 x 30 cm, eluted with a 30 to 100% water-methanol gradient (120 ml). One hundred fractions (1.2 ml each) were collected and counted.

RESULTS

Fractionation of Mouse and Calf DNA by Ag^+-Cs_2SO_4 Isopycnic Density Gradient Centrifugation

The successful enrichment and purification of eukaryotic DNA which codes for ribosomal RNA (21), 5S RNA (6), $tRNA_1^{Met}$ (7), silk fibroin mRNA (31) and histone mRNA (2,3), as well as a variety of satellite DNA (8,9,11,23,32,33), was facilitated by the differential interaction of heavy ions (Ag^+ and Hg^{++}) with DNA and the centifugation of these metal-DNA complexes in Cs_2SO_4 isopycnic density-gradient centrifugation. It has been shown in earlier studies that the interaction of Ag^+ with (G+C)-rich and (A+T)-rich sequences is dependent on pH (17,28). We have used this pH-dependent reaction to fractionate mouse liver and calf thymus DNA on Cs_2SO_4 gradients (Papaconstantinou, DuBrul and Triplett, in preparation). The patterns in Figure 1A and B show that DNA from mouse liver and calf thymus can be spread throughout the entire gradient at pH 7 and $r_f = 0.27$ (mouse), and pH 7 and $r_f = 0.35$ (calf), where r_f represents the molar ratio of Ag^+:DNA-P. Analysis of the various fractions from these gradients by isopycnic CsCl density-gradient centrifugation and thermal denaturation (T_m) have shown that under these experimental conditions a gradient of (A+T)-rich to (G+C)-rich sequences is set up from the top (or low-density region) to the bottom (or high-density region) of the gradient. (The details of the fractionation of eukaryotic DNA by this procedure will be published elsewhere.) In view of the availability of a variety of (G+C)-rich and (A+T)-rich DNA sequences, studies were initiated to determine the characteristics of the interaction of anti-BPDE with these fractions. Our major purpose in these experiments was to determine if there are regions within a single genome which show a differential interaction with anti-BPDE.

Interaction of Anti-BPDE with Unfractionated Mouse Liver and Calf Thymus DNA

The rates of reaction of anti-BPDE with unfractionated Sencar mouse liver DNA and calf thymus DNA are shown in Figure 2. The conditions of the reaction are such that the dNTP of the DNA are in excess, *i.e.*, the reactants are present in a molar ratio of 1 anti-BPDE:100 dNTP. In both cases there is a rapid initial rate of binding of anti-BPDE to the DNA for approximately 10 min, after which the reaction slows down. It should be noted that the initial rate of reaction and the final amount of anti-BPDE bound to calf thymus DNA is higher than with Sencar mouse liver DNA. Although there is

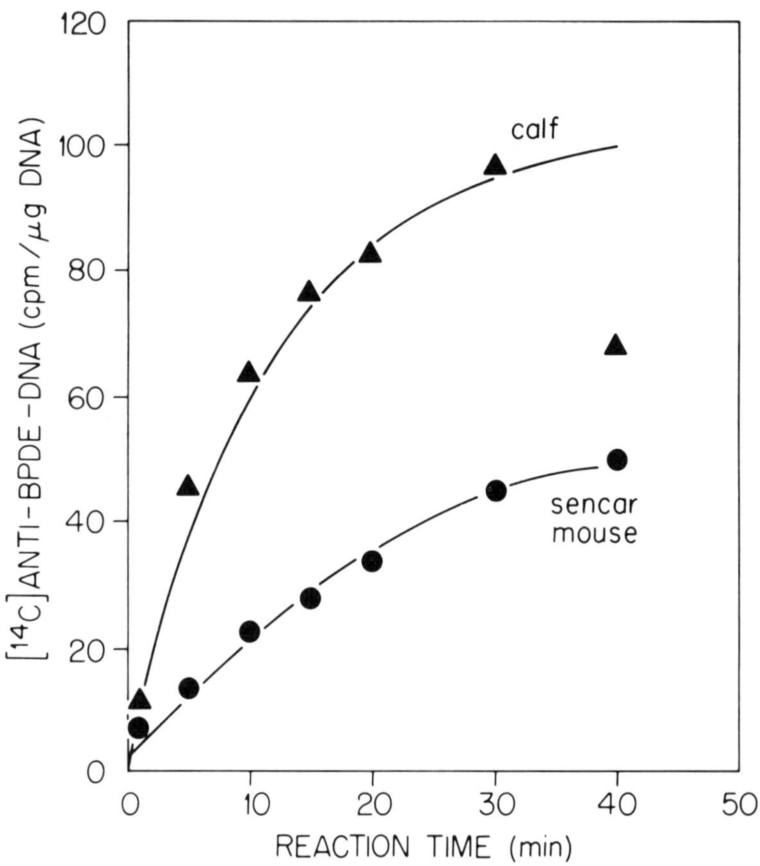

FIGURE 2. The rate of reaction of Sencar mouse DNA and calf thymus DNA with [^{14}C] anti-BPDE.

no significant difference in the total (G+C) content of calf thymus DNA (39% G+C) and mouse liver DNA (40% G+C) that might explain this difference, it has been shown that the calf thymus genome contains several (G+C)-rich satellites that are not found in the mouse genome (11,23,32). Thus, in addition to (G+C) content, the significant difference in the reaction of anti-BPDE with calf and mouse DNA may also be due to these satellites as well as specific sequences within the satellites.

Interaction of Anti-BPDE with Ag^+-Cs_2SO_4 Gradient Fractions of Mouse Liver and Calf Thymus DNA

In view of the differences in the initial rates of reaction of mouse liver and calf thymus DNA, the (G+C)-rich and (A+T)-rich fractions from Ag^+-Cs_2SO_4 gradients were used to answer the following questions: (1) Is there a difference in the initial rate of reaction among the various Ag^+-C_2SO_4 gradient fractions? (2) Is there a difference in the amount of anti-BPDE that can be bound by these fractions? (3) Is there a relationship between the (G+C) content of the fraction and the rate of reaction? (4) Is there a relationship between the (G+C) content of the fractions and the amount of anti-BPDE bound?

Sencar Mouse Liver DNA

The data in Figure 3A and B show the rates of reaction of anti-BPDE with Sencar mouse DNA fractions from Ag^+-Cs_2SO_4 gradients (Figure 1). The (G+C)-rich fractions 2 and 3 show the highest rate of reaction and also show that more anti-BPDE is bound to the nucleotide at the end of the reaction. Fractions 5, 6 and 7 represent the regions of the Ag^+-Cs_2SO_4 gradient where the (A+T)-rich fractions sediment. The rate of reaction of these fractions with anti-BPDE is less, and the amount of anti-BPDE bound at the end of the reaction is approximately half that bound to the (G+C)-rich fractions. Our calculations indicate that approximately 1 anti-BPDE is bound per 750 dNTP in (G+C)-rich fraction 2, whereas 1 anti-BPDE is bound per 1,500 dNTP in the (A+T)-rich fractions 5, 6 and 7. The (G+C) contents of fractions 2 and 3 range from 41% to 50%, whereas the (G+C) contents of fractions 5, 6 and 7 range from 32% to 42%. We conclude that the (G+C)-rich fractions of Sencar DNA react more rapidly with anti-BPDE and bind more of the carcinogen at the end of the reaction. However, the reaction with (A+T)-rich fractions ranging from the 32% (G+C) (fraction 6) to 40% (G+C) (fraction 4) do not show significant differences. Furthermore, since dGMP, which is the major reactant with anti-BPDE, is not rate-limiting, the difference in reaction and binding by these fractions is not directly related to (G+C) content alone. We postulate that the nucleotide sequences also play a role in the extent of reaction and binding of anti-BPDE to DNA.

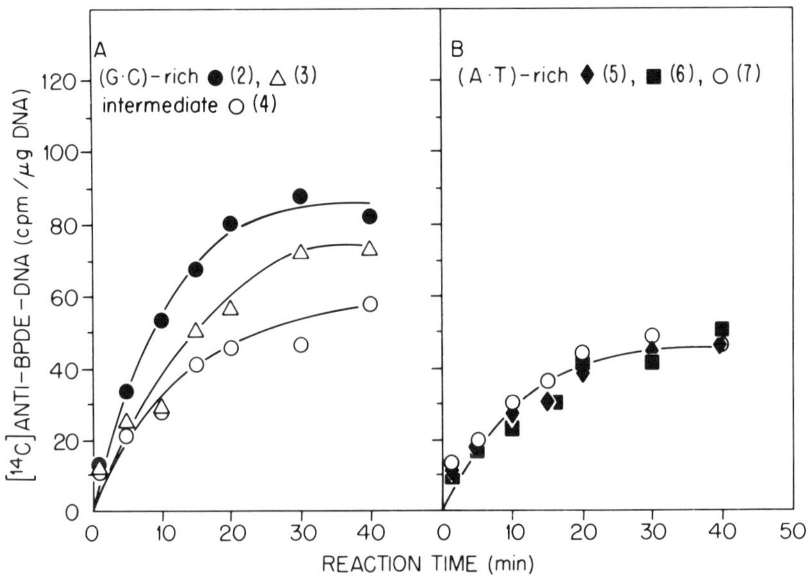

FIGURE 3. The rate of reaction of Sencar mouse DNA fractions from Ag^{+}-Cs_2SO_4 gradients with [^{14}C] anti-BPDE. (A) Fraction 2 (●—●) and fraction 3 (△—△) are (G+C)-rich; fraction 4 (○—○) is intermediate in (G+C) content. (B) Fraction 5 (X—X), fraction 6 (■—■) and fraction (○—○) are (A+T)-rich.

BALB/c Mouse Liver DNA

A similar series of reactions was carried out with Ag^{+}-Cs_2SO_4 gradient fractions of BALB/c mouse liver DNA. These data (Figure 4) show that the initial rate of reaction is greatest with the (G+C)-rich fractions 2 and 3, as was seen with the Sencar DNA. Fractions 5 and 6, which represent the (A+T)-rich fractions as well as the (A+T)-rich satellite DNA, show a similar rapid initial rate of reaction but bind only half the amount of anti-BPDE at the end of the reaction. Thus, both BALB/c and Sencar mouse liver DNA show significant differences between the (G+C)-rich and (A+T)-rich fractions of the genome.

A note should be made here with respect to the differences in the reaction of fraction 4 in Sencar versus BALB/c mouse liver DNA. This fraction shows a twofold greater ability to bind anti-BPDE in Sencar DNA; the Sencar mouse shows a significantly increased sensitivity to carcinogens in comparison to BALB/c. Although there does not appear to be significant differences in the (G+C) content of fraction 4 from one experiment to the other, this reaction may be pointing out subtle differences in the DNA sequences of Sencar vs BALB/c DNA that could not be seen in the Ag^{+}-Cs_2SO_4 or CsCl gradients. In fact, a high degree of heterogeneity of this fraction is indicated

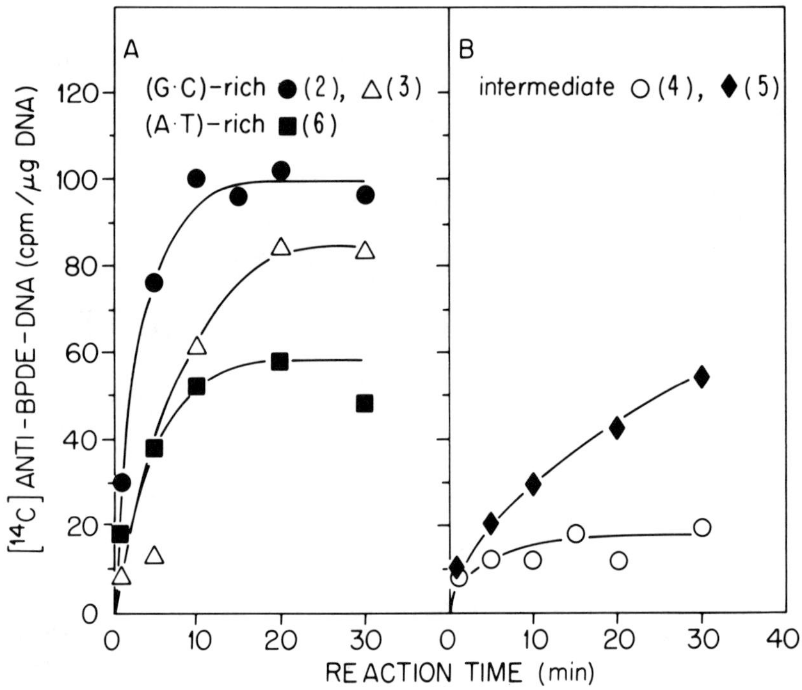

FIGURE 4. The rate of reaction of BALB/c mouse DNA fractions from Ag^+-Cs_2SO_4 gradients with [^{14}C] anti-BPDE. (A) Fractions 2 (●—●) and 3 (△—△) are (G+C)-rich; fraction 6 (■—■) is (A+T)-rich. (B) Fraction 4 (○—○) is intermediate, and fraction 5 (◆—◆) is (A+T)-rich.

by the broadness of the peak (ρ = 1.702 g/cm^3) in a CsCl gradient (Papaconstantinou, DuBrul and Triplett, in preparation). The significance of this difference in binding must await further studies involving fractionation and sequence analysis.

Calf Thymus DNA

Calf thymus DNA was chosen for this comparative study because of the presence of (G+C)-rich satellites in the genome and because these (G+C)-rich satellites are easily isolated by Ag^+-Cs_2SO_4 gradients (11,23,32). In addition, the calf genome does not contain (A+T)-rich satellites typical of the mouse genome, which further facilitates comparison between naturally occurring (G+C)-rich and (A+T)-rich sequences.

Ag^+-Cs_2SO_4 fractions (Figure 1B) of calf thymus DNA were reacted with anti-BPDE under the same conditions used with the mouse DNA fractions.

The data in Figure 5 show that the initial rate of reaction is greatest with the (G+C)-rich fractions 2, 3 and 4, which represent the (G+C)-rich satellites as well as the ribosomal DNA sequences (Papaconstantinou, DuBrul and Triplett, in preparation). Furthermore, fractions 5, 6, 7 and 8, which represent the major fraction of the genome, show a significantly lower rate of reaction and also show that the amount of anti-BPDE bound at the end of the reaction is 1.5- to 2-fold greater with the (G+C)-rich sequences. We conclude from these studies that there is a preferential reaction of anti-BPDE with (G+C)-rich sequences of both calf thymus and mouse liver DNA. In addition, even though there is a significant difference in the (G+C) content of fractions 5, 6, 7 and 8, which make up the bulk of the genomic DNA, these fractions react similarly and bind the same amount of anti-BPDE.

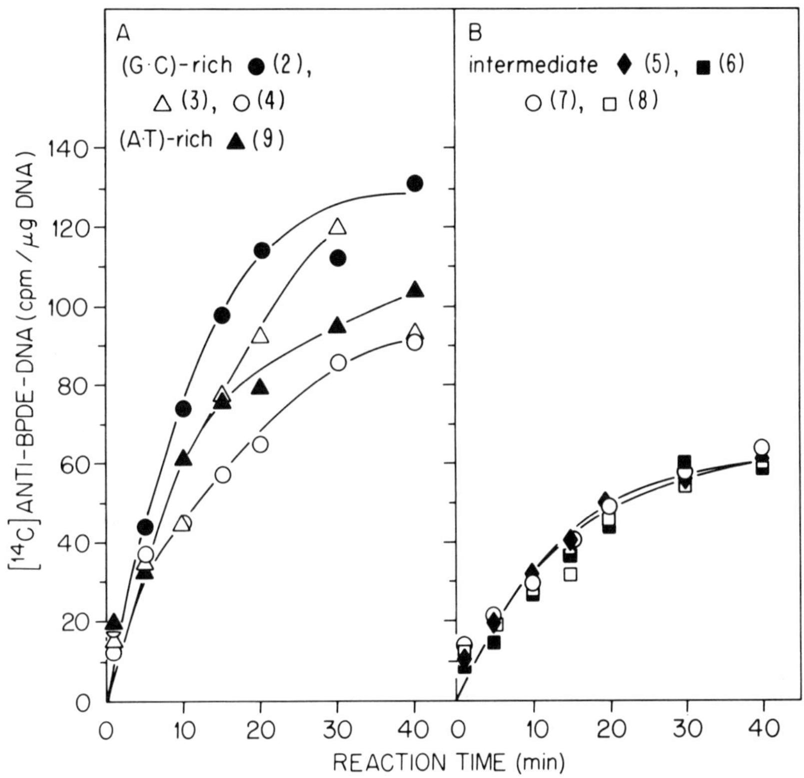

FIGURE 5. The rate of reaction of calf thymus DNA fractions from Ag^+-Cs_2SO_4 gradients with [^{14}C] anti-BPDE. (A) Fractions 2 (●–●), 3 (△–△) and 4 (○–○) are (G+C)-rich; fraction 9 (▲–▲) is (A+T)-rich. (B) Fractions 5 (◆–◆), 6 (■–■), 7 (○–○) and 8 (□–□) are intermediate and (A+T)-rich.

An interesting outcome of this study (as with the mouse) is that fraction 9, an (A+T)-rich fraction, also shows a high capacity to react with and bind anti-BPDE. The data in Figure 5A show that fraction 9 has an initial rate of reaction comparable to that for the other (G+C)-rich fractions 2 and 3 but binds slightly less anti-BPDE than these fractions. This DNA has an A_{260}/A_{280} = 1.806, and we cannot at this time rule out the possibility that a small amount of tightly bound protein exists in this fraction. Further characterization of this DNA is in progress. In addition, studies with the highly purified (G+C)-rich sequences of fractions 2 and 3 are being done.

Determination of the Formation of DNA-BPDE Covalent Linkage

Experiments were done to determine if the DNA-BPDE formed represents covalently bound material. Calf DNA-[^{14}C]anti-BPDE was enzymatically degraded to deoxynucleotides and chromatographed on an LH-20 column (Figure 6). Appproximately 85-90% of the radioactivity was eluted between

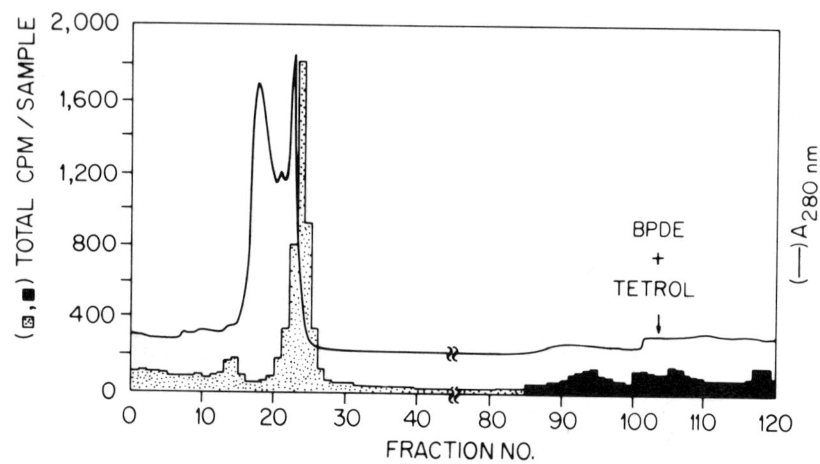

FIGURE 6. Sephadex LH-20 chromatographic elution profile of DNA containing bound BPDE(II). The DNA was enzymatically degraded to deoxynucleotides placed on the column and eluted with a 30-100% water:methanol linear gradient. The volume of the fraction applied was 0.8 ml. The flow rate was 0.6 ml/min.

fractions 20-30, which is the region of elution of the nucleotide-adduct (1). The unreacted deoxynucleotides elute between fractions 10-20; elution of reference samples, *i.e.*, unreacted [^{14}C]anti-BPDE and [^{14}C]BP-tetrol, occurs between fractions 90-100. Approximately 10-15% of the radioactivity in the digested DNA fraction eluted in this region. We believe this represents unreacted and possibly intercalated carcinogen. We interpet these studies to indicate that the anti-BPDE is covalently bound to the DNA. Identification of these adducts is in progress.

DISCUSSION

In the present investigation we examined the binding of anti-BPDE with fractionated eukaryotic DNA. Using Ag^{+}-Cs_2SO_4 isopycnic density-gradient centrifugation, it was possible to generate a gradient of (G+C)-rich to (A+T)-rich DNA sequences which we divided into fractions for use in these studies (see Figure 1A and B). Since recent observations from other laboratories have indicated that guanine is the major adduct formed in the reaction of DNA with anti-BPDE, we decided to use these Ag^{+}-Cs_2SO_4 fractions to study this reaction with fractionated mouse and calf DNA. We chose to use both BALB/c and Sencar mouse DNA because of the relative resistance and sensitivity of these mice to carcinogen treatment. Calf thymus DNA was also used to enable us to compare it with mouse DNA because (1) mouse DNA contains a major (A+T)-rich satellite DNA, and (2) calf DNA contains a variety of (G+C)-rich satellite. This would permit us to study the characteristics of this reaction with naturally occurring DNA having significantly different base composition.

Our studies suggest that under conditions in which anti-BPDE is rate-limiting, there are regions in the eukaryotic genome that preferentially bind this carcinogen. As would be expected, the (G+C)-rich fractions of BALB/c, Sencar and calf DNA interact strongest with anti-BPDE. In addition to having a rapid initial rate of reaction, the amount of anti-BPDE bound at the end of the reaction is greater for the (G+C)-rich fractions. Since the nucleotide (and guanine) residues are not rate-limiting, the final amount of anti-BPDE bound should be similar for all of the fractions, *i.e.*, for both (G+C)-rich and (A+T)-rich DNA. This is obviously not the case, and we interpret these data to indicate that in addition to the (G+C) content some other factor, possibly base-sequence, is also important in determining the amount of anti-BPDE bound to eukaryotic DNA.

An interesting outcome of the studies with BALB/c and Sencar DNA is seen in the amount of anti-BPDE bound by fraction 4 of the Ag^{+}-Cs_2SO_4 gradient. This DNA has an intermediate (G+C) content, as determined by T_m and isopycnic density-gradient centrifugation in CsCl (Papaconstantinou, DuBrul and Triplett, in preparation). However, the Sencar fraction is at least twofold more reactive with anti-BPDE than the corresponding BALB/c fraction. As far as we can tell, there is no obvious difference between these fractions. In fact, we postulate that this difference in reaction may represent a difference in (G+C) content or sequence that is below the limits of detection of the Ag^+-Cs_2SO_4 gradient. On the other hand, a very subtle difference in (G+C) content or sequence might be detected by the sensitivity of the reaction with small amounts of high specific activity anti-BPDE, *i.e.*, a reaction run under our conditions of 1 BPDE:100 dNTP. The significance of this difference is not understood, although it is interesting that the increased binding occurs with DNA derived from the carcinogen-sensitive Sencar mouse. Further characterization of this DNA is in progress.

Another interesting observation is the relatively high level of binding of anti-BPDE with the (A+T)-rich satellite of mouse DNA and the (A+T)-rich fraction of calf thymus DNA. Assuming that there is no tightly bound protein in these DNA fractions, these observations indicate that factors other than (G+C) content are important in the reaction and in the degree of binding of anti-BPDE with eukaryotic DNA. Although we do not understand the functional role of the (A+T)-rich or (G+C)-rich satellite DNA, it is known that they are localized in the centromeric region of the chromosome and are a component of the heterochromatin (18,22). With the rapid development of restriction enzyme and base-sequence technology, we propose to use these satellites in our future studies on the question of the role of specific base sequences in the reaction of anti-BPDE and the diol epoxides of other PAH with DNA. Studies of this nature are presently in progress.

ACKNOWLEDGMENTS

This research was sponsored by the Division of Biomedical and Environmental Research, U.S. Department of Energy, under Contract W-7405-eng-26 with the Union Carbide Corporation.

REFERENCES

1. Baird, W. M., and Brooks, P. (1973): Isolation of the hydrocarbon-deoxynucleoside products from the DNA of mouse embryo cells treated in culture with 7-methylbenz[a]anthracene-^{3}H. *Cancer Res.* 33:2378-2385.
2. Birnstiel, M., Telford, J., Weinberg, E., and Stafford, D. (1974): Isolation and some properties of the genes coding for histone proteins. *Proc. Nat. Acad. Sci., U.S.* 71:2900-2904.
3. Birnstiel, M. L., Wallace, H., Sirlin, J., and Fischberg, M. (1966): Localization of the ribosomal DNA complements in the nucleolar

organizer regions of *Xenopus laevis. Nat. Cancer Inst. Monograph.* 23:431-444.

4. Borgen, A., Darvey, H., Castagnoli, N., Crocker, T. T., Rasmussen, R. E., and Wang, I. Y. (1973): Metabolic conversion of benzo[a]pyrene by Syrian hamster liver microsomes and binding of metabolites to DNA. *J. Med. Chem.* 16:502-506.
5. Brookes, P., and Lawley, P. D. (1964): Evidence for the binding of polynuclear aromatic hydrocarbons to the nucleic acids of mouse skin: Relation between carcinogenic power of hydrocarbons and their binding to deoxyribonucleic acid. *Nature* 202:781-784.
6. Brown, D. D., Wensink, P. C., and Jordan, E. (1971): Purification and some characteristics of 5S DNA from *Xenopus laevis. Proc. Nat. Acad. Sci., U.S.* 68:3175-3179.
7. Clarkson, S. G. and Kurer, V. (1976): Isolation and some properties of DNA coding for $tRHA_1^{Met}$ from *Xenopus laevis. Cell* 8:183-195.
8. Corneo, G., Ginelli, E., and Polli, E. (1970): Different satellite deoxyribonucleic acids of guinea pig and ox. *Biochemistry* 9:1565-1571.
9. Corneo, G., Ginelli, E., Soave, C., and Bernardi, G. (1968): Isolation and characterization of mouse guinea pig satellite deoxynucleic acids. *Biochemistry* 7:4373-4379.
10. Dandel, P., Duquesne, M., Vigny, P., Grover, P. L., and Sims, P. (1975): Fluorescence spectral evidence that benzo[a]pyrene-DNA products in mouse skin arise from diol-epoxides. *FEBS Lett.* 57:250-253.
11. Filipski, J., Thiery, J. P., and Bernardi, G. (1973): Analysis of the bovine genome by cesium sulfate-silver(I) ion density gradient centrifugation. *J. Mol. Biol.* 80:177-197.
12. Fu, P. P., and Harvey, R. G. (1977): Synthesis of the diols and diol-epoxidesof carcinogenic hydrocarbons. *Tetrahedron Lett.* 24:2059-2062.
13. Gelboin, H. V. (1969): A microsome-dependent binding of benzo[a]-pyrene to DNA. *Cancer Res.* 29:1272-1276.
14. Heidelberger, C. (1975): Chemical carcinogenesis. *Ann. Rev. Biochem.* 44:79-121.
15. Huberman, E., Sachs, L., Yang, S. K., and Gelboin, H. V. (1976): Identification of mutagenic metabolites of benzo[a]pyrene in mammalian cells. *Proc. Nat. Acad. Sci., U.S.* 73:607-612.
16. Ivanovic, V., Geacintov, N. E., and Weinstein, I. B. (1976): Cellular binding of benzo[a]pyrene to DNA characterized by low temperature fluorescence. *Biochem. Biophys. Res. Commun.* 70:1172-1179.
17. Jensen, R. H., and Davidson, N. (1966): Spectrophotometric, potentiometric, and density gradient ultracentrifugation studies of the binding of silver ion by DNA. *Biopolymers* 4:17-32.
18. Jones, K. W. (1970): Chromosomal and nuclear location of mouse satellite DNA in inidividual cells. *Nature* 225:912-915.
19. Kapitulnik, J., Levin, W., Conney, A. H., Yagi, H., and Jerina, D. M. (1977): Benzo[a]pyrene 7,8-dihydrodiol is more carcinogenic than benzo[a]pyrene in newborn mice. *Nature* 266:378-380.

20. Kapitulnik, J., Wislocki, P. G., Levin, W., Yagi, H., Jerina, D. M., and Conney, A. H. (1978): Tumorigenicity of diol-epoxides of benzo-[a]pyrene: Evidence that (±)-trans-7β,8α-dihydroxy-9α-10α-epoxy-7,8,9,10-tetrahydrobenzo[a]pyrene is an ultimate carcinogen in newborn mice. *Cancer Res.* 38:354-358.
21. Karijord, O., and Tordis, B. O. (1975): Study of haploid yeast strain with an unusually high rDNA content. II. Fractionation of the DNA in preparative Ag^+-Cs_2SO_4 density gradients. *Biochim. Biophys. Acta.* 383:255-265.
22. Kurnit, D. M., Shafit, B., and Maio, J. J. (1973): Multiple satellite deoxynucleic acids in the calf and their relation to the sex chromosomes. *J. Mol. Biol.* 81:273-284.
23. Macaya, G., Theiry, J. -P., and Bernardi, G. (1976): An approach to the organization of eukaryotic genomes at a macromolecular level. *J. Mol. Biol.* 108:237-254.
24. Marmur, J. (1961): A procedure for the isolation of deoxynucleic acid from micro-organisms. *J. Mol. Biol.* 3:208-218.
25. Meehan, T., Straub, K., and Calvin, M. (1976): Benzo[a]pyrene diol-epoxide covalently binds to deoxyguanosine and deoxyadenosine in DNA. *Nature* 269:725-727.
26. Miller, E. C., and Miller, J. A. (1974): Biochemical mechanisms of chemical carcinogenesis. In: *Molecular Biology of Cancer,* H. Busch, Ed., pp. 377-402, Academic Press, New York.
27. Newbold, R. F., and Brookes, P. (1976): Mutagenicity of benzo[a]-pyrene diol-epoxides in cultured mammalian cells. *Nature* 261:52-54.
28. Rinehart, F. P., and Schmid, C. W. (1976): The effect of silver ion binding and pH on the buoyant density of DNA and its use in fractionating heterogeneous DNA. *Biochim. Biophys. Acta.* 425:451-462.
29. Sims, P., Grover, P. L., Swaisland, A., Pal, L., and Hewer, A. (1974): Metabolic activation of benzo[a]pyrene proceeds by a diol-epoxide. *Nature* 252:326-328.
30. Slaga, T. J., Bracken, W. M., Viaje, A., Levin, W., Yagi, H., Jerina, D. M., and Conney, A. H. (1977): Comparison of the tumor-initiating activities of benzo[a]pyrene oxides and diol-epoxides. *Cancer Res.* 37:4130-4133.
31. Suzuki, Y., Gage, P. L., and Brown, D. D. (1972): The genes for silk fibroin in *Bombyx mori. J. Mol. Biol.* 70:637-649.
32. Theiry, J. -P., Macaya, G., and Bernardi, G. (1976): An analysis of eukaryotic genomes by density gradient centrifugation. *J. Mol. Biol.* 108:219-235.
33. Wilson, V. L., Rinehart, F. P., and Schmid, C. W. (1976): Comparison of fractionation methods for DNA of varying base composition by equilibrium density-gradient centrifugation. *Anal. Biochem.* 73:350-362.
34. Yagi, H., Hernandez, O., and Jerina, D. M. (1975): Synthesis of (±)-7β-8α-hydroxy-9β,10β-epoxy-7,8,9,10-tetrahydrobenzo[a]pyrene. A potential metabolite of the carcinogen benzo[a]pyrene with stereochemistry related to the antileukemic triptolides. *J. Am. Chem. Soc.* 97:6881-6883.

Polynuclear Aromatic Hydrocarbons, edited by P. W. Jones and P. Leber. Ann Arbor Science Publishers, Inc. Ann Arbor, MI ©1979.

STUDIES ON THE BINDING OF B[a]P DIOL EPOXIDE TO DNA AND CHROMATIN

A. Kootstra, T. J. Slaga and D. E. Olins

The University of Tennessee–
Oak Ridge Graduate School of Biomedical Sciences

Biology Division
Oak Ridge National Laboratory
Oak Ridge, Tennessee 37830

INTRODUCTION

Benzo[a]pyrene (B[a]P) is a ubiquitous environmental pollutant and its toxic, mutagenic and carcinogenic effects are a consequence of the metabolic conversion to reactive intermediates by the mixed-function oxidase system (8,12,23,35). These reactive metabolites of B[a]P are capable of interacting with nucleic acids and cellular proteins (3,9,17,31,36). Although the interaction of these carcinogens with any of the cellular components may eventually lead to tumor formation, it is currently thought that a covalent DNA-B[a]P adduct may be one of the primary events in the initiation process of carcinogenesis (3,36).

In recent studies the ultimate carcinogenic form of B[a]P appears to be the (±)7β,8α-dihydroxy-9α,10α-epoxy-7,8,9,10-tetrahydro benzo[a]pyrene [(±B[a]P diol epoxide-anti] (10,13,25,28,37,38,40). B[a]P diol epoxide has been shown to form covalent adducts with RNA and DNA *in vitro* and *in vivo*, and the major adduct isolated was N^2(7,8,9-dihydroxy-7,8,9,10-tetrahydro benzo[a]pyrene-10-yL) guanine nucleoside. Other covalent adducts have now been isolated and characterized involving deoxyadenine, deoxycytidine (21) and the N^7 position of deoxyguanine (29). Covalent binding of carcinogens to DNA has been shown to destablize the DNA and interfere with DNA-dependent transcription *in vitro* (18,33).

However, since eukarotic DNA is associated with histones, which are responsible for packaging the DNA into discrete subunits called nucleosomes (5,6,15,16,26,41), the binding of carcinogens to chromosomal DNA may be subject to additional constraints. Nucleosomes in all eukaryotes examined to date contain 140 base pairs of DNA which are wrapped around a histone core, containing two of each H2A, H2B, H3, H4. At the primary level of organization the nucleosomes are arranged like beads on a string (26), each nucleosome is connected to the next by a spacer region, consisting of DNA which is variable in length and seems to have histone H1 and some non-histones associated with it (19). Recent evidence has shown that the binding of carcinogens to chromatin is not uniformly distributed (11,22). However, since current data suggest that nuclease digestion may differentially hydrolyse DNA of "active" and "inactive" chromatin (5,19,39), these results should be carefully evaluated. We have therefore studied the *in vitro* interaction of B[a]P diol epoxide (anti) with chicken erythrocyte chromatin. Chicken erythrocyte chromatin was chosen since it has been well characterized, it is an example of inactive genome and the internucleosomal spacer DNA is rather constant containing 60-70 base pairs (20). Consequently, the binding of a compound to nucleosomal and internucleosomal DNA can be more precisely quantitated.

Our studies have revealed that the binding of B[a]P diol epoxide (anti) with chromosomal DNA appears to involve intercalation and covalent binding. This reaction can be differentially inhibited by a competing nucleophile such as cysteine. Analysis of the distribution of the carcinogen to mononucleosomes (ν_1) and dinucleosomes (ν_2) revealed that the internucleosomal DNA bound three to four times more carcinogen than the nucleosomal DNA. Furthermore, dissociation of labeled chromatin showed that this ratio is no longer detectable after reconstitution.

MATERIALS AND METHODS

Preparation of Carcinogen-Labeled Nuclei, Chromatin and Nucleosomes

Chicken erythrocyte nuclei were isolated from either fresh or frozen chicken blood (27). The nuclear suspension was diluted with isolation buffer [10 m*M* tris (*p*H 7.0), 10 m*M* NaCl, 3 m*M* $MgCl_2$, 0.5% NP-40] to contain 1.4×10^9 nuclei/ml, in 1.0 m*M* $CaCl_2$, 0.1 m*M* phenylmethanesulfonyl-fluoride (PMSF). Micrococcal nuclease (60 units/ml) was added, and the digestion at 37°C was stopped after 10 min by the addition of EDTA (10 m*M* final concentration). The digested nuclear suspension was dialyzed for 24 hr with three changes of 0.2 m*M* EDTA (pH 7.0), 0.1 m*M* PMSF. The lysed nuclear suspension was centrifuged at 10,000 *g* for 10 min at 4°C. The

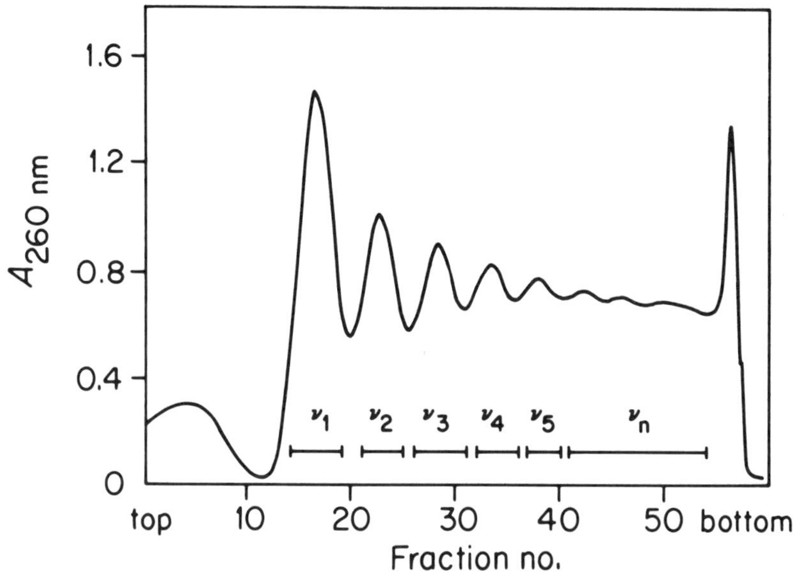

FIGURE 1. Optical density profile of sucrose gradient. 2.5 ml chromatin (A_{260} = 83.0) was layered on a 5-25% linear sucrose gradient containing 0.2 m*M* EDTA (pH 7.0), 20 m*M* NaCl and was centrifuged for 14 hr at 23,000 rpm in an SW 25.2 rotor at 4°C. The gradients were fractionated with an ISCO density-gradient collector (Model 640) equipped with an ISCO UV-5 absorbance monitor. Nucleosome monomers, dimers and oligomers are shown as ν_1, ν_2 and ν_n.

supernatant containing solubilized chromatin was concentrated on an Amicon UM 10 membrane, and further purification was achieved by centrifugation through sucrose gradients (Figure 1). Chromatin from fractions 41-54 (Figure 1) was pooled and dialyzed against 10 m*M* tris (pH 7.0), 0.1 m*M* PMSF at 4°C for 12 hr. It was then reacted with [^{14}C] (±)*trans*-7β,8α-dihydroxy-9α,10α-epoxy-7,8,9,10-tetrahydrobenzo[a]pyrene (specific activity, 29.4 mCi/mmol) at 37°C for 30 min under yellow light. The reaction mixture was subsequently dialyzed against 0.2 m*M* EDTA (pH 7.0), 0.1 m*M* PMSF at 4°C for 48 hr with four buffer changes. The final dialysis step was against 10 m*M* tris (pH 7.0), 0.1 m*M* PMSF. The dialysate containing oligonucleosomes (ν_n) was made 3 m*M* $CaCl_2$ and redigested with micrococcal nuclease (60 μ/ml) for 5 min at 37°C. Nucleosome monomers (ν_1) and dimers (ν_2) were isolated after centrifugation through sucrose gradients (Figure 2). The purified ν_1 and ν_2 were dialyzed against 0.2 m*M* EDTA (pH 7.0), 0.1 m*M* PMSF, and their specific activity was determined.

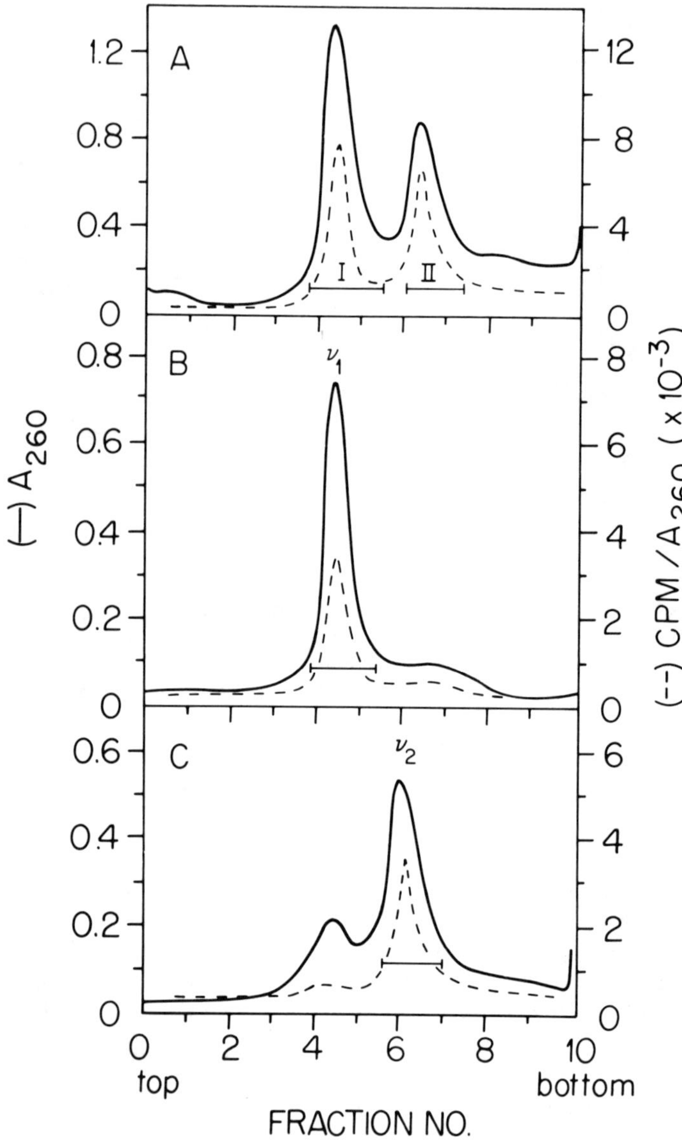

FIGURE 2. Chromatin obtained from pool ν_n (Figure 1) was reacted with B[a]P diol epoxide and redigested with micrococcal nuclease as described in the text. ν_1 and ν_2 were purified by ultracentrifugation through a 5-25% sucrose gradient containing 0.2 m*M* EDTA (pH 7.0), 20 m*M* NaCl by use of an SW 41 Ti rotor at 39,000 rpm for 12 hr at 4°C. The fractions containing ν_1 and ν_2 were pooled, concentrated and subjected to another ultracentrifugation run under the same conditions to yield purified ν_1 (B) and ν_2 (C).

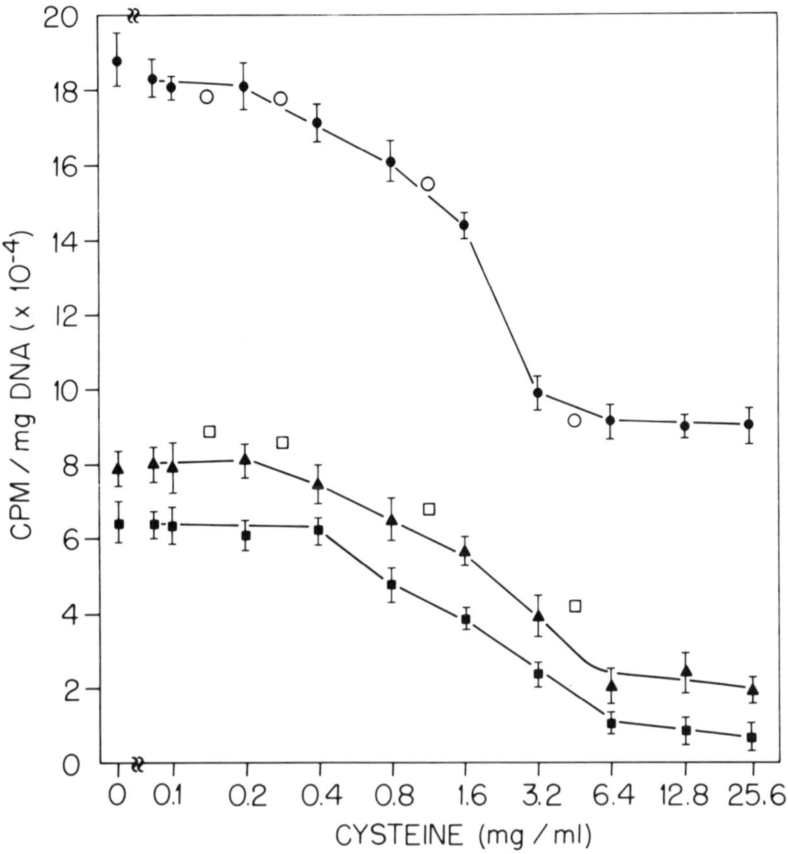

FIGURE 3. Incorporation of B[a]P diol epoxide into DNA, chromatin and nucleosomes as a function of cysteine concentration. Chicken erythrocyte DNA, chromatin and nucleosomes (A_{260} = 1.0) was reacted with 1.0 μg B[a]P diol epoxide (anti) in the presence of cysteine. Closed symbols represent buffer containing 0.14 *M* KCl, open symbols represent buffer without KCl, DNA (●, ○), mononucleosomes (▲), chromatin (■, □).

Binding of B[a]P in the Presence of Cysteine

Chicken erythrocyte DNA, chromatin and mononucleosomes were dissolved in 100 m*M* tris (pH 7.0), 0.1 m*M* PMSF and 0.14 *M* KCl containing a variable amount of cysteine. (±) B[a]P diol epoxide (anti) was added and the reaction mixture was dialysed as described above. The specific activities of nondialysable material are shown in Figure 3 (closed symbols). Since chromatin precipitated in the presence of 0.14 *M* KCl, the above reaction was repeated in the absence of KCl. Under these conditions chromatin remained soluble (Figure 3, open squares).

Purification of DNA

Chromatin, ν_1 and ν_2 were repeatedly extracted with phenol:chloroform (1:1 v/v) followed by chloroform and ether extractions. The purified DNA was precipitated with ethanol at -20°C. The dried DNA pellet was dissolved in 0.2 m*M* EDTA (pH 7.0). A known volume (0.1 ml) was routinely dissolved in 10 ml Aquasol (New England Nuclear) and the radioactivity was determined, using a Packard (Model 3385) scintillation counter.

Purification of Histone Core Complex

For the isolation of histone core complex, purified chicken erythrocyte nuclei were labeled with (±) B[a]P diol epoxide in the isolation buffer (see above). This reaction mixture was extensively dialysed against 0.2 m*M* EDTA (pH 7.0), 0.1 m*M* PMSF. The subsequent chromatin suspension was made 2.0 *M* NaCl by slow addition of solid NaCl with gentle mixing. The resulting viscous solution was left for 12 hr at 4.0°C and centrifuged in a Beckman 50 Ti rotor at 48,000 rpm for 12 hr. The pellet (i.e., DNA and residual proteins) was solubilized in 0.2 m*M* EDTA (pH 7.0), and the DNA was purified as described above. The supernatant solution containing histones, and some nonhistone proteins, was concentrated by vacuum dialysis. The core histone proteins (H2A, H2B, H3 and H4) were purified by centrifugation through a 5-20% linear sucrose gradient containing 2.0 *M* NaCl, 10 m*M* tris (pH 7.2) and 0.1 m*M* dithiothreitol (DTT) for 44 hr at 50,000 rpm in an SW 50.1 rotor at 4°C (4). The histone core was isolated and the specific activity was determined by use of the value $E^{0.1\%}_{230(1cm)} = 4.2$.

Reconstitution of Carcinogen Chromatin Complex

Chromatin was isolated, labeled and dialysed as described above except that in this experiment [^{3}H](±)B[a]P diol epoxide (anti) (specific activity 269 mCi/mmole) was used. Aliquots of labeled chromatin were made, 0.65 *M* and 2.0 *M* in NaCl concentration, by slow addition of solid NaCl. Reconstitution was achieved by stepwise dialysis for at least 4 hr against: (A) 2.0 *M* NaCl, 0.1 m*M* DTT, 10 m*M* tris (pH 7.0); (B) 1.4 *M* NaCl, 0.1 m*M* DTT, 10 m*M* tris (pH 7.0); (C) 1.0 *M* NaCl, 0.1 m*M* DTT, 10 m*M* tris (pH 7.0); (D) 0.3 *M* NaCl, 10 m*M* tris (pH 7.0); (E) 0.1 *M* NaCl, 10 m*M* tris (pH 7.0); (F) 10 m*M* tris (pH 7.0). All solutions contained 0.1 m*M* PMSF and 0.2 m*M* EDTA (pH 7.0), and final dialysis for 12 hr was against 0.2 m*M* EDTA (pH 7.0) and 0.1 m*M* PMSF. The specific activities of reconstituted chromatin are shown in Table I. Further digestion of this labeled chromatin to mononucleosomes and oligonucleosomes was obtained by redigestion of reconstituted chromatin (Figure 4), using the parameters described in Figure 5. Aliquots of each

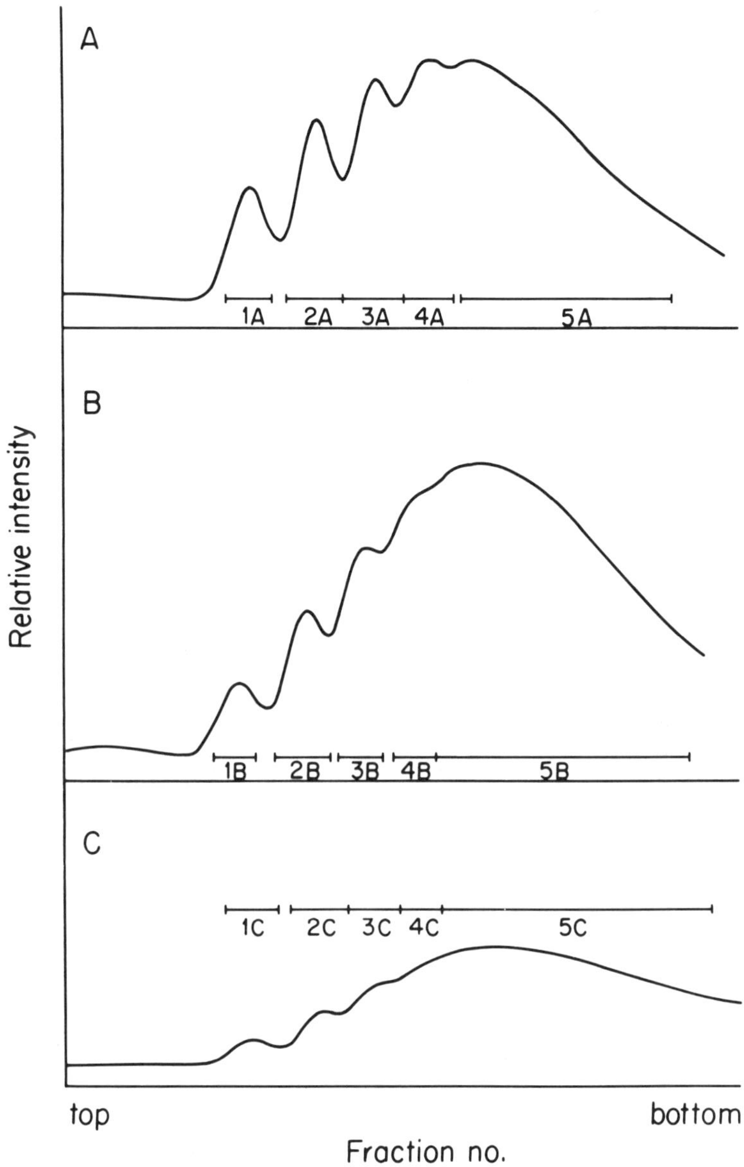

FIGURE 4. Optical density profiles of sucrose gradients. The conditions are as those described in Figure 1: (A) control chromatin; (B) reconstituted 0.65 *M* dissociated chromatin; (C) reconstituted 2.0 *M* dissociated chromatin. Pool 1, 2, 3, 4 and 5 $\nu_1, \nu_2, \nu_3, \nu_4$ and $\geqslant \nu_5$.

TABLE I. Specific Activity of Reconstituted Chromatin

Chromatin	Dissociation Condition	cpm/A_{260}
ν_n	0.00 *M* NaCl (control)	14,814.9
ν_n	0.65 *M* NaCl	13,993.9
ν_n	2.00 *M* NaCl	14,940.0

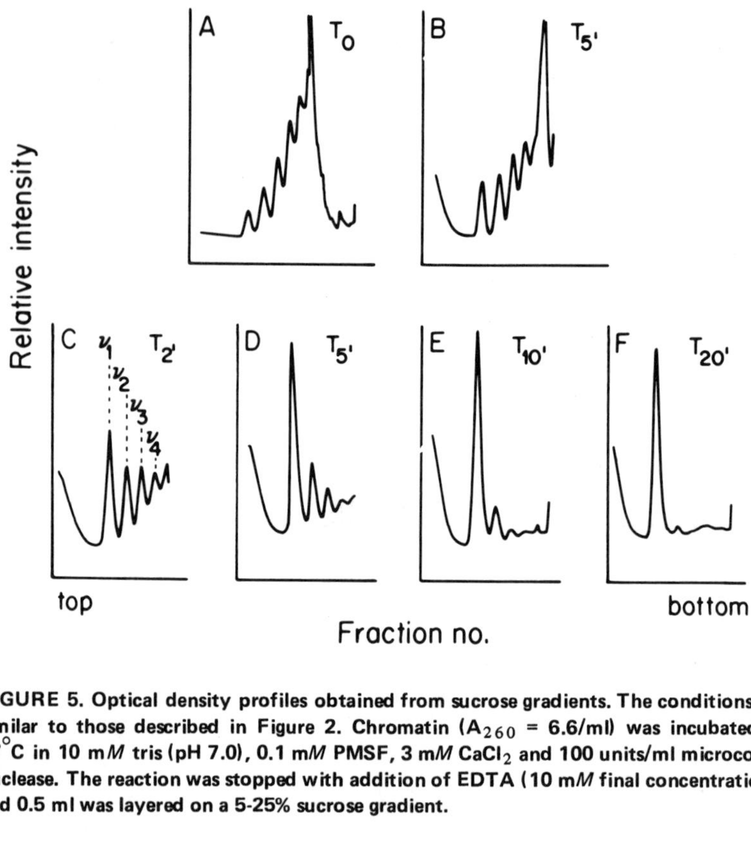

FIGURE 5. Optical density profiles obtained from sucrose gradients. The conditions are similar to those described in Figure 2. Chromatin (A_{260} = 6.6/ml) was incubated at 37°C in 10 m*M* tris (pH 7.0), 0.1 m*M* PMSF, 3 m*M* $CaCl_2$ and 100 units/ml micrococcal nuclease. The reaction was stopped with addition of EDTA (10 m*M* final concentration), and 0.5 ml was layered on a 5-25% sucrose gradient.

individual pool (A, B and C) were further digested to mononucleosomes, and the specific activities are shown in Table II.

Fluorescence Spectrophotometry

All fluorescence measurements were made with a conventional fluorescence spectrophotometer (Perkin Elmer MPF44) at room temperature. The

TABLE II. Specific Activities of Mono- and Oligonucleosomes Obtained from Reconstituted Chromatin

Pool	cpm/A_{260}	cpm/$A_{260}(\nu_1)$[a]
1A	21,514.1	6,569.3
2A	15,886.4	7,358.5
3A	14,789.3	7,307.6
4A	14,508.2	–
5A	13,170.2	6,168.8
1B	21,704.5	8,433.8
2B	17,564.1	9,158.0
3B	15,103.4	7,878.0
4B	14,189.2	–
5B	13,424.5	7,388.5
1C	27,000.0	16,490.1
2C	18,205.0	16,363.6
3C	15,315.8	15,595.7
4C	14,727.3	–
5C	15,396.2	15,079.4

[a]Each individual pool was redigested to mononucleosome and the specific activity of purified ν_1 was determined. Pool, 1 = ν_1; 2 = ν_2; 3 = ν_3; 4 = ν_4; 5 $\geqslant$ ν_5.

Note: A–control chromatin; B–reconstituted 0.65 *M* NaCl dissociated chromatin; C–reconstituted 2.0 *M* NaCl dissociated chromatin.

reaction buffer was 10 m*M* tris (pH 7.0), containing a variable amount of DNA or 200 μg of mononucleosomes. The reaction mixture was equilibrated at room temperature and the reaction in each experiment was started by the addition of 10 μl (8.6 μg) B[a]P diol epoxide in tetrahydrofuran:triethylamine (19:1 v/v) which was mixed by rapid inversion. The excitation wavelength (λex)was 327 nm, the emission wavelength (λem) was either maintained at 379 nm for continuous measurement (Figure 6), and/or scanned from 300 nm to 475 nm (Figure 6, inset).

RESULTS AND DISCUSSION

In order to establish initial conditions and parameters for the reaction of B[a]P diol epoxide (anti) with various nucleophiles, we have studied the rate of breakdown of B[a]P diol epoxide in buffer and buffer containing different amounts of DNA, either as free DNA or as nucleohistone complex.

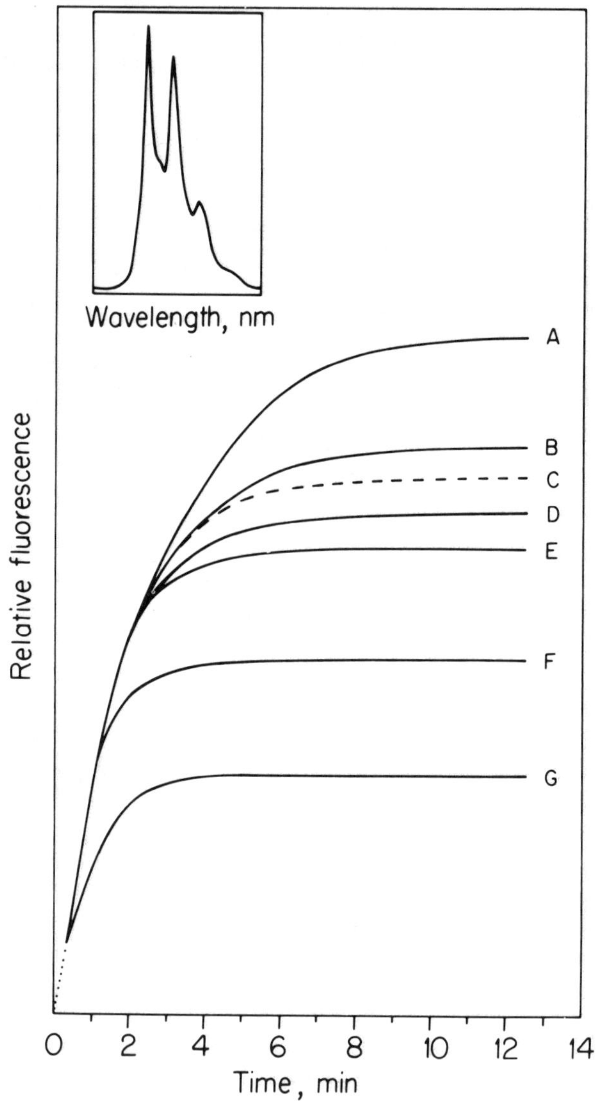

FIGURE 6. Kinetics of product formation from B[a]P diol epoxide measured by fluorescence spectrophotometry; λex = 327 nm, λem = 378 nm. B[a]P diol epoxide (8.6 μg) was added to each reaction mixture. A, 10 m*M* tris (pH 7.0); B, 0.5 μg DNA/ml; C, 100 μg DNA/ml as mononucleosomes; D, 25 μg DNA/ml; E, 100 μg DNA/ml; F, 500 μg DNA/ml; G, 1,000 μg DNA/ml.

Inset shows typical spectrum obtained at any point during the reaction. This spectrum is identical to the spectrum of tetraoles having emission maxima at 378 nm, 397 nm and 420 nm.

Fluorescence

It was found that the rate of breakdown of B[a]P diol epoxide to products could be monitored by fluorescence spectrophotometry, since B[a]P diol epoxide has no detectable fluorescence spectrum in aqueous medium, while the products show a fluorescence spectrum analogous to one obtained by pyrene adducts (32) (Figure 6). The rate of product formation from B[a]P diol epoxide in solution at fixed emission and excitation wavelengths can therefore be monitored. Discontinuous monitoring of the spectrum at time intervals during this reaction revealed that the increase in relative fluorescence intensity was due to an increase in the magnitude of the pyrene-like spectrum (Figure 6, inset), and not due to any other form of spectra. Free tetraoles added to the reaction buffer produced an identical spectrum which did not change with time. In the case of solvolysis (Figure 6 A) an initial fast reaction was observed while a slower reaction continued for about 35 min; after this no further changes were observed. The presence of a nucleophile such as DNA, or mononucleosomes appeared to abolish the slow reaction observed in A. It should be pointed out that the relative intensity decreased with increasing amounts of nucleophile. Fluorescence quenching of a variety of dyes such as acriflavine, proflavine and quinacrine, in the presence of DNA, has been observed by a number of investigators (1,30,34). This phenomenon was identified with intercalation into G-C rich regions of DNA and the mechanism of quenching has been related to the formation of a charge-transfer complex in the excited state involving π electron interactions. Consequently, the fluorescence quenching observed when B[a]P diol epoxide is reacted with DNA would appear to indicate that the interaction involves not only covalent binding to the exocyclic amino group of guanine, but also intercalation (7). The latter may be responsible for the decrease in fluorescence intensity. Furthermore, the reaction in the presence of a nucleophile was complete within 10 min, and thus an incubation time of 30 min at 37°C would be more than adequate to ensure completion of the reaction.

In the reaction of the diol epoxide with purified mononucleosomes, the decrease in relative intensity would suggest that this reaction is not of the same magnitude in the nucleohistone complex as that found with free DNA. This may indicate that the presence of histones decreases the availability of binding to chromosomal DNA. In other words, it may well be that the presence of histones in eukaryotic chromatin confers an additional level of specificity on the reaction of carcinogens with DNA.

The Effect of Cysteine on the Binding of B[a]P Diol Epoxide

To inhibit the reaction of the diol epoxide with DNA, chromatin and mononucleosomes we have used a competing nucleophile, cysteine. The effects of cysteine on adduct formation are shown in Figure 3. Free DNA contained almost twice the number of nondialysable molecules as ν_1 and chromatin. The lower incorporation of radioactivity into chromatin, compared with ν_1 (closed symbols), could be attributed to a solvent effect, since in the presence of 0.14 *M* KCl chromatin precipitated. By changing the ionic strength of the medium (in the absence of KCl) chromatin remained soluble and the incorporation of adducts was increased (open squares, Figure 3). The presence of cysteine therefore would largely abolish the covalent-type interaction and at [cys] $>$ 3.2 mg/ml, 85% of the radioactivity associated with DNA could be removed by repeated phenol extraction. Repeated extraction of chromatin or ν_1 with phenol removed only 20% of the bound carcinogen. This data would therefore suggest that under our *in vitro* condition, DNA incorporated a substantial amount of noncovalently bound material. In chromatin or mononucleosomes about 80% of the carcinogen seems to be covalently bound to the DNA.

Distribution of B[a]P Adducts in Chromosomal Components

In order to examine the distribution of the carcinogen with respect to chromosomal DNA and histone proteins in detail, isolated chicken erythrocyte nuclei were labeled with B[a]P diol epoxide. The labeled nuclei were lysed and the resulting chromatin component, DNA and proteins were purified as described in Materials and Methods. The phenol-purified DNA had a specific activity of 29,629 cpm/mg DNA and the purified histone core proteins (H2A, H2B, H3 and H4) had a specific activity of 4,476.5 cpm/mg histone. Consequently, only 15% of the radioactivity was associated with the core histone while the remainder was associated with chromosomal DNA. The binding of B[a]P diol epoxide to the core histone complex is thus a minor component, in contrast with other studies in which the parent hydrocarbon has been used (2,31). It would therefore be important to know if some other metabolites of B[a]P are involved in binding to chromosomal proteins.

Distribution of B[a]P Adducts in Nucleosomal and Internucleosomal DNA

Since it is now apparent that most of the *in vitro* binding of the diol epoxide to chromatin involves the formation of covalent adducts with DNA, we have examined the distribution of adducts in chromosomal DNA, with respect to nucleosomal and internucleosomal regions. The isolation of chromatin (ν_n) (Figure 1), labeling and redigestion has been described in Materials and Methods. The isolated ν_1 and ν_2 (Figure 2) were pooled, concentrated, and the specific activities of ν_1 and ν_2 from three independent experiments

TABLE III. Distribution of Labeled Carcinogen in Chromatin

Fraction	Experiment I Specific Activity	Experiment II Specific Activity	Experiment III Specific Activity
	(cpm/A_{260})	(cpm/A_{260})	(cpm/A_{260})
Chromatin	11,805.0	9,641.0	6,461.6
ν_1	4,443.0	4,213.0	2.452.0
ν_2	6,250.0	6,690.0	3,580.0

Note: Chromatin was reacted with B[a] P diol epoxide as described in the text. The ratio of DNA to B[a] P diol epoxide in experiments I and II was 100:1 and in experiment III, 200:1. Assuming that $\nu_2 = 2\nu_1$ + spacer, the spacer region contained 3.8 ± 0.5 times more carcinogen than ν_1.

are shown in Table III. These experiments showed that ν_2 contained more bound carcinogen per unit DNA than did ν_1.

Furthermore, phenol extraction of ν_n, ν_1 and ν_2 revealed that in each case 75 ± 5% of the carcinogen was associated with the DNA. Since in chicken erythrocyte chromatin the internucleosomal region appears rather constant at 60-70 base pairs of DNA and the core ν_1 contains approximately 140 base pairs (20,24), we have estimated that the internucleosomal region bound 3.8 ± 0.5 times more carcinogen than the core nucleosome. This increase in binding to the internucleosomal region appears to be due to the increased accessibility of the internucleosomal DNA, since phenol extraction does not change the ratio of specific activities. Furthermore, since chicken erythrocyte chromatin may be viewed as a 200-210 base pair repeat containing a 140 base pair ν_1 plus a 60-70 base pair spacer region, the specific activity of chromatin calculated from our experimental findings was within 10% of the experimental value observed for chromatin.

Reconstitution of Chromatin Containing B[a]P Adducts

We have dissociated and reconstituted labeled chromatin to examine whether the presence of B[a]P adducts would interfere with the assembly process of nucleosomes and if the distribution of adducts remains similar after reconsititution. In this manner one might be able to use the distribution of adducts as a probe to evaluate the fidelity of reconsitituted chromatin *in vitro*. Furthermore, some of the answers could then be related to the *in vivo* situation of chromosomal assembly in the presence of carcinogens during the S-phase of the cell.

Chromatin was isolated as previously described, treated with [^{3}H] B[a]P diol epoxide (anti) and extensively dialysed to a specific activity of 14,520.0 cpm/A_{260}. Aliquots of this labeled chromatin were made 0.65 *M* NaCl, to dissociate histones H1 and H5, and 2.0 *M* NaCl, to dissociate all histones and some nonhistone protein, from chromosomal DNA. Reassociation was achieved as described in Materials and Methods. The specific activities of the reconstituted chromatin are shown in Table I.

In order to establish parameters for redigestion with micrococcal nuclease, labeled chromatin was redigested, as shown in Figure 5. This revealed that chromatin had been subjected to slow digestion in the absence of added enzyme, presumably by residual micrococcal nuclease activity, and that the digestion to multimers was optimum in 3 min, while complete digestion to mononucleosomes took place in 15-20 min (Figure 5). The control and reconstituted chromatin solutions were redigested to monomers and oligomers of nucleosomes (Figure 4 A, B and C). Each pool (pools 1-5) was isolated and the specific activities were determined (Table II). Aliquots of each pool were redigested to mononucleosomes using the digestion parameters shown in Figure 5F. The specific activities of the purified ν_1 from each pool are shown in Table II.

One of the striking features of this analysis showed that the mononucleosomes of the 2.0 *M* NaCl dissociated and reassociated chromatin (Table II C) have the same specific activity as chromatin, while the control (Table II A), showed the typical nonuniform distribution of the carcinogen with respect to nucleosome vs internucleosomal region. This would indicate that in the 2.0 *M* NaCl dissociated chromatin, reconstitution leads to random distribution of histones. This suggests that the presence of B[a]P adducts does not hinder the formation of nucleosomes and that a compound such as B[a]P diol epoxide can be used to ascertain the fidelity of reconstituted chromatin. A second feature that we have observed consistently is the high specific activity of mononucleosomes obtained from short digestion of labeled nuclei, or chromatin with micrococcal nuclease. This could be attributed to either a subpopulation of ν_1 that contain spacer regions, which are much more susceptible to binding of the carcinogen or to some other macromolecule (i.e., RNA or nuclear protein) associated with the internucleosomal region that can be removed by micrococcal nuclease digestion.

CONCLUSION

The data presented here suggests that the interaction of B[a]P diol epoxide (anti) with free DNA or chromosomal DNA involved, apart from covalent binding, intercalation or some form of stacking interaction of the covalent and noncovalent-bound carcinogen and that in the case of chromatin or mononucleosomes binding appeared to be decreased. This decrease may be a function of changes in DNA conformation, steric hindrance or both due to the presence of chromosomal proteins. Furthermore, in the presence of a competing nucleophile such as cysteine, the covalent binding of B[a]P diol epoxide could be greatly diminished, and our data suggest that little if any noncovalent association occurred in chromatin.

Analysis of the distribution of carcinogen with respect to chromosomal components revealed that most of the adducts were associated with chromosomal DNA and that the internucleosomal DNA contained three to four times more carcinogen than nucleosomal DNA. Reconstitution of labeled chromatin showed that removal of the very lysine-rich histones by 0.65 *M* NaCl did not affect the distribution of the carcinogen, while "complete" dissociation using 2.0 *M* NaCl and reassociation revealed randomization of the bound carcinogen.

The reconstitution experiments showed that the presence of carcinogen did not interfere with the reassociation of histones to DNA, and that B[a]P diol epoxide could be used to evaluate the fidelity of reconstituted chromatin *in vitro*.

ACKNOWLEDGMENTS

We are grateful to Ms. E. B. Wright for technical assistance. We also wish to thank P. N. Bryan and Drs. A. L. Olins, A. P. Butler, and G. J. Bunick for suggestions and discussions during the course of this investigation. This research was jointly sponsored by NIH Grants CA 20076 and GM 19334 and the Office of Health and Environmental Research, U.S. Department of Energy, under Contract W-7405-eng-26 with the Union Carbide Corporation.

Parts of this work were presented in preliminary form in the abstracts of the 22nd Annual Meeting of the Biophysical Society (14).

REFERENCES

1. Bontemps, J., and Fredicq, E. (1974): *Biophys. Chem.* 2:1-22.
2. Bresnick, E. *et al.* (1977): *Arch. Biochem. Biophys.* 181:257-269.
3. Brookes, P., and Lawley, P. D. (1964): *Nature* 202:781-784.
4. Butler, A. P. *et al.* (1978): *Nucleic Acids Res.* (submitted).
5. Felsenfeld, G. (1978): *Nature* 271:115-122.
6. Finch, J. T. *et al.* (1977): *Nature* 269:29-36.
7. Frenkel, K. *et al.* (1978): *Biochemistry* 17(7):1278-1282.
8. Gelboin, H. V. *et al.* (1972): *Fed. Proc.* 31:1298-1309.
9. Heidelberger, C. A. (1975): *Rev. Biochem.* 44:79-121.
10. Huberman, E. *et al.* (1976): *Proc. Nat. Acad. Sci. U.S.* 73:607-611.

11. Jahn, C. L., and Litman, G. W. (1977): *Biochem. Biophys. Res. Commun.* 76:534-540.
12. Jerina, D. M., and Daly, J. W. (1974): *Science* 185:573-582.
13. Kapitulnik, J. *et al.* (1977): *Nature* 266:378-380.
14. Kootstra, A. *et al.* (1978): *Biophys. J.* 21:67a.
15. Kornberg, R. D. (1974): *Science* 184:868-871.
16. Kornberg, R. D., and Thomas, J. O. (1974): *Science* 184:865-868.
17. Kuroki. T., and Heidelberger, C. (1971): *Cancer Res.* 31:2168-2175.
18. Leffler, S. *et al.*(1977): *Biochemistry* 16:3133-3136.
19. Levy, W. B. *et al.* (1978): *J. Biol. Chem.* (in press).
20. Lohr, D. *et al.* (1977): *Proc. Nat. Acad. Sci. U.S.* 74:79-83.
21. Meehan, T. *et al.*(1977): *Nature* 269:725-727.
22. Metzger, G. *et al.* (1977): *Biochem. Biophys. Res. Commun.* 75:703-710.
23. Miller, E. C., and Miller, J. A. (1974): In: *The Molecular Biology of Cancer,* H. Bush, Ed., pp. 377-403, Academic Press, New York.
24. Morris, R. N. (1976): *Cell* 9:627-632.
25. Newbold, R. F, and Brookes, P. (1976): *Nature* 261:52-54.
26. Olins, A. L., and Olins, D. E. (1974): *Science* 183:330-332.
27. Olins, A. L. *et al.* (1976): *Nucleic Acids Res.* 3:3271-3291.
28. Osborne, M. R. *et al.* (1976): *Int. J. Cancer* 18:362-368.
29. Osborne, M. R. *et al.* (1978): *Chem.-Biol. Interact.* 20:123-130.
30. Pachmann, U., and Rigler, R. (1972): *Exp. Cell Res.* 72:602-608.
31. Pezzuto, J. M. *et al.* (1978): *Cancer Res.* 36:3647-3654.
32. Pranis, R. A., and Klotz, I. M. (1977): *Biopolymers* 16:299-316.
33. Pulkrabek, P. *et al.* (1977): *Biochemistry* 16:3127-3132.
34. Selander, R. K. (1973): *Biochem. J.* 131:749-755.
35. Sims, P., and Grover, P. L. (1974): *Adv. Cancer Res.* 20:165-274.
36. Slaga, T. J. *et al.* (1977): *Cancer Res.* 37:3126-3131.
37. Slaga, T. J. *et al.* (1977): *Cancer Res.* 37:4130-4133.
38. Weinstein, I. B. *et al.* (1976): *Science* 193:592-595.
39. Weintraub, H., and Groudine, H. (1976): *Cancer Res.* 193:848-856.
40. Wood, A. W. *et al.* (1976): *Cancer Res.* 36:3358-3366.
41. Woodcock, C. J. (1973): *Cell Biol.* 59:368a.

Polynuclear Aromatic Hydrocarbons, edited by P. W. Jones and P. Leber. Ann Arbor Science Publishers, Inc. Ann Arbor, MI ©1979.

ASBESTOS-FACILITATED MEMBRANE UPTAKE OF POLYNUCLEAR AROMATIC HYDROCARBONS STUDIED BY FLUORESCENCE SPECTROSCOPY: A POSSIBLE EXPLANATION OF THE COCARCINOGENIC EFFECTS OF PARTICULATES AND PAH

J. R. Lakowicz,[1] J. L. Hylden, F. Englund, A. Hidmark and M. McNamara

Freshwater Biological Institute and
Department of Biochemistry
University of Minnesota
P.O. Box 100
Navarre, Minnesota 55392

INTRODUCTION

Cocarcinogenic Effects of Particulate Matter and Chemical Carcinogens

Recent epidemiological evidence indicates that human exposure to asbestos in the workplace is a major cause of occupational cancers, primarily cancers of the lungs and the intestinal tract (9). Earlier studies have indicated that asbestos insulation workers have a sevenfold increased incidence of bronchogenic carcinoma as compared to persons not exposed to asbestos (4,15,38). When the smoking habits of these asbestos workers were examined it was found that the smokers had a 92-fold increased risk of lung cancer, whereas the nonsmoking asbestos workers had little or no increased risk of cancer. These data indicate that the cocarcinogenic effects of asbestos inhalation and cigarette smoking are responsible for approximately 90% of the cancers resulting from asbestos exposure. Similar synergistic or cocarcinogenic effects with cigarette smoking have been observed among other workers who are exposed to other particulate matter, these being coke oven (29) and uranium mine workers (2). In light of the realization that cocarcinogenesis of asbestos with cigarette smoke is a major cause of human

[1]Author to whom inquiries should be sent.

cancer, the Johns-Mansville Corporation, a major producer of mineral fibers, has recently banned cigarette smoking in work environments where respirable asbestos is present and has stopped hiring persons who smoke to work in these environments (23). Similar restrictions were not placed on workers exposed to particulates such as fibrous glass which is not known to be co-carcinogenic in humans (13), although fibrous glass is known to be a respiratory irritant (31). These actions of the Johns-Mansville corporation represent a pioneering effort in preventive environmental health in a industrial environment.

Cocarcinogenic effects of polynuclear aromatic hydrocarbons (PAH), known to be present in cigarette smoke and thought to be major contributors to its carcinogenic effects, have also been observed in studies with experimental animals. In fact, in the absence of particulate matter, intratracheal instillation of benzo[a]pyrene (B[a]P) results in only a low incidence of lung cancer, even at high dosages (36). However, a wide variety of particulates, when coinstilled with PAH, result in a high tumor yield. These particulates include hematite (35), asbestos (34), carbon particles (41), india ink (33) and aluminum and titanium oxide (41). Thus particulates in general, irrespective of chemical composition, appear to act as cocarcinogens. However, human epidemiological evidence points to asbestos as being the most significant cocarcinogen.

When are inhaled particulates and PAH cocarcinogenic? In the induction of lung cancer the following processes seem essential:

1. The inhaled carcinogen must be retained by the lungs. Adsorption of PAH to particulates which are themselves efficiently retained can increase the total exposure level to PAH (19).
2. The carcinogens must enter the cells which are subsequently transformed. This transport can be facilitated by particles which release adsorbed carcinogens rapidly in comparison with the clearance rates of these particles from the lungs, by particles which are rapidly phagocytized or by those which penetrate cell membranes as a result of their shape or surface properties.
3. The PAH must be metabolized to more active compounds, presumably diol epoxides (22), which react with cellular macromolecules. At present we do not know if particulates which increase cellular exposure to PAH will alter the amounts of these reactive species.

In this chapter we demonstrate that the adsorption of PAH to particulates greatly enhances the rate of PAH uptake into model membranes. We also demonstrate that asbestos, in particular the amosite form, is superior to silica in its ability to adsorb B[a]P in a monomeric state and to transport this carcinogen into membranes (process 2 above). Other studies in this laboratory have shown that many other particulates, such as hematite, talc and titanium oxide, also enhance B[a]P uptake into model membranes and that Canadian chrysotile and anthophyllite are more effective in facilitating B[a]P transport than are crocidolite, amosite and the nonfibrous minerals (26). Similar results were obtained with rat liver microsomes and 3T3 cells (24), indicating the validity of extrapolating results obtained with model membranes to natural membranes and to the microsomal fraction of cells. It should be noted that our model membranes consist of dipalmitoyl-L-α-phosphatidylcholine (DPPC). When particulates lodge in the lungs they initially contact a layer of lung surfactant, which is about 70% DPPC (30). Hence, DPPC vesicles appear to be a reasonable model for investigating the elution of carcinogens from particulates.

Thus it appears likely that the cocarcinogenic effects of particulates and PAH could be a result of the ability of particulates to adsorb PAH and then release these carcinogens for cellular uptake prior to clearance of these particulates by ciliary action. The following observations support this hypothesis. Asbestos is known to adsorb B[a]P out of benzene solution (34), to adsorb the natural oils from the jute bags in which it is often shipped (16), and to adsorb machining oils in the process of mining and milling (42). In addition, smoking-related lung cancers appear to be localized in regions where particles concentrate during ciliary clearance, such as vocal cords and the larger bronchi where other bronchial openings block ciliary flow. These cancers are less prevalent in the trachea where ciliary flow is uninterrupted (21). It appears likely that interruptions of ciliary flow in the lungs at a specific point would result in an increased carcinogen dose at this site. This speculation is supported by the observation that B[a]P is removed from soot particles which are present in human lungs (10).

However, we hasten to point out that mechanisms other than enhanced transport are also likely to be operative in the process of cocarcinogenesis. For example, mesothelioma can result from exposure to high levels of asbestos, but this form of human cancer does not appear to be related to cigarette smoking (12,14,37). Others have demonstrated that if an animal is exposed to a subthreshold level of coal tar, and if this site is subsequently subjected to surgical incisions or scarification, tumors develop at the irritated sites (8). In fact most types of cell trauma are known to be cocarcinogenic, or to predispose the scarred area to cancer (3). Implantation of foreign bodies into organisms also results in cancer after the formation of scar tissue around the implant (6). Particulate matter is known to cause cell damage (1,17,18, 32), cell proliferation (41), hyperplasia and fibrosis (7). The shape of the particulate also appears to have an effect on its ability to induce mesothelioma (40). Since all experimental animals are exposed to carcinogens to some extent, whether from external sources or internal metabolic processes, it is impossible to determine whether cell trauma alone causes cancer or whether the particles cause the expression of previously silent preneoplastic cells. Moreover, particulates can enhance the tumor incidence resulting from exposure to systemic carcinogens such as diethylnitrosamine (11). In this instance particle-enhanced uptake of the carcinogen appears unlikely. Thus, particle-enhanced cellular uptake of PAH appears to be a possible explanation of the known cocarcinogenic effects of these materials, but this mechanism is unlikely to explain all the known effects of particulates on the incidence of tumors in humans and animals.

In the present paper we describe how fluorescence spectroscopy may be utilized to investigate the surface distributions of PAH on particulates and how this spectroscopic method can be used to quantitate the membrane uptake of carcinogens off particulates. These methods should be valuable in investigations designed to reveal the mechanism or mechanisms of cocarcinogenesis.

MATERIALS AND METHODS

Fluorescence Spectral Data

All fluorescence spectral data were obtained on a computerized photon-counting spectrofluorometer (SLM Instruments, Inc., Urbana, IL). The following instrumental conditions were used for BA: excitation wavelength, 343 nm; excitation and emission bandpasses, 8 and 2 nm, respectively; excitation filter, Corning 7-54. Spectral data from samples containing silica were obtained using a 2-cm x 2-cm cuvette, 4 cm high, which was thermostated and positioned on a magnetic stirrer so that the stirring bar remained below the light path. Illumination of the cuvette was perpendicular to one of its sides. In spite of the turbidity of samples which contained silica, the scattered light did not interfere with our measurements. Continual stirring kept the particulates suspended and the resulting fluorescence intensity of the silica-adsorbed BA did not fluctuate or lose intensity over the course of the spectral scans. Membrane uptake of BA was followed by the increase in fluorescence intensity at 389 nm (BA, 1,2-benzanthracene).

Observations of B[a]P were obtained using the following instrumental conditions: excitation wavelength, 296 nm; excitation and emission bandpasses 8 and 4 nm, respectively; excitation filter, Corning 7-54; emission filters, 0-52 and 2 mm of 1 *M* $NaNO_2$. B[a]P uptake was measured by the intensity increase at 405 nm. The emission filters do not transmit light below 395 nm. Without these filters we do observe a moderate background signal from the asbestos at wavelengths below 395 nm. Some background signal was also seen above 395 nm, typically less than 10% of the total intensity. These backgrounds were quantitated using particles without B[a]P, and subtracted from the spectra shown.

Spectral data for B[a]P were also obtained using the 2-cm x 2-cm cuvette. However, in this instance the exciting beam impinged upon the sample at an angle of 20°, and the emission was observed from this illuminated surface at 90° to the exciting light.

Phospholipid Vesicles and Vesicle Integrity

Phospholipid vesicles of DPPC were prepared by ultrasonic irradiation of the lipid suspended in buffer at a lipid concentration of 10 mg/ml. The buffer was 0.01 *M* tris/0.05 *M* KCl, pH 7.5. Sonication was performed with a Heat Systems Model 350 for 10 min at about 50°C. The power input was near 200 W. The vesicles were then annealed at 50°C for 1 hr (28), followed by centrifugation at 54,000 x g for 90 min.

The structural integrity of phospholipid vesicles was investigated by their permeability to a fluorogenic probe, umbelliferyl phosphate (39). DPPC vesicles containing umbelliferyl phosphate in the internal aqueous volume were formed by cosonication of DPPC and the probe at final concentrations of 13 and 1 m*M*, respectively. External umbelliferyl phosphate was removed by dialysis. Alkaline phosphatase was added to these preformed vesicles. Upon disruption of the vesicles the umbelliferyl phosphate is released into the external volume where it is converted to the highly fluorescent hydrolysis product, umbelliferone. The vesicle integrity assay mixture contained 10 ml of buffer, 0.5 ml of the labeled DPPC vesicles, and 8.8 μg of *Escherichia coli* Type 3 alkaline phosphatase.

Umbelliferone fluorescence was observed at 445 nm, using an excitation wavelength of 370 nm and a Corning 7-54 excitation filter. In order to calibrate this assay for complete disruption, the vesicles were heated to 100°C for 7 min. A second addition of alkaline phosphatase was made after cooling to the experimental temperature.

Preparation of Particulates with Adsorbed PAH

BA concentrations were determined by optical density measurements at 343 nm with the use of an extinction coefficient of 6,920 M^{-1} cm^{-1} in benzene. Silica-adsorbed BA was prepared by chromatography on silica gel G preparative thin-layer plates (20-cm x 20-cm, 1,000 μ thickness of silica, obtained from Analabs, Inc., North Haven, CT). BA in benzene was streaked on the plate, which was subsequently developed in benzene/cyclohexane, 80/20. The band containing BA was removed from the plate and air-dried. A small fraction of this sample was extracted six times with benzene. The optical density of these extracts revealed a loading density of the BA-silica sample. After preparation these samples were stored in the dark under argon. The fluorescence emission spectrum of the extracted BA was identical to that of the starting material, which indicates that both the starting material and the adsorbed BA did not contain any significant additional fluorescent compounds. In addition, the BA migrated as a single band on the chromatographic plate.

The silica used in these studies had a surface area of 381 m^2/g, as determined by N_2 adsorption; 95% of the particles were between 1 and 15 μ in diameter, and most were about 2 μ in diameter (25).

Microcrystalline dispersions of BA in aqueous buffer were prepared in two ways. First 6 μg of BA were suspended in buffer and sonicated with a Heat Systems Model 350 sonicator (Heat Systems–Ultrasonics, Inc., Plainview, NY) for 20 min at about 200 W-sec. To obtain a more finely divided suspension of BA in buffer, a benzene solution containing the required amount of BA was evaporated to dryness in a beaker under an N_2 flow. In this way the BA was coated evenly on the surface of the vessel. After aqueous buffer was added, the sample was sonicated as before.

Benzo[a]pyrene (B[a]P), 99+% purity, was obtained from Aldrich Lot #101847 and used without further purification. B[a]P concentrations were obtained using a molar extinction coefficient of 57,500 $M^{-1} cm^{-1}$ at 299.4 nm. Particulate samples containing B[a]P were prepared by mixing the particulate with a benzene solution of B[a]P, followed by evaporation of the benzene under reduced pressure. For the studies reported here 0.1 to 3.0 mg of B[a]P was added for each gram of particulate. The sample was rotated continuously during the benzene evaporation. After the sample appeared dry it was kept under vacuum for 30-45 min at 85°C. At the higher weight ratios B[a]P crystals could be seen on the walls of the glass vessel. Any material which clung to the walls was discarded. After removal from the vessel the particulates were stored in the dark under an argon atmosphere.

The amount of B[a]P present on the particles was determined by solvent extraction using a Soxhlet extractor. Typically the extraction was continued for 4.5 hr, which constituted about 50 individual extractions. All extractions were done in duplicate. The fluorescence emission spectra of the extracted material were identical to that of the starting material. Thus there did not appear to be any degradation of B[a]P on this particulate which interfered with our measurements. The amounts of B[a]P adsorbed to silica and asbestos were found to be 85 ± 15% of the amount added. Aqueous dispersions of B[a]P crystals were prepared by evaporation of a benzene solution of B[a]P to dryness, addition of buffer, followed by sonication for 30 min at 40 W using a Cole-Parmer Model 8845-2 water-bath-type sonicator. We will refer to this preparation as being microcrystalline.

For all the B[a]P uptake kinetics reported here we used 5 μg of B[a]P and 10 mg of DPPC. An amount of particulate containing 5 μg B[a]P, or 5 μg

B[a]P crystals, was suspended in 10 ml of buffer. After measurement of the initial intensity, 10 mg of DPPC were added in 1 ml of buffer to initiate the reaction. Complete B[a]P transfer to the bilayers was obtained by heating the sample to 55°C for 30 min. The final fluorescence intensity was measured after reequilibration at the experimental temperature of 25°C.

For measurements of membrane uptake of both BA and B[a]P, the lipid-to-PAH weight ratios were always in excess of 500 to 1. Otherwise, the final fluorescence spectra indicate the presence of PAH crystals and/or excimer formation in the lipid bilayers.

RESULTS

Fluorescence Emission Spectra of 1,2-Benzanthracene (BA)

The fluorescence emission spectra and apparent quantum yields of the PAH are dependent on their immediate local environments. For example, the emission spectrum of BA in ethanol is distinct from the dispersion of BA microcrystals in aqueous buffer (Figure 1). The emission spectrum of BA adsorbed to silica is similar, but not identical, to that observed for the BA microcrystals (Figure 2 A). The emission spectrum of BA in DPPC vesicles is very similar to that observed for the dilute ethanol solution. The differences in the crystalline, silica-adsorbed and membrane-bound emission spectra of BA suggests that these changes can be used to quantitate membrane uptake of BA.

Membrane Uptake of 1,2-Benzanthracene

When silica with adsorbed BA is dispersed in aqueous buffer, both the fluorescence spectral distribution and intensity is constant over a time period from hours to days. On addition of phospholipid vesicles there occurs a rapid increase in fluorescence intensity at 389 nm (Figure 3) and a shift of the emission spectrum to the solvent-like form (Figure 2). These spectral changes are essentially complete after 30 min at 31°C at which time the fluorescence spectrum has lost any observable crystal-like component.

The percentage of BA which is transferred to the bilayer at time t is calculated from the fluorescence intensity at this time, I(t), and the initial and final fluorescence intensities at 389 nm, I_0 and I_∞, respectively.

$$\text{\% Carcinogen transferred to membrane} = \frac{I(t) - I_0}{I_\infty - I_0} \times 100 \qquad (1)$$

The final intensity is found by heating the samples to 41°C for 2 hr, followed by reequilibration to the experimental temperature. No spectral changes occurred upon heating the silica-adsorbed BA in buffer in the absence of lipid.

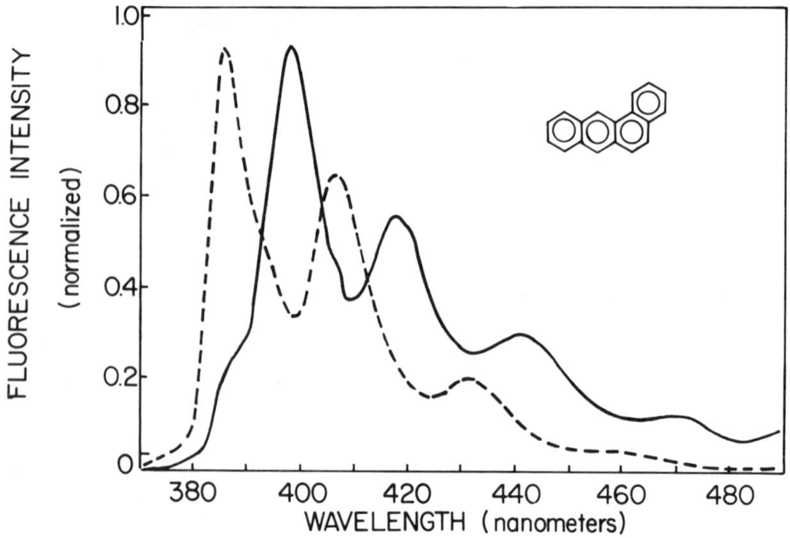

FIGURE 1. Fluorescence emission spectra of 1,2-benzanthracene. Emission spectra are shown in ethanol (- - - -, 1.7 x 10^{-5} *M*), and for a dispersion of sonicated BA crystals in aqueous buffer (——, 3.7 μg BA per ml of buffer).

We used these spectral changes to quantitate the membrane uptake of BA from crystals, sonicated crystals, sonicated crystals after benzene evaporation and the silica-adsorbed state (Figure 4). The silica-adsorbed state is more readily transported to DPPC vesicles by almost two orders of magnitude when compared with crystalline BA. The adsorbed state of the carcinogen is more available than for any crystals we could prepare by either solvent evaporation or sonication. Silica, added after sonication of the crystals, may cause a slight increase in the carcinogen availability, but given the difficulty in obtaining reproducible crystal sizes for BA by sonication, the effect shown in Figure 4 due to added silica is within experimental limits.

Data presented in Figure 4 demonstrate that adsorption of BA to silica greatly enhances its rate of transport into DPPC vesicles. This result may be of significance in explaining the enhanced carcinogenicity of PAH in the presence of particulate matter.

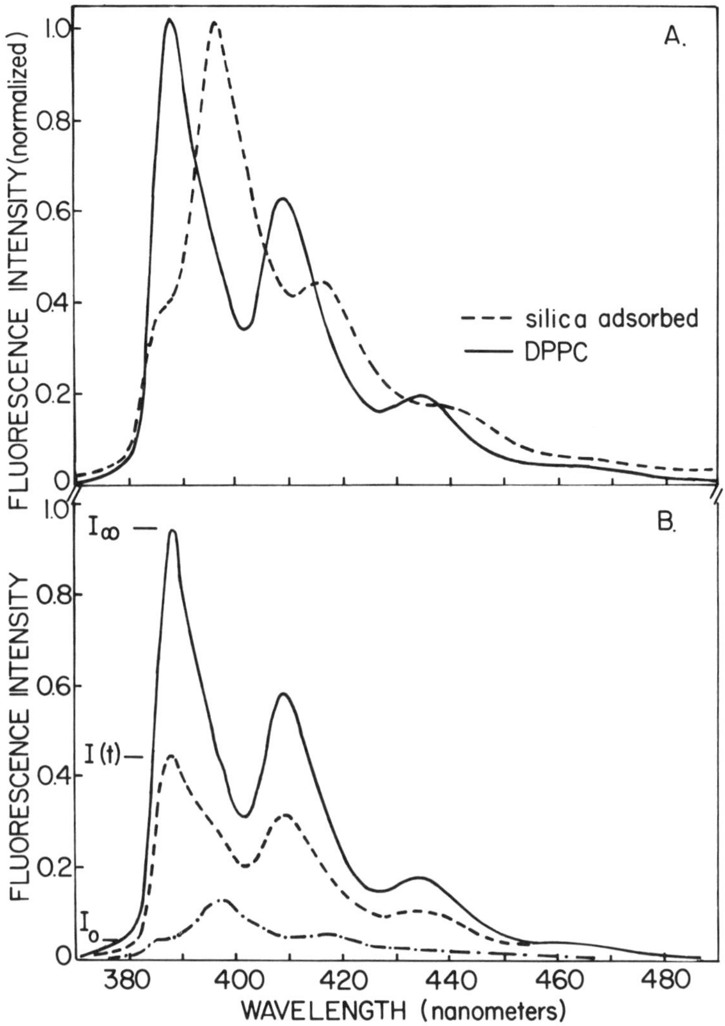

FIGURE 2. Fluorescence emission spectra of 1,2-benzanthracene. (A) Normalized emission spectra are shown for silica-adsorbed BA (– – – –), 1.1 mg BA/g silica, and the same sample after addition of and equilibration with 3 mg of DPPC vesicles (———). (B) 5.5 mg of the silica sample was suspended in 10 ml of buffer. These spectra were recorded at the same instrumental gain settings before addition of DPPC (•-•-•), 3 min after DPPC addition (- - - -), and after 16 hr at 40°C, followed by reequilibration to the experimental temperature of 31°C (——). About 3 min are required to record an emission spectrum. Reprinted with permission, see Reference 25.

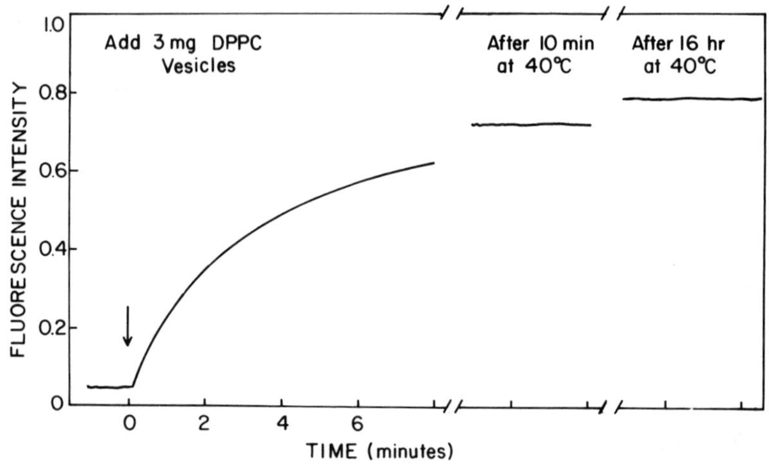

FIGURE 3. Fluorescence intensity of 1,2-benzanthracene during transfer from silica to DPPC vesicles. 5.5 mg of the 1.1 mg/g silica sample was suspended in 10 ml of buffer, T = 31°C. The fluorescence intensity was recorded at 389 nm. At t = 0, 3 mg of DPPC in 0.3 ml of buffer was added. In order to effect complete membrane uptake of BA the sample was heated to 40°C for the times indicated. Reprinted with permission, see Reference 25.

Membrane-Uptake of Benzo[a]Pyrene from a Fibrous and a Nonfibrous Mineral Particulate

Figure 5 shows fluorescence emission spectra of B[a]P resulting from a dilute benzene solution, and from a suspension of B[a]P crystals in an aqueous buffer. When B[a]P is dissolved in benzene, a structured emission is observed, whereas the suspension of B[a]P crystals yields a broad structureless emission at longer wavelengths. This latter emission results from formation of an excited state charge-transfer complex between two or more adjacent B[a]P molecules (5) and is called the excimer emission. The fluorescence emission spectra may therefore be used to reveal the proximity of B[a]P molecules to one another. Also shown in Figure 5 is the fluorescence emission spectrum of B[a]P which is partitioned into lipid bilayers after uptake from asbestos fibers. Only the structured emission is observed which is indicative of widely dispersed B[a]P molecules.

Figure 6 shows the fluorescence emission spectra of B[a]P when present as microcrystals which are suspended in aqueous buffer, and when the B[a]P is adsorbed to silica or amosite. The emission spectra of the microcrystals and of the silica-adsorbed B[a]P are similar, indicating the presence of B[a]P microcrystals on the silica surface, or the presence of B[a]P crystals which did not adsorb to silica during the solvent evaporation process. The emission spectra of the asbestos-adsorbed B[a]P is dominated by the structured emission, indicating the presence of monomeric adsorbed B[a]P. The superior ability of amosite to disperse B[a]P is especially surprising when the surface areas of these two particulates are compared. As determined by nitrogen adsorption these surface areas are 5.7 and 381 m^2/g for asbestos and silica, respectively (25,42).

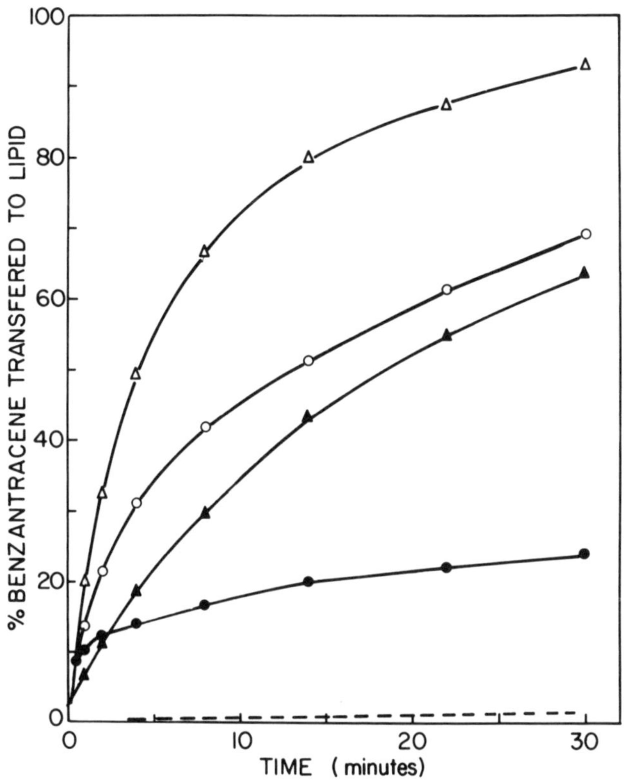

FIGURE 4. Membrane uptake of 1,2-benzanthracene. The percentage of BA transferred is shown for unsonicated crystals (- - - -), sonicated crystals (●——●), sonicated crystals after benzene evaporation (▲——▲) and silica-adsorbed BA (△——△). The data shown for silica-adsorbed BA is an average of the values observed for BA to silica weight ratios ranging from 1.2 to 0.08 mg BA per gram of silica. The BA uptake kinetics were not sensitive to its surface density on the silica. Also shown is the uptake data for sonicated crystals formed from benzene evaporation to which 10 mg of silica was added after sonication (o——o). The amount of BA varied from 6 to 30 μg, and the amounts of DPPC vesicles from 4 to 15 mg. The DPPC to BA weight ratio was 500. T = 31°C. Reprinted with permission, see Reference 25.

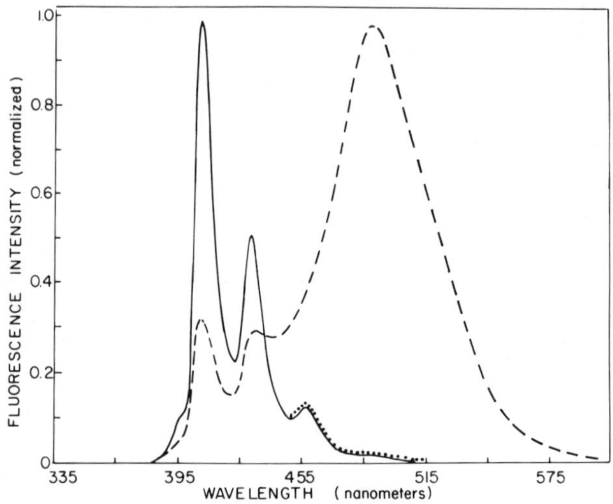

FIGURE 5. Fluorescence emission spectra of benzo[a]pyrene. Emission spectra are shown for B[a]P in benzene (———), and as a suspension of 0.5 mg B[a]P crystals in 10 ml of aqueous buffer (- - - -). After addition of DPPC vesicles to asbestos containing adsorbed B[a]P (• • •) the emission spectra closely resembles that of B[a]P in benzene. Reprinted with permission, see Reference 26.

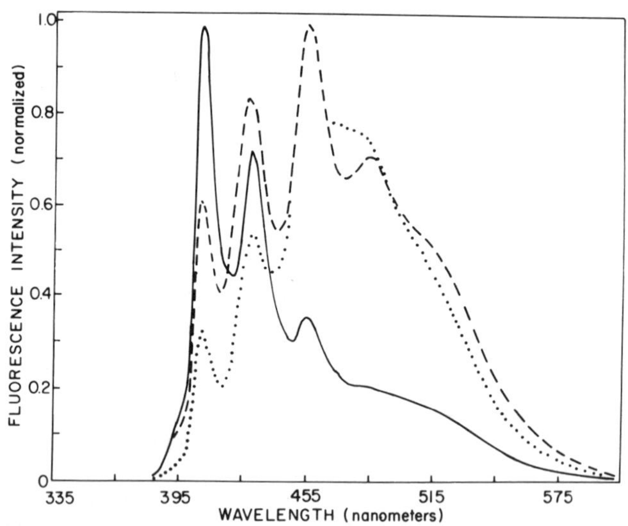

FIGURE 6. Fluorescence emission spectra of benzo[a]pyrene. Normalized emission spectra are shown for B[a]P microcrystals (• • •), and for B[a]P adsorbed to 8.6 mg silica (————) and 9.5 mg asbestos (———). In each case the sample was suspended in 10 ml of buffer and contained 5 μg B[a]P.

The fluorescence emission spectra of B[a]P are sensitive to both the weight percentage of carcinogen which is on the particulate and the age of the samples. Over a period of weeks after preparation, the emission spectra of B[a]P on the silica samples become more structured and show less excimer emission. At present we do not know what factors control this process, but temperature of sample storage and the presence of residual solvent do not appear to affect the rate of sample aging. Less significant changes occur with asbestos-adsorbed B[a]P. Irrespective of these considerations we consistently observed more B[a]P excimer emission from the silica than from the asbestos samples. These observations include B[a]P-to-particle weight ratios ranging from 3 to 0.1 mg B[a]P per gram of particulate. For both particulates the excimer emission is less pronounced at the lower weight ratios (Table I).

TABLE I. Benzo[a] Pyrene Uptake by Phospholipid Vesicles from Particulate Adsorbed and Microcrystalline States

mg BP / g Particulate	Time for 50% Uptake (min) Silica	Time for 50% Uptake (min) Asbestos	(Monomer Emission)[a] / (Total Emission) Silica	(Monomer Emission)[a] / (Total Emission) Asbestos
0.1	4.5	3	0.32	0.34
0.3	24	6.5	0.24	0.34
1.0	30	12	0.19	0.28
3.0	34	17	0.23	0.22

BP Microcrystals	Time for 15% Uptake (min)	(Monomer Emission) / (Total Emission)
no particles	23	0.17
+ amosite[b]	24	0.21
+ silica[b]	24	0.19

[a]This value was estimated from the ratio of area under the fluorescence emission spectrum from 380 to 440 nm to the total area from 380 to 605 nm. It should be noted that the emission spectra of B[a] P adsorbed to silica appears to lose eximer emission over a period of weeks after preparation. After complete uptake of BP by vesicles this ratio is 0.47. Reprinted with permission from *Nature* (Lakowicz and Hylden, 1978).
[b]Particles were added after formation of the microcrystals.

As was observed for adsorbed BA, the emission spectra of aqueous suspensions of particles which contain B[a]P retain a constant intensity and shape over a period of hours. Upon addition of phospholipid vesicles, an increase in fluorescence intensity occurs at 405 nm. Simultaneously the excimer emission disappears and the emission spectra of all the samples listed in Table I become identical to that representative of B[a]P in a hydrophobic environment, indicating transfer of B[a]P from the surface of the particle to the lipid bilayer. At present we do not know whether the B[a]P undergoes significant decomposition when adsorbed to these particulates. However, the final emission spectra indicate that any such decomposition products do not interfere with our spectral measurements of B[a]P uptake.

In Figure 7 we compare the rates of membrane uptake of B[a]P from the microcrystalline, silica-adsorbed and asbestos-adsorbed states. The percentage of B[a]P transferred was calculated according to Equation 1, except that the fluorescence intensity at 405 nm was utilized. Asbestos-adsorbed B[a]P shows the most rapid rate of membrane uptake, and the microcrystals of B[a]P the slowest rate. In Table I we list the time required for 50% transfer of B[a]P for particulate samples with a range of B[a]P-to-particulate weight ratios. Asbestos is seen to be more effective than silica at all weight ratios, with the most dramatic effect being observed at the lower ratios. These rates increased as surface density of B[a]P on the particulates decreased. Asbestos or silica, when added to preformed microcrystals of B[a]P in buffer, did not increase the rate of B[a]P membrane uptake (Table I) or alter the shape of the fluorescence emission spectra of the suspended microcrystals. Therefore, only B[a]P which is adsorbed to the particles displays an increased rate of membrane uptake. These studies are in agreement with animal studies which indicate that B[a]P is more highly carcinogenic when adsorbed to asbestos (34) or hematite (20) than when simply mixed with these particulates.

Vesicle Integrity in the Presence of Silica and Asbestos

In the preceding paragraphs we assumed that the fluorescence spectra which appeared upon addition of phospholipid vesicles was the spectrum of the PAH present in the nonpolar region of the bilayers. However, we questioned whether similar spectral changes could be a result of adsorption of the vesicles to the surface of the particulates without the actual transfer of the PAH into the bilayer.

As a result of the nonpolar nature of BA and B[a]P we expected these compounds to partition strongly into the hydrophobic region of lipid bilayers. We investigated the emission spectra of BA in DPPC vesicles which were formed by cosonicating the DPPC in the presence of the BA but in the absence of either silica or asbestos and the emission spectrum of B[a]P after uptake from microcrystals, again in the absence of particulates. These emission spectra are identical to the spectra which result from equilibration of the vesicles with the particulate-adsorbed PAH. Thus it appears likely that the PAH are located in a hydrophobic environment. We would not expect such spectral changes to result from binding of the phospholipid vesicles to the silica surface unless this binding resulted in contact of the nonpolar alkyl side chains with the polar surface of the particulates. Such binding is expected to be weak, but nonetheless we investigated this possibility.

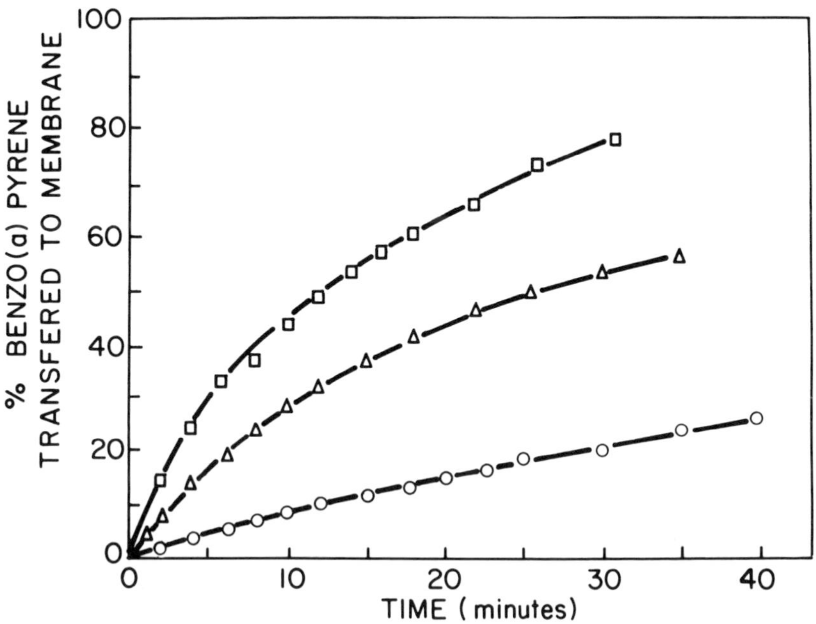

FIGURE 7. Rate of benzo[a]pyrene uptake into lipid bilayers. The percentage of benzo[a]pyrene transferred to DPPC vesicles is shown for microcrystals of B[a]P (0), silica-adsorbed B[a]P (△), and amosite-adsorbed B[a]P (□). See the legend to Figure 6 for additional details. Reprinted with permission from *Nature* (Lackowicz and Hylden, 1978).

Contact of the alkyl side chains of the vesicles with the surface of the particulates could only occur with disruption of the phospholipid vesicles. We investigated the integrity of the DPPC vesicles by their impermeability to the fluorogenic probe umbelliferyl phosphate as described in the Materials and Methods. Under the conditions of our experiments neither silica nor amosite caused release of umbelliferyl phosphate, and we therefore conclude that these particulates did not cause disruption of the vesicles. We note that no disruption was observed under conditions which would reveal even small amounts of disruption, this being a 25-fold increase in the weight ratio of particulate to phospholipid as compared to the normal experimental ratio.

Solubilization in the Aqueous Phase as a Rate-Limiting Step for Membrane Uptake of PAH

At present we do not know if BA and B[a]P are being transferred into vesicles which are adsorbed to the surface of the particulate or to those which are free in solution. Under our experimental conditions with silica, 30% of the vesicles were bound to the particulate. In addition, we do not know if the extent of binding is important in the transfer process.

If binding of the vesicles to the particulates is important we expect an increased lipid concentration to result in an increased rate of membrane uptake of the PAH. Similarly, if collisional encounters between the vesicles and the particles are important, we again expect an increased vesicle concentration to result in an increased uptake rate. Contrary to these expectations we find that the membrane uptake rates of both BA (25) and B[a]P (26) to be independent of lipid concentration and independent of particle concentration at the same PAH to particulate weight ratio. These observations strongly indicate that solubilization of these PAH in the aqueous phase is the rate-limiting step in their membrane uptake.

CONCLUSIONS

1. The fluorescence emission spectra and apparent quantum yields of the PAH, when adsorbed to particulates, are dependent on the surface density of the PAH on the particulate, the chemical composition of the particulate, and the method by which the PAH and the particulate are complexed. As a result these spectral properties provide a means to investigate the nature of the adsorbed state.

2. These spectral properties of the PAH undergo dramatic changes upon transfer from the surface of the particulate to lipid bilayers. These spectral changes can be conveniently used to monitor the rates of membrane uptake of the PAH.

3. In general, PAH which are adsorbed to particulates are more rapidly transported into membranes than are PAH which are in the crystalline or microcrystalline state. The rate-limiting step for membrane uptake of the PAH appears to be their rate of solubilization into the aqueous phase.

4. Most importantly, B[a]P which is adsorbed to the amosite form of asbestos shows a higher membrane availability than B[a]P which is adsorbed to silica.

These observations may provide a partial explanation of particle-PAH cocarcinogenesis.

ACKNOWLEDGMENTS

We thank the Freshwater Biological Research Foundation, and especially its founder, Mr. Richard Gray, Sr., without whose assistance this work would not have been possible. In addition, we acknowledge the generous support of the American Cancer Society, Grant No. BC-261, the National Institutes of Health, Grant No. ES-1283 and a Grant-in-Aid from the American Heart Association. J. R. Lakowicz is an Established Investigator of the American Heart Association.

In addition we thank Dr. David Bevan for his assistance in the vesicle integrity studies and his critical review of this manuscript.

REFERENCES

1. Allison, A. C. (1971): Lysosome and the toxicity of particulate pollutants. *Arch. Intern. Med.* 128:131-139.
2. Arcker, V. E., Wagoner, J. K., and Lundin, R. E. (1973): Lung cancer among uranium miners in the United States. *Health Phys.* 25:351-371.
3. Berenblum, I. (1944): Irritation and carcinogenesis. *Arch. Pathol.* 38: 233-244.
4. Berry, G., Newhouse, M. I., and Turok, M. (1971): Combined effect of asbestos exposure and smoking on mortality from lung cancer in factory workers. *Lancet* 2:476-479.
5. Birks, J. B., and Cameron, J, W. (1959): *Proc. R. Soc. London* Ser. A249:297-317.
6. Brand, K. G., Buoen, L. C., and Brand, I. (1975): Foreign body tumorigenesis induced by glass and smooth and rough plastic: comparative study of preneoplastic events. *J. Nat. Cancer Inst.* 55:319-322.
7. Creasia, D. A., and Netteshein, P. (1974): Respiratory cocarcinogenesis studies with ferric oxide: A test of current experimental models. In: *Experimental Lung Cancer*, E. Karbe and J. F. Park, Eds., pp. 234-243, Springer-Verlag, New York.
8. Deelman, H. T. (1927): The part played by injury and repair in the development of cancer, with some remarks on the growth of experimental cancers. *Proc. Roy. Soc. Med.* 20:19-20.
9. Department of Housing, Education and Welfare (1978): Estimation of the Fraction of Cancer in the United States Related to Asbestos. September 15, 1978, National Cancer Institute.
10. Falk, H. L., Kotin, P., and Markul, I. (1958): The disappearance of carcinogens from soot in human lungs. *Cancer* 11:483-489.
11. Feron, V. J., Emmelot, P., and Vossenarr, T. (1972): Lower respiratory tract tumors in Syrian golden hamsters after intratracheal instillations of diethylnitrosamine alone and with ferric oxide. *Eur. J. Cancer* 8:445-449.
12. Godwin, M. C., and Jagatic, J. (1970): Asbestos and mesotheliomas. *Environ. Res.* 3:391-416.
13. Gross, P., Tuma, J., and deTrevill, R. T. P. (1971): Lungs of worker exposed to fiber glass: A study of their pathological changes and their dust content. *Arch. Environ. Health* 23:67-76.

14. Haley, T. J. (1975): Asbestosis: A reassessment of the overall problem. *J. Pharm. Sci.* 64:435-448.
15. Hammond, E. C., and Selikoff, I. J. (1973): Relation of cigarette smoking to risk of death of asbestos-associated disease among insulation workers in the United States. In: *Biological Effects of Asbestos.* IARC Sci. Publ., 8:312-317.
16. Harrington, J. S. (1965): Chemical studies of asbestos. *Ann. N.Y. Acad. Sci.* 132:31-47.
17. Harrington, J. S. (1976): The biological effects of mineral fibers, especially asbestos, as seen from *in-vitro* and *in-vivo* studies. *Ann. Anat. Pathol.* (Paris) 21:155-198.
18. Harrington, J. S., Miller, K., and Magnab, G. (1971): Hemolysis by asbestos. *Environ. Res.* 4:95-117.
19. Hatch, T. F., and Gross, P. (1964): *Pulmonary Deposition and Retention of Inhaled Aerosols*, Chap. 4, Academic Press, New York.
20. Henry, M. C., Port, C. D., and Kaufman, D. G. (1975): Importance of physical properties of benzo[a]pyrene-ferric oxide mixtures in lung tumor induction. *Cancer Res.* 35:207-217.
21. Hilding, A. C. (1957): Ciliary streaming in the bronchial tree and the time element in carcinogenesis. *New England J. Med.* 256:634-640.
22. Jerina, D. M., and Daly, J. W. (1974): Arene oxides: A new aspect of drug metabolism. *Science* 185:573-585.
23. Johns Mansville Corporation (1978): Banning Employee Smoking at Asbestos Plants. *Johns-Mansville News*, January 31, Denver, CO.
24. Lakowicz, J. R., and Bevan, D. (1978): Unpublished observations.
25. Lakowicz, J. R., Englund, F., and Hidmark, A. (1978): Particle enhanced membrane uptake of 1,2-benzanthracene observed by fluorescence spectroscopy: A possible role in co-carcinogenesis. *Biochim. Biophys. Acta* 543:202-216.
26. Lakowicz, J. R., and Hylden, J. I. (1978): *Nature* 275:446-448.
27. Lakowicz, J. R., McNamara, M., and Steenson, L. (1978): Particle mediated membrane uptake of chemical carcinogens studied by fluorescence spectroscopy. *Science* 199:305-307.
28. Lawaczeck, R., Kainosho, M., and Chan, S. I. (1976): The formation and annealing of structural defects in lipid bilayers and vesicles. *Biochim. Biophys. Acta* 443:313-330.
29. Lloyd, J. W. (1972): Long-term mortality study of steelworkers. Respiratory cancer in coke plant workers. *J. Occup. Med.* 14:621-629.
30. McDermott, M., Wagner, J. C., Tetley, T., Harwood, J., and Richards, R. J. (1975): The effects of inhaled silica and chrysotile on the elastic properties of rat lungs; Physiological, physical and biochemical studies of lung surfactant. *Inhaled Particulates* 4:415-427.
31. Milby, T. H., and Wolf, C. R. (1969): Respiratory tract irritation from fibrous glass inhalation. *J. Occup. Medicine* 11:409-410.
32. Miller, K., and Harrington, J. S. (1972): Some biochemical effects of asbestos on macrophages. *Brt. J. Exp. Path.* 53:397-405.
33. Pylev, L. N. (1961): Experimental induction of lung cancer in rats by intratracheal administration of 9,10-dimethyl-1,2-benzanthracene. *Bull. Exp. Biol. Med.* (USSR) 52:1316-1319.
34. Pylev, L. N., and Shabad, I. M. (1973): Some results of experimental studies in asbestos carcinogenesis. In: *Biological Effects of Asbestos,* IARC Sci. Publ., 8:99-106.

35. Saffiotti, U., Cefis, F., and Kolb, L. H. (1968): A method for the experimental induction of bronchogenic carcinoma. *Cancer Res.* 28:104-124.
36. Saffiotti, U., Cefis, F., Kolb, L. H., and Shubik, P. (1965): Experimental studies of the conditions of exposure to carcinogens for lung cancer induction. *J. Air Poll. Control Assoc.* 15:23-25.
37. Selikoff, I. J., and Hammond, E. C. (1968): Community effects of monoccupational environmental asbestos exposure. *Am. J. Public Health* 58:1658-1666.
38. Selikoff, I. J., Hammond, E. C., and Chung, J. (1968): Asbestos exposure, smoking, and neoplasia. *J. Am. Med. Assoc.* 204:104-110.
39. Six, H. R., Young, W. W., Ulmura, K., *et al.* (1974): Effect of antibody-complement on a multiple vs. single compartment liposomes. Application of fluorometric assay for following changes in liposomal permeability. *Biochemistry* 13:4050-4057.
40. Stanton, M. F., and Wrench, C. (1972): Mechanism of mesothelioma induction with asbestos and fibrous glass. *J. Natl. Cancer Inst.* 48:797-821.
41. Stenback, F., Rowland, J., and Sellakumar, A. (1976): Carcinogenicity of benzo[a]pyrene and dusts in hamster lung (instilled intratracheally with titanium oxide, carbon and ferric oxide). *Oncology* 33:29-34.
42. Timbrell, V. (1969): *Pneumoconiosis*, Proc. Intl. Conf., Johannesburg, pp. 24-37, Oxford University Press, Oxford.
43. Wagner, J. C., Berry, G., Cooke, T. J., Hill, R. J. Pooley, F. D., and Skidmore, J. W. (1975): Animal experiments with talc. *Inhaled Particles* 4:647-654.

Polynuclear Aromatic Hydrocarbons, edited by P. W. Jones and P. Leber. Ann Arbor Science Publishers, Inc. Ann Arbor, MI © 1979.

BIOCHEMICAL STUDIES ON THE METABOLISM AND DNA-BINDING OF DMBA AND SOME OF ITS MONOFLUORO DERIVATIVES OF VARYING CARCINOGENICITY

F. B. Daniel,[1,2,3] L. K. Wong,[1] C. T. Oravec,[1]
F. D. Cazer,[3] C'L. A. Wang,[1] S. M. D'Ambrosio,[1,2]
R. W. Hart[1,2,3] and D. T. Witiak[3]

Departments of Pharmacology[1] and Radiology[2]
College of Medicine
Division of Medicinal Chemistry[3]
College of Pharmacy
The Ohio State University
Columbus, Ohio 43210

Recent years have witnessed an increasing awareness in the possible biological consequences of environmental chemicals. Epidemiological studies have indicated that environmental factors may well be causative components in a sizeable proportion of human cancer (14). Correspondingly, research in the field of chemical carcinogenesis has been expanding rapidly in recent years. Much of the credit for this expansion of knowledge can be traced to the postulate of the Millers that chemical carcinogens either are electrophiles or can be metabolically converted to such species (30-32). Thus, on the molecular level, much of the effort in chemical carcinogenesis has centered on elucidating the method by which cells convert precarcinogenic chemicals to electrophilic species and the subsequent reaction of the activated species with cellular nucleophiles.

Numerous authors have postulated that cellular DNA is the critical target for chemical carcinogenesis (5,16,18,23). Thus, this paper reports on the metabolism and DNA-binding potential of a series of monofluoro derivatives of the polycyclic aromatic hydrocarbon (PAH) 7,12-dimethylbenz[a]-anthracene (DMBA). These compounds have widely varying degrees of carcinogenicity and yet are structurally similar. They thus provide a convenient model system for evaluating the importance of metabolism and DNA binding on the oncogenic process.

DMBA (Figure 1), one of the earliest pure chemicals to be proven carcinogenic (27,37), is the most powerful PAH carcinogen yet studied on mouse

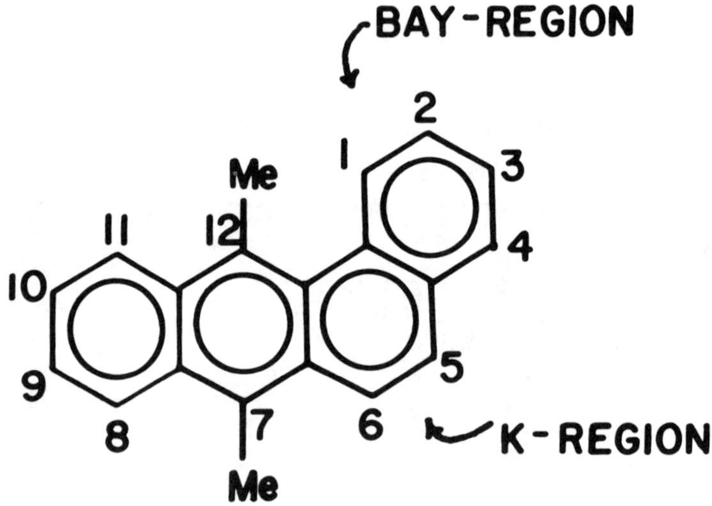

FIGURE 1. Structural formula of 7,12-dimethylbenz[a] anthracene.

skin (3,4). A single intravenous or intragastric dose of DMBA can induce mammary tumors (21,38), lung tumors (42,43), ovarian tumors (26,41), leukemias (41) and tumors of Zymbal's gland (19). As little as three micrograms of DMBA applied to the skin of a mouse and followed by an appropriate promoter will result in epidermal carcinomas (4). At larger doses (50-75 μg) DMBA will induce skin tumors in the absence of promoters (3,4). Additionally, in the proper test system, DMBA is a potent mutagen (20).

DMBA can be induced to bond to DNA in a variety of *in vivo* (6,15,24), *in vitro* (1,11,13,33) and cell-free systems (2,7,11). Similarly, DMBA has been shown to bond to the DNA of mouse skin (target tissue) more extensively than other less carcinogenic PAH (6). Similar results were obtained in a cell-free system that employed mouse epidermal microsomes to activate the DMBA (7).

In a collaborative study Professor M. S. Newman prepared a series of monofluoro derivatives of 7-methylbenz[a] anthracene and these compounds were tested for their carcinogenicity by the Millers (34). The rationale behind these experiments was that the aryl-fluorine bond, which is of greater bond energy than the aryl-hydrogen bond [128 kcal/mole fluoro-benzene vs 112 kcal/mole benzene (8)], would block or reduce carcinogenicity if it was inserted in a position that is critical in the actual bonding to the cellular target. Alternatively, fluorine could block or reduce carcinogenicity if it were substituted into a position critical for the activation of the hydrocarbon. Of the compounds synthesized and tested (*i.e.,* 1F-, 2F-, 3F-, 4F-, 5F-, 6F-,

7F-, 8F-, 9F- and 10F-) only the 5-fluoro-7-methylbenz[a]anthracene was inactive as a carcinogen (34). Newman and the Millers (34) also synthesized and tested four monofluoro derivatives of DMBA (*i.e.*, the 4F-, 5F-, 8F- and 11F-) and of these only the 5F-DMBA was inactive. Similar results were obtained by Huggins (34). The recent results of Harvey and Dunne (17) demonstate that the 1F-DMBA and 2F-DMBA do not produce sarcomas in Long-Evans rats under conditions where DMBA and its 4F-, 8F and 11F-derivatives are active. The recent work of Slaga and Huberman *et al.* (40) has shown that as a tumor initiator 11F-DMBA is comparable to DMBA itself whereas 5F-DMBA is only about one one-hundredth as effective. In the same system, 1F-DMBA and 2F-DMBA are essentially inactive (40). These same investigators have shown that the mutagenicity of these compounds (in V79 cells employing a Syrian hamster feeder layer) parallels their tumor-initiating power with the 2F- and 1F- being essentially inactive, the 5F- weakly active and the 11F- only slightly less active than DMBA itself (40).

The relative lack of biological activity observed with 1F- and 2F-DMBA is consistent with the molecular orbital calculations of Jerina and Daly (25) which postulate that metabolites of the "bay region" (for DMBA the A-ring or the 1-, 2-, 3- and 4- positions; Figure 1) are responsible for the carcinogenic and/or mutagenic properties of PAH. The elegant study of Moschel *et al.* (33) seems to corroborate these predictions. They have shown that the DMBA-DNA adducts formed when mouse embryo cells are treated with DMBA exhibit an excitation and emission spectrum very similar to 9,10-dimethylanthracene. Additional studies by Dipple and Nebzydoski (13) have indicated that the DNA binding metabolite of DMBA is an A-ring diol epoxide.

MATERIALS AND METHODS

Hydrocarbons

[G-^{3}H]-DMBA was purchased from Amersham Corporation and purified on the day of usage by thin-layer chromatography (TLC) (double development, silica, cyclohexane:benzene, 80:20 v/v). All work on this and other hydrocarbons was carried out under gold fluorescent lamps (Westinghouse F40GO). 2F-DMBA and 11F-DMBA, generously supplied by Professor M. S. Newman, The Ohio State University, were catalytically tritiated by Amersham, and were purified by high-pressure liquid chromatography (HPLC). The 5F-(*bis*-7,12-[^{14}C]methyl)benz[a]anthracene ([^{14}C]5F-DMBA) was prepared in our laboratory by published procedures (9) and purified by TLC on the day of usage.

Cell-Free Metabolic Oxidation

Liver microsomal (S10) preparations were prepared from male Sprague-Dawley rats (200-300 g) that were injected two days prior to sacrifice with 20 mg/kg 3-methylcholanthrene in corn oil. Following homogenization in 0.1 *M* potassium phosphate, pH 7.45 (Waring blender, 30 sec), the homogenate was centrifuged at 10,000 *g* for 20 min. The supernate (S10) was divided and frozen at -90°C until usage.

The microsomal metabolites were produced by addition of 3 μmoles of the hydrocarbon (0.75 ml acetone) to a 30-ml reaction mixture containing 120 μmoles *DL*-isocitrate (Sigma), 40 μmoles NADP (Sigma), 5 μmoles $MgCl_2$, 3.6 international units isocitrate dehydrogenase, 177 μmoles tris buffer pH 7.45. The reaction was carried out at 37°C in open flasks on a metabolic shaker under reduced yellow light. The reactions were stopped at 30 min by the addition of one volume of acetone. After the addition of two volumes of ethyl acetate the vessels were shaken, centrifuged and the organic layer passed through anhydrous Na_2SO_4. The solvent was then evaporated on a roto-evaporator and the residue (metabolites) dissolved in 50 μl of HPLC-grade (Burdick and Jackson) methanol.

Isolation and Spectral Analysis of Hydrocarbon Metabolites

An aliquot (25 μl) of the metabolite mixture was injected on a Laboratory Data Control (LDC) HPLC fitted with an LDC Spherisorb 5μ octadecyltrimethoxysilane (ODS) column and eluted with a 60-min water-to-methanol gradient (25% to 100%), 1.5 ml/min. The eluate was monitored by ultraviolet light (UV) (254 nm) with an LDC UV detector and fluorescence (exitation: 297-nm bandpass filter Baird-Atomic; emission: 415-nm Kodak Wratten 2A) with an Aminco fluoromonitor.

The resulting metabolite peaks were collected manually, dried on a roto-evaporator and redissolved in 2 ml of HPLC-grade methanol. The UV spectra were recorded on a Beckman Acta II spectrophotometer and the fluorescence data collected on an Aminco SPF-500 spectrofluorimeter.

Gas Chromatographic and Mass Spectral (GC/MS) Analysis of Hydrocarbon Metabolites

After the spectral data was collected the metabolite fractions were subject to GC/MS analysis. Each fraction was evaporated to dryness under a stream of N_2. The residue was trimethylsilylated by reaction with 30 μl of N_1O-*bis*(trimethylsilyl)-trifluoroacetamide at 60°C for 30 min. The volume of the solution was reduced to 10 μl under N_2 and an aliquot (3 μl) of this mixture was mixed with 1 μl of solution containing gas chromatographic reference standards (C_{24}, C_{28}, C_{32} and C_{36}) of aliphatic hydrocarbons.

The entire mixture was subjected to GC/MS analysis on a Hewlett-Packard Model 5985 system. The gas chromatograph component of the system consists of an HP 5840 gas chromatograph equipped with 6-ft x 1/8-in. glass column of 3% OV-17 on gas-chrom (100-120 mesh). The oven was temperature-programmed from 220 to 310° at 6°/min. The mass spectrometer was operated at the scanning mode and mass spectra were recorded continuously during an entire GC run. The ionization voltage of the mass spectrometer was 70 eV. The ion source temperature was held constant at 150°C.

DNA-Binding Analysis

Syrian hamster embryo (SHE) cells, at passage 2, were subdivided at a 1:2 ratio and grown in 20 ml of minimal essential medium (Earle's salts) [MEM(E)] with 10% fetal calf serum (FCS) on 150-mm Corning plastic culture plates to 60-80% confluency (1 day). The monolayers (10-20 plates) were treated with 50 μl of a 0.2 m*M* acetone solution of the radiolabeled hydrocarbon (final hydrocarbon concentration = 0.5 μ*M*). For tritium-labeled hydrocarbons the specific activity was 500-1,000 μCi/μmole; for the ^{14}C-labeled compounds, 100-120 μCi/μmole.

Aliquots of the cell media were treated at various times with one volume of 0.5 *N* NaOH-80% ethyl alcohol (EtOH) and the resulting solution extracted with cyclohexane as an index of the extent of metabolism (12).

After 24-hr hydrocarbon exposure, the cell media was removed and the monolayer was rinsed two times with 25 ml of phosphate-buffered saline (PBS), treated with 4 ml of lysing solution [8 *M* urea, 0.035 *M* sodium dodecyl sulfate (SDS), 0.01 *M* ethylenediaminetetraacetic acid (EDTA), 0.24 *M* sodium phosphate, pH 6.8] and the viscous lysate removed, treated with isoamyl alcohol (2 ml/50 ml lysate), homogenized (Waring, 30 sec) and loaded onto a 6-g hydroxylapatite (Bio-Rad HTP-DNA-grade) column. The DNA was eluted with 0.48 *M* sodium phsophate, pH 7.5, after removal of the protein and RNA (29). These DNA preparations contained negligible amounts of RNA (orcinol assay) or protein (Bio-Rad assay). The extent of hydrocarbon binding was measured by measuring the absorbance at 260 nm (E_{260} = 6,825 liter/mole–deoxynucleotide) and the specific activity of the hydrocarbon. The DNA preparation (0.5-2 mg in 50 ml) was dialysed twice against one liter distilled water (in the dark, 5°C) and the hydrocarbon binding redetermined by analysis of the radioactivity and the DNA content [3,5-diaminobenzoic acid (27)]. The resulting dialysate is dried on a roto-evaporator and the DNA residue was hydrolysed in 2 ml of 0.2 *M* tris, pH 7.5, 0.01 *M* $MgCl_2$ with DNase I, snake-venom phosphodiesterase and alkaline phosphatase (Worthington) in the dark at 37°C for 20-24 hr. Each hydrolysate was checked by TLC (Eastman cellulose, 1 *M* ammonium acetate; ethanol, 3:7, v/v) to assure hydrolysis was completely to the nucleoside level.

Sephadex LH-20 Chromatography

The hydrolysates were loaded on to 0.9 x 33-cm Sephadex LH20 columns and eluted with a 240-ml water-to-methanol gradient (0 to 100% methanol). The column effluent was monitored at 254 nm with an ISCO type 6 monitor and 2.0-ml fractions collected on an ISCO Model 328 fraction collector fitted with a Model 400 volumeter. Fractions were collected in 6-ml mini-vials and the solvent was evaporated in an oven at 70°C overnight. The radioactivity was analysed by adding 4 ml of Amersham PCS counting fluid and counting in a Beckman 8000 or Packard mini-vial spectrometer.

Analysis of Cellular Metabolites

Monolayers of Syrian hamster embryo cells (see DNA binding methods) were treated with 50 μl of 0.2-m*M* solution of radiolabeled hydrocarbon. This resulted in a 0.5-μM concentration of either DMBA, 2F-DMBA, 5F-DMBA or 11F-DMBA. At various times, the media was removed and treated with one volume of acetone. The resulting solution was extracted with two volumes of ethyl acetate and the organic layer dried over sodium sulfate. The solvent was removed and the ethyl acetate-soluble metabolites were dissolved in 50 μl of methanol and injected onto the HPLC system used for the microsomal metabolites. Fractions were collected at 30-sec intervals (0.75 ml) into 5-ml mini-vials and analysed on a Beckman 8000 spectrometer after addition of 4 ml of Amersham PCS.

The cell monolayer was rinsed four times with PBS, lysed with distilled water, extracted with ethyl acetate and analysed in the same manner.

RESULTS AND DISCUSSION

DMBA and its monofluoro derivatives were all readily metabolized by the liver homogenate preparations (44,45). Figure 2 shows the fluorescence profile obtained when metabolites from a 30-min oxidation of the various hydrocarbons were injected on the HPLC ODS column. In control oxidations, in which microsomes are heated to 100°C for 10 min before the reaction, only the parent compounds were eluted. Each metabolite fraction was collected and subjected to UV and fluorescence analysis and the percent methanol at which each metabolite eluted was recorded and compared to the parent hydrocarbon (HPLC retention index). Tables I and II detail the mass spectral and spectrophotometric data, respectively, of the DMBA metabolites isolated by HPLC (Figure 3).

The mass spectral analysis confirms the UV and fluorescence data and indicates that peak 14 (Figure 3, DMBA trace) is DMBA as it has a molecular ion m/e 256 as the base peak. Peaks 13 and 12 gave rise to very similar mass

Likewise, peak 9 did not give an interpretable mass spectrum but radioactivity studies indicate that this material is a DMBA derivative. Spectrophotometric data indicate a fully aromatic DMBA ring system (Table I). Peak 8 is highly interesting since this metabolite appears to be largely the methanolysis product of the DMBA-5,6-oxide (45). The fluorescence, UV and retention data (HPLC and GC) are consistent with this assignment, but UV and fluorescence data also indicate the presence of a DMBA phenol. Thus, this peak probably contains another hydroxylated phenol in addition to the DMBA-5,6-hydroxymethoxyl compound. Peak 7 may not be a DMBA derivative. Peak 6 is a dihydrodiol of DMBA (m/e 434 for the TMS derivative) since on GC it yields two minor products, one of which is identical to the DMBA-9-ol. Further, its spectrophotometric data are consistent with those reported for the 8,9-dihydro-8,9-dihydroxy-7,12-dimethylbenz[a]anthracene (DMBA-8,9-DHD) (39,45). Peak 5 elutes in the position of the *trans*-DMBA-5,6-DHD and the mass spectral analysis of the TMS derivative is consistent with a DMBA dihydrodiol (DHD). Further, when subjected to GC analysis this compound yields two thermolysis products, one being the DMBA-5-ol (the other apparently being the DMBA-6-ol). Thus, on the basis of this evidence and its UV fluorescence and HPLC retention time, this product is identified as *trans*-DMBA-5,6-DHD. By radiometric analysis, peak 4 is a DMBA product but is as yet not characterized. The fluorescence, however, strongly suggests that peak 4 is a metabolite resulting from additional oxidation of the DMBA-8,9-DHD. Peak 3 did not provide a mass spectrum suitable for analysis, but its UV properties are consistent with those reported by Sims (39) for a DMBA-10,11-DHD. Thus, it is the most likely product resulting from the further oxidation of that compound. Peaks 1 and 2 have not yet been characterized.

Since chemical standards of most of the monofluoro-DMBA metabolites are not available, tentative identification has been made by comparison with the analogous DMBA metabolites on the basis of HPLC retention, UV and fluorescence spectroscopy, GC retention and mass spectral analysis. Such comparisons are greatly facilitated by the fact that metabolites of the monofluoro derivatives have spectral and retention properties similar to the analogous DMBA compound (*e.g.*, 11F-DMBA-8,9-DHD has a fluorescence spectrum similar to DMBA-8,9-DHD) (45). Table III summarizes the comparative spectrophotometric data for the 8,9-DHD and 7OHM-metabolites from the various DMBA monofluoro derivatives (45). Thus, by such comparative studies tentative identification on a great number of the metabolites of the fluoro-DMBA derivatives has been made. The mass spectral, UV and fluorescence data, as well as the GC and HPLC retention times were used in identifying the numerous metabolites shown in Figure 2. In the case of 9F-DMBA no metabolite exhibiting an 8,9-DHD spectrum was found (44). In an analogous manner no metabolite exhibiting a 5,6-DHD spectrum was found in

TABLE III. Comparative Data on a DMBA Metabolite and Analogous Monofluoro Compounds

Parent	Derivative	Retention Index (HPLC)[a]	UV Maxima[b]	Emission Maxima[b]
DMBA	8,9-DHD	0.761	*265*,318,332	*387*,405
1F-DMBA	8,9-DHD	0.760	*261*,312,326	*390*,410
2F-DMBA	8,9-DHD	0.759	*269*,314,329	*389*,407
4F-DMBA	8,9-DHD	0.768	*265*,315,330	*388*,405
5F-DMBA	8,9-DHD	0.759	*268*,311,332	*390*,410
9F-DMBA	8,9-DHD	–	–	–
11F-DMBA	8,9-DHD	0.764	*260*,317,331	*384*,403
DMBA	70 HM	0.889	274,284,*294*	408,427
1F-DMBA	70HM	0.890	265,*286*,296,360	*408*,428
2F-DMBA	70HM	0.893	268(sh),285,*293*,361	412,432
4-F-DMBA	70HM	0.888	not analyzed	407,430
5F-DMBA	70HM	0.889	265,285,295,360	407,428
9F-DMBA	70HM	0.881	274,283,*294*,361	410,425
11F-DMBA	70HM	0.897	274,*286*,296,364	412,431

[a]Retention Index = $\frac{\text{\% methanol metabolite elutes}}{\text{\% methanol parent hydrocarbon elutes}}$

[b]Data collected as for Table I.

the rat liver metabolites of 5F-DMBA (44). Thus, it appears that the original assumption of Newman and the Millers (34), *i.e.*, that substitution of fluorine in a PAH will block or greatly reduce metabolism at that position, seems correct.

We next investigated the metabolism of radiolabeled DMBA, 2F-DMBA, 5F-DMBA and 11F-DMBA by cultured Syrian hamster cells since these cells were to be used for the DNA-binding studies (35). In this study we analyzed the ethyl acetate-extractable metabolites that are present both in the media and inside the cells at the end of the incubation. The flow diagram for these operations are shown in Figure 4. In addition, we have looked at the radioactivity in the media that is non-ethyl acetate-extractable. The HPLC profiles of these metabolite fractions for DMBA are shown in Figure 5.

With all of the DMBA derivative more than 95% of the radioactivity was present in the media at the end of the incubation. In the case of DMBA itself about 70-72% of the total activity found in the media was not extractable in ethyl acetate. HPLC analysis of the non-ethyl acetate-extractable radioactivity reveals that it is a very polar material and elutes before the 56% methanol portion on the gradient. We are currently conducting analysis on this material. In the case of all the hydrocarbons, the more polar metabolites (*e.g.*, dihydrodiols) were predominantly in the media whereas the less polar products (*e.g.*, phenols) and the parent hydrocarbons represented the majority of the ethyl acetate-extractable radioactivity found in the cells, except in the case of 2F-DMBA where substantial quantities of 8,9-DHD is found in the cells as well as the media (Figures 5, 6, 7 and 8).

Interestingly, for DMBA itself the major portion of the radioactivity extracted from the lysed-cell monolayer was found to elute in the position of DMBA phenols on the HPLC (35). The radioactivity of the cell-generated DMBA phenols cochromatographed with the fluorescence from microsomally generated phenols on the HPLC system. However, the cell-generated radioactivity eluting at fraction 58 to 60 (78 to 80% MeOH) did not exactly cochromatograph with the fluorescence of the microsomally produced metabolite we previously identified as 7OHM-12MBA (Figure 5). The most likely DMBA metabolite eluting this close to 7OHM-12MBA would probably be the isomeric 7-methyl-12-hydroxymethylbenz[a]anthracene (7M-12OHMBA). Final confirmation awaits the synthesis of an appropriate standard.

The major ethyl acetate-extractable metabolite found in the media of cells treated with [^{3}H]DMBA is postulated to be the DMBA-8,9-DHD since it cochromatographs (HPLC) with the microsomally generated product (Figure 5).

Figures 6, 7 and 8 show HPLC profiles resulting from identical ethyl acetate extractions of media and cells after incubation with [^{3}H]2F-DMBA, [^{3}H]11F-DMBA and [^{14}C]5F-DMBA. As expected, the more polar

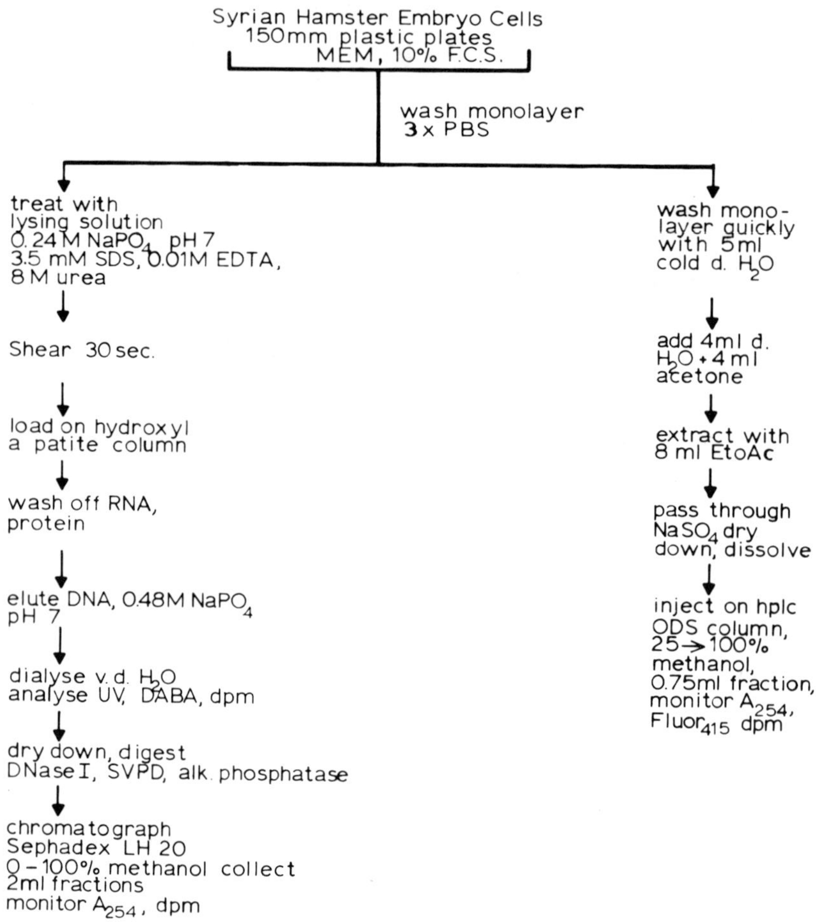

FIGURE 4. Flow diagram for Syrian hamster embryo DNA-binding and metabolism studies (see Materials and Methods).

metabolites are distributed predominantly in the media and the less polar products and parent hydrocarbons are in the cells. With monofluoro-DMBA compounds, a decidedly smaller portion of the extractable radioactivity found in the cells elutes in the phenol region. We are currently working on the identification of these cell-generated monofluoro-DMBA metabolites. The SHE cells seem to produce relatively more phenolic DMBA derivatives than do mouse embryo cells (2) or macrophages (36).

The comparative DNA-binding experiments were carried out in SHE cells and the flow chart for these experiments is detailed in Figure 4. The dose of

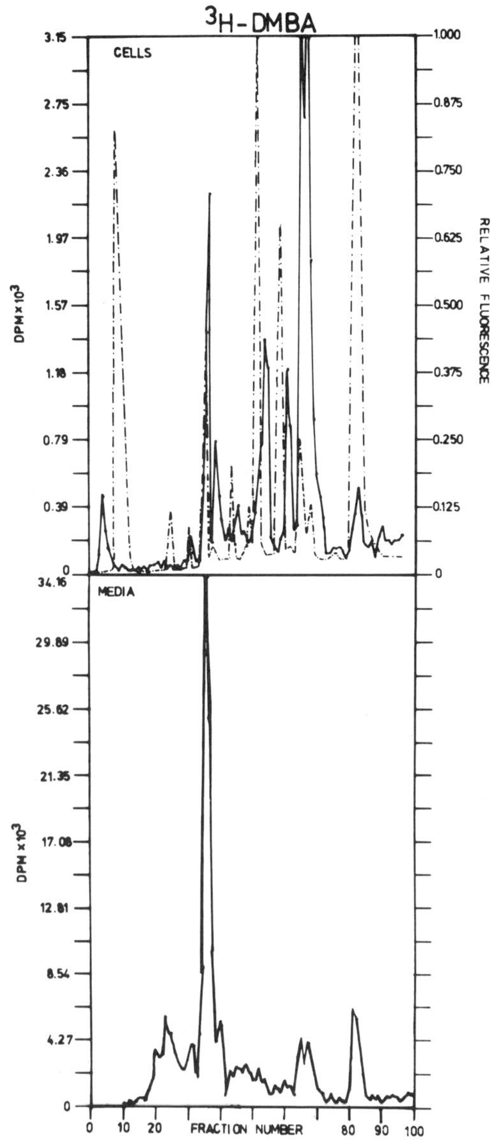

FIGURE 5. HPLC separation of [^{3}H]DMBA metabolites produced by SHE cultures. Upper: [^{3}H]metabolites found inside the cells after 24 hr (——); fluorescence trace of unlabeled DMBA metabolites generated via a liver microsomal enzyme system (—·—·—). (The fluorescence centered at fraction 53 is due to an unidentified ethyl acetate-extractable material from SHE cells). Lower: ethyl acetate [^{3}H]metabolites from the cell media at 24 hr. The major peak at fraction 34 is the DMBA-8,9-DHD. The DMBA phenols and DMBA elute at fractions 62-72 and 80-85, respectively.

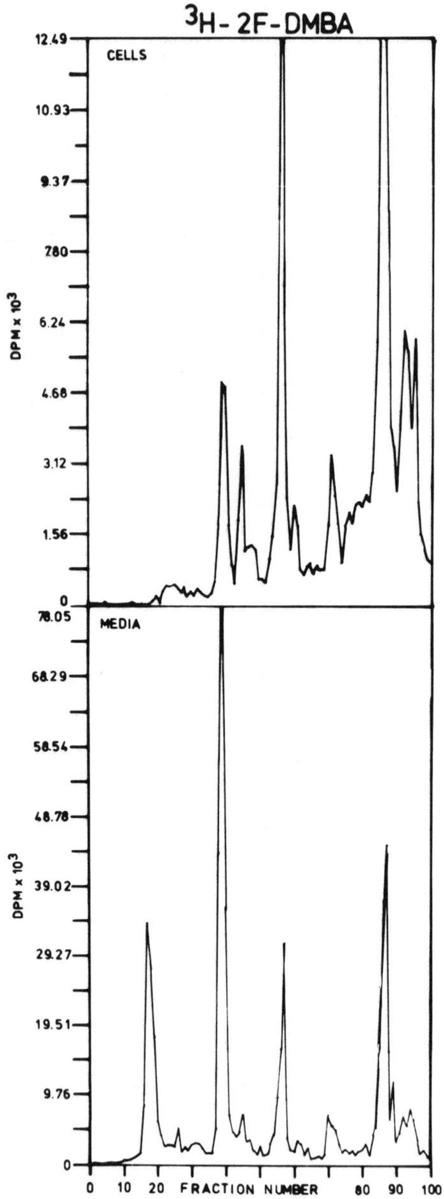

FIGURE 6. HPLC analysis of [^{3}H] 2-F-DMBA metabolites produced by SHE cultures. Upper: [^{3}H] metabolites extracted from the cells at 24 hr. Lower: [^{3}H] metabolites derived from the cell media. 2F-DMBA-8,9-DHD elutes at fraction 39.

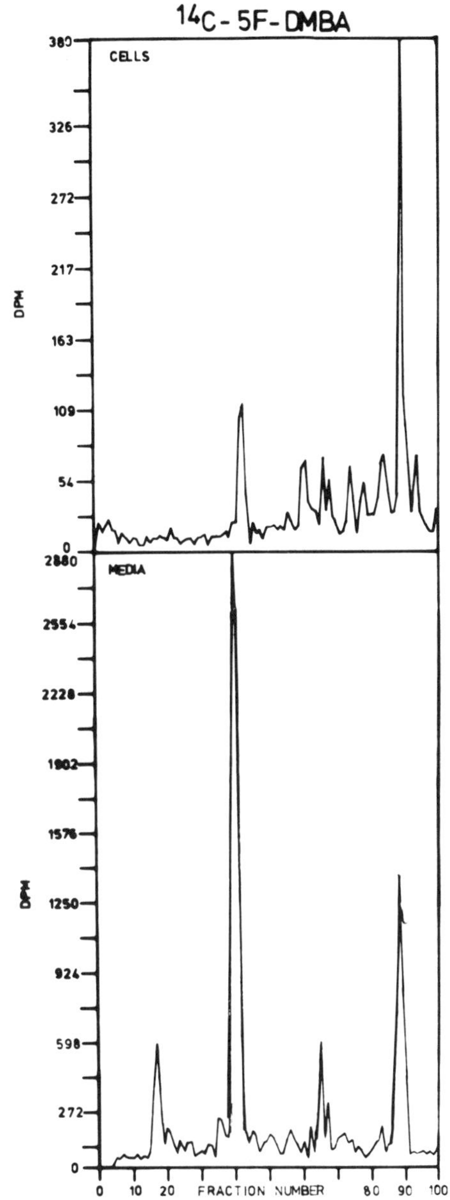

FIGURE 7. HPLC analysis of [^{14}C]-5F-DMBA metabolites produced by SHE cultures. Upper: [^{14}C] metabolites extracted from cells at 24 hr. Lower: [^{14}C] metabolites extracted from the cell media at 24 hr. The 5F-DMBA-8,9-DHD elutes at fraction 40 and the parent hydrocarbon elutes at fraction 88.

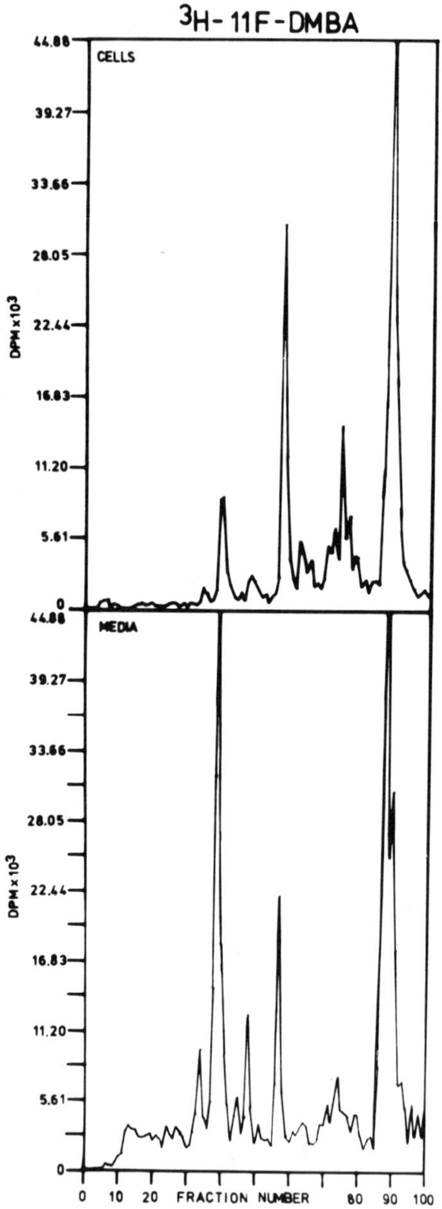

FIGURE 8. HPLC analysis of the [^{3}H]-11F-DMBA metabolites produced by SHE cells. Upper: [^{3}H] metabolites extracted from cells at 24 hr. Lower: [^{3}H] metabolites extracted from the cell media at 24 hr. The 11F-DMBA-8,9-DHD and 11F-DMBA elute at fractions 39 and 89, respectively.

0.5 μM was used as the treatment concentration since for DMBA this concentration was not noticeably cytotoxic and gave good binding levels (Figure 9). In addition, 24 hr was chosen as a convenient treatment time since metabolism of DMBA (as measured by the percent of radioactivity rendered non-cyclohexane soluble) was largely complete (Figure 10). In each set

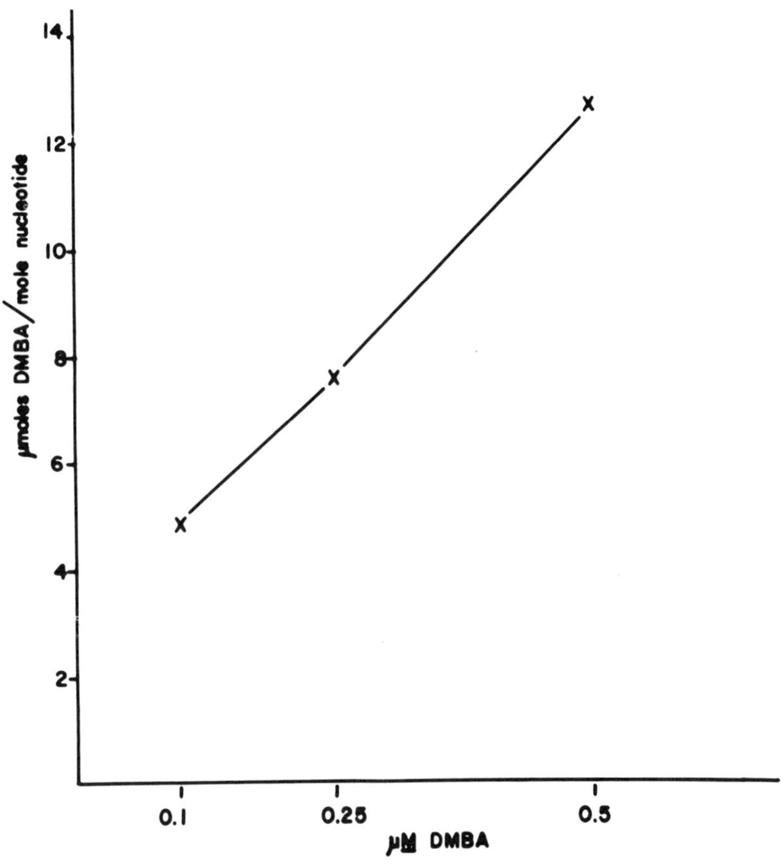

FIGURE 9. Binding level of [^{3}H]DMBA to SHE cell DNA as a function of treatment concentration. Treatment time was 24 hr.

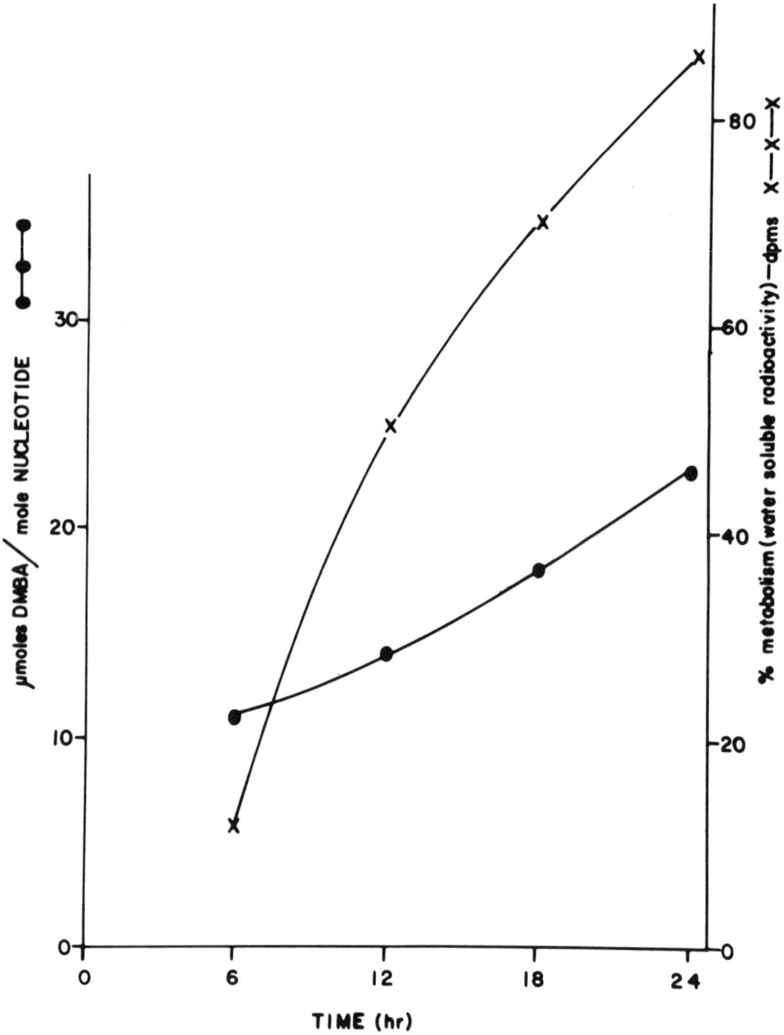

FIGURE 10. Extent of metabolism (x–x) and DNA binding (●–●) of [^{3}H]DMBA in SHE cells as a function of time. Treatment concentration was 5 µM [^{3}H]DMBA.

of experiments involving the testing of a monofluoro derivative, the parent compound DMBA was administered to an identical set of SHE cells grown up at the same time. This was necessary so as to eliminate the batch-to-batch variation observed with the SHE cells with respect to their potential to activate PAH to DNA binding forms. The results of direct comparative binding studies are summarized in Table IV. After determination of the

TABLE IV. Direct Comparative Binding of DMBA and Selcted Monofluoro Derivatives to SHE Cell DNA

Experiment	Fluoro-Derivative	Fluoro-DMBA Binding Level (μmole HC/mole nucleotide)	DMBA Binding Level (μmole HC/mole nucleotide)	Binding Ratio Fluoro-DMBA/DMBA	Average
1	2F	0.95	23.00	0.041	
2	2F	0.53	15.3	0.034	0.039
3	2F	0.75	17.69	0.042	
4	5F	3.71	10.56	0.35	
5	5F	2.28	6.29	0.36	0.34
6	5F	8.34	25.38	0.32	
7	11F	40.36	18.18	2.22	
8	11F	31.48	12.85	2.45	2.34

binding level, the DNA preparations were hydrolysed enzymatically and chromatographed on Sephadex LH-20 columns. The profiles obtained from this chromatography are shown in Figures 11 and 12. Thus, it is clear that all the fluoro derivatives tested can be activated by SHE cells to metabolites that bind firmly to the cellular DNA.

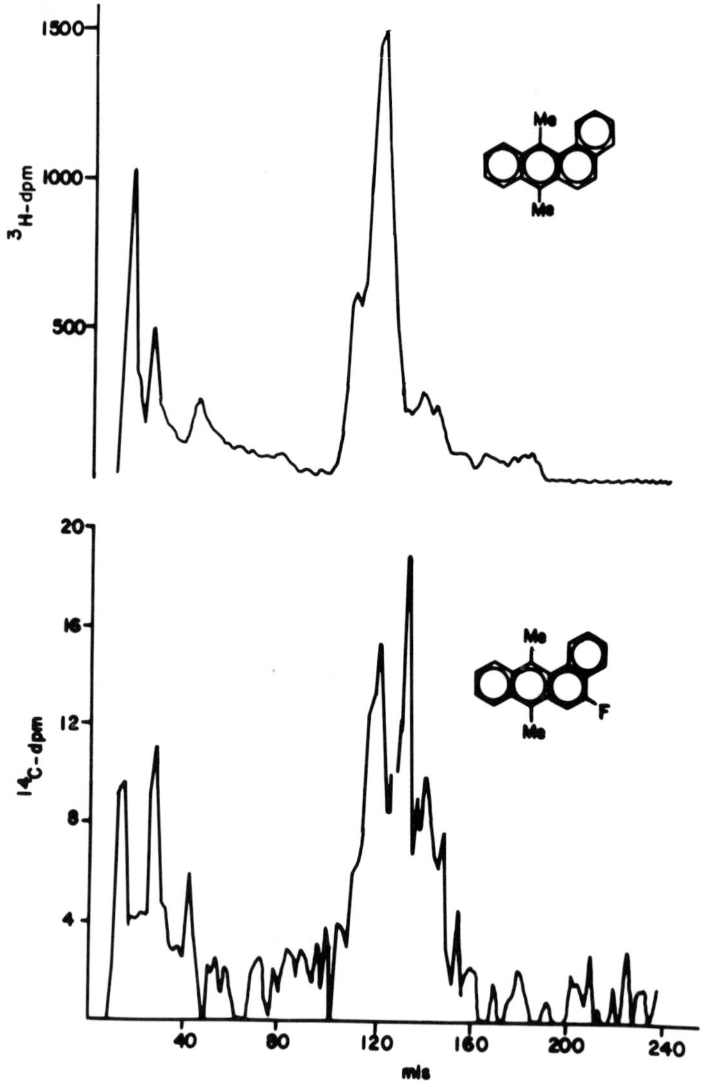

FIGURE 11. Sephadex LH-20 analysis of deoxynucleoside hydrocarbon adducts formed after 24 hr in SHE cells. Upper: [^{3}H] DMBA. Lower: [^{14}C] 5F-DMBA. The UV marker, 4-(*p*-nitrobenzyl)pyridine, elutes at 96-100 ml on this system.

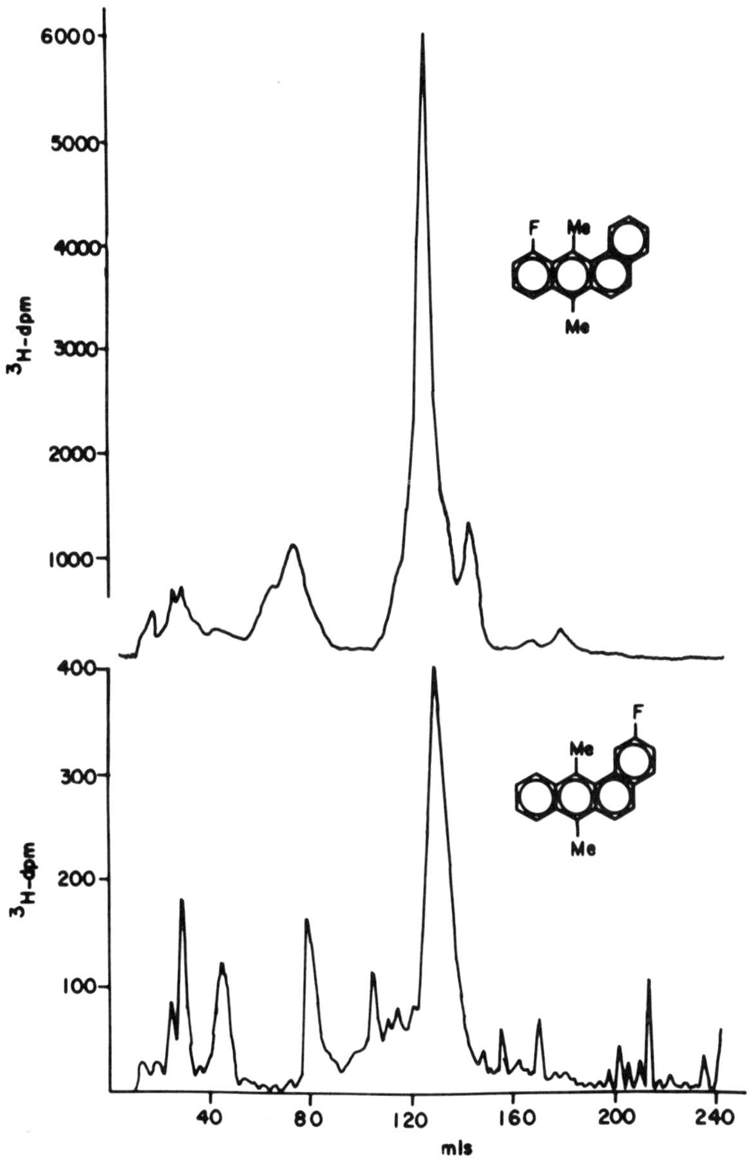

FIGURE 12. Sephadex LH-20 analysis of deoxynucleoside hydrocarbon adducts formed after 24 hr in SHE cells. Upper: [^{3}H] 11F-DMBA. Lower: [^{3}H] 2F-DMBA. The UV marker, 4-(*p*-nitrobenzyl)pyridine, elutes at 96-100 ml.

The binding profiles obtained for DMBA (Figure 11a) and 11F-DMBA (Figure 12a) are nearly identical and both compounds bind to SHE DNA at high levels. The radioactivity eluting in the 55- to 85-ml range on the 11F-DMBA gradient apparently represents the photodecomposition of the deoxynucleoside-DMBA (DN-DMBA) adducts (110-140 ml). This material is routinely found in the DMBA profiles as well and was present even when the DN-DMBA adducts (90-130 ml) were rechromatographed on the same LH-20 system (Figure 13). Such photodecomposition was reported by Moschel *et al.* (33) and others (1,13) after prolonged (> 4 hr) exposure to white fluorescence light.

Since Baird *et al.* (1) have reported that such decomposition is characteristic of DMBA derivatives with a saturated A-ring, we interpret these observations to mean that the carcinogenic 11F-DMBA, like DMBA, binds to DNA through the bay region (A-ring). Such photodecomposition has not been observed with 5F-DMBA and 2F-DMBA and thus may indicate that these weakly carcinogenic and noncarcinogenic compounds do not bind through the bay region. (The radioactivity at 80 ml on the 2F-DMBA profile shown in Figure 12 did not appear on three previous runs and thus we believe it represents a radiochemical contamination and not a DN-2F-DMBA decomposition product.)

These observations, in summary, can most easily be interpreted as indicating that the 2F-DMBA and 5F-DMBA are noncarcinogenic because insertion of fluorine into those positions blocks requisite bay-region activation and subsequent bay-region DNA binding.

Recent studies by Slaga and Huberman *et al.* (40) have indicated that the monofluoro derivatives of DMBA have a widely varying range of mutagenic and carcinogenic potency. Their investigations indicate that 11F-DMBA is a potent carcinogen/mutagen comparable to DMBA itself, while 2F-DMBA is virtually devoid of carcinogenic/mutagenic properties (40). 5F-DMBA, in the hands of these investigators, is a relatively weak carcinogen/mutagen (40). Similar data on 5F-DMBA had been obtained previously by the Millers (11). Thus it appears that the DNA-binding potential of these compounds falls in the same rank orders as their carcinogenic potency. In agreement with these observations, postreplication studies indicate that the most carcinogenic derivatives (DMBA and 11F-DMBA) produce the smallest molecular weight DNA (9,10).

These investigations support the bay-region theory of PAH carcinogens advanced by Jerina and Daly (25). It is unclear, however, how substitution of fluorine at the 5-position of DMBA perturbs metabolism at the bay region. Recent results of Harvey and Dunne (17) have shown that 4F-DMBA is carcinogenic, an observation in agreement with earlier studies by the Millers (34). It is also interesting that in the 7-methylbenz[a] anthracene (7MBA)

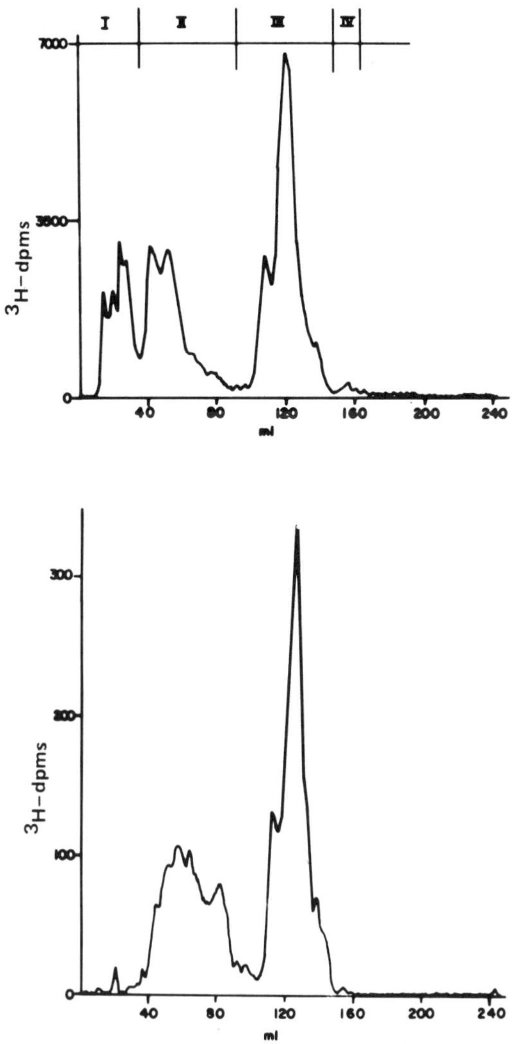

FIGURE 13. Sephadex LH-20 analysis of DN-[^{3}H]DMBA adducts from SHE cells. Upper: LH-20 chromatograph of [^{3}H]DMBA-DNA from SHE cells. Lower: Rechromatograph (same system) of a portion of fraction III.

series the A-ring monofluoro derivatives are carcinogenic whereas 5F-7MBA is inactive like 5F-DMBA (34).

CONCLUSION

Our data indicate that:

1. Substitution of fluorine for hydrogen in the DMBA nucleus may block or substantially reduce metabolism at that position.

2. Substitution of fluorine for hydrogen in certain positions of DMBA (*e.g.*, 2- and 5-) reduces DNA binding in mammalian cultured cells relative to DMBA, whereas, substitution at other positions (*e.g.*, 11-) does not reduce binding.

3. The rank order of the various DMBA derivatives for DNA-binding potential is the same as their reported oncogenic potential.

4. The binding studies tentatively indicate that the DN-DMBA adducts of the carcinogenic derivatives (*e.g.*, DMBA and 11F-DMBA) bind via the bay region whereas the noncarcinogenic and weakly carcinogenic derivatives (*e.g.*, 2F-DMBA and 5F-DMBA) do not.

ACKNOWLEDGMENTS

We gratefully acknowledge the excellent technical assistance of M. A. Drum, S. Dhawale, P. Sedmak, N. J. Joyce and D. K. Allison in these studies.

We thank Professor M. S. Newman for the DMBA monofluoro derivatives as well as numerous DMBA phenols. We thank the National Cancer Institute Chemical Repository for the 7OHM-12MBA.

Finally, we acknowledge the support for this work from the National Cancer Institute Public Health Service Grant No. CA-21371, U.S. Environmental Protection Agency Grant R-804201 and American Cancer Society Grant No. IN-16-0.

REFERENCES

1. Baird, W. M., and Dipple, A. (1977): Photosensitivity of DNA-bound 7, 12-dimethylbenz[a] anthracene. *Int. J. Cancer* 20:427-431.
2. Bigger, C. A. H., Tomaszewski, J. E., and Dipple, A. (1978): Differences between products of binding of 7,12-dimethylbenz[a] anthracene to DNA in mouse skin and in a rat liver microsomal system. *Biochem. Biophys. Res. Comm.* 80:229-235.
3. Bock, F. G. (1964): Early effects of hydrocarbons on mammalian skin. *Prog. Exp. Tumor Res.* 4:126-168.
4. Boutwell, R. G. (1967): Some biological aspects of skin carcinogenesis. *Prog. Exp. Tumor Res.* 4:207-250.
5. Brookes, P. (1975): Covalent interaction of carcinogens with DNA. *Life Sciences* 16:331-334.

6. Brookes, P., and Lawley, P. D. (1964): Evidence for the binding of polynuclear aromatic hydrocarbons to nucleic acid in mouse skin: Relation between carcinogenic power of hydrocarbons and their binding to deoxyribonucleic acid. *Nature* 202:781-784.
7. Buty, S. G., Thompson, S., and Slaga, T. J. (1976): The role of epidermal aryl hydrocarbon hydroxylase in the covalent binding of polycyclic hydrocarbon to DNA and its relationship to tumor initiation. *Biochem. Biophys. Res. Comm.* 70:1102-1108.
8. Cox, J. O., and Pilcher, G. (1970): In: *Thermochemistry of Organic and Organometallic Compounds,* Academic Press, New York.
9. D'Ambrosio, S. M., Daniel, F. B. and Hart, R. W. (1978): Cellular repair of DNA damage induced by 7,12-dimethylbenz[a]anthracene and its fluoro analogues *in vitro.* Chapter 52, this volume.
10. D'Ambrosio, S. M., Daniel, F. B., Hart, R. W., Cazer, F. D., and Witiak, D. T. (1978): DNA repair in Syrian hamster embryo cells treated with 7,12-dimethylbenz[a]anthracene and its weakly carcinogenic 5-fluoro analogue. *Cancer Lett.* (submitted).
11. Daniel, F. B., Cazer, F. D., D'Ambrosio, S. M., Hart, R. W., Kim, W. H., and Witiak, D. T. (1978): Comparative metabolism and DNA binding of 7,12-dimethylbenz[a]anthracene and its weakly carcinogenic 5-fluoro analogue. *Cancer Lett.* (submitted).
12. DePierre, J. W., Moron, M. S., Johannesen, K. A. M., and Enster, L. (1975): A reliable, sensitive, and convenient radioactive assay for benzopyrene monooxygenase. *Anal. Biochem.* 63:470-484.
13. Dipple, A., and Nebzydoski, J. A. (1978): Evidence for the involvement for a diol-epoxide in the binding of 7,12-dimethylbenz[a]anthracene to DNA in cells in culture. *Chem. Biol. Interact.* 20:17-26.
14. Doll, R., and Vodopija, I., Eds. (1973): *Host Environment Interactions in the Etiology of Cancer in Man,* IARC Scientific Publication No. 7, Lyon, France.
15. Goshman, L. M., and Heidelberger, C. (1967): Binding of tritium labeled polycyclic hydrocarbons to DNA of mouse skin. *Cancer Res.* 27: 1678-1688.
16. Hart, R. W., Daniel, F. B., Davis, M. T., and Lewis, N. J. (1978): A rational evaluation of the use of DNA damage as an environmental carcinogenesis/mutagenesis prescreen in the proceeding of the Interagency Collaborative Research Meeting on environmental carcinogenesis and mutagenesis. (in press).
17. Harvey, R. B., and Dunne, F. B. (1978): Multiple regions of metabolic activation of carcinogenic hydrocarbons. *Nature* 273:566-568.
18. Heidelberger, C. (1976): Chemically and metabolically induced DNA adduct: Relationship to chemical carcinogenesis. In: *Aging, Carcinogenesis and Radiation Biology: The role of nucleic acid addition reactions,* K. C. Smith, Ed., pp. 341-371, Plenum Press, New York.
19. Heimann, R., Heuson, J. C., and Coune, A. (1968): Tumors developing in oophorectomized Sprague-Dawley rats after a single gastric instillation of 7,12-dimethylbenz[a]anthracene. *Cancer Res.* 28:309-313.

20. Huberman, E., and Sachs, L. (1976): Mutability of different genetic loci in mammalian cells by metabolically activated carcinogenic polycyclic hydrocarbons. *Proc. Nat. Acad. Sci., U.S.* 73:188-192.
21. Huggins, C., Grand, L. C., and Brillantes, F. B. (1961): Mammary cancer induced by a single feeding of polynuclear hydrocarbons and its suppression. *Nature* 189:204-207.
22. Huggins, C. B. (1975): Communicated through Professor M. S. Newman, Ohio State University.
23. Irving, C. C. (1973): Interaction of chemical carcinogens with DNA. In: *New Methods in Environmental Toxicology,* F. Coulston, F. Korte and M. Goto, Eds., pp. 99-112, International Academic Print Co., Tokyo.
24. Janss, D. H., and Ben, T. L. (1978): Age-related modification of 7,12-dimethylbenz[a]anthracene binding to rat mammary gland DNA. *J. Nat. Cancer Inst.* 60:173-177.
25. Jerina, D. M., and Daly, J. W. (1976): Oxidation at carbon. In: *Drug Metabolism,* D. V. Parke and R. L. Smith, Eds., pp. 15-33, Taylor and Francis, London.
26. Jull, J. W., Hawryluk, A., and Russell, A. (1968): Mechanism of induction of ovarian tumors in the mouse by 7,12-dimethylbenz[a]anthracene. Tumor induction in organ culture. *J. Nat. Cancer Inst.* 40:687-706.
27. Kennaway, E. L. (1930): Further experiments on cancer producing substances. *Biochem. J.* 24:497-504.
28. Kissane, J. M., and Robbins, E. (1958): The fluorometric measurement of deoxyribonucleic acid in animal tissues with special reference to the central nervous system. *J. Biol. Chem.* 233:184-188.
29. Meinke, W., Goldstein, D. A., and Hall, M. R. (1974): Rapid isolation of mouse DNA from cells in tissue culture. *Anal. Biochem.* 58:82-88.
30. Miller, J. A. (1970): Carcinogenesis by chemicals: An overview, G. H. A. Clowes Memorial Lecture. *Cancer Res.* 30:559-576.
31. Miller, J. A., and Miller, E. C. (1966): Mechanisms of chemical carcinogens: Nature of proximate carcinogens and interactions with macromolecules. *Pharmacol. Rev.* 18:806-838.
32. Miller, J. A., and Miller, E. C. (1974): Biochemical mechanisms of chemical carcinogenesis. In: *The Molecular Biology of Cancer,* H. Busch, Ed., pp. 377-402, Academic Press, New York.
33. Moschel, R. C., Baird, W. M., and Dipple, A. (1977): Metabolic activation of the carciongen 7,12-dimethyl[a]anthracene for DNA binding. *Biochem. Biophys. Res. Comm.* 76:1092-1098.
34. Newman, M. S. (1976): Carcinogenic activity of benz[a]anthracenes. In: *Carcinogenesis, Vol. 1,* R. I. Freudenthal and P. W. Jones, Eds., pp. 203-207, Raven Press, New York.
35. Oravec, C., Daniel, F. B., Wong, L., Wang, A., and Hart, R. W. (1978): Metabolism of 7,12-dimethylbenz[a]anthracene by cultured Syrian hamster embryo cells. *Drug Metabol. Dispos.* (in preparation).

36. Palmer, W. G., Allan, T. J., and Tomaszewski, J. E. (1978): Metabolism of 7,12-dimethylbenz[a]anthracene by macrophages and uptake of macrophage derived metabolites by respiratory tissues *in vitro. Cancer Res.* 38:1079-1084.
37. Shear, M. J. (1938): Studies in carcinogenesis; methyl derivatives of 1:2-benzanthracene. *Am. J. Cancer* 33:499-537.
38. Shellabarger, C. J. (1976): Modifying factors in rat mammary gland carcinogenesis. In: *Biology of Radiation Carcinogenesis,* J. M. Yuhas, R. W. Tennant and J. D. Regan, Eds., pp. 31-42, Raven Press, New York.
39. Sims, P. (1970): Studies on the metabolism of 7-methylbenz[a]anthracene and 7,12-dimethylbenz[a]anthracene and its hydroxymethyl derivatives in rat liver and adrenal homogenates. *Biochem. Pharmacol.* 19:2261-2275.
40. Slaga, T. J., Bracken, M., Gleason, G. L., DiGiovanni, J., Berry, D. L., Juchau, M. R., and Harvey, R. G. (1978): Tumor initiating activities of various derivatives of benz[a]anthracene and 7,12-dimethylbenz[a]-anthracene in mouse skin. Chapter 49, this volume.
41. Uematsu, K., and Huggins, C. (1968): Induction of leukemia and ovarian tumors in mice by pulse-dose of polycyclic aromatic hydrocarbons. *Mol. Pharmacol.* 4:411-426.
42. Walters, M. A. (1966): The induction of lung tumors by the injection of 9,10-dimethyl-1,2-benzanthracene (DMBA) into newborn suckling and young adult mice. A dose response study. *Brit. J. Cancer* 20:148-160.
43. Walters, M. A., and Roe, F. J. C. (1966): The time appearance of lung tumors in mice injected when newly born with 9,10-dimethyl-1,2,-benzanthracene (DMBA). *Brit. J. Cancer* 20:161-167.
44. Wong, L., Wang, A., and Daniel, F. B. (1978): Unpublished observations, The Ohio State University.
45. Wong, L., Wang, A., and Daniel, F. B. (1978): Analysis of microsomally generated metabolites of 7,12-dimethylbenz[a]anthracene via high pressure liquid chroamtography, mass-spectral, and spectrofluorometric analysis. (in preparation).
46. Yang, S. K., and Dower, W. V. (1975): Metabolic pathways of 7,12-dimethylbenz[a]anthracene in hepatic microsomes. *Proc. Nat. Acad. Sci., U.S.* 72:2601-2605.

INDEX

MORE VALUABLE BOOKS